This book is a graduate text devoted to the main aspects of the physics of recombination in semiconductors. It is the first book to deal exclusively and comprehensively with the subject, and as such is a self-contained volume, introducing the concepts and mechanisms of recombination from a fundamental point of view. Professor Landsberg is an internationally acknowledged expert in this field and, while not neglecting the occasional historical insights, he takes the reader to the frontiers of current research.

Following initial chapters on semiconductor statistics and recombination statistics, the text moves on to examine the main recombination mechanisms: Auger effects, impact ionization, radiative recombination, defect and multiphonon recombination. A final chapter deals with the topical subject of quantum wells and low-dimensional structures. Altogether the book covers a remarkably wide area of semiconductor physics.

The book will be of importance to scientists and engineers who are studying the physics and applications of semiconductors. Much of the book, particularly the first half, has been used in graduate courses and the explanations throughout are adequate for this purpose.

Recombination in semiconductors

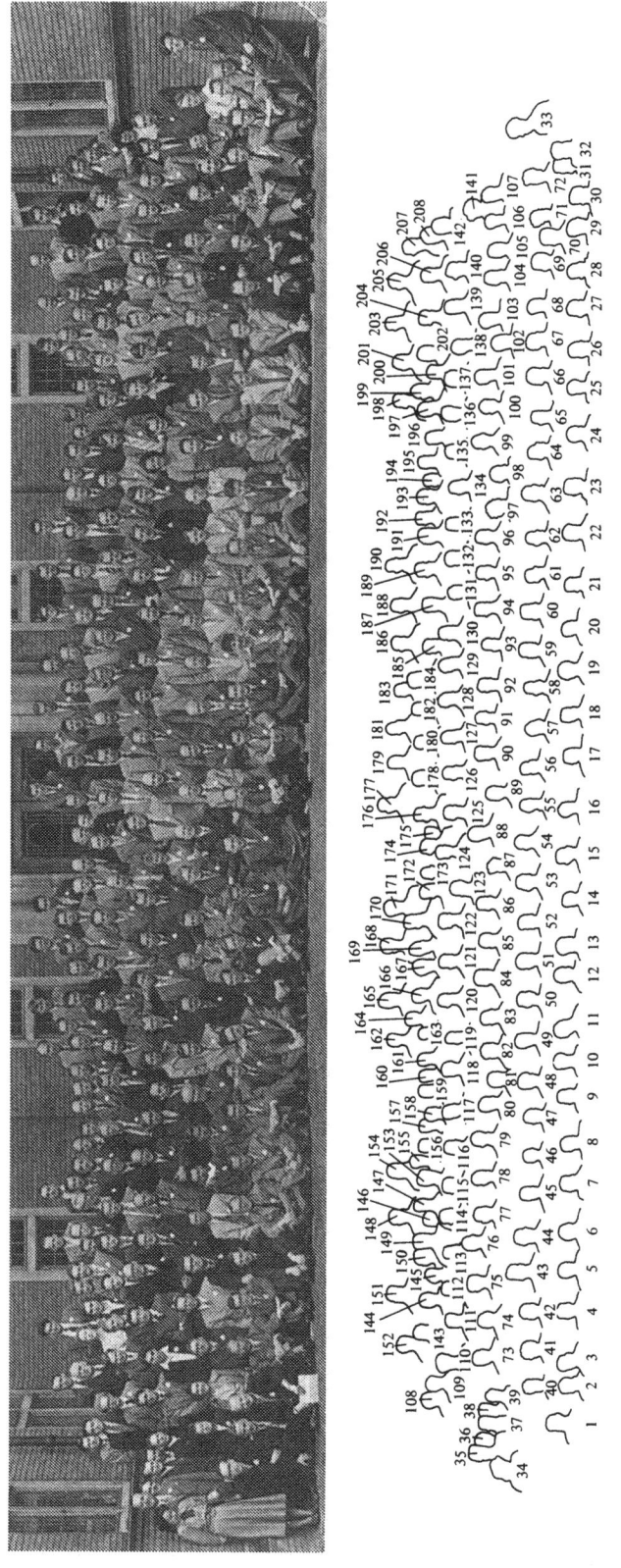

The first International Conference on the Physics of Semiconductors, Reading, England, 1950.

This is the official photograph which was taken at the time.

It shows most of the participants.

1 John Hodgson (Keele). 9 R.G. Breckenridge, National Bureau of Standards, Washington, †4.9.86. 11 Dr Forsberg, Washington. 16 Dr Howard Etzel, US Naval Research Lab. (1950–6), Deputy Director National Science Foundation, 1971. 18 Dr E.W. (Ted) Elcock, Aberdeen. 20 Dr Ron Cooper, AEI Laboratories, Aldermaston. Later Professor of Electrical Engineering at the University of Manchester. 21 Dr Peter T. Landsberg, AEI Laboratories, Aldermaston. Later Professor of Applied Mathematics at Universities of Wales and Southampton. 22 Dr Trevor Moss, RRE, later RAE, 1953–78, then Deputy Director RSRE 1978–81. 23 Dr Dennis Sciama, Fellow, Trinity College, Cambridge 1952–6. Later Professor of Astrophysics, Trieste 1983, FRS (1983). 24 Dr A.F. Gibson, RRE. Later Professor of Physics at the University of Essex, FRS (1978). †1988(?). 25 Ron Bloomer, AEI Laboratories, Aldermaston (1947–59), Harlow (1959–63), CEGB (1964–86). 30 Dr P. Jutsum, University of Reading, later University of the West Indies. 33 O. Heavens, University of Reading. Later Professor of Physics at the University of York. 36 Dr T.B. Rymer, University of Reading. 38 Dr Pierre Baruch, Ecole Normale Superieure; later Professor at the University of Paris. 41 Mlle M. Francois, Universities of Reading and Paris. 42 Dr Koop, Philips Research Laboratories †1955(?). 44 Dr C.A. Hogarth, University of Reading. Later Professor of Physics at Brunel University. 45 Dr Park H. Miller, Professor, University of Pennsylvania, Philadelphia. Later at General Dynamics. 46 R.A. Smith, REE, later Principal Herriot-Watt University, FRS (1962). †1980. 47 Dr J.W. Mitchell, Reader, University of Bristol (1945–59). Professor of Physics at the University of Virginia since 1959. FRS (1956). 48 Prof. Perruci. 49 Prof. A.E. Sandström, Sweden. 50 Mr T.R. Scott, Standard Telecommunications Laboratories, London. 51 Prof. K. Lark-Horowitz, University of Purdue. 52 Dr E.J.W. Verwey, Philips Research Laboratories. 53 Prof. N.F. Mott, FRS (1936), N.L. (1977). Professor of Physics at the Universities of Bristol, 1933–54, and Cambridge, 1955–66. 54 Prof. Fleury, Conservatoire National des Arts et Métiers. Later Director of L'Institute d'Optique, Paris. †1976. 55 Prof. R.W. Ditchburn, Professor of Physics at the University of Reading, FRS (1962). †1987. 56 W. Shockley, N.L. (1956). Bell Laboratories (1936–42, 1945–54). Professor at Stanford University (1963–75). †13.8.89. 57 W.H. Brattain, N.L. (1956). Bell Laboratories (1929–67). †13.10.87. 58 H.Y. Fan, Professor at MIT (1948–63) and Purdue University. (1963+). 59 Prof. R.L. Sproull, Professor at Cornell University (1946–63), President of University of Rochester (1975). 60 Prof. R.W. Pohl, University of Göttingen, 1916–53. †1976. 61 Prof. G. Busch, ETH Zurich (1949–78). 1954 supervisor of Nobel prize winner Alex Müller. 62 Dr L.P. Smith, Cornell University 1904. †17.6.88. 64 Pierre Aigrain, Research & Study Centre, French Navy (1948–50). Later Secretary of State in charge of research (1978–81), Scientific Advisor Thomson Group (1983+). 65 Dr C.R. Dugas, Paris. 68 Dr Meltzer. 69 Dr W. Ehrenberg, Birkbeck College, University of London. He and R.E. Siday found in 1949 a precursor of the Aharanov–Bohm effect. 70 Dr P.C. Banbury, University of Reading. 71 Dr Vera Daniel, Electrical Research Association (ERA). 72 Dr Stella Mayne, later Rymer, University of Reading. 77 Dr W.W. Tyler. 87 Dr Ruth Warminsky, later Broser, Berlin. 88 H.K. Henisch, University of Reading. Later Professor at Pennsylvania State University. 89 Dr N. Mostovetch, Paris. 90 Dr O. Klemperer. 96 Dr H. Labhart. Later Professor of Physical Chemistry, University of Zürich (1964–77). †1977. 97 Dr Tatjana Kousmine, University of Lausanne. 99 Dr J.B. Gunn (of the Gunn effect), Elliott Bros. Ltd (1948–53). Later Professor at the University of British Columbia (1956–71), IBM (1971). 101 Dr J.R. Drabble, GEC Research Laboratories. Later Professor of Physics at the University of Reading. Later Dr A. Hogarth. 106 E.W.J. Mitchell, University of Bristol. Later Professor at the Universities of Reading (1961–78) and Oxford (1978–88), Chairman SERC (1985–1990). FRS (1986). 108 Dr Torgesen, Norway. 109 Dr C.G. Kuper, later Prof. Technion, Haifa. 112 Donald Avery, RRE. Later Atomic Energy Establishment Director, Sellafield? 114 Dr K.W. Plessner, AEI Laboratories, Aldermaston. Later at Dielectrics Ltd. 119 Dr L. Pincherle, RRE. Later Reader Bedford College. 120 J.W. Granville, University of Reading. Later at RRE. 121 Dr E.H. Putley, RRE. 122 Dr B. Vodar, Paris? 123 Dr R.P. Chasmar, RRE. Later AEI? 124 Dr George G. McFarlane, TRE 1941–60. Director RRE 1962–7, Kt. 1971, Controller R&D MOD 1971–5, Board member British Telecom 1981–4, Corporate Director British Telecom 1984–7. 129 D.A. Wright, GEC Laboratories. Later Professor at Durham University. 134 Prof. F.H. Stieltjes, Philips Research Laboratories. †1986. 140 Dr S.E. Mayer, Standard Telecommunications Research Laboratories, London. †1986(?). 144 Dr Audrey Jones, University of Reading. Later Dr A. Hogarth. 149 H. Fritzsche, Purdue University (1954–63). Professor at Chicago University (1963). 150 Jacques Friedal, Ecole des Mines, Paris. Later Professor at the Univeristy of Paris-Sud (Orsay), Academie de France 1977. 151 Dr John Hirsch, Birkbeck College, London. 152 A.H. Gebbie, National Physical Laboratories. 158 Dr Fred Ansbacher, University of Aberdeen. 159 Dr George A.P. Wyllie, University of Glasgow. 160 Dr Charles W. McCombie, University of Aberdeen. Later Professor of Physics at the University of Reading. 161 Dr M.J.O. Strutt, Philips Research Laboratories. Professor at E.T.H. Zürich (1948–74). 163 Dr A. Fairweather, Post Office Research Laboratories. 165 Dr M. Wise, Philips Research Laboratories. 167 Dr P.W. Haayman, Philips Research Laboratories. 168 Dr W. Morton Jones, AEI Aldermaston (1949–61), Culham (1962), Saclay (1963–5), University of Strathclyde (1966+). 169 Dr J. Volger, Philips Research Laboratories. †1984. 171 Dr J. Ewels, University of Purdue. 172 Dr H. Krebs, Germany. 173 Dr W. Meyer, Philips Research Laboratories, of the Meyer-Neldel rule. 175 Dr R.W.H. Stevenson, University of Aberdeen. 183 Dr H.J. Vink, Philips Research Laboratories. 190 Dr Philip Rhodes, University of Leeds. 198 Dr Richard Mansfield, Bedford College. Later at Brunel. 202 Dr R.W. Sillars, AEI Laboratories, Trafford Park, Manchester. 206 Dr Peter Myers, University of Aberdeen. Later Professor at the Chalmers University of Technology, Göteburg, Sweden.

Recombination in semiconductors

PETER T. LANDSBERG

University of Southampton, UK

The right of the
University of Cambridge
to print and sell
all manner of books
was granted by
Henry VIII in 1534.
The University has printed
and published continuously
since 1584.

Cambridge University Press

Cambridge

New York   Port Chester

Melbourne   Sydney

PUBLISHED BY THE PRESS SYNDICATE OF THE UNIVERSITY OF CAMBRIDGE
The Pitt Building, Trumpington Street, Cambridge, United Kingdom

CAMBRIDGE UNIVERSITY PRESS
The Edinburgh Building, Cambridge CB2 2RU, UK
40 West 20th Street, New York NY 10011–4211, USA
477 Williamstown Road, Port Melbourne, VIC 3207, Australia
Ruiz de Alarcón 13, 28014 Madrid, Spain
Dock House, The Waterfront, Cape Town 8001, South Africa

http://www.cambridge.org

First published 1991
First paperback edition 2003

A catalogue record for this book is available from the British Library

Library of Congress cataloguing in publication data

Landsberg, Peter Theodore.
Recombination in semiconductors/P.T. Landsberg.
p. cm.
ISBN 0 521 36122 2 hardback
1. Semiconductors—Recombination. I. Title.
QC611.6.R43L36 1991
537.6′226—dc20 90-45623 CIP

ISBN 0 521 36122 2 hardback
ISBN 0 521 54343 6 paperback

Contents

Main symbols

* References are to equations unless otherwise stated.

$E_t \equiv (m^*/m\varepsilon^2) t^{-2}(me^4/2\hbar^2)$ hydrogen-like ionization energy from orbit with principal quantum number t (1.8.4), $me^4/2\hbar^2 = 13.6$ eV; $E_{At} = \hbar^2/2m_A \hat{r}_{At}^2$ acceptor ionization energy (5.2.19); $E_t = \hbar^2 t^2/2mr_t^2$ corresponding result for Bohr atom (5.2.20)

eq suffix for equilibrium conditions; suffix 0 is also used

f_c, f_v activity coefficients (1.11.4)

$f(E)$ Fermi–Dirac distribution function, section 1.6.3

F Helmholtz free energy (1.1.15)

$F = B^s + B_1 n + B_2 p$ combination of band–band recombination parameters (2.2.11)

F overlap integral (3.4.15)

F_e, F_h reduced quasi-Fermi levels (i.e. divided by kT) (1.12.25), (3.5.19); γ_e, γ_h are also sometimes used

$F_s(\gamma)$ Fermi integral (1.6.4)

g degeneracy due to spin (1.3.9)

g momentum transfer, section 3.8.2

g_n degeneracy of the ground state of an n-electron defect, section 1.7.1, (1.7.15)

G Gibbs free energy (1.1.10)

$G = T_1^s + T_1 n + T_2 p = c_n$ band–trap recombination parameters (2.2.11), (2.4.22)

G_e, G_n, G_v dimensionless superposition function (3.4.24), (3.4.27), (5.2.1); Fourier coefficient of hydrogen-like function, e.g. (5.2.21)

h momentum transfer, section 3.8.2

H enthalpy (1.1.15), (1.7.33)

$H = T_2^s + T_3 n + T_4 p = c_p$ band–trap recombination parameters (2.2.11), (2.4.23)

H Hamiltonian operator, sections 3.1, 6.2

$I_0 = me^4/2\hbar^2 \simeq 13.6$ eV ionization energy of the hydrogen atom (2.6.34), (3.7.17), (5.2.71)

j current density, section 1.2.2, (2.5.38)

$\mathbf{k}$ wavevector (1.3.9)

$\mathbf{K}$ vectors in $\mathbf{k}$-space, possibly extending beyond the first zone (3.6.24)

K_n second modified Bessel function (1.6.23)

K_D equilibrium constant (1.12.15)

l number of directions of polarization (4.4.13)

l_{Deb} Debye screening radius (1.8.22), (5.4.3)

l_j integers (1.4.1)

$\mathbf{L}$ vector between lattice points in $\mathbf{k}$-space (3.4.9)

L_n diffusion length of electrons (2.2.24)

m free electron rest mass

$m_j, m_\parallel, m_\perp^*, m_c, m_v, m_s, m_1, m_h, m^*$ effective masses (1.4.2), (1.4.9), (1.4.11), (1.4.12), (1.8.2); m_1 and m_h refer to a light-hole and a heavy-hole band, respectively

$\left(\dfrac{1}{m}\right)_{ij}$ reciprocal effective mass tensor (1.5.6)

M maximum number of detachable electrons per defect (1.7.1)

M_D, M_E general matrix element for direct and exchange transitions (3.3.21)

$M_s (s = 0, 1, \ldots)$ critical points in **k**-space, section 1.5.1, Table 3.7.1

n electron concentration (1.6.3)

n_0 or n_{eq} thermal equilibrium concentration, p. 104

$n(r - \frac{1}{2})$, $n_1 = n(\frac{1}{2})$ special electron concentration, section 2.2.4

$n_g = e_{ng}/c_{ng}$, $n_e = e_{ne}/c_{ne}$ (2.4.16)

n_i intrinsic concentration (1.12.10)

n_θ intrinsic concentration away from equilibrium (2.2.13)

N number of particles (1.1.1); concentration of defects (1.8.27)

N photon number (4.2.3)

$\mathcal{N}$ density of states of dimension (energy)$^{-1}$ (1.3.12), (1.4.4)

N_c 'effective' number of states per unit volume (1.6.7)

N_D, N_A number of donors, acceptors (1.7.24) (1.11.8) normally per unit volume

N_e mean number of trapped electrons (1.7.5), (1.7.17)

N_t number of traps (1.7.24)

N_v 'effective' number of states per unit volume (1.6.9)

$N(\boldsymbol{\mu}, 1, 1')$ supplementary matrix element (3.4.30), (5.3.4)

0 suffix for equilibrium conditions, also denoted by 'eq'

O_{12} overlap integral of vibrational wavefunctions (6.4.20)

p pressure (1.1.1)

p concentration of holes (1.6.9)

p_0 or p_{eq} thermal equilibrium concentration, p. 104

$\mathbf{p}$ momentum, section 1.3

$p(r - \frac{1}{2})$, $p_1 = p(\frac{1}{2})$ special hole concentration, section 2.2.4

$p_g = e_p/c_p$ (2.4.17)

p_I probability of occupation of state I, section 1.11

P probability (1.2.1), (1.6.1)

P transition rate per unit volume (3.7.4), (4.3.15)

q general charge (1.1.2)

q normal coordinates (6.2.5)

q magnitude of phonon wavevector (4.5.15)

q_J probability of vacancy of state J, section 1.11

Q constant pressure partition function, section 1.7.3

Q general charge (1.8.5)

Q configuration coordinates (6.2.12)

$\mathbf{r}$ position vector, section 1.2.2

$r_{ij}^{sp}(E)$, $r_{ij}^{st}(E)$ rates of spontaneous and stimulated emission per unit volume per unit energy range (4.4.15)

$\tilde{r}_t \equiv (\varepsilon m/m^*) r_1 t^2$ radius of Bohr orbit of principal quantum number t in semiconductor (1.8.2), (4.5.18), (5.2.18)

$r_t = t^2 r_1$ radius of Bohr orbit of principal quantum number t, $r_1 \equiv \hbar^2/me^2 = 0.528$ Å (1.8.2), (3.7.17), (4.5.18)

R recombination rate, usually per unit volume (2.1.1)

s density of states exponent (1.3.2)

s velocity of sound, section 2.6.1

S entropy (1.1.1)

S Huang–Rhys factor (6.2.3)

S_{IJ} transition probability per unit time (2.2.1)

t principal quantum number (1.8.2)

$t_0 = \hbar^3/e^4 m \sim 2.42 \times 10^{-17}$ s, section 3.7.1, (3.7.17), (5.2.71)

T absolute temperature (1.1.1)

T_{el}, T_l electron and lattice kinetic energy (6.2.4), (6.2.8)

T_1^s, T_2^s, T_1, T_2, T_3, T_4 band–trap recombination coefficients, Fig. 2.1.1, p. 103

$T(E)$ transmission coefficient (2.5.35)

T_{if} probability of a transition $i \to f$ in time t (3.7.1)

u recombination rate per unit volume (2.2.1)

u ion displacement (6.2.24)

$u_a(\mathbf{k}, \mathbf{r})$ dimensionless modulating part of Bloch function (3.4.9)

$u_{cr-\frac{1}{2}}$, $u_{r-\frac{1}{2}v}$ recombination rate into or out of traps (2.2.12b,c)

u_{cv} rate between bands (2.2.9)

U mean energy (1.1.1)

U_{el}, U_l electron and lattice potential energy (6.2.5), (6.2.7)

v mobility (1.2.10)

v saturation velocity $(E_0/m_0^*)^{\frac{1}{2}}$ in a nonparabolic band (1.6.31)

v thermal velocity of electrons

$v_a(\mathbf{k}, \mathbf{L})$ dimensionless Fourier coefficient of modulating part of Bloch function (3.4.10), (3.4.16), (3.4.22)

V volume (1.1.1)

V electronic potential energy, section 2.5.3

W combinatorial factor (1.7.11), (1.9.7)

W transition rate (6.3.21)

x coordinate (1.4.1)

X entropy factor, $\exp(\Delta S/k)$ (1.7.34)

X^s, X_1, X_2 band–trap ionization coefficients, Fig. 2.1.1

$X_{IJ} \equiv S_{IJ} p_I q_J / S_{JI} p_J q_I$ (2.2.2a,b)

Y^s, Y_1, Y_2 band–band ionization coefficients, Fig. 2.1.1

Z canonical partition function (1.2.9)

Z_d canonical partition function for a defect (1.12.34)

Z_D canonical partition function for a donor (1.12.37)

Z_s canonical partition function for an s-electron defect (1.7.1)

Greek

α absorption coefficient (4.5.1)

$\alpha_0 = e^2/\hbar c = 1/137$ (5.2.71), (5.4.6)

β covers uncertainty in value of like-spin matrix element (3.3.27)

$\gamma \equiv (q/|q|)(E_j - \mu)/kT$ (1.2.16), also μ/kT (1.7.24)

γ_e, γ_h reduced quasi-Fermi levels $\mu_{e,h}/kT$ (1.11.1); F_e, F_h is also used (3.5.19). γ_n and
γ_p are used in section 2.4 for γ_e and γ_h, respectively, to release γ_e for the
excited state

Γ gamma function (1.3.4)

Γ width of an energy level (6.5.4)

δn, δp electron and hole concentration departures from equilibrium (2.1.2)

Δ split-off band gap, Fig. 1.6.3(a)

$\Delta = \Delta_c + \Delta_v$ (1.11.5a)

Δ_c, Δ_v band edge shifts due to band-gap narrowing (1.11.5)

Δ_{eff} effective band-gap narrowing (1.12.23)

ε particle kinetic energy (1.2.15)

ε dielectric constant (1.8.2)

$\varepsilon(\mathbf{k}, \omega)$ dielectric function (1.8.10)

$\zeta(x) \equiv \exp[e\varphi(x)/kT]$ (2.7.10)

η reduced energy E/kT (1.7.23), (1.7.28)

η quantum efficiency (2.1.10)

η spin wavefunction (3.3.1)

$(\eta_D, \eta_A) \equiv (E_D/kT, E_A/kT)$

κ wavevector in tunneling region (6.4.1)

$\lambda \equiv \exp(\mu/kT)$ (1.7.1)

λ mean free path (2.5.17), (2.6.1)

λ screening parameter $[L^{-1}]$ of the Coulomb potential (3.4.19)

λ subscript specifying polarization (2.6.15), (4.3.17)

Λ^{-1} Debye screening radius (1.8.7), (1.8.22) ($\equiv l_{Deb}$)

μ electrochemical potential (1.1.3), Fermi level

μ refractive index (4.3.32)

μ m_c/m_v (3.8.6)

μ_0 chemical potential (1.1.1)

$\mu_e = \gamma_e kT$, $\mu_h = \gamma_h kT$ quasi-electrochemical potentials (1.11.1)

μ_i chemical potential for intrinsic sample (1.12.11)

v general volume *concentration* of particles or defects (1.2.10), (1.8.16)

v photon frequency (4.2.9)

v_r sometimes the *number* of r-electron defects (1.7.4)

ξ variance of random potential (5.2.36)

Ξ grand partition function (1.2.2)

$\rho(\mathbf{r}, t)$ volume density of particles (1.8.5)

σ (recombination coefficient in cm^3 s^{-1}/thermal velocity) cross section (2.3.31)

$\sigma_i = \pm 1$ parameters for energy surfaces (1.5.10)

σ_{ix}, σ_{iy}, σ_{iz} Pauli spin operators of particle i (3.3.1)

τ_n, τ_p lifetime of electrons or holes (2.1.4)

τ_r radiative life time, Table 4.1.1

φ electrostatic potential (1.1.2)

φ wavefunction (3.1.7)

$\varphi_{\mathbf{k}}$ Fourier coefficient of electrostatic potential (1.8.8)

$\chi = e_n(\mathscr{E})/e_n(\mathscr{E} = 0)$ for ground and excited state (2.5.6), (2.5.23)

$\chi_i(Q)$ vibrational wavefunction (6.4.1)

ψ wavefunction (electrons) (1.4.1), section 3.1

ω_p angular frequency of plasma oscillations (1.8.26)

$\omega(k)$ normal mode frequencies (6.2.2), (6.2.11)

Ω volume of unit cell, Appendix B

Note on units

Some important books on solid-state physics use Gaussian units, and these units have also been adopted in this book in preference to SI units. Thus the dielectric constant ε is a number just like the relative dielectric constant ε_r in the SI system, in which ε_0 is the permittivity of the vacuum and is given by 8.854×10^{-12} farads per meter.

Some examples in terms of the two systems of units are given in the table.

	Gaussian units	SI units
Poisson's equation for the electrostatic potential φ in terms of the charge density ρ	$\varphi'' = -\dfrac{4\pi}{\varepsilon}\rho$	$\varphi'' = -\dfrac{\rho}{\varepsilon_0 \varepsilon_r}$
Coulomb's law for the potential energy of two charges q_1, q_2 a distance r apart	$\dfrac{q_1 q_2}{\varepsilon r}$	$\dfrac{q_1 q_2}{4\pi\varepsilon_0 \varepsilon_r r}$
Poynting's vector when the displacement is proportional to the electric field	$\dfrac{\varepsilon\mathscr{E}^2}{8\pi}$	$\dfrac{\varepsilon_0 \varepsilon_r \mathscr{E}^2}{2}$
For vacuum	$\varepsilon = 1$	$\varepsilon_r = 1$

Atomic units are employed occasionally (see pp. 280, 404):

$$a_0 \equiv \frac{\hbar^2}{e^2 m} = 0.529 \times 10^{-8} \text{ cm}$$

$$I_0 \equiv \frac{\hbar^2}{2ma_0^2} = \frac{me^4}{2\hbar^2} = 13.6 \text{ eV}$$

$$t_0 = \frac{\hbar^3}{e^4 m} = 2.419 \times 10^{-17} \text{ s}$$

$\alpha_0 \equiv e^2/\hbar c = 1/137 = $ (fine-structure constant)$^{-1}$

Atomic units for recombination:

$(\alpha_0 a_0/t_0)^3 = 2.4 \times 10^{-15} \text{ cm}^3 \text{ s}^{-1}$, $a_0^6/t_0 = 0.91 \times 10^{-33} \text{ cm}^6 \text{ s}^{-1}$

Note also that the Ångstrom unit is used: $1\text{Å} = 10^{-10}$ m.

Introduction

A systematic interest in recombination in semiconductors dates roughly from 1950 and gave rise for example to the Shockley–Read–Hall statistics in 1952 and the application of detailed balance to radiative processes in semiconductors in 1954. But our story really started with quantum mechanics and its application to solids. In contrast to its junior cousin, the black hole, the hole of semiconductor physics was first seen experimentally (in the anomalous Hall effect) and quantum mechanics was used to elucidate it [1]. Quantum mechanics was also used later to propose the band model of a semiconductor [2]. Actually, the copper oxide plate rectifier had been a useful solid-state device since the early days of quantum mechanics in the 1920s, but its action came to be understood only just before the war using electrons, holes and the band model [3]. This work has been reviewed for example by Mott and Gurney [4] and by Henisch [5]. Solid-state electronics was already in the air then, and rudimentary solid-state amplification had been proposed by Lilienfeld [6] in the late 1920s and established by Hilsch and Pohl [7] in 1938, using potassium bromide crystals. A useful semiconductor 'triode' was clearly ready to be born in 1938/9. But the war intervened. Still, solid-state detection was now important for radar, and programs to study silicon and germanium were initiated partly with government funding, notably at Purdue University under K. Lark-Horovitz. At the end of the war one could claim that what Shockley called 'one-current theories' of metal-semiconductor rectification [8] were in reasonable shape, apart from a little tinkering here and there, for example, by the present author (see [5]). Of course, the study of surfaces was then, and continues to be, a very active area of research.

The two-current theories, and with them recombination, came a few years after the war when in 1947 Bardeen and Brattain discovered the point contact transistor along with minority carrier injection [9]. Electron–hole pair generation was also studied and was rather like another cousin, this time a senior one, namely electron–positron pair creation. (This requires energy $2m_0 c$ which corresponds in the solid state to $E_G = 2E_0 = 2m_0^* v^2$, where E_G is the energy gap, E_0 is the nonparabolicity parameter (1.6.16) and v is a saturation velocity (1.6.31) which indeed corresponds to c.) The junction transistor followed, and was discussed by Shockley in a post-deadline paper at an international conference held in Reading, England, in July 1950 [10]. (It did not appear in the proceedings [11].)

Other developments followed, as is well known: the onward march of silicon, many new devices, including solar cells which enabled space exploration to proceed, GaAs and other III–V compounds, integrated circuits, microminiaturization, ever faster and more compact computers, etc. With vacuum microelectronics the wheel may come full circle. The story is well outlined elsewhere ([12]–[14]).

If we fasten our attention on the year 1950 for the systematic beginning of our subject, we can think of the Reading Conference of July 1950 as the appropriate event. Indeed, seven of the people mentioned above were present, as is clear from the group photograph given as the Frontispiece. (Additions and corrections to the identifications are needed and will be welcomed.) It is a historical document, particularly since the Reading Conference came to be regarded as the first of a series of international conferences on the physics of semiconductors which runs as follows: (2) Amsterdam, 1954; (3) Garmisch, 1956; (4) Rochester, 1958; (5) Prague, 1960; (6) Exeter, 1962; (7) Paris, 1964; (8) Kyoto, 1966; (9) Moscow, 1968; (10) Cambridge USA, 1970; (11) Warsaw, 1972; (12) Stuttgart, 1974; (13) Rome, 1976; (14) Edinburgh, 1978; (15) Kyoto, 1980; (16) Montpellier, 1982; (17) San Francisco, 1984; (18) Stockholm, 1986; (19) Warsaw, 1988; (20) Thessaloniki, 1990. If someone exists who attended them all, he should be given a medal for devotion to the subject, longevity, and willingness to travel.

This book, then, is devoted to the main aspects of the physics of recombination in semiconductors, omitting related topics that are well covered elsewhere, such as band theory, resonances, details of phonon effects, and amorphous systems. The concepts are introduced so that graduate students who are beginning research can follow the argument. Some things may *look* complicated, but they *are* explained, so that some of the work (chapter 1, sections 2.1 to 2.3, and the beginnings of chapters 3 and 4) can be studied already in an undergraduate course. My idea was to make the book almost completely self-contained with an emphasis on general principles. This should enable a reader to make new applications while it will also detach the book a little from the precise contemporary state of research, so that, with luck, it will not go out of date too quickly. In a first study of the book the portions in small print can be omitted. A short guide to the book is given at the end of this Introduction.

It will be seen that an attempt to cover the main topics in the recombination area has been made. However, not all relevant subjects could be discussed, and I regret the disappointment this may cause some readers. In general, these omissions are due to my lack of competence in these areas, coupled with the need to keep the book from becoming too long. The knowledgeable reader will, perhaps, be recompensed by finding relatively new material already early in the book and in unexpected places.

Imbedded in chapters 1 and 2 is the material for a possible slim volume called *The Partition Function Approach to Semiconductor Kinetics*. For, on looking through this book, it became clear to me that this rather useful tool, largely developed and used by my collaborators and myself over the last decades, is not as widely known as it should be. Perhaps it has not been popularized with sufficient skill. Hence, the possibility of the above mentioned slim volume. This is no 'plug' for an (unwritten) book. On the contrary, the careful reader of the present book will already know all there is to be taught by me, and therefore will not need to have it, should it ever materialize.

I have tried to give an indication of current thought with regard to most topics, but at the same time I wanted to preserve a sense of the historical development of our subject. For example the paper [1.8.10] by Debye and Conwell (1954) is not often cited now, but it was very influential at the time. I have also cited papers which, while good and relevant, have had a rather low citation count, thus helping to 'save' them for possible future work – [5.2.57] is an example. Thus many papers are cited. This is done by number (for example [1.8.10] is reference 10 in section 1.8), for which I apologize, but it has saved much space, as many papers have three authors and some have more, and it has helped to keep the continuity of argument.

For help with the identification of the persons in the frontispiece I thank Dr P.C. Banbury (Reading), Dr R.N. Bloomer (Locksheath, Southampton), Prof. G. Busch (Zürich), Prof. H.K. Henisch (State College, Pa.), Sir George MacFarlane (Esher, UK), Dr T.S. Moss (Malvern, UK), and Dr A.J. Vink (V.D. Waalre, The Netherlands).

It is a pleasure to thank several colleagues for comments on part of the manuscript, notably Dr R.A. Abram (Durham), Prof. M. Jaros (Newcastle), Dr Jerzy M. Langer (Warsaw), Dr T. Markvart (Southampton), Dr D. J. Robbins, Dr R.I. Taylor (Plessey Research Laboratories)* and Leendert Verhoef (Amsterdam). Dr Markvart kindly wrote chapter 6, and Dr Taylor wrote chapter 7, and I am indebted to them for these important contributions. R.I. Taylor acknowledges tenure of an SERC CASE studentship in collaboration with British Telecom

* Now GEC-Marconi Materials Technology Ltd.

A short guide to the contents of this book

References are to sections unless stated otherwise.

	Band–band	Band–trap
Statistics	Statistics involving recombination coefficients: Figs 2.1.1, 2.1.2	
	Statistics for combining recombination channels: 2.2.4, 2.3.7	
	Surface recombination and grain boundary barrier heights: 2.7	
	Recombination at dislocations: 2.8	
	Recombination–generation induced phase transitions and chaos: 4.6	
	Recombination-enhanced reactions 5.4.4	
	Reaction constants: 1.11.1	Reaction constants: 1.11.2
		Statistics for trap spectra: 2.3.6
	Quantum efficiency in an intrinsic semiconductor: 2.1.3	Cascade recombination kinetics: 2.4.2
		Capture and emission coefficients: 2.4.6
		Their field dependence: 2.5.3, 2.5.4
		Thermionic and field emission: 2.5.5
		Geminate recombination: 2.5.6
		Cascade capture: 2.6
Radiative processes	Radiative emission: 4.4, 4.5	Radiative recombination: 5.2.1
	Radiative absorption: 4.5	Photoionization: 5.2.4
		Donor–acceptor radiative transition: 5.2.6
		Excitons in radiative recombination: 5.2.7
Nonradiative processes	Impact ionization thresholds: 3.5	Impact ionization and Auger recombination: 5.3
	Impact ionization and Auger probabilities: 3.7	Auger quenching: 5.3.4
	Auger lifetimes: 3.8	More involved Auger effects: 5.4
		Multiphonon effects: chapter 6
		Tunneling: 2.5.5, 2.5.6
Mixture of above	Recombination in low-dimensional structures (chapter 7)	

Research Laboratories during the course of part of the work on chapter 7. R.I. Taylor would also like to thank Drs R.A. Abram, M.G. Burt and C. Smith for useful discussions. Thanks are also due to D. Harding (Southampton) for his help with the index, and to Irene Pizzie for her careful reading of the manuscript.

I am also indebted to the Leverhulme Trust for the award of an emeritus fellowship; to the Department of Electrical Engineering of the University of Florida in Gainesville for the opportunity to give some of this material in graduate courses; and to my home University of Southampton for providing facilities.

It is common for an author to thank his family for suffering his absences during the writing of his masterpiece. The present case is slightly different. It is my family who are thanking me for keeping away from under their feet by undertaking the task, the result of which I have now pleasure to present.

I

Semiconductor statistics

1.1 The electrochemical potential

The movement of electrons and holes in space and their quantum transitions between states govern the behavior of semiconductor devices. The understanding of these processes involves thermodynamic and statistical concepts, and these are introduced in this chapter.

In the following sections we shall denote by μ the electrochemical potential for charged particles. In this section, however, it is convenient to introduce also the chemical potential μ_0 of a component in a system. It will be recalled from thermodynamics that μ_0 plays the same role as regards equilibrium with respect to particle diffusion or chemical equilibrium, as temperature, T, does with respect to heat conduction or thermal equilibrium. That T, μ_0 and also pressure p are on the same footing may be seen from the Gibbs equation for the mean energy U of a system of N neutral particles, which is

$$dU = T\,dS - p\,dV + \mu_0\,dN \tag{1.1.1}$$

(where S, V are entropy and volume, respectively). For a system of particles of charge q (positive or negative) a change in the number of particles leads to an *additional* energy term $q\varphi\,dN$, where $\varphi(\mathbf{r})$ is the electrostatic potential at the position $\mathbf{r}$ considered. Hence the analogue of eq. (1.1.1) for a system of charged particles is

$$dU = T\,dS - p\,dV + (\mu_0 + q\varphi)\,dN \tag{1.1.2}$$

Many thermodynamic functions need to be amended by virtue of this change, and μ_0 has to be replaced by the electrochemical potential

$$\mu = \mu_0 + q\varphi \tag{1.1.3}$$

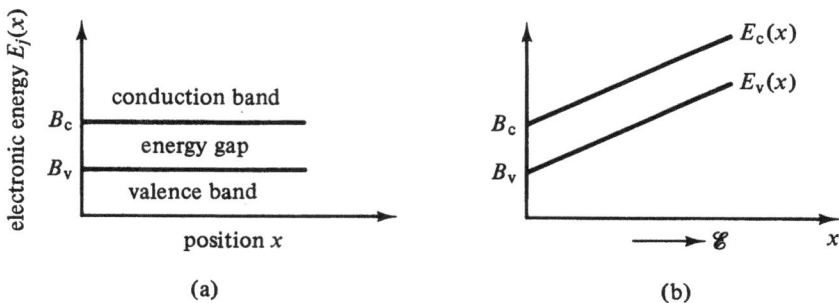

Fig. 1.1.1. Semiconductor bands (a) in the absence of an electric field, (b) in the presence of a constant electric field acting to the right. Typically $\varphi(x) = A - \mathscr{E}x$, $E_j(x) = B_j - |q| A + |q| \mathscr{E}x$. A has been taken to be zero.

Key uses of μ include (i) its occurrence in the expression for the mean occupation number in thermal equilibrium at temperature T of particles in a quantum state j (say) at energy e_j

$$\bar{N}_j = \frac{1}{\exp\left[(e_j - \mu)/kT\right] + 1} \sim \exp\frac{\mu - e_j}{kT} \tag{1.1.4}$$

The last expression holds if $\bar{N}_j \ll 1$, i.e. when μ has a large negative value. This result is derived as eq. (1.2.8), below. (ii) Just as a temperature gradient is proportional to the force which drives a thermal current, so a gradient in electrochemical potential is proportional to the force which drives an electrical current. This is discussed in section 1.2.2, below.

For complete thermodynamic equilibrium then, we must have thermal equilibrium, mechanical equilibrium and chemical equilibrium. Thermal equilibrium requires that the temperature must be the same throughout, while for mechanical equilibrium the pressure must be the same throughout the system. If a component in the system is to be stable to diffusion from one part to another, it may be shown from the condition of chemical equilibrium $\Sigma\mu_i \, dN_i \leqslant 0$ that the electrochemical potential must have the same value throughout the system:

$$\mu_j(\mathbf{r}_1) = \mu_j(\mathbf{r}_2) = \mu_{0j}(\mathbf{r}) + q_j\, \varphi(\mathbf{r}) \quad (j \text{ labels a component}) \tag{1.1.5}$$

The above relations for complete thermodynamic equilibrium may be derived by considering the equilibrium between two parts of an isolated system and using the condition that according to the second law of thermodynamics $dS \geqslant 0$ for any process occurring in such a system. If eq. (1.1.5) is not valid, the system is not in equilibrium and particles will tend to move to regions of lower electrochemical potential. The process terminates when μ has the same value throughout. In applying eq. (1.1.5) we are considering volumes of the system which are small enough for μ and φ to have roughly constant values. Yet these volumes must be large enough for thermodynamics to be applicable to them, so that μ, φ and T can be defined. This rules out very large gradients in these quantities since eq. (1.1.5) would then become invalid [1.1.1].

We now investigate the effect of the electrostatic potential on the energies of the electronic states. Let B_c and B_v be the energies of the conduction and valence band edges in the absence of an electric field $\mathscr{E} = -\,\mathrm{grad}\,\varphi(x)$ (Fig. 1.1.1). Since these are electronic energies both B_c and B_v are shifted by $-|q|\,\varphi$ to

$$E_c(\mathbf{r}) = B_c - |q|\,\varphi(\mathbf{r}), \qquad E_v(\mathbf{r}) = B_v - |q|\,\varphi(\mathbf{r}) \tag{1.1.6}$$

The energy gap $E_g = E_c(\mathbf{r}) - E_v(\mathbf{r})$ is normally constant in space. Thus we can write generally with $j = c$ or v

$$E_j(\mathbf{r}) = B_j - |q|\,\varphi(\mathbf{r}) \tag{1.1.7}$$

From eqs. (1.1.5) and (1.1.6) one sees that [1.1.2]:

$$-|q|\,\varphi(\mathbf{r}) = E_j(\mathbf{r}) - B_j = \mu_j(\mathbf{r}) - \mu_{0j}(\mathbf{r}) \tag{1.1.8}$$

In thermal equilibrium $\mu_j(\mathbf{r})$ is independent of $\mathbf{r}$, and E_j and $-\mu_{0j}$ vary linearly with φ. If in addition $\nabla\varphi(\mathbf{r}) = 0$, then these quantities are constant in space.

In semiconductor work we deal with electronic states in the conduction band and vacancies in electronic states in the valence bands. At equilibrium, both of these can be treated in terms of a single electrochemical potential μ (cf. sections 1.6 and 1.11). This chemical potential pertains to electronic states so that q is negative. Consequently the change of μ with φ is given by

$$\mu = \mu_0 - |q|\,\varphi \tag{1.1.9}$$

An alternative approach, which we shall not follow in this chapter, is to treat electron vacancies in the valence band as positively charged particles. This interpretation is suggested by the formula describing the distribution of holes (cf. eq. (1.6.9), below). The energy of the holes is measured from the top of the valence band energy E_v, with the added qualification that there is a minus sign: thus a vacancy in a state of energy E corresponds, in this scheme, to a hole of energy $-(E - E_v)$. The corresponding electrochemical potential for the hole in this scheme is $-(\mu - E_v)$. The electrochemical potential of the hole will be shifted by $+|q|\,\varphi$ as a result of an electrostatic potential φ, as is to be expected for a positively charged particle.

A useful way of looking at the electrochemical potential is in terms of the Gibbs free energy

$$G = \mu N = U - TS + pV \tag{1.1.10}$$

so that

$$dG = \mu\,dN + N\,d\mu = dU - T\,dS - S\,dT + p\,dV + V\,dp \tag{1.1.11}$$

Adding eqs. (1.1.1) and (1.1.11) yields the Gibbs–Duhem equation

$$S\,dT - V\,dp + N\,d\mu = 0 \tag{1.1.12}$$

By eqs. (1.1.11) and (1.1.12)

$$dG = \mu\,dN - S\,dT + V\,dp \tag{1.1.13}$$

Thus one sees from eqs. (1.1.2) and (1.1.13) that

$$\mu = \left(\frac{\partial U}{\partial N}\right)_{S,V} = \left(\frac{\partial G}{\partial N}\right)_{p,T} \tag{1.1.14}$$

The second form is particularly suitable since experiments are often performed at constant temperature and pressure. Thus μ is the Gibbs free energy per particle as may be seen from eq. (1.1.10) and, in differential form, from eq. (1.1.14).

The thermodynamic results

$$\mu = (\partial H/\partial N)_{S,p} = (\partial F/\partial N)_{V,T} \tag{1.1.15}$$

and eq. (1.1.14) suggest that under appropriate conditions μ may also be regarded as an energy U or an enthalpy $H = U + pV$ or a Helmholtz free energy $F = U - TS$ per particle. These results will be found useful when entropy factors of impurity levels are considered in section 1.7.3.

1.2 A shorter statistical mechanics

1.2.1 The distribution laws

The discussion which follows utilizes the grand canonical distribution. This applies if a system may be regarded as exchanging both energy and particles with its surroundings. For a gas of identical particles it is therefore assumed to be in contact with two reservoirs. One can provide and take up energy and is large enough to remain at constant temperature under these exchanges; the other provides and takes up molecules and is large enough to remain at constant chemical potential under these exchanges. The walls of the enclosure which are kept at constant temperature and on which molecules can be adsorbed, and then get free again, can be thought of as a physical realization of this concept. For a canonical distribution one withdraws the facility of exchanging particles and their number is then constant. For a microcanonical distribution one withdraws the remaining facility of exchanging energy and the system energy is then a constant.

The quantities T and μ occur automatically if a grand distribution is used. For a canonical distribution, μ has to be introduced by a special device and for a microcanonical distribution both T and μ have to be introduced by a special device. The results are often very similar whichever of these three distributions is used, but this is not always true, as seen in section 7.1. If a grand canonical distribution is used, one expects the energy and the particle number to be variables while the volume V and also μ and T are given.

We consider next how μ enters statistical mechanics, and derive the Fermi distribution function. It is shown in statistical mechanics that for an equilibrium system of given volume V, electrochemical potential μ and temperature T the probability of finding it with N indistinguishable particles and with energy E_i is

$$P(V, \mu, T; N, E_i) = \Xi^{-1} \exp\left[\mu N - E_i\right)/kT] \tag{1.2.1}$$

When there is no chance of confusion, it will sometimes be abbreviated as P below. Here the normalizing factor is the grand partition function

$$\Xi \equiv \sum_{N,i} \exp\left[(\mu N - E_i)/kT\right] \tag{1.2.2}$$

the sum being, for each particle number N, over all states i of the system. Relation (1.2.2) ensures that the probabilities (1.2.1) sum to unity. This is necessary since it is certain that the system is in some state i, corresponding to some number N. A second key formula is that the statistical mechanical analogue of the entropy is

$$S = -k \sum_{N,i} P(V, \mu, T; N, E_i) \ln P(V, \mu, T; N, E_i) \tag{1.2.3}$$

(The notation 'ln' means the natural logarithm.) Since the probabilities lie between 0 and 1, each term in eq. (1.2.3) is positive (or zero).

To gain some familiarity with these equations let us substitute eq. (1.2.1) into eq. (1.2.3) and multiply by T. One finds

$$TS = -kT \sum_{N,i} \left[\left(\frac{\mu N - E_i}{kT} - \ln \Xi\right) P\right] = -\mu \bar{N} + \bar{E} + kT \ln \Xi \tag{1.2.4}$$

where it has been noted that the P's are normalized, and the definition of an average over the distribution P,

$$\bar{x} = \sum_{N,i} x(N, i) P$$

has been used.

Consider now, from the point of view of thermodynamics, a system of N identical particles in thermal equilibrium at pressure p, absolute temperature T, electrochemical potential μ, mean energy U and entropy S. Comparing eq. (1.1.10) with eq. (1.2.4), statistical mechanics and thermodynamics are consistent if we make the following identification between thermodynamic and statistical mechanical quantities:

$$\text{thermodynamics} \quad N \quad U \quad pV \tag{1.2.5}$$

$$\text{statistical mechanics} \quad \bar{N} \quad \bar{E} \quad kT\ln\Xi$$

Thus one sees that statistical mechanics peers more deeply into the fine structure of matter: it arrives at distribution functions such as (1.2.1), while thermodynamics tends to deal with the macroscopic, and therefore average, quantities.

Although the 'particles' we consider are idealized as points, we still envisage collisions between them. For a system of particles which interact only when they collide (i.e. they have *point interactions*) one can consider the different quantum states $j = 1, 2, \ldots$ in which they can be, and label the number in state j by N_j and the energy of the *single-particle* state by e_j; E_i denotes the *system* energies. One can simply sum over all permitted integers $N_1, N_2 \ldots$ etc. instead of summing over all N and all i, since this procedure covers all possible states of the system if the particles involved are indistinguishable. For simplicity we shall consider a system of two single-particle states. Thus eq. (1.2.2) becomes, using only simple algebra,

$$\Xi = \sum_{N_1=0} \sum_{N_2=0} \exp\left[\frac{\mu(N_1+N_2)-e_1 N_1 - e_2 N_2}{kT}\right]$$

$$= \left(\sum_{N_1=0} t_1^{N_1}\right)\left(\sum_{N_2=0} t_2^{N_2}\right) \left[t_j \equiv \exp\frac{\mu-e_j}{kT}, j=1,2\right] \tag{1.2.6}$$

The upper limits in the summations are either unity, so that a quantum state is either empty or full, or infinity, so that a quantum state can accommodate any number of particles. The first case (fermions) applies to electrons, the second case (bosons) applies, for example, to gas molecules and also to radiation (photons).

Returning to eq. (1.2.1), using eq. (1.2.6), the probability can be written in a different way:

$$P(V,\mu,T;N_1,N_2) = \frac{t_1^{N_1} t_2^{N_2}}{\left(\sum_{N_1=0} t_1^{N_1}\right)\left(\sum_{N_2=0} t_2^{N_2}\right)}$$

$$= \frac{t_1^{N_1}}{\sum_{N_1=0} t_1^{N_1}} \frac{t_2^{N_2}}{\sum_{N_2=0} t_2^{N_2}} \tag{1.2.7}$$

This gives the probability that an equilibrium system of particles with point interactions and of given volume, electrochemical potential and temperature is in a state such that N_1 particles are in the first single-particle quantum state, and N_2 in the second.

The mean occupation number for fermions in the quantum state 1 is

$$\bar{N}_1 = \sum_{N_1}\sum_{N_2} N_1 \left[\frac{t_1^{N_1}}{\sum\limits_{N_1=0} t_1^{N_1}}\right]\left[\frac{t_2^{N_2}}{\sum\limits_{N_2=0} t_2^{N_2}}\right]$$

$$= \sum_{N_1=0}^{1} N_1 \frac{t_1^N}{\sum\limits_{l=0}^{1} t_1^l} = \frac{0t_1^0+1t_1}{t_1^0+t_1} = \frac{t_1}{1+t_1} = \frac{1}{\exp[(e_1-\mu/kT]+1}$$

It is not hard to extend the above argument to show that for an arbitrary number of states

$$\bar{N}_j = \frac{1}{\exp[(e_j-\mu)/kT]+1} \quad (j=1,2,\ldots)$$

For bosons we sum to infinity and find similarly

$$\bar{N}_j = \frac{t_j+2t_j^2+\ldots}{1+t_j+t_j^2+\ldots} = \frac{t_j/(1-t_j)^2}{1/(1-t_j)}$$

since the denominator is a geometrical progression, and the numerator is t_j times the derivative with respect to t_j of the denominator. Hence

$$\bar{N}_j = \frac{1}{\exp[(e_j-\mu)/kT]+a} \tag{1.2.8}$$

$a = 1$ for fermions,
$a = -1$ for bosons,
a can be neglected for large negative μ.

These are the famous distribution functions of Fermi–Dirac, Bose–Einstein and Maxwell–Boltzmann. For electrons and holes in semiconductors one requires the first and the third distribution functions. But the second one is also needed when interactions with excitons, photons or phonons are considered.

The above argument, apart from being based on the grand canonical distribution through eqs. (1.2.1) and (1.2.3), uses only the assumption of identical particles with point interaction. All alternative derivations have to rely on spurious assumptions of a mathematical nature. The *method of the most probable distribution* relies on the existence of large numbers so that Stirling's approximation can be used. Also integer variables must be replaced by continuous variables so that differentiations can be performed. The *canonical ensemble* treats N as fixed rather than as an average. Consequently, instead of eq. (1.2.8), one finds the mean occupation number $\bar{N}_{N,j}$ of state j (as discussed also on p. 482),

$$\bar{N}_{N,j} = \frac{1}{(Z_{N+1}\bar{N}_{N+1,j}/Z_N\bar{N}_{N,j})\exp(e_j/kT)+a} \tag{1.2.9}$$

where Z_N is the canonical partition function for a system of N particles. It has recently been shown that the grand canonical and the canonical ensemble give significantly different values for the factor multiplying $\exp(e_j/kT)$ in the limit $T\to 0$ [1.2.1]–[1.2.3]. It is only by an approximation that the first factor in the denominator can be regarded as independent of the quantum state label j. The method of steepest descents, used in connection with the canonical or microcanonical ensemble, also requires the neglect of certain terms. Thus all these methods are *mathematically* less direct. We return to this point at the end of section 1.10.

1.2.2 The Einstein diffusion–mobility ratio

Suppose now that μ_0 depends on position $\mathbf{r}(x,y,z)$ through the $\mathbf{r}$-dependence of a general carrier concentration v and the temperature T. We then have, using (1.1.9), that the electric field is given by

$$\mathscr{E} = -\nabla\varphi = \frac{1}{|q|}\nabla(\mu-\mu_0) = \frac{1}{|q|}\left\{\nabla\mu-\left(\frac{\partial\mu_0}{\partial v}\right)_{T,V}\nabla v-\left(\frac{\partial\mu_0}{\partial T}\right)_{v,T}\nabla T\right\} \tag{1.2.10}$$

Now, the current density j due to the current carriers of concentration v is the sum of a conduction current density $|q|vv\mathscr{E}$ and a diffusion current density of $-qD\nabla v$. Here q is the charge on the current carriers, D is their diffusion coefficient and v is their mobility*. The dimensions are

$$[D] = [\mathrm{L^2T^{-1}}], \quad [v\mathscr{E}] = [\mathrm{LT^{-1}}]$$

$v\mathscr{E}$ being the 'drift velocity' of the carriers in the electric field. Thus

$$[qD/v] = [q^2\mathrm{L^{-1}}] = [\mathrm{ML^2T^{-2}}] \tag{1.2.11}$$

since $[\mathscr{E}] = [q/r^2]$ and so $[q^2] = [\mathrm{ML^{-1}T^{-2}}]$. We shall next identify the energy term (1.2.11). For uniform temperature

$$\mathbf{j} = |q|vv\mathscr{E} - qD\nabla v$$

$$= vv\left[\nabla\mu-\frac{\partial\mu_0}{\partial v}\nabla v\right]-qD\nabla v \tag{1.2.12}$$

---

* We have solved a notational problem by using V, v, v, μ for volume, mobility, carrier concentration, electrochemical potential, respectively. Sometimes μ is used for mobility, v for volume, etc.

We also note that from irreversible thermodynamics

$$j = v v \nabla \mu \qquad (1.2.13)$$

so that the coefficient of ∇v must vanish in eq. (1.2.12). *One need not appeal to eq.* *(1.2.13)* if one observes that *in thermal equilibrium* $\nabla \mu = 0$ from thermodynamics, that $j = 0$ by hypothesis, so that the coefficient of ∇v in eq. (1.2.12) must vanish. The relation obtained may then be expected to be valid also away from, but in the neighborhood of, equilibrium. Hence

$$\frac{|q| D}{v} = -\frac{|q|}{q} \left(\frac{\partial \mu_0}{\partial \ln v} \right)_{T, V}$$

Since μ and μ_0 differ by a quantity ($|q| \varphi$) which can normally be regarded as independent of v, one finally finds

$$\frac{|q| D}{v k T} = -\frac{|q|}{q} \left(\frac{\partial (\mu / kT)}{\partial \ln v} \right)_{T, V} \qquad (1.2.14)$$

Note that for a Maxwell–Boltzmann distribution of electrons or holes of concentrations n and p, respectively, and with A_e, A_h the appropriate functions of temperature and volume

$$v \rightarrow \begin{cases} n = A_e \exp (\mu / kT) \\ p = A_h \exp (-\mu / kT) \end{cases} \quad \text{so that one has} \quad \frac{|q| D}{v k T} = 1$$

The Einstein ratio $|q| D / v$ is therefore just the typical thermal energy kT of a carrier in this simple case. This identifies the energy (1.2.11).

   Relation (1.2.14) is a generalized form of Einstein's diffusion–mobility relation believed to go back to [1.2.4]. A simpler form, $|q| D / vkT = 1$ was first derived by Einstein in the context of Brownian motion (see [1.2.5] and [1.2.6] for historical reviews). For large departures from equilibrium, however, eqs. (1.2.12) to (1.2.14) are modified because of the occurrence of nonlinear terms [1.2.7]; for the case of gases, see [1.2.8].

   A slightly different form of eq. (1.2.14) is sometimes convenient. The total energy of a current carrier of charge q in an electrostatic potential field φ is in generalization of eq. (1.1.7)

$$E = E_j - \frac{q}{|q|} \varepsilon \qquad (1.2.15)$$

on the electron energy scale. Here ε is the kinetic energy which is added for electrons ($q < 0$) and subtracted for holes ($q > 0$). The energy shift $-|q| \varphi$ for both types of carrier was discussed in eq. (1.1.9) and is included in E_j (the band edge for some constant electrostatic potential). In the Fermi–Dirac distribution (1.2.8) the exponent is for electrons

$$E - \mu = E_c + \varepsilon - \mu = \varepsilon - (\mu - E_c)$$

For electron vacancies, or holes, the exponent is obtained from

$$1 - \bar{N} = 1 - \frac{1}{\exp[(E - \mu)/kT] + 1} = \frac{1}{\exp[(\mu - E)/kT] + 1}$$

as discussed more fully in connection with eq. (1.6.8), below. It is therefore

$$\mu - E = \mu - E_v + \varepsilon$$

This leads one to the introduction of a parameter which is relevant to electrons *and* holes:

$$\gamma \equiv \frac{q}{|q|} \frac{E_j - \mu}{kT} \tag{1.2.16}$$

This makes the mean occupation number of a quantum state

$$\left[\exp\left(\frac{\varepsilon}{kT} - \gamma \right) + 1 \right]^{-1} \tag{1.2.17}$$

for both electrons and holes. Insert now eq. (1.2.16) into eq. (1.2.14) to find

$$\frac{|q| D}{v} = kT v \left(\frac{\partial \gamma}{\partial v} \right)_{T, V} \tag{1.2.18}$$

This is another form of the generalized Einstein relation to which we shall return in eq. (1.8.18). The relation also exists in other contexts, for example for thermal currents [1.2.9] and for membranes [1.2.10]. Other examples will be found in section 1.11.3.

1.3 The classical gas theory

In this section we show how one can advance from sums such as those occurring in eqs. (1.2.2) and (1.2.3) to more convenient expressions. This will be done by choosing the simple example of an ideal gas, defined by the equation of state $pV = \bar{N}kT$. The procedure adopted will be seen to lead to just this result. We will thus be encouraged to use the same procedure later for the electron and hole gases in semiconductors. In particular we introduce here the essential concept of *density of states*. The simplest formula, and the derivation of it, is given in eq. (1.3.12). In sections 1.4 and 1.5 the reader will find more advanced discussions to which this section is therefore an introduction.

One can give a simple statistical mechanical theory by assuming the electrochemical potential of such a system to be large and negative, i.e. by putting $a = 0$ in eq. (1.2.8). The average number of particles is then

$$\bar{N} = \sum_i \bar{N}_i = \exp(\mu/kT) \sum_i \exp(-e_i/kT) \tag{1.3.1}$$

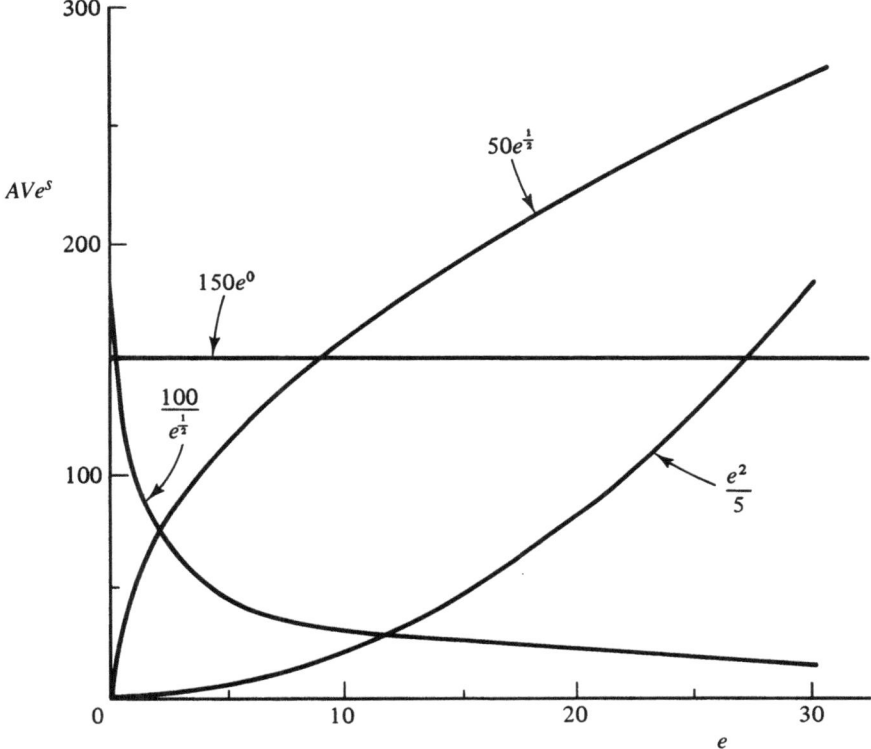

Fig. 1.3.1. Density-of-states curves for $s = -\frac{1}{2}, 0, \frac{1}{2}$ and 2.

Suppose that the energies e_i are distributed so that, with A and s constants (whose values are found in eqs. (1.3.13) below),

$$AVe^s\,de \text{ quantum states lie in single-particle energy range } (e, e+de)\quad(1.3.2)$$

For electron motion in d dimensions one finds $s = d/2 - 1$. So $s = -\frac{1}{2}, 0, \frac{1}{2}$ for one, two and three dimensions. This includes the case of quantum wires ($d = 1$), quantum wells ($d = 2$), and normal motion. The shapes of the density of states curves are illustrated in Fig. 1.3.1. For radiation in d dimensions $s = d-1$ and the normal case $s = 2$ which arises is also illustrated.

Replacing the sum (1.3.1) by an integral:

$$\bar{N} = AV\exp(\mu/kT)\int_0^\infty e^s\exp(-e/kT)\,de$$

$$= AV(kT)^{s+1}\exp(\mu/kT)\int_0^\infty x^s\exp(-x)\,dx$$

$$= AV(kT)^{s+1}\,\Gamma(s+1)\exp(\mu/kT) \qquad (1.3.3)$$

Here $x \equiv e/kT$ is a dimensionless variable and the definition of the gamma function

$$\Gamma(s+1) \equiv \int_0^\infty x^s \exp(-x)\,dx \quad (s > -1)$$

has been used. It is shown in books on the calculus that:

(i) $\Gamma(s+1) = s\Gamma(s)$ (1.3.4)

(ii) $\Gamma(s+1) = s![\equiv 1\cdot 2\cdot\ldots\cdot(s-1)\cdot s]$ for positive integers

(iii) $\Gamma(1/2) = \sqrt{\pi}$ (1.3.5)

In order to understand the significance of large and negative μ rewrite eq. (1.3.3) in the form

$$\frac{\mu}{kT} = \ln\frac{\bar{N}}{AV(kT)^{s+1}\Gamma(s+1)}$$ (1.3.6)

where eq. (1.2.5) has been used. Thus $\bar{N}/V$ must be small, i.e. the gas must be sufficiently dilute, if $\mu/kT \leqslant -1$, s being a fixed quantity.

Carrying out the sums in eq. (1.2.6), one has for fermions and bosons, respectively,

$$\Xi = \prod_j (1+t_j) \text{ or } \Xi = \prod_j (1-t_j)^{-1}$$

One finds from eq. (1.2.5)

$$pV = kT\ln\Xi = \pm kT\sum_j \ln(1\pm t_j) \simeq kT\sum_j t_j$$ (1.3.7)

For $\mu/kT \leqslant -1$, t_j is small compared with unity and hence $\ln(1\pm t_j) \approx \pm t_j$. Using (1.3.2) in eq. (1.3.7)

$$pV = AVkT\exp(\mu/kT)\int_0^\infty e^s \exp(-e/kT)\,de$$

$$= AV(kT)^{s+2}\exp(\mu/kT)\Gamma(s+1) = \bar{N}kT$$ (1.3.8)

where (1.3.3) has been used. Thus we have confirmed that (1.3.2) can describe a classical ideal gas.

We now derive a general result for estimating the density of states for particles which are free except for the point interactions noted in section 1.2. If the box confining the particles is a cube of side L, then the uncertainty in the position of a particle $\Delta x = \Delta y = \Delta z = L$. Since $\hbar\mathbf{k} = \mathbf{p}$ is the momentum and $\Delta x \Delta p \geqslant h$, the uncertainties in the components of the so-called wavevector $\mathbf{k}$ are

$$\Delta k_1 = \Delta k_2 = \Delta k_3 = \frac{\Delta p_1}{\hbar} = \frac{1}{\hbar}\frac{h}{\Delta x} = \frac{2\pi}{L}$$

This is the length of the side of a cube in k-space such that one can associate just one single translational quantum state with it. A greater number would violate the uncertainty principle. However, each allowed spin state contributes separately. If the spin degeneracy is g, the number of quantum states in an element $\mathbf{dk}$ of k-space is

$$g\,\mathbf{dk}/(\Delta k_1)^3 = (gV/8\pi^3)\,\mathbf{dk}$$

and is just $\mathscr{N}(e)\,de$ in terms of particle energies. Note that the quantum states are uniformly distributed in $\mathbf{k}$-space. One finds a general formula which can deal with three-dimensional energy–wavevector dispersion relations:

$$\mathscr{N}(e)\,de = \frac{gV}{8\pi^3}\,\mathbf{dk} = \frac{gV}{2\pi^2}k^2\,dk \tag{1.3.9}$$

It will be used for nonparabolic bands later (eq. (1.6.17)).

For free electrons the dispersion relation is

$$k = (2me)^{\frac{1}{2}}/\hbar \tag{1.3.10}$$

so that

$$k^2\,dk/de = \tfrac{1}{2}(2m/\hbar^2)^{\frac{3}{2}}e^{\frac{1}{2}} \tag{1.3.11}$$

Hence eqs. (1.3.9) and (1.3.11) yield a so-called parabolic band

$$\mathscr{N}(e)\,de = \frac{gV}{4\pi^2}\left(\frac{2m}{\hbar^2}\right)^{\frac{3}{2}} e^{\frac{1}{2}}\,de \tag{1.3.12}$$

Comparing eqs. (1.3.12) and (1.3.2) one sees that particles of spin degeneracy g have:

$$s = \tfrac{1}{2}, \quad A = \frac{gV}{4\pi^2}\left(\frac{2m}{\hbar^2}\right)^{\frac{3}{2}} \tag{1.3.13}$$

For electrons one can take $g = 2$.

Other thermodynamic quantities of the ideal classical gas can readily be calculated (without using the special form of eqs. (1.3.13)). This will now be shown. Following eqs. (1.3.1) and (1.3.3) the energy turns out to be

$$U = \bar{E} = \sum_i e_i \bar{N}_i = \exp(\mu/kT) \sum_i e_i \exp(-e_i/kT)$$
$$= \exp(\mu/kT)\,AV \int e^{s+1}\exp(-e/kT)\,de$$

Thus

$$U = AV(kT)^{s+2}\exp(\mu/kT)\,\Gamma(s+2) = (s+1)pV \tag{1.3.14}$$

The entropy is obtainable from eq. (1.1.10),

$$TS = U + pV - \mu N = (s+1)NkT + NkT - \mu N$$

i.e.

$$S = (s + 2 - \mu/kT)kN \tag{1.3.15}$$

Also from section 1.1

$$F \equiv U - TS = (\mu - kT)N \tag{1.3.16}$$

and

$$H \equiv U + pV = (s+2)NkT \tag{1.3.17}$$

The heat capacities are

$$C_V = \left(\frac{\partial U}{\partial T}\right)_{V,N} = (s+1)Nk, \quad C_p = \left(\frac{\partial H}{\partial T}\right)_{p,N} = C_V + Nk \tag{1.3.18}$$

It follows that the specific heat ratio is

$$\frac{C_p}{C_V} = 1 + \frac{kN}{C_V} = 1 + \frac{1}{s+1} = \frac{s+2}{s+1} \tag{1.3.19}$$

1.4 Density-of-states expressions

A generalized theory of the density of states in a field-free region of space will now be given. Assume a d-dimensional rectangular box of sides $D_1, D_2, \ldots, D_d$ and volume $V_d = D_1 D_2 \ldots D_d$, so that an electron meets infinitely high potential walls at these surfaces. If a corner of the box is taken as origin of a Cartesian coordinate system, an eigenfunction of the Hamiltonian is

$$\psi_{\mathbf{l}}(x_1, x_2, \ldots, x_d) = B \prod_{j=1}^{d} \sin(\pi l_j x_j/D_j), \quad \mathbf{l} \equiv (l_1, l_2, \ldots, l_d) \tag{1.4.1}$$

where the l_j are positive or negative integers. Equation (1.4.1) satisfies $\psi = 0$ at the surfaces $x_1 = 0$, $x_2 = 0, \ldots, x_1 = D_1, \ldots, x_d = D_d$. The Schrödinger equation is

$$\left[\sum_{j=1}^{d}(p_j^2/2m_j) + E_0\right]\psi_{\mathbf{l}} = E_{\mathbf{l}}\psi_{\mathbf{l}} \tag{1.4.2}$$

where the p_j are electron momentum component operators, E_0 is the constant potential energy and the m_j are *effective* masses for the different directions. These masses incorporate the effect of the periodic potential in the crystal in an approximate manner. Substitution of eqs. (1.4.1) into eq. (1.4.2) yields with $p_j \rightarrow (\hbar/i)\,\partial/\partial x_j$:

$$\tfrac{1}{2}\pi^2 \hbar^2 \sum_{j=1}^{d}[(l_j/D_j)^2(1/m_j)] = E_{\mathbf{l}} - E_0 \tag{1.4.3}$$

Now count the sets of integers l for energies E_0 to E by constructing a d-dimensional lattice with coordinates $l_1, \ldots, l_d$. Each translational electron quantum state is represented by a lattice point; as we wish to count only linearly independent wavefunctions, attention will now be confined to lattice points having all $l_j \geqslant 0$. The maximum number of points l_k along the l_k-axis is found if all other l_j's vanish, so that eq. (1.4.3) yields

$$l_k < \hat{l}_k \equiv [2(E-E_0)m_k]^{\frac{1}{2}} D_k/\pi\hbar \text{ and so } \sum_{k=1}^{d} \left(\frac{l_k}{\hat{l}_k}\right)^2 = 1$$

Since each state belongs to a volume of size unity, the number of points is given by the appropriate volume in l-space. This is a half-line for $d = 1$, a quarter of an ellipse for $d = 2$, an eighth of an ellipsoid for $d = 3$. In general the number of states is given by the formula for a hyper-ellipsoid of semiaxes $\hat{l}_1, \hat{l}_2, \ldots, \hat{l}_d$. This is obtained in the appendix to this section. One finds

$$N_d(E) = \frac{\pi^{d/2}\hat{l}_1 \ldots \hat{l}_d}{2^d \Gamma(1+d/2)} = \frac{\pi^{d/2}(2^d m_1 \ldots m_d)^{\frac{1}{2}}}{2^d(d/2)\Gamma(d/2)\pi^d\hbar^d} V_d(E-E_0)^{d/2}$$

For $\hat{l}_1 = \hat{l}_2 = \hat{l}_3 = r$ and $d = 3$ one finds the volume $\pi^{\frac{3}{2}}r^3/\Gamma(5/2) = 4\pi r^3/3$ of a sphere from the general formula $\pi^{d/2}\hat{l}_1 \ldots \hat{l}_d/\Gamma(1+d/2)$. The density of translational states $dN_d(E)/dE$ per unit energy range has to be multiplied by a factor g for spin degeneracy

$$\mathcal{N}_d(E) = \frac{gV_d(m_1 m_2 \ldots m_d)^{\frac{1}{2}}}{\Gamma(d/2)(2\pi\hbar^2)^{d/2}}(E-E_0)^{d/2-1} \tag{1.4.4}$$

This yields, on using $\Gamma(1/2) = \sqrt{\pi}$,

$$\mathcal{N}_3(E) = \left(\frac{g}{2}\right)\frac{(2m_1 m_2 m_3)^{\frac{1}{2}}}{\pi^2\hbar^3} V_3(E-E_0)^{\frac{1}{2}} \tag{1.4.5}$$

$$\mathcal{N}_2(E) = \left(\frac{g}{2}\right)\frac{(m_1 m_2)^{\frac{1}{2}} V_2}{\pi\hbar^2} \tag{1.4.6}$$

$$\mathcal{N}_1(E) = \left(\frac{g}{2}\right)\frac{(2m_1)^{\frac{1}{2}} V_1}{\pi\hbar}(E-E_0)^{-\frac{1}{2}} \tag{1.4.7}$$

Here V_1 is the length, V_2 is the area and V_3 the volume of the system considered. For $g = 2$ and $m_1 = m_2 = m_3 = m$, eq. (1.4.5) reproduces eq. (1.3.13), but we now have also the results (1.4.6) and (1.4.7) for a surface film (or a quantum well) and for a quantum wire.

   Applying eq. (1.4.5) to the conduction band in a semiconductor, E_0 is the energy at the bottom of the band where $\mathcal{N}(E)$ vanishes, and it can be interpreted as the energy at the minimum $E(k_0)$ of the conduction band. There may be n_c *equivalent* minima (which result from crystal symmetry), whence eq. (1.4.5) may be written:

$$\mathcal{N}(E) = n_c \frac{g}{2} \frac{(2m_1 m_2 m_3)^{\frac{1}{2}}}{\pi^2 \hbar^3} V(E - E_0)^{\frac{1}{2}} \qquad (1.4.8)$$

It is usual to write E_c (instead of E_0) for the bottom of a conduction band. The surfaces of constant energy in k-space can be given the general form

$$E - E_c = \frac{\hbar^2}{2} \left(\frac{k_1^2}{m_1} + \frac{k_2^2}{m_2} + \frac{k_3^2}{m_3} \right)$$

if one is in the neighborhood of E_c by choosing appropriate (i.e. 'principal') axes in k-space. In the case of Ge, Si and GaAs the symmetry of the crystal requires that at least two of the effective masses are equal. If for example $m_2 = m_3$ the surface is an ellipsoid of revolution about the k_1-axis. Thus, m_1 is called the longitudinal effective mass $m_\parallel$, while $m_2 = m_3$ is called the transverse effective mass $m_\perp$. It is associated with the plane at right angles to the axis. One then finds

$$E - E_c = \frac{\hbar^2}{2} \left(\frac{k_1^2}{m_\parallel} + \frac{k_2^2 + k_3^2}{m_\perp} \right) \qquad (1.4.9)$$

The number n_c is obtained from band theory. Thus the lowest conduction band of Si lies at roughly $\frac{3}{4}$ of the distance from the zone center (usually labelled Γ) to the center (labelled X) of the square face perpendicular to the k_x-axis. There are six ($n_c = 6$) symmetrically placed such minima. In Ge there are eight symmetrically placed minima, but only half of each lies in the first Brillouin zone, so that $n_c = 4$.

One can write (1.4.8) as

$$\mathcal{N}(E) = \frac{g}{2} \frac{(2m_c^3)^{\frac{1}{2}}}{\pi^2 \hbar^3} V(E - E_c)^{\frac{1}{2}} \qquad (1.4.10)$$

and m_c is then called the density-of-states mass. Comparing with eq. (1.4.8)

$$m_c = n_c^{\frac{2}{3}} (m_1 m_2 m_3)^{\frac{1}{3}} \text{ or } n_c^{\frac{2}{3}} (m_\parallel m_\perp^2)^{\frac{1}{3}} \qquad (1.4.11)$$

This gives the density-of-states mass in terms of the effective masses, for example, for Si and Ge. For GaAs one can put $n_c = 1$, $m_1 = m_2 = m_3$ near E_c since the constant energy surface is there nearly spherical. Hence $m_c = m_1$ in this case.

The valence bands for these materials (diamond and zinc blend structures) have four subbands, three of which are degenerate at $k = 0$. The spin–orbit interaction causes a splitting of the bands which consist of a heavy-hole, a light-hole and a 'split-off' band. For further details a specialized treatment of band theory should be consulted. Assuming spherical surfaces for the light-hole and heavy-hole bands of masses m_1 and m_h, respectively, one finds

$$\mathcal{N}(E) = \frac{g}{2} \frac{(2m_v^3)^{\frac{1}{2}}}{\pi^2 \hbar^3} V(E_v - E)^{\frac{1}{2}} = \frac{g}{2} \frac{2^{\frac{1}{2}}}{\pi^2 \hbar^3} (E_v - E)^{\frac{1}{2}} [m_h^{\frac{3}{2}} + m_1^{\frac{3}{2}}] V \qquad (1.4.12)$$

so that in that case

$$m_v = [m_h^{\frac{3}{2}} + m_1^{\frac{3}{2}}]^{\frac{2}{3}}$$

Some numerical values are given in Table 1.4.1.

Appendix: The volume V_n of an n-dimensional sphere

Suppose $V_d(r) = C_d r^d$, where r is the radius. We then work out the integral of $\exp(-r^2)$ over all space by two methods: (i) using Cartesian coordinates $x_1, \ldots, x_d$, and (ii) using spherical coordinates

$$\int \cdots \int_{-\infty}^{\infty} \exp(-r^2) \, dx_1 \ldots dx_d = \int_0^{\infty} \exp(-r^2) \, dV_d(r) \qquad (1.4.13)$$

The left-hand side is with $r^2 = x_1^2 + \ldots + x_d^2$

$$\left[\int_{-\infty}^{\infty} \exp(-x^2) \, dx \right]^d = \left[2 \int_0^{\infty} \exp(-x^2) \, dx \right]^d = \left[\int_0^{\infty} y^{-\frac{1}{2}} e^{-y} \, dy \right]^d$$

$$= [\Gamma(\tfrac{1}{2})]^d = \pi^{d/2} \qquad (1.4.14)$$

where eq. (1.3.5) has been used. The right-hand side of eq. (1.4.13) is with $y = r^2$, $dr = dy/2y^{\frac{1}{2}}$,

$$C_d d \int_0^{\infty} \exp(-r^2) r^{d-1} \, dr$$

$$= \frac{d}{2} C_d \int_0^{\infty} y^{d/2-1} \exp(-y) \, dy$$

$$= \frac{d}{2} C_d \Gamma\left(\frac{d}{2}\right) = C_d \Gamma\left(1 + \frac{d}{2}\right) \qquad (1.4.15)$$

Equating eqs. (1.4.14) and (1.4.15),

$$C_d = \frac{\pi^{d/2}}{\Gamma\left(1 + \dfrac{d}{2}\right)}$$

Table 1.4.1. *Some numerical data*

(Values are approximate and are liable to be adjusted at various times.)

		Ge	Si	GaAs
(1) Main energy gap at 300 K (eV)		0.66	1.12	1.42
(2) Effective densities of states (cm^{-3}) at 300 K (see eqs. (1.6.3) and (1.6.9))	N_c	1.03×10^{19}	3.22×10^{19}	4.21×10^{17}
	N_v	5.35×10^{18}	1.83×10^{19}	9.52×10^{18}
(3) Effective electron masses in units of the electron rest mass (see eq. (1.4.9)) at 4 K	$m_\parallel$	1.59	0.92	
	$m_\perp$	0.082	0.19	
Parabolic mass near band bottom				0.067
(4) Effective hole masses in units of the electron rest mass at 4 K				
heavy-hole band (m_h)		0.35	0.54	0.51
light-hole band (m_l)		0.043	0.15	0.082
split-off band (m_s)		0.077	0.23	0.15
(5) Density-of-states effective masses at 4 K in units of the electron rest mass (see eqs. (1.4.11) and (1.4.12))	m_c	0.55	1.06	0.067
	m_v	0.36	0.59	0.53
(6) Dielectric constant		16	11.8	13.2

For a hyper-ellipsoid C_d is retained and r^d is replaced by the product of the d semiaxes.

1.5 Density-of-states expressions in the neighborhood of various points in k-space*

1.5.1 General theory

The result (1.4.4) can be seen in a yet more general setting, without appeal to wavefunctions, by integrating over a surface of constant energy E as follows. Let A and B be two points picked out by a common normal to two neighboring surfaces of constant energy (Fig. 1.5.1). The number of k-vectors in a volume element dk of k-space is for d-dimensional vectors (cf. eq. (1.3.9)):

$$dv = [V_d/(2\pi)^d] dk \tag{1.5.1}$$

Let Δk be the change in k on passing along AB. The rate of change of E normal to the surface being $\nabla_k E(k)$, the energy change is

$$\Delta E = \nabla_k E(k) \cdot \Delta k = |\nabla_k E(k)| \Delta k$$

* This section may be omitted at first reading.

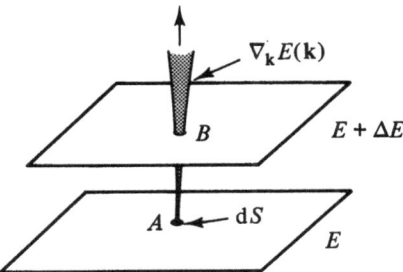

Fig. 1.5.1. Neighboring elements of distinct surfaces of constant energy in k-space.

Let dS be an incremental area around A on the surface of constant energy E in k-space. Then the number of states in an incremental volume $dS\,\Delta k$ between the surfaces is by (1.5.1)

$$dv = \frac{V_d}{(2\pi)^d}dS\,\Delta k \xrightarrow{(d-3)} \frac{V_3}{(2\pi)^3}\frac{dS\Delta E}{|\nabla_k E(k)|} \tag{1.5.2}$$

Thus in an energy interval the number of states is given by an integral over the surface S of constant energy E

$$\mathcal{N}(E)\,dE = 2\int dv = \frac{2V_d\,dE}{(2\pi)^d}\int_{\text{all } k}\delta[E-E(k)]\,dk \tag{1.5.3}$$

$$\xrightarrow{(d-3)} \frac{2V_3\,dE}{(2\pi)^3}\int_S \frac{dS}{|\nabla_k E(k)|}$$

We have multiplied by two to allow for spin. Also the Dirac δ-function is defined to have the property (see eq. (A.7) of Appendix A):

$$\int_{-a}^{b} f(x)\,\delta(x)\,dx = f(0), \quad [f(x) \text{ any continuous function}; a,b>0]$$

It looks after the confinement of the integration to the particular energy E.
    The expression (1.5.3) will be used, with the aid of a lemma.

Lemma

If **y** is a d-dimensional vector of length $y = (\sum_{j=1}^{d} y_j^2)^{\frac{1}{2}}$, then

$$I = \int_{\text{all } y}\delta(A-y^2)\,dy = \begin{cases} (\pi^{d/2}A^{d/2-1})/\Gamma(d/2) & (A\geqslant 0) \\ 0 & (A<0) \end{cases} \tag{1.5.4}$$

Proof
The surface area of a d-dimensional sphere of radius y has the form $C_d dy^{d-1}$, where $3C_3 = 4\pi$ and, more generally from the appendix to section 1.4,

$$C_d d = 2\pi^{d/2}/\Gamma(d/2)$$

Since the integrand in I is independent of angles, the volume element dy is $\mathrm{d}y$ multiplied by $C_d\,\mathrm{d}y^{d-1}$. Hence with $y^2 = z$

$$I = C_d d \int_0^\infty \delta(A-y^2)\,y^{d-1}\,\mathrm{d}y = \frac{d}{2} C_d \int_0^\infty \delta(A-z)\,z^{(d-2)/2}\,\mathrm{d}z$$

This vanishes if A lies outside the range of the z-integration, i.e. if $A < 0$. If it lies inside the range of integration, one has $\dfrac{d}{2} C_d A^{(d-2)/2}$, as required.

Returning to eq. (1.5.3), useful results can be obtained for many band structures, even if they are complicated, by considering the neighborhood of an extremum in $\mathbf{k}$-space, at $\mathbf{k}_0$ say. Then

$$E(\mathbf{k}_0) \equiv E_0 \tag{1.5.5}$$

denotes a kind of potential energy of an electron in that region of $\mathbf{k}$-space. In that region one can write the Taylor expansion

$$E(\mathbf{k}) = E_0 + \sum_{i=1}^d (k_i - k_{0i})\left(\frac{\partial E}{\partial k_i}\right)_{\mathbf{k}_0} + \frac{1}{2}\sum_{i,j=1}^d \left(\frac{\partial^2 E}{\partial k_i\,\partial k_j}\right)_{\mathbf{k}_0} (k_i - k_{i0})(k_j - k_{j0}) + \dots$$

One may define an effective mass tensor by

$$\left[\left(\frac{1}{m}\right)_{ij}\right]_{\mathbf{k}_0} \equiv \frac{1}{\hbar^2}\left(\frac{\partial^2 E}{\partial k_i\,\partial k_j}\right)_{\mathbf{k}_0} \tag{1.5.6}$$

Then, since $(\partial E/\partial k_i)_{\mathbf{k}_0} = 0$,

$$E(\mathbf{k}) - E_0 = \tfrac{1}{2}\hbar^2 \sum_{i,j=1}^d (k_i - k_{0i})(k_j - k_{0j})(1/m)_{ij,\mathbf{k}_0} + \dots \tag{1.5.7}$$

If the axes are chosen parallel to the principal axes of this hyper-ellipsoid in $\mathbf{k}$-space, the effective mass tensor becomes diagonal and

$$E(\mathbf{k}) - E_0 = \tfrac{1}{2}\hbar \sum_{i=1}^d (k_i - k_{0i})^2\,(1/m)_{ij,\mathbf{k}_0} + \dots \tag{1.5.8}$$

We now use eq. (1.5.8) in eq. (1.5.3) to obtain the $\mathcal{N}(E)$-function. In order to deal with positive masses m_i the parameters σ_i will be introduced:

$$(1/m)_{ii,\mathbf{k}_0} \equiv \sigma_i/m_i, \quad m_i > 0, \quad \sigma_i = +1 \text{ or } -1$$

and

$$y_i^2 \equiv (\hbar^2/2m_i)(k_i - k_{0i})^2, \quad \mathrm{d}k_i = [(2m_i)^{\frac{1}{2}}/\hbar]\,\mathrm{d}y_i \tag{1.5.9}$$

It follows that

$$\mathbf{dk} = \mathrm{d}k_1 \dots \mathrm{d}k_d = [2^d m_1 \dots m_d)^{\frac{1}{2}}/\hbar^d]\,\mathbf{dy}$$

Hence, using eq. (1.5.3)

$$\mathcal{N}(E) = \frac{2V_d}{(2\pi)^d}\frac{(2^d m_1 m_2 \dots m_d)^{\frac{1}{2}}}{\hbar^d}\int \delta\left(E - E_0 - \sum_{i=1}^d \sigma_i y_i^2\right)\mathbf{dy} \tag{1.5.10}$$

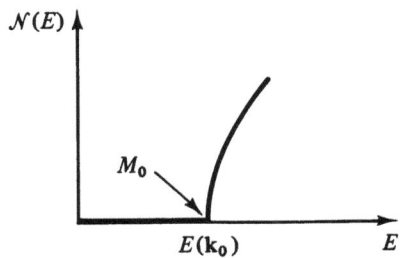

Fig. 1.5.2. Schematic diagram of the density-of-states equations (1.5.11) and (1.5.12) for $d = 3$ if there is a minimum at $\mathbf{k} = \mathbf{k}_0$. The point $\mathbf{k} = \mathbf{k}_0$ is of type M_0.

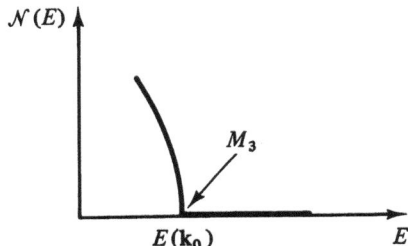

Fig. 1.5.3. As Fig. 1.5.2, but there is a maximum at $\mathbf{k} = \mathbf{k}_0$. For $d = 3$ this is a point of type M_3.

One can discuss these integrals in a general way, allowing for various distributions of σ's over $+1$ or -1 [1.5.1]. Thus $\mathbf{k}_0$ can be defined to be a point in k-space of type $M_0, M_1, \dots,$ M_d depending on the number of σ_i which are equal to -1. If E_0 is a minimum, eq. (1.5.8) shows that $\mathbf{k}_0$ is of type M_0, since all effective mass components are positive. If they are all negative, we have a maximum and require all σ's in (1.5.10) to have the values -1 and the point is of type M_d. In all other cases the energy surface has a saddle point in k-space. These matters are of interest in connection with vibrational spectra [1.5.2], radiative transitions [1.5.3] and impact ionization in semiconductors [1.5.4] (see section 3.7.3).

1.5.2 Conditions near extrema and saddle points

If E_0 is a minimum, all σ's are $+1$ and the lemma applied to (1.5.10) yields

$$\mathcal{N}(E) = \begin{cases} \dfrac{2V_d(m_1 \dots m_d)^{\frac{1}{2}}}{\Gamma(d/2)(2\pi\hbar^2)^{d/2}}(E-E_0)^{d/2-1} & [E \geqslant E_0] \qquad (1.5.11) \\[2ex] 0 & [E < E_0] \qquad (1.5.12) \end{cases}$$

Thus eq. (1.5.11) is again the result (1.4.4) and it is illustrated in Fig. 1.5.2. We can also see how the lemma and these results can be extended to other types of points $\mathbf{k}_0$. For a maximum this is particularly easy (Fig. 1.5.3), and densities near saddle points are illustrated in Fig. 1.5.4. The conditions near a conduction band minimum and a valence band maximum are as depicted in Figs. 1.5.2 and 1.5.3, respectively.

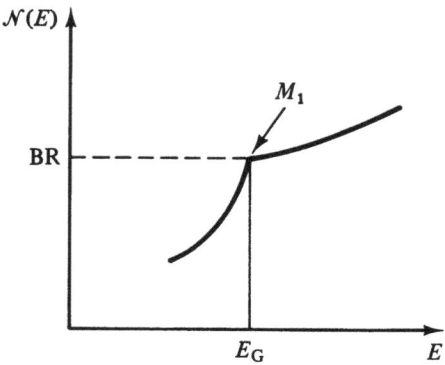

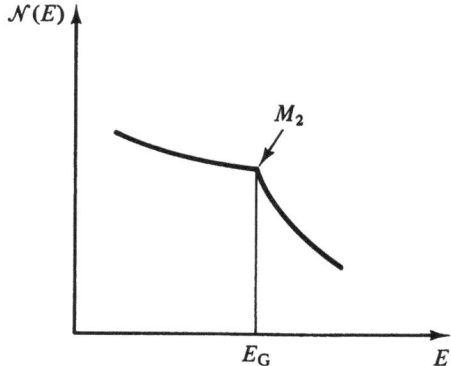

Fig. 1.5.4. As Fig. 1.5.2 for saddle points M_1 and M_2. For M_1 the calculation leading to eq. (1.5.15) applies.

We obtain the density of states near a saddle point of type M_1 next. Suppose therefore

$$\sigma_1 = \sigma_2 = +1, \quad \sigma_3 = -1 \tag{1.5.13}$$

Introduce cylindrical coordinates into (1.5.10) as follows:

$$y_1^2 + y_2^2 \rightarrow r^2, \quad y_3 \rightarrow z, \quad \mathrm{d}y \rightarrow r\,\mathrm{d}r\,\mathrm{d}\theta\,\mathrm{d}z \tag{1.5.14}$$

where θ is defined by $\sin\theta = y_2(y_1^2 + y_2^2)^{-\frac{1}{2}}$ and runs from 0 to 2π. Then

$$\mathcal{N}(E) = \frac{V}{4\pi^3} \frac{(8m_1 m_2 m_3)^{\frac{1}{2}}}{\hbar^3} 2\pi \int r\,\mathrm{d}r \int_{-\infty}^{\infty} \delta(E_0 - E + r^2 - z^2)\,\mathrm{d}z$$

$$= \frac{V}{\pi^2} \frac{(2m_1 m_2 m_3)^{\frac{1}{2}}}{\hbar^3} \int r\,\mathrm{d}r \int_0^{\infty} \delta(E_0 - E + r^2 - u)\,u^{-\frac{1}{2}}\,\mathrm{d}u$$

If $E > E_G$, we have $u = r^2 - (E - E_G)$ and a contribution arises only if $r^2 > E - E_G$. If $E < E_G$, then $r^2 > E - E_G$ and a contribution arises for all r values. A logarithmic divergence

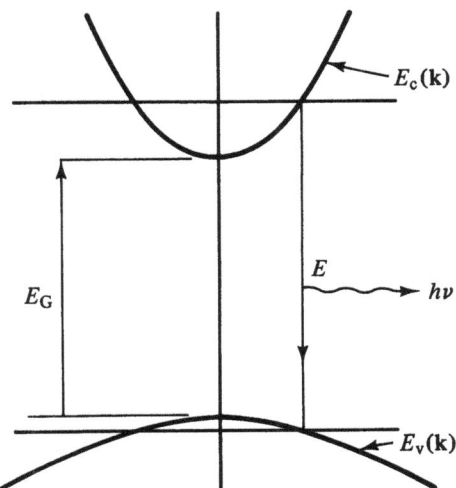

Fig. 1.5.5. A radiative transition.

arises in both cases for $r = \infty$. The r integration must therefore be limited to a value $r \leqslant R$, beyond which the assumed band structure becomes a poor approximation. Hence

$$\mathcal{N}(E) = \begin{cases} B(R^2 + E_G - E)^{\frac{1}{2}} & (E \geqslant E_G) \\ B|(R^2 + E_G - E)^{\frac{1}{2}} - (E_G - E)^{\frac{1}{2}}| & (E \leqslant E_G) \end{cases} \qquad (1.5.15)$$

where

$$B \equiv \frac{V(2m_1 m_2 m_3)^{\frac{1}{2}}}{\pi^2 \hbar^3}$$

This type of curve is shown in Fig. 1.5.4 and is of importance in the theory of optical transitions (section 4.4.3).

The simple example (1.5.13) arises from writing

$$E_c(\mathbf{k}) - E_v(\mathbf{k}) = E_G + \sum_{i=1}^{3} \left(\frac{\hbar^2 k_i^2}{2m_{ci}} + \frac{\hbar^2 k_i^2}{2m_{vi}} \right)$$

$$= E_G + \sum_{i=1}^{3} \sigma_i \frac{\hbar^2 k_i^2}{2m_i}$$

where

$$\frac{1}{m_{ci}} + \frac{1}{m_{vi}} \equiv \frac{\sigma_i}{m_i} \qquad (\sigma_i = \pm 1, m_i > 0, i = 1, 2, 3) \qquad (1.5.16)$$

In eq. (1.5.13) one is considering a band structure in which m_{v3}, say, is negative but less in magnitude than m_{c3}, while the other four masses are all positive. For instance if $m_{c3} = 0.2m$ and $m_{v3} = -0.1m$ then $\sigma_1 = \sigma_2 = 1$, $\sigma_3 = -1$, and $m_3 = 0.2m$. The k-conservation shown in Fig. 1.5.5 is characteristic of a certain type of radiative transition (see section 4.4.2). Use of the m_i's in the expression for the density of states gives one the two-band (or 'joint') density of states.

1.6 The number of electrons and holes in bands

1.6.1 Standard or parabolic bands

For the conduction band of a semiconductor, assumed to extend to infinite energy, the mean number of electrons in thermal equilibrium at temperature T and chemical potential μ is

$$\bar{N} = \int_{E_c}^{\infty} P(E)\,\mathcal{N}(E)\,\mathrm{d}E \tag{1.6.1}$$

where E_c is the energy at the bottom of the band and $P(E)$ is the occupation probability of a quantum state of energy E. This drops to zero exponentially for large energies, by eq. (1.2.8). Hence the finite extent of a band is well simulated by the limit $E = \infty$ in (1.6.1). Using eqs. (1.2.8) and (1.4.8):

$$\bar{N} = AV \int_{E_c}^{\infty} \frac{(E-E_c)^{\frac{1}{2}}\,\mathrm{d}E}{1+\exp\left[(E-\mu)/kT\right]}$$

$$= AV(kT)^{\frac{3}{2}} \int_{0}^{\infty} \frac{x^{\frac{1}{2}}\,\mathrm{d}x}{1+\exp\left[x-(\mu-E_c)/kT\right]} \tag{1.6.2}$$

where, with m_c given by eq. (1.4.11),

$$A \equiv \frac{1}{2\pi^2}\left(\frac{2m_c}{\hbar^2}\right)^{\frac{3}{2}}$$

One finds for the electron concentration [1.6.1–1.6.3]

$$n \equiv \frac{\bar{N}}{V} = N_c\,F_{\frac{1}{2}}[(\mu-E_c)/kT] \tag{1.6.3}$$

The parameter in the $F_{\frac{1}{2}}$ function is, by eq. (1.1.8), in general a function of position. We define the Fermi–Dirac integrals by

$$F_s(\gamma) \equiv \frac{1}{\Gamma(s+1)} \int_{0}^{\infty} \frac{x^s\,\mathrm{d}x}{1+\exp(x-\gamma)} \tag{1.6.4}$$

where $\Gamma(s)$ is defined in section 1.3. The value of N_c is, from eqs. (1.6.2) and (1.6.3),

$$N_c = A(kT)^{\frac{3}{2}}\Gamma\left(\frac{3}{2}\right) = \frac{1}{2\pi^2}\left(\frac{2m_c}{\hbar^2}\right)^{\frac{3}{2}}(kT)^{\frac{3}{2}}\frac{\sqrt{\pi}}{2} \tag{1.6.5}$$

$$= 2(m_c kT/2\pi\hbar^2)^{\frac{3}{2}} \tag{1.6.6}$$

Hence

$$N_c = 2(2\pi m_c kT/h^2)^{\frac{3}{2}} \tag{1.6.7}$$

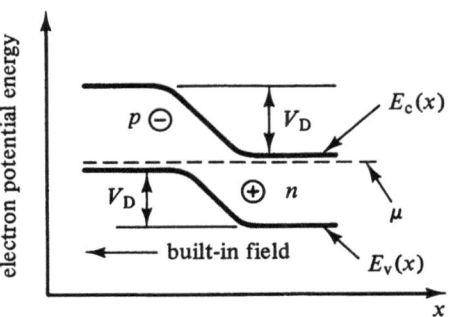

Fig. 1.6.1. A *p–n* junction in thermal equilibrium. The built-in potential has been denoted by V_D and the space charge has also been indicated. Fig. 1.13.1 gives additional details.

The valence band, being almost full of electrons, is usually better described in terms of electron vacancies [1.6.1]. The probability of an electron vacancy in a quantum state of energy E is

$$1 - \frac{1}{1 + \exp\left[(E-\mu)/kT\right]} = \frac{1}{1 + \exp\left[(\mu-E)/kT\right]} \tag{1.6.8}$$

Assume a valence band which extends from energy $-\infty$ to an energy E_v. Then the concentration of electron vacancies in a valence band is, in analogy with eq. (1.6.3),

$$p = \frac{\bar{P}}{V} = A \int_{-\infty}^{E_v} \frac{(E_v - E)^{\frac{1}{2}}\,dE}{1 + \exp\left[(\mu-E)/kT\right]}$$

$$= N_v F_{\frac{1}{2}}[(E_v - \mu)/kT] \tag{1.6.9}$$

Here

$$N_v = 2(m_v kT/2\pi\hbar^2)^{\frac{3}{2}}, \quad m_v \equiv (n_v^2 m_1 m_2 m_3)^{\frac{1}{3}}$$

where m_v is the density-of-states effective mass, based here on the valence band effective masses (not distinguished notationally from the conduction band effective masses). Also n_v is the number of equivalent valence band maxima.

The results (1.6.3) and (1.6.9) are fundamental in all simple semiconductor calculations of equilibrium properties like heat capacities, transport properties like electrical conductivity or generation–recombination processes. As seen in section 1.1, E_c, E_v and μ, and therefore $n \equiv \bar{N}/V$ and $p \equiv \bar{P}/V$, can be functions of position. We omit the bars on n and p and assume that average concentrations are intended. Thus as one goes from left to right in a simple one-dimensional *p–n* junction in thermal equilibrium (Fig. 1.6.1), the hole concentration decreases as $E_v(x) - \mu$ becomes more negative, while the electron concentration increases as $\mu - E_c(x)$ becomes less negative. On the left, where holes dominate, we have the *p*-type material, on the right we have the *n*-type material. Donors, which on losing

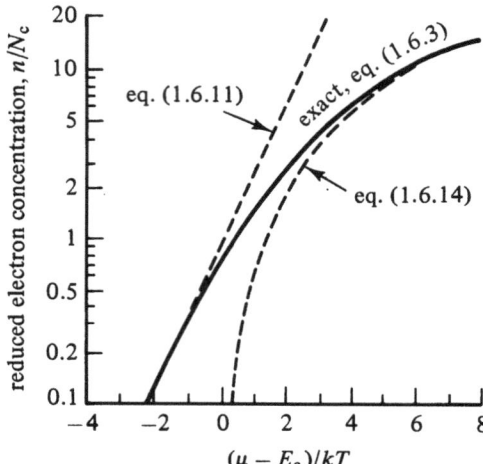

Fig. 1.6.2. The concentration of electrons in a parabolic conduction band in units of N_c.

electrons become positively charged, are the majority impurity on the right; acceptors, which become negatively charged on accepting electrons from the valence band, are the majority impurity on the left.

We note two properties. The first is that for large and negative chemical potential, as discussed in section 1.3,

$$F_s(\gamma) \rightarrow \frac{1}{\Gamma(s+1)} \int_0^\infty x^s e^{-x+\gamma} \, dx = e^\gamma \quad \text{(all } s) \tag{1.6.10}$$

Hence (with $E_c = 0$, $s = \frac{1}{2}$), eq. (1.6.3) yields the classical result (1.3.3), if eq. (1.6.7) is used:

$$n = N_c \exp[(\mu - E_c)/kT], \quad N_c \equiv 2(m_c kT / 2\pi\hbar^2)^{\frac{3}{2}} \tag{1.6.11}$$

Also eq. (1.6.9) yields

$$p = N_v \exp[(E_v - \mu)/kT] \tag{1.6.12}$$

These approximations are called classical or nondegenerate limits or Maxwell–Boltzmann statistics. If the Fermi integral has to be taken into account, one speaks of *Fermi degeneracy*. (This has to be distinguished from the *degeneracy* when several linearly independent wavefunctions belong to the same eigenvalue.) The electrochemical potential μ is more usually referred to as the Fermi level, and this will often be done in the sequel.

Secondly, for highly (Fermi) degenerate material the parameter in the integrals (1.6.3) and (1.6.9) is positive, i.e. the Fermi level has risen into the conduction band or dropped into the valence band. In such situations a rough approximation is to

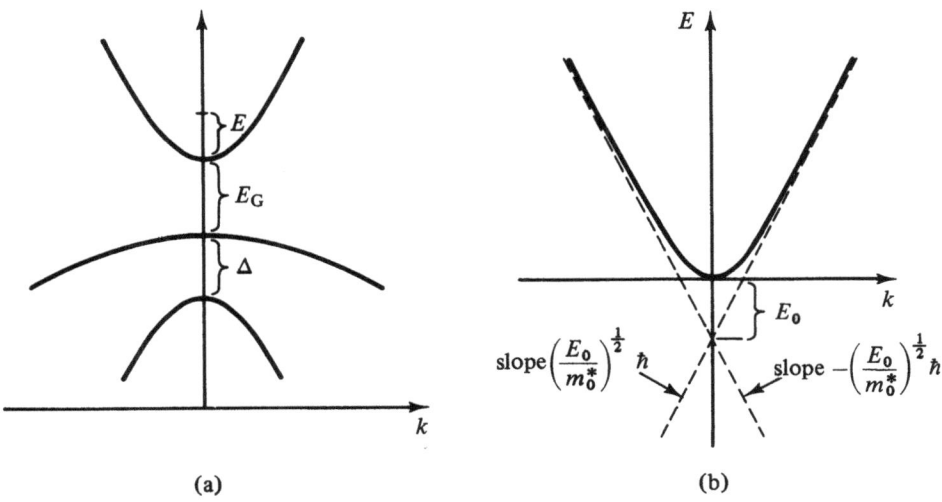

Fig. 1.6.3. (a) A common band structure near $k = 0$ for III–V compound semiconductors. (b) Dispersion relation (1.6.16) showing also the asymptotes.

replace the Fermi function by unity for energies up to μ and by zero for energies beyond μ. Then

$$F_s(\gamma) \to \frac{1}{\Gamma(s+1)} \int_0^\gamma x^s \, \mathrm{d}x = \frac{\gamma^{s+1}}{(s+1)\,\Gamma(s+1)} \qquad (1.6.13)$$

In particular

$$F_{\frac{1}{2}}(\gamma) \to 4\gamma^{\frac{3}{2}}/3\sqrt{\pi} \qquad (1.6.14)$$

The accurate result and the classical and degenerate approximations are illustrated in Fig. 1.6.2.

Other density-of-states functions are needed for disordered or amorphous materials, for nonparabolic bands, etc. As an illustration, a case of nonparabolic bands is considered below.

1.6.2 Nonparabolicity

We consider briefly the case of nonparabolic bands. Let m_0^* be the effective mass at a conduction band minimum $k = 0$ and suppose that there is also a valence band and a split-off band as shown in Fig. 1.6.3(a). Then, with the notation of that figure, it can be shown from reference [1.6.4] that the relation between the energy E and the wavevector k is

$$\frac{\hbar^2 k^2}{2m_0^*} = \frac{(3E_G + 2\Delta)(E + E_G)(E + E_G + \Delta)}{3E_G(E_G + \Delta)(E_G + E + \frac{2}{3}\Delta)} E \qquad (1.6.15)$$

It is readily seen that both for large Δ and also for $\Delta \to 0$ the above dispersion relation becomes (Fig. 1.6.3(b))

$$\frac{\hbar^2 k^2}{2m_0^*} = \left(1 + \frac{E}{2E_0}\right) E, \quad (2E_0 \sim E_G) \tag{1.6.16}$$

In fact one can show from eq. (1.6.15) that

$$E\left(1 + \frac{E}{2E_0}\right) > \frac{\hbar^2 k^2}{2m_0^*} > \tfrac{2}{3}E\left(1 + \frac{E}{2E_0}\right)$$

so that eq. (1.6.16) is often a fair approximation. The density of states based on eq. (1.6.16) is, from eq. (1.3.9)

$$\mathcal{N}(E) = \frac{gV}{2\pi^2} k^2 \frac{\mathrm{d}k}{\mathrm{d}E} = \frac{gV}{4\pi^2}\left(\frac{2m^*}{\hbar^2}\right)^{\frac{3}{2}} [E(1 + E/2E_0)]^{\frac{1}{2}}(1 + E/E_0) \tag{1.6.17}$$

For $g = 2$ and $E_0 = \infty$ this goes over into the parabolic case of eq. (1.4.5).

The result (1.6.16) has been applied quite widely, and we give here some relevant references. In silicon the piezo-resistance was studied in this way recently [1.6.5]; in InSb the metal–insulator–semiconductor (MIS) capacitance [1.6.6], surface waves [1.6.7] and the heat capacity of thin films [1.6.8] could be discussed with the aid of the dispersion relation (1.6.16); it was used for GaAs to study the Boltzmann equation for the interaction of energetic electrons with polar optical phonons [1.6.9]. One may also note studies of nonparabolicity by the measurement of thermoelectric power in a strong magnetic field, applied for example to lead compounds [1.6.10].

One can give discussions analogous to those based on eqs. (1.6.1) to (1.6.4) and Fig. 1.6.3, by using eq. (1.6.16). The integrals turn out to be more complicated. The statistical thermodynamics of such a 'Kane gas' has not been used in the past, but is developed below because of its intrinsic interest and for possible future use.

1.6.3 The 'Kane gas'

We again apply the pV-formula (1.3.7) for the pressure p of the gas. The sum is replaced by an integration, using eq. (1.3.11). The number of states in a small volume $\mathrm{d}\mathbf{k}$ of phase space is

$$\frac{Vg}{8\pi^3} \mathrm{d}\mathbf{k} = \frac{Vg}{h^3} \mathrm{d}\mathbf{p} \tag{1.6.18}$$

where V is the volume of the material, g the spin degeneracy, and $\mathbf{p} = \hbar\mathbf{k}$ is the momentum. Hence

$$p = \frac{4\pi g kT}{h^3} \int \ln\left[1 + t(p')\right] (p')^2 \, dp'$$

$$= \frac{4\pi g}{3h^3} \int_0^\infty [p'(E)]^3 f(E) \, dE$$

where

$$t(p') \equiv \exp\left[\frac{\mu - E(p')}{kT}\right]$$

by a partial integration. Here p' denotes the magnitude of the momentum (the symbol p might cause confusion with the pressure here), and f is the Fermi–Dirac distribution function

$$f(E) = (1 + t^{-1})^{-1} = \left\{1 + \exp\frac{E - E_c - (\mu - E_c)}{kT}\right\}^{-1}$$

The energy is now explicitly referred to the band edge energy E_c, so that $x \equiv (E - E_c)/E_0$ is a measure of the kinetic energy of the electron. Defining a velocity v by $m_0^* v^2 = E_0$ and using eq. (1.6.16)

$$p = \frac{4\pi g}{3h^3} (m_0^*)^4 v^5 \int_0^\infty (x^2 + 2x)^{\frac{3}{2}} f \, dx$$

The substitution $x = \cosh\theta - 1$ leads to $x^2 + 2x = \sinh^2\theta$ so that the pressure of a Kane gas is

$$p = B \int_0^\infty \frac{\sinh^4\theta \, d\theta}{1 + \exp\left\{\frac{1}{kT}[m_0^* v^2 (\cosh\theta - 1) - (\mu - E_c)]\right\}} \tag{1.6.19}$$

where

$$B \equiv 4\pi g (m_0^*)^4 v^5 / 3h^3 \tag{1.6.20}$$

One finds also the mean number of particles to be

$$N \equiv V\left(\frac{\partial p}{\partial \mu}\right)_{V,T} = \frac{3BV}{m_0^* v^2} \int_0^\infty \frac{\sinh^2\theta \cosh\theta \, d\theta}{1 + \exp\left\{\frac{1}{kT}[m_0^* v^2 (\cosh\theta - 1) - (\mu - E_c)]\right\}} \tag{1.6.21}$$

The average energy is obtained by multiplying the integrand in N by $E = xE_0 = (\cosh\theta - 1) m_0^* v^2$. Hence

$$U = 3BV \int_0^\infty \frac{\sinh^2\theta \cosh\theta \, (\cosh\theta - 1) \, d\theta}{1 + \exp\left\{\frac{1}{kT}[m^* v^2 (\cosh\theta - 1) - (\mu - E_c)]\right\}} \tag{1.6.22}$$

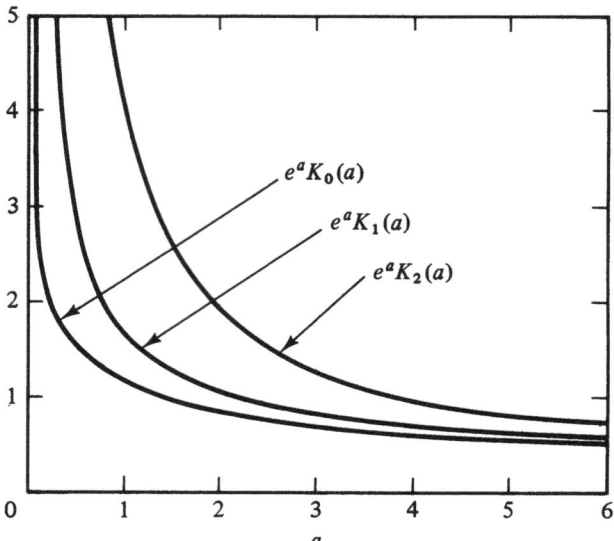

Fig. 1.6.4. The functions $e^a K_n(a)$ for $n = 0, 1, 2$ which occur in the expressions for N and for U in eqs. (1.6.25) and (1.6.27).

These general results can be evaluated in the *nondegenerate* limit. Consider e.g. (1.6.21) first. We have

$$N = \frac{4\pi g (m_0^* v)^3 V}{h^3} \exp\left(\frac{\mu + m_0^* v^2 - E_c}{kT}\right) \int_0^\infty \sinh^2 \theta \cosh \theta$$

$$\exp\left(-\frac{m_0^* v^2}{kT} \cosh \theta\right) d\theta \quad (1.6.21')$$

Now

$$\sinh^2 \theta \cosh \theta = \tfrac{1}{4}(\cosh 3\theta - \cosh \theta)$$

and

$$\int_0^\infty \cosh n\theta \exp\left(-z \cosh \theta\right) d\theta = K_n(z) \qquad (1.6.23)$$

where K_n is the second modified Bessel function. This gives for the integral

$$\frac{1}{4}\left[K_3\left(\frac{m_0^* v^2}{kT}\right) - K_1\left(\frac{m_0^* v^2}{kT}\right)\right] = \frac{1}{4} \frac{4kT}{m_0^* v^2} K_2\left(\frac{m^* v^2}{kT}\right)$$

where the recurrence relation for the K's has been used:

$$K_{n+1}(a) - K_{n-1}(a) = \frac{2n}{a} K_n(a) \qquad (1.6.24)$$

Hence (see Fig. 1.6.4)

$$N = \frac{4\pi g v V m_0^{*2} kT}{h^3} \exp\left(\frac{\mu + m_0^* v^2 - E_c}{kT}\right) K_2\left(\frac{m_0^* v^2}{kT}\right) \qquad (1.6.25)$$

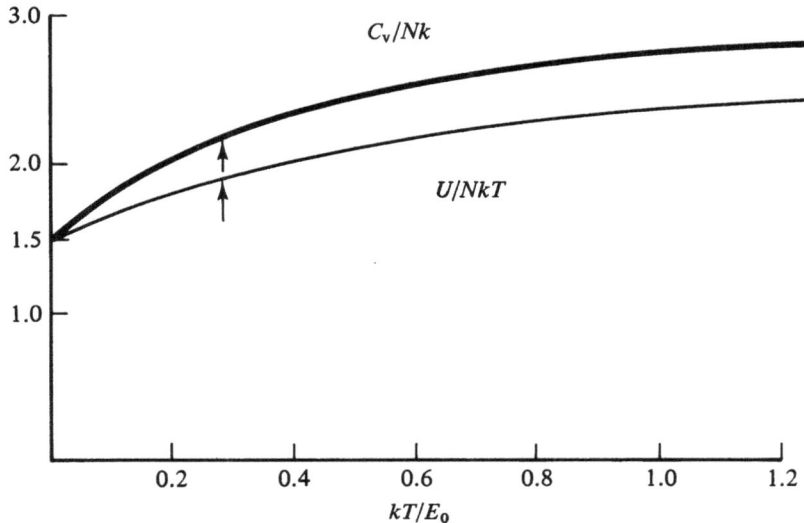

Fig. 1.6.5. The ratios U/NkT and C_v/NK of a nondegenerate electron gas as a function of the nonparabolicity parameter E_0^{-1}. The numerical data of Table 24 from [1.6.11] has been used. Arrows refer to the case of InSb.

One sees at once that in the nondegenerate limit

$$pV = NkT$$

and, using eq. (1.6.25), one can find an expression for p.

Next, consider U. We have

$$U = \frac{4\pi g(m_0^*)^4 v^5}{h^3} V \exp\left(\frac{\mu + m_0^* v^2 - E_c}{kT}\right) \int_0^\infty \sinh^2 \theta \cosh \theta (\cosh \theta - 1)$$

$$\exp\left(\frac{m_0^* v^2}{kT} \cosh \theta\right) d\theta \quad (1.6.26)$$

The hyperbolic functions in the integrand reduce to

$$\tfrac{1}{4}\{\tfrac{1}{2}\cosh 4\theta - \tfrac{1}{2} - \cosh 3\theta + \cosh \theta\}$$

so that eq. (1.6.23) can be used to integrate them. Using the recurrence relation (1.6.24) twice, to eliminate K_4 and the sum of K_2 and K_0, one finds

$$U = \frac{4\pi g(m_0^* v)^3 kT}{h^3} V \exp\left(\frac{\mu + m_0^* v^2 - E_c}{kT}\right)$$

$$\times \left\{ \frac{3}{4} K_3\left(\frac{m_0^* v^2}{kT}\right) + \frac{1}{4} K_1\left(\frac{m_0^* v^2}{kT}\right) - K_2\left(\frac{m_0^* v^2}{kT}\right) \right\} \quad (1.6.27)$$

Writing $a = m_0^* v^2/kT$ for brevity,

$$\frac{U}{NkT} = a \frac{\frac{3}{4}K_3(a) + \frac{1}{4}K_1(a) - K_2(a)}{K_2(a)} \quad (1.6.28)$$

This quantity is shown in Fig. 1.6.5 and rises from 1.5 to 3 as the nonparabolicity is increased.

The limiting behavior of eq. (1.6.28) is readily worked out as follows. For vanishing nonparabolicity we should obtain $U = \frac{3}{2}NkT$. This can be verified using

$$K_n(a) \sim \left(\frac{\pi}{2a}\right)^{\frac{1}{2}} \exp(-a)\left[1 + \frac{n^2 - \frac{1}{4}}{2a} + \dots\right] \quad (a \gg 1) \tag{1.6.29}$$

The other limit $a \to 0$ of very large nonparabolicity is not usually of interest, though future experiments may perhaps identify it. It corresponds to

$$K_n(a) \sim 2^{n-1}(n-1)!/a^n \tag{1.6.30}$$

It leads to $U = 3NkT$, as shown in Fig. 1.6.5, where the heat capacity per particle, C_v/NK, is also shown.

The case of InSb is also marked on the figure. Using $2E_0 = E_G = 0.18$ eV this yields at room temperature $kT/E_0 \sim 0.278$. Using also $m^* = 0.013m$, where m is the normal electron rest mass, one finds $v \sim 1.1 \times 10^8$ cm s^{-1} or 0.37% of the velocity of light.

The velocity of light is relevant for the following reason. For an isotropic but nonparabolic energy band the electron velocity $\mathbf{w}$ is parallel to the wavevector $\mathbf{k}$ and is given by $\mathbf{w} = \hbar^{-1}\nabla_{\mathbf{k}} E(\mathbf{k})$. Its magnitude is obtained by first solving the quadratic equation (1.6.16) for E:

$$E(k) = E_0\left\{1 \pm \left(1 + \frac{\hbar^2 k^2}{E_0 m_0^*}\right)^{\frac{1}{2}} - 1\right\}$$

The energy zero is at the band minimum and the positive sign is taken. Now, using eq. (1.6.16),

$$w = \hbar^{-1}\frac{dE}{dk} = \frac{\hbar k}{m_0^*}\left(1 + \frac{\hbar^2 k^2}{E_0 m_0^*}\right)^{-\frac{1}{2}} = \left(\frac{2E}{m_0^*}\right)^{\frac{1}{2}}\left(1 + \frac{E}{2E_0}\right)^{\frac{1}{2}}\left(1 + \frac{\hbar^2 k^2}{E_0 m_0^*}\right)^{-\frac{1}{2}}$$

and, using (1.6.16) again,

$$\left(1 + \frac{\hbar^2 k^2}{E_0 m_0^*}\right)^{\frac{1}{2}} = \left(1 + \frac{\hbar^2 k^2}{2m_0^*}\frac{2}{E_0}\right)^{\frac{1}{2}} = \left[1 + \frac{2}{E_0}\left(E + \frac{E^2}{2E_0}\right)\right]^{\frac{1}{2}} = 1 + \frac{E}{E_0}$$

Hence

$$w = \frac{(1 + 2E_0/E)^{\frac{1}{2}}}{1 + E_0/E}v \quad (v^2 \equiv E_0/m_0^*)$$

This shows that as the kinetic energy is increased, the velocity saturates at v, rather like the velocity of a relativistically moving particle saturates at c. In fact, the momentum–mass is

$$m^*(k) = \frac{\hbar k}{w} = \frac{E + E_0}{E_0}m_0^* = \left(1 - \frac{w^2}{v^2}\right)^{-\frac{1}{2}}m_0^* \tag{1.6.31}$$

This is the familiar formula from special relativity, except that the velocity of light has been replaced by v [1.6.12]. Some verification of this analogy has been obtained by the study of electrons in InSb in crossed electric and magnetic fields [1.6.13]. A key point is that the above effective mass (and others that could be defined) is increased as the electron gains energy.

The density-of-states formula (1.6.17) has not been used explicitly here. However, it provides an alternative way of obtaining the thermodynamic functions. This will be illustrated by re-deriving eq. (1.6.21). If $f(E)$ is the Fermi–Dirac distribution function

$$\left\{1+\exp\left[\frac{E_0}{kT}\frac{E}{E_0}-\frac{\mu-E_c}{kT}\right]\right\}^{-1},$$

$$N = \int \mathcal{N}(E)f(E)\,\mathrm{d}E = \frac{gV}{4\pi^2}\left(\frac{2m_0^*}{\hbar^2}\right)^{\frac{3}{2}}\int_0^\infty \left(1+\frac{E}{E_0}\right)(E+E^2/2E_0)^{\frac{1}{2}}f(E)\,\mathrm{d}E$$

$$= \frac{4\pi gV}{h^3}m_0^* v^3 \int_0^\infty (1+x)(2x+x^2)^{\frac{1}{2}}f\,\mathrm{d}x$$

The integral is then identical with that which occurs in eq. (1.6.21) and so is the pre-factor. The electron concentration may be written in a form analogous to (1.6.3):

$$n = N_c I, \quad I \equiv \frac{2}{\sqrt{\pi}}\int_0^\infty \frac{(1+\eta/\eta_0)(\eta+\eta^2/2\eta_0)^{\frac{1}{2}}}{1+\exp\left(\eta-\dfrac{\mu-E_c}{kT}\right)}\,\mathrm{d}\eta \tag{1.6.32}$$

where $\eta \equiv E/kT$ and $\eta_0 \equiv E_0/kT$. In the expression for N_c, m_c is now the effective mass at the bottom of the conduction band.

If one wants to work out the internal energy, one simply notes that in

$$U = \int E\mathcal{N}(E)f(E)\,\mathrm{d}E$$

one has now an additional factor, which is

$$E = E_0 x = m_0^* v^2(\cosh\Theta - 1)$$

This gives eq. (1.6.22).

1.6.4 Additional remarks

The properties of bands can also be formulated in greater generality, allowing for both nonparabolicity and for Fermi degeneracy [1.6.14] and by allowing explicitly for electric fields [1.6.15] and screening [1.6.16], but the resulting formulae are too involved to be given here.

The effect of electron–electron interactions have been neglected here, except in so far as an averaged effect is included in a band scheme. They lead to complications which can be treated only approximately. For full discussions see [1.6.17] and [1.6.18]. For a less detailed treatment see [1.6.19]. For simple introductions see [1.6.20] and [1.6.21].

The Kane model holds approximately for several III–V compounds (InP, InSb, GaAs, etc.) whose properties are widely reviewed. For example, see the review [1.6.22] of GaAs properties.

A slow drop by a factor of about 2×10^{-5} K^{-1} in the conduction band edge effective mass ratio m_c/m of n-type GaAs, InSb and InP with rise in temperature between 70 K and 220 K has been noted in [1.6.23] by means of magnetophonon resonance measurements. This is due to dilational changes in the band gap, see also [1.6.24] and [1.6.25].

Fig. 1.7.1. Illustrating donors and acceptors.

1.7 The number of electrons in localized states

1.7.1 General theory

If the electron and hole concentrations are of the same order, the semiconductor is called *intrinsic*. This situation is encouraged by mechanical perfection (few vacancies, dislocations, interstitial atoms, etc. in the lattice) and chemical purity (few foreign atoms). However, some mechanical and chemical imperfections are always present, and if their effect is important the semiconductor is called *extrinsic*. Chemical impurities, called *dopants*, are often implanted or diffused to tailor the material appropriately. Electrons occupying these defects tend to have localized wavefunctions and they give rise to the localized states. Dopants can be neutral when inserted and then can give off electrons (*donors*), or they can be electrically neutral when inserted and then capture an electron, normally from the valence band where this leaves a hole (*acceptors*). Some chemical impurities can act either as donors or as acceptors; they are called *amphoteric*.

The occupation probabilities for electrons in localized states are important throughout semiconductor physics and device design. They are therefore derived here by two alternative methods. The first is based on the grand partition function (1.2.2). However, for readers who have concentrated only on the *results* of section 1.2.1, rather than the derivations, we add an alternative argument at the end of this section, which is based on a simple free energy minimization and is independent of the grand canonical ensemble. The result of this work is, in the simplest case, that the occupation probability of a defect center is

$$\left[1 + \frac{g_0}{g_1} \exp \frac{E - \mu}{kT}\right]^{-1}$$

where g_0 and g_1 are the degeneracies of the ground states of the unoccupied center and of the occupied center, respectively, and E is an effective energy of the center. However, the precise meaning of E, the effect of the excited states of the unoccupied and the occupied center, and the modification of the formula for centers which can be in different charge conditions, have all to be explained. The result for the general case is given in (1.7.3), below, which is still basically a very simple result, as will be seen from the special cases studied in section 1.7.2.

We now seek to derive the occupation probabilities of localized states, basing ourselves on eq. (1.2.2). Suppose a localized lattice defect or chemical impurity or vacancy is able to capture and lose electrons. The state that has the lowest positive (or largest negative) charge that is likely under usual conditions will be called 'M'. The change from 0 to M is effected by the capture of M electrons. For any charge state r ($r = 0, 1, \ldots, M$) we sum over all states l of the center. By allowing different energy spectra for different values of r the effect of interaction between different electrons on the same center can be taken into account. Denote the appropriate sum of terms $\sum_l \exp[-E(l,r)/kT]$ by Z_r, the canonical partition function for an r-electron center. Thus from eq. (1.2.2), μ being the electrochemical potential,

$$\Xi = \sum_{r=0}^{M} \lambda^r Z_r, \quad \lambda \equiv \exp(\mu/kT) \tag{1.7.1}$$

The probability of finding a center in a quantum state (l, r) whose energy is $E(l, r)$ is

$$P(l, r) = -\frac{\partial \ln \Xi}{\partial[E(l, r)/kT]} = \frac{\lambda^r \exp[-E(l, r)/kT]}{\sum_{s=0}^{M} \lambda^s Z_s} \tag{1.7.2}$$

If this state is degenerate, this result has to be multiplied by the degeneracy of this state, and l then refers to the energy level of the center rather than the quantum state. Summing over l, one finds the probability of finding a center in charge state r [1.7.1]–[1.7.3]

$$P(r) = \lambda^r Z_r / \sum_{s=0}^{M} \lambda^s Z_s \tag{1.7.3}$$

If there are N noninteracting centers, the number of centers in charge state r is

$$\nu_r = NP(r) \tag{1.7.4}$$

The mean number of electrons trapped in these centers at temperature T and electrochemical potential μ is

$$N_e = \sum_{r=0}^{M} rNP(r) = \left[\sum_{r=0}^{M} r\lambda^r Z_r \middle/ \sum_{s=0}^{M} \lambda^s Z_s \right] N \tag{1.7.5}$$

This matter has been reviewed in a number of books where experimental details may also be found. See [1.7.4]–[1.7.6].

A simplification may be made by neglecting the excited states of a center for each charge condition r. If g_r is the appropriate degeneracy of the ground state, one can then put

$$\frac{Z_r}{Z_s} = \frac{g_r}{g_s} \exp\left(\frac{E_s - E_r}{kT}\right) \tag{1.7.6}$$

where E_r is the energy of the r-electron center in its ground state. The exponential involves therefore a difference between an s-electron and an r-electron energy, and this is difficult to introduce into a one-electron energy scheme such as has been considered in sections 1.2.3 to 1.2.6. It is therefore more convenient to work with quantities [1.7.7]

$$E(r-\tfrac{1}{2}) \equiv E_r - E_{r-1} \quad (r = 1, 2, \ldots, M) \tag{1.7.7}$$

This is the energy needed to take the least strongly bound electron from an r-electron center and to deposit it at infinity at the energy level which is taken to be the zero of energy. This is, as required, a single-electron energy. One can then use

$$\frac{Z_r}{Z_s} = \begin{cases} \dfrac{g_r}{g_s} \exp\left[\dfrac{E(s-\tfrac{1}{2}) + E(s-\tfrac{3}{2}) + \ldots + E(r+\tfrac{1}{2})}{kT} \right] (r < s) \\[2ex] \dfrac{g_r}{g_s} \exp\left[-\dfrac{E(s+\tfrac{1}{2}) + E(s+\tfrac{3}{2}) + \ldots + E(r-\tfrac{1}{2})}{kT} \right] (r > s) \end{cases} \tag{1.7.8}$$

One can of course use eqs. (1.7.6) to (1.7.8) *even if the excited states are not neglected*. But the E_r's are then *effective* energy levels, which one would not expect to see precisely in optical experiments. Also the E_r's would become (at least weakly) temperature-dependent. The neglect of the excited states is normally satisfactory if they lie a few kT above the ground state. This matter has been studied in connection with acceptors in Ge [1.7.8].

A number of quite involved discussions exist in the literature designed to derive what turn out to be merely special cases of eq. (1.7.3) by means of free energy arguments. We here give an improved, brief and quite general argument of this type. Instead of deriving special cases, we shall again obtain the full result (1.7.3). Consider N centers in charge states $r = 0, 1, 2, \ldots, M$. Hence

$$\sum_{r=0}^{M} P(r) = 1, \quad \text{i.e.} \sum_{r=0}^{M} v_r = N \tag{1.7.9}$$

and

$$\sum_{r=0}^{M} r v_r = N_e \tag{1.7.10}$$

Now, as shown in statistical mechanics, the free energy corresponding to the canonical partition function Z_r is $-kT \ln Z_r$. Also the number of ways of choosing v_0 empty centers, v_1 single-electron centers, etc., up to $r = M$, given that there are N centers, is, if one regards $(v_0, v_1, \ldots, v_M)$ as the vector $\mathbf{v}$

$$W(\mathbf{v}) \equiv N!/v_0! v_1! \ldots v_M! \tag{1.7.11}$$

Now $-kT \ln [W(\mathbf{v}) Z_0^{v_0} Z_1^{v_1} \ldots Z_M^{v_M}]$ is the corresponding free energy $F(\mathbf{v})$. Using Stirling's approximation, and denoting by 'e' the base of the natural logarithm,

$$v! \sim (v/e)^v, \quad (v \gg 1) \tag{1.7.11a}$$

one finds for the free energy of the system

$$F(\mathbf{v}) = -kT \sum_r v_r \ln Z_r - kTN \ln N + kT \sum_r v_r \ln v_r \tag{1.7.12}$$

Allowing for the conservation conditions (1.7.9) and (1.7.10) by Lagrangian multipliers $\alpha'kT$ and γkT, we can minimize the free energy subject to these constraints, by minimizing in fact

$$L(\mathbf{v}) \equiv -kT\left\{N\ln N - \sum_{r=0}^{M} v_r[\ln Z_r - \ln v_r + \alpha' + r\gamma]\right\}$$

with respect to each of $v_0, v_1, \ldots, v_M$ separately, treating N as a constant. This will yield the equilibrium expressions for the v_r according to standard statistical mechanical procedures. Thus keeping all v_r fixed, except for v_j,

$$\partial L/\partial v_j = -kT[\ln Z_j - \ln v_j + \alpha' + j\gamma] + kT = 0$$

The equilibrium expressions are, writing $\alpha \equiv \alpha' - 1$,

$$v_{j0}/Z_j = \exp(\alpha + j\gamma) \tag{1.7.13}$$

In order to interpret the Lagrangian multipliers, note that by putting eq. (1.7.13) into eq. (1.7.9),

$$e^{\alpha} \sum_{r=0}^{M} Z_r \exp(r\gamma) = N$$

This enables one to write eq. (1.7.13) as

$$v_{j0}/N = \lambda^j Z_j / \sum_{r=0}^{M} \lambda^r Z_r \quad (\lambda \equiv \exp\gamma) \tag{1.7.14}$$

In order that this agrees with the results (1.7.3) and (1.7.4), one has to check that $\gamma = \mu/kT$. This can be done by noting from eqs. (1.7.12), (1.7.13) together with (1.7.9) and (1.7.10) that

$$F(\mathbf{v}_0) = -kTN\ln N + kT\sum_{r=0}^{M} v_{r0}(\alpha + r\gamma)$$

$$= kTN\ln N + \alpha kTN + \gamma kTN_e$$

However, from general thermodynamics, the chemical potential of an electron system is given by $\partial F/\partial N_e$, whence $\gamma kT = \mu$ as required.

Note that the approach of section 1.2, based on the grand canonical ensemble, is exact, while the free energy approach utilizes Stirling's approximation. This situation is very much as already indicated at the end of section 1.2. In all these cases the Stirling approximation is rather spurious. Our procedure via the grand canonical ensemble is therefore more direct in yielding the desired mean, rather than most probable, values.

1.7.2 Some special cases of occupation probabilities for localized states

The simplest special case arises for $M = 1$, when eq. (1.7.4) gives

$$\frac{v_0}{N} = \frac{1}{1 + \lambda Z_1/Z_0} = \left\{1 + \frac{g_1}{g_0}\exp\left[\frac{\mu - E(\tfrac{1}{2})}{kT}\right]\right\}^{-1} \tag{1.7.15}$$

where the substitution (1.7.6) and the notation (1.7.7) have been used. For a hydrogen-like atom the ($r = 0$)-state might correspond to a proton-like particle which has therefore $g_0 = 1$. A captured electron may have either of two spins so that $g_1 = 2$ in this case. This is the case of the so-called 'unpaired' spin. If the $(r = 0)$-state has an electron in an s-state, then the $(r = 1)$-state must have the matching spin. In that case $g_0 = 2$, $g_1 = 1$. This is the case of 'paired' spins. The number of trapped electrons is for $r = 1$

$$N_t = v_1 = N - v_0 = N \left\{ 1 + \frac{g_0}{g_1} \exp\left[\frac{E(\frac{1}{2}) - \mu}{kT} \right] \right\}^{-1} \tag{1.7.16}$$

where $g_0/g_1 = \frac{1}{2}$ if the spins are unpaired, and $g_0/g_1 = 2$ if the spins are paired. If the center has excited states, then for this, and also for other reasons, $E(\frac{1}{2})$ can depend on temperature.

Some additional examples can be based on eq. (1.7.4). Thus

$$N_e = \sum_{j=0}^{M} j\lambda^j Z_j \Big/ \sum_{r=0}^{M} \lambda^r Z_r = \frac{M}{1 + \sum_{j=0}^{M-1} (M-j)\lambda^j Z_j \Big/ \sum_{r=1}^{M} r\lambda^r Z_r} \tag{1.7.17}$$

which leads to the following expressions for N_e:

$$1/(1 + Z_0/\lambda Z_1) \qquad\qquad (M = 1) \tag{1.7.18}$$

$$2 \Big/ \left(1 + \frac{2Z_0 + \lambda Z_1}{\lambda Z_1 + 2\lambda^2 Z_2} \right) \qquad (M = 2) \tag{1.7.19}$$

$$3 \Big/ \left(1 + \frac{3Z_0 + 2\lambda Z_1 + \lambda^2 Z_2}{\lambda Z_1 + 2\lambda^2 Z_2 + 3\lambda^3 Z_3} \right) \quad (M = 3) \tag{1.7.20}$$

These results have been reviewed (with additional details) elsewhere in [1.7.9] and [1.7.10]. They apply also in principle to vacancies or interstitials (see section 1.10).

In the important case of Au in Si, $r = 0, 1, 2$ refer to the positively, neutral and negatively charged Au atoms, respectively. The three-component vector $\mathbf{v}$ is

$$(v_0, v_1, v_2) = (Z_0/\lambda Z_1, 1, \lambda Z_2/Z_1) v_1 \tag{1.7.21}$$

$$= \left(\frac{g_0}{g_1} \exp\left[\frac{E(\frac{1}{2}) - \mu}{kT} \right], \quad 1, \quad \frac{g_2}{g_1} \exp\left[\frac{\mu - E(\frac{3}{2})}{kT} \right] \right) v_1 \tag{1.7.22}$$

where the substitution (1.7.6), (1.7.7) has been used. The energy $E(\frac{1}{2})$ is the donor energy level E_D and $E(\frac{3}{2})$ is the acceptor energy level, both regarded as located on a single-electron energy level scheme. The energy zero is arbitrary and does not occur in eq. (1.7.22), which features only single-electron energy differences. Also v_0 and v_2 have been expressed in terms of the number of neutral centers. This interpretation of v_1 is, however, not essential and the argument is independent of which of the charge states is neutral. A careful analysis of carrier equilibrium effects as a function of temperature has been made for Cr-doped GaAs, for example. the Cr levels follow the valence band as the temperature is changed [1.7.11].

As another simple example, consider N_D singly-ionizable donors under neglect of excited states. Let g_0 and g_1 be the degeneracies of the ground states when no electron is trapped and when one electron is trapped, respectively, so that

$$\frac{Z_0}{Z_1} = \frac{g_0}{g_1} \exp \eta_D, \quad \eta_D \equiv \frac{E(\frac{1}{2})}{kT} \equiv \frac{E_D}{kT} \tag{1.7.23}$$

Then E_D is the energy level of the donor in the one-electron energy scheme. The number of trapped electrons is given, using eq. (1.7.16) with a slight change of notation, by

$$N_t = \frac{N_D}{(g_0/g_1) \exp (\eta_D - \gamma) + 1} \tag{1.7.24}$$

Suppose now N_{Du} neutral donors with unpaired electrons and N_{Dp} neutral donors with paired electrons are introduced into a pure semiconductor at absolute temperature T and that they have ground state levels which lie at the *same* energy E_D. Then the number of electrons in the conduction band which come from the paired sites is, with $a \equiv \exp(\eta_D - \gamma)$,

$$N_p \equiv N_{Dp} - \frac{N_{Dp}}{1+2a} = \frac{2aN_{Dp}}{1+2a} \tag{1.7.25}$$

The number from unpaired sites is

$$N_u \equiv N_{Du} - \frac{N_{Du}}{1+\frac{1}{2}a} = \frac{aN_{Du}}{2+a} \tag{1.7.26}$$

Hence the fraction f of electrons in the conduction band from paired sites is given by

$$\frac{1}{f} = \frac{N_p + N_u}{N_p} = 1 + \frac{1+2a}{4+2a} \frac{N_{Du}}{N_{Dp}}$$

Suppose now that one does not know if the donors introduced into the sample were paired or unpaired and that we simply regard them as $N_{Dp} + N_{Du}$ donors with energy level E_D and uncertain degeneracy factor $g \left[\equiv \left(\frac{g_0}{g_1}\right)_{\text{effective}} \right]$. We wish to calculate g. Its equation is given by

$$\frac{N_{Du}}{1+\frac{1}{2}a} + \frac{N_{Dp}}{1+2a} = \frac{N_{Du} + N_{Dp}}{1+ga}$$

Solving for g,

$$g = \frac{(1+2a)(N_{Du}/N_{Dp}) + 4 + 2a}{(2+4a)(N_{Du}/N_{Dp}) + 2 + a}$$

Substituting for N_{Du}/N_{Dp} in terms of f, finally gives

$$g = \frac{2}{4-3f} \tag{1.7.27}$$

Thus for $N_{\mathrm{Dp}} = 0$ only unpaired electrons play a part, so that $f = 0$ and $g = \frac{1}{2}$. If $N_{\mathrm{Du}} = 0$ only paired electrons play a part, so that $f = 1$ and $g = 2$. Intermediate values are found in other cases. As N_{Dp} rises from zero, g rises from one-half to a maximum value of two.

More generally, let E, E' denote *effective* energy levels which incorporate the effect of degeneracy and excited states

$$Z_1/Z_0 \equiv \exp\left(-\frac{E}{kT}\right), \quad Z_1'/Z_0' \equiv \exp\left(-\frac{E'}{kT}\right)$$

Let the number of levels be N at E and N' at E'. Then an effective energy level $E_{\mathrm{eff}} = \eta_{\mathrm{eff}} kT$ may be defined by $[a \equiv \exp(\eta - \gamma), \; a' \equiv \exp(\eta' - \gamma), \; a_{\mathrm{eff}} \equiv \exp(\eta_{\mathrm{eff}} - \gamma)]$

$$\frac{N}{1+a} + \frac{N'}{1+a'} = \frac{N+N'}{1+a_{\mathrm{eff}}}$$

Then one can show that the fraction of conduction band electrons $f \equiv n'/(n+n')$ which come from the N' levels is in the absence of other levels and bands

$$f = \frac{\exp(-\eta) - \exp(-\eta_{\mathrm{eff}})}{\exp(-\eta) - \exp(-\eta')}$$

This can be established by first deriving

$$a_{\mathrm{eff}} = \frac{a(a'+1)(N/N') + a'(a+1)}{1+a+(a'+1)(N/N')}$$

Into this one can substitute the following expression derived from the f-equation:

$$\frac{N}{N'} = \left(\frac{1}{f} - 1\right)\frac{a'(a+1)}{a(a'+1)}$$

This gives the desired relation for f.

As a first application of the f-formula note that if $E = E' (= E_0$ say) then E_{eff} is also E_0, as one would expect.

As a second application suppose that, when occupied, the N' levels are paired and the N levels are unpaired, both at the same energy E_0. Then

$$\exp(-\eta) = 2\exp(-\eta_0), \quad \exp(-\eta') = \tfrac{1}{2}\exp(-\eta_0)$$

and the f-relation gives at once

$$\exp(\eta_{\mathrm{eff}}) = [2/(4-3f)]\exp(\eta_0)$$

One sees from

$$(Z_0/Z_1)_{\mathrm{eff}} = g \exp \eta_0$$

that the effective energy level lies at E_0 with effective degeneracy (1.7.27).

The effective energy level concept is useful when one wishes to ignore either degeneracies of energy levels or ignore excited states, or ignore both. For example, if excited states are neglected so that eq. (1.7.6) holds

$$\frac{Z_r}{Z_{r+1}} = \frac{g_r}{g_{r+1}}\exp\frac{E_{r+1}-E_r}{kT} = \frac{g_r}{g_{r+1}}\exp[\eta(r+\tfrac{1}{2})] \equiv \exp[\eta_{\mathrm{eff}}(r+\tfrac{1}{2})] \qquad (1.7.28)$$

where η_{eff} incorporates the effect of degeneracies. This implies in general only a numerically small change because of the occurrence of the logarithm:

$$\eta_{eff}(r+\tfrac{1}{2}) - \eta(r+\tfrac{1}{2}) = \ln(g_r/g_{r+1})$$

There is thus an entropy contribution $k\ln(g_r/g_{r+1})$ to $E_{eff}(r+\tfrac{1}{2})$ due to transitions between an r- and an $(r+1)$-electron centre. An additional term arises from the changes in atomic vibrations due to lattice relaxation. These terms must be regarded as formally included in our later results for E_{eff}, ΔG, etc. in equations (1.7.29) to (1.7.36) and in Fig. 1.7.2.

More generally one can put

$$\frac{Z_r}{Z_{r+1}} = \frac{\sum_i g_{r,i}\exp(-\eta_{r,i})}{\sum_i g_{r+1,i}\exp(-\eta_{r+1,i})} = \exp[\eta_{eff}(r+\tfrac{1}{2})] \qquad (1.7.29)$$

to include the effect of degeneracy *and* excited states. Such effective energies are not solutions of the Schrödinger equation. A kind of averaging has occurred which brings in the Boltzmann factor, and hence a temperature-dependence of the effective energy.

1.7.3 Gibbs free energies and entropy factors

Let us consider a special case $r = 0$ of relation (1.7.28). It tells us that the degree of ionization of a donor is

$$\frac{v_0}{N_D} = \frac{v_0}{v_0+v_1} = \frac{1}{1+\lambda Z_1/Z_0} = \frac{1}{1+\exp[\gamma-\eta_{eff}(\tfrac{1}{2})]} \qquad (1.7.30)$$

Since $\mu = kT\gamma$ can represent different thermodynamic functions depending on the variables which are kept constant (and put in brackets below):

$$U(S,V), \quad H(S,p), \quad F(V,T), \quad G(p,T) \qquad (1.7.31)$$

one now has a choice already foreshadowed in eqs. (1.1.14) and (1.1.15). The last form is the most practical from the experimental point of view. It is therefore becoming accepted that

$$\Delta G \equiv E_{eff}(\tfrac{1}{2}) - \mu \qquad (1.7.32)$$

can be regarded as a Gibbs free energy which is in this case associated with the ionization of a donor. If we go back to 1954 when J.A. Burton gave his key review of impurity centers in Ge and Si at the International Conference on Semiconductors in Amsterdam, the talk was about energies and not yet about free energies [1.7.12].

It is fortunate that different ensembles yield normally similar results for large systems, since the change to a basis of constant pressure from constant volume (the latter is implied by use of the grand canonical ensemble) really requires one to switch the basis of the theory to the so-called constant pressure ensemble. This is not greatly loved (even by statistical mechanicians), because it has a certain arbitrariness associated with it; see, for example, [1.7.13]. Fortunately this need not worry us here, and we shall often continue to speak about energy rather than free energy.

The basic idea in the pressure ensemble is to suppose that the probability of finding a certain volume V and energy E when pressure and temperature are given is

$$P_i \equiv P(N; E_{iN}, V_{iN}) = Q^{-1} \exp\left(-\frac{E_{iN} + pV_{iN}}{kT}\right)$$

Here the number N of indistinguishable particles is also given, and Q is a normalizing factor – the constant pressure partition function. The suffix i specifies a state. Hence

$$TS = -kT \sum_i P_i \ln P_i = kT \sum_i \left[\frac{E_{iN} + pV_{iN}}{kT}\right] P_i + kT \ln Q$$

$$= H + kT \ln Q$$

where H is the mean (thermodynamic) enthalpy. One sees that for consistency with thermodynamics

$$P_i = \exp\frac{G}{kT} \exp\left(-\frac{H_{iN}}{kT}\right)$$

where H_{iN} is the enthalpy $E_{iN} + pV_{iN}$ of state (i, N). Thus the ratio of probabilities yields an *enthalpy* difference, i.e. an enthalpy of activation:

$$\frac{P_i}{P_j} = \exp\frac{H_{jN} - H_{iN}}{kT}$$

rather than the *energy* difference, as expected from a canonical ensemble.

We know from eq. (1.1.10) that $G = H - TS$ so that

$$\Delta G = \Delta H - T\Delta S \tag{1.7.33}$$

at constant temperature. Using eqs. (1.7.30), (1.7.32) and (1.7.33)

$$\frac{v_0}{N_D} = \left\{1 + X \exp\left(-\frac{\Delta H}{kT}\right)\right\}^{-1}, \quad X \equiv \exp\left(\frac{\Delta S}{k}\right) \tag{1.7.34}$$

The 'entropy factor', X, for the interaction of the impurity and the conduction band can be estimated from

$$\Delta S = -\left(\frac{\partial \Delta G}{\partial T}\right)_{p,N} \tag{1.7.35}$$

which follows at once from eq. (1.1.13). Also

$$\Delta H = \left(\frac{\partial (\Delta G/T)}{\partial (1/T)}\right)_{p,N} \tag{1.7.36}$$

since

$$H = G + TS = G - T(\partial G/\partial T)_{p,N}$$

The heat capacity at constant pressure is

$$\Delta C = (\partial \Delta H/\partial T)_{p,N} \tag{1.7.37}$$

For any energy gap ΔG, therefore, its temperature dependence furnishes the key ingredients, ΔS and ΔH, of equations like (1.7.34). Some results for the main energy gap (i.e. for the formation of electrons and holes) are shown in Fig. 1.7.2. If ΔS goes up to six Boltzmann constants, X rises to a value of 403. It has lower values for the interaction of a defect level with a band; see section 2.4.6, Table 2.4.6, where some physical interpretation of (1.7.35) is also given.

For the main energy gap electron–hole pair creation can be represented by a reaction with an equilibrium condition:

$$0 \rightleftharpoons e + h, \quad \mu_{eq} + \mu_{heq} = 0 \tag{1.7.38}$$

One can, in a chemical analogy, associate standard states with electrons and holes for which $(n, p, \mu_e, \mu_h) = (n^0, p^0, \mu_e^0, \mu_h^0)$. With Boltzmann statistics one then has for low concentrations 'compositional fractions'

$$x_n \equiv \frac{n_{eq}}{n^0} = \exp\frac{\mu_{eeq} - \mu_e^0}{kT}, \quad x_p \equiv \frac{p_{eq}}{p^0} = \exp\frac{\mu_{heq} - \mu_h^0}{kT} \tag{1.7.39}$$

so that the equilibrium product of concentrations is

$$(np)_{eq} = n^0 p^0 \exp\left(-\frac{\mu_e^0 + \mu_h^0}{kT}\right)$$

The obvious standard states are those at the band extrema, given by $(\mu_e^0, \mu_h^0) = (E_c, -E_v)$, whence from eqs. (1.6.11) and (1.6.12)

$$(n^0, p^0) = (N_c, N_v)$$

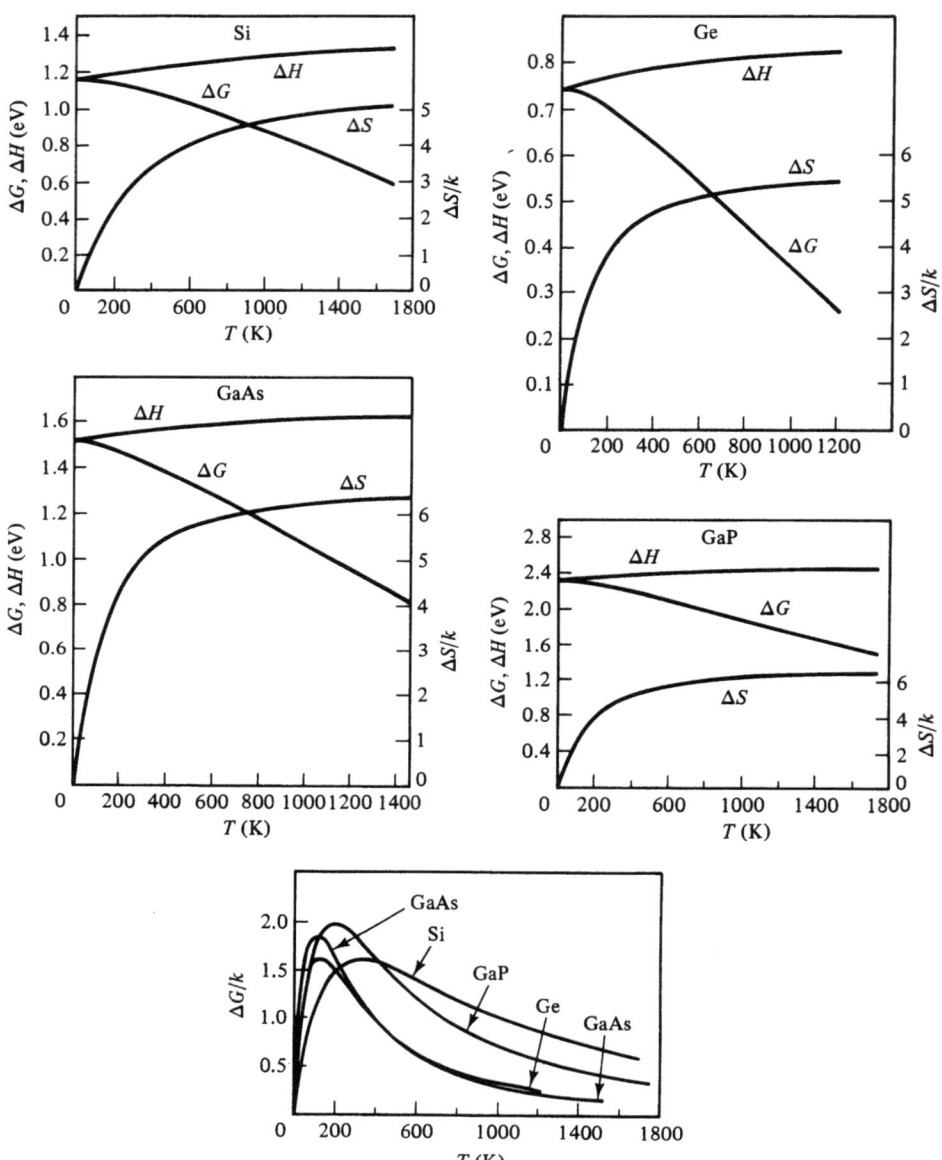

Fig. 1.7.2. The quantities (1.7.35)–(1.7.37) for the forbidden energy gap between the conduction and valence bands as a function of absolute temperature for four important semiconductors [1.7.14].

The negative sign for E_v arises from the fact that on the usually adopted energy scale, electronic energies are measured upwards and hole energies downwards.

1.7.4 Points from the literature

(1) An operator method of dealing with the statistics of localized impurities has also been presented [1.7.15].

(2) The theory has been applied in other areas, for example in the determination of capture cross sections by means of an amplitude-modulated electron beam [1.7.16].

(3) It should be noted that an overlap of localized states with a continuum of states, such as a conduction band, is not only theoretically possible, but has in fact been observed experimentally, for instance in CdF_2 [1.7.17].

(4) The experimental methods of characterizing the properties of semiconductors with special reference to impurities are reviewed in [1.7.18].

(5) Some additional early papers in which the subject of this section was explored are [1.7.19], [1.7.20], [1.7.21] and [1.7.22].

(6) The importance of Au in Si, referred to below eq. (1.7.20) is due to the fact that with its energy levels near midgap it can be an efficient recombination center. This, together with its large diffusion coefficient, makes it useful for controlling the lifetimes of Si devices. Au and Pt are used, for example, to decrease the switching time of transistors. The solubility of Au in Si was studied for instance in [1.7.23].

1.8 Interaction effects from impurities, including screening

1.8.1 Interaction effects (impurity bands, Mott and Anderson transitions)

At low temperatures weakly doped semiconductors have a d.c. electrical conductivity which goes exponentially to zero, since the number of conduction band electrons does. This can be seen, for example, from eq. (1.7.26). We have $\mu > E_D$ so that all donor states are occupied near $T = 0$. Therefore $a \ll 1$ and

$$N \sim \frac{g_0}{g_1} N_D \exp(\eta_D - \gamma) \to 0 \tag{1.8.1}$$

Electrical d.c. conductivity is thus of an activated type. At sufficient impurity concentration there is a *significant* overlap of electron wavefunctions for neighboring impurities so as to enable electrons to hop from impurity to impurity and so contribute significantly to the current. When this happens, one expects a positive but small conductivity as $T \to 0$, just as in the case of metals (where it is larger). One can take the concentration $N_D = N_{Dcrit} \sim 10^7$ to 3×10^{18} cm^{-3}. One also talks of impurity-band conduction. This type of metal–insulator transition depends on the effect of electron–electron interactions and is often referred to as the Mott transition [1.8.1]; for a simple account see [1.8.2]. A large literature exists on this subject, but the nature of the transition is not fully understood; in some cases it can be due to a transition to a new crystal lattice, for example. What is certain is that the impurities form a band which covers an energy range and leads to a decrease in impurity ionization energy as doping increases (see section 1.8.2).

A related transition occurs already in a model which neglects the Coulomb interactions among electrons. Instead of impurities consider a random arrangement of potential wells with an electron occupying the appropriate level in each well. Under what conditions are the electrons localized so that the conductivity again drops to zero as $T \to 0$? This is the problem of Anderson localization [1.8.3], [1.8.4]. This is a many-scatterer wave coherence effect which leads to trapping of a wave excitation in a region with many scatterers or inhomogeneities. It applies to electron, optical and acoustic waves, and does not need to involve local trapping.

It can also be shown that the density of states in an impurity band has a minimum at the Fermi level, should it lie in the band. This means that there is effectively an energy gap between the occupied and the empty states. It is brought about by the Coulomb interaction, and the energy gap is sometimes called the Coulomb gap.

The calculation of the density of states in a heavily doped material is difficult. Several approximational schemes have been used (e.g. [1.8.5]–[1.8.8]), and we return to this problem in section 5.2.3, following the introductory remarks in this section.

For a lightly doped semiconductor the resistivity, following (1.8.1), may be expected to behave as

$$\rho \propto \frac{1}{N} \propto \exp(\eta_c - \eta_D) \equiv \exp(E/kT)$$

where E is the ionization energy of the relevant impurity. It was found early in the work on semiconductors that E depends both on doping and on *compensation* of donors by acceptors, or acceptors by donors ([1.8.9], [1.8.10]). Such effects have to be borne in mind when doping is varied. They are discussed semiquantitatively in the next subsection. For work in this area see [1.8.11] and [1.8.12] and for detailed reviews see [1.8.13] and [1.8.14].

1.8.2 Concentration-dependent activation energies

A rough semiquantitative model will show that the activation energy of an impurity may be expected to decrease as doping increases. Consider a hydrogen-like impurity in a medium of dielectric constant ε. The radius of the orbit with principal quantum number t is

$$\tilde{r}_t = \frac{\varepsilon m}{m^*} r_1 t^2 \quad \left(r_1 \equiv \frac{\hbar^2}{me^2} = 0.528 \text{ Å} \right) \tag{1.8.2}$$

where m is the electron rest mass and m^* is the effective mass. If N_D is the number of donors per unit volume, then the tth orbits will just begin to overlap when

$$\frac{4\pi}{3} \tilde{r}_t^3 N_D = 1 \tag{1.8.3}$$

Let us define t by this condition. To be in the continuum of levels a bound electron need therefore be promoted only to the 'ionization' level t which comes down by (1.8.3) as N_D goes up. In fact, the ionization energy is for an initial electron state of principal quantum number s:

$$E = A(s^{-2} - t^{-2}), \quad A \equiv (m^*/m\varepsilon^2)(me^4/2\hbar^2) = 13.6m^*/m\varepsilon^2 \text{ eV}$$

$$= \frac{A}{s^2}[1 - (BN_D)^{\frac{1}{3}}], \quad B \equiv \frac{4\pi}{3} \tilde{r}_s^3 \tag{1.8.4}$$

Using Table 1.4.1, this gives for germanium

$$E = (0.029 - 0.72 \times 10^{-8} N_D^{\frac{1}{3}}) \text{ eV} \quad (N_D \text{ in cm}^{-3})$$

which is in broad agreement with experiments on arsenic donors in germanium (see Fig. 1.8.1 [1.8.10])

$$E = (0.0125 - 2.35 \times 10^{-8} N_D^{+\frac{1}{3}}) \text{ eV} \tag{1.8.4a}$$

Here N_D^+ is the concentration of ionized donors. The approach to $E = 0$ later gave rise to the suggestion [1.8.15] that E should vanish abruptly at a critical impurity concentration. This group of phenomena, in which activated electrical conduction goes over to metallic conduction as the impurity content is increased, is another aspect of the Mott transition. For more recent experiments on germanium see [1.8.16]. If both donor and acceptors are present in a material, the phenomenon of compensation occurs. Since donor electrons can fill the acceptors, only a small concentration of current carriers may result. The simple model (1.8.4) would have to be extended to allow for compensation.

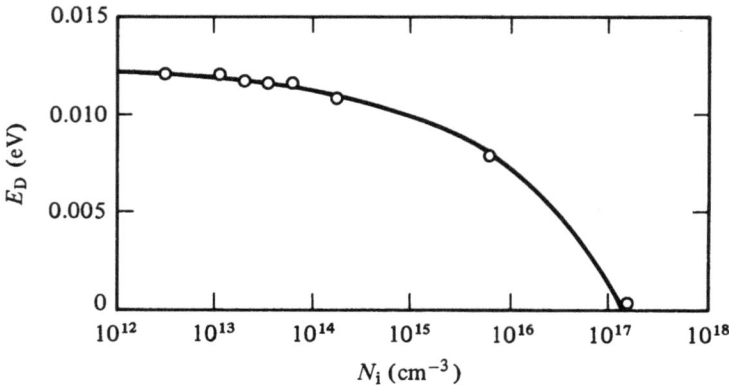

Fig. 1.8.1. The ionization energy for As donors in Ge, as a function of the average density N_i of ionized donors at low temperature. The curve is given by eq. (1.8.4a). The experimental points are from [1.8.10].

1.8.3 Screening

The Coulomb interactions among charge carriers, both fixed and mobile, in any given volume leads to very complicated correlations among the motions of the particles which present us with a truly many-body effect. Any simplified, if partial, view which one can gain of this situation is therefore welcome, and we present one such view here. The repulsion of like and the attraction of unlike charges in the neighborhood of a fixed charged impurity is represented by an electrostatic potential $\varphi(\mathbf{r}, t)$ which has in principle a long range, falling off only as r^{-1} with distance. However, mobile carriers of predominantly opposite charge tend to be near the fixed charge, and as a result the potential reaches a constant value in a shorter distance. The incredibly complicated dynamics thus leads to a comparatively simple law as a first approximation. This will now be derived.

Suppose a charge Q, inserted at $\mathbf{r} = 0$ at time $t = 0$ moves with velocity $\mathbf{v}$. This produces a change in the charge concentration due to electrons by an amount $q\rho(\mathbf{r}, t)$ which is a function of space and time. If $q = -|q|$ is the charge on an electron, the electrostatic potential $\varphi(\mathbf{r}, t)$ due to the charge Q is related to $\rho(\mathbf{r}, t)$ by Poisson's equation

$$\nabla^2\varphi(\mathbf{r}, t) = -4\pi[Q\delta(\mathbf{r} - \mathbf{v}t) + q\rho(\mathbf{r}, t)] \tag{1.8.5}$$

It follows that

$$[\nabla^2 - \Lambda^2(\mathbf{r}, t)]\varphi(\mathbf{r}, t) = -4\pi Q\delta(\mathbf{r} - \mathbf{v}t) \tag{1.8.6}$$

where

$$\Lambda^2(\mathbf{r}, t) \equiv 4\pi|q|\rho(\mathbf{r}, t)/\varphi(\mathbf{r}, t) \tag{1.8.7}$$

This rather general result lends itself to the study of three special cases. The first is used to lead naturally to the more important second and third special cases.

(i) The simplest case arises in the absence of polarizable material, $\rho(\mathbf{r}, t) = \Lambda(\mathbf{r}, t) = 0$, and a static charge. Then using a Fourier expansion

$$\varphi(\mathbf{r}) = \sum_{\mathbf{k}} \varphi_{\mathbf{k}} \exp(i\mathbf{k} \cdot \mathbf{r}) \tag{1.8.8}$$

of $\varphi(r)$ and $\delta(r)$, eqs. (1.8.6) and (1.8.7) yield

$$-\sum_{\mathbf{k}} k^2 \varphi_{\mathbf{k}} \exp(i\mathbf{k} \cdot \mathbf{r}) = -\frac{4\pi Q}{V} \sum_{\mathbf{k}} \exp(i\mathbf{k} \cdot \mathbf{r})$$

It follows at once that for all $\mathbf{k}$

$$\varphi_{\mathbf{k}} = \frac{4\pi Q}{Vk^2} \quad \text{i.e. } \varphi(\mathbf{r}) = Q/|r| \tag{1.8.9}$$

It is from this argument that one can expect *nonzero* $\Lambda(\mathbf{r}, t)$ and *nonzero* velocity $\mathbf{v}$ to yield in the presence of polarizable material

$$\varphi_{\mathbf{k},\omega} = \frac{4\pi Q}{Vk^2 \varepsilon(\mathbf{k}, \omega)} \tag{1.8.10}$$

It is now no longer possible to write down the expression for $\varphi(\mathbf{r}, t)$ in closed form: the dielectric function (it is hardly a 'constant') depends on the wavevector and the frequency ω. The latter is introduced by a Fourier analysis in time. The quantity $\varepsilon(\mathbf{k}, 0)$ is the static dielectric constant, and $\varepsilon(0, \omega)$ is the long-wavelength dielectric constant. The dielectric response function yields a great deal of insight into the behavior of many-body charged systems [1.8.17].

(ii) The form of eq. (1.8.7) suggests the next approximation of 'linear screening' in which $\rho(\mathbf{r}, t)$ is proportional to $\varphi(\mathbf{r}, t)$. Again neglect the time dependence, put $\mathbf{v} = 0$ and assume $\Lambda(\mathbf{r}, t)$ to be a constant. The Fourier component (1.8.10) is then

$$\varphi_{\mathbf{k}} = \frac{4\pi Q}{Vk^2(1 + \Lambda^2/k^2)} \quad \text{i.e. } \varepsilon(\mathbf{k}, 0) = 1 + \Lambda^2/k^2 \tag{1.8.11}$$

In this case the potential can be found by (C.14) of Appendix C, and is

$$\varphi(r) = \frac{Q}{r} \exp(-\Lambda r)$$

It is sometimes convenient to insert an effective dielectric constant to take account of the effect of interband transitions [1.8.17], the polarization of the ions, etc., in an *ad hoc* manner to find

$$\varphi(r) = \frac{Q}{\varepsilon r} \exp(-\Lambda r) \tag{1.8.12}$$

Effectively one has here a static charge Q inserted at $r = 0$ and finds that it is screened out at distances of order Λ^{-1}. This is the case of static linear screening. It can give only a rough guide for the case of dynamic linear screening when all the interacting charges are in motion. It may be reasonable for slowly moving electrons, but if one considers a fast electron, shot through an electron gas, the remaining electrons have insufficient time to follow the motion. Accordingly the polarization cloud cannot form fully round the fast electron. In this case one would expect the screening effect to be slight. This apparent wavevector dependence of the screening parameter Λ merely reflects the fact that the passage from eq. (1.8.10) to eq. (1.8.11) is no longer justified. The detailed theory confirms this conclusion ([1.8.18] and [1.8.19]).

(iii) Here it will be adequate to confine our attention to the approximations embodied in eq. (1.8.11). A slightly more general formulation considers groups of carriers distinguished by a subscript j. The linear screening approximation is then

$$\rho_j(\mathbf{r}) = f_j\,\varphi(\mathbf{r})$$
$$(j = 1, 2, \ldots; f_j \text{ independent of } \mathbf{r}). \tag{1.8.13}$$

The Poisson equation for the disturbance due to a charge Q at $r = 0$ is, in extension of eq. (1.8.5),

$$\nabla^2\varphi(\mathbf{r}) = -\frac{4\pi}{\varepsilon}\left(Q\delta(\mathbf{r}) + \sum_j q_j\,\rho_j(\mathbf{r})\right)$$

where an effective dielectric constant has been introduced. Hence

$$(\nabla^2 - \Lambda^2)\,\varphi(\mathbf{r}) = -\frac{4\pi}{\varepsilon}\,Q\delta(\mathbf{r})$$

and one recovers eqs. (1.8.11) and (1.8.12). However, the screening parameter is now given by

$$\Lambda^2 = -\frac{4\pi}{\varepsilon}\sum_j f_j q_j \tag{1.8.14}$$

It is the object of specific theories of linear screening to find expressions for the constants f_j which occur in eq. (1.8.13) and hence to determine Λ via eq. (1.8.14).

The above results show that the potential near a charged impurity falls off exponentially. In order to evaluate Λ, note from eqs. (1.2.16) and (1.1.7) that the parameter enabling us to treat electrons and holes by similar integrals is

$$\gamma = \frac{|q_j|}{q_j}\frac{E_j - \mu_j}{kT} = \frac{|q_j|}{q_j}\frac{B_j - |q|\,\varphi - \mu_j}{kT} \tag{1.8.15}$$

Now let $\nu_j(\gamma_j)$ be the carrier concentration (electrons or holes) in a group j of carriers when the quasi-Fermi level for the group has the value μ_j. A disturbance occurs and there is a small change from a constant electrostatic potential, taken to be zero, to the additional potential $\varphi(r)$ produced by the charge. Then the change $\delta\gamma_j$ in γ_j is by eq. (1.8.15) $- q_j\varphi/kT_j$, so that, assuming temperature and volume to be kept constant,

$$\rho_j = v_j(\gamma_j + \delta\gamma_j) - v_j(\gamma_j) = \delta\gamma_j\left(\frac{\partial v_j}{\partial\gamma_j}\right)_{T_j, V} = -\frac{q_j\varphi}{kT_j}\left(\frac{\partial v_j}{\partial\gamma_j}\right)_{T_j, V} \qquad (1.8.16)$$

This leads by eq. (1.8.13) to

$$f_j = \frac{q_j}{kT_j}\left(\frac{\partial v_j}{\partial\gamma_j}\right)_{T_j, V} \qquad (1.8.17)$$

and finally by eq. (1.8.14) to ([1.8.20])

$$\Lambda^2 = \frac{4\pi}{\varepsilon}\sum_j \frac{q_j^2}{kT_j}\left(\frac{\partial v_j}{\partial\gamma_j}\right)_{T_j, V} = \frac{4\pi}{\varepsilon}\sum_j \frac{|q_j|\,v_j v_j}{D_j} \qquad (1.8.18)$$

This is our main result in this subsection showing that the screening radius is related to the Einstein ratio (1.2.18). One would indeed expect intuitively an easy response of carriers to an electric field (large v_j) to lead to an efficient screening of the Coulomb field (large Λ).

For parabolic bands eq. (1.6.3) enables us to put

$$v_j = v_{0j} F_{\frac{1}{2}}(\gamma_j) \qquad (1.8.19)$$

Hence one can verify by a partial integration that

$$\left(\frac{\partial v_j}{\partial\gamma_j}\right)_{T_j, V} = v_{0j} F_{-\frac{1}{2}}(\gamma_j) \qquad (1.8.20)$$

The formula (1.8.18) then becomes

$$\Lambda^2 = \frac{4\pi}{\varepsilon}\sum_j \frac{q_j^2 v_{0j}}{kT_j} F_{-\frac{1}{2}}(\gamma_j)$$

For example, for one conduction band and one valence band one finds [1.8.21]

$$\Lambda^2 = \frac{4\pi q^2}{k\varepsilon}\left[\frac{v_{0c}}{T_c}F_{-\frac{1}{2}}\left(\frac{\mu_c - E_c}{kT_c}\right) + \frac{v_{0v}}{T_v}F_{-\frac{1}{2}}\left(\frac{E_v - \mu_h}{kT_v}\right)\right] \qquad (1.8.21)$$

As $F_s(a) \sim e^a$ for nondegenerate materials for all $s > -1$, eq. (1.8.21) gives the so-called Debye screening length in that limit (with $v_{0c} \equiv n$):

$$l_{\text{Deb}} \equiv \Lambda^{-1} = \left(\frac{\varepsilon kT}{4\pi q^2 n}\right)^{\frac{1}{2}} \qquad (1.8.22)$$

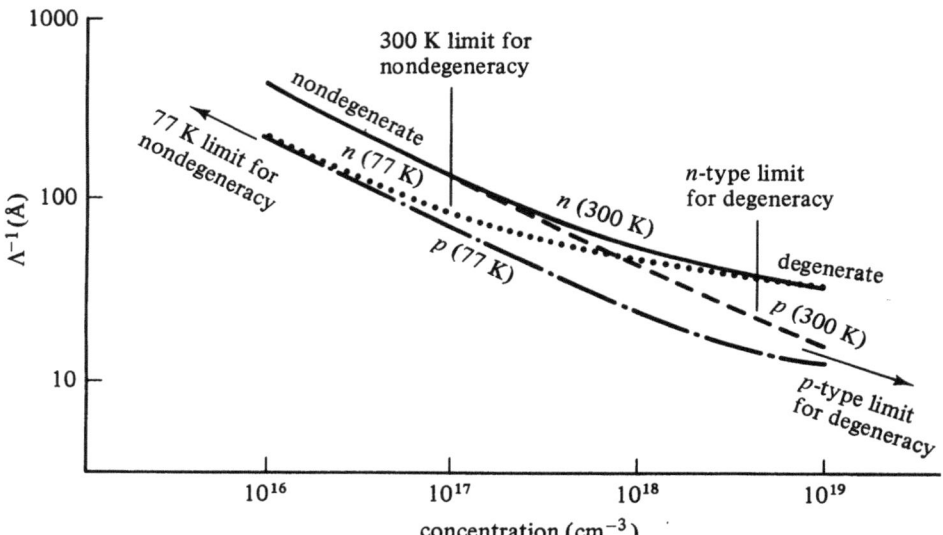

Fig. 1.8.2. Screening lengths in GaAs, according to eq. (1.8.21), for electrons and holes in GaAs at 300 K and 77 K using $\varepsilon = 12.6$, $m_c = 0.072m$, $m_v = 0.5m$.

It is believed to be a good approximation to the self-consistent screening length in the presence of band tails [1.8.22]. In the case of electrons, eq. (1.6.3) gives

$$n = 2\left(\frac{2\pi m^* kT}{h^2}\right)^{\frac{3}{2}} F_{\frac{1}{2}}(\gamma) = N_c F_{\frac{1}{2}}(\gamma) \tag{1.8.23}$$

and in the degenerate limit we can use relation (1.6.14). Hence in that limit

$$\gamma = \left(\frac{3}{4}\sqrt{\pi}\frac{n}{N_c}\right)^{\frac{2}{3}} \tag{1.8.24}$$

so that

$$\Lambda^{-1} = \left[\frac{\varepsilon \hbar^2}{4q^2 m^*}\left(\frac{\pi}{3n}\right)^{\frac{1}{3}}\right]^{\frac{1}{2}} \tag{1.8.25}$$

For GaAs at room temperature with $n = 10^{16}$ cm^{-3} and in c.g.s. units, eq. (1.8.22) gives

$$l_{\text{Deb}} = \left[\frac{12.5(\frac{1}{40}\times 1.602 \times 10^{-12})^{\frac{1}{2}}}{4\pi(4.8 \times 10^{-10})^2\, 10^{16}}\right]^{\frac{1}{2}} \sim 416\ \text{Å}$$

Fig. 1.8.2 gives a plot based on eq. (1.8.21) ([1.8.20]). Fig. 1.8.3 gives the results of a computation using calculated density of states curves for heavily doped specimens [1.8.23]. The screening length decreases with doping, i.e. as the potential due to a fixed or moving charge is more rapidly damped out, the more mobile current carriers are available.

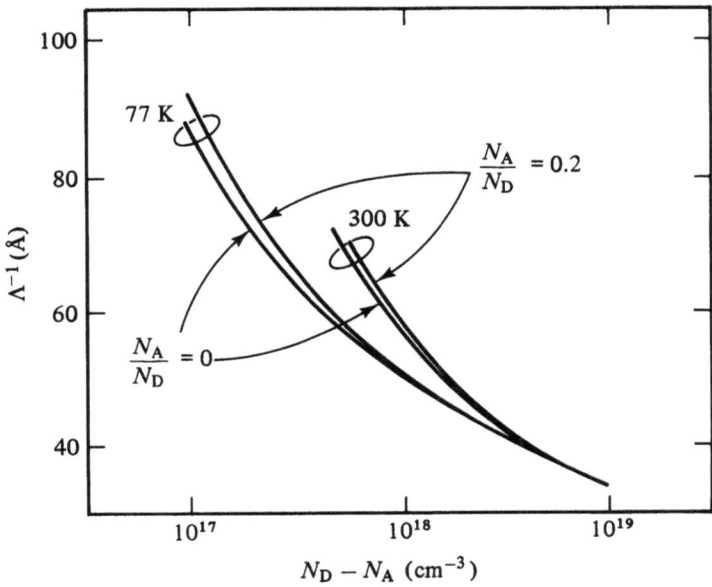

Fig. 1.8.3. Dependence of the electron screening length in GaAs on net carrier concentration $N_D - N_A$.

An imbalance of charge set up purposely or by a fluctuation causes mobile carrier of opposite charge to flow into the region, overshoot and be pulled back by the force field so induced, and then to overshoot again. This is the origin of the plasma oscillations whose angular frequency in the long-wavelength limit will be denoted by ω_p. This frequency can be estimated for one group of nondegenerate carriers by associating a travel distance Λ_j^{-1} with the periodic time ω_{pj}^{-1} of the oscillation. Attributing to this oscillation also an energy kT_j, i.e. a velocity $(kT_j/m_j^*)^{\frac{1}{2}}$, one finds

$$\frac{\omega_{pj}}{\Lambda_j} = \left(\frac{kT_j}{m_j^*}\right)^{\frac{1}{2}} \tag{1.8.26}$$

Using eqs. (1.8.25) and (1.8.26) one finds, if several groups of carriers are involved,

$$\omega_p^2 = \sum_j \frac{4\pi q_j^2 \nu_j}{\varepsilon m_j^*}$$

This is a key parameter entering the theory of plasma oscillations. The plasmon energy $\hbar\omega_p$ is of order 10 meV at $\nu \sim 10^{18}$ cm^{-3}. For a modern introduction to this type of theory see [1.8.24].

Other theories of screening are available. That which is associated with the Thomas–Fermi method yields also eq. (1.8.25) and is discussed in many books.

The improved theory of screening in [1.8.25] is widely used. It has singularities for disturbances whose wavevectors have a length equal to the diameter of the Fermi sphere in **k**-space. The singularity is responsible for weak kinks in the phonon spectrum, an effect predicted by W. Kohn and called the Kohn anomaly. There is, however, no need to go into this matter here.

The *static* dielectric screening was introduced fifty years ago in connection with the theory of electrolytes [1.8.26]. It was applied to the determination of the field surrounding a dissolved atom in a metal in [1.8.27] and [1.8.28], using the Thomas–Fermi approximation. The *dynamic* screening parameter was first used in calculations concerning soft X-ray spectra of metals [1.8.29] and was related to plasma oscillations in metals by [1.8.30]. Steadily improving theories of the dynamical screening effects in a wide variety of physical situations were developed in the last twenty years, [1.8.18] and [1.8.19]. However, the empirical and theoretical estimations of the screening parameters are still subject to some serious uncertainties [1.8.31].

1.8.4 Negative-U centers

The elegance of the partition function approach must not blind one to the fact that it holds for *any* interaction energy which arises when an electron is captured by a defect, whereas *specific* interactions occur in reality. Suppose as an illustration that a defect in its ground state has an unpaired valence electron or, equivalently, a partially occupied highest energy orbital. The defect energy is now expected to increase if an additional electron is captured. This is due to the repulsive nature of the Coulomb interaction energy which is however decreased somewhat by the lattice relaxation consequent upon capture. This converts the original interaction energy (associated with the name of J. Hubbard [1.8.32]) to an effective interaction energy (associated with P.W. Anderson [1.8.33]) here to be denoted by U. If U is negative due to a strong defect–lattice interaction, as seems to be the case in important examples, the two-electron state of the defect has a lower energy than the one-electron orbit. The latter is then never stable and the ground state of the defect cannot contain an odd number of spins. Thus no electron spin paramagnetism can arise from such defects. It is in fact this lack in amorphous chalcogenide glasses that led to the model in the 1970s in the first place.

In the meantime it has been suggested that the self-interstitial in Si [1.8.34], the Si vacancy [1.8.35], interstitial B in Si [1.8.36] as well as certain thermal donors in Si [1.8.37] are negative-U centers (this list is not complete). These identifications depend to some extent on using the correct statistics. This turns out to be equivalent to analyses developed 15 years earlier since the partition function approach holds for positive- as well as negative-U centers. It is described in section 1.7.

As an example consider the case $M = 2$ for a p-type semiconductor. Then the total concentration of defects is

$$N \equiv v_0 + v_1 + v_2 = \frac{pv_1}{p_1} + v_1 + \frac{p_2 v_1}{p} = \left(\frac{p}{p_1} + 1 + \frac{p_2}{p}\right)v_1 \qquad (1.8.27)$$

where p is the hole concentration, and for $i = 1, 2$

$$p_i \equiv pv_i/v_{i-1} = N_v Z_i \exp \eta_v/Z_{i-1}$$

The concentration of bound holes is, with eq. (1.8.27),

$$v_1 + 2v_0 = \left(1 + \frac{2p}{p_1}\right)v_1 = \frac{(1 + 2p/p_1)2N}{2p/p_1 + 2 + 2p_2/p} = \frac{2N}{1 + (1 + 2p_2/p)/(1 + 2p/p_1)}$$

Let p_T be the doping level, i.e. the concentration of free and bound holes distributed over the valence band and localized states. Then the neutrality condition is

$$p_T = p + v_1 + 2v_0 = p + \left[1 + \frac{1 + 2p_2/p}{1 + 2p/p_1}\right]^{-1} 2N \qquad (1.8.28)$$

This gives the doping level as a function of p, as required for the analysis of experiments. If excited states are neglected, as in eq. (1.7.22), then

$$p_i = (g_i/g_{i-1}) N_v \exp[-\eta(i - \tfrac{1}{2})] \qquad (1.8.29)$$

and into these relations one substitutes

$$E(\tfrac{3}{2}) = E(\tfrac{1}{2}) + U$$

so that the value of U and its sign will affect the relation between p_T and p. Thus one obtains, as a special case, a key result, viz. equation (5) of a paper by Hoffmann [1.8.38]. See also [1.8.39]. Some additional temperature dependence from the excited states may be expected when the approximation (1.8.29) is *not* appropriate. However, the link with partition functions is here made for the first time. Note that the v_0 defects can be doubly ionized donors D^{++}, singly ionized donors D^+, neutral donors $D^\times$ or acceptors, etc.

If one were to change the value of U at will, one would find that as U became negative the defects with one electron would decompose into equal numbers of defects which had captured no electrons and defects which had captured two electrons:

$$v_0 = v_2 = \tfrac{1}{2}v_1$$

A phase diagram could then be drawn with a phase transition at $U = 0$. Lattice relaxation effects are discussed further in section 5.4.4 and chapter 6.

The conventional notation in the negative-U literature is to write

$E(\tfrac{1}{2})$ as $E(+/++)$

$E(\tfrac{3}{2})$ as $E(0/+)$

suggesting, respectively, the capture or emission of a second hole by a vacancy and the capture or emission of a first hole by a vacancy. In the notation of this book $++$ is in this case the state of zero electron captured (which corresponds to a vacancy which has captured two holes), $+$ refers to the state of one electron captured, and 0 to the state of two electrons captured by the vacancy.

1.9 Saturation solubilities of impurities

Lifetimes of excess carriers in very pure materials are often limited by small residual impurity concentrations of order 10^{12} cm^{-3} or less, which are hard or impossible to measure. For cases like these it is useful to have an understanding of the principles governing the solubility of defects in semiconductors. Such theoretical considerations can provide clues as to the nature of these residual defects. The theory of section 1.7 can be applied to a situation which does depend on the charge state, s say, corresponding to electrically neutral centers. Its elucidation depends on two observations: (1) Electric fields, for example in a p–n junction, do not affect the distribution of such centers so that their concentration v_s has a constant value throughout the material in thermal equilibrium. (2) The impurities will in general diffuse from a source and in thermal equilibrium diffusion has effectively ceased.

In the case of mechanical defects such as vacancies the value of v_s is determined by the history of the sample, the equilibrium temperature, etc. In any case there are important situations when the value of v_s is independent of the Fermi level [1.9.1]. This holds, for example, for diffusion at sufficiently high temperature so that the impurity in its neutral charge state attains its maximum solubility, from which the maximum solubility of the defect in other charge states then follows by arguments based on eq. (1.7.6). In the theory which follows, ion pairing and the formation of new compounds are neglected. It is found that the concentration v_s of electrically neutral centers can be determined experimentally and so furnishes an important base line for the other concentrations which can be obtained from eq. (1.7.3). Fig. 1.9.1 illustrates schematically the situation for $M = 3$ and $s = 1$. The equations for the lines are:

$$y_0 \equiv \ln \frac{v_0}{v_1} = \ln \frac{Z_0}{Z_1} - \frac{\mu}{kT}$$

$$y_1 \equiv \ln (v_1/v_1) = 0$$

$$y_2 \equiv \ln \frac{v_2}{v_1} = \ln \frac{Z_2}{Z_1} + \frac{\mu}{kT}$$

$$y_3 \equiv \ln \frac{v_3}{v_1} = \ln \frac{Z_3}{Z_1} + \frac{2\mu}{kT}$$

Note that if μ/kT is varied at constant temperature, the lines drawn are exactly straight within the model. If T is allowed to vary, then the ratio of the partition functions becomes

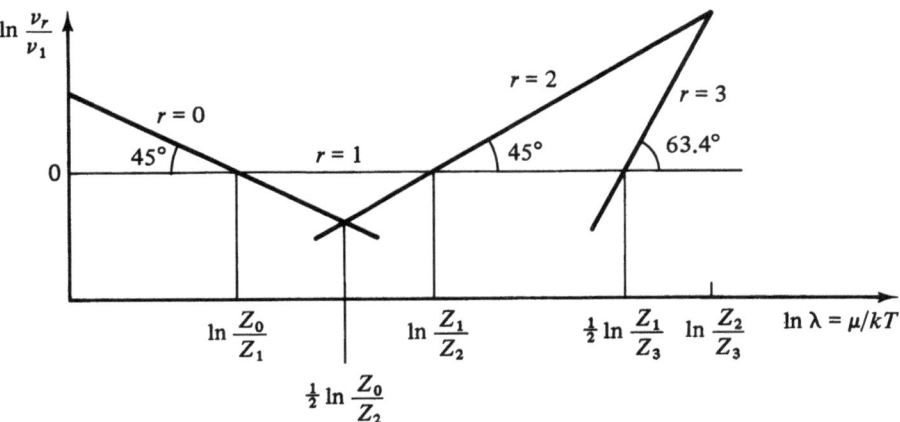

Fig. 1.9.1. Concentration of r-electron centers for $r = 0, 1, 2, 3$ as a function of Fermi level divided by kT. Saturation conditions and thermal equilibrium at a fixed temperature have been assumed.

temperature dependent and one ceases to have straight lines. If one uses *effective* energies as in eq. (1.7.28) with unit degeneracies, then the five values of μ/kT shown in the figure become

$$\frac{E(\tfrac{1}{2})}{kT}, \frac{E(\tfrac{1}{2})+E(\tfrac{3}{2})}{2kT}, \frac{E(\tfrac{3}{2})}{kT}, \frac{E(\tfrac{3}{2})+E(\tfrac{5}{2})}{2kT}, \frac{E(\tfrac{5}{2})}{kT} \qquad (1.9.1)$$

They are arranged correctly in the figure provided $E(\tfrac{1}{2}) < E(\tfrac{3}{2}) < E(\tfrac{5}{2})$.

The variation of μ in Fig. 1.9.1 can be assumed to be due to variable dopings by some other majority impurity. One sees that this affects the take-up by the material of the multicharge impurity under investigation, even if the source of this impurity is maintained at constant strength. This can be seen from the expression for the total number of such impurities to be found in the material in thermal equilibrium. This is

$$N = \sum_{r=0}^{M} v_r = \left(\sum_{r=0}^{M} \lambda^{r-s} Z_r \right) v_s / Z_s \qquad (1.9.2)$$

Note that

$$\frac{\lambda}{N}\left(\frac{\partial N}{\partial \lambda}\right)_{v_s} = \sum_{r=0}^{M} (r-s)\lambda^{r-s} Z_r v_s / N Z_s = \frac{1}{N}\sum_{r=0}^{M} (r-s) v_r$$

$$= \overline{r-s} \qquad (1.9.3)$$

is the average number of negative electronic charges on the impurities. The average charge is $-\overline{(r-s)}|q|$. It is positive for low Fermi levels and becomes increasingly negative for higher Fermi levels as the impurities fill up. For small λ the inverse powers of λ dominate in eq. (1.9.2), provided $s > 0$, and the take-up (N) decreases as the Fermi level rises. For large enough λ, N increases with the Fermi level. We then have an enhancement of solubility due to doping. This is illustrated in Fig. 1.9.2. N is also called the saturation solubility.

There are many experiments on solubility effects. Note, for example, studies [1.9.2], [1.9.3] and [1.9.4] of the increased solubility of gold in silicon as a result of boron doping. If excited

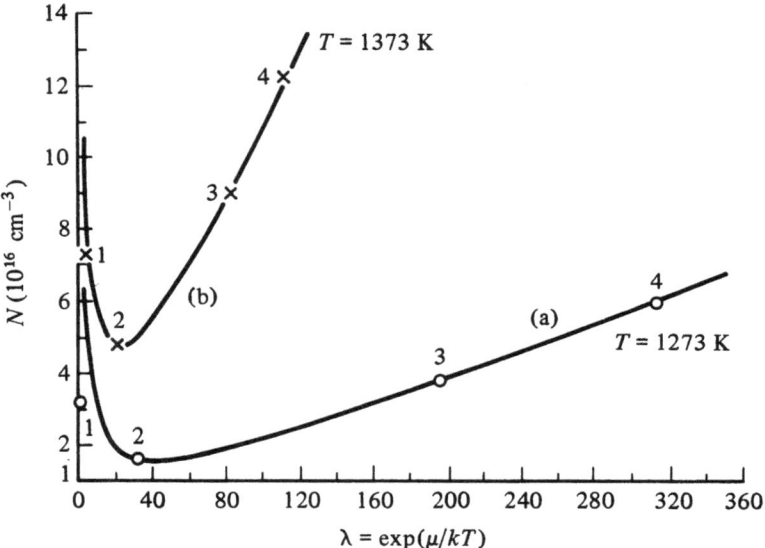

Fig. 1.9.2. Saturation solubilities of Au in Si for different dopings. The curves are theoretical for case 4 of Table 1.9.2, using: (a) $(Z_0/Z_1)\nu_1 = 30.6 \times 10^{16}$ cm^{-3}, $(Z_2/Z_1)\nu_1 = 0.019 \times 10^{16}$ cm^{-3}; (b) $(Z_0/Z_1)\nu_1 = 50.0 \times 10^{16}$ cm^{-3}, $(Z_2/Z_1)\nu_1 = 0.107 \times 10^{16}$ cm^{-3}. The points are experimental as follows: (1) [1.9.2] for 9×10^{19} cm^{-3} B-doped Si; (2) [1.9.2] for intrinsic Si; (3) [1.9.11] for 4×10^{19} cm^{-3} p-doped Si; (4) [1.9.11] for 6×10^{19} cm^{-3} p-doped Si.

states are neglected, these experiments yield some indication of the degeneracy factors which appear in eq. (1.7.6).

A method of arriving at these factors [1.9.5] may be explained by using superfix $l = 1$, $2, \ldots, D$ to distinguish D different heavy shallow dopings which set the Fermi level at $\mu^{(l)}$ for the temperature T. The additional dopings by centers which can be in different charge conditions are assumed small enough *not to affect* $\mu^{(l)}$. Using eq. (1.9.2) one finds

$$N^{(l)} = \sum_{r=0}^{M} \nu_r^{(l)},$$

$$\nu_r^{(l)} = (\lambda^{(l)})^{r-s} Z_r \nu_s/Z_s \quad (r = 0, 1, \ldots, M; l = 1, 2, \ldots, D) \qquad (1.9.4)$$

We have not attached a superscript l to ν_s because the neutral centers have the same concentration for all Fermi levels, as discussed above. We have *known*: $N^{(l)}$ (e.g. from radioactive tracer technique) and $\mu^{(l)}$ from Hall effect and conductivity or from the charge balance equation;

$$\text{unknown:} \quad \frac{Z_0}{Z_s}, \frac{Z_1}{Z_s}, \ldots, \frac{Z_{s-1}}{Z_s}, \frac{Z_{s+1}}{Z_s}, \ldots, \frac{Z_M}{Z_s}, \nu_s$$

The $M+1$ unknowns can be identified for a given temperature T from eqs. (1.9.4) provided $D = M+1$. Repeating the double-doping procedure for other temperatures furnishes curves of quantities (1.7.8) as a function of $x \equiv 1/kT$:

$$y \equiv \ln\frac{Z_r}{Z_s} = \ln\frac{g_r}{g_s} + \begin{cases} [E(s-\tfrac{1}{2}) + \ldots + E(r+\tfrac{1}{2})]x & (r < s) \\ -[E(s+\tfrac{1}{2}) + \ldots + E(r-\tfrac{1}{2})]x & (r > s) \end{cases} \qquad (1.9.5)$$

Table 1.9.1. *Theoretical estimates of degeneracy factors*

r	e	p	g_r given by eq. (1.9.7)
0	2	4	6
1	3	4	4
2	4	4	1

This yields the g_r/g_s from the intercepts, and the energies from the slopes. The method is, however, complicated by the temperature dependence of energy levels which are often of the form

$$E(T) = E(0) - \alpha k T \qquad (1.9.6)$$

whence the intercept is $\ln(g_r/g_s) \pm \alpha$ and the slope yields the energy at the absolute zero of temperature. For more details see the appendix to this section.

At the theoretical level one might argue that the degeneracies can be calculated from the number of ways W of filling p distinguishable places by e indistinguishable electrons:

$$W = p!/e!(p-e)! \qquad (1.9.7)$$

For trivalent Au in Si the results of Table 1.9.1 are possible. The values of e are due to the fact that the $r = 0$ state corresponds to two 5d electrons in the ions. The value $p = 4$ arises from the four positions of the unsatisfied valency. Of course, one could have monovalent Au in Ge and the degeneracy factors could then be $g_0 = 1$, $g_1 = 4$, $g_2 = 6$ [1.9.6]. The question of the correct degeneracy factors is hard to resolve [1.9.7] even for Au in Si.

An attempt to identify experimentally degeneracy factors as well as energy levels has been made for Mg in Si [1.9.8] and for Cr in Si [1.9.9].

If one is not interested in separating energy levels and degeneracies and so lumps these together in the partition functions, eq. (1.9.2) gives the take up N per unit volume of a minority impurity in terms of the concentration v_s of the neutral impurity at a given temperature. It will be assumed that v_s is not affected by the heavy majority doping. As already explained, v_s is taken as independent of the amount of the heavy majority doping, which is assumed to exceed about 100 N. Thus v_s is a constant *for a given temperature*. Table 1.9.2 gives a number of possibilities which can arise.

We know of only one case where experimental points of N as a function of λ are available, and they are marked in Fig. 1.9.2. The theory which yields the curves in the figure is very simple. One just uses eq. (1.9.2) and puts

$$N = v_0 + v_1 + v_2 = \frac{a}{\lambda} + v_1 + b\lambda \quad \left(a \equiv \frac{Z_0}{Z_1} v_1, b \equiv \frac{Z_2}{Z_1} v_1 \right)$$

One then looks for the a and b values of best fit. This was done in Fig. 1.9.2 for each temperature. The minimum occurs at $\lambda^2 = Z_0/Z_2$ when $v_0 = v_2$. The agreement with the experimental points is remarkably good [1.9.10].

The concepts introduced in this section have had wide applications which are, however, largely outside the present scope, and only some will be briefly described. (a) For example,

Table 1.9.2. *Saturation solubility of minority impurity as a function of*
$\lambda = \exp(\mu/kT)$, *which is assumed fixed by heavy primary doping*

Case number	M	Number of electrons on neutral center	Equation for N/v_s
1	1	0	$1 + \dfrac{\lambda Z_1}{Z_0}$
2	1	1	$\dfrac{Z_0}{Z_1}\lambda^{-1} + 1$
3	2	0	$1 + \dfrac{Z_1}{Z_0}\lambda + \dfrac{Z_2}{Z_0}\lambda^2$
4	2	1	$\dfrac{Z_0}{Z_1}\lambda^{-1} + 1 + \dfrac{Z_2}{Z_1}\lambda$
5	2	2	$\dfrac{Z_0}{Z_2}\lambda^{-2} + \dfrac{Z_1}{Z_2}\lambda^{-1} + 1$

the solubility of an acceptor in n-type extrinsic material should exceed its solubility in intrinsic material. For, in the simplest case, we then have $M = 2$ and $s = 0$. So in a suitably modified Fig. 1.9.1 the horizontal line represents $r = s = 0$, and the line rising at 45° represents $\ln(v_1/v_0) = \ln(Z_1/Z_0) + \mu/kT$ indicating an increase in solubility with μ/kT. This has been applied for example to Si$^+$-implanted InP [1.9.12]. (b) Conversely, donor-type defects should have an enhanced solubility in a strongly p-doped crystal. One has two charge conditions $r = 0, 1 (M = 2)$ in the simplest case and one can use the lines marked $r = 0$ and $r = 1$ in Fig. 1.9.1. As the Fermi level is lowered by p-type doping, the solubility of the donor increases by virtue of an increase in v_0. This consideration has been applied to Column III interstitials (donors) in AlGaAs heterostructures [1.9.13]. (c) An impurity or defect may be an acceptor in a given semiconductor when it is on one type of site (α), and it may act as a donor when it is on another type of site (β). Hence the Fermi level position in the band may decide which of the sites is stable in the sense that it leads to the lowest total energy of the system. An example is provided by Li in ZnSe where α is an interstitial site and β a Zn substitutional position [1.9.14]. For an introduction to this problem of the metastability of defects see [1.9.15].

Appendix: Solubilities of multivalent impurities

A more formal presentation of the method of inferring solubilities of multivalent impurities in the presence of majority impurity doping is as follows. We use the notation of section 1.9. Let $x_j \equiv Z_j/Z_s$, where the neutral impurity contains s detachable electrons, and let

$$a_{ij} \equiv [\lambda^{(l)}]^{j-s} = \exp[(j-s)\,\mu^{(l)}/kT] \quad \text{(assumed known at given } T) \tag{1.9.8}$$

$$\alpha_l \equiv N^{(l)}/v_s = \left[\sum_{r=0}^{M} v_r^{(l)}\right] \Big/ v_s \quad (N^{(l)} \text{ assumed known}) \tag{1.9.9}$$

Then eq. (1.9.4) is the set of $D(=M+1)$ equations, one for each doping,

$$\sum_{j=0}^{M} a_{lj} x_j = \alpha_l \ (l = 1, 2, \ldots, D) \tag{1.9.10}$$

The unknowns are $x_0, x_1, \ldots, x_M$. In the matrix a_{lj} strike out the ith row and the kth column and denote $(-1)^{i+k}$ times the determinant of the matrix which has resulted by A_{ik}. Then from the theory of linear equations the solution of eq. (1.9.10) is

$$x_j = \frac{Z_j}{Z_s} = \frac{A_{1j}\alpha_1 + A_{2j}\alpha_2 + \ldots + A_{Dj}\alpha_D}{|a|} \quad (j = 0, 1, \ldots, M) \tag{1.9.11}$$

The case $j = s$ is somewhat special as the left-hand side yields unity. So it is convenient to rewrite this particular solution as

$$\frac{A_{1s} N^{(1)} + A_{2s} N^{(2)} + \ldots + A_{Ds} N^{(D)}}{|a|} = v_s$$

This equation thus furnishes the solution for v_s. The $(M+1)$ relations (1.9.11) give the $M+1$ unknowns noted in the main text. From the x_j one can then pass to the quantities $y_j = \ln x_j$, as considered in eq. (1.9.5) and plot these for various temperatures.

1.10 The equilibrium of simple lattice defects

1.10.1 General theory based on the grand canonical ensemble

A lattice defect is typically a vacancy arising from the migration of an atom from a lattice site to the surface. Consider a compound of type CA_r, where C denotes the (positively charged) cation and A the (negatively charged) anion. Then a *Schottky defect* consists of a C vacancy and r A vacancies, as shown in Fig. 1.10.1 for $r = 1$. The energy of formation (W_s) of such a defect relative to the ideal crystal is the energy required to extract the ions and to deposit them on the surface of the crystal. A *Frenkel pair*, on the other hand, consists of a vacant lattice site and a nearby ion in an interstitial position (Fig. 1.10.1 and Table 1.10.1). It exists in elements and does not require a compound. The energy of formation of a Frenkel defect will be denoted by W_F. At the absolute zero there is ideally no such defect present in a crystal. But this number increases with temperature. With W_s or $W_F \sim 1$ eV one would expect 10^{10} cm^{-3} defects at room temperature and about 10^{16} cm^{-3} at 300 °C, as is readily calculated from the formulae below. As the creation of defects requires the supply of energy, they can be studied experimentally by looking for specific heat anomalies.

It will be assumed in Example 1 that vacancies, in Example 2 that Schottky pairs, and in Example 3 that Frenkel pairs are the only defects present, and that they are present only in small numbers so that the interaction between them can be neglected. Furthermore the grand canonical partition function will be used in this context [1.10.1]. It has the merits of (a) expeditiousness and (b) absence of

Table 1.10.1. *Elementary defects*

			Energy range in alkali halides (eV)
Schottky defect	$V_c V_a$	*	1.8–2.5
Frenkel defect	$V_c I_c$	*	2.0–4.3
Frenkel defect	$V_a I_a$		2.6–4.6
Unnamed defect	$I_a I_c$		?

* Illustrated in Fig. 1.10.1.
V stands for vacancy, I stands for interstitial, c and a stand for cation and anion.

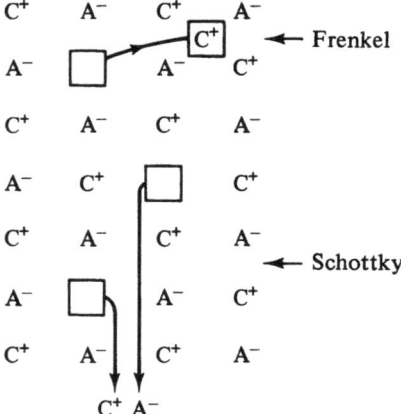

Fig. 1.10.1. A Frenkel and a Schottky defect are illustrated at the top and bottom of the figure, respectively.

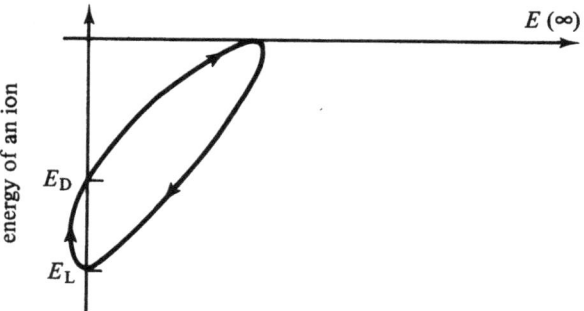

Fig. 1.10.2. It requires less energy to remove an ion from a defect site to infinity than it does to remove an ion from a lattice site.

mathematical approximations, over the free energy arguments normally employed. It presumes thermal equilibrium of the system at temperature T, volume V, and given chemical potential μ. One can imagine this situation to be brought about by regarding (for each lattice site) the surrounding crystal as a reservoir at temperature T and chemical potential μ.

Consider now the reaction

$$L^\circ + D^\bullet \rightleftarrows L^\bullet + D^\circ \qquad (1.10.1)$$

where L° and $L^\bullet$ are, respectively, vacant and occupied lattice sites, while D° and $D^\bullet$ refer to defect sites. The latter are surface sites for Schottky defects and interstitial sites for Frenkel defects. These sites have *grand* partition functions

$$\Xi_A \equiv Z_{A0} + \lambda_A Z_{A1} \qquad (1.10.2a)$$

$$Z_{A1}/Z_{A0} \equiv \exp(-E_A/kT) \qquad (1.10.2b)$$

$$\lambda_A \equiv \exp(\mu_A/kT) \qquad (1.10.2c)$$

Here $A = L$ or D and Z_{A0}, Z_{A1} are the canonical partition functions for an empty site A° and an occupied site $A^\bullet$. Let N_A, n_A denote, respectively, the number of sites A and the number of occupied sites A, respectively. Then

$$\frac{n_A}{N_A - n_A} = \frac{\lambda_A Z_{A1}}{Z_{A0}} = \exp\left(\frac{\mu_A - E_A}{kT}\right) \quad (A = L, D) \qquad (1.10.3)$$

and this corresponds to the reaction $A^\bullet \rightleftarrows A^\circ$. Using $\mu_L = \mu_D$, the equilibrium constant for the reaction (1.10.1) is, on using eq. (1.10.3) twice,

$$\frac{(N_L - n_L) n_D}{n_L (N_D - n_D)} = \exp\left(-\frac{E_D - E_L}{kT}\right) \qquad (1.10.4)$$

Note that eq. (1.10.2b) is analogous to eq. (1.7.23) and E_A is the effective energy required to take one ion from site A, and to deposit it at infinity at energy $E(\infty)$, which can be taken to be the energy zero. Since it requires less energy to remove an ion from a defect site to infinity than it does to remove an ion from a regular lattice site,

$$E(\infty) - E_D < E(\infty) - E_L$$

i.e. E_D lies above E_L on the energy scale. $E(\infty)$ can be regarded as lying above E_D.

1.10.2 Examples

Example 1

Removal of an ion to the surface (σ), leaving a vacancy (V). In this case:

$$D \to \sigma, N_L - n_L = n_\sigma (\equiv n_V)$$

Then eq. (1.10.3) with $A \to \sigma$, L yields:

$$\frac{n_v}{N_\sigma - n_v} = \exp\left(\frac{\mu_\sigma - E_\sigma}{kT}\right) \tag{1.10.5a}$$

$$\frac{n_v}{N_L - n_v} = \exp\left(-\frac{W_v}{kT}\right) \tag{1.10.5b}$$

$$W_v \equiv \mu_L - E_L \tag{1.10.5c}$$

Hence eq. (1.10.4) with $\mu_L = \mu_\sigma$ takes the form

$$\frac{n_v^2}{(N_\sigma - n_v)(N_L - n_v)} = \exp\left(-\frac{W_v'}{kT}\right) \tag{1.10.6a}$$

$$W_v' \equiv E_\sigma - E_L \tag{1.10.6b}$$

Equation (1.10.5b) is a usual result as derived by free energy arguments; it assumes the number of surface sites to be unlimited, as is usual. Equation (1.10.5a) is not usually given. But eqs. (1.10.5a) and (1.10.5b) together are equivalent to eq. (1.10.6a). Equation (1.10.6a), being based on the grand canonical ensemble, appears to be relatively new, since the fact that the number of surface sites is finite is normally neglected. W_v and W_v' are the vacancy formation energies for these two theories, respectively. Note that the new formation energy W_v' is independent of chemical potentials. The value of n_v can be estimated from density measurements of the solid using X-ray data. Vacancy formation energies in Ge and Si are of order 2.5 eV.

Example 2

For a Schottky defect in a material of type CA, where C stands for cation and A for anion, two 'moves' (1.10.1) are needed. One finds on replacing n_v by n_s and W_v by W_s in eqs. (1.10.5b) and (1.10.6a, b) and putting $W_s \equiv 2(\mu_L - E_L)$, $W_s' \equiv 2(E_\sigma - E_L)$ for the two theories, respectively:

$$\frac{n_s^2}{(N_L - n_s)^2} = \exp\left(-\frac{W_s}{kT}\right) \text{ or } \frac{n_s^4}{(N_L - n_s)^2 (N_\sigma - n_s)^2} = \exp\left(-\frac{W_s'}{kT}\right)$$

whence

$$\frac{n_s}{N_L - n_s} = \exp\left(-\frac{W_s}{2kT}\right) \tag{1.10.7a}$$

or

$$\frac{n_s^2}{(N_L - n_s)(N_\sigma - n_s)} = \exp\left(-\frac{W_s'}{2kT}\right) \tag{1.10.7b}$$

(If the cation and anion energy levels are different, one would have

$$W_s = 2\mu_L - E_{LC} - E_{LA}, W_s' = E_{\sigma C} + E_{\sigma A} - E_{LC} - E_{LA})$$

The energy to create a Schottky defect relative to the ideal lattice is thus W_s (number of surface sites unlimited) or W_s' (surface sites limited in number). Schottky pairs can also be studied using density measurements.

Example 3

For Frenkel defects

$$N_L - n_L = n_D (\equiv n_F, \text{ say})$$

and eq. (1.10.4) becomes

$$\frac{n_F^2}{(N_L - n_F)(N_I - n_F)} = \exp\left(-\frac{W_F}{kT}\right), \quad (W_F \equiv E_I - E_L) \tag{1.10.8}$$

where D has been replaced by I (for 'interstitial') in E_D and N_D. As the number of interstitial sites is of the same order as the number of lattice sites ($N_L \sim N_I$), one finds the special case

$$n_F = N_L \exp(-W_F/2kT) \quad (N_L \sim N_I \gg n_F)$$

as given in [1.10.2].

1.10.3 The free energy argument

We now give the free energy argument for Schottky defects to show up the difference in the treatments. The entropy of mixing of two species of n_s and $N - n_s$ indistinguishable members is

$$S = 2k \ln[N!/(N - n_s)! \, n_s!]$$

where k is Boltzmann's constant. We have put

$$S = k \ln P^2 \text{ (Boltzmann's equation)}$$

where the number of different ways in which n_s vacant cation sites and $N - n_s$ occupied cation sites can be distributed on N cation sites is

$$P = \frac{N!}{(N - n_s)! \, n_s!}$$

The second factor P arises from the anion sites. The change in Helmholtz free energy due to the creation of n_s Schottky pairs is

$$F = W_s n_s - 2kT[\ln N! - \ln(N - n_s)! - \ln n_s!]$$

For the equilibrium concentration $(\partial F/\partial n_{\mathrm{s}})_{T,V} = 0$. Introducing Gauss's ψ-function,

$$\psi(n) \equiv \frac{\mathrm{d}}{\mathrm{d}n}[\ln \Gamma(1+n)]$$

and replacing the integer-valued functions $n!$ by the continuous functions

$$n! \to \Gamma(1+n) \tag{1.10.9}$$

one finds for thermal equilibrium

$$W_{\mathrm{s}} - 2kT\psi(N-n_{\mathrm{s}}) + 2kT\psi(n_{\mathrm{s}}) = 0$$

Here relation (1.10.9) is the only assumption. Let us put [1.10.3]

$$\psi(n) \simeq \ln(n+a) \tag{1.10.10}$$

Then $a = 0$ corresponds to the Stirling approximation and $a = \frac{1}{2}$ corresponds to an improved approximation. One finds

$$\exp\left(-\frac{W_{\mathrm{s}}}{2kT}\right) = \exp\left[\psi(n_{\mathrm{s}}) - \psi(N-n_{\mathrm{s}})\right] \simeq \frac{n_{\mathrm{s}}+a}{N-n_{\mathrm{s}}+a} \tag{1.10.11}$$

as the exact result, subject to relation (1.10.9), and as the approximate result respectively. One recovers eq. (1.10.7a) (if $N = N_{\mathrm{L}}$) for $a = 0$, *whereas the grand partition function reaches this result without approximation.* The difference is normally of no consequence, but as a matter of principle, eq. (1.10.6a, b) must be accepted as exact within the model, whereas its derivation via eq. (1.10.10) has introduced approximations which may be considered to be spurious. They arise from the fact that an average and a most probable value must be expected to differ, the average being closer to what one actually measures. The same point arises in other contexts when most probable distributions are investigated [1.10.4].

The application of defect statistics to typical semiconductor problems which are more involved than those usually considered (including those in this section) was carried out in [1.10.5]. This work was extended by developing the grand partition function approach [1.10.6].

Calculations covering more complex situations exist (for example [1.10.7]–[1.10.9]) but have not been applied very extensively. Other literature includes [1.10.10]–[1.10.13].

For a review of the general area of the entropy and the enthalpy of defects, we refer to a comprehensive review [1.10.14].

1.11 Quasi-Fermi levels, reaction kinetics, activity coefficients and Einstein relation

1.11.1 Reaction constants for band–band processes

If a semiconductor is steadily irradiated with photons whose energies exceed the band-gap energy, then electrons from the valence band can be promoted to the conduction band by the absorption of energy. The radiation thus creates new electrons and holes. These come rapidly into equilibrium with the other current carriers (electrons or holes) in their respective band. Excess electrons will, however, tend to drop back into the valence band in a recombination process. The energy liberated may be given to the lattice or it may be emitted as radiation. In this way a steady state can be set up in which carriers recombine at the same rate as they are generated by the external source, leaving the number of electrons and holes constant in time.

One would expect the rate of recombination to be proportional to the concentrations of electrons (n) and holes (p), since this governs the probability of an electron 'meeting' a hole with the possibility of a recombination act. This leads us to consider a product of eq. (1.6.3) and eq. (1.6.9):

$$np = N_c N_v F_{\frac{1}{2}}(\gamma_e - \eta_c) F_{\frac{1}{2}}(\eta_v - \gamma_h) \qquad (1.11.1)$$

We have here replaced the electrochemical potential by two distinct ones: μ_e and μ_h for electrons and holes respectively. This takes account of the fact that the carriers in each band can equilibriate rapidly among themselves, while the electrons and holes are not in equilibrium with each other. Complete equilibrium would occur only if $\mu_e = \mu_h (\equiv \mu_{eq})$. The quantities μ_e and μ_h are called quasi-Fermi levels and provide a means of describing recombination away from thermal equilibrium.

The existence of a quasi-Fermi level leads to the following generalization of eq. (1.2.8). The probability of occupation, p_I, and of a vacancy, q_I, of a state I in a conduction band are, respectively,

$$p_I = 1/[1 + \exp(\eta_I - \gamma_e)]$$
$$q_I = \exp(\eta_I - \gamma_e)/[1 + \exp(\eta_I - \gamma_e)]$$

Here the energy of state I is E_I and $\eta_I \equiv E_I/kT$. It follows that

$$\frac{q_I p_{I\mathrm{eq}}}{p_I q_{I\mathrm{eq}}} = \exp(\gamma_{eq} - \gamma_e) \qquad (1.11.1a)$$

where the suffix 'eq' denotes equilibrium conditions. Similarly for a state J in a valence band of quasi-Fermi level $\mu_h = kT\gamma_h$,

$$\frac{q_J p_{J\mathrm{eq}}}{p_J q_{J\mathrm{eq}}} = \exp(\gamma_{eq} - \gamma_h) \qquad (1.11.1b)$$

Consider now the recombination of a hole (h) and an electron (e) to yield an electron in the valence band (e$_{vb}$) as a chemical reaction,

$$e + h \rightleftarrows e_{vb} \qquad (1.11.2)$$

The mass action law would lead one to expect a concentration-independent quantity np/n_{vb} where n_{vb} is the concentration of valence band electrons. This latter quantity is, however, so great that it is essentially constant when electron and hole numbers vary, provided we assume

$$n, p \ll n_{vb}$$

This enables us to confine attention to nondegenerate semiconductors, whence eq. (1.11.1) yields

$$np = N_c N_v \exp(\gamma_e - \eta_c + \eta_v - \gamma_h)$$
$$= N_c N_v \exp(\gamma_e - \gamma_h - \eta_G) \tag{1.11.3}$$

In the limit of a near-equilibrium situation the right-hand side is indeed independent of electron and hole concentration, showing that the model of the process as a reaction is satisfactory. If the electron or hole gas becomes degenerate, one can put

$$(f_c^* n)(f_v^* p) = N_c N_v \exp(\gamma_e - \gamma_h - \eta_G)$$

where

$$\left. \begin{array}{l} f_c^* \equiv \exp(\gamma_e - \eta_c)(N_c/n) \\ f_v^* \equiv \exp(\eta_v - \gamma_h)(N_v/p) \end{array} \right\} \tag{1.11.4}$$

The concentrations are here generalized to effective concentrations or activities $f_c^* n, f_v^* p$, where the f_c^*, f_v^* can be called activity coefficients [1.11.1]. Note that f_c^* plotted against n/N_c gives a curve which starts from unity for $n \ll N_c$ and rises rapidly with n/N_c (Fig. 1.11.1, curve (a)). With the aid of the f's a form of mass action law thus survives.

The other curves in Fig. 1.11.1 are theoretically derived by taking into account exchange and Coulomb interactions. The effects are believed to be somewhat overestimated in the curves [1.11.3]. In any event, they contribute terms to the Hamiltonian of the system which are complicated so that only approximations can be employed. The attraction between carriers and oppositely charged ionized defects contribute a negative term. On the single-electron energy scale this leads to a band-gap shrinkage. But there are other effects which have to be considered, as discussed at the end of section 1.12. Here we merely take account of this effect by introducing reduced energy shifts Δ_c, Δ_v for the band edges. Each is regarded as positive if the shift is in the direction of band-gap narrowing, so that

$$\left. \begin{array}{l} n \equiv N_c F_{\frac{1}{2}}(\gamma_e - \eta_c) \exp \Delta_c \equiv N_c \exp(\gamma_e - \eta_c) \exp \Delta_c/f_c \\ p \equiv N_v F_{\frac{1}{2}}(\eta_v - \gamma_h) \exp \Delta_v \equiv N_v \exp(\eta_v - \gamma_h) \exp \Delta_v/f_v \end{array} \right\} \tag{1.11.5}$$

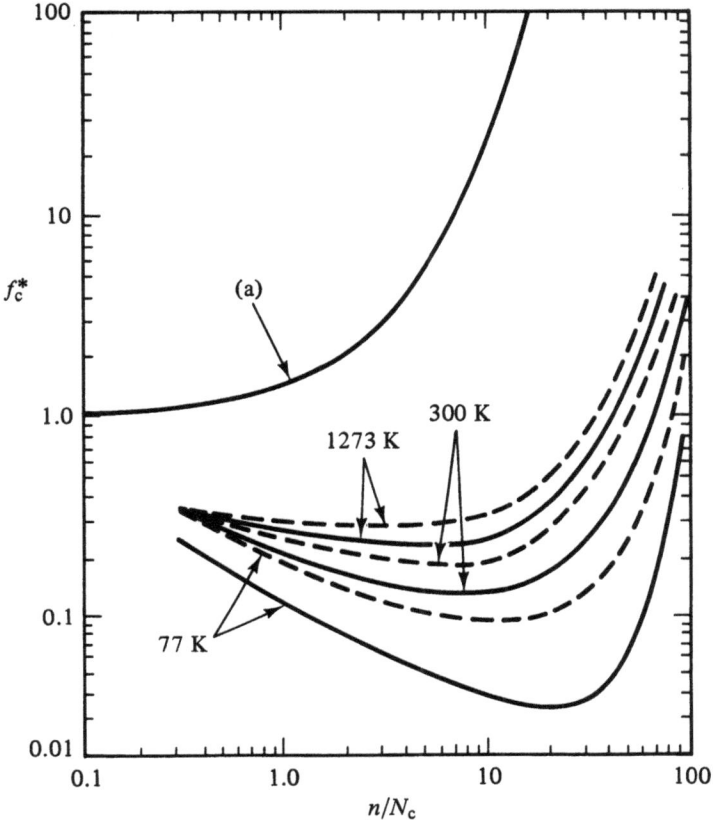

Fig. 1.11.1. The activity coefficient of electrons. Curve (a) is according to eq. (1.11.4) for almost free electrons. The remaining cases are from [1.11.2] and allow for band shift due to Coulomb and exchange interactions. Solid curves: GaAs (dielectric constant 12.5, effective mass $= 0.072m$). Dashed curves: InSb (dielectric constant 17, effective mass $= 0.014m$).

The new activity coefficients f_c, f_v can alternatively be defined by

$$f_c = f_c^* \exp(\Delta_c), \quad f_v = f_v^* \exp \Delta_v, \quad \Delta = \Delta_c + \Delta_v \qquad (1.11.5a)$$

Δ is the total band-gap shrinkage. Thus f_c^*, f_v^* take account of interactions and degeneracy effects; f_c and f_v take account of degeneracy effects only. The effect of the many-body interactions adds immense complications to the interpretation and calculation of activity coefficients such as those shown in the lower curves of Fig. 1.11.1. Then interactions affect the density of states and this effect can often be split up into (a) a rigid shift of the density of states on the one-electron energy scale, and (b) a distortion of the function itself. These effects are particularly marked for heavily doped semiconductors ($\geqslant 10^{19}$ cm^{-3}) as this increases the various Coulomb and exchange interactions.

1.11.2 Reaction constants for processes involving traps

The recombination process may of course take place between the conduction band and impurities. Suppose the impurities are initially neutral ($D^\times$). If they can donate electrons to the conduction band, they become positively charged (D^+). Let the concentration of such impurities (donors) be N_D^+. The recombination process is then the forward reaction, and the carrier generation process is the reverse reaction in

$$e + D^+ \rightleftarrows D^\times \tag{1.11.6}$$

If N_D is the total concentration of donors, one needs to consider the concentration of occupied and empty donors, which are, using eq. (1.7.15)

$$\left.\begin{array}{l}
N_D^\times \equiv v_1 = \dfrac{\lambda_D Z_{1D}}{Z_{0D} + \lambda_D Z_{1D}} N_D \quad \left(\lambda_D \equiv \exp\dfrac{\mu_D}{kT} \equiv \exp\gamma_D\right) \\[3mm]
N_D^+ \equiv v_0 = \dfrac{Z_{0D}}{Z_{0D} + \lambda_D Z_{1D}} N_D
\end{array}\right\} \tag{1.11.6a}$$

Here μ_D is the quasi-Fermi level for the donors. For a nondegenerate band we have

$$\frac{n N_D^+}{N_D^\times} = N_c \exp(\gamma_e - \eta_c) \frac{Z_{0D} N_D}{Z_{0D} + \lambda_D Z_{1D}} \frac{Z_{0D} + \lambda_D Z_{1D}}{\lambda_D Z_{1D} N_D}$$

$$= (Z_{0D}/Z_{1D}) N_c \exp(\gamma_e - \gamma_D - \eta_c) \tag{1.11.7}$$

One sees that the mass action law is now satisfied at equilibrium when $\gamma_e = \gamma_D$, since the right-hand side of eq. (1.11.7) is then independent of Fermi levels, and so independent of the concentrations which occur on the left-hand side [1.11.4].

The example of an acceptor, analogous to reaction (1.11.6), is

$$e + A^\times \rightleftarrows A^-$$

and it leads (in an obvious notation) to

$$n N_A^\times = N_c F_{\frac{1}{2}}(\gamma_e - \eta_c) \frac{Z_{0A}}{\lambda_A Z_{1A}}$$

$$\rightarrow N_c \frac{Z_{0A}}{Z_{1A}} \exp(\gamma_e - \gamma_A - \eta_c)$$

$$\rightarrow N_c \frac{Z_{0A}}{Z_{1A}} \exp(-\eta_c) \tag{1.11.8}$$

In the first step nondegeneracy has been assumed, and in the second step thermal equilibrium has been assumed.

If one neglects the excited states, then by eq. (1.7.8):

$$Z_1/Z_0 = (g_1/g_0)\exp[-\eta(\tfrac{1}{2})] \tag{1.11.9}$$

If this neglect is, however, not justified for Z_1, then the energy level introduced in eq. (1.11.9) has to be given the following interpretation:

$$E(\tfrac{1}{2}) = kT\ln\frac{g_1}{\sum_i g_{1i}\exp[(E_0 - E_{1i})kT]} \tag{1.11.10}$$

where the g_{1i} and E_{1i} are the degeneracies and energies of the one-electron center. Thus $E(\tfrac{1}{2})$ is not then a normal eigenvalue of the Hamiltonian, but it is a temperature-dependent theoretically introduced quantity.

Other reactions are [1.11.4]:

$$h + D^\times \rightleftarrows D^+, \quad h + A^- \rightleftarrows A^\times$$

The reaction constants are

$$pN_D^\times/n_D^+ = N_v F_{\frac{1}{2}}(\eta_v - \gamma_h)\lambda_D Z_{1D}/Z_{0D} \tag{1.11.11}$$

$$pN_A^-/N_A^\times = N_v F_{\frac{1}{2}}(\eta_v - \gamma_h)\lambda_A Z_{1A}/Z_{0A} \tag{1.11.11a}$$

Given the concentration of neutral centers, the concentration of positively charged centers is proportional to the p-type doping, and the concentration of negatively charged centers is inversely proportional to the p-type doping. More generally, one has

$$nv_j/v_{j+1} = N_c F_{\frac{1}{2}}(\gamma_e - \eta_c)Z_j/\lambda Z_{j+1} \quad (e + j \rightleftarrows j + 1)$$

$$pv_{j+1}/v_j = N_v F_{\frac{1}{2}}(\eta_v - \gamma_h)\lambda Z_{j+1}/Z_j \quad (h + (j+1) \rightleftarrows j)$$

The right-hand sides are again independent of the number of particles if the bands are nondegenerate and $\lambda = \exp\gamma_e = \exp\gamma_h$ (equilibrium).

Lastly, if all electrons are supplied by N_D donors, the valence band being neglected, one finds for $N_A + n < N_D$ that

$$N_D^+ = N_A + n, \quad N_D^\times = N_D - N_A - n$$

The reaction

$$e + D^+ \rightleftarrows D^\times$$

has therefore the reaction constant

$$\frac{nN_D^+}{N_D^\times} = \frac{n(N_A + n)}{N_D - N_A - n} = N_c F_{\frac{1}{2}}(\gamma_e - \eta_c)\frac{Z_{0D}}{\lambda_D Z_{1D}} \tag{1.11.11b}$$

This is a generalization of an early result [1.11.5].

If all donors lose their electrons to acceptors and only the valence band needs to be considered, then the mass action rate for

$$h + A^- \rightleftarrows A^\times$$

becomes

$$\frac{pN_A^-}{N_A^\times} = \frac{p(N_D+p)}{N_A-N_D-p} = N_v F_{\frac{1}{2}}(\eta_v-\gamma_h)\frac{\lambda_A Z_{1A}}{Z_{0A}} = N_v F_{\frac{1}{2}}(\eta_v-\gamma_h)\,e^{\gamma_1-\eta(\frac{1}{2})}$$

where the quasi-Fermi level of the trap divided by kT has been denoted by γ_1 and $kT\eta(\frac{1}{2})$ is an *effective* energy level (as explained on p. 42) incorporating of course the degeneracy factors. In equilibrium and for nondegeneracy we obtain a special case of eq. (1.11.11a)

$$\left(\frac{pN_A^-}{N_A^+}\right)_{eq} = N_v \exp\left[\eta_v-\eta(\tfrac{1}{2})\right] = N_v \exp\left(-\eta_A\right) \qquad (1.11.11c)$$

where

$$\eta_A \equiv \eta(\tfrac{1}{2})-\eta_v$$

is the effective trap excitation energy divided by kT.

Equations such as (1.11.11) to (1.11.11c) are widely used in the analysis of data relating thermal carrier densities and temperature, see for example [1.11.6]. The reinterpretation (1.7.30) → (1.7.34), i.e.

$$\exp\left(\gamma-\eta_{\mathrm{eff}}\right) \to X \exp\left(-\frac{\Delta H}{kT}\right) \qquad (1.11.11d)$$

can also be applied to the above mass action ratios.

For a center which has captured r detachable electrons one simply generalizes eqs. (1.11.6a) from $r = 1$ to general r. The concentration of such r-electron centers is then

$$v_r = \left[Z_r \exp\left(r\gamma_r\right) \middle/ \sum_{s=0}^{M} Z_s \exp\left(s\gamma_s\right)\right] n_t \qquad (1.11.12)$$

where the γ_r are quasi-Fermi levels divided by kT and n_t is the total concentration of centers: $\sum_{r=0}^{M} v_r = n_t$. In equilibrium one recovers eq. (1.7.3):

$$P(r) = \frac{v_r}{n_t} = \frac{Z_r \exp\left(r\gamma_{eq}\right)}{\displaystyle\sum_{s=0}^{M} Z_s \exp\left(s\gamma_{eq}\right)}$$

1.11.3 Einstein relation and activity coefficient

We now establish a relation between the Einstein diffusion–mobility ratio (1.2.14) and the activity coefficients [1.11.7]. If the density of states for electrons is $\mathcal{N}_e(E)$ and for holes $\mathcal{N}_h(E)$, the activity coefficients may be defined alternatively as

$$f_c^* \equiv \int_{E_c}^{\infty} \mathcal{N}_e(E) e^{\gamma_e - \eta} dE \bigg/ \int_{E_c}^{\infty} \frac{\mathcal{N}_e(E) dE}{\exp(\eta - \gamma_e) + 1} = \frac{N_c}{n} e^{\gamma_e - \eta_c} \qquad (1.11.13)$$

$$f_v^* \equiv \int_{-\infty}^{E_v} \mathcal{N}_h(E) e^{\eta - \gamma_h} dE \bigg/ \int_{-\infty}^{E_v} \frac{\mathcal{N}_h(E) dE}{\exp(\gamma_h - \eta) + 1} = \frac{N_v}{p} e^{\eta_v - \gamma_h} \qquad (1.11.14)$$

Here γ_e, γ_h are the reduced quasi-Fermi levels, η_c, η_v reduced band-gap energies, and

$$N_c \equiv \int_{E_c}^{\infty} \mathcal{N}_e(E) e^{\eta_c - \eta} dE, \quad N_v \equiv \int_{-\infty}^{E_v} \mathcal{N}_h(E) e^{\eta - \eta_v} dE \qquad (1.11.15)$$

which has been elaborated in [1.11.8]. Thus, although the form of eqs. (1.11.4) is recovered, the density of states involved is seen to be arbitrary. N_c and N_v are an *effective* number of levels for the conduction and valence band, respectively, and can be used for any density of states, as seen already in eq. (1.6.32). For parabolic bands, $\mathcal{N}_e(E) = AV(E - E_c)^{\frac{1}{2}}$, where A is a constant, one recovers eq. (1.6.7):

$$N_c = AV \int_{E_c}^{\infty} (E - E_c)^{\frac{1}{2}} e^{\eta_c - \eta} dE = \frac{\sqrt{\pi}}{2} AV(kT)^{\frac{3}{2}}$$

Observe now from eqs. (1.11.5a), (1.11.13) and (1.11.14) that

$$\gamma_e = \ln n + \ln f_c - \ln N_c - \Delta_c + \eta_c$$
$$\gamma_h = -\ln p - \ln f_v + \ln N_v + \Delta_v + \eta_v$$

so that eq. (1.2.14) gives

$$\frac{|q| D}{vkT} = \begin{cases} \left(\dfrac{\partial \gamma_e}{\partial \ln n} \right)_{T,V} = 1 + \left(\dfrac{\partial \ln f_c}{\partial \ln n} - \dfrac{\partial \Delta_c}{\partial \ln n} \right)_{T,V} = 1 + \left(\dfrac{\partial \ln f_c^*}{\partial \ln n} \right)_{T,V} \\[3mm] -\left(\dfrac{\partial \gamma_h}{\partial \ln p} \right)_{T,V} = 1 + \left(\dfrac{\partial \ln f_v}{\partial \ln p} - \dfrac{\partial \Delta_v}{\partial \ln p} \right)_{T,V} = 1 + \left(\dfrac{\partial \ln f_v^*}{\partial \ln p} \right)_{T,V} \end{cases} \qquad (1.11.16)$$

Consider, as an example, a band with a constant density of states. Then eqs. (1.11.15) and (1.11.13) yield

$$N_c = kT\mathcal{N}_e, \quad n = kT\mathcal{N}_e \ln[1 + e^{\gamma_e - \eta_c}] \qquad (1.11.17)$$

so that

$$\exp\left(\frac{n}{N_c} \right) = 1 + \exp(\gamma_e - \eta_c) \qquad (1.11.18)$$

and

$$f_c^* = \left(\frac{N_c}{n} \right) \left[\exp\left(\frac{n}{N_c} \right) - 1 \right] \qquad (1.11.19)$$

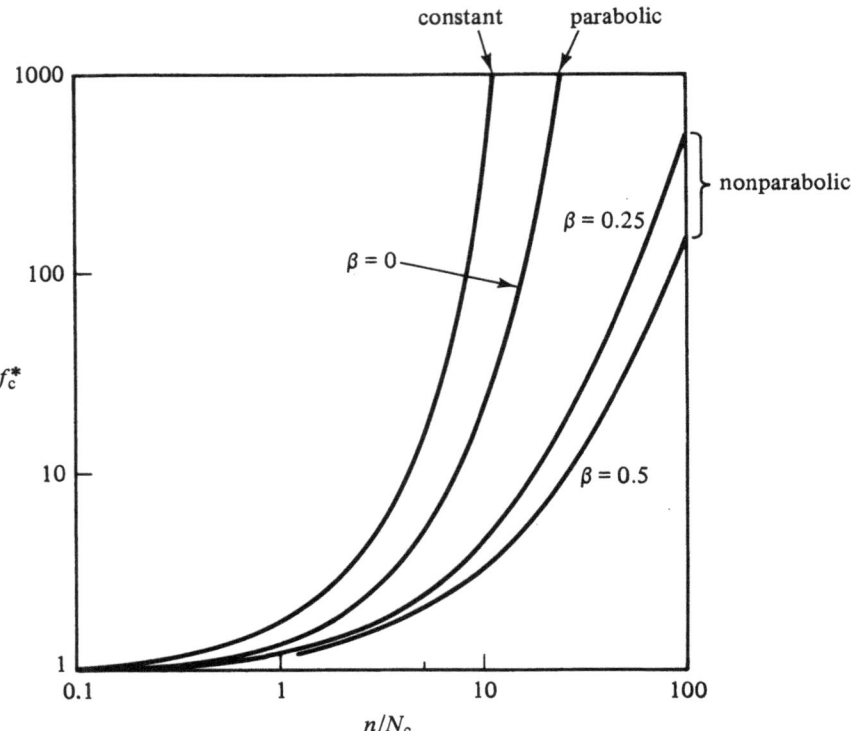

Fig. 1.11.2. The activity coefficient f_c^* is reduced by increased nonparabolicity $\beta \equiv kT/2E_0$ for given n/N_c.

From eq. (1.2.14) for electrons in a band of constant density of states

$$\frac{|q|D_e}{v_e kT} = \frac{n}{\partial n/\partial \gamma_e} = \frac{n/N_c}{1-\exp(-n/N_c)} = \frac{\exp(n/N_c)}{f_c^*} \qquad (1.11.20)$$

Thus the activity coefficient, as well as the ratio (1.11.20), rises from unity as n increases from zero. One can also verify eq. (1.11.16) in this case [1.11.7], [1.11.9].

In the case of a nonparabolic band similar calculations can be made. Using eqs. (1.6.17) and (1.11.13), we show the activity coefficient for this more complicated case in Fig. 1.11.2. Using eqs. (1.2.14) and (1.6.17) the Einstein diffusion–mobility ratio has also been calculated; see Fig. 1.11.3 [1.11.9]. For other relevant calculations see [1.11.10].

Another simple case arises for a narrow band at energy E_t, e.g. for an impurity band. This may be modelled by

$$\mathcal{N}_e(E) = N_t \delta(E - E_t) \qquad (1.11.21)$$

where N_t is the number of levels. In this case from eq. (1.11.13)

$$f_c^* = \frac{|q|D_e}{v_e kT} = \frac{1}{1-n/N_t} \qquad (1.11.22)$$

and these quantities rise as the N_t levels fill up [1.11.11].

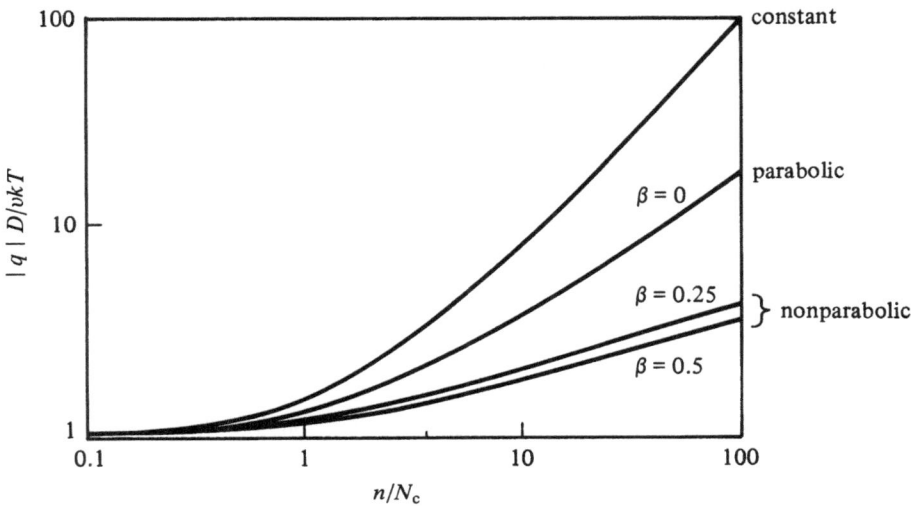

Fig. 1.11.3. The Einstein ratio is decreased by increased nonparabolicity $\beta \equiv kT/2E_0$ for given n/N_c.

For a normal parabolic band one finds

$$f_c^* = e^{\gamma_e - \eta_c}/F_{\frac{1}{2}}(\gamma_e - \eta_c), \quad f_v^* = e^{\eta_v - \gamma_h}/F_{\frac{1}{2}}(\eta_v - \gamma_h) \tag{1.11.23}$$

$$\frac{|q|D_e}{v_e kT} = \frac{F_{\frac{1}{2}}(\gamma_e - \eta_c)}{F_{-\frac{1}{2}}(\gamma_e - \eta_c)}, \quad \frac{|q|D_h}{v_h kT} = \frac{F_{\frac{1}{2}}(\eta_v - \gamma_h)}{F_{-\frac{1}{2}}(\eta_v - \gamma_h)} \tag{1.11.24}$$

[1.11.12]. Fig. 1.11.4 shows the ratios (1.11.24), allowance for some interaction effects having been made [1.11.13]. Some additional results are given in Table 1.11.1.

We give a simple proof of the last entry in the Table. Since the Fermi function drops steeply to zero at low temperatures, a density of states $A \exp(E/E_1)$, where A and E_1 are constants, leads at low temperatures to a concentration of electrons

$$n \simeq A \int_0^\mu \exp\left(\frac{E}{E_1}\right) dE = AE_1 \left(\exp\frac{\mu}{E_1} - 1\right) \simeq AE_1 \exp\frac{\mu}{E_1}$$

Thus eq. (1.2.14) gives

$$\frac{|q|D}{v} = \left(\frac{\partial(\mu/kT)}{\partial \ln n}\right)_{T,V} \quad kT = E_1$$

and $|q|D/v$ is again a typical carrier energy.

Additional discussions of Einstein relations can be found as follows:

heavily doped semiconductors [1.11.16],
experimental justification [1.11.17],
two-dimensional systems [1.11.18], [1.11.19],
multi-band properties [1.11.20],
the effect of writing the diffusion current as $\nabla(Dn)$ instead of $D\nabla n$ [1.11.21]–[1.11.23],

Table 1.11.1. *Einstein relation for conduction bands of semiconductors under simple conditions* $[a \equiv (\mu - E_c)/kT]$

Physical system	$\dfrac{|q| D_e}{v_e}$ as given by eqs. (1.11.16)	(i) Nondegenerate limit (ii) Degenerate limit	Reference
Parabolic band, extending from energy E_c	$kTF_{\frac{1}{2}}(a)/F_{-\frac{1}{2}}(a)$	(i) kT (ii) $\frac{2}{3}(\mu - E_c)$	[1.11.12] [1.11.14]
As above but nonparabolic band derived from $\hbar^2 k^2/2m^* = E[1 + E/2E_0]$ where $E/2E_0 \ll 1$	$kT\dfrac{F_{\frac{1}{2}}(a) + \dfrac{15}{4}\dfrac{kT}{2E_0}F_{\frac{3}{2}}(a)}{F_{-\frac{1}{2}}(a) + \dfrac{15}{4}\dfrac{kT}{2E_0}F_{\frac{1}{2}}(a)}$	(i) kT (ii) $\frac{2}{3}(\mu - E_c)\left(1 - \dfrac{\mu - E_c}{2E_0}\right)$	[1.11.10]
Constant density of states $\mathcal{N}_e(E) = \mathcal{N}_e$	$kT(1 + e^{-a})\ln(1 + e^a)$	(i) kT (ii) $\mu - E_c$	
Exponential band tail (with $A\exp(E/E_1)$ as density of states) at low temperature	E_1	Degenerate case only	[1.11.15]

quantum corrections [1.11.24],
numerical estimates for the degenerate case [1.11.25], [1.11.26],
nonparabolicity and superlattice [1.11.10],
multi-band degenerate semiconductors [1.11.20],
anisotropic parabolic bands and size quantization [1.11.27].

1.12 Fermi level identifications, intrinsic carrier concentrations and band-gap shrinkage

1.12.1 Fermi level identifications

The Fermi level at temperature T in a simple equilibrium semiconductor having N_D donor and N_A acceptors can be identified from an equation which asserts electron conservation. Suppose that at $T = 0$, N_e electrons are available in localized states and in bands above the valence bands ($V = 1, 2, \ldots$), and that the valence bands are fully occupied. Then at an elevated temperature extra electrons are added to N_e by the thermal creation of holes (p_V in number) in the valence bands. The resulting $N_e + \sum_V p_V$ electrons are distributed over the conduction bands $C = 1, 2, \ldots$ and the

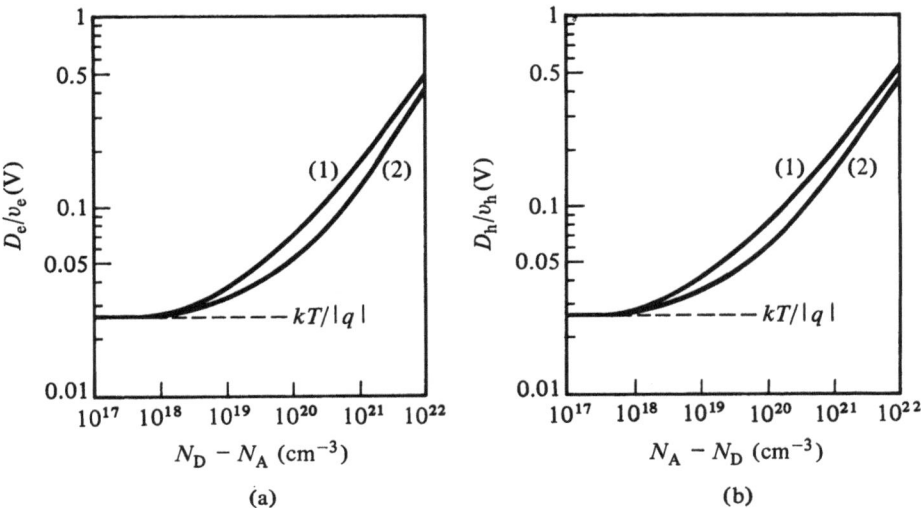

Fig. 1.11.4. (a) Theoretical plots of the ratio of the electron diffusivity to mobility as a function of the net donor concentration in (1) Ge and (2) Si at 300 K including some heavy doping effects; $N_A = 10^{17}$ atoms cm^{-3}. (b) Theoretical plots of the ratio of the hole diffusivity to mobility as a function of the net acceptor concentration in (1) Ge and (2) Si at 300 K including some heavy doping effects; $N_D = 10^{17}$ atoms cm^{-3}.

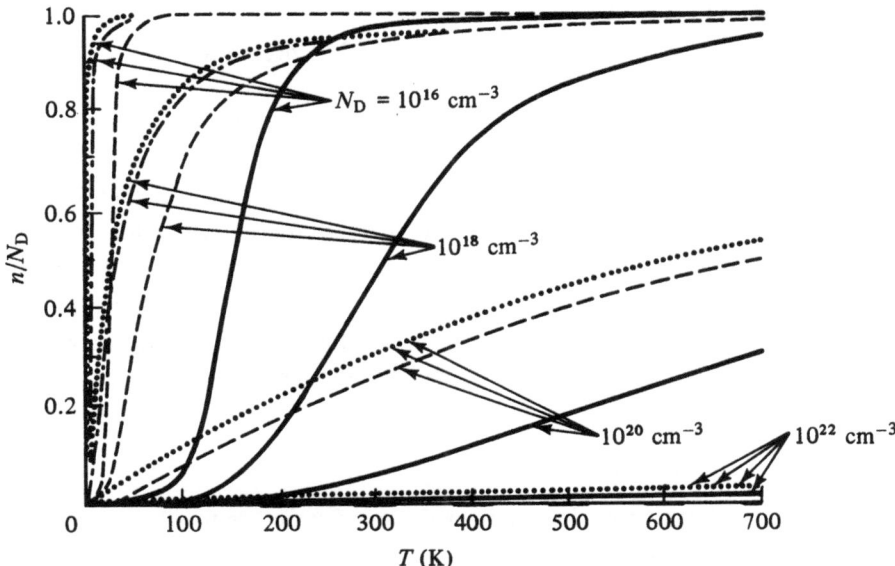

Fig. 1.12.1. Relation between the degree of ionization of a donor and the temperature for different donor concentrations and energy levels. $E_c - E_D$ (eV): ······· 0.0001; — · — · — 0.001; — — — — 0.01; —— 0.1.

localized states or traps, of which we assume that there are $N^{(t)}$ of type t ($t = 1$, $2, \ldots$). Hence from eqs. (1.6.3), (1.6.9) and (1.7.14)

$$\sum_C N_e^{(C)} F_{\frac{1}{2}}(\gamma_c^{(C)} - \eta_c^{(C)}) + \sum_t \frac{N(t) Z_1^{(t)} \exp \gamma^{(t)}}{Z_0^{(t)} + Z_1^{(t)} \exp \gamma^{(t)}}$$

$$= N_e + \sum_V N_v^{(V)} F_{\frac{1}{2}}(\eta_v^{(V)} - \gamma_h^{(V)}) \quad (1.12.1)$$

N_e is thus the number of ionizable electrons in all the donor atoms plus the number of electrons in the conduction bands at $T = 0$. The relation links quasi-Fermi levels $\mu_e^{(C)} (\equiv kT\gamma_e^{(C)})$ of electrons in conduction bands C, the quasi-Fermi levels $\mu_h^{(V)}$ of holes in valence bands V and the quasi-Fermi levels $\mu(t)$ of traps $t = 1, 2, \ldots$. The traps are assumed to be either occupied or empty, the modification to multiple charge conditions of traps being obvious. Assuming quasi-Fermi levels to exist, eq. (1.12.1) is a reasonable relation for a semiconductor in which equilibrium is disturbed by the injection of charge carriers or by incident radiation. In true thermal equilibrium,

$$\mu_e^{(C)} = \mu_h^{(V)} = \mu^{(t)} (\equiv \mu_{eq}) \quad (C, V, t = 1, 2, \ldots) \quad (1.12.2)$$

In that case eq. (1.12.1) determines the equilibrium Fermi level μ_{eq}.

Using an *effective* energy level $E^{(t)}$ of a trap (as in eq. (1.7.29)),

$$Z_1^{(t)}/Z_0^{(t)} \equiv \exp(-\eta^{(t)}), \quad \eta^{(t)} \equiv E^{(t)}/kT \quad (1.12.3)$$

and suppressing the sums over bands, eq. (1.12.1) becomes

$$N_c F_{\frac{1}{2}}(\gamma_e - \eta_c) + \sum_t \frac{N^{(t)}}{1 + \exp[\eta^{(t)} - \gamma^{(t)}]} = N_e + N_v F_{\frac{1}{2}}(\eta_v - \gamma_h) \quad (1.12.4)$$

With eq. (1.12.2) this is an equation for $\gamma = \mu/kT$. In the case of thermal equilibrium and nondegeneracy one has a quadratic equation for $\exp \gamma$. These situations are analyzed in the literature [1.12.1] and [1.12.2] and will not be rehearsed here. However, because of the importance of equations such as (1.12.4), it is worth recalling some approximation schemes for the Fermi integral which have been considered from time to time. We note here that the approximation

$$F_{\frac{1}{2}}(\gamma) = \frac{\exp \gamma}{1 + a \exp \gamma} \quad (1.12.5)$$

has sometimes been used, since it gives the same structure to all terms other than N_e in eq. (1.12.4). The normal classical approximation is $a = 0$; $a = 2^{-\frac{2}{3}}$ is an improved classical approximation [1.12.3], and $a = 0.25$ [1.12.4] is known empirically to be accurate to within 5% for most semiconductor situations. Tabulated values have also been used [1.12.5] in the 1950s when these matters were elucidated. For typical references see [1.12.6] and [1.12.7]; the question of

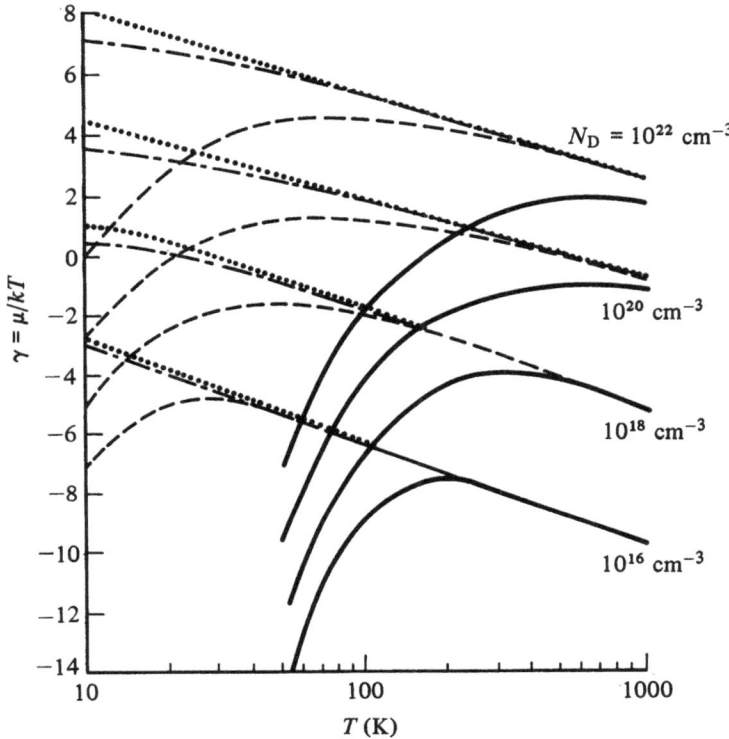

Fig. 1.12.2. Relation between the reduced Fermi level and the temperature for the cases covered in Fig. 1.12.1, $E_c - E_D$ (eV): ...... 0.0001; — – — 0.001; – – – – 0.01; —— 0.1.

improving the mathematical expansions involved has also been raised again recently; see [1.12.8]–[1.12.12]. There is also a recent review [1.12.13].

Suppose that N_D is the concentration of donors of one type only, that the valence band can be neglected, and that $N^{(t)} = N_e = N_D$. The concentration of conduction band electrons is then given by

$$(n =) N_c F_{\frac{1}{2}}(\gamma - \eta_c) = N_D / [1 + \exp(\gamma - \eta_D)] \qquad (1.12.6)$$

Given T, N_D, N_c (which depends on the effective mass) and the trapping level at energy $E_D (= kT\eta_D)$, eq. (1.12.6) determines the Fermi level. The concentration of conduction electrons, n, and the degree of ionization of the donors, n/N_D, can then also be worked out. Typical results are illustrated in Figs. 1.12.1 and 1.12.2.

For two bands, the energy gap $E_G = E_c - E_v$ enters the argument, and it depends on temperature roughly as

$$E_G(T) \simeq E_G(0) - \alpha T \qquad (1.12.7)$$

for most semiconductors. The situation for Ge is illustrated in Fig. 1.12.3 [1.12.14], assuming $N^{(t)} = N_e = N_D$, a conduction and a valence band, and either one type

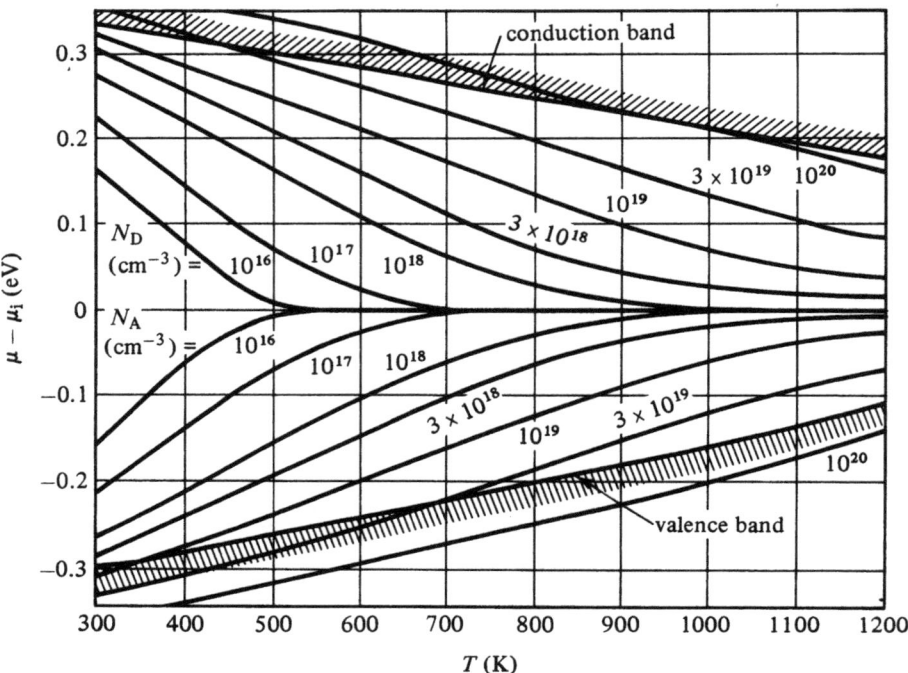

Fig. 1.12.3. Fermi levels and band edges for intrinsic Ge as a function of temperature (see text).

of donor level close to the energy value E_c, or one type of acceptor level close to the energy value E_v. In the nondegenerate limit the equation for donors can be written as

$$xN_c \exp(-\eta_G) - \frac{N_D}{1 + x\exp(\eta_v - \eta_D)} - \frac{N_v}{x} = 0, \quad x \equiv \exp(\gamma - \eta_v) \quad (1.12.8)$$

At high temperatures $x \to 1$, and the N_D term attains a value of order $N_D/2$, which is normally negligible. Hence eq. (1.12.8) yields

$$x^2 \simeq (N_v/N_c)\exp\eta_G \quad (1.12.9)$$

It follows that the concentration of conduction electrons, n, and the concentration of holes, p, are equal:

$$n = (N_c N_v)^{\frac{1}{2}}\exp(-\eta_G/2) = p (\equiv n_i) \quad (1.12.10)$$
$$n_i^2 = 32\pi^2(m_c m_i)^{\frac{3}{2}}(kT)^3 h^{-6}\exp(-\eta_G)$$

The number (n_c, n_v) of extrema are by eqs. (1.4.11) and (1.6.9) included in the definition of m_c and m_v. The quantity (1.12.10) is called the *intrinsic* electron or hole concentration n_i. In this condition the impurity concentration has become unimportant (the intrinsic condition also holds good at lower temperatures if the impurity concentration is small enough). The temperature dependence of n_i-values

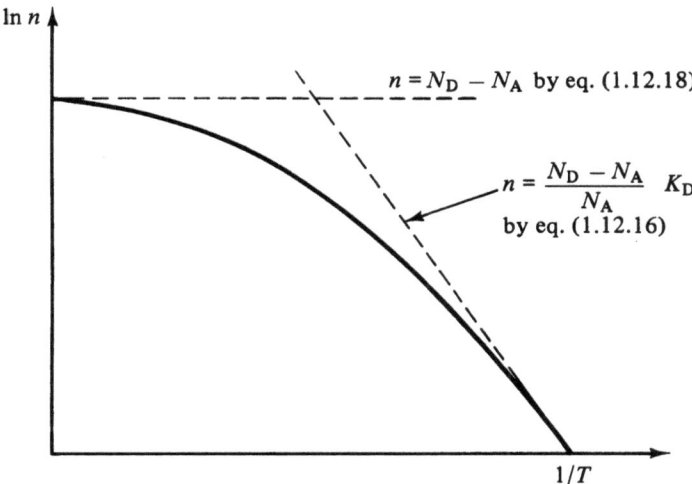

Fig. 1.12.4. The eléctron concentration in an n-type compensated semi-conductor.

for Ge, Si and GaAs is given in [1.12.15], p. 19. A few expressions for the Fermi level are given below.

(i) Intrinsic Fermi level

The intrinsic Fermi level lies by eq. (1.12.9) at a value given by

$$\mu = \tfrac{1}{2}(E_c + E_v) + \tfrac{1}{2}kT\ln(N_v/N_c)(\equiv \mu_i) \qquad (1.12.11)$$

This value is at each temperature taken as the origin in Fig. 1.12.3. The last term in eq. (1.12.11) can by eqs. (1.6.7) and (1.6.9) be rewritten as

$$\tfrac{3}{4}kT\ln(m_v/m_c)$$

(ii) Weak ionization of donors and acceptors, two nondegenerate bands

In this case $\eta_D \ll \gamma \ll \eta_A$ and eq. (1.12.4) becomes with $N^{(t)} = N_e = N_D$

$$N_c\exp(\gamma-\eta_c)+N_A\exp(\gamma-\eta_A) = N_v\exp(\eta_v-\gamma)+N_D\exp(\eta_D-\gamma)$$

whence, on solving for γ,

$$\mu = \tfrac{1}{2}(E_D + E_A) + \tfrac{1}{2}kT\ln\frac{N_D+N_v\exp(\eta_v-\eta_D)}{N_A+N_c\exp(\eta_A-\eta_c)} \qquad (1.12.12)$$

If N_D and N_A are small enough, eq. (1.12.12) reduces to eq. (1.12.11), as expected.

(iii) Donors, two nondegenerate bands

In this case eq. (1.12.4) is

$$N_c\exp(\gamma-\eta_c) = N_D/[\exp(\gamma-\eta_D)+1]$$

and this is a quadratic equation in $\exp \gamma$. It leads to

$$\exp\left(\frac{\mu}{kT}\right) = \frac{1}{2}\left[\left\{1 + \frac{4N_{\mathrm{D}}}{N_{\mathrm{c}}}\exp(\eta_{\mathrm{c}} - \eta_{\mathrm{D}})\right\}^{\frac{1}{2}} - 1\right]\exp \eta_{\mathrm{D}}$$

Thus for small $N_{\mathrm{D}}/N_{\mathrm{c}}$,

$$\mu \simeq E_{\mathrm{c}} + kT \ln(N_{\mathrm{D}}/N_{\mathrm{c}})$$

This holds for an n-type semiconductor, whose Fermi level lies normally below the conduction band (if $N_{\mathrm{D}} < N_{\mathrm{c}}$).

(iv) Acceptors, two nondegenerate bands

In this case one finds:

$$\mu \simeq E_{\mathrm{v}} - kT \ln(N_{\mathrm{A}}/N_{\mathrm{v}})$$

This holds for a p-type semiconductor, whose Fermi level lies above the valence band. The more complete expression is

$$\exp\left(-\frac{\mu}{kT}\right) = \frac{1}{2}\left\{\left[1 + \frac{4N_{\mathrm{A}}}{N_{\mathrm{v}}}\exp(\eta_{\mathrm{A}} - \eta_{\mathrm{v}})\right]^{\frac{1}{2}} - 1\right\}\exp(-\eta_{\mathrm{A}})$$

(v) Compensated semiconductor, one nondegenerate band

Assume that the acceptors have all trapped donor electrons and that the concentration of donors and acceptors is N_{D} and N_{A}. Then eq. (1.12.4) with $N^{(t)} = N_{\mathrm{e}} = N_{\mathrm{D}}$ yields

$$N_{\mathrm{c}}\exp(\gamma - \eta_{\mathrm{c}}) + N_{\mathrm{A}} + \frac{N_{\mathrm{D}}}{1 + \exp(\eta_{\mathrm{D}} - \gamma)} = N_{\mathrm{D}} \quad (N_{\mathrm{D}} > N_{\mathrm{A}}) \tag{1.12.13}$$

The quadratic equation for e^{γ} can be solved in the form

$$n = \tfrac{1}{2}(N_{\mathrm{A}} + K_{\mathrm{D}})[(B+1)^{\frac{1}{2}} - 1] \tag{1.12.14}$$

where

$$B \equiv \frac{4(N_{\mathrm{D}} - N_{\mathrm{A}})K_{\mathrm{D}}}{(N_{\mathrm{A}} + K_{\mathrm{D}})^2}, \quad K_{\mathrm{D}} \equiv N_{\mathrm{c}}\exp[-(\eta_{\mathrm{c}} - \eta_{\mathrm{D}})] \tag{1.12.15}$$

If N_{A}, N_{D} are constants, then the temperature dependence resides in K_{D}, which can be interpreted as an equilibrium constant already familiar from eq. (1.11.11b):

$$K_{\mathrm{D}} = \frac{nN_{\mathrm{D}}^{+}}{N_{\mathrm{D}}^{\times}} = \frac{n(N_{\mathrm{A}} + n)}{N_{\mathrm{D}} - N_{\mathrm{A}} - n} \tag{1.12.16}$$

$$= N_{\mathrm{c}}\exp(\gamma - \eta_{\mathrm{c}})\frac{Z_{\mathrm{D0}}}{Z_{\mathrm{D1}}\exp\gamma} = N_{\mathrm{c}}\exp[-(\eta_{\mathrm{c}} - \eta_{\mathrm{D}})] \tag{1.12.17}$$

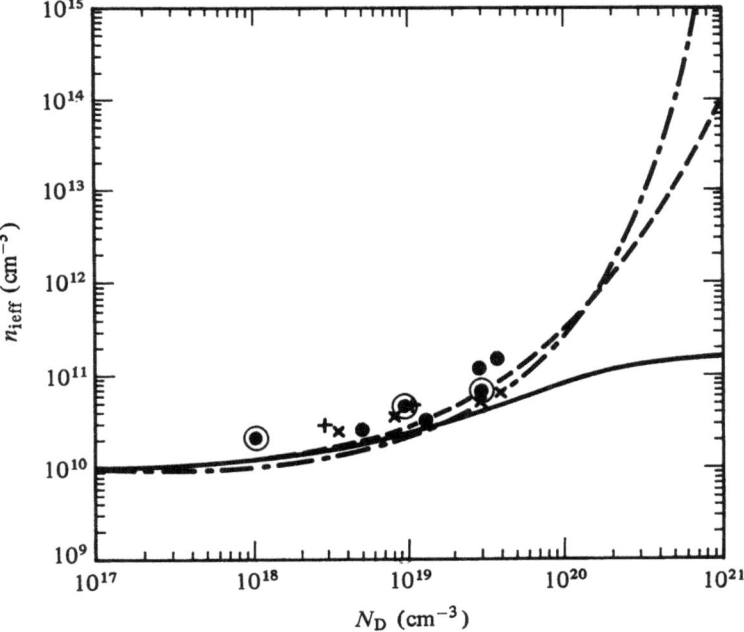

Fig. 1.12.5. The effective intrinsic concentration versus donor concentration in silicon at 300 K temperature [1.12.24]. *Points* are experimental values: ●, [1.12.25]; ⊙, [1.12.26]; ×, [1.12.27]; +, [1.12.28]. Curves are theoretically derived: ——, [1.12.29]; – – –, [1.12.26]; —·—, [1.12.30].

The increase in the conduction band electron concentration with temperature follows approximate exponential laws in two cases (Fig. 1.12.4).

(a) $K_D \ll N_A$, favored by low temperatures; alternatively $K_D \gg N_D - N_A$. In these cases $B \ll 1$ and

$$n \to n_1 = K_D(N_D - N_A)/(N_A + K_D)$$

(b) $K_D \ll N_D$. In this case $B \gg 1$ and

$$n \to n_2 = [(N_D - N_A) K_D]^{\frac{1}{2}}$$

The activation energy $E_c - E_D$ of the curve of $\ln n$ against $1/kT$ is halved in the second case.

At the highest temperatures excitation from the valence band can no longer be neglected. Before this situation is reached, one approaches the largest conduction band electron concentration $n \to N_D - N_A$ which is possible under neglect of the valence band, when

$$K_D \sim N_c \sim n \sim (N_D - N_A) \tag{1.12.18}$$

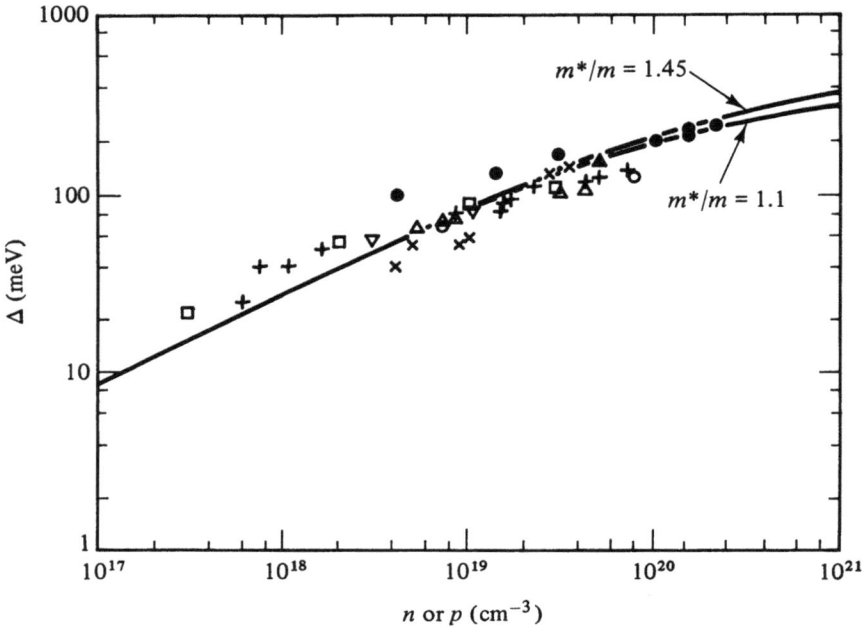

Fig. 1.12.6. Gap shrinkage Δ as inferred from transport measurements for n-type layers of Si from various sources at a mean temperature of $\simeq 340$ K. The curves are based on eq. (1.12.24). The upper curve is for $m^*/m = 1.45$ [1.12.31]; the lower curve is for $m^*/m = 1.1$ [1.12.18]; and $\varepsilon = 11.7$ (Si) has also been used. The horizontal axis is the majority-carrier concentration. Points are experimental, as follows: $\square$, [1.12.32]; $+$, [1.12.33]; $\triangle$, [1.12.27]; $\bigcirc$, [1.12.34]; $\times$, [1.12.25]; $\triangledown$, [1.12.28]; $\bullet$, [1.12.35]; $\blacktriangle$, [1.12.18].

For a p-type semiconductor the above results hold provided the following replacements are made:

$$n \rightarrow p, N_A \rightarrow N_D, N_D \rightarrow N_A, K_D \rightarrow K_A \equiv pN_A^-/N_A^\times \qquad (1.12.19)$$

Various ways of characterizing impurities starting from thermal carrier measurements have recently been compared, using as an example In and Hg acceptors in Ge [1.12.16].

1.12.2 Heavy doping effects

There is considerable interest in heavy doping phenomena. They arise from the use of thin highly doped diffused and implanted layers in bipolar transistors, as used in Si integrated circuits, where a highly doped emitter is desirable. Heavy doping is also employed in other semiconductor devices such as solar cells. This brings into play the complicated phenomena outlined qualitatively in section 1.8.2. In particular, it brings about band-gap shrinkage already noted on p. 74.

As observed in section 1.8.2, the heavy doping effects arise from Coulomb interactions among current carriers, and between carriers and charged impurity ions, and also from the spatial fluctuations in the electrostatic potential. The latter is due to the random distribution of impurities and the electron–phonon interactions. These complicated effects can be assumed to give rise to the band-gap shrinkage parameter (1.11.5a). We shall now make use of this assumption.

The *equilibrium* value of the *np*-product can be written in various ways

$$(np)_{eq} = N_c N_v F_{\frac{1}{2}}(\gamma_{eq} - \eta_c) F_{\frac{1}{2}}(\eta_v - \gamma_{eq}) \exp \Delta \quad [\text{parabolic } \mathcal{N}(E)] \quad (1.12.20)$$

$$= (N_c N_v / f_c f_v) \exp(-\eta_G) \exp \Delta = (n_i^2 / f_c f_v) \exp \Delta \quad (1.12.21)$$

$$= (N_c N_v / f_c^* f_v^*) \exp(-\eta_G) \quad (1.12.22)$$

$$\equiv n_i^2 \exp \Delta_{eff} \equiv n_{ieff}^2 \quad (1.12.23)$$

The first three forms arise from eqs. (1.11.5). The form (1.12.23) is more conventional, although eq. (1.12.21) is also used [1.12.17]. Equation (1.12.23) lumps together the effect of degeneracy and density of states shifts and distortions into an effective band-gap narrowing reduced energy parameter Δ_{eff}. It then gives rise to an effective intrinsic carrier concentration n_{ieff}. Figs. 1.12.5 and 1.12.6 give numerical values of n_{ieff} and Δ. In much of the literature Boltzmann statistics have been used to infer the band-gap narrowing even when the semiconductor was degenerate. These data have been recalculated so as to isolate band-gap shrinkage and remove the effect of degeneracy, assuming parabolic bands. Thus Fig. 1.12.6 gives Δ rather than Δ_{eff}. Note that eqs. (1.12.21) to (1.12.23) can be used for arbitrary density of states, but it is possible to separate degeneracy effects only if a density-of-states function is *assumed*. The curves in Fig. 1.12.6 [1.12.18] are based on the eq.

$$\Delta = \frac{e^2}{\varepsilon} \left\{ \frac{4\pi e^2}{\varepsilon kT} \left[N_c F_{-\frac{1}{2}}(\gamma_{eq} - \eta_c) + N_v F_{-\frac{1}{2}}(\eta_v - \gamma_{eq}) \right] \right\} \quad (1.12.24)$$

where ε is the dielectric constant. Reviews of the heavy doping effects are given in [1.12.19]–[1.12.23]. The interpretation of these experiments is not yet reliable enough to warrant a very detailed theory, and we give in section 1.12.3 merely a simple argument leading to the qualitative result (1.12.24).

Away from equilibrium, but if quasi-Fermi levels exist, one has to replace in eq. (1.12.20)

$$\gamma_{eq} - \eta_c \rightarrow F_e - \eta_c, \quad \eta_v - \gamma_{eq} \rightarrow \eta_v - F_h \quad (1.12.25)$$

and eq. (1.12.23) becomes

$$np = N_c N_v F_{\frac{1}{2}}(\gamma_e - \eta_c) F_{\frac{1}{2}}(\eta_v - \gamma_h) \exp \Delta \quad [\text{parabolic } \mathcal{N}(E)]$$

$$= N_c N_v \exp(-\eta_G + \Delta_{eff} + F_e - F_h) = n_i^2 \exp(\Delta_{eff} + F_e - F_h)$$

$$= n_{ieff}^2 \exp(F_e - F_h) \quad (1.12.26)$$

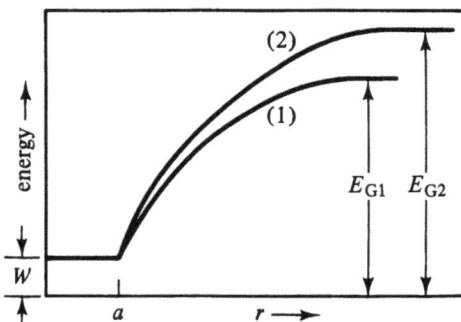

Fig. 1.12.7. Schematic diagram showing the decomposition of the band-gap energy into W and the work done against attraction. More carriers are assumed present for curve (1) than for curve (2).

The experimental data is not easy to analyze, and the gap shrinkage can be plotted somewhat differently. For a precise definition of 'apparent band-gap shrinkage' the original papers [1.12.36] should be consulted.

1.12.3 A simple model of band-gap shrinkage

The experimental data on the reduced band-gap shrinkage Δ discussed in the last subsection depend on whether the experiments have been based on transport properties (typically electrical conductivity and Hall effect) or on optical measurements. The magnitude of Δ depends on temperature and doping. The explanations in terms of many-body theory involve the effects discussed in section 1.8, but they are complicated and beyond the present scope. These theories must ultimately furnish satisfactory results. For steps in this direction see [1.12.23] and [1.12.37]–[1.12.39]. A naive picture of some of the effects involved is based simply on the Coulomb interaction and leads to formula (1.12.24) which gives a fair account of the experimental results, and will now be established.

In the present model the semiconductor is regarded to some extent as a neutral dielectric continuum in which positive and negative charges are smeared out. However, the particulate structure is not entirely neglected, as will be seen. The first step is to create an electron–hole pair which is in a bound state for a very short time. The distance, a say, between the maxima of their wave packets will be a few angstroms only, so that the effect of the smeared out electron and hole densities will not affect the energy, W say, to create the pair. The normal Coulomb potential acts between the particles for $r > a$ and is cut off at $r = a$. Such cut-offs are often needed for small r as the Coulomb potential diverges as $r \to 0$. Imagine now the hole to be trapped at a defect at $r = 0$ and the electron to be removed to infinity against the Coulomb attraction starting at the cut-off distance. Bearing in mind that the 'continuum' consists of particles which are in very complicated and correlated motion, one cannot just use the bare Coulomb potential. The correlated motions of the many-body problem incorporate approximately the effect of the long-range part of the Coulomb potentials of the electrons, leaving short-range, or screened, potentials acting between largely independent particles. The screening parameter is approximated as a constant (it really depends on the wavevectors involved in the Coulombic collisions). With these approximations the total energy supplied to create the pair, and to separate it, is (Fig. 1.12.7)

$$E_G(n,p) \equiv W + (e^2/\varepsilon a)\exp(-\Lambda a) \tag{1.12.27}$$

This quantity is interpreted as the energy gap if n, p are the carrier concentrations. If the semiconductor is highly nondegenerate, then screening can be neglected and eq. (1.12.27) yields

$$E_G(0,0) = W + (e^2/\varepsilon a) \tag{1.12.28}$$

As already explained, W and a are to be approximated as concentration independent. By subtraction, and with the notation (1.11.5a),

$$\Delta \equiv E_G(0,0) - E_G(n,p) = (e^2/\varepsilon a)[1 - \exp(-\Lambda a)] \tag{1.12.29}$$

Using Debye or Thomas–Fermi screening gives the same result in the limit of extreme degeneracy, as given for n^+-material by eq. (1.8.25), so that

$$\Lambda = \left(\frac{m^*/m}{\varepsilon/10}\right)^{\frac{1}{2}}\left[\frac{n}{10^{18}}\right]^{\frac{1}{6}} \times 0.0863 \text{ Å}^{-1} \tag{1.12.30}$$

$$\left(\simeq 0.08\left[\frac{n}{10^{18}}\right]^{\frac{1}{6}} \text{Å for Si}\right.$$

with $\varepsilon = 11.7$ and for $m^* = m$, where m^* is the density-of-states effective mass for electrons.)

The approximation

$$\exp(-\Lambda a) \sim 1 - \Lambda a \tag{1.12.31}$$

requires the constraint $\Lambda a < z$, where z is a number of order of $\frac{1}{4}$. From this and eq. (1.12.30) one has

$$a \lesssim \left[\frac{\varepsilon/10}{m^*/m}\right]^{\frac{1}{2}}\left[\frac{10^{18}}{n}\right]^{\frac{1}{6}} 11.59z \text{ Å}$$

$$\left(= 12.53z\left[\frac{10^{18}}{n}\right]^{\frac{1}{6}} \text{Å for Si}\right).$$

For $n \sim 10^{20}$ cm^{-3} in Si, one has $\Lambda = 0.17$ Å^{-1} and $a \leqslant 1.45$ Å if $z \sim \frac{1}{4}$. Hence from eqs. (1.12.29) and (1.12.31)

$$\Delta = e^2\Lambda/\varepsilon \tag{1.12.32}$$

Numerically, eqs. (1.12.30) and (1.12.32) yield

$$\Delta = 126.6 \left[\frac{m^*/m}{(\varepsilon/10)^3}\right]^{\frac{1}{2}} \left[\frac{n}{10^{18}}\right]^{\frac{1}{6}} \text{meV}$$

where n is in cm^{-3}. For Si, with $\varepsilon \simeq 11.7$ and $m^* \simeq m$, this gives $\Delta \sim 215$ meV at $n \sim 10^{20}$ cm^{-3}. In fact, one can put, using $\varepsilon = 11.7$ for Si,

$$\Delta = 215(m^*/m)^{\frac{1}{2}} \left[\frac{n}{10^{20}}\right]^{\frac{1}{6}} \text{meV}$$

Although the band-gap shrinkage has been determined in eq. (1.12.29) in terms of a and Λ, a more sophisticated theory is needed to estimate a. The importance of a arises from the fact that it determines the relative contributions of the two terms in eq. (1.12.27). The beauty of the present treatment is that no commitment needs to be made as regards the numerical value of a. The reason is that we need only eq. (1.12.29) from which a cancels if eq. (1.12.31) holds, so that the actual value of a, which enters only eqs. (1.12.27) and (1.12.28), is not required.

The above simple theory begins to fail unless by eq. (1.12.31) $\Lambda a < z$ where $z \sim \frac{1}{4}$ and also, by eq. (1.12.28) we need $E_G(0,0) > e^2/\varepsilon a$, so that

$$\Lambda < z\varepsilon E_G(0,0)/e^2$$

We saw in eq. (1.8.25) that $\Lambda \propto n^{\frac{1}{3}}$ and so one sees that n must lie below a critical concentration, n_{cr} say. This turns out to be

$$n_{\text{cr}} = \frac{(\varepsilon/10)^9}{(m_c/m)^3} (4z)^6 E_G^6(0,0) \times 0.659 \times 10^{20} \text{ cm} \rightarrow 5 \times 10^{20} \text{ cm}^{-3} \qquad (1.12.33)$$

where E_G is expressed in eV. We have assumed one band and degeneracy, and the last expression holds for silicon ($\varepsilon \sim 11.7$, $\varepsilon_G \sim 1.1$ eV).

An alternative derivation of a somewhat smaller band-gap shrinkage

$$\Delta = \frac{3}{4}\frac{e^2\lambda}{\varepsilon} \qquad (1.12.32a)$$

can be given [1.12.40]. Like most simple treatments it only gives the flavor of the argument and it can be criticized [1.12.38]. This derivation attributes the shrinkage to the difference in the electrostatic energy which resides in the unscreened field $\mathscr{E}_1$ and the screened field $\mathscr{E}_2$

$$\Delta = \frac{\varepsilon}{8\pi} \int [\mathscr{E}_1(r)^2 - \mathscr{E}_2(r)^2] \, d\tau$$

Now

$$\mathscr{E}_1(r) = \frac{e}{\varepsilon r^2}$$

$$\mathscr{E}_2(r) = -\text{grad}\left[\frac{e}{\varepsilon r}\exp(-\lambda r)\right] = \frac{e}{\varepsilon r}e^{-\lambda r}\left(\frac{1}{r}+\lambda\right)$$

$$d\tau = 4\pi r^2 \, dr$$

Hence

$$\Delta = \frac{e^2}{2\varepsilon} I$$

To show that the integral I has the value $3\lambda/2$, one can carry out the spatial integration between $r = a$ and $r = b$, and then let $a \to 0$, $b \to \infty$. One finds

$$I \equiv \int_a^b [r^{-2} - r^{-2}\exp(-2\lambda r) - 2\lambda r^{-1}\exp(-2\lambda r) - \lambda^2\exp(-2\lambda r)]\,dr$$

Integrating the second and third terms together,

$$I = [-r^{-1} + \tfrac{1}{2}\lambda\exp(-2\lambda r) + r^{-1}\exp(-2\lambda r)]_a^b$$

For $b \to \infty$ the upper limit does not contribute and one finds

$$I = \frac{1}{a} - \frac{1}{2}\lambda\exp(-2\lambda a) - a^{-1}\exp(-2\lambda a)$$

For $a \to 0$ the last term yields $-a^{-1} + 2\lambda$ so that

$$I = 2\lambda - \tfrac{1}{2}\lambda = \tfrac{3}{2}\lambda$$

as required.

1.12.4 The relation between doping and solubility of defects

Consider a nondegenerate semiconductor containing *donors* of known concentration N_D and mechanical or other lattice *defects* of neutral concentration $N_d^\times$. This can be regarded as also 'known' since special considerations apply to the solubility of neutral defects in a semiconductor (see section 1.9). We shall assume that the defects are of the acceptor type, so that they can also be negatively charged by one electron charge. With these assumptions the concentration of negatively charged defects is given by

$$\frac{N_d^-}{N_d^\times} = \frac{Z_{d1}}{Z_{d0}}\lambda = \frac{Z_{d1}}{Z_{d0}} \times \frac{n}{N_c}\exp\eta_c \equiv \frac{n}{n_{1d}} \tag{1.12.34}$$

using the notation of section 1.7. Hence

$$n_{1d} \equiv N_c Z_{d0}/Z_{d1}\exp\eta_c \tag{1.12.35}$$

is independent of electron concentration. The total defect concentration is

$$N_d = N_d^\times\left(1 + \frac{n}{n_{1d}}\right) \tag{1.12.36}$$

and is unknown. By virtue of the separability assumption (see p. 127) the main doping N_D determines it by fixing the Fermi level.

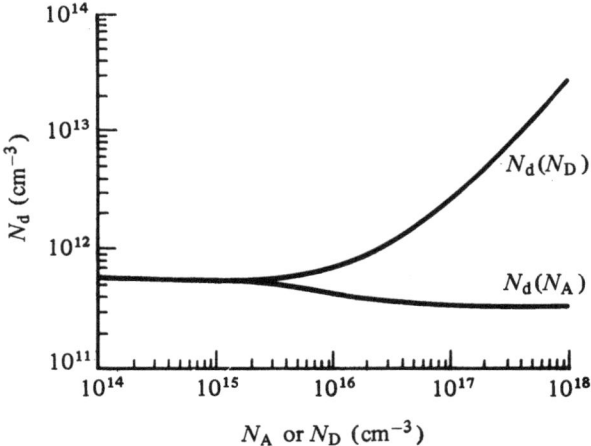

Fig. 1.12.8. Defect density (acceptor type) as function of impurity concentration in n-type $[N_d(N_D)]$ and p-type $[N_d(N_A)]$ nondegenerate Si at $T_f = 620$ K.

Similarly one can obtain N_D^+, the concentration of ionized donors, from

$$N_D^+ = \frac{N_D}{1 + \lambda Z_{D1}/Z_{D0}} = \frac{N_D}{1 + n/n_{1D}} \tag{1.12.37}$$

where in analogy with eq. (1.12.35)

$$n_{1D} \equiv N_c Z_{D0}/Z_{D1} \exp \eta_c \tag{1.12.38}$$

The equation for the Fermi level is

$$p - n + N_D^+ = N_d^- \tag{1.12.39}$$

Here the concentration p can be eliminated by

$$p = n_i^2/n = N_d^\times n_i^2/N_d^- n_{1d} \tag{1.12.40}$$

where eq. (1.12.34) has been used. After some algebra one finds a cubic equation for N_d^-:

$$[(N_d^-)^2 - cb^2][N_d^- - b] = db^2 N_d^- \tag{1.12.41}$$

$$b \equiv \frac{n_{1D}}{n_{1d}} N_d^\times, \quad c \equiv \frac{(n_i/N_{1D})^2}{1 + N_d^\times/n_{1d}}, \quad d \equiv \frac{N_D n_{1D}}{1 + N_d^\times/n_{1d}} \tag{1.12.42}$$

Assuming b, c, d to be known in addition to $N_d^\times$, eq. (1.12.41) yields N_d^-, eq. (1.12.34) yields λ and n, and eq. (1.12.40) yields p. Hence one can find N_d of eq. (1.12.36) as a function of the doping concentration N_D (see Fig. 1.12.8). The looked-for link between N_d and N_D is due to the fact that as the Fermi level is raised by doping, N_d^- increases for given $N_d^\times$, and hence N_d *increases* [1.12.41].

Table 1.12.1. *Data used for Fig.* 1.12.8

(1) Donor level in *n*-type silicon: 44 meV below conduction band
(2) Acceptor level in *p*-type silicon: 45 meV above valence band
(3) $N_d^\times = 1.39 \times 10^{11}$ cm^{-3}
(4) $T = 620$ K

In a *p*-type semiconductor the situation is rather different. Acceptor doping lowers the Fermi level, and hence N_d^- and N_d are *decreased*. In fact, N_d^- is again given by an equation of the form (1.12.41), except that in eq. (1.12.42) all D's have to be replaced by A's; everything else remains the same. Fig. 1.12.8 gives the expected relationship.

In [1.12.42] and [1.12.41] it was conjectured that

$$N_d^\times = M \exp(-E_a/kT) \tag{1.12.43}$$

with $M = 5 \times 10^{22}$ cm^{-3} representing the concentration of host lattice sites, E_a the activation energy of the defect, and T the temperature at which the defects can be regarded as becoming practically immobile. It was found by comparison with experimental results, as explained in connection with eqs. (2.3.30), that $E_a \sim 1.375$ eV, $T \sim 620$ K. These are the items (3) and (4) in Table 1.12.1.

Suppose that in the production of silicon crystals, the freeze-in temperature $T \equiv T_f \sim 620$ K has the significance discussed above. Then the mechanical defect density N_d is determined as in Fig. 1.12.8 also for $T < T_f$. When the crystal is used at such a lower temperature T, one finds new values of N_d^- and N_D^+:

$$\frac{N_d^-}{N_d} = \frac{\lambda Z_{d1}}{Z_{d0} + \lambda Z_{d1}} = \left[1 + \frac{Z_{d0}}{\lambda Z_{d1}}\right]^{-1} = \left[1 + \frac{n_{1d}}{n}\right]^{-1} \tag{1.12.44}$$

$$\frac{N_D^+}{N_D} = \frac{Z_{D1}}{Z_{D0} + \lambda Z_{D1}} = \left[1 + \frac{\lambda Z_{D1}}{Z_{D0}}\right]^{-1} = \left[1 + \frac{n}{n_{1D}}\right]^{-1} \tag{1.12.45}$$

where N_d is given by the known value of $N_d^\times$ and N_d^- at T_f through eq. (1.12.41). One can now use these results in eq. (1.12.39) to find a quartic equation for the electron concentration n in an *n*-type material at the new temperature. It is

$$(n^2 - n_i^2)(n + n_{1D})(n + n_{1d}) + (n + n_{1D}) N_d n^2$$
$$- (n + n_{1d}) N_D n_{1D} n = 0 \tag{1.12.46}$$

An analogous relation holds for *p*-type material. One can use the solution of this equation to discuss recombination traffic and lifetimes (see section 2.3.5).

1.12.5 *Points from the literature*

(i) For a review of doping effects, light as well as heavy, see [1.12.43].

(ii) The grand partition function has been used to study equilibrium effects in transition-metal-doped semiconductors, notably Cr-doped GaAs [1.12.44].

(iii) A review of temperature-dependent electronic conduction in semiconductors is available [1.12.45].

1.13 Quantities referred to the intrinsic level

Semiconductor parameters are sometimes referred to the intrinsic Fermi level (1.12.11), in bulk materials, in p–n junctions and in other devices. This referral is never essential, and sometimes in fact interferes with a simple appreciation of the formulae. Because of the popularity of this procedure, some of the relevant results are derived in this section, without, however, including band shrinkage.

For nondegenerate materials, eq. (1.12.11) yields

$$E_c - \mu_i = \tfrac{1}{2}E_G + \tfrac{1}{2}kT \ln (N_c/N_v) \tag{1.13.1}$$

$$\mu_i - E_v = \tfrac{1}{2}E_G + \tfrac{1}{2}kT \ln (N_v/N_c) \tag{1.13.2}$$

We shall apply these results in the first place to a p–n junction. We shall model it in the simple form of Fig. 1.13.1, the band edges E_c, E_v, and hence also μ_e, μ_h become functions of position x. We shall assume the temperature T, and therefore N_c, N_v, to be independent of x. Thus a one-dimensional model is adopted. *In the transition region* of width W, the band edges and μ_i are assumed to be sloping linearly, while the quasi-Fermi levels are assumed constant and parallel [1.13.1]. These assumptions are expressed by

$$\gamma_e - \gamma_h = \gamma = \text{const} \tag{1.13.3}$$

$$\gamma_i(x) = \gamma_i(0) - (\Theta/W)x \quad (-\tfrac{1}{2}W \leqslant x \leqslant \tfrac{1}{2}W) \tag{1.13.4}$$

where

$$\gamma_i \equiv \mu_i/kT \text{ and } \Theta \equiv (V_D - U)/kT$$

Here V_D is the built-in potential energy difference and $U/|e|$ is the applied (forward) potential, which is negative for a reverse applied potential. The assumptions encoded in Fig. 1.13.1 are widely used, but they are approximations.

The results (1.6.11) and (1.6.12) can now be rewritten

$$n = n_i \exp (\gamma_e - \gamma_i) \tag{1.13.5}$$

$$p = n_i \exp (\gamma_i - \gamma_h) \tag{1.13.6}$$

since the right-hand side of eq. (1.13.5) is, using eqs. (1.12.10) and (1.12.11),

$$\left[(N_c N_v)^{\frac{1}{2}} \exp\left(\frac{\eta_v - \eta_c}{2}\right) \right] \exp \gamma_e \left[\exp\left(\frac{-\eta_c - \eta_v}{2}\right) (N_c/N_v)^{\frac{1}{2}} \right]$$

$$= N_c \exp (\gamma_e - \eta_c)$$

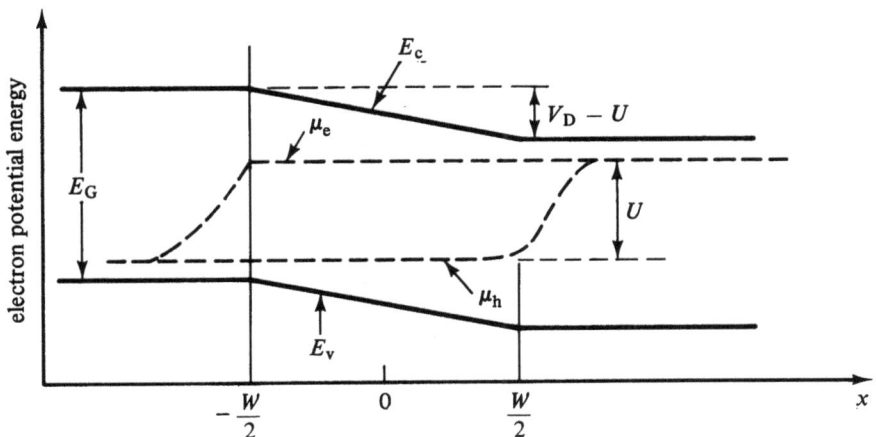

Fig. 1.13.1. A simple model of a p–n junction (see also Fig. 1.6.1).

A similar argument holds for eq. (1.13.6). Applying results (1.13.1) and (1.13.6) to a p–n junction, and using suffices p and n for situations at the left-hand and the right-hand sides of the junction,

$$\left.\begin{aligned}
n_n/n_i &= \exp\left[\gamma_e - \gamma_i(W/2)\right] = \exp\left[\gamma_e - \gamma_i(0) + \tfrac{1}{2}\Theta\right] \\
&= (N_c/N_v)^{\frac{1}{2}}\exp(\eta_D/2) \\
p_p/n_i &= \exp\left[\gamma_i(-W/2) - \gamma_h\right] = \exp\left[\gamma_i(0) + \tfrac{1}{2}\Theta - \gamma_h\right] \\
&= (N_v/N_c)^{\frac{1}{2}}\exp(\eta_D/2)
\end{aligned}\right\} \tag{1.13.7}$$

where

$$\eta_D \equiv V_D/kT = \Theta + U/kT \tag{1.13.8}$$

Thus if $\mu_e - \mu_h$ denotes the Fermi level separation in the transition region

$$n_n p_p = n_i^2 \exp\eta_D = n_i^2 \exp\left[(V_D - U)/kT + (\mu_e - \mu_h)/kT\right] \tag{1.13.9}$$

All these results are useful in device work.

Consider next a bulk semiconductor with quasi-Fermi levels μ_e, μ_h. Using Boltzmann statistics, one can introduce standard concentrations

$$n_0 = \exp(\gamma_e^0 - \eta_c) \quad p_0 = \exp(\eta_v - \gamma_h^0)$$

which generalize eqs. (1.13.5) and (1.13.6) in a simple way. Also

$$np = n_0 p_0 \exp(\gamma_e - \gamma_h - \gamma_e^0 + \gamma_h^0)$$

Normally one chooses

$$\mu_e^0 (\equiv kT\lambda_e^0) = E_c, \quad \text{whence } n_0 = N_c$$

and

$$\mu_h^0 = E_v, \quad \text{whence } p_0 = N_v$$

Hence

$$np = N_c N_v \exp(-\eta_G)\exp(\gamma_e - \gamma_h) \tag{1.13.10}$$

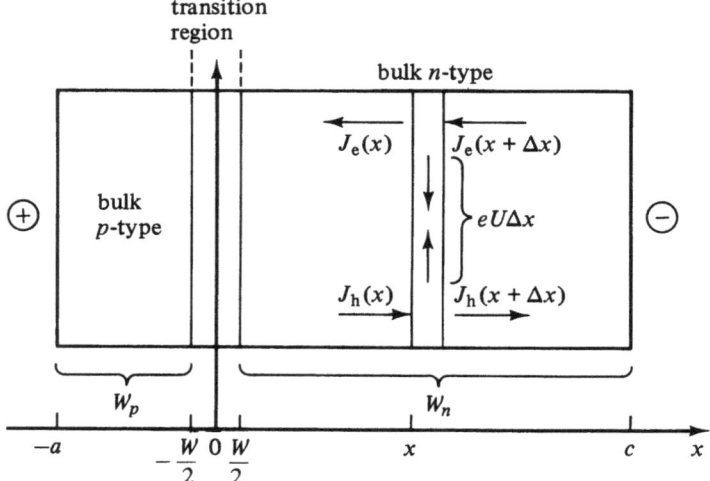

Fig. 1.14.1. Main x-coordinates of a p–n junction. Charge conservation diagram for forward bias.

If one chooses to interpret the μ's as Gibbs free energies per particle as suggested by eqs. (1.1.10) and (1.1.14) then E_c, E_v and E_G should also be regarded in that way.

A nonequilibrium intrinsic carrier density can be defined by

$$n_\theta \equiv [np \exp{(\gamma_h - \gamma_e)}]^{\frac{1}{2}} \tag{1.13.11}$$

whence

$$n_\theta^2 = N_c N_v \exp{(-\eta_G)}$$

in generalization of eq. (1.12.10).

1.14 Junction currents as recombination currents

Let a semiconductor pass a current in the x direction only, so that a one-dimensional model can be used. Consider any narrow region lying between x and $x + \Delta x$. Then electron and hole current densities flow in and out of this region and there is also a recombination current density $eU\Delta x$ (where $e \equiv |e|$). The illustration (Fig. 1.14.1) of the charge conservation condition is drawn with the current directions appropriate for a p–n junction under forward bias. One sees that

$$J_h(x) = eU\Delta x + J_h(x + \Delta x)$$
$$J_e(x + \Delta x) = J_e(x) + eU\Delta x$$

whence

$$\frac{\mathrm{d}J_e(x)}{\mathrm{d}x} = -\frac{\mathrm{d}J_h(x)}{\mathrm{d}x} = eU \tag{1.14.1}$$

The total current density at any plane x is

$$J(x) = J_e(x) + J_h(x) \tag{1.14.2}$$

With the notation $x = -a$, $x = c$ established in Fig. 1.14.1, integration of eq. (1.14.1) yields

$$J_h(-a) - J_h(c) = e \int_{-a}^{c} U \, dx \tag{1.14.3}$$

The last term in eq. (1.14.3) can be split into recombination currents for the bulk p-type material, the transition region and the bulk n-type material. Putting eq. (1.14.3) into eq. (1.14.2) we have, for instance at $x = -a$,

$$J = J_e(-a) + J_h(c) + J_{bp} + J_{tr} + J_{bn} \tag{1.14.4}$$

where

$$J_{bp} \equiv e \int_{-a}^{-W/2} U \, dx, \quad J_{tr} \equiv e \int_{-W/2}^{W/2} U \, dx, \quad J_{bn} \equiv e \int_{W/2}^{c} U \, dx \tag{1.14.5}$$

Now $J_e(-a)$, $J_h(c)$ are the minority-current densities at the contacts and they will in many cases be small. In any case, relation (1.14.4) reveals a considerable part of the *current density in a p–n junction as due to recombination or generation*. The current densities at the edges of the transition region, but just in the bulk, are the so-called 'diffusion current densities'

$$J_e(-a) + J_{bp}, \quad J_h(c) + J_{bn}$$

Hence eq. (1.14.4) can also be written as

$$J = \text{sum of diffusion current densities} + J_{tr}$$

This does not however fully emphasize the importance of recombination–generation processes for junction properties.

The above considerations hold whenever the bulk regions and the transition region can be defined by (possibly voltage-dependent) planes. To proceed further, however, one must know, or assume, something about the variation of quasi-Fermi levels and electrical potential within the different regions. The accurate determination of these quantities, by the solution of Poisson's equation and the drift–diffusion–recombination equations, poses a complicated problem. In order to obtain analytical results, one may base a theory on simple assumptions concerning the shapes of the quasi-Fermi levels and band edges.

It should be pointed out, however, that general integrations of the basic equations can be done numerically. This then leads to interesting results to be found in the literature on semiconductor devices.

2

Recombination statistics

2.1 Basic assumptions for recombination statistics

2.1.1 General orientation

A transition in which an electron jumps from a high energy to a low energy state which is located in a different group of levels is called *recombination*. One can think of the electron vacancy in the lower states as a trapped hole. The transition is then an electron–hole recombination, in which both 'particles' disappear and energy is released, for example in the form of photons or phonons. The different groups of levels are normally different energy bands or valleys in **k**-space or different groups of localized states. Transitions within a band are not normally considered as recombination events as they take place very quickly.

Figure 2.1.1 gives examples of recombination transitions. As in chemical kinetics, the rate per unit volume is regarded as proportional to the frequency of collision (which is itself proportional to the appropriate product of concentrations) and a recombination coefficient. In part (a) of the figure, for example, the frequency of collision is np and the coefficient, B^s say, has the dimension $[L^3 T^{-1}]$. The other recombination rate expressions are obtained similarly, v_0 and v_1 being the concentration of empty and occupied defect centers. In the fourth process, for example, one may suppose that an electron recombines with a hole trapped on a defect. In writing down the rates one must remember that full states in the valence band and unoccupied states in the conduction band 'do not count'. There are usually enough of them so that they do not affect the rate. This rule assumes nondegenerate bands. For degenerate bands the situation is more complicated (see section 2.2).

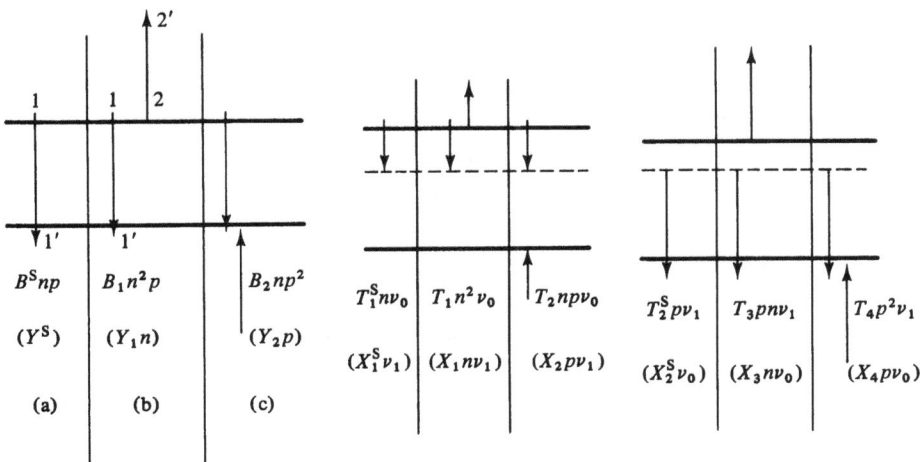

Fig. 2.1.1. Definitions of recombination coefficients. Transition rates per unit volume are stated with each process, and in brackets for the reverse process. B_1, B_2, T_1 to T_4 refer to Auger processes; B^S, T_1^S, T_2^S refer to single-electron recombination; Y^S, X_1^S, X_2^S refer to carrier generation processes; Y_1, Y_2, X_1 to X_4 refer to impact ionization processes.

The reverse process to recombination is *generation* and the same rules apply. For example, in process (a) the rate of recombination per unit volume, Y^S say, is independent of concentration. This remark has again to be modified if either or both bands are degenerate.

When a semiconductor is excited optically or by carrier injection, then the *unavoidable* recombination processes [2.1.1] leading to the decay of electrons and holes involve the energy bands. Physical defects such as vacancies, divacancies, dislocations, grain boundaries and the like, as well as chemical defects in the form of impurities, also lead to recombination. The latter can in principle be avoided by using a perfect sample. The unavoidable band–band processes (Fig. 2.1.1a–c) will however remain, and we shall now discuss these.

The processes (a) to (c) are unavoidable because they simply depend on the existence of energy bands. Process (a) can take place by emission or absorption of photons and phonons. If there are no phonons involved, it is said to be a purely radiative process. If no photons are involved, one has a nonradiative multiphonon process. The reverse process, upward in energy, is normally due to thermal agitation in a crystal and proceeds by the absorption of phonons or/and photons whose energy then promotes the electron to a higher energy level. Processes (b) and (c) are due to electron collisions: the recombining electron gives up its energy to a second electron in the conduction or valence band. These so-called Auger processes (first discovered in atomic systems by Pierre Auger; soft g, please, the

gentleman is French not German!) increase in importance with carrier concentration, as can be seen from the rate expressions given in the figure. For large
enough electron concentration $B_1 n^2 p$ will exceed $B^s np$.

Two sets of three avoidable processes remain. Each set of three avoidable
processes mirror the significance of the three band–band processes just discussed,
one band being replaced by impurity levels in each case. There are in these two sets
a total of two direct and four Auger-type avoidable processes.

Consider some recombination rate $A n^r p^s$, which is proportional to a simple
power of the number of electrons and holes, and to a recombination coefficient A.
One can then write for the recombination rate due to this type of process

$$R = A n^r p^s + B n^s p^r \qquad (2.1.1)$$

where the contribution from the partner process, arising from an interchange of
electrons with holes, has also been included. The two coefficients A, B will in
general be different. Thus, if the quantities δn, δp measure the departure from
equilibrium,

$$R = A n_0^r p_0^s \left(1 + \frac{\delta n}{n_0}\right)^r \left(1 + \frac{\delta p}{p_0}\right)^s + B n_0^s p_0^r \left(1 + \frac{\delta n}{n_0}\right)^s \left(1 + \frac{\delta p}{p_0}\right)^r \qquad (2.1.2)$$

In thermal equilibrium the concentrations are n_0, p_0 and excitation is in balance
with the following equilibrium recombination rate:

$$R_0 = A n_0^r p_0^s + B n_0^s p_0^r \qquad (2.1.3)$$

The excess concentrations of electrons and holes will decay in times of order

$$\tau_n \equiv \frac{\delta n}{R - R_0}, \quad \tau_p \equiv \frac{\delta p}{R - R_0} \qquad (2.1.4)$$

This assumes that the decay rate (= recombination rate − generation rate, away
from equilibrium) behaves as follows: the recombination rate obeys eq. (2.1.1).
The generation rate is due to lattice agitation appropriate to the lattice temperature
and remains at R_0. One can also define an *equilibrium lifetime*

$$\tau_{n0} \equiv n_0 / R_0, \quad \tau_{p0} \equiv p_0 / R_0 \qquad (2.1.5)$$

One now finds from eqs. (2.1.1) to (2.1.5) that for small excursions from
equilibrium

$$\delta n / n_0 \sim \delta p / p_0 \ll 1 \qquad (2.1.6)$$

$$\tau_n \sim \frac{\delta n}{(r+s)(\delta n / n_0) R_0} \sim \frac{\tau_{n0}}{r+s}, \quad \tau_p \sim \frac{\tau_{p0}}{r+s} \qquad (2.1.7)$$

In cases where relation (2.1.6) holds the thermal equilibrium values τ_{n0}, τ_{p0} thus give a good guide to the values of τ_n, τ_p [2.1.2].

2.1.2 The assumptions needed

The recombination problem in semiconductors is greatly complicated by the interaction of the electrons with each other, which allows one to speak only of the quantum states of the semiconductor crystal as a whole. However, a simplified picture is successful. In this the electron interactions are first neglected but are later taken into account as a perturbation. These perturbations lead necessarily to changes of state of electron pairs. They can be pictured as transitions due to electron collisions, and can still be described within the framework of the single-particle states found in the unperturbed problem.

The *first approximation* is to neglect most (but not all) electron interactions; two-electron transitions are, however, retained in our recombination–generation schemes. In addition, electrons interact with the radiation and lattice fields and emit or absorb photons or phonons. But these electron–boson interactions are represented by the transitions of single electrons in a single-electron energy band scheme. We have attached a superscript S to the recombination coefficients for such processes. These are denoted by B_i^S or T_i^S, depending on whether only bands are involved or also traps. These recombination coefficients are illustrated in Fig. 2.1.1 and are

$$B^S, B_1, B_2; \quad T_1^S, T_1, T_2; \quad T_2^S, T_3, T_4 \tag{2.1.8}$$

Note the basic distinction between single-electron and two-electron recombination coefficients: The former (B^S, T_1^S, T_2^S) have dimension [$L^3 T^{-1}$], the latter (B_1, B_2, T_1 to T_4) have dimensions [$L^6 T^{-1}$].

Electron–electron interactions can at least formally be taken into account in connection with electrons trapped in a center, by taking its energy spectrum as a function of the number, r, of electrons captured (see section 1.7). The irremovable electrons are included with the ion core. One can simply associate a set of quantum states $(l, r, \mathbf{R})$ $(l = 1, 2, \ldots)$ with a center located at position $\mathbf{R}$. There then exist also a set of energies and a partition function Z_r, as in eq. (1.7.1).

Suppose we can arrange for the equilibrium Fermi levels to rise from the valence band to the conduction band. At first most centers (assumed here to be of positive U (see p. 60)) are in a state $r = 0$. As the Fermi level rises, the states corresponding to $r = 1$ begin to appear according to eq. (1.7.2). In equilibrium the ground states are always more highly populated than the excited states so that we can confine attention to them. As the Fermi level rises, the ground states for

$r = 2$ may be more important, and there may now be hardly any centers which have captured less electrons. If a larger number of electrons cannot be captured by the time the Fermi level reaches the conduction band, then states of the center with $r > 2$ can be neglected as unstable, i.e. $M = 2$ in the notation of section 1.7.1. That this is a satisfactory picture is our *second approximation*.

The recombination problem is still complex because of the many states available to electrons in bands and centers. A key simplification arises from the fact that it is often possible to talk about a small number of *groups* of quantum states: I labels quantum states in group i, J labels quantum states in group j, etc.

Within each group it is supposed that the transitions are much more rapid than they are between groups. In a sense, therefore, electrons in each group are in equilibrium among themselves and can be assigned a quasi-Fermi level (divided by kT) γ_i, γ_j, etc. This is our *third assumption* (see section 1.11). With this assumption, recombination problems can be discussed by neglecting *transitions confined to one group*, because they proceed at exactly the same rate as the reverse transitions. The number of transition types to be considered is thereby greatly reduced.

Assuming the main groups of states to be conduction band, valence band, and one-electron, two-electron, ..., M-electron centers, appropriate quasi-Fermi levels divided by kT [2.1.3] are denoted by

$$\gamma_e, \gamma_h, \gamma_1, \ldots, \gamma_M$$

Away from equilibrium we shall then assume in generalization of the equilibrium equation (1.7.2),

$$P(l,r) = \frac{\exp[r\gamma_r - \eta(l,r)]}{Z_0 + Z_1 e^{\gamma_1} + Z_2 e^{2\gamma_2} + \ldots} \tag{2.1.9}$$

The definition of the γ_r is largely a matter of convention. Each γ_r is defined by the set $\{P(l,r)\}$ for fixed r and it is assumed that $P(l,r) \propto \exp[-\eta(l,r)]$. The centers will be assumed to be all of the same type, so that our results do not depend on location $\mathbf{R}$, and this label can be omitted. The assumption of a quasi-Fermi level for each state of charge of a center implies that all the excited states of an r-electron center are populated according to

$$P(l,r) \propto \exp[-\eta(l,r)] \text{ for given } r \quad (r = 1, 2, \ldots, M)$$

as the quantum state label l is varied. A further point is that eq. (2.1.9) will also be valid for a state I in a band i, since for $Z_0 = 1$, $\eta(i,r) = \eta$, and $r = M = 1$ the usual Fermi–Dirac probability factor is obtained. However, eq. (2.1.9) is not always correct and this arises from the quasi-Fermi level assumption. Improved theories include the cascade theories (see section 2.4) and others.

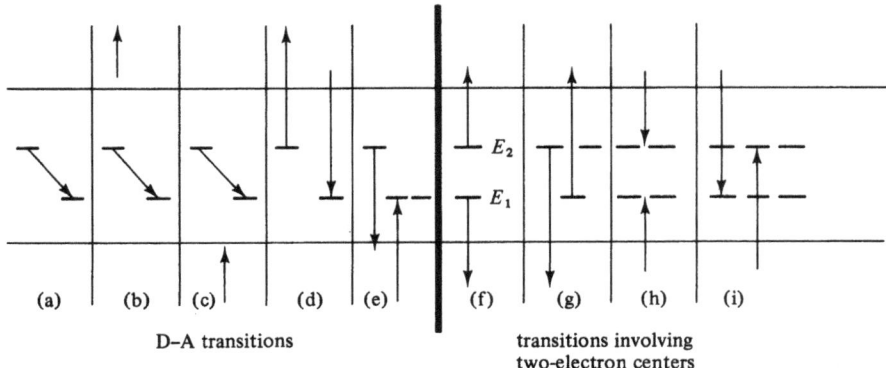

(a)   (b)   (c)   (d)   (e)   (f)   (g)   (h)   (i)

D–A transitions   transitions involving
                                   two-electron centers

Fig. 2.1.2. Some other recombination processes.

Equation (2.1.9) goes over into the equilibrium equation (1.7.2) if the replacements

$$(e^{\gamma_1} \to \lambda), \ldots, (e^{\gamma_M} \to \lambda)$$

are made, i.e. if all quasi-Fermi levels coincide.

Fig. 2.1.1 does not exhaust the catalog of possible transitions, and Fig. 2.1.2 gives some additional ones. Donor–acceptor or 'pair' transitions were much investigated in the 1960s and 1970s, and are reviewed in [2.1.4]. It was found possible to distinguish radiation from both close and widely separated pairs, as discussed in section 5.2.6. As always, nonradiative transitions are in competition with the radiative ones. For example, the energy released radiatively as in (a) may be taken up by a second (Auger) particle as in (b) or (c) (see section 5.4.2). Their exchange processes are (d) and (e). A two-electron center in a ground state of energy E_2 can lose an electron to become a one-electron center in a ground state of energy E_1. It can lose both electrons by Auger effect as in (f), and in its exchange process (g), thus effectively converting a hole into an electron, keeping the total number of current carriers constant. The reverse processes of (f) and (g) convert an electron into a hole as in (h) and (i). Whereas (b) to (e) have (a) as the radiative analogue, processes (f) to (i) have no radiative analogue. The processes (f) to (i) are briefly discussed in section 5.3.5. For further elaborations see [2.1.5]. An entirely new set of processes are added to the list if one wishes to consider excitons, see Table 5.4.1.

A first review of the whole of our subject [2.1.6] already boasted 131 references. Eight years later there was a review with 362 references [2.1.7]. But thereafter reviews tended to deal with specific aspects of recombination; for example radiative decay in compound semiconductors [2.1.8], Auger effects [2.1.9], [2.1.10], Si [2.1.11], and narrow-gap semiconductors [2.1.12].

2.1.3 Example: Quantum efficiency in an intrinsic semiconductor

If we add to the recombination rates already discussed a nonradiative one, An, proportional to the injected carrier density $n \sim p$ (when doping is negligible), one can make an instructive calculation already at this stage. The radiative band–band recombination is $B^S n^2$. Hence the rate of radiative recombination as a fraction of the total recombination rate, i.e. the quantum efficiency, for this intrinsic semiconductor is

$$\eta = \frac{B^S n^2}{An + (B_1 + B_2) n^3 + B^S n^2} \tag{2.1.10}$$

The denominator contains the radiative and the nonradiative terms which includes the band–band Auger effects. Traps are neglected. Since $\eta \to 0$ as $(B^S/A)n$ for small n, and as $B^S/(B_1 + B_2)n$ for large n, it goes through a maximum. This occurs for

$$n = n_1 \equiv [A/(B_1 + B_2)]^{\frac{1}{2}}$$

and is

$$\eta_{\mathrm{max}} = \left\{ 1 + \frac{2}{B^S} A^{\frac{1}{2}} (B_1 + B_2)^{\frac{1}{2}} \right\}^{-1} \tag{2.1.11}$$

For a high quality epitaxial AlGaAs/GaAs double heterostructure of great purity we may take [2.1.13]

$$A \sim \tfrac{1}{2} 10^6 \ \mathrm{s^{-1}}, \quad B^S \sim 10^{-10} \ \mathrm{cm^3 \ s^{-1}}, \quad B_1 + B_2 \sim 10^{-29} \ \mathrm{cm^6 \ s^{-1}}$$

whence

$$n_1 \sim 2 \times 10^{17} \ \mathrm{cm^{-3}}, \quad \eta_{\mathrm{max}} \sim 0.96$$

Thus for sufficiently pure materials spontaneous emission ($\sim B^S$) can dominate.

This coefficient is not a material constant. It is due to the interaction between the relevant atom or crystal and the infinity of vacuum states. These can be manipulated by placing the excited atom between mirrors or in a cavity, and, in the case of solids, by periodic spatial modulation which introduces a *photonic* band gap, analogous to the *electronic* band gap. If these two gaps overlap, the spontaneous emission is inhibited, since there is then a little wave propagation in the direction considered. This previously leaked energy could in lasers go into stimulated emission, which is desirable. The light-trapping which results could also be useful in solar cells. Thus the subject of *cavity quantum electrodynamics* has recently been born. The precise conditions under which a photonic energy gap will be exhibited by periodic dielectric structures are still being refined [2.1.13]. The inhibition of spontaneous emission by the removal of appropriate electromagnetic modes has also been investigated for atomic systems.

In the above considerations the generation processes specified by Y^s, Y_1, Y_2 in Fig. 2.1.1 could be neglected since the system is far from equilibrium. Near equilibrium one needs the reverse processes as well since detailed balance holds at equilibrium.

2.2 The main recombination rates

2.2.1 Basic theory

Recombination and its converse, generation, consist of a transition of an electron from one state to another. The observed rate is the net rate of recombination and is the algebraic sum of the recombination and generation processes. During these processes both total energy and total momentum must be conserved. This is achieved by creation or absorption of photons, or phonons, excitation of secondary electrons, etc.

To obtain the general expression for the net recombination we proceed as follows. The transition probability per unit time, S_{IJ}, for a single-electron transition from state I to J depends on no states other than I and J. For this transition to occur, state I should be 'occupied', with probability p_I say, and state J should be 'vacant' with probability q_J say. The general expression for the average rate of the transition from I to J then takes the form $p_I S_{IJ} q_J$. The reverse process is the transition of an electron from state J to state I. State J will have to be occupied and state I vacant. Thus the rate of this reverse process is $p_J S_{JI} q_I$. The net recombination rate per unit volume of the process $I \rightarrow J$ can then be written as

$$u_{IJ} = (p_I S_{IJ} q_J - p_J S_{JI} q_I) V^{-1}$$

By the principle of detailed balance, this expression vanishes at equilibrium. We have

$$u_{IJ} = p_J S_{JI} q_I (X_{JI} - 1) V^{-1} = p_I S_{IJ} q_J (1 - X_{JI}^{-1}) V^{-1} \qquad (2.2.1)$$

where

$$X_{JI} \equiv \begin{cases} \dfrac{S_{IJ} p_I q_J}{S_{JI} p_J q_I}, & (2.2.2a) \\[2ex] \dfrac{S_{IJ}/S_{JI}}{(S_{IJ}/S_{JI})_0} \dfrac{p_I q_{I0}}{p_{I0} q_I} \dfrac{q_J p_{J0}}{q_{J0} p_J} & (2.2.2b) \end{cases} \qquad (X_{JI})_0 = 1$$

For recombination by a single mechanism (e.g. radiative recombination, or capture by traps, etc.) the overall rate of recombination will be the sum of expressions like eq. (2.2.1) over allowed states I in some group i and over allowed states J in some group j. The experimentally observed rate of recombination will be the sum of the overall rates for the different mechanisms; here we assume that the different mechanisms do not interfere with each other.

2.2.2 Identification of the X_{JI}

We illustrate the evaluation of X_{JI} for three cases.

(i) When I and J refer to states in bands i, j, then p_I is the probability that state I is occupied and q_I is the probability that state I is vacant. Clearly $p_I + q_I = 1$, and using eq. (1.11.1a), we have

$$\frac{p_I q_{I0}}{q_I p_{I0}} = \exp(\gamma_i - \eta_I)\exp(\eta_I - \gamma_0) = \exp(\gamma_i - \gamma_0) \qquad (2.2.3)$$

Also with $F \equiv (\mu_i - \mu_j)/kT$ and $\eta_G \equiv (E_I - E_J)/kT \equiv E_G/kT$

$$X_{JI} = \begin{cases} \dfrac{S_{IJ}}{S_{JI}}\exp(F - \eta_G) & (2.2.4a) \\[3mm] \dfrac{S_{IJ}/S_{JI}}{(S_{IJ}/S_{JI})_0}\exp F & (2.2.4b) \end{cases}$$

These are alternative expressions. One can go from the first to the second by using detailed balance in the form

$$(X_{JI})_0 = 1, \quad \text{or} \quad (S_{IJ}/S_{JI})_0 = \exp\eta_G \qquad (2.2.4c)$$

F is a measure of the departure from equilibrium. One says that the system is *pumped* in order to produce such a departure.

If the pumping of the system does not affect the transition probability ratio, i.e. if S_{IJ}/S_{JI} is *excitation-independent*, then one can simplify eq. (2.2.4b) and generalize eq. (2.2.4c):

$$X_{JI} = \exp\frac{\mu_i - \mu_j}{kT} \equiv \exp F \qquad (2.2.4d)$$

$$\frac{S_{IJ}}{S_{JI}} = \exp\eta_G \equiv \exp\frac{h\nu}{kT} \qquad (2.2.4e)$$

The photon frequency to excite an electron from level I to level J has been denoted by ν. These results have the important consequence that if one sums eq. (2.2.1) over the groups i and j of states, a factor comes out which depends on a typical voltage $\varphi \equiv (\mu_i - \mu_j)/|e| \equiv kTF/|e|$

$$u_{ij} \equiv \left[\sum_{I\in i}\sum_{J\in j} p_J S_{JI} q_I\right]\left[\exp\frac{|e|\varphi}{kT} - 1\right]V^{-1} \qquad (2.2.4f)$$

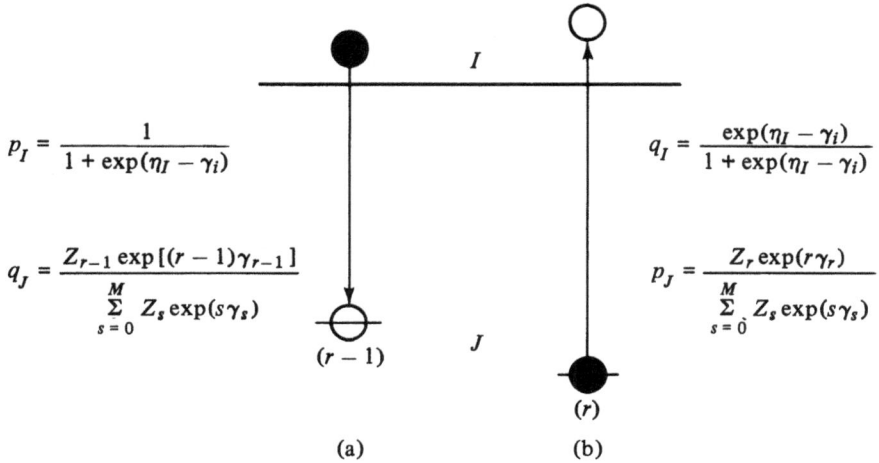

Fig. 2.2.1. Examples of factors p and q for single-electron transitions into traps (a) and their inverses (b).

Now a transition rate, when multiplied by the charge of current carriers, is a current, and when divided by the area of the surface involved, is a current density. If i and j denote the states of the conduction and valence bands of a semiconductor, excitation independence may often be assumed, and one then expects a current density proportional to

$$\exp\left(\frac{|e|\,\varphi}{kT}\right)-1$$

This is characteristic of the current through $p-n$ junctions, metal semiconductor junctions, etc. In these configurations the Fermi level difference between the ends of the device determines the voltage across it [2.2.1]. When radiation is involved, however, excitation dependence tends to spoil this simple story.

(ii) When a trap is involved, matters are rather different. Because of the interactions among the electrons on a center it is not possible to talk of the same level being occupied or vacant. Consequently, identification of forward and reverse processes in terms of levels becomes impossible. Instead, it is sufficient to deal with a center, say an r-electron center, as a whole; we then need the probability that a given center is an r-electron center. For the reverse transition, where the capture of an electron converts the center back again into an r-electron center, the capturing center must be an $(r-1)$-electron center. Thus the p's and q's must be replaced as shown in Fig. 2.2.1.

Consider the capture of an electron into an $(r-1)$-electron center. Then for a typical transition (as shown in Fig. 2.2.1) eq. (2.2.2b) applies. The first factor is formally unchanged, the second factor is given by eq. (2.2.3) and the third factor is with eq. (2.1.9)

$$\frac{q_J p_{J0}}{p_J q_{J0}} = \frac{Z_{r-1}}{Z_r} \exp\left[(r-1)\gamma_{r-1} - r\gamma_r\right] \frac{Z_r}{Z_{r-1}} \exp(\gamma_0)$$

$$= \exp\left[(r-1)\gamma_{r-1} - r\gamma_r + \gamma_0\right]$$

Define a reduced quasi-Fermi level for the $[(r-1) \to r]$-electron center transitions by

$$r\gamma_r - (r-1)\gamma_{r-1} \equiv \gamma_{r-\frac{1}{2}} \tag{2.2.5}$$

Then one finds in the case of electron capture by an $(r-1)$-electron center

$$X_{JI} = \frac{S_{IJ}/S_{JI}}{(S_{IJ}/S_{JI})_0} \exp(\gamma_i - \gamma_{r-\frac{1}{2}}) \quad (i = \text{e or h}) \tag{2.2.6}$$

which is again of the form (2.2.4b), except that γ_J is replaced by $\gamma_{r-\frac{1}{2}}$. (Note that $\gamma_1 = \gamma_{\frac{1}{2}}$.)

(iii) For single-boson transitions eqs. (2.2.1)–(2.2.2) are still valid. One has to note that the occupation probability of a quantum state is for fermions exactly the same as the mean number of particles in that state. It is the latter interpretation that is more basic, since it alone holds also for bosons. In this case one has

$$p_I = \frac{1}{\exp(\eta_I - \gamma_I) - 1}, \quad q_I = 1 + p_I \text{ (bosons)} \tag{2.2.7}$$

in contrast to

$$p_I = \frac{1}{\exp(\eta_I - \gamma_I) + 1}, \quad q_I = 1 - p_I \text{ (fermions)}$$

Equation (2.2.7) expresses the increased attraction an occupied boson state has for additional bosons. However, for both fermions and bosons

$$q_I/p_I = \exp(\eta_I - \gamma_I) \tag{2.2.8}$$

2.2.3 Recombination coefficients as state averages

We now proceed to consider some overall recombination rates. The simplest special case arises for single-electron transitions from conduction to valence band under neglect of the excitation dependence of the S-ratios. This is done for simplicity of the resulting formulae, and we return to the explicit possibility of excitation dependence at the end of the section (see eq. (2.2.23)). Thus assuming excitation-independent S-ratios, eqs. (2.2.1) and (2.2.4d) yield:

$$u_{cv} = \left(\sum_{\substack{I \in c \\ J \in v}} p_I S_{IJ} q_J \right) [1 - \exp(\gamma_h - \gamma_e)] V^{-1}$$

$$= B^s np [1 - \exp(\gamma_h - \gamma_e)] \equiv B^s np - Y^s \quad \text{(using Fig. 2.1.1)} \tag{2.2.9}$$

Noting $n = \left(\sum_C p_C \right) / V$ and $p = \left(\sum_V q_V \right) / V$, we have introduced reaction constants

$$B^s \equiv \frac{V^{-1} \sum_{C \in c} \sum_{V \in v} p_C S_{CV} q_V}{V^{-2} \sum_C p_C \sum_V q_V} \quad [\text{L}^3\text{T}^{-1}] \tag{2.2.10a}$$

$$Y^s \equiv V^{-1} \sum_{C \in c} \sum_{V \in v} p_C e^{-\gamma_e} S_{CV} q_V e^{\gamma_h} \quad [\text{L}^{-3}\text{T}^{-1}] \tag{2.2.10b}$$

The dimensions are also given. Here B^s is an average of S_{CV} over the electrons in the conduction band and the holes in the valence band. Both are needed for the transitions of rate $B^s np$. For the average transition rate Y^s one would have to divide by all the electrons in the valence band and all the holes in the conduction band. These are large numbers and this is not done. The various forms of eq. (2.2.9) immediately yield the thermal excitation coefficient Y^s. However, its general form can be obtained independently from first principles as follows. The average excitation probability per unit time per unit volume (without the divisions, as just explained) is

$$Y^s = V^{-1} \sum_{V,C} p_V S_{VC} q_C = V^{-1} \sum_{V,C} X_{VC}^{-1} p_C S_{CV} q_V \tag{2.2.10c}$$

using eq. (2.2.2a) with $I, J \to C, V$. Upon using eq. (2.2.4b) in the form

$$X_{VC} = \exp(\gamma_e - \gamma_h)$$

one arrives again at eq. (2.2.10b), as required. These results are valid even for degenerate bands and for any density of states.

Two special cases are of interest. (i) For nondegenerate bands

$$p_c = \exp(\gamma_e - \eta_c), \quad q_v = \exp(\eta_v - \gamma_h)$$

so that B^s and Y^s become independent of γ_e and γ_h and hence of carrier concentrations, as would be expected of mass action constants:

$$B^s = V \frac{\sum_C \sum_V S_{CV} \exp[-(\eta_C - \eta_V)]}{\sum_C \sum_V \exp[-(\eta_C - \eta_V)]}$$

$$Y^s = V^{-1} \sum_C \sum_V S_{CV} \exp[-(\eta_C - \eta_V)]$$

(ii) If one does not make the assumption of nondegeneracy, but assumes S_{CV} to be independent of the states C and V in eq. (2.2.10), then B^S, but not Y^S, is again independent of the quasi-Fermi levels:

$$B^S = V^{-1} S_{CV}$$

$$Y^S = V^{-1} S_{CV} \exp[-(\gamma_e - \gamma_h)] \sum_C \sum_V p_C q_V$$

$$= V S_{CV} \exp[-(\gamma_e - \gamma_h)] np$$

Generalizations of these arguments proceed in an analogous way for single-electron processes with a participation of a trap and for the various collision processes involving just two electrons. To discuss the coefficient B_1 of Fig. 2.1.1, for example, one needs four states instead of only two in eq. (2.2.10a). Denoting the initial states by C, $C' \in c$, the final states by $C'' \in c$ and $V \in v$ and the transition probability per unit time by T, the contribution of these processes to u_{cv} is

$$\left[\sum_{C, C', C'', V} p_C p_{C'} T_{CC', C''V} q_{C''} q_V \right] [1 - \exp(\gamma_h - \gamma_e)] V^{-1}$$

Regarding the *second* square bracket, the ratio in X_{JI}^{-1} for the single-electron process of eq. (2.2.2b), namely (cf. eq. (2.2.3))

$$\frac{p_{C0} q_C}{p_C q_{C0}} \cdot \frac{p_V q_{V0}}{p_{V0} q_V} = \exp(\gamma_0 - \gamma_e) \cdot \exp(\gamma_h - \gamma_0) = \exp(\gamma_h - \gamma_e)$$

has now to be multiplied by an analogous ratio for the two new states, namely

$$\frac{p_{C'0} q_{C'}}{p_{C'} q_{C'0}} \cdot \frac{p_{C''} q_{C''0}}{p_{C''0} q_{C''}} = \exp(\gamma_0 - \gamma_e) \cdot \exp(\gamma_e - \gamma_0) = 1$$

Because the second electron remains in the same band, the second square bracket is thus the same as in the single-electron processes. The contribution to u_{cv} is therefore (compare eq. (2.2.9))

$$B_1 n^2 p[1 - \exp(\gamma_h - \gamma_e)]$$

where

$$B_1 \equiv \frac{V^{-1} \sum_{CC'C''V} p_C p_{C'} T_{CC', C''V} q_C q_V}{V^{-3} \sum_C p_C \sum_{C'} p_{C'} \sum_V q_V}$$

Thus the processes (a) and (b) of Fig. 2.1.1 together contribute

$$(B^S + B_1 n) np[1 - \exp(\gamma_h - \gamma_e)]$$

to u_{cv}. This shows clearly how the required generalizations must proceed.

2.2.4 Joint effects of distinct recombination channels

Regarding rates determined by more than one mechanism, Fig. 2.1.1 shows three sets of three processes each. It leads one to expect three transition rates, each incorporating the effect of three transition types. We shall use eq. (2.2.1)

$$u_{ij} = \sum_{I \in i} \sum_{J \in i} p_I S_{IJ} q_J (1 - X_{JI}^{-1}) V^{-1}, \quad X_{JI}^{-1} = \exp(\gamma_j - \gamma_i) \tag{2.2.10d}$$

Introducing

$$\left. \begin{aligned} F &\equiv B^S + B_1 n + B_2 p \\ G(r-\tfrac{1}{2}) &\equiv T_1^S(r-\tfrac{1}{2}) + T_1(r-\tfrac{1}{2})n + T_2(r-\tfrac{1}{2})p \quad [\mathrm{L^3 T^{-1}}] \\ H(r-\tfrac{1}{2}) &\equiv T_2^S(r-\tfrac{1}{2}) + T_3(r-\tfrac{1}{2})n + T_4(r-\tfrac{1}{2})p \end{aligned} \right\} \tag{2.2.11}$$

one can express the three transition rates per unit volume in the generalized form indicated above

$$u_{cv} = Fnp[1 - \exp(\gamma_h - \gamma_e)] \tag{2.2.12a}$$

$$u_{c\,r-\frac{1}{2}} = G(r-\tfrac{1}{2})nv_{r-1}[1 - \exp(\gamma_{r-\frac{1}{2}} - \gamma_e)] \tag{2.2.12b}$$

$$u_{r-\frac{1}{2}v} = H(r-\tfrac{1}{2})pv_r[1 - \exp(\gamma_h - \gamma_{r-\frac{1}{2}})] \tag{2.2.12c}$$

v_r is here the concentration of r-electron centers. Equation (2.2.12b) gives the transition rate per unit volume from the conduction band into $(r-1)$-electron centers; eq. (2.2.12c) gives the rate from r-electron centers to the valence band. The first terms in eq. (2.2.11) are the usual ones and refer to single-electron processes. The terms in n may be thought of as arising from the partner in the electron collision which makes a transition which is confined to the conduction band. The terms in p arise similarly from the collision partner which makes a transition in the valence band. Each of the nine processes adds to the rate conduction → valence band, conduction band → trap, trap → valence band as if the other processes were absent. This leads to the sums occurring in F, G and H. With proper definitions, analogous to eq. (2.2.10), these expressions hold even if the bands are degenerate. However, in the case of degeneracy the recombination coefficients are themselves concentration-dependent. (Capture and emission coefficients of electrons and holes are discussed in section 2.4.6.)

By defining new concentrations

$$\left. \begin{aligned} n_\theta &\equiv [np \exp(\gamma_h - \gamma_e)]^{\frac{1}{2}} \\ n(r-\tfrac{1}{2}) &\equiv \frac{nv_{r-1}}{v_r} \exp(\gamma_{r-\frac{1}{2}} - \gamma_e) = \frac{Z_{r-1}}{Z_r} n \exp(-\gamma_e) \\ p(r-\tfrac{1}{2}) &\equiv \frac{pv_r}{v_{r-1}} \exp(\gamma_h - \gamma_{r-\frac{1}{2}}) = \frac{Z_r}{Z_{r-1}} p \exp \gamma_h \end{aligned} \right\} \tag{2.2.13}$$

we can cast eqs. (2.2.12) into the equivalent forms

$$u_{cv} = F(np - n_\theta^2) \tag{2.2.14a}$$

$$u_{cr-\frac{1}{2}} = G(r - \tfrac{1}{2})[nv_{r-1} - n(r - \tfrac{1}{2})v_r] \tag{2.2.14b}$$

$$u_{r-\frac{1}{2}v} = H(r - \tfrac{1}{2})[pv_r - p(r - \tfrac{1}{2})v_{r-1}] \tag{2.2.14c}$$

If the quasi-Fermi levels lie such that

$$\frac{v_{r-1}}{v_r} \exp(\gamma_{r-\frac{1}{2}} - \gamma_e) = 1 \quad \text{then } n = n(r - \tfrac{1}{2})$$

Also if

$$\frac{v_r}{v_{r-1}} \exp(\gamma_h - \gamma_{r-\frac{1}{2}}) = 1 \quad \text{then } p = p(r - \tfrac{1}{2})$$

This gives an interpretation to the new concentrations (2.2.13). In particular in thermal equilibrium we see that the single condition $v_{r-1} = v_r$ is sufficient to ensure that $n = n(r - \tfrac{1}{2})$ and $p = p(r - \tfrac{1}{2})$. Thus the new concentrations can be interpreted as special values of n and p. If $r = 0, 1$ are the only charge states of the center then the condition simplifies further to

$$v_0 = v_1 = \tfrac{1}{2}N_t$$

The new concentrations $n(r - \tfrac{1}{2})$, $p(r - \tfrac{1}{2})$ of eqs. (2.2.13) are *Fermi level-dependent* if the bands are degenerate. In that case they cannot be identified by noting that expressions (2.2.14) vanish in thermal equilibrium, since the passage to equilibrium alters these concentrations. The forms (2.2.12) are therefore less likely to lead to error than the forms (2.2.14). In the case of nondegeneracy there is no problem since the quasi-Fermi level cancels out of eqs. (2.2.13). One finds, for example, that $(n_\theta^2)_0 = n_i^2$. Furthermore, on introducing eqs. (1.13.5) and (1.13.6) one finds, using (1.7.28),

$$n(r - \tfrac{1}{2}) = (Z_{r-1}/Z_r)n_i \exp(-\gamma_i) \sim n_i \exp[\eta_{\text{eff}}(r - \tfrac{1}{2}) - \gamma_i]$$

$$p(r - \tfrac{1}{2}) = (Z_r/Z_{r-1})n_i \exp\gamma_i \sim n_i \exp[\gamma_i - \eta_{\text{eff}}(r - \tfrac{1}{2})]$$

These are familiar formulae for the quantities n_1, p_1 frequently introduced into the simpler forms of "SRH" (i.e. Shockley–Read–Hall) statistics, and we find them here as special cases of our more flexible partition function notation. They give respectively the electron and hole concentrations when the Fermi level is at the trap level.

For *steady-state* recombination via traps, one can identify the quasi-Fermi level $\gamma_{r-\frac{1}{2}}$ from the steady-state condition

$$u_{cr-\frac{1}{2}} = u_{r-\frac{1}{2}v} \quad (r = 1, 2, \dots, M) \tag{2.2.15}$$

Their substitution in the rate

$$u = u_{cv} + \sum_{r=1}^{M} u_{cr-\frac{1}{2}} \qquad (2.2.16)$$

then yields the final steady-state recombination rate. One finds the following characteristics of recombination involving traps:

$$\exp\left(\gamma_{r-\frac{1}{2}}\right) = \frac{Z_{r-1} G(r-\frac{1}{2}) n + Z_r H(r-\frac{1}{2}) p \exp\left(\gamma_h\right)}{Z_{r-1} G(r-\frac{1}{2}) n \exp\left(-\gamma_e\right) + Z_r H(r-\frac{1}{2}) p}$$

$$u_{cr-\frac{1}{2}} = \frac{np[1 - \exp\left(\gamma_h - \gamma_e\right)]}{\tau_n(r-\frac{1}{2})[p + p(r-\frac{1}{2})] + \tau_p(r-\frac{1}{2})[n + n(r-\frac{1}{2})]} \qquad (2.2.17)$$

where the $n(r-\frac{1}{2})$, $p(r-\frac{1}{2})$ are the quantities introduced in eqs. (2.2.13) and

$$\left.\begin{aligned} \frac{1}{\tau_n(r-\frac{1}{2})} &\equiv G(r-\frac{1}{2})(v_{r-1} + v_r) \\ \frac{1}{\tau_p(r-\frac{1}{2})} &\equiv H(r-\frac{1}{2})(v_{r-1} + v_r) \end{aligned}\right\} \qquad (2.2.18)$$

These τ's can be interpreted as lifetimes. Suppose the smallest $\tau_n(r-\frac{1}{2})$ occurs for $r = r_1$. Then for strongly p-type material, with $p \sim p_0$ and band–band recombination negligible,

$$u \sim \frac{np - n_0 p_0}{p \tau_n(r_1-\frac{1}{2})} \sim \frac{n - n_0}{\tau_n(r_1-\frac{1}{2})} \qquad (2.2.19)$$

Thus $\tau_n(r_1-\frac{1}{2})$ acts like a minority carrier lifetime in this case. Also $\tau_p(r_1-\frac{1}{2})$ acts similarly in n-type material.

A more detailed derivation of eq. (2.2.17) is given in section 2.3.1.

These results hold for all density-of-states formulae and they are valid for degeneracy. Although the generalized SRH statistics (2.2.17) were introduced in a paper published by David Evans and the present author already in 1963, and has been much used by the present author in the intervening years, it is sometimes not fully understood. A brief explanation of it is therefore given here. The normal Shockley–Read process consists of the transition T_1^s followed by T_2^s; let us denote it by (T_1^s, T_2^s). However, one single-electron capture may also go together with an Auger process to effect an electron–hole recombination:

$$(T_1^s, T_3), (T_1^s, T_4) \quad \text{and also} \quad (T_1, T_2^s), (T_2, T_2^s)$$

In addition, one has combinations of Auger processes which also lead to electron–hole recombination, namely

$$(T_1, T_3), (T_1, T_4); \quad (T_2, T_3), (T_2, T_4)$$

These eight additional processes are taken into account in the generalized statistics. Any given electron has, therefore, a number of transition paths available to it.

They are constrained by the imposition of the steady-state condition for the electron and hole concentrations. The advantage of our procedure is precisely that it covers these various possibilities in one argument.

Let us invent a new symbol, $\blacktriangleleft$, for 'generalizes to' or, equivalently, 'is a special case of'. Then we have shown here, with c_n, c_p taken from eqs. (2.4.67) and (2.4.68) below, that

$$\tau_n^{-1} = c_n N_t \blacktriangleleft GN_t = G(v_0+v_1) \blacktriangleleft G(r-\tfrac{1}{2})(v_{r-1}+v_r)$$

$$\tau_p^{-1} = c_p N_t \blacktriangleleft HN_t = H(v_0+v_1) \blacktriangleleft H(r-\tfrac{1}{2})(v_{r-1}+v_r)$$

$$n_1 = N_c e^{\eta_t-\eta_c} \blacktriangleleft \frac{Z_0}{Z_1} n e^{-\gamma_e} \blacktriangleleft \frac{Z_{r-1}}{Z_r} n e^{-\gamma_e} = n(r-\tfrac{1}{2})$$

$$p_1 = N_v e^{\eta_v-\eta_t} \blacktriangleleft \frac{Z_1}{Z_0} N_v e^{\eta_v} \blacktriangleleft \frac{Z_1}{Z_0} p e^{\gamma_h} \blacktriangleleft \frac{Z_r}{Z_{r-1}} p e^{\gamma_h} = p(r-\tfrac{1}{2})$$

as clear from eqs. (2.2.18) and (2.2.13) respectively.

2.2.5 Special cases

Noteworthy special cases are:

(i) No electron collisions. Equations (2.2.11) lose the terms in n and in p.

(ii) $r = M = 1$; nondegeneracy. Equation (2.2.17) can still exhibit the effect of electron collisions.

(iii) $r = M = 1$; no electron collisions. Equation (2.2.17) can still exhibit the effect of degeneracy.

(iv) $r = M = 1$; nondegeneracy; no electron collisions. This yields the standard SRH formula [2.2.2] in which $n(\tfrac{1}{2})$, $p(\tfrac{1}{2})$ are often denoted by n_1, p_1.

In the simplified form (iv), eq. (2.2.17) becomes

$$u_{c\frac{1}{2}} = \frac{np-n_i^2}{\tau_n(\tfrac{1}{2})[p+p(\tfrac{1}{2})]+\tau_p(\tfrac{1}{2})[n+n(\tfrac{1}{2})]} \tag{2.2.20}$$

where (with total impurity concentration $N_t = v_0+v_1$)

$$\tau_n(\tfrac{1}{2}) = \frac{1}{N_t T_1^s(\tfrac{1}{2})}, \quad \tau_p(\tfrac{1}{2}) = \frac{1}{N_t T_2^s(\tfrac{1}{2})} \tag{2.2.21}$$

hence, if $n = n_0+\Delta p$, $p = p_0+\Delta p$, and the suffix 0 distinguishes equilibrium,

$$u_{c\frac{1}{2}} = \frac{n_0 \Delta p+p_0 \Delta p+(\Delta p)^2}{\tau_n(\tfrac{1}{2})[p_0+p(\tfrac{1}{2})+\Delta p]+\tau_p(\tfrac{1}{2})[n_0+n(\tfrac{1}{2})+\Delta p]}$$

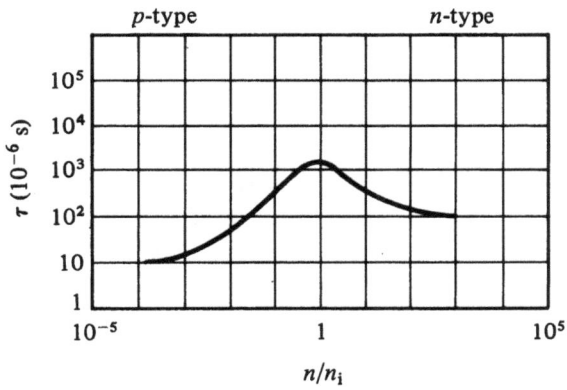

Fig. 2.2.2. Minority-carrier lifetime as a function of electron concentration for Ge [2.2.2]. It is in agreement with eq. (2.2.22) using $\tau_n(\frac{1}{2}) = 10$ μs, $\tau_p(\frac{1}{2}) = 100$ μs.

The lifetime τ for *small* disturbance is given by

$$\tau = \frac{\Delta p}{u_{c\frac{1}{2}}} = \tau_p(\tfrac{1}{2})\frac{n_0 + n(\frac{1}{2})}{n_0 + p_0} + \tau_n(\tfrac{1}{2})\frac{p_0 + p(\frac{1}{2})}{n_0 + p_0} \qquad (2.2.22)$$

As the Fermi level is raised from valence to conduction band τ changes from $\tau_n(\frac{1}{2})$ ($p_0 \gg n_0$) to $\tau_p(\frac{1}{2})$ ($p_0 \ll n_0$) as shown in Fig. 2.2.2. Auger effects would add a carrier concentration dependence in eqs. (2.2.21) as indicated in eqs. (2.2.11). Recombination via bound excitons follows somewhat similar laws.

The concentration dependence of the minority lifetime $\tau_p(\frac{1}{2})$ in gold-doped germanium at 300 K has been used to give the empirical estimates $T_2^s \sim 6.7 \times 10^{-10}$ cm³ s⁻¹ and $T_3 \sim 10^{-26}$ cm⁶ s⁻¹ [2.2.3] experimentally by using variable donor and acceptor concentrations whereby carrier concentrations are varied, but which do not act as recombination centers (see Fig. 2.2.3). A similar test which involves both band–band and band–trap recombination is discussed in section 2.3.4.

2.2.6 Brief survey of selected literature on recombination *(mainly to about 1980)*

Some additional useful identifications of Auger–trap coefficients include experimental values for GaP (Zn, 0) [2.2.5], results from Ge and Si under laser irradiation [2.2.6], impact ionization from Zn in Ge [2.2.7], theoretical studies of two-level centers [2.2.8], and luminescence investigations [2.2.9]. Theoretical work using Bloch functions [2.2.10] and Bloch and Coulomb functions [2.2.11] has also recently been published in connection with the trap Auger effect, discussed more fully in section 5.3.

In the case of band–band Auger processes, discussed more fully in chapter 3, one can refer to reviews [2.2.12]–[2.2.15] and to work on the indirect Auger effect [2.2.16]–[2.2.21], as well as on the normal effect [2.2.22]–[2.2.26]. The relation between the work of Haug *et al.* [2.2.24] and that of Hill and Landsberg [2.2.16] was discussed by Robbins and Young

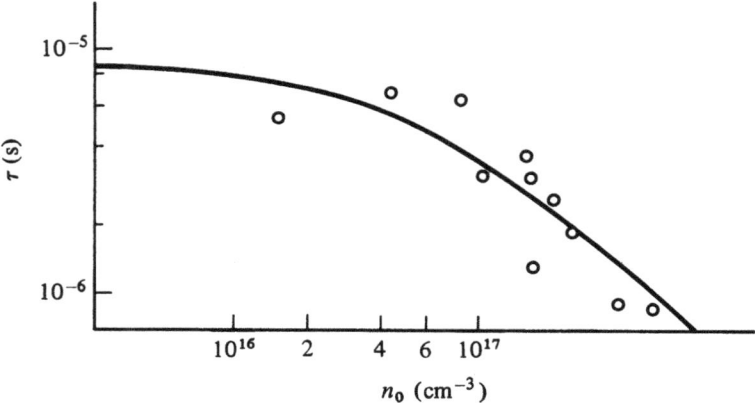

Fig. 2.2.3. Hole lifetime in n-type Ge. Experimental points are from [2.2.3].
The curve is from [2.2.4] and is based on the second eq. (2.2.18) for $r = 1$.

[2.2.27]. Studies also exist of the effect of electric [2.2.28] and magnetic [2.2.29] fields, of
degeneracy [2.2.30], [2.2.31], of screening [2.2.32], and of their effect on electron–hole drops
[2.2.33], [2.2.34]. Experimental work should be noted [2.2.35]–[2.2.37] as well as a review
[2.2.38]. Experimental identifications exist of single-electron transitions, Shockley–Read
processes etc., and they lead to values for the coefficients which have a superfix S in (2.1.8).
Some data collections exist [2.2.39]–[2.2.42] but they are somewhat out of date, and the
literature contains a vast amount of information which has not been systematized. A review
of capture by attractive centers is of interest in this connection [2.2.31]. The whole
recombination area was reviewed in [2.2.43].

That recombination statistics for centers in multiple charge conditions are needed is well
known from the study of Au, Cu, Ni and Mn in Ge or Si [2.2.44] (see sections 1.7–1.9).
Excited states are needed (for example) to account for the photoconductivity of Ga Sb
[2.2.45] and for cascade models ([2.2.46]–[2.2.49], [2.2.42] and section 2.4 below), in which
electrons hop from state to state, emitting a phonon at each step. If several phonons are
involved at the same time, configuration coordinate diagrams are needed as discussed in
chapter 6; see also [2.2.50]. The combination of Auger and multiphonon recombinations is
also important [2.2.51], [2.2.52]; see p. 434 for more details.

A comparative neglect over the years has been the systematic study of Fermi degeneracy
effects in recombination statistics. Although formally included here, more detailed formulae
are required [2.2.53], [2.2.54]. The effect of degeneracy in recombination statistics is
particularly important now in connection with the heavy doping needed for certain devices,
and referred to in section 1.12.

Donor–acceptor transitions, with which we deal in section 5.2.6, have been reviewed
[2.2.55] and both radiative [2.2.56] and nonradiative transitions [2.2.57]–[2.2.59] and their
interrelation [2.2.60] have been studied. At least in CdS and GaAs the Auger competition
is unlikely to be important under normal conditions. The drop in the radiative
recombination rate as the separation between pairs increases has also been considered both
theoretically [2.2.56], [2.2.57] and experimentally [2.2.61], [2.2.62].

Note that the recombination statistics for traps can also be applied to problems of
chemisorption on semiconductors [2.2.63].

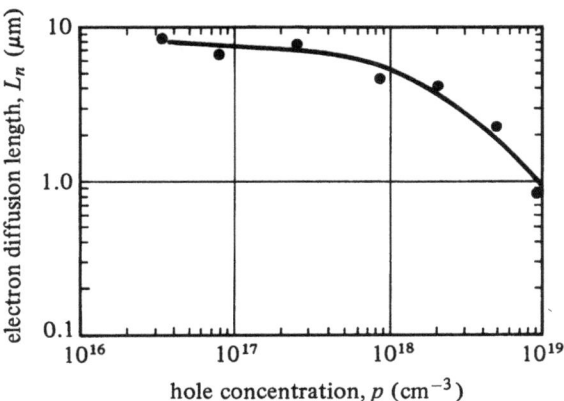

Fig. 2.2.4. Electron diffusion length in Ge-doped p-type GaAs layers as a function of hole concentration [2.2.64].

The minority carrier lifetime determines the minority-carrier diffusion length via $L = (D\tau)^{\frac{1}{2}}$ in the simplest cases. Thus for thermal band–band pair generation Y^s and band–band radiative recombination B^s one has

$$u_{cv} = B^s np - Y^s = \frac{Y^s}{n_0}\left[\frac{B^s n_0}{Y^s}np - n_0\right] = \frac{1}{\tau}\left[\frac{B^s Y_0^s}{B_0^s Y^s}\frac{np}{p_0} - n_0\right] \qquad (2.2.23)$$

This is an alternative approach to eq. (2.2.12a). The band–band recombination rate is for p-type material (allowing for excitation dependence of B^s and Y^s):

$$u_{cv} = \frac{1}{\tau_n}\left[\frac{B^s Y_0^s}{B_0^s Y^s}n - n_0\right] \sim \frac{n - n_0}{\tau_n}, \quad \frac{1}{\tau_n} = \frac{Y^s}{n_0} = \frac{Y^s}{Y_0^s}p_0 B_0^s$$

Hence for p-type layers having $p = p_0$

$$L_n = (D\tau_n)^{\frac{1}{2}} = (Dn_0/Y^s)^{\frac{1}{2}} = (DY_0^s/B_0^s Y^s p_0)^{\frac{1}{2}} \qquad (2.2.24)$$

It has been found that L_n decreases as p_0 rises for majority carrier concentration $p_0 \gtrsim 10^{18}$ cm^{-3} while the luminescent efficiency still rises, Ge-doped GaAs being an example (Fig. 2.2.4). This has been interpreted as implying, via eq. (2.2.24), that radiative recombination commences to influence the total recombination traffic only at $p_0 \gtrsim 10^{18}$ cm^{-3}, while nonradiative recombination is important below this value [2.2.64]. An analogous sharp drop of τ_p in GaP with increased impurity concentration has also been observed [2.2.58], [2.2.59], [2.2.65]. Some lifetime limitation in these materials is due to dislocations, see [2.2.66] and section 2.8.

The recent improvement in epitaxial growth techniques has focused interest on the effect of thin layers on devices. This has led to the manufacture of multilayer and graded band-gap devices, and to electrons flowing in approximate confinement to two dimensions. In the case of semiconductor lasers, for example, it has raised the question of the improvements obtainable by the use of these so-called low-dimensional structures which include quantum wires. Can the threshold current and its temperature dependence be lowered by going to the quantum wells? Is the deleterious nonradiative transition rate as a fraction of the total radiative plus nonradiative rate less for the low-dimensional structures? This matter has been discussed, and the present consensus is that the last question is to be answered in the negative [2.2.67], [2.2.68]; see also chapter 7.

2.3 Shockley–Read–Hall (SRH) statistics: additional topics

2.3.1 A derivation of the main formulae (any degeneracy, any density of states)

Since the Shockley–Read–Hall statistics (2.2.20) for carrier recombination via traps has played an important role in comparisons with experiment, we shall return to it in this section. First we establish in more detail the result (2.2.17), using Fig. 2.1.1. Regard $G(r-\frac{1}{2})$, $H(r-\frac{1}{2})$ to be the mass action coefficients for the downward transitions conduction band $\rightarrow$ $(r-1)$-electron trap and r-electron trap $\rightarrow$ valence band. They are given by eqs. (2.2.11). Let $\lambda G(r-\frac{1}{2})$, $\mu H(r-\frac{1}{2})$ be the corresponding coefficients for the upward transitions; one then has the following transition rates per unit volume:

$$u_{cr-\frac{1}{2}} = G(r-\tfrac{1}{2})[nv_{r-1} - \lambda v_r] \tag{2.3.1}$$

$$u_{r-\frac{1}{2}v} = H(r-\tfrac{1}{2})[pv_r - \mu v_{r-1}] \tag{2.3.2}$$

In a steady state these two rates are equal and this leads to

$$\frac{v_{r-1}}{v_r} = \frac{\lambda G(r-\tfrac{1}{2}) + pH(r-\tfrac{1}{2})}{nG(r-\tfrac{1}{2}) + \mu H(r-\tfrac{1}{2})} = \frac{Z_{r-1}}{Z_r} \exp\left[-\gamma(r-\tfrac{1}{2})\right] \tag{2.3.3}$$

where eqs. (1.11.12) and (2.2.5) have been used in the last step. The steady-state trap quasi-Fermi level for the transition $\{(r-1)\text{-electron center} \rightleftarrows r\text{-electron center}\}$ has thus been identified:

$$\exp\left[\gamma(r-\tfrac{1}{2})\right] = \left(\frac{Z_{r-1}}{Z_r}\right)\left[\frac{nG(r-\tfrac{1}{2}) + \mu H(r-\tfrac{1}{2})}{\lambda G(r-\tfrac{1}{2}) + pH(r-\tfrac{1}{2})}\right] \tag{2.3.4}$$

This holds for $r = 1, 2, \ldots, M$, the most important case being $r = 1$. The steady-state recombination rate per unit volume involving these centers is

$$
\begin{aligned}
u_{cr-\frac{1}{2}} &= (v_{r-1} + v_r)\, G(r-\tfrac{1}{2})\left[n\frac{v_{r-1}}{v_{r-1}+v_r} - \lambda\frac{v_r}{v_{r-1}+v_r}\right] \\[4pt]
&= (v_{r-1}+v_r)\,G(r-\tfrac{1}{2})\frac{n[\lambda G(r-\tfrac{1}{2}) + pH(r-\tfrac{1}{2})] - \lambda[nG(r-\tfrac{1}{2}) + \mu H(r-\tfrac{1}{2})]}{(n+\lambda)\,G(r-\tfrac{1}{2}) + (p+\mu)\,H(r-\tfrac{1}{2})} \\[4pt]
&= \frac{G(r-\tfrac{1}{2})\,H(r-\tfrac{1}{2})\,(v_{r-1}+v_r)\,(np-\lambda\mu)}{(n+\lambda)\,G(r-\tfrac{1}{2}) + (p+\mu)\,H(r-\tfrac{1}{2})} \\[4pt]
&= \frac{np - \lambda\mu}{\tau_p(r-\tfrac{1}{2})(n+\lambda) + \tau_n(r-\tfrac{1}{2})(p+\mu)}
\end{aligned}
\tag{2.3.5}
$$

which is of the form (2.2.17), (2.2.18). It remains to identify λ and μ.

The general theorem (2.2.10d) applied to eq. (2.3.1) yields

$$u_{cr-\frac{1}{2}} = G(r-\tfrac{1}{2})\,nv_{r-1}\left(1 - \lambda\frac{v_r}{nv_{r-1}}\right) = G(r-\tfrac{1}{2})\,nv_{r-1}[1 - \exp(\gamma_{r-\frac{1}{2}} - \gamma_e)]$$

$\gamma_{r-\frac{1}{2}}$ and $\gamma(r-\frac{1}{2})$ being equivalent notations. It follows that

$$\lambda = \frac{v_{r-1}}{v_r} n \exp(\gamma_{r-\frac{1}{2}}-\gamma_e) = \frac{Z_{r-1} \exp[(r-1)\gamma_{r-1}]}{Z_r \exp[r\gamma_r]} n \exp(\gamma_{r-\frac{1}{2}}-\gamma_e)$$

$$= \frac{Z_{r-1}}{Z_r} n \exp(-\gamma_e) = n(r-\tfrac{1}{2}) \tag{2.3.6}$$

where eqs. (2.2.5) and (2.2.13) have been used. Similarly

$$\mu = \frac{v_r}{v_{r-1}} p \exp(\gamma_h - \gamma_{r-\frac{1}{2}}) = \frac{Z_r}{Z_{r-1}} p \exp\gamma_h = p(r-\tfrac{1}{2}) \tag{2.3.7}$$

so that

$$n(r-\tfrac{1}{2})p(r-\tfrac{1}{2}) = np\exp[-(\gamma_e-\gamma_h)] \to n_0 p_0 \tag{2.3.8}$$

The last step is justified if the bands are nondegenerate. Thus the important result (2.2.17) has been rederived. It is sometimes convenient to re-express it as

$$u_{cr-\frac{1}{2}} = \frac{1}{2}\left[\frac{np}{\tau_n(r-\frac{1}{2})\tau_p(r-\frac{1}{2})}\right]^{\frac{1}{2}} \frac{1-\exp(\gamma_h-\gamma_e)}{\cosh\alpha + \exp[\frac{1}{2}(\gamma_h-\gamma_e)]\cosh\beta} \tag{2.3.9}$$

where

$$\exp(2\alpha) \equiv \frac{n\tau_p(r-\frac{1}{2})}{p\tau_n(r-\frac{1}{2})}, \quad \exp(2\beta) \equiv \frac{n(r-\frac{1}{2})\tau_p(r-\frac{1}{2})}{p(r-\frac{1}{2})\tau_n(r-\frac{1}{2})} \tag{2.3.10}$$

As regards the steady-state trap quasi-Fermi level (2.3.4) it is seen to be given by

$$\exp[\gamma(r-\tfrac{1}{2})] = \frac{Z_{r-1}}{Z_r} \frac{nG(r-\frac{1}{2})+p(r-\frac{1}{2})H(r-\frac{1}{2})}{n(r-\frac{1}{2})G(r-\frac{1}{2})+pH(r-\frac{1}{2})} \tag{2.3.4a}$$

2.3.2 The addition of reciprocal lifetimes

Consider first band–band recombination. The electron lifetime may be defined, using eqs. (2.2.11), by

$$\frac{1}{\tau_{n,bb}} \equiv \frac{u_{cv}}{n-n_0} = (B^s + B_1 n + B_2 p)\frac{np - np\exp(\gamma_h-\gamma_e)}{n-n_0} \tag{2.3.11}$$

There is no straightforward way of writing this as an addition of reciprocal lifetimes. However, if the assumption of nondegeneracy is made, together with

$$\Delta n \equiv n-n_0 = \Delta p \equiv p-p_0 \tag{2.3.12}$$

the last factor is then

$$\frac{(n_0+\Delta n)(p_0+\Delta n)-n_0 p_0}{\Delta n} = n_0+p_0+\Delta n$$

Hence the addition of reciprocal lifetimes is possible:

$$\frac{1}{\tau_{n,\,\mathrm{bb}}} = \sum_{j=1}^{3} \frac{1}{\tau_{n,\,\mathrm{bb},\,j}} \tag{2.3.13}$$

where

$$\left(\frac{1}{\tau_{n,\,\mathrm{bb},\,1}}, \frac{1}{\tau_{n,\,\mathrm{bb},\,2}}, \frac{1}{\tau_{n,\,\mathrm{bb},\,3}}\right) \equiv (n_0 + p_0 + \Delta n)(B^{\mathrm{S}}, B_1 n, B_2 p) \tag{2.3.14}$$

The lifetimes refer respectively to direct band–band transitions (notably by photon emission), and band–band Auger transitions with the second electron in the conduction band or the valence band. Only $\tau_{n,\,\mathrm{bb},\,1}$ is independent of the nonequilibrium concentrations, and that only if $\Delta n \ll n_0 + p_0$.

The same assumptions, nondegeneracy and formula (2.3.12), enable one to write the SRH lifetime (2.2.17), for electrons and for $r = 1$,

$$\frac{1}{\tau_{n,\,\mathrm{SRH}}} = \frac{1}{a\tau_n(\tfrac{1}{2}) + b\tau_p(\tfrac{1}{2})} \tag{2.3.15}$$

$$a \equiv \frac{p_0 + p(\tfrac{1}{2}) + \Delta n}{n_0 + p_0 + \Delta n}, \quad b \equiv \frac{n_0 + n(\tfrac{1}{2}) + \Delta n}{n_0 + p_0 + \Delta n}$$

where

$$\frac{1}{\tau_n(\tfrac{1}{2})} = G(\tfrac{1}{2})\, N_{\mathrm{t}} = [T_1^{\mathrm{S}}(\tfrac{1}{2}) + T_1(\tfrac{1}{2})\, n + T_2(\tfrac{1}{2})\, p]\, N_{\mathrm{t}} \tag{2.3.16}$$

N_{t} being the concentration of traps, $v_0 + v_1$ in this instance. Also

$$\frac{1}{\tau_p(\tfrac{1}{2})} = H(\tfrac{1}{2})\, N_{\mathrm{t}} = [T_2^{\mathrm{S}}(\tfrac{1}{2}) + T_3(\tfrac{1}{2})\, n + T_4(\tfrac{1}{2})\, p]\, N_{\mathrm{t}} \tag{2.3.17}$$

Again only the single-electron transition component is concentration-independent, and that only if $\Delta n \ll n_0 + p_0$. The Auger effects always produce a dependence on carrier concentration, see eqs. (2.3.14), (2.3.16) and (2.3.17).

It would appear that one cannot in general split up eq. (2.3.15) into a sum of constituent reciprocal lifetimes. However, using eq. (2.2.16), one sees that the electron lifetime resulting from band–band *and* SRH transitions is

$$\frac{1}{\tau_n} = \frac{u}{\Delta n} = \frac{u_{\mathrm{cv}} + u_{\mathrm{c}\frac{1}{2}}}{\Delta n} = \frac{1}{\tau_{n,\,\mathrm{bb}}} + \frac{1}{\tau_{n,\,\mathrm{SRH}}} \tag{2.3.18}$$

Thus one sees that the situation is relatively straightforward, and reciprocal addition of lifetimes is possible. But constituent lifetimes will in general depend on carrier concentration. This effect can be neglected only in special cases: nondegeneracy, small excitations Δn, no Auger effects.

2.3.3 Reduction of SRH to Fermi–Dirac occupation probabilities

The steady-state trap occupation probability (2.3.3) is with eqs. (2.3.6) and (2.3.7) for $r = 1$ given by

$$\frac{v_1}{v_0 + v_1} = \frac{v_1}{N_t} = \frac{nG(\tfrac{1}{2}) + p(\tfrac{1}{2})\,H(\tfrac{1}{2})}{[n + n(\tfrac{1}{2})]\,G(\tfrac{1}{2}) + [p + p(\tfrac{1}{2})]\,H(\tfrac{1}{2})} \tag{2.3.19}$$

One must inquire how this probability reduces in equilibrium to the Fermi–Dirac type of occupation probability as given for example in eq. (1.7.16). Using an *effective* energy level which absorbs the degeneracy factor and excited states,

$$\left(\frac{v_1}{N_t}\right)_0 = \frac{1}{1 + \exp(\eta_t - \gamma_0)} \tag{2.3.20}$$

where η_t is the reduced energy of the trapping level and γ_0 is the reduced equilibrium Fermi level. We accordingly write eq. (2.3.19)

$$\frac{v_1}{N_t} = [1 + g\exp(\eta_t - \gamma_0)]^{-1} \tag{2.3.21}$$

where g has to be identified.

Assume again nondegeneracy, equate eqs. (2.3.19) and (2.3.21) and take reciprocals to find

$$g = \frac{n(\tfrac{1}{2})\,G(\tfrac{1}{2}) + pH(\tfrac{1}{2})}{nG(\tfrac{1}{2}) + p(\tfrac{1}{2})\,H(\tfrac{1}{2})}\exp(\gamma_0 - \eta_t) \tag{2.3.22}$$

Simple manipulations then show that

$$g\exp(\eta_t - \gamma_0) = \frac{\cosh[\theta + \tfrac{1}{2}(\gamma_h - \gamma_0)]}{\cosh[\theta + \tfrac{1}{2}(\gamma_e - \gamma_0)]}\exp[\eta_t - \tfrac{1}{2}(\gamma_e + \gamma_h)] \tag{2.3.23}$$

where

$$\exp\theta = G(\tfrac{1}{2})\,n(\tfrac{1}{2})/H(\tfrac{1}{2})\,p_0$$

An alternative way of writing this is

$$g = \frac{\cosh[\ln(a\exp\gamma_h)^{\frac{1}{2}}]}{\cosh[\ln(a\exp\gamma_e)^{\frac{1}{2}}]}\exp[\gamma_0 - \tfrac{1}{2}(\gamma_e + \gamma_h)] \tag{2.3.24}$$

where

$$a \equiv \frac{n_0\,G(\tfrac{1}{2})}{p_0\,H(\tfrac{1}{2})}\exp(\eta_t - 2\gamma_0)$$

This solves the problem posed, and in thermal equilibrium $g = 1$ as required [2.3.1]. In a manner of speaking Shockley–Read–Hall embrace Fermi–Dirac.

Table 2.3.1. *Inferred recombination parameters for Si*

Temperature of Au diffusion (°C)	Temperature of measurement (K)	$\dfrac{1}{T_i^S N_t}$ (10^{-9} s)	$T_j N_t$ (10^{-12} cm^3 s^{-1})	B_k (10^{-31} cm^6 s^{-1})	
p-type		$i = 1$	$j = 2$	$k = 2$	
850	400	2580	0	1.2	
850	300	2253	0	0.99	
850	77	69	0	0.78	
920	300	57	0	0.99	
920	77	5.3	0	0.78	
n-type		$i = 2$	$j = 3$	$k = 1$	T_3/T_2^S (10^{-19} cm^3)
850	400	157	3.2	2.8	5.02
850	300	150	5.3	2.8	7.95
850	77	65	6.5	2.3	4.22
920	300	13	82	2.8	10.7
920	77	7.1	82	2.3	5.82

2.3.4 An application of the formalism to Si and Hg$_{1-x}$Cd$_x$Te

One can make eqs. (2.2.18) the basis for the comparison of lifetimes with experiment. One then has for minority carriers in *p*-type material and in *n*-type material, respectively, and for $r = 1$,

$$\frac{1}{\tau_n} = pF + G(\tfrac{1}{2}) N_t = p(B_1^S + B_1 n + \underline{B_2 p}) + (\underline{T_1^S} + T_1 n + \underline{T_2 p}) N_t \qquad (2.3.25)$$

$$\frac{1}{\tau_p} = nF + H(\tfrac{1}{2}) N_t = n(B_1^S + \underline{B_1 n} + B_2 p) + (\underline{T_2^S} + \underline{T_3 n} + T_4 p) N_t \qquad (2.3.26)$$

The underlined terms are clearly the most important ones in studies of minority-carrier lifetimes. Consider Au-diffused Si at high carrier concentration for a number of different temperatures [2.3.2]. Using T_1^S, T_2, B_2 as fitting parameters for *p*-type Si, and T_2^S, T_3, B_1 for *n*-type Si, the results given in Table 2.3.1 were found.

It is seen that a trap–Auger effect had to be invoked only for *n*-type Si, though one might have expected *p*-type samples to be similar in this respect. The experiment suggests

$$T_2 < 10^{-27} \text{ cm}^6 \text{ s}^{-1}$$
$$T_3 \sim 5 \times 10^{-19} T_2^S \sim 10^{-26} \text{ cm}^6 \text{ s}^{-1} \qquad (2.3.27)$$

These results are illustrated in Fig. 2.3.1.

A *separability* assumption has been made here, and it is often justified. It says that the dopant helps to set the Fermi level, but it does not participate in the recombination traffic which limits the lifetime. The recombination defects on the other hand, even if of low concentration, are included in the Fermi level equation. In the experiment under review, the validity of the assumption was established experimentally.

Curves of lifetimes which fall as the equilibrium number of carriers (or the doping) is increased are shown in Fig. 2.2.3 (*n*-type Ge) and 2.3.1 (*n*-type Si). They occur in many other materials. Figure 2.3.2 shows an analogous curve for the material mercury cadmium telluride, $Hg_{1-x}Cd_xTe$, which is widely used for photodetectors. The agreement with experiment is again good. The curves are theoretical and have been obtained not by the method of [2.3.2], which is favored here, but by the reciprocal addition of lifetimes as in eq. (2.3.18). τ_{bb} was calculated from

$$\frac{1}{\tau_{bb}} = \frac{1}{\tau_r} + \frac{1}{\tau_A} \qquad\qquad (2.3.28)$$

which refers to radiative and Auger lifetimes for band–band transitions [2.3.3].

2.3.5 The residual defect in Si

If one looks at measured Si lifetimes as a function of doping, one finds the jumble of points shown in Fig. 2.3.3. However, one may consider only the *best* lifetimes for given doping on the argument that these crystals have attained some ideal lifetime, limited only by a particular, but unknown, defect. This defect could be mechanical (e.g. an interstitial), chemical, or an association of several of these. What are the characteristics of this 'residual' lifetime-limiting defect? To answer this question add to the separability assumption of section 2.3.4, secondly, the hypothesis that the defect has only one recombination level and that it is negatively charged when occupied; otherwise it is neutral. A third assumption is that the concentration of the neutral variant of the residual defect, 'frozen in' at temperature T_f, has a concentration given by eq. (1.12.43) with

$$M = 5 \times 10^{22} \text{ cm}^{-3} \qquad\qquad (2.3.29)$$

the concentration of host lattice sites of silicon. As explained in section 1.9, the maximum solubility N_d^x of the neutral defects at temperature T_f can be regarded as in equilibrium independent of position in the material and also as independent of Fermi level. Because of the statistical link (1.12.34) between N_d^- and N_d^x an increase in donor-doping raises N_d, while acceptor-doping decreases it, as already explained

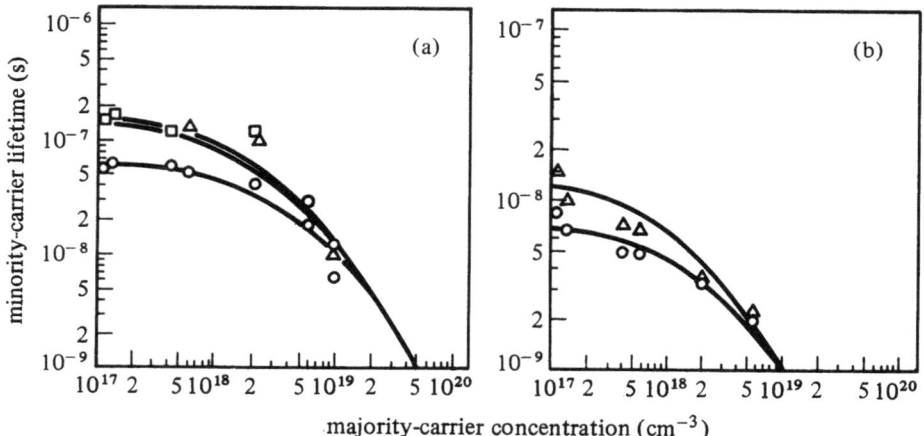

Fig. 2.3.1. Minority-carrier lifetime as a function of the majority-carrier concentration in n-Si at different temperatures. The samples were diffused with Au at (a) 850 °C and (b) 920 °C, respectively. The lines are fits of eq. (2.3.26) to the data with $B_1^s = B_2 = T_4 = 0$. $\square$, 400 K; $\triangle$, 300 K; $\bigcirc$, 77 K.

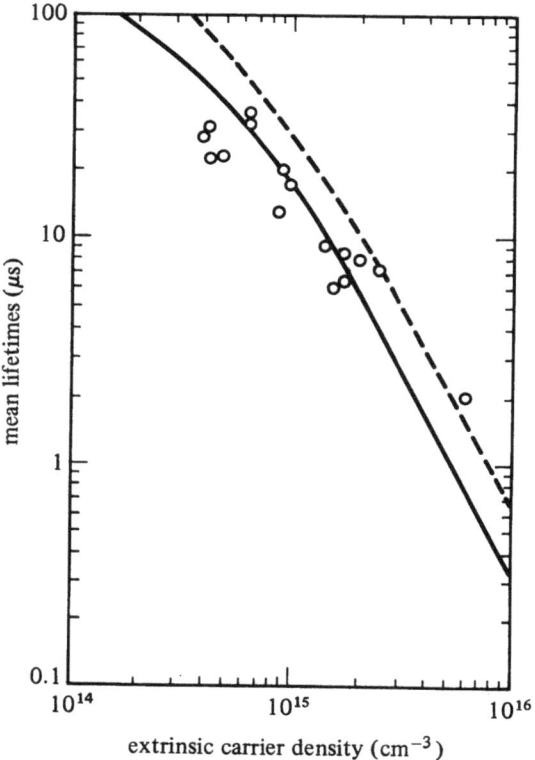

Fig. 2.3.2. Mean minority-carrier lifetime in slices of n-type $Hg_{1-x}Cd_x$Te. The curves are theoretical, based on eqs. (2.3.18) and (2.3.28) [2.3.3]. $T = 193$ K; $\bigcirc$, measured; ——, $x = 0.30$; -----, $x = 0.35$.

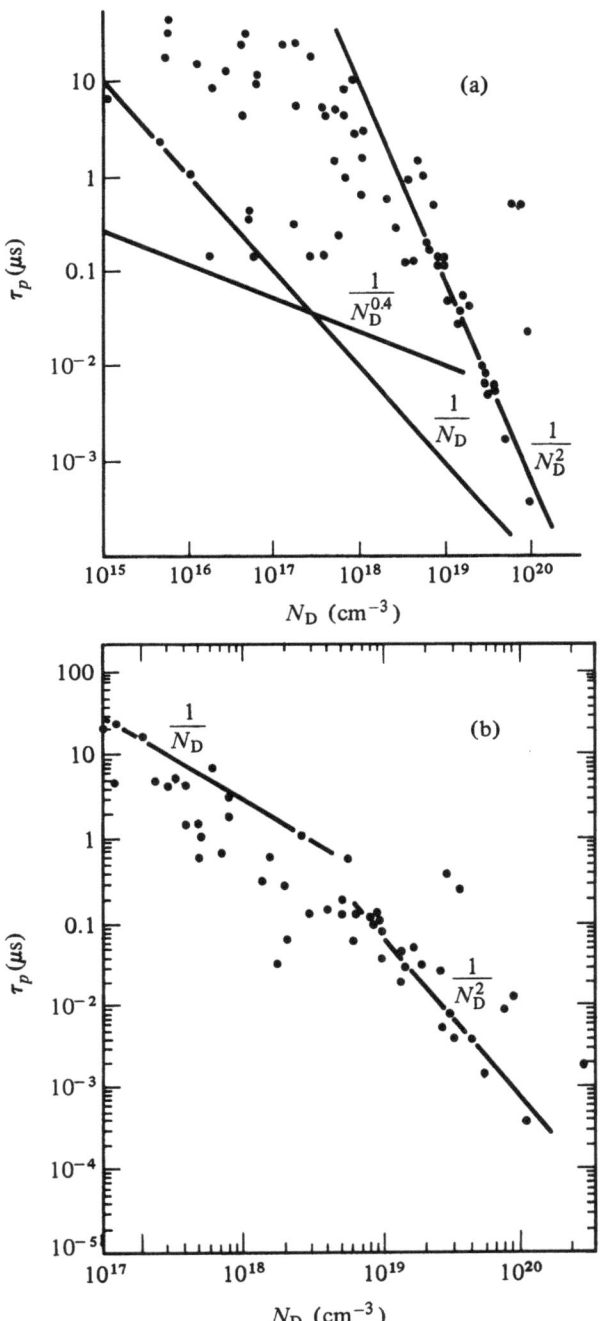

Fig. 2.3.3. Some experimental minority-hole lifetimes in n-type Si. (a) From [2.3.4]; (b) from [2.3.5].

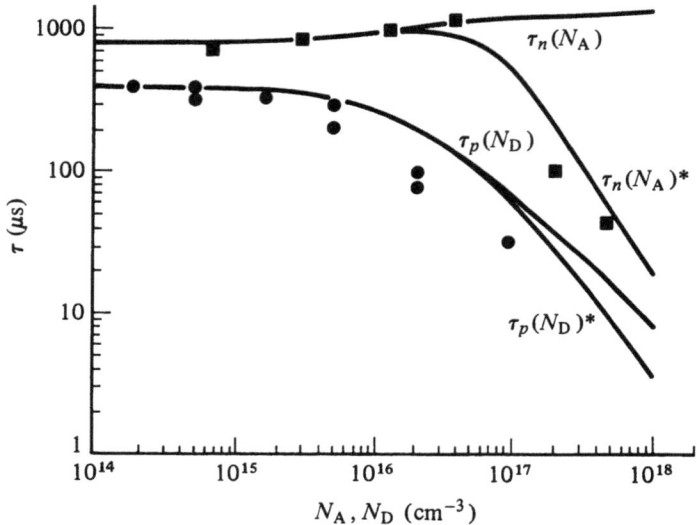

Fig. 2.3.4. Doping dependence of the best room temperature minority-carrier lifetime in Si according to experiments (points). Asterisks indicate that the band–Auger process is included along with the normal Shockley–Read process (unasterisked). Circles (for holes) and squares (for electrons) represent experimental points [2.3.6].

in connection with Fig. 1.12.7. This leads to *lower* minority lifetimes $\tau_p(N_D)$ with doping, as expected, but to *longer* lifetimes $\tau_n(N_A)$ with doping, until these lifetimes are pulled down again by the band–band Auger effect as shown in Fig. 2.3.4 [2.3.6]. The second assumption concerning the charge on the unknown defect therefore enables the model to reproduce the asymmetric behavior observed experimentally as regards τ_n compared with τ_p.

The lifetime curves for the correct concentration N_d of defects, as calculated at T_f, can be used at the lower measurement temperature T, assuming a generalized Shockley–Read mechanism [2.3.7]. They follow roughly the law $(N_d)^{-1}$ with doping, until they are both pulled down by band–band Auger effects which are strong at high concentrations. The details of the Fermi level calculation at the temperatures T_f and T are given in section 1.12.4.

One can neglect T_1, T_2, T_3, T_4 and B^s and adopt [2.3.5] $B_1 \sim 2B_2 \sim 2 \times 10^{-31}$ cm^6 s^{-1} from Table 2.3.1. As to T_1^s, T_2^s, one may regard them as fitting parameters, along with T_f and E_a. The inferred values from the best fit (Fig. 2.3.4) are then found to be:

$$\left. \begin{aligned} 2T_1^s = T_2^s &\sim 5 \times 10^{-5} \text{ cm}^3 \text{ s}^{-1} \\ E_a = 1.375 \text{ eV}, \quad T_f &= 620 \text{ K} \end{aligned} \right\} \tag{2.3.30}$$

Position of defect level: 45 meV above mid-gap

Note that the SRH mechanism is invoked only for the calculation at the measurement temperature $T < T_f$ when a relation of the type (1.12.46), of fourth power in the electron concentration n, has to be used. The defect concentration N_d entering this calculation is derived from a cubic equation in N_d^- which holds at temperature T_f.

We are left with two matters of interest: (1) What is T_f? (2) What is the nature of the inferred defect specified in eqs. (2.3.30)?

As to the first question, recall the early quenching experiments on silicon which led to a relation of the type [2.3.8]

$$\frac{1}{\tau} = C \exp\left(-E_a/kT_q\right)$$

where the activation energy was found to be 0.6 eV, τ was the minority-carrier lifetime and T_q was the temperature from which the sample was quenched. Data enabling one to find C was given later for these thermally generated recombination centers:

$$C \sim 2.13 \times 10^{13}\ \mathrm{s}^{-1}, \quad E_a = 0.9\ \mathrm{eV}$$

([2.3.9]; note that the captions of Figures 8 and 10 should be interchanged.) More recently a thermally generated donor density

$$v = C' \exp\left(-E_a/kT_q\right)$$

was found with $C' \sim 8 \times 10^{23}\ \mathrm{cm}^{-3}$, $E_a = 2.5\ \mathrm{eV}$ in 'pure' p-type Si. The appropriate level was located 0.37 eV above the valence band edge [2.3.10]. These results suggest that eq. (2.3.29) is a reasonable assumption and that the freezing-in temperature T_f may be identified as the quenching temperature for infinitely rapid cooling at least for some heat treatment histories. This corresponds to the 'perfect' quench. Departure from the perfect quench by slower cooling should lead to $T_f < T_q$. This relation between T_f and T_q needs further study.

The second question is made difficult by the variety of levels found by different methods in the forbidden gap of Si. In particular we cite nine relevant pre-1980 papers on thermally generated and/or quenched-in centers in Si [2.3.11]–[2.3.19]. Thus a donor level at $E_v + 0.4$ eV was found in p-type Si in [2.3.10], [2.3.11] and in B-doped Si in [2.3.12], but not in [2.3.13], where the B concentration was heavier. It was again found in [2.3.15] as a complicated defect. The thermally generated defects were found to be hard to anneal out in [2.3.16] and in later work.

In a series of later papers fast ('s') and slow ('r, r', r″') thermal recombination centers were found and characterized. They have formation energies of 1.0 eV, 1.2 eV and 2.5 eV [2.3.17], the slower centers being less soluble. The high binding energy and the consequent difficulty of annealing out thermal centers was confirmed [2.3.18], [2.3.19]. The slow centers were attributed to vacancy-copper complexes and later to vacancy-oxygen complexes [2.3.20]. The fast centers were attributed to native defects ([2.3.20], Figure 3).

As regards energy level structure, many inconsistencies remain. Some of the discrepancies between the various experiments have been attributed to electrically active defects connected with traces of Fe in Si which may have been present in varying amounts [2.3.21]. They can be kept down to below 10^{14} cm^{-3} by special treatment. Fe-related deep levels have, in fact, been studied separately [2.3.22] as has the level at 0.45 eV above the valence band edge [2.3.23].

Swirl defects (due to point defect agglomerates, presumably interstitial) of formation energy 1.3–1.4 eV were also noted in p-type floating zone grown heat-treated Si [2.3.24], and their annealing characteristics differ from those of divacancies of a similar formation energy (1.3 eV).

Possible interpretations of the defect inferred and characterized in eqs. (2.3.30) will now be proposed. Note that a defect similar to the one inferred in relations (2.3.30) appears to have been found in swirl and dislocation-free float zone grown Si by deep level transient spectroscopy and derivative surface photovoltage [2.3.25]. This dominant recombination level was located at $E_v + 0.56$ eV with a capture cross section for holes equal to twice the capture cross section for electrons:

$$\sigma_2^s = 2\sigma_1^s = 10^{-14} \text{ cm}^2 \qquad\qquad (2.3.31)$$

in fair agreement with the specification (2.3.30). If one puts

$$T_2^s = fv\sigma_2^s \ (v \sim 10^7 \text{ cm s}^{-1} \sim \text{thermal velocity})$$

and inserts $T_2^s \sim 5 \times 10^{-5}$ cm^3 s^{-1} into relation (2.3.31), the factor f (giving a recombination efficiency) turns out to be

$$f = 0.05$$

The same result is found if T_1^s and σ_1^s are used. The defect may be a self-interstitial or a cluster of these [2.3.25] – this is a first interpretation of the residual defect.

In view of the importance of controlling Si wafers for device work, two secondary suggestions can be made to interpret relations (2.3.30). The first suggestion is that it is a swirl. The A-type swirl, believed to consist of dislocation loops, loop clusters, etc., occurs in concentrations of typically 10^6–10^7 cm^{-3}, and is therefore not a serious candidate. B-type swirls are smaller and are found in concentrations up to 10^{11} cm^{-3} or so [2.3.26]. This is of the order (10^{11}–10^{13} cm^{-3}) of defect densities implied by Fig. 1.12.8. The formation activation energy of 1.3–1.4 eV [2.3.24] is also of the right order. If such swirls can supply an acceptor level near mid-gap (their energy level structure does not seem to be well known

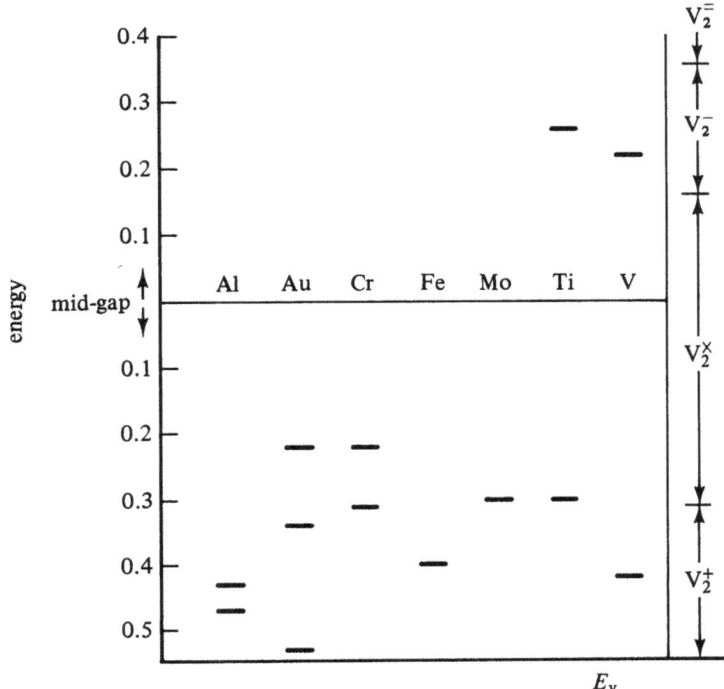

Fig. 2.3.5. Some deep levels in Si due to divacancies V_2 [2.3.28] and due to metallic ions [2.3.29].

yet), the swirl B would be a serious candidate. This interpretation of the 'residual' defect in Si as used for semiconductor work, if correct, would be of importance for two reasons: In the first place swirl defects are known to have detrimental effects on Si, and secondly the elimination of swirl defects is under active study. One can use slow or fast crystal pulling rates, inert ambients during growth, or annealing after growth to reduce their occurrence.

A second candidate is the 's' (native, fast) recombination center [2.3.20]. The slow centers ('r, r′, r″') have levels which lie too close to the band edges, whereas the 's' center has a level near mid-gap. A recombination coefficient for minority carriers of $\sim 10^{-7}$ cm^3 s^{-1} has been suggested [2.3.14] which is 100 times larger than the inferred values of $T_1^s \sim T_2^s \sim 10^{-9}$ cm^3 s^{-1}. This could, however, be understood in terms of different thermal histories. It is, of course possible that the 's'-center and the swirl B center are the same defect. Even a recent study [2.3.27] on the relation between recombination mechanisms and doping density leaves these matters unresolved.

Deep level spectra are now well known, and some are shown in Fig. 2.3.5. It will be seen that they do *not* seem to apply to the residual defect. For a discussion of these topics see [2.3.30].

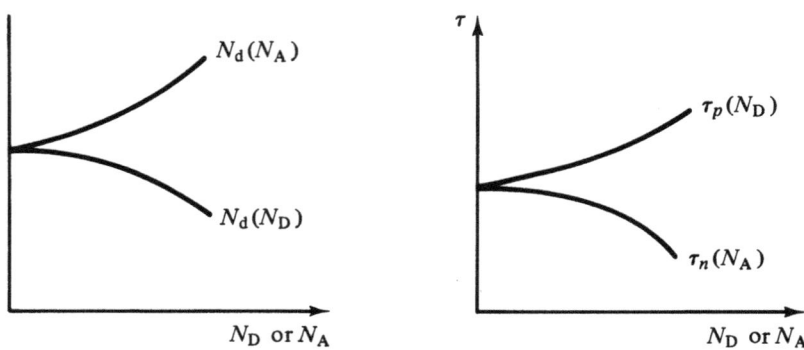

Fig. 2.3.6. Schematic representation of the expected defect concentration N_d and minority-carrier lifetime τ on N_D and N_A. It is assumed that the residual defect is a donor.

It has been assumed in the preceding discussion that the residual defect is of acceptor-type, i.e. is negatively charged when occupied. If it is positively charged when empty, it is a type of donor. The situation reverses relative to Figs. 1.12.8 and 2.3.4 as shown in Fig. 2.3.6. The lowering of the Fermi level increases the defect concentration N_d (because the solubility of N_d^+ increases) while raising of the Fermi level decreases N_d. In brief, X-doping (X = D or A) encourages the solubility of a non-X (A or D) type of defect while reducing the solubility of an X-type defect. The system is in a sense self-compensating.

2.3.6 SRH recombination via a spectrum of trapping levels

Let us consider centers which can be only empty or singly occupied [$r = 1$ in eq. (2.3.5)], and let us replace $n(\frac{1}{2})$, $p(\frac{1}{2})$ by n_1, p_1 and $G(\frac{1}{2})$, $H(\frac{1}{2})$ by G and H, which is a more usual notation in this case. Assuming also nondegeneracy, the recombination rate per unit volume into a spectrum $D(E_t)$ of trapping levels per unit volume per unit energy range at energy E_t is by eq. (2.2.20) (but see also eq. (2.7.4), below)

$$u = GH(np - n_i^2) \int \frac{D(E_t)\,dE_t}{G(n+n_1) + H(p+p_1)} \qquad (2.3.32)$$

where the trap concentration is $N_t = \int D(E_t)\,dE_t$. The integrations are performed over the trap energy E_t. A more convenient variable is, by eqs. (2.2.13),

$$n(r - \tfrac{1}{2}) \to n(\tfrac{1}{2}) \to n_1 = n \exp(\eta_t - \gamma_e) = N_c \exp(\eta_t - \eta_c)$$

Its lower and upper values will be denoted by n_{tl} and n_{tu}, respectively. Put

$$dn_1 = d(N_c e^{\eta_t - \eta_c}) = \frac{n_1}{kT} dE_t, \quad p_1 = n_i^2/n_1, \quad \eta_t - \eta_c = \ln(n_1/N_c)$$

so that with $n_{11} \equiv N_c \exp(\eta_{t1} - \eta_c)$ and $n_{1u} \equiv N_c \exp(\eta_{tu} - \eta_c)$

$$u = GH(np - n_i^2)kT \int_{n_{11}}^{n_{1u}} \frac{D \, dn_1}{a + bn_1 + Hn_1^2} \qquad (2.3.33)$$

where $a = Hn_i^2$, $b \equiv Gn + Hp$. This gives the SRH recombination rate per unit volume into a *spectrum* of trapping states. We wish to inquire how this rate compares with the rate into an equal number of trapping levels located at a *single* energy half-way between E_{t1} and E_{tu}.

For a spectrum of constant D eq. (2.3.33) can be integrated and yields

$$u_{sp} = \frac{GHkTD(np - n_i^2)}{[(Gn + Hp)^2 - 4GHn_i^2]^{\frac{1}{2}}} \ln \left[\frac{1 - \theta(n_{1u})}{1 - \theta(n_{11})} \frac{1 + \theta(n_{11})}{1 + \theta(n_{1u})} \right]$$

where

$$\theta(a) \equiv \frac{[(Gn + Hp)^2 - 4GHn_i^2]^{\frac{1}{2}}}{Gn + Hp + 2Ga}$$

The total concentration of trapping levels involved is

$$N_t = (E_{tu} - E_{t1}) D = DkT \ln \lambda$$

where

$$\lambda \equiv n_{1u}/n_{11} = \exp(\eta_{tu} - \eta_{t1})$$

For recombination via N_t traps at a single energy E_t, assumed to be placed half-way in the band

$$\eta_t \equiv \tfrac{1}{2}(\eta_{tu} + \eta_{t1}) = \eta_c + \ln \frac{(n_{1u} n_{11})^{\frac{1}{2}}}{N_c} = \eta_c + \ln \frac{n_{11}}{N_c} \lambda^{\frac{1}{2}} \qquad (2.3.34)$$

The SRH recombination rate per unit volume through a single level is then

$$u_1 = \frac{GHN_t(np - n_i^2)}{G(n + n_{1t}) + H(p_i + n_i^2/n_{1t})} \quad [n_{1t} \equiv N_c \exp(\eta_t - \eta_c)]$$

$$= \frac{GHDkT(np - n_i^2) \ln \lambda}{Gn + Hp + Gn_{11} \lambda^{\frac{1}{2}} + Hn_i^2/n_{11} \lambda^{\frac{1}{2}}}$$

The ratio of interest is, after some algebra,

$$R \equiv \frac{u_1}{u_{sp}} = \frac{\lambda Gn_{11}^2 + \tfrac{1}{2}(\lambda + 1) n_{11}(Gn + Hp) + Hn_i^2}{(Gn_{1t} + Gn + Hp + Hn_i^2/n_{1t}) n_{11}} \cdot \frac{\ln \lambda}{\lambda - 1} \cdot \frac{2C}{\ln[(1 + C)/(1 - C)]} \qquad (2.3.35)$$

where

$$C \equiv \frac{\tfrac{1}{2}n_{11}(\lambda - 1)[(Gn + Hp)^2 - 4GHn_i^2]^{\frac{1}{2}}}{\tfrac{1}{2}n_{11}(\lambda + 1)(Gn + Hp) + \lambda Gn_{11}^2 + Hn_i^2}$$

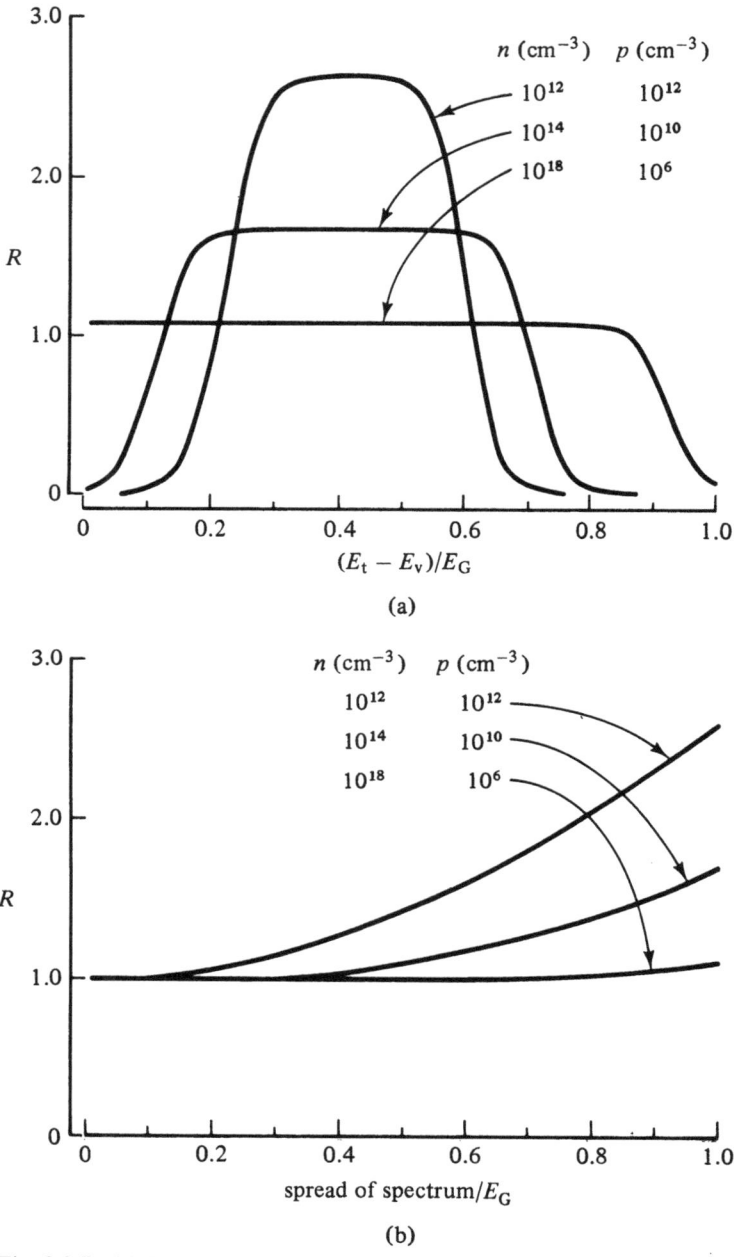

Fig. 2.3.7. (a) Dependence of $R = u_1/u_{sp}$ on the energy E_t of the single level. A constant density of states per unit energy range has been assumed, with this constant the same for each curve, and the spectrum extending throughout the energy gap. Energy independent transition probabilities have been assumed and eq. (2.3.35) has been used. (b) The dependence of R on the width of the spectrum, the latter being normalized by dividing it by E_G. A constant density of states per unit energy range has been assumed, and cancels out of R; $G = 10^{-8}$ cm^3 s^{-1}, $H = 10^{-11}$ cm^3 s^{-1}.

The results of these calculations are illustrated in Figs. 2.3.7 and 2.3.8. Figure 2.3.7 assumes a uniform energy spectrum which extends throughout the energy gap. In Fig. 2.3.7(b) the total number of states was kept fixed, the spectrum being centred on the mid-gap energy. Energy-independent transition probabilities have been assumed and eq. (2.3.35) has been used. In Fig. 2.3.8 this uniform spectrum has been cut off at 0.01 eV from the band edges which has only a negligible effect on u_{sp}. Whereas an energy-independent transition probability per unit time between a level of energy E in one of the bands and the trapping level at energy E_t has been assumed for Fig. 2.3.7, this transition probability has been taken as proportional to $|E - E_t|^{-1}$ for Fig. 2.3.8. The cut-off mentioned above is therefore made for computational reasons. The energy dependence distorts the curves of R toward the valence band, for a strongly n-type material. If the probability of a transition between valence band and trap level *decreased* as the single trapping level moved toward the valence band edge, then we could expect a distortion toward the conduction band edge for strongly n-type material.

As regards the shapes of the curves in Figs. 2.3.7(a) and 2.3.8, note that there always exists a level $E_t = E_{t\,max}$ at which R is a maximum. The spectrum of states is uniformly distributed over the gap and so includes levels which are far less recombination-efficient than the level at $E_{t\,max}$. One would therefore expect the curve of R versus $E_t - E_v$ to have a maximum. Similarly if all the N_t levels are placed at a single level away from $E_{t\,max}$, then this *single* level will be recombination-inefficient whereas the *spectrum* will include levels which are more recombination-efficient. Hence $R \ll 1$ when E_t lies close to either band.

In Fig. 2.3.7(b) one sees the obvious effect that $R \sim 1$ if the spectrum of levels is very narrow at mid-gap. As it broadens out, the single level stays at mid-gap whereas the spectrum acquires recombination-inefficient levels. So one would expect R to rise with Δ. For further details the original paper should be consulted [2.3.31].

2.3.7 The joint action of band–band and SRH recombination

The action of all three terms (2.2.12) will now be considered for the simple case $r = 1$. Thus the band–band recombination rate $F(np - n_0 p_0)$ has to be added to eq. (2.2.20). The steady-state recombination rate involving bands and traps is then

$$u = \left[F + \frac{GHN_t}{G(n + n_1) + H(p + p_1)} \right](np - n_0 p_0) \tag{2.3.36}$$

where we have written G, H, n_1, p_1 for $G(\tfrac{1}{2})$, $H(\tfrac{1}{2})$, $n(\tfrac{1}{2})$, $p(\tfrac{1}{2})$. Such an expression is useful for the sensitive experiments with relatively low impurity concentrations

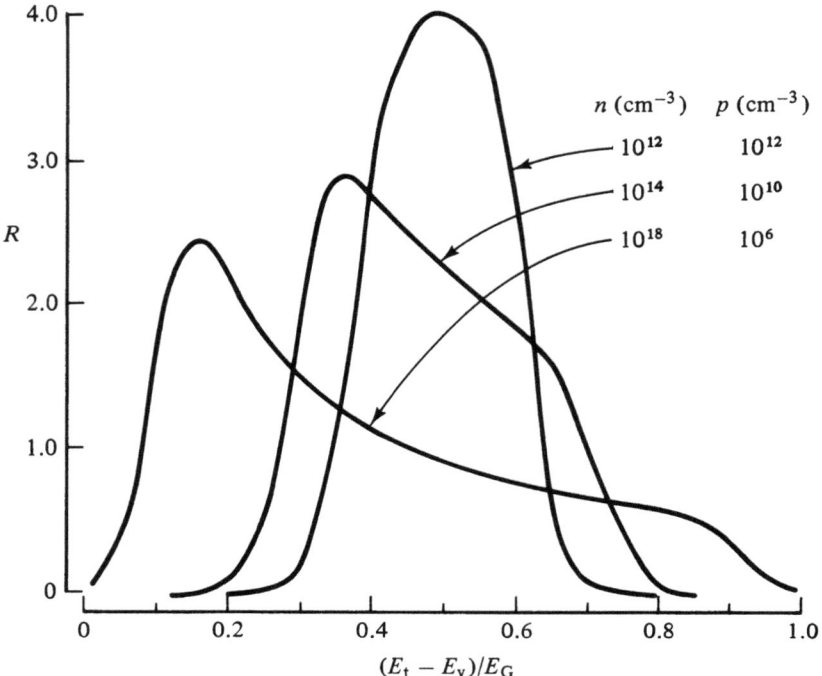

Fig. 2.3.8. Dependence of R on the energy E_t of the single level. The single-electron and single-hole capture coefficients are assumed inversely proportional to the energy bridged by a captured carrier. A constant density of states per unit energy range has been assumed.

which have recently become possible [2.3.32]. They can pick up both band–band and band–impurity recombination. We shall merely consider the case

$$n = n_0 + \Delta n, \quad p = p_0 + \Delta n$$

so that the excess carrier concentrations Δn may actually dominate the equilibrium concentration.

The experimentally accessible variation of the lifetime with Δn

$$\frac{1}{\tau} = u/\Delta n \equiv X(\Delta n) + Y(\Delta n) \tag{2.3.37}$$

is quite involved when considered from the theoretical point of view. Thus the contribution of the band–band processes alone consists of fifteen terms which arise from multiplying

$$F = B^s + (B_1 + B_2)\,\Delta n + B_1 n_0 + B_2 p_0$$

by

$$np - n_0 p_0 = (n_0 + p_0)\,\Delta n + (\Delta n)^2$$

One obtains the contribution $X(\Delta n)$ which is

$$
\begin{aligned}
X = {} & (B_1 + B_2)(\Delta n)^2 + [(B_1 + B_2)(n_0 + p_0) + B^s + B_1 n_0 + B_2 p_0]\,\Delta n \\
& + (B^s + B_1 n_0 + B_2 p_0)(n_0 + p_0)
\end{aligned}
\tag{2.3.38}
$$

The first term is readily recognized to depend only on Auger effects, whereas the other terms depend on single-electron transitions as well.

The contribution from the band–trap processes is represented by Y in eq. (2.3.37) and is algebraically more involved. Because of the widespread use of Shockley–Read statistics, it is given here in full:

$$
Y = \frac{N_0 + N_1\Delta n + N_2(\Delta n)^2 + N_3(\Delta n)^3}{D_0 + D_1\Delta n + D_2(\Delta n)^2}
\tag{2.3.39}
$$

$$
\left[\rightarrow \frac{N_3}{D_2}\Delta n + \frac{N_2}{D_2} - \frac{D_1 N_3}{D_2^2} \right]
$$

The expression in square brackets holds in the case of large Δn. We have also used, with the notation of eqs. (2.2.11),

$$
\left.
\begin{aligned}
N_0 &\equiv (T_1^s + T_1 n_0 + T_2 p_0)(T_2^s + T_3 n_0 + T_4 p_0)(n_0 + p_0)\,N_t \equiv G_0 H_0(n_0 + p_0)\,N_t \\
N_1 &\equiv [H_0(T_1 + T_2) + G_0(T_3 + T_4)](n_0 + p_0)\,N_t + G_0 H_0 N_t \\
N_2 &\equiv [H_0(T_1 + T_2) + G_0(T_3 + T_4) + (n_0 + p_0)(T_1 + T_2)(T_3 + T_4)]\,N_t \\
N_3 &\equiv (T_1 + T_2)(T_3 + T_4)\,N_t \\
D_0 &\equiv G_0(n_0 + n_1) + H_0(p_0 + p_1) \\
D_1 &\equiv (T_1 + T_2)(n_0 + n_1) + (T_3 + T_4)(p_0 + p_1) + G_0 + H_0 \\
D_2 &\equiv T_1 + T_2 + T_3 + T_4
\end{aligned}
\right\}
\tag{2.3.40}
$$

One observes that the terms multiplying the highest powers of Δn, namely N_3 and D_2, depend only on Auger effects, whereas the other terms involve both Auger processes and single-electron transitions.

We proceed with the case of large Δn. Substitution of eqs. (2.3.38) and (2.3.39) into eq. (2.3.37) yields

$$
\frac{1}{\tau} = \alpha + \beta\Delta n + \gamma(\Delta n)^2
\tag{2.3.41}
$$

where α and β are still rather complicated [2.3.33]:

$$
\gamma \equiv B_1 + B_2
$$

$$
\beta \equiv (B_1 + B_2)(n_0 + p_0) + B^s + B_1 n_0 + B_2 p_0 + \frac{(T_1 + T_2)(T_3 + T_4)}{T_1 + T_2 + T_3 + T_4}\,N_t
$$

$$
\alpha \equiv \alpha_1 \frac{N_t}{T_1 + T_2 + T_3 + T_4} + \alpha_2 \frac{(T_1 + T_2)(T_3 + T_4)\,N_t}{(T_1 + T_2 + T_3 + T_4)^2} + \alpha_3
$$

$$
\begin{aligned}
\alpha_1 \equiv {} & (T_2^s + T_3 n_0 + T_4 p_0)(T_1 + T_2) + (T_1^s + T_1 n_0 + T_2 p_0)(T_3 + T_4) \\
& + (n_0 + p_0)(T_1 + T_2)(T_3 + T_4)
\end{aligned}
$$

$$
\alpha_2 \equiv (T_1 + T_2)(n_0 + n_1) + (T_3 + T_4)(p_0 + p_1) + T_1^s + T_1 n_0 + T_2 p_0 + T_2^s + T_3 n_0 + T_4 p_0
$$

$$
\alpha_3 \equiv (B^s + B_1 n_0 + B_2 p_0)(n_0 + p_0)
$$

Thus γ represents band–band Auger coefficients. Turning to β, the terms for the band–band processes are small, except possibly B^s ($\sim 2 \times 10^{-15}$ cm^3 s^{-1} in the case of Si), which covers the radiative band–band recombination coefficient. The T's refer to the band–trap Auger processes which fall at room temperatures within the range [2.3.34], [2.3.35] $10^{-29} \lesssim T_i \lesssim 10^{-23}$ cm^3 s^{-1}. Assuming all T_i of the same order,

$$\beta \sim B^s + T_1 N_t \tag{2.3.42}$$

Possible values are $T_1 \sim 10^{-24}$ cm^6 s^{-1} and $N_t \sim 10^{10}$ cm^{-3}. They yield $\beta \sim 10^{14}$ cm^3 s^{-1} with B^s presumably almost negligible. The coefficient α is largely due to trapping effects. Applying the same procedure to α, together with $n_0 \sim p_0$,

$$\alpha \sim [\tfrac{3}{4}(T_1^s + T_2^s) + 6n_0 T_1 + \tfrac{1}{2}(n_1 + p_1) T_1] N_t \tag{2.3.43}$$

The SRH process with coefficients T_1^s, T_2^s is liable to dominate α. However, some very small cross sections (σ) are known ($\sigma \sim 10^{-21}$ cm^2) and σ values can extend over ten decades (see, for example, [2.3.36], which applies to GaAs and GaP). Hence, it is not easy to simplify α any further as it would lead to serious loss of generality.

One can fit the lifetimes found for Si in [2.3.32] by use of eq. (2.3.41) with the results shown in figs. 2.3.9 and Table 2.3.2. One can regard the numbers in the last column of Table 2.3.2 as empirical data which requires interpretation in terms of the parameters in the second column of Table 2.3.2.

As regards γ, a widely accepted value for the band–band Auger coefficient in Si at room temperature [2.3.37] is

$$B_1 + B_2 \sim (2.8 + 0.99)\, 10^{-31} \sim 3.79 \times 10^{-31} \text{ cm}^6 \text{ s}^{-1}$$

This is roughly a third of the value inferred in Table 2.3.2. This discrepancy has not been explained.

The β-value and the α-value (interpreted as $T_1^s N_t$) suggest:

$\langle 100 \rangle$ sample: $T_1 \sim 10^{-25}$ cm^6 s^{-1}, $\quad T_1^s \sim 10^{-9}$ cm^3 s^{-1}, $\quad N_t \sim 10^{11}$ cm^{-3}

$\langle 111 \rangle$ sample: $T_1 \sim 10^{-24}$ cm^6 s^{-1}, $\quad T_1^s \sim 10^{-9}$ cm^3 s^{-1}, $\quad N_t \sim 10^{10}$ cm^{-3}

These are reasonable, and they show that the trap Auger effect can make a significant contribution to the recombination traffic even in relatively pure material. This has also been observed in the context of solar cells [2.3.38], to which we turn next.

The situation is here quite simple. In the n-type part of the solar cell (the emitter, in the present case, which receives the light on its front surface) one can put

$$n \sim n_0 \sim N_D \gg p, \quad \Delta p \equiv p - p_0$$

and neglect B^s, and the terms in H which involve p. The neglect of p_1 is valid for trap levels near mid-gap. Hence the minority-carrier lifetime is

$$\tau_p^{-1} \equiv \frac{u}{\Delta p} \sim B_1 N_D^2 + (T_2^s + T_3 N_D) N_t \tag{2.3.44}$$

Table 2.3.2. *Numerical data concerning* α, β, γ *in eq.* (2.3.41)

	Approximate theoretical expression	From fit of experimental curve to Si data	
		⟨100⟩ Si	⟨111⟩ Si
α (s⁻¹)	$\frac{3}{4}(T_1^s + T_2^s)\,N_t$	191	18.6
β (10⁻¹⁴ cm³ s⁻¹)	$B^s + T_1\,N_t$	2.46	1.35
γ (10⁻³⁰ cm⁶ s⁻¹)	$B_1 + B_2$	1.19	1.52

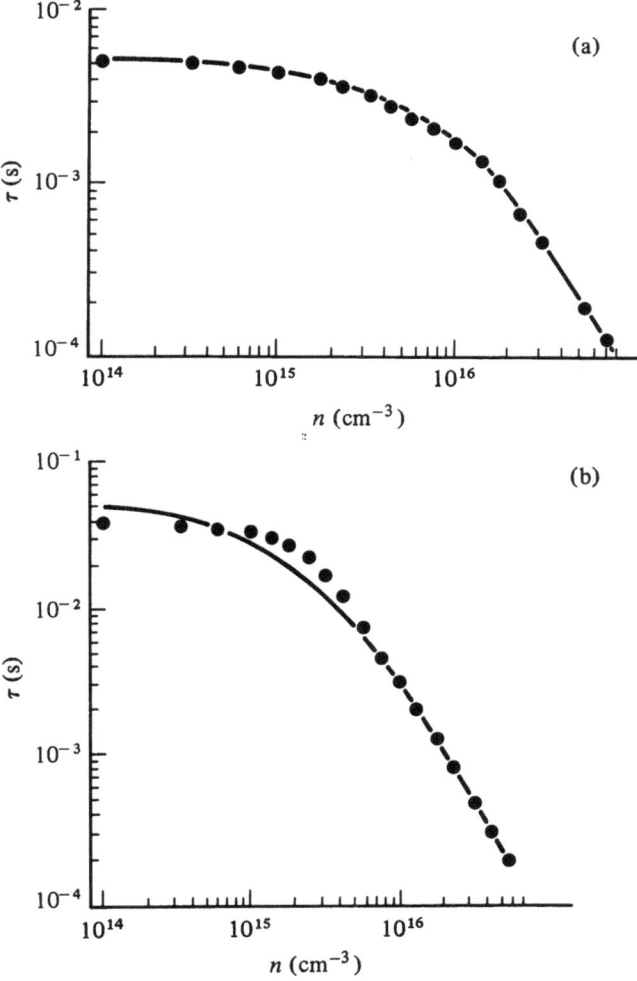

Fig. 2.3.9 [2.3.33]. Room-temperature lifetime of Si. The points represent experimental results from [2.3.32] for (a) ⟨100⟩ Si and (b) ⟨111⟩ Si. The curves are based on eq. (2.3.41) and Table 2.3.2.

Table 2.3.3. *Three scenarios for lifetime calculations for a Si solar cell*

Scenario no.	Curve	Concentration	RR
$1 : \tau - N_A$	dash-dot	$N_t = 10^{11}$ cm^{-3}	w
$(N_D = 5 \times 10^{19}$ cm$^{-3})$	solid	$\begin{cases} 10^{11} \\ 10^{14} \end{cases}$	s w
	dashed	10^{14}	s
$2 : \tau - N_D$	dashed		w
$(N_A = 1.3 \times 10^{15}$ cm^{-3}	solid		s
$N_t = 10^{13}$ cm$^{-3})$			
$3 : \tau - N_t$	dash-dot	$N_A = 1.3 \times 10^{15}$ cm^{-3}	w
$(N_D = 5 \times 10^{19}$ cm$^{-3})$	dotted	1.85×10^{17}	w
	solid	1.3×10^{15}	s
	dashed	1.85×10^{17}	s

In the *p*-region (the base of the solar cell) we find similarly for the minority-carrier lifetime

$$\tau_n^{-1} \equiv \frac{u}{\Delta n} \sim B_2 N_A^2 + (T_1^S + T_2 N_A) N_t \qquad (2.3.45)$$

The lifetimes now follow by taking reasonable numerical values, N_t being assumed uniform throughout.

We shall utilize two recombination regimes ('RR'): weak ('w') and strong ('s'). These are specified respectively by

$$\text{w} : T_1^S = T_2^S = 10^{-10} \text{ cm}^3 \text{ s}^{-1}, \quad T_2 = T_3 = 10^{-28} \text{ cm}^6 \text{ s}^{-1}$$

$$\text{s} : T_1^S = T_2^S = 10^{-7} \text{ cm}^3 \text{ s}^{-1}, \quad T_2 = T_3 = 10^{-25} \text{ cm}^6 \text{ s}^{-1}$$

Consider now three scenarios appropriate for a Si solar cell (Table 2.3.3).

The resulting lifetimes are shown in Figs. 2.3.10–2.3.12 with the dominant recombination mechanism marked at the appropriate parts of the curves [2.3.39]. Fig. 2.3.10 shows that τ_p is constant since it is independent of N_A while τ_n decreases due to the Auger processes B_2 and T_2 as N_A is increased. We here assume $B_1 \sim B_2 \sim 10^{-31}$ cm^6 s^{-1} and τ_p lies below τ_n since N_D exceeds N_A. If the *emitter* doping N_D is varied, it is now τ_n which is constant by virtue of eq. (2.3.45) and is determined by normal capture T_1^S. The factor 10^3 between regimes s and w yields the factor of 10^3 between the two values of τ_n (Fig. 2.3.11). For τ_p the dominant mechanism is normal capture for low N_D, and B_1 (or B_1 and T_3) for high N_D. If N_t is varied one finds that τ_n is greatest for the w regime and for the lower of the two values of N_A (Fig. 2.3.12). In order to understand why one of the four τ_n is limited

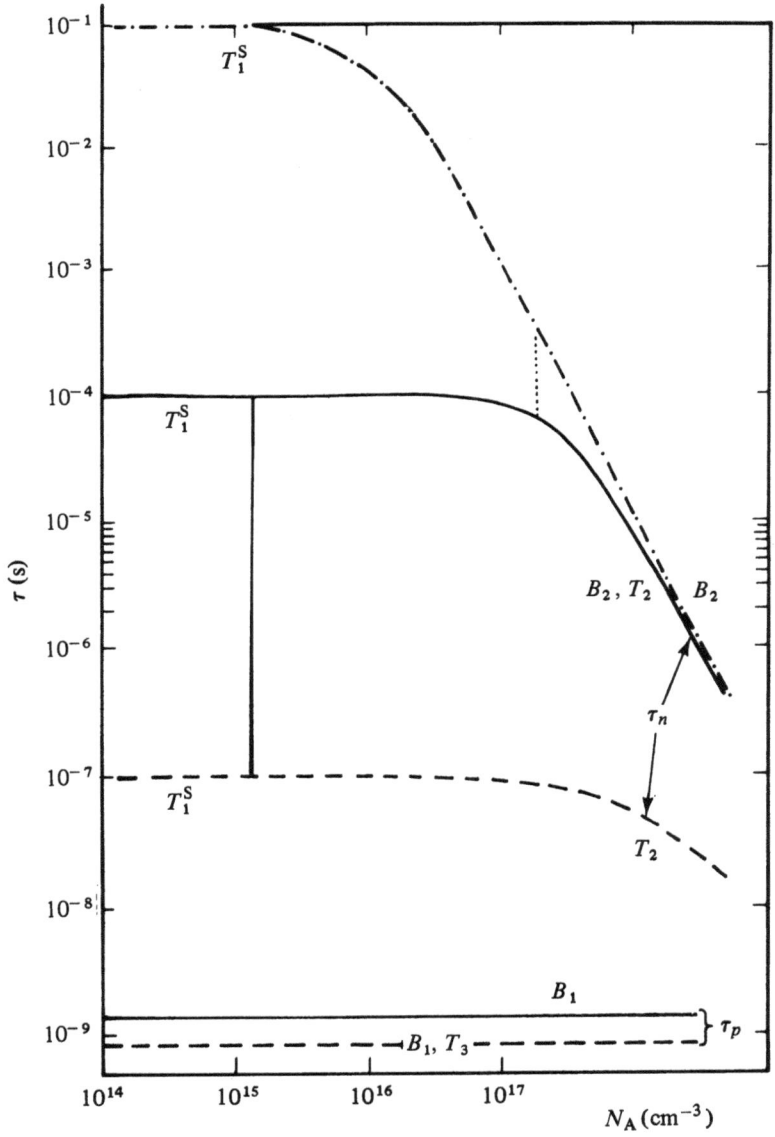

Fig. 2.3.10. Calculated minority-carrier lifetimes in the base (τ_n) and in the emitter (τ_p) of a Si solar cell for scenario 1.

by B_2 at the lower N_t values the solid and dotted curves are shown as vertical solid and dotted lines in Fig. 2.3.10. The limitation of τ_n by B_2 is then seen as due to the comparatively large value of N_A used for the dotted curve which brings it into the Auger regime for $N_t \sim 10^{11}$ cm^{-3} but almost out of it for $N_t \sim 10^{14}$ cm^{-3}.

The fractional contributions of five recombination mechanisms in Si point-contact concentrator solar cells have been given in [2.3.40]. The mechanisms

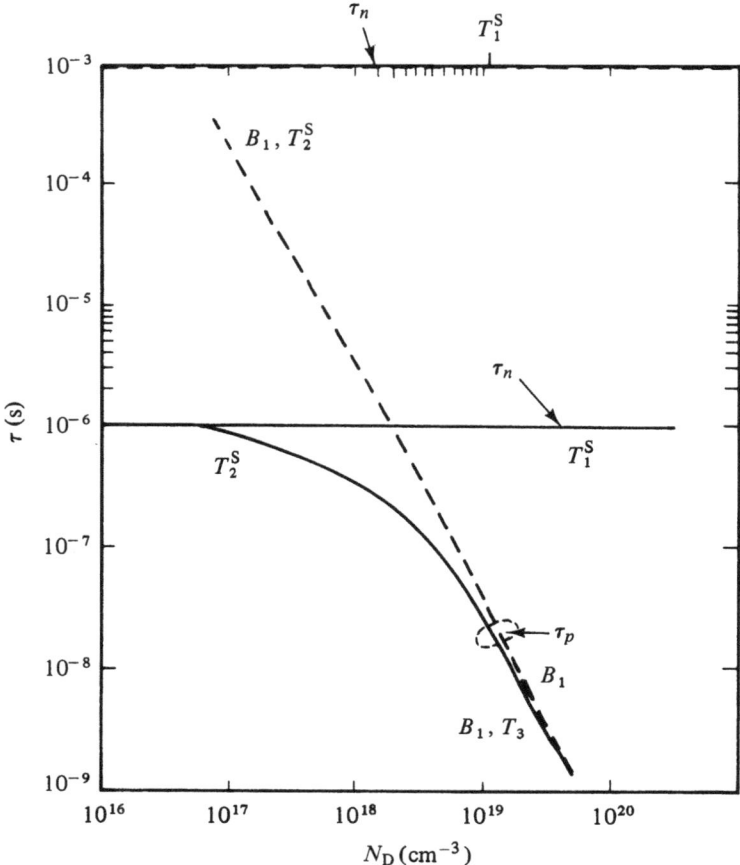

Fig. 2.3.11. As Fig. 2.3.10 for scenario 2.

involved are bulk SRH, surface, emitter and bulk Auger recombination. Between 0.5 and 20 W cm^{-2} recombination at the maximum power point is dominated by emitter recombination. Auger recombination dominates above 20 W cm^{-2}.

2.4 Cascade recombination

2.4.1 Introduction

So far the existence of excited states of traps or defects has been acknowledged, but they have not been examined with a view to finding new physical effects which are entirely due to them. The whole procedure of making extensive use of partition functions, for example already in eq. (1.2.2), was designed to lump together ground state and excited states of a system by summing over them and then using the partition function which results. While this is reasonable, there is evidence that excited states can play an important physical role. For example, a conduction band

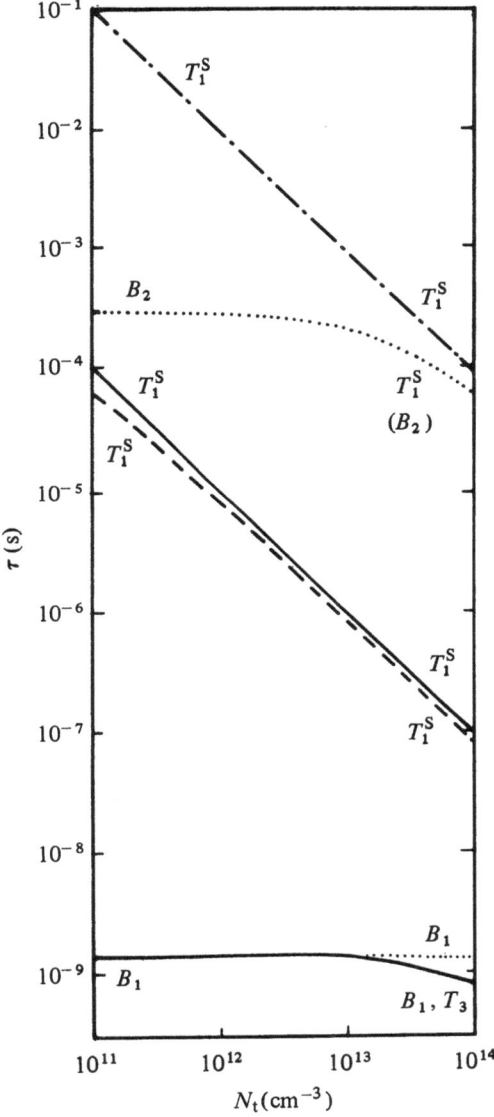

Fig. 2.3.12. As Fig. 2.3.10 for scenario 3. The top four curves are for τ_n, the bottom two curves are for τ_p.

electron can be trapped into high excited state of a trap. It can then be re-emitted into the conduction band with a certain probability or it can make a transition into a lower excited state. From this state the electron can be emitted into the conduction band, make a transition back into the higher level, or it can fall further into a lower level. One can see that the description of this kind of cascading down a ladder of excited states into a ground state can become enormously complicated.

In fact it *has* been pursued in connection with both astrophysical plasmas and semiconductors. However, a conceptual simplification is possible by associating a probability p_i with each excited level i that an electron in this level will in fact reach the ground state without being first re-emitted into the conduction band. The difficulty of the problem is now largely transferred to the calculation of these so-called *sticking probabilities*. However, simplifying approximations can make the problem tractable. The theoretically calculated capture cross section of a trap or a defect can be greatly increased by taking the effect of the excited states into account.

The subject was largely pioneered by Lax [2.4.1], [2.4.2] who coined the term 'giant traps' in this connection. Later work refined and developed the theory, as reviewed by Stoneham [2.4.3], see also [2.4.4]–[2.4.6]. For full cascade treatments which dispenses with the idea of a sticking probability, see [2.4.7] and [2.4.8].

So far we have considered *trapping*, and indeed the mechanisms considered do not involve the valence band. In that respect they are more specialized than the SRH process of section 2.3. In [2.4.6] only *one* effective excited state was considered and this simplified the problem. A further development is to stick to only one effective excited state for electron trapping, but to add also one excited state for hole trapping and facilitate the establishment of genuine steady-state *recombination* by adding the valence band. This model has been explored in [2.4.9] and [2.4.10]. In fact these papers considered also the effect of finite relaxation times of the traps immediately after capture of an electron or hole into the excited states. Later, the excited hole state was neglected for simplicity but the possibility of a *direct transition* between the ground state of the trap and the bands was also allowed. This transition had been neglected in cascade trapping and cascade recombination. Yet it can be important at elevated temperatures when excited states can hold electrons only with difficulty. In section 2.4 this type of theory will be outlined. It can be called truncated cascade recombination (TCR) since only one excited level ('truncation') but both bands ('recombination') are considered [2.4.11], [2.4.12]. Fig. 2.4.1 shows the model and introduces the notation which will be used for the transition rates.

2.4.2 *The kinetics of the model* [2.4.12]

In the present model of TCR v_0, v_g, v_e are, respectively, the concentration of empty traps (e.g. ionized donors), traps with electrons in their ground state and in an excited state. The capture probability per unit time of a conduction band electron into the donor ground state is taken to be $C_{ng}(E) \mathcal{N}(E) \mathrm{d}E$. Here the electron is assumed to come from an energy range $(E, E + \mathrm{d}E)$ for which the density of states per unit volume is $\mathcal{N}(E) \mathrm{d}E$. The emission probability per unit time in the reverse

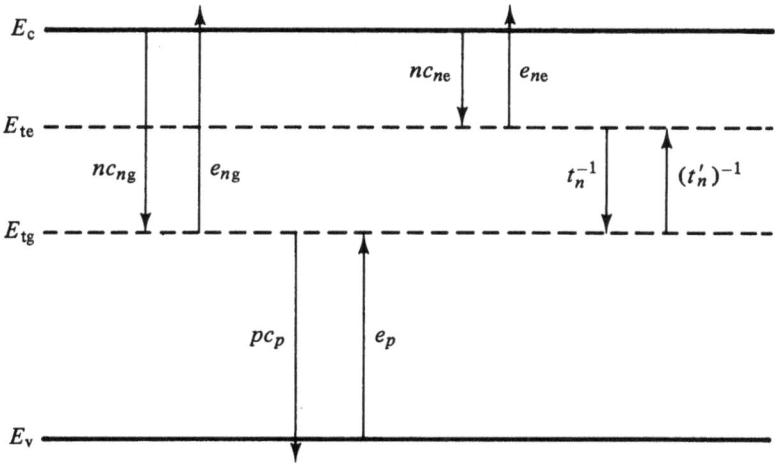

Fig. 2.4.1. The transition rates considered here. These rates are *not* per unit volume.

direction is denoted by $C'_{ng}(E)\mathcal{N}(E)\,dE$. Both probabilities presume the donor center empty or occupied in its ground state so that the actual net transition rate per unit volume from conduction band to trap ground state is

$$R_{ng} = v_0\,nc_{ng} - v_g\,e_{ng} \tag{2.4.1}$$

With the conduction band Fermi function denoted by f_n, two transition rates (averaged over the conduction band) have been defined

$$nc_{ng} \equiv \int_{\text{c.b.}} C_{ng}(E)\,\mathcal{N}(E) f_n(E)\,dE \tag{2.4.2}$$

$$(S_c - n)\,c'_{ng} \equiv \int_{\text{c.b.}} C'_{ng}(E)\,\mathcal{N}(E)[1 - f_n(E)]\,dE \equiv e_{ng} \tag{2.4.3}$$

where $S_c - n$ is the concentration of unoccupied conduction band states. Here e_{ng} is the conventionally used 'emission rate' from the trap in its ground state. For the excited states one can write similarly after integrating over the conduction band

$$R_{ne} = v_0\,nc_{ne} - v_e\,e_{ne} \tag{2.4.4}$$

The assumed isotropy of the distribution function limits the validity of the theory to low fields.

The equilibrium concentrations are

$$(v_0, v_g, v_e)_{\text{eq}} = (Z_0, e^{\gamma_{\text{eq}}}Z_{1g}, e^{\gamma_{\text{eq}}}Z_{1e}) \frac{N_T}{Z_0 + (\exp\gamma_{\text{eq}})(Z_{1g} + Z_{1e})} \tag{2.4.5}$$

where γ_{eq} is the equilibrium Fermi level divided by kT, N_T is the trap concentration, Z_0 is the partition function of the empty trap and $Z_1 = Z_{1g} + Z_{1e}$ is the partition

function for the filled trap, split into ground state and excited state terms. By
detailed balance for an energy range dE (and writing $\eta \equiv E/kT$),

$$\left[\frac{C'_{ng}(E)}{C_{ng}(E)}\right]_{eq} = \left[\frac{f_n(E)}{1-f_n(E)}\frac{v_0}{v_g}\right]_{eq} = \frac{Z_0}{Z_{1g}}e^{-\eta} = \frac{C'_{ng}(E)}{C_{ng}(E)} \tag{2.4.6}$$

The last step assumes excitation independence of the capture coefficients. Using
eqs. (2.4.3), (2.4.6) and

$$e^{-\eta}[1-f_n(E)] = e^{-\gamma_n}f_n(E) \tag{2.4.7}$$

where γ_n is the quasi-Fermi level (divided by kT) for the conduction band,

$$e_{ng} = \frac{Z_0}{Z_{1g}}\int e^{-\eta}C_{ng}(E)[1-f_n(E)]\,\mathcal{N}(E)\,dE$$

$$= \frac{Z_0}{Z_{1g}}e^{-\gamma_n}nc_{ng} \tag{2.4.8}$$

(Note that γ_e is used in this section 2.4 for the quasi-Fermi level of the excited state,
so that γ_n has to serve for the conduction band. In section 2.2 we used γ_e, γ_h for
conduction and valence band. Here we use γ_n, γ_p. It is hoped this will not cause
confusion.) Let us generalize eq. (2.4.5) by the use of quasi-Fermi levels $kT\gamma_g$ and
$kT\gamma_e$ for ground and excited states of the trap

$$(v_0, v_g, v_e) = (Z_0, e^{\gamma_g}Z_{1g}, e^{\gamma_e}Z_{1e})\frac{N_T}{Z_0 + (\exp\gamma_g)Z_{1g} + (\exp\gamma_e)Z_{1e}} \tag{2.4.9}$$

Inserting eqs. (2.4.8) and (2.4.9) into (2.4.1)

$$R_{ng} = [1 - \exp(\gamma_g - \gamma_n)]\,v_0\,nc_{ng} \tag{2.4.10}$$

In thermal equilibrium $R_{ng} = 0$. A similar argument shows that eq. (2.4.4) yields

$$R_{ne} = [1 - \exp(\gamma_e - \gamma_n)]\,v_0\,nc_{ng} \tag{2.4.11}$$

Next, the electron transition rate per unit volume from the ground state to the
valence band is

$$R_{gp} = v_g\,pc_p - v_0\,e_p$$

Here c_p is an average, corresponding to eq. (2.4.2), but now over the valence band:

$$pc_p \equiv \int_{v.b.} C_p(E)[1-f_n(E)]\,\mathcal{N}(E)\,dE$$

$$(S_v - p)c'_p \equiv \int_{v.b.} C'_p(E)f_n(E)\,\mathcal{N}(E)\,dE \equiv e_p$$

By detailed balance and assuming excitation independence [2.4.13], [2.4.14]

$$\left[\frac{C_p'(E)}{C_p(E)}\right]_{eq} = \frac{Z_{1g}}{Z_0} e^\eta = \frac{C_p'(E)}{C_p(E)}$$

so that analogously to eq. (2.4.8)

$$e_p = \frac{Z_{1g}}{Z_0} e^{\gamma_p} p c_p \tag{2.4.12}$$

and

$$R_{gp} = [1 - \exp(\gamma_p - \gamma_g)] v_g p c_p \tag{2.4.13}$$

where γ_p is the quasi-Fermi level (divided by kT) for the valence band.

The net transition rate per unit volume from the excited states to the ground state of a trap will be written in the form

$$R_{eg} = v_e/t_n - v_g/t_n' \tag{2.4.14}$$

where by detailed balance

$$\left(\frac{t_n'}{t_n}\right)_{eq} = \left(\frac{v_g}{v_e}\right)_{eq} = \frac{Z_{1g}}{Z_{1e}} \tag{2.4.15}$$

It is convenient to define certain new concentrations as ratio of averages of the type (2.4.2), (2.4.3):

$$n_g \equiv \frac{e_{ng}}{c_{ng}} = \frac{Z_0}{Z_{1g}} n \exp(-\gamma_n), \quad n_e \equiv \frac{e_{ne}}{c_{ne}} = \frac{Z_0}{Z_{1e}} n \exp(-\gamma_n) \tag{2.4.16}$$

$$p_g \equiv \frac{e_p}{c_p} = \frac{Z_{1g}}{Z_0} p \exp \gamma_p \tag{2.4.17}$$

These relations, along with others in this section, hold also if the bands are Fermi-degenerate. If they are, the concentrations (2.4.16), (2.4.17) depend on the Fermi levels. In the nondegenerate case the new concentrations are Fermi-level-independent and correspond to the usual Shockley–Read parameters (η_T = trap energy level divided by kT and we take $r = 1$ in eqs. (2.2.13)) $n(\frac{1}{2}) = n_1$ or $p(\frac{1}{2}) = p_1$, see p. 118,

$$n_g \to n \exp(\eta_{Tg} - \gamma_n), \quad n_e \to n \exp(\eta_{Te} - \gamma_n), \quad p_g \to p \exp(\gamma_p - \eta_{Tg}) \tag{2.4.18}$$

The latter can be obtained from eqs. (2.4.16), (2.4.17) by introducing effective trap levels of energy $E_{Tg} = kT\eta_{Tg}$ and $E_{Te} \equiv kT\eta_{Te}$ by

$$Z_0/Z_{1g} = \exp \eta_{Tg}, \quad Z_0/Z_{1e} = \exp \eta_{Te} \tag{2.4.19}$$

If several excited states exist, E_{Te} must be expected to be temperature-dependent.

The net recombination rates per unit volume can be written in the form

$$R_{ng} = (v_0 n - v_g n_g) c_{ng}, \quad R_{ne} = (v_0 n - v_e n_e) c_{ne} \tag{2.4.20}$$

$$R_{gp} = (v_g p - v_0 p_g) c_p \tag{2.4.21}$$

It should also be noted that the quantity c_{ng} of relation (2.4.2) need not describe only a single-electron transition given by a recombination coefficient $T_1^s(E)$. Trap–Auger effects involving the conduction band or the valence band for the second electron can be regarded as included. In that case one has to interpret

$$c_{ng} \text{ as } T_{1g}^s + T_{1g} n + T_{2g} p \tag{2.4.22}$$

Similarly one can interpret

$$c_p \text{ as } T_2^s + T_3 n + T_4 p \tag{2.4.23}$$

$$c_{ne} \text{ as } T_{1e}^s + T_{1e} n + T_{2e} p \tag{2.4.24}$$

as is clear by analogy with eqs. (2.2.11).

2.4.3 The transient properties of the concentrations [2.4.12]

If one eliminates the concentration $v_0 \ (= N_t - v_g - v_e)$ one finds, as the differential equations of the model, the coupled pair,

$$\dot{v}_e = R_{ne} - R_{eg} = a - b v_e - c v_g \tag{2.4.25}$$

$$\dot{v}_g = R_{eg} + R_{ng} - R_{gp} = d - h v_e - l v_g \tag{2.4.26}$$

The constants are given in Table 2.4.1. The above equations have to be solved together with the electron and hole continuity equations,

$$\dot{n} = -(R_{ne} + R_{ng}) + G + \frac{1}{q} \nabla \cdot \mathbf{j}_n \tag{2.4.27}$$

$$\dot{p} = -R_{pg} + G - \frac{1}{q} \nabla \cdot \mathbf{j}_p \tag{2.4.28}$$

where $\mathbf{j}_n$ and $\mathbf{j}_p$ are respectively the electron and hole current densities and G is the generation rate per unit volume. However, for experimental conditions in which n and p are kept *time-independent* one finds

$$\ddot{v}_i + P \dot{v}_i + Q v_i + S_i = 0 \quad (i = 0, g, e) \tag{2.4.29}$$

Similar equations have been obtained by Rees et al. [2.4.15] by neglecting electron transitions to the ground state and hole transitions and by Pickin [2.4.16] for the electron–donor recombination.

If one uses

$$\mu_i \equiv v_i + S_i / Q$$

then

$$\ddot{\mu}_i + P \dot{\mu}_i + Q \mu_i = 0$$

Table 2.4.1. *Expressions for constants*

Defining equation	Constant	Dimension	Expression for constant	
(2.4.25)	a	$L^{-3}T^{-1}$	$nc_{ne}N_t$	
	b	T^{-1}	$nc_{ne}+e_{ne}+1/t_n$	
	c	T^{-1}	$nc_{ne}-1/t_n'$	
(2.4.26)	d	$L^{-3}T^{-1}$	$[nc_{ng}+e_p]N_t$	
	h	T^{-1}	$nc_{ng}+e_p-1/t_n$	
	l	T^{-1}	$nc_{ng}+e_{ng}+pc_p+e_p+1/t_n'$	
(2.4.29)	P	T^{-1}	$b+l=n(c_{ng}+c_{ne})+pc_p+e_{ng}+e_{ne}+e_p$ $+1/t_n+1/t_n'$	(2.4.33)
(2.4.29)	Q	T^{-2}	$bl-ch=c_{ne}c_pnp+[(n_e+n_g)c_{ng}c_{ne}+(c_{ng}+c_{ne})$ $(1/t_n+1/t_n')]n+(n_ec_{ne}+1/t_n)c_pp$ $+(n_gc_{ng}+p_gc_p)(n_ec_{ne}+1/t_n)$ $+(c_{ne}+p_gc_p/n_e)(n_g/t_n)$	(2.4.34)
(2.4.29)	S_g	$L^{-3}T^{-2}$	$ah-bd=-(nc_{ng}+p_gc_p)(1/t_n+n_ec_{ne})N_t$ $-nc_{ne}N_tt_n^{-1}$	(2.4.35)
(2.4.29)	S_e	$L^{-3}T^{-2}$	$cd-al=-(n_gc_{ng}+pc_p)nc_{ne}N_t$ $-(nc_{ng}+p_gc_p+nc_{ne})N_t/t_n'$	(2.4.36)
(2.4.29)	S_0	$L^{-3}T^{-2}$	$-S_e-S_g-QN_t$	(2.4.37)

Table 2.4.2. *Values of parameters used in the present section (Si)*

σ_{ne}	10^{-13} cm^2
$\sigma_{ng}=\sigma_p$	10^{-16} cm^2
v_{th} (for electrons and holes)	$10^7\left(\dfrac{T}{300}\right)^{\frac{1}{2}}$ cm s^{-1}
N_T	10^{13} cm^{-3}
E_c-E_{Te}	0.05 eV
$E_c-E_{Tg}=E_{Tg}-E_v$	0.55 eV
t_n	10^{-10} s
$N_c=N_v$ (at 300 K)	2×10^{19} cm^{-3}

Parameters inferred from the above at T = 300 K

$n_e=N_c\exp[-(\eta_c-\eta_{Te})]$	2.89×10^{18} cm^{-3}
$\eta_g=N_c\exp[-(\eta_c-\eta_{Tg})]$	1.15×10^{10} cm^{-3}
$p_g=N_v\exp[-(\eta_{Tg}-\eta_v)]$	1.15×10^{10} cm^{-3}
$t_n'=t_nn_e/n_g$	0.0251 s

Table 2.4.3. *Steady-state occupation probabilities* $f_i = v_i/N_T$ (a) *for a depleted*
semiconductor and (b) *under a uniform injection of electron concentration*

Condition	T (K)	p	n	f_g	f_e	f_0
Depletion	300	0	0	0.184	2.52×10^{-12}	0.82
	200	0	0	0.045	2.29×20^{-16}	0.95
	100	0	0	0.002	6.28×10^{-29}	1.00
Uniform injection	300	0	10^{12} cm^{-3}	0.99	3.92×10^{-9}	1.13×10^{-2}
of electrons	200	0	10^{12} cm^{-3}	1.00	2.51×10^{-13}	1.50×10^{-7}
	100	0	10^{12} cm^{-3}	1.00	6.30×10^{-26}	0.00

This is a homogeneous linear differential equation with constant coefficients so that

$$\mu_i(t) = v_i(t) + S_i/Q = A_i e^{-\omega_1 t} + B_i e^{-\omega_2 t}$$

where

$$\omega_1 \equiv \tau_1^{-1} = \frac{P}{2}\left[1 + \left(1 - \frac{4Q}{P^2}\right)^{\frac{1}{2}}\right], \quad \omega_2 \equiv \tau_2^{-1} = \frac{P}{2}\left[1 - \left(1 - \frac{4Q}{P^2}\right)^{\frac{1}{2}}\right] \tag{2.4.30}$$

Inserting some initial conditions by assuming $v_i(0)$ and $\dot{v}_i(0)$ to be given and, noting that for positive $\omega_{1,2}$

$$v_i(\infty) = -S_i/Q, \quad (i = 0, g, e) \tag{2.4.31}$$

one finds the transient solutions for $i = 0, g, e$:

$$(\omega_1 - \omega_2)[v_i(t) - v_i(\infty)] = \{\dot{v}_i(0) + \omega_1[v_i(0) - v_i(\infty)]\} e^{-\omega_2 t}$$
$$- \{\dot{v}_i(0) + \omega_2[v_i(0) - v_i(\infty)]\} e^{-\omega_1 t} \tag{2.4.32}$$

The explicit and exact expressions for τ_1, τ_2 and $v_i(\infty)$ can be obtained from Table 2.4.1 using eqs. (2.4.25) and (2.4.31).

For a numerical evaluation use the parameters given in Table 2.4.2. The capture cross section σ_{ne} has been assumed to be of the order of the largest measured cross sections [2.4.17], and more typical values have been chosen for σ_{ng} and σ_p. They have been assumed to be independent of temperature. The resulting effective electron capture cross section, which is defined below in eq. (2.4.49), does however show a temperature dependence. The excited states have been assumed to have an effective energy level at 50 meV below the conduction band edge. This is in accordance with the arguments given by Lax [2.4.2], Abakumov *et al.* [2.4.18] and others who assume that electrons are captured by levels with binding energy greater than kT. As to traps, Rees *et al.* [2.4.15] have obtained from the measured emission rates [2.4.19] for singly charged S and Se center in Si a value of $t_n \simeq 0.5 \times 10^{-10}$ s. We have kept to the same order of magnitude here (Table 2.4.2).

In Table 2.4.3 typical values of the steady-state occupation probabilities $f_i = v_i(\infty)/N_T$ are given for two areas of practical interest. In the first case the semiconductor is depleted of carriers ($p = n = 0$), for example by reverse biasing a p–n junction, whereas in the second

Table 2.4.4A. *Calculated values of parameters involved in a transient for the measurement of emission rates ($n = p = 0$), for the semiconductor parameters given in Table 2.4.2*

T (K)	P (s^{-1})	Q (s^{-2})	$\tau_1 \equiv \omega_1^{-1}$ (s)	$\tau_2 \equiv \omega_2^{-1}$ (s)
300	2.90×10^{12}	1.81×10^{14}	3.44×10^{-13}	1.59×10^{-2}
260	1.62×10^{12}	4.33×10^{12}	6.16×10^{-13}	3.74×10^{-1}
220	7.79×10^{11}	3.11×10^{10}	1.28×10^{-12}	2.49×10^{1}

Table 2.4.4B. *Calculated values of parameters involved in a transient for the measurement of capture rates ($p = 0$, $n = 10^{12}$ cm^{-3}) for the semiconductor parameters given in Table 2.4.2*

T (K)	P (s^{-1})	Q (s^{-2})	$\tau_1 = \omega_1^{-1}$ (s)	$\tau_2 = \omega_2^{-1}$ (s)
240	1.15×10^{12}	9.97×10^{15}	8.69×10^{-13}	1.15×10^{-4}
180	2.96×10^{11}	7.97×10^{15}	3.37×10^{-12}	3.71×10^{-4}
120	3.50×10^{10}	6.34×10^{15}	2.82×10^{-11}	5.57×10^{-6}

case a uniform flux of electrons is injected in the otherwise depleted semiconductor. In either case, because n is small, f_e is very small. Thus, excited states, though important for the capture of electrons, do not hold them (as will be seen later, such an assumption is violated under a high injection condition). Usually for a depleted semiconductor most of the traps are empty. Thus $f_0 \simeq 1$, whereas $f_g \ll 1$. These conditions are best met at low temperatures. Similarly for uniform electron injection all the traps are filled with electrons, giving $f_g \simeq 1$ and $f_0 \ll 1$.

Two types of experiments usually performed to determine the emission and capture rates correspond to switching the semiconductor from one to the other of the above conditions [2.4.20]. In the first set of experiments, traps (donors here) are first filled with electrons and then, at some time $t = 0$, electrons and holes are swept out of the conduction and valence bands by applying a large reverse bias. In the depletion layer we then have $n = p \simeq 0$ at $t > 0$. Therefore the condition of constant n and p is met and eq. (2.4.32) applies.

Under constant reverse bias, decay of $f_g(t)$ from near unity to a small value occurs as electrons are emitted from traps. The values of the time constants are given in Tables 2.4.4A and B using the values of the parameters given in Table 2.4.2. It is seen that

$$\tau_1(\equiv \omega_1^{-1}) \ll \tau_2(\equiv \omega_2^{-1}) \text{ and also } 4Q/P^2 \ll 1 \tag{2.4.38}$$

Hence from eqs. (2.4.30)

$$\tau_1^{-1} \simeq P, \quad \tau_2^{-1} \simeq Q/P \tag{2.4.39}$$

Consider now the emission experiment, denoted by a suffix e attached to the time constants, and assume $n_e c_{ne} \gg n_g c_{ng}$, $p_g c_p$. Then an approximate relation for the readjustment time τ_1 of the population of the excited states is obtained from eq. (2.4.33) by noting that only the last five terms contribute and, of these, the terms e_{ne} and $1/t_n$ dominate. Hence

$$\tau_1 \rightarrow \tau_{1e} = t_n/(1 + n_e c_{ne} t_n) \tag{2.4.40}$$

Similarly for τ_{2e} we have

$$\tau_{2e}^{-1} = \left(\frac{Q}{P}\right)_{n=p=0} = \left(c_{ng} + \frac{c_{ne}}{1 + n_e c_{ne} t_n}\right) n_g + c_p p_g = e_{n\,\text{eff}} + e_p \tag{2.4.41}$$

where

$$e_{n\,\text{eff}} \equiv \left(c_{ng} + \frac{c_{ne}}{1 + n_e c_{ne} t_n}\right) n_g \tag{2.4.42}$$

is an effective emission rate for electrons.

For Tables 2.4.4A and B $\tau_{2e} \gg \tau_{1e}$, so that after a short initial time the transient is dominated by τ_{2e} and from eq. (2.4.32)

$$v(t) - v(\infty) = [v(0) - v(\infty)] \exp(-t/\tau_{2e}) \tag{2.4.43}$$

where $v \equiv v_g + v_e$ is the concentration of trapped electrons. In a transient capacitance experiment, the above equation is used to determine τ_{2e}, whose variation with temperature gives the activation energy of the trap. From eq. (2.4.41) we find that τ_{2e} involves three different activation energies which cannot always be distinguished experimentally. The assumption that only one activation energy dominates, which was made by Gibb *et al.* [2.4.6], can be derived if the following approximations are made:

$$e_p \ll e_{n\,\text{eff}}, \quad c_{ng} \ll \frac{c_{ne}}{1 + n_e c_{ne} t_n}, \quad 1 \ll n_e c_{ne} t_n$$

Then

$$\tau_{2e}^{-1} \simeq \frac{n_g}{t_n n_e} = \frac{1}{t_n} \exp[-(\eta_{\text{Te}} - \eta_{\text{Tg}})] \tag{2.4.44}$$

Although the interpretation based on this relation appears justified in the case of Gibb *et al.*, the present analysis shows its limitations. In particular, the last approximation becomes poorer at lower temperatures.

In Fig. 2.4.2 we have plotted τ_{2e}^{-1} against $1/T$ for these parameters. The curve shown is only approximately straight with activation energy 513 meV. In the approximation (2.4.44) the straight line is exact with activation energy of 500 meV.

The second set of transient experiments involving v_t is for direct measurement of the effective capture rate of electrons at empty traps, usually by using a reverse-biased junction. At $t \leqslant 0$, we assume $v_0 \simeq N_T$. Since for $t \leqslant 0$ we have a steady state, we can write $\dot{v}_t(0) = 0$. Now at $t = 0$, electron–hole pairs are injected, say by light. Electrons, which are minority carriers, flow steadily into the depletion layer with a drift velocity v_{dn} such that $\Delta n = j_n/v_{dn}$ is constant with respect to time. Thus, in the transient phase both n and p are again constant and eq. (2.4.32) applies. Values of the time constants for this case are given in Tables 2.4.4A

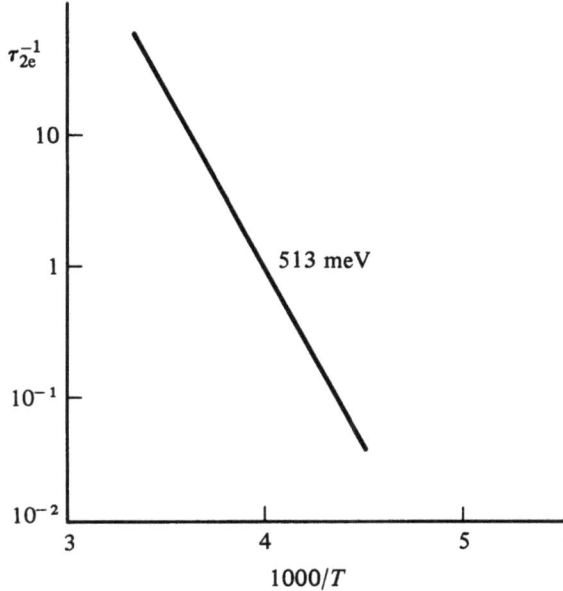

Fig. 2.4.2. A plot of τ_{2e}^{-1} against $1000/T$ according to eq. (2.4.41) using the parameters of Table 2.4.2.

and **B** and show that relations (2.4.38) and (2.4.39) hold. The low-injection capture time constants can therefore be defined by

$$\tau_{1c}^{-1} \equiv (P)_{p=0,\,n-\Delta n \ll n_e} \simeq \tau_{1e}^{-1}$$

and

$$\tau_{2c}^{-1} = \left(\frac{Q}{P}\right)_{p=0,\,\Delta n \ll n_e} = \left(c_{ng} + \frac{c_{ne}}{1 + n_e c_{ne} t_n}\right)\Delta n + e_{n\,\mathrm{eff}} + e_p \qquad (2.4.45)$$

A subscript c has been used with τ to denote a *capture* experiment. Also note that at low temperatures the first term on the right of eq. (2.4.45) dominates. Since it clearly deals only with capture it can be used to define an effective capture coefficient,

$$c_{n\,\mathrm{eff}} \equiv \left(\frac{1}{\tau_{2c}\,\Delta n}\right)_{\mathrm{low\,temp}} = c_{ng} + \frac{c_{ne}}{1 + n_e c_{ne} t_n} \qquad (2.4.46)$$

By eqs. (2.4.42) and (2.4.46)

$$e_{n\,\mathrm{eff}}/c_{n\,\mathrm{eff}} = n_g = \left(\frac{\tau_{2c}\,\Delta n}{\tau_{2e}}\right)_{\mathrm{low\,temp}} \qquad (2.4.47)$$

A simple deduction of the key results (2.4.42) and (2.4.46) is possible for steady-state conditions. For this one need not go through the analysis of transient effects given here. Because of the importance of these results we give two additional proofs. One is fairly intuitive (section 2.5.2), the other utilizes steady-state kinetics (appendix to section 2.5.2).

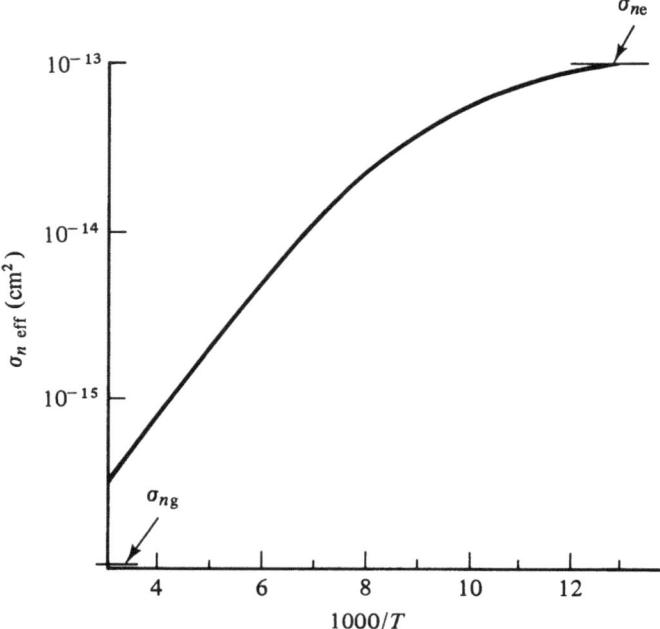

Fig. 2.4.3. A plot for the present theory of $\sigma_{n\,\text{eff}}$ against $1000/T$ showing that the cross sections for electron capture into the ground state and into the effective excited state represent its limiting values. The parameters of Table 2.4.2 have been used.

In capture experiments the trapped electron concentration behaves as

$$v(t) = v(\infty)[1 - \exp(-t/\tau_{2c})] \quad (t \gg \tau_{1c}) \tag{2.4.48}$$

An effective electron capture cross section is

$$\sigma_{n\,\text{eff}} \equiv c_{n\,\text{eff}}/v_{\text{th}} \tag{2.4.49}$$

In Fig. 2.4.3 we have plotted $\sigma_{n\,\text{eff}}$ against $1/T$ and find that at low temperature $\sigma_{n\,\text{eff}}$ is close to σ_{ne}, whereas at high temperature it tends to σ_{ng}. Thus, direct capture into the ground state, even if not of interest at low temperatures, may become significant at room temperature.

Equation (2.4.46) can be viewed as an approximation to a cascade theory [2.4.1]–[2.4.3], [2.4.21] in which all the excited states have been lumped together into a single effective energy level E_{Te} defined by eqs. (2.4.19). Sticking probabilities P_i that an electron from a state i will reach the ground state can be identified by writing eq. (2.4.46) as

$$c_{n\,\text{eff}} = c_{ng} + c_{ne}\,P_{\text{e}}$$

whence

$$P_{\text{e}} = \frac{1}{1 + n_{\text{e}}\,c_{ne}\,t_n} = \frac{t_n^{-1}}{t_n^{-1} + e_{ne}} \tag{2.4.50}$$

This shows that the present approach is consistent with the cascade theory, simplified to have one effective excited level. Equations (2.4.46) and (2.4.50) can be obtained by a simple steady-state argument (p. 168).

If one neglects direct transitions to the ground state ($c_{ng} = 0$) and assumes $n_e c_{ne} t_n \gg 1$, one arrives at

$$\tau_{2c}^{-1} = \Delta n / n_e t_n$$

(equation (14) of [2.4.6]). These authors obtain eq. (2.4.47) (their equation (11)) even though their individual expressions (denoted by ε_T and $\sigma_T v_{th}$ in their paper) refer to their special case. By these assumptions ([2.4.6], [2.4.15]) one can get $\sigma_{n\,eff} \simeq (v_{th} n_e t_n)^{-1}$. Since $v_{th} \sim T^{\frac{1}{2}}$ and $n_e \sim T^{\frac{3}{2}} \exp(\eta_{Te} - \eta_c)$ a plot of $\log(\sigma_{n\,eff} T^2)$ against T^{-1} should then give a straight line. For the present theory one finds Fig. 2.4.4 which it is difficult to approximate by a straight line. If such an approximation is made for a limited range of temperatures one is liable to obtain a misleading value of the activation energy. Note that these experiments are performed for the depletion layer where a large electric field exists which modifies capture and emission rates [2.4.22]–[2.4.24].

2.4.4 The steady-state recombination rate [2.4.12]

On substituting the expressions $v_i(\infty)$ in a net recombination rate formula such as eq. (2.4.21) one finds the steady-state recombination rate u per unit volume. The expression normally found in SRH statistics will be denoted by u_{SRH}. One then finds

$$u_{SRH} = \frac{np[1 - \exp(\gamma_p - \gamma_n)] c_{ng} c_p N_T}{(n + n_g) c_{ng} + (p + p_g) c_p} \tag{2.4.51}$$

$$u = \frac{1 + n_e c_{ne} t_n + c_{ne}/c_{ng}}{1 + n_e c_{ne} t_n + A} u_{SRH} \equiv F u_{SRH} \tag{2.4.52}$$

where

$$A \equiv \frac{n_e c_{ne}(pc_p t_n + n_g c_{ng} t_n + 1) + (t_n/t'_n)[(n + n_e) c_{ne} + n c_{ng} + p_g c_p]}{(n + n_g) c_{ng} + (p + p_g) c_p} \tag{2.4.53}$$

Note that one can put $c_{ne} \sim t_n \sim 0$ on neglecting the excited states, whence $u \to u_{SRH}$.

Now for *any* electron lifetime one has

$$\tau_n = \frac{\Delta n}{u} = \frac{\Delta n}{F u_{SRH}} = \frac{1}{F} \tau_{SRH} \tag{2.4.54}$$

This turns out to be the product of three factors:

$$\tau_n = \frac{\alpha np + \beta n + \gamma p + \delta}{\lambda n + \mu p + \rho} \cdot \frac{n_e c_{ng} + pc_p + n_g c_{ng} + p_g c_p}{pc_{ng} c_p N_T} \cdot \frac{\Delta n}{n\{1 - \exp[-(\gamma_n - \gamma_p)]\}} \tag{2.4.55}$$

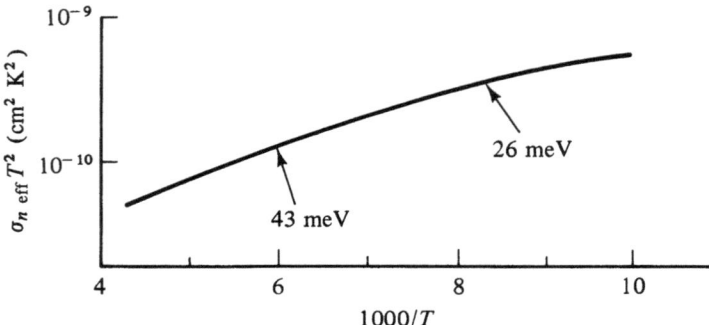

Fig. 2.4.4. A plot for the present theory of $\sigma_{n\,\text{eff}}\,T^2$ against $1000/T$ based on eqs. (2.4.46) and (2.4.49) showing that one does not find a straight line. The parameters of Table 2.4.2 have been used and two activation energies are shown which apply for a small range of temperatures.

The first factor is F^{-1}, the last factor is usually of order unity and is part of τ_{SRH}. The new parameters (see Table 2.4.5) α, β, γ, δ, λ, μ, ρ depend on c_{ng}, c_{ne}, c_p, t_n, t'_n, n_g, n_e, c_p, and can readily be obtained from eqs. (2.4.51)–(2.4.53). If the last factor of eq. (2.4.55) is put equal to unity, one finds for the low level ($p \gg n, n_g, p_g$) and high injection level ($\Delta n \sim n \sim p \gg n_g, p_g$) lifetimes respectively

$$\tau_{nl} = \frac{\gamma}{\mu c_{ng} N_t} = \frac{1 + n_e c_{ne} t_n}{c_{ng}(1 + n_e c_{ne} t_n) + c_{ne}} \frac{1}{N_T} = \frac{1}{c_{n\,\text{eff}} N_T} \rightarrow \frac{1}{c_{ng} N_T} \tag{2.4.56}$$

$$\tau_{nh} = \frac{\alpha^2 n^2 + (\beta + \gamma) n}{(\lambda + \mu) n} \frac{(c_{ng} + c_p) n}{p c_{ng} c_p N_T} \tag{2.4.57}$$

$$= \frac{n c_{ne} c_p t_n + (1 + n_e c_{ne} t_n)(c_{ng} + c_p) + c_{ne}(n_g c_{ng} t_n + 1 + t_n/t'_n) + c_{ng} t_n/t'_n}{[(1 + n_e c_{ne} t_n) c_{ng} + c_{ne}] c_p N_T}$$

$$\rightarrow \frac{c_{ng} + c_p}{c_{ng} c_p N_T} \tag{2.4.58}$$

Note that the $c_{n\,\text{eff}}$ of the transient problem occurs again here in the steady-state case. The arrows indicate the results in the limit in which one goes to SRH statistics. In eq. (2.4.57) one need not retain the term $(\beta + \gamma) n$ for very large n, but without it one would obtain zero in the Shockley–Read limit of eq. (2.4.58). One also has

$$\tau_{nh} = \tau_{nl} + \frac{1 + t_n/t'_n}{c_p N_T} + \frac{n c_{ne} t_n}{[(1 + n_e c_{ne} t_n) c_{ng} + c_{ne}] N_T} \rightarrow \frac{1}{c_{ng} N_T} + \frac{1}{c_p N_T} \tag{2.4.59}$$

Thus τ_{nh} consists of three terms. The first deals with electron capture and dominates in p-type material at low injection. The second term is a contribution

Table 2.4.5. *Parameters in eq. (2.4.55)*

$$\alpha = c_{ne} c_p t_n$$
$$\beta = c_{ng}(1 + n_g c_{ne} t_n + n_e c_{ne} t_n + t_n/t_n') + c_{ne}(1 + t_n/t_n')$$
$$\gamma = c_p(1 + n_e c_{ne} t_n)$$
$$\delta = n_g c_{ng} + p_g c_p + n_g n_e c_{ng} c_{ne} t_n + n_e p_g c_{ne} c_p t_n + (t_n/t_n')(n_e c_{ne} + p_g c_p)$$
$$\lambda = c_{ng}(1 + n_e c_{ne} t_n) + c_{ne}$$
$$\mu = c_p(1 + n_e c_{ne} t_n + c_{ne}/c_{ng})$$
$$\rho = n_g c_{ng}(1 + n_e c_{ne} t_n) + p_g c_p(1 + n_e c_{ne} t_n) + (c_{ne}/c_{ng})(n_g c_{ng} + p_g c_p)$$

due to hole capture, which is for many cases of practical interest (in which $t_n/t_n' \ll 1$) the same as in the SRH theory. The last term is new and shows that there can be a significant increase in the lifetime at high injection. This is due to the bottleneck produced at the excited levels whose delay time t_n leads to them being largely occupied thus withdrawing their availability for electron capture. In this last stage of a high injection

$$\tau_{nh} \simeq \frac{n t_n / N_T}{1 + (c_{ng}/c_{ne})(1 + n_e c_{ne} t_n)} \tag{2.4.60}$$

The proportionality of τ and $n = \Delta n$ implies that at this level the steady-state recombination rate $u = \Delta n/\tau_n$ reaches a constant plateau as a function of Δn.

To understand the effect arising due to the level of injection we have calculated the fractions

$$f_i = v_i/N_T, \quad i = 0, g, e \tag{2.4.61}$$

for empty traps and for traps filled with electrons in the ground and excited states, respectively. These have been plotted against Δn in Fig. 2.4.5. The semiconductor has been assumed to be of p-type Si with $p_0 = 10^{15}$ cm^{-3}.

At *low level* of injection ($\Delta n \sim 10^{13}$ cm^{-3}) f_0 is close to unity whereas f_g and f_e are very small. Thus the deciding factor for recombination is the minority-carrier capture (into the excited or ground state of the trap). The lifetime of excess electron–hole pairs is then given by eq. (2.4.56). As Δn is increased, more and more traps are filled, thereby reducing the concentration $v_0 = f_0 N_T$ of the traps available for electron capture. At *high level* of injection (10^{15} cm$^{-3} < \Delta n < 10^{18}$ cm^{-3}) the recombination is mostly determined by the hole capture ($f_g > f_0$) and the lifetime becomes nearly constant with respect to Δn. In Fig. 2.4.5 we have plotted f_i at two different temperatures and find that the ratio f_g/f_0 is higher at lower temperature. In the high-level range, though f_e is very small, it keeps on increasing almost

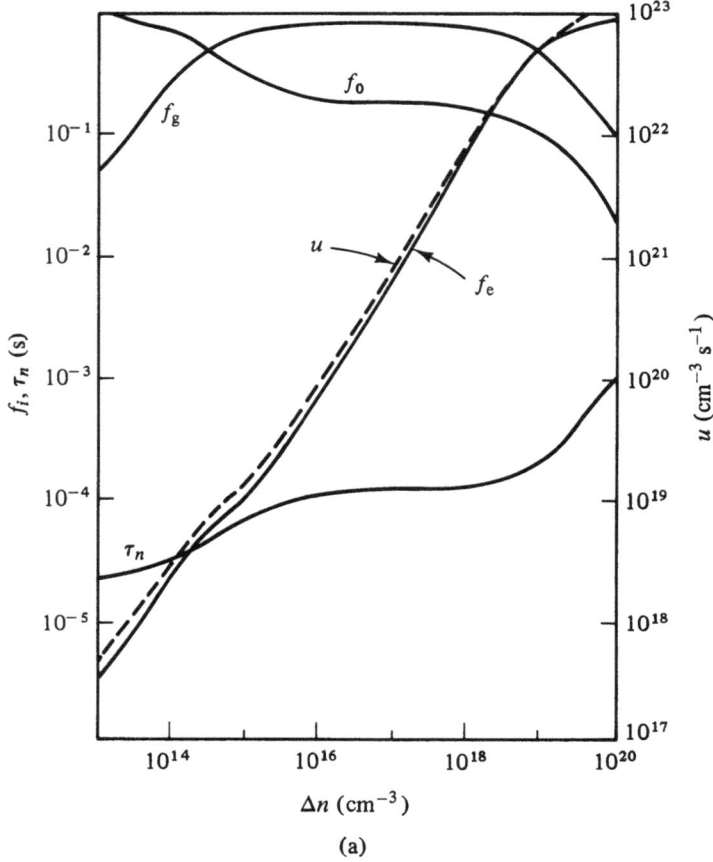

(a)

Fig. 2.4.5.a For legend see facing page.

proportionally to Δn. Since $f_0 + f_g + f_e = 1$ this increase in f_e is at the cost of f_0 and f_g and they start decreasing. As shown in Fig. 2.4.5, u follows f_e very closely, the small difference being due to direct capture into the ground state. An approximate value of u is given by

$$u = R_{ng} + R_{eg} = f_0 N_t c_{ng} \Delta n + \frac{N_t f_e}{t_n} \qquad (2.4.62)$$

where the reverse process of thermal emission has been neglected at high levels of injection. As Δn increases $f_e \rightarrow 1$ and $f_0 \rightarrow 0$ and a maximum value of u may be approximated as [2.4.9], [2.4.10]

$$(u)_{\Delta n \rightarrow \infty} \simeq \frac{N_t}{t_n}$$

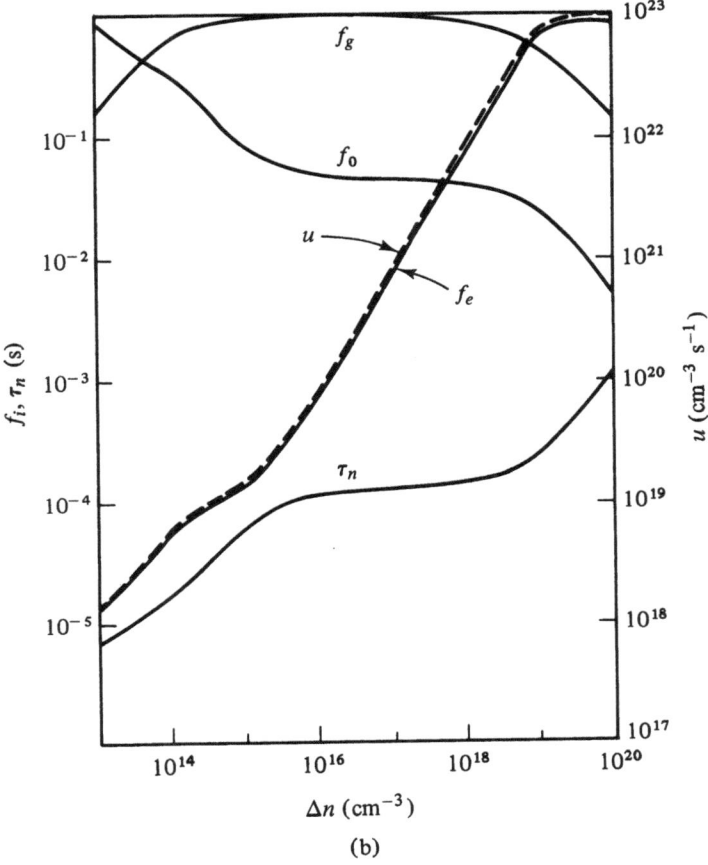

Fig. 2.4.5. Steady-state occupation probabilities f_i, minority-carrier lifetime τ_n and steady-state recombination rate per unit volume u as functions of excess minority-carrier concentration n at (a) $T = 300$ K, (b) $T = 200$ K. Other parameters are given in Table 2.4.2.

However, the more accurate expression (2.4.52) for u has a limiting value

$$(u)_{\Delta n \to \infty} = \frac{N_t}{t_n}\left[1 + \frac{c_{ng}}{c_{ne}}(1 + n_e c_{ne} t_n)\right] \qquad (2.4.63)$$

Such a saturation of u has been reported in the literature ([2.4.25]; also [2.4.26], Figure 6). It is interesting to note that even when a direct transition into the ground state has been allowed, the trap saturation occurs because of the readjustment of the electron population between the ground state and the excited states.

Because u approaches the constant in eq. (2.4.63) the lifetime (2.4.54) increases in proportion to Δn at high levels of injection. Such an effect may however not be observable in some cases because of other recombination processes. For example,

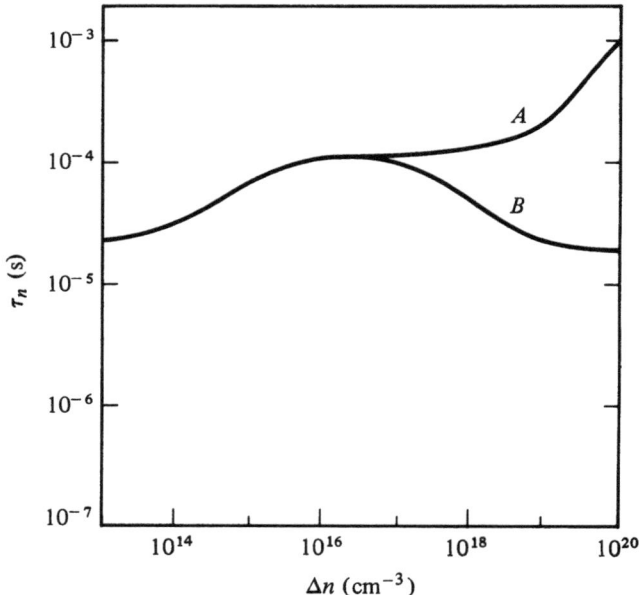

Fig. 2.4.6. Curve A shows the lifetime curve of Fig. 2.4.5(a). Curve B includes Auger recombination coefficients $T_{1g} = T_{2g} = T_3 = T_4 = 10^{-27}$ cm^6 s^{-1} and $T_{1e} = T_{2e} = 10^{-24}$ cm^6 s^{-1}. An effect of Auger recombination is to reduce the minority-carrier lifetime at high injections.

if the Auger coefficients in eqs. (2.4.22) and (2.4.23) became important the lifetime may decrease and then become constant. This has been shown in Fig. 2.4.6.

Note that additional experimental studies on the dependence of τ_n on Δn exist and tend to give somewhat conflicting results [2.4.27]–[2.4.36].

2.4.5 Generation lifetime

We add some remarks about the rate of generation of thermal carriers in the steady state. The condition corresponds typically to a reverse biased p–n junction and the resulting current is called the reverse saturation current. In the absence of electrons and holes ($n = p = 0$), eq. (2.4.52) gives the rate of generation per unit volume

$$G = -(u)_{n=p=0} = \frac{(1 + n_e c_{ne} t_n + c_{ne}/c_{ng}) c_{ng} c_p N_T n_i^2}{(1 + n_e c_{ne} t_n)(n_g c_{ng} + p_g c_p) + (t_n/t_n')(n_e c_{ne} + p_g c_p)}$$

$$= \frac{N_t e_{n\,\mathrm{eff}} e_p}{e_{n\,\mathrm{eff}} + e_p} \qquad (2.4.64)$$

where $e_{n\,\mathrm{eff}}$ has been defined in eq. (2.4.42), p. 154.

A generation lifetime [2.4.37], [2.4.38] is an average time needed to produce an additional minority carrier in a depletion region. It can be defined by

$$\tau_g \equiv \frac{n_i}{G} = \frac{n_i}{N_t}\left[\frac{1}{e_p} + \frac{1}{e_{n\,\text{eff}}}\right] \tag{2.4.65}$$

(A minority-carrier recombination lifetime is in contrast an average time a carrier can 'live' in a material.) In the steady state the generation rates e_p and $e_{n\,\text{eff}}$ form two paths *in series*, and the slower of the two decides the rate of generation. This is in contrast to the time constant involved in the transient emission experiment, τ_{2e} (eq. (2.4.41)), which is due to a *parallel* combination of e_p and $e_{n\,\text{eff}}$. Thus, measurements of τ_g and τ_{2e} give complementary information. (This is not the case with τ_{n1} (eq. (2.4.56)) and τ_{2c} (eq. (2.4.45)). They are related by

$$c_{n\,\text{eff}} = \frac{1}{\Delta n \tau_{2c}} = \frac{1}{N_t \tau_{n1}}\Big)$$

2.4.6 Capture and emission coefficients in general

The alternative notation, foreshadowed on p. 118, involving capture and emission coefficients, has been extensively used in this section. In general this notation arises from the convention sometimes made to write for the electron capture rates by $(r-1)$-electron centers and for the electron emission rate from r-electron centers, both per unit volume,

$$c_n(r-\tfrac{1}{2})\,nv_{r-1}, \quad e_n(r-\tfrac{1}{2})\,v_r$$

Also for the hole capture rate by r-electron centers and for the hole emission rate by $(r-1)$-electron centers one puts

$$c_p(r-\tfrac{1}{2})\,pv_r, \quad e_p(r-\tfrac{1}{2})\,v_{r-1}$$

Note that dimensionally

$$[c_n(r-\tfrac{1}{2})] = [c_p(r-\tfrac{1}{2})] = [L^3 T^{-1}], \quad [e_n(r-\tfrac{1}{2})] = [e_p(r-\tfrac{1}{2})] = [T^{-1}] \tag{2.4.66}$$

Comparison with eqs. (2.2.12) then yields the interpretation

$$c_n(r-\tfrac{1}{2}) = G(r-\tfrac{1}{2}), \quad e_n(r-\tfrac{1}{2}) = G(r-\tfrac{1}{2})\,n(r-\tfrac{1}{2}) \tag{2.4.67}$$

$$c_p(r-\tfrac{1}{2}) = H(r-\tfrac{1}{2}), \quad e_p(r-\tfrac{1}{2}) = H(r-\tfrac{1}{2})\,p(r-\tfrac{1}{2}) \tag{2.4.68}$$

where $n(r-\tfrac{1}{2})$, $p(r-\tfrac{1}{2})$ are concentrations which were defined in terms of partition functions in eqs. (2.2.13). Thus in the present development it may seem that the use of c_n, c_p, e_n, e_p represents an unnecessary multiplication of symbols. However, in the usual case $r = 1$ it is quite expeditious and widely used.

Note from eqs. (2.4.67) and (2.4.68) that

$$\frac{e_n(r-\frac{1}{2})}{c_n(r-\frac{1}{2})} = n(r-\tfrac{1}{2}) = \frac{Z_{r-1}}{Z_r} n\, e^{-\gamma_e}$$

$$\frac{e_p(r-\frac{1}{2})}{c_p(r-\frac{1}{2})} = p(r-\tfrac{1}{2}) = \frac{Z_r}{Z_{r-1}} p\, e^{\gamma_h} \tag{2.4.69}$$

and also from eqs. (2.2.13)

$$\frac{e_n(r-\frac{1}{2})\, e_p(r-\frac{1}{2})}{c_n(r-\frac{1}{2})\, c_p(r-\frac{1}{2})} = n(r-\tfrac{1}{2})\, p(r-\tfrac{1}{2}) = np \exp(\gamma_h - \gamma_e) = n_\theta^2 \tag{2.4.70}$$

This notation, and the results (2.4.69) and (2.4.70), do not appear in the early Shockley papers, e.g. [2.4.39], [2.4.40], but make their appearance first in 1957 ([2.4.41], equation (2.10)) and 1958 [2.4.42] for the case of thermal equilibrium. We see that the above results hold also away from equilibrium, provided quasi-Fermi levels F_h, F_e exist for holes and electrons [2.4.43].

The identification (2.4.67) and (2.4.68) also shows that one can take the ratio of single electron/hole transition terms (the usual interpretation of the e's and the c's) or, more generally, one can take the ratio of terms such as $G(r-\frac{1}{2})$, $H(r-\frac{1}{2})$ which include effects of Auger transitions, and still eqs. (2.4.69) and (2.4.70) hold [2.4.43].

If one puts $Z_{r-1}/Z_r = \exp[\eta(r-\frac{1}{2})]$, where $E(r-\frac{1}{2}) \equiv kT\eta(r-\frac{1}{2})$ is an effective energy level, then one has from eqs. (2.2.13) and (2.4.69) for non-degeneracy

$$e_n(r-\tfrac{1}{2}) = c_n(r-\tfrac{1}{2}) N_c \exp[\eta(r-\tfrac{1}{2}) - \eta_c] \tag{2.4.71}$$

$$e_p(r-\tfrac{1}{2}) = c_p(r-\tfrac{1}{2}) N_v \exp[\eta_v - \eta(r-\tfrac{1}{2})] \tag{2.4.72}$$

Interpreted as Gibbs free energies divided by kT, i.e. envisaging experiments at constant pressure and temperature, the exponent in eq. (2.4.71) can be written as

$$-\frac{1}{kT} G_n(r-\tfrac{1}{2}) = -\frac{1}{kT}[H_n(r-\tfrac{1}{2}) - TS_n(r-\tfrac{1}{2})]$$

where H_n is the enthalpy and S_n the entropy. They both refer to changes in these quantities resulting from electron excitation from the trap to the conduction band due to lattice relaxation, changes in degeneracies, etc. One finds

$$e_n(r-\tfrac{1}{2}) = c_n(r-\tfrac{1}{2}) N_c \exp\left(\frac{\Delta S_n(r-\frac{1}{2})}{k}\right) \exp\left(-\frac{\Delta H_n(r-\frac{1}{2})}{kT}\right) \tag{2.4.73}$$

$$e_p(r-\tfrac{1}{2}) = c_p(r-\tfrac{1}{2}) N_v \exp\left(\frac{\Delta S_p(r-\frac{1}{2})}{k}\right) \exp\left(-\frac{\Delta H_p(r-\frac{1}{2})}{kT}\right) \tag{2.4.74}$$

The slope of an 'Arrhenius plot' of the logarithm of $e_n(r-\frac{1}{2})/N_c c_n(r-\frac{1}{2})$ against $1/T$ is thus the appropriate enthalpy (note that N_c and c_n can depend on

Table 2.4.6. *Some enthalpies and entropy factors for impurities*

Material	T (K)	σ (10^{-16} cm²)	ΔH (eV)	$X = e^{\Delta S/k}$	Experimental method	Reference
Gold donor in *p*-type Si	1243		0.35	28	Rutherford back scattering	[2.4.46]
	250–300	110	0.340	36	pulsed junction capacitance	[2.4.47]
	250–300	55	0.353	20	DLTS	[2.4.48]
Gold acceptor in *p*-type Si	250	0.69	0.553	48	DLTS	[2.4.49]

temperature as well and hence have to be divided out). The equations show us again the entropy factor of eq. (1.7.34), this time applied to the interaction between a band and a defect. Equations (2.4.73) and (2.4.74) go back to [2.4.44]. The discussion given above shows rather effortlessly that the results given there hold also for general charge conditions and with the inclusion of Auger effects [2.4.45]. Table 2.4.6 gives some numerical estimates of H and S. Many other systems have been investigated, e.g. sulphur-related defects in Si [2.4.19] and neutral nickel acceptors in GaP [2.4.50].

It is hard to obtain separately degeneracy factors and true energy levels from such data as they represent basically a measurement of an effective energy level, i.e. a partition function ratio, and this incorporates both these quantities. This agrees with the applications of eq. (2.4.73) to interface traps [2.4.51], [2.4.52]. These have also been studied (via eq. (2.4.73)) with field effect transistors in [2.4.53], whose authors gave from their experiments 'the first direct and unambiguous demonstration of the necessity of incorporating entropy factors when relating capture and emission rates at interface states'.

If one incorporates the empirical proposal [2.4.54]

$$\Delta G = \alpha - \frac{\beta T^2}{T + \gamma} \tag{2.4.75}$$

where α, β, γ are constants, for the temperature-dependence of ΔG, then one finds upon using eqs. (1.7.35) and (1.7.36) [2.4.55]

$$\Delta S = \frac{\beta T(T + 2\gamma)}{(T + \gamma)^2} \tag{2.4.76}$$

$$\Delta H = \alpha + \frac{\beta \gamma T^2}{(T + \gamma)^2} \tag{2.4.77}$$

Hence eq. (2.4.73) gives (we write $r = 1$ for simplicity)

$$n(\tfrac{1}{2}) = \frac{e_n(\tfrac{1}{2})}{c_n(\tfrac{1}{2})} = N_c \exp\left[\frac{\beta T}{k(T+\gamma)}\right]\exp\left[-\frac{\alpha}{kT}\right] \qquad (2.4.78)$$

The first exponential derives from both entropy and enthalpy factors.

As an application of these ideas consider Cr-doped GaAs. This has been used to produce semi-insulating material which is difficult to manufacture by purification owing to residual defects. For device modeling and in order to understand the physics of the material the intrinsic carrier concentration is needed as a function of temperature. Of course this can be done from eq. (1.12.10) if one knows the effective masses and the energy gap:

$$n_i = 2\left[\frac{2\pi(m_c m_v)^{\frac{1}{2}}}{h^2}kT\right]^{\frac{3}{2}}\exp\left[-\frac{E_G}{2kT}\right]$$

But one must include effects of temperature on effective masses, the effect of higher bands and nonparabolicity, and the temperature-dependence of E_G. In any case, a procedure based on recombination phenomena, utilizing eq. (2.4.70), can also be pursued. It has been applied with reference to Cr acceptors (settling substitutionally on Ga-sites) and it has led to an inferred $n_i(T)$ function in good agreement with the procedure using the effective mass–energy gap approach [2.4.55]. As a result of a careful survey, Blakemore suggests semi-empirically

$$e_n(\tfrac{1}{2}) = 7.9 \times 10^4 T^2 \exp(-0.883/kT) \text{ s}^{-1}$$

$$c_n(\tfrac{1}{2}) = 3.1 \times 10^{-11} T^{\frac{1}{2}} \exp(-0.117/kT) \text{ cm}^3 \text{ s}^{-1}$$

$$e_p(\tfrac{1}{2}) = 4.25 \times 10^6 T^2 \exp(-0.858/kT) \text{ s}^{-1}$$

$$c_p(\tfrac{1}{2}) = 9.8 \times 10^{-11} T^{\frac{1}{2}} \exp(-0.020/kT) \text{ cm}^3 \text{ s}^{-1}$$

for a range of temperatures $T = 300\text{–}475$ K. Hence using eq. (2.4.70) one has, near equilibrium,

$$n(\tfrac{1}{2}) = 2.55 \times 10^{15} T^{\frac{3}{2}} \exp(-0.766/kT) \text{ cm}^{-3}$$

$$p(\tfrac{1}{2}) = 4.34 \times 10^{16} T^{\frac{3}{2}} \exp(-0.838/kT) \text{ cm}^{-3} \qquad (2.4.79)$$

$$n_i = [n(\tfrac{1}{2})p(\tfrac{1}{2})]^{\frac{1}{2}} = 1.05 \times 10^{16} T^{\frac{3}{2}} \exp(-0.802/kT) \text{ cm}^{-3}$$

One sees that the temperature-dependence of the important intrinsic carrier concentration of a semiconductor can be derived by such a procedure. The result for $n(\tfrac{1}{2})$ is probably more reliable than eq. (2.4.78). The capture and emission rates are seen in (2.4.79) to be thermally activated.

If the electron–ion interaction is weak, one can subdivide the exponent of the partition function $\exp(-G/kT)$ of section 1.7.3, and hence also S and H, into one component due to the electronic system and one component due to the lattice. In particular, the electronic contribution to the entropy depends on the small term arising from the change in degeneracy (from g_i to g_f, say) of the defect state: $\Delta S_e = k \ln (g_f/g_i)$. The remaining term, ΔS_l, say, is then due to the lattice relaxation and the change in the vibrational frequency. For strong coupling this subdivision is not possible.

2.5 Field dependence of capture and emission

2.5.1 Introduction

The kinetics of trapping and detrapping by impurities in solid-state devices has always to be formulated in the presence of an electric field. The effect of the field is sometimes neglected but in fact it facilitates electron emission from traps and it is often important, e.g. [2.5.1], [2.5.2]. It is for this reason that a great deal of work has been done in this area. It started broadly in 1938 when Frenkel [2.5.3] put forward his paper on what has become known as the Poole–Frenkel effect, but accurate comparison with experiment was difficult. For a brief review of various models see [2.5.4], where earlier references can also be found. The results of these calculations can be given by a factor, usually denoted by χ. While Frenkel considered a simple one-dimensional situation (eq. (2.5.27), below), an approximate three-dimensional calculation of the effect was made thirty years later (eq. (2.5.5), below) [2.5.5], [2.5.6]. These papers gave a fairly simple result, significantly different from Frenkel's, and [2.5.6] has been widely cited. Nonetheless, experiments sometimes yielded good agreement with Frenkel's formula even though the three-dimensional argument should apply [2.5.7], [2.5.8]. We give in section 2.5.4 details of a recent improved argument [2.5.9].

In section 2.5.2 we present a simplified derivation of the key kinetic results (2.4.42) and (2.4.46), and in section 2.5.3 we give an introduction to the theory of the Poole–Frenkel effect. We then give in section 2.5.4 an improved three-dimensional calculation of the factor χ. It is still reasonably elementary, but it takes into account that after an emitted electron has entered the conduction band the density of states which it encounters there is a function of position (see Fig. 2.5.3) until it is scattered. The resulting formula involves the mean free path λ, together with the variables occurring in the earlier calculation (eq. (2.5.23), below). A characteristic of the result is that for high electric fields one obtains a Frenkel-type formula as an approximation. This could explain the good agreements which have occasionally been obtained with the Frenkel formula when one would have expected the Jonscher–Hartke three-dimensional formula to perform better.

A spatially averaged density of states is needed for the analysis in order to obtain an effective density of states for the emitted electron (section 2.5.4).

The exposition is confined to the Poole–Frenkel effect and additional mechanisms such as tunneling and phonon-assisted tunneling are neglected. They are in any case found to be unimportant except at low temperature [2.5.10].

2.5.2 A simple intuitive treatment of the steady-state TCR model (with application to electric field effects)

In replacing the spectrum of excited states by a single, effective level, one again arrives at Fig. 2.4.1 as representing the truncated cascade recombination (TCR) model. The sticking probability for electrons in this excited state is

$$P_{e} = t_{n}^{-1}/[e_{ne} + t_{n}^{-1}] = (1 + e_{ne} t_{n})^{-1} \tag{2.5.1}$$

where the notation of the figure has been used. This leads to an effective capture coefficient

$$c_{n\,\mathrm{eff}} = c_{ng} + c_{ne} P_{e} = c_{ng} + c_{ne}/(1 + e_{ne} t_{n}) \tag{2.5.2}$$

The recombination rate per unit volume is

$$c_{n\,\mathrm{eff}}\, n v_{0} - e_{n\,\mathrm{eff}}\, v_{g} \tag{2.5.3}$$

where v_{0} is the concentration of centers without captured electrons and v_{g} the concentration of centers with a captured electron in the ground state. Detailed balance yields for the effective emission coefficient

$$e_{n\,\mathrm{eff}} = \left(\frac{n v_{0}}{v_{g}}\right)_{\mathrm{eq}} c_{n\,\mathrm{eff}} \equiv n_{g} c_{n\,\mathrm{eff}} = n_{g}\left(c_{ng} + \frac{c_{ne}}{1 + e_{ne} t_{n}}\right) \tag{2.5.4}$$

Here n_{g} corresponds to the concentration, usually denoted by n_{1} in the SRH statistics, which gives the conduction band electron concentration when the Fermi level is at ground level of the trap. The term $n_{g} c_{ng}$ is the emission rate e_{ng} directly from the ground state.

In a three-dimensional Poole–Frenkel effect the trap activation energy is reduced by $\ln \chi$ where, using for the moment the Jonscher–Hartke result of section 2.5.3,

$$\chi = \tfrac{1}{2} + \xi^{-2}[1 + (\xi - 1)\,e^{\xi}], \quad \xi \equiv \frac{2}{kT}\left(\frac{z q^{3} \mathscr{E}}{\varepsilon}\right)^{\frac{1}{2}} \tag{2.5.5}$$

Here $\mathscr{E}$ is the electric field, ε the dielectric permittivity of the semiconductor, and zq is the charge on the center which is here assumed positive. The factor χ has to be applied to the emission coefficients e_{ng}, e_{ne}, but not to the capture coefficients

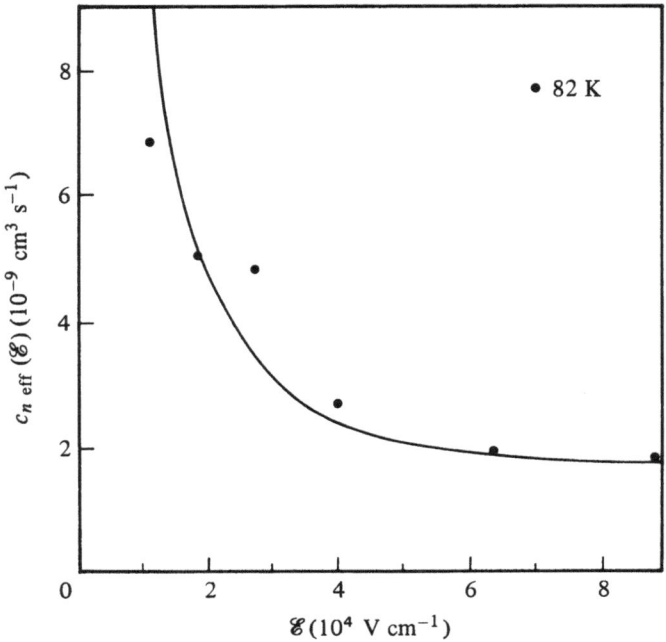

Fig. 2.5.1. The effect of an electric field on $c_{n\,\text{eff}}$ using eq. (2.5.6) and the data of Table 2.5.1. The experimental points apply to S in Si at 82 K [2.5.12].

c_{ng}, c_{ne}. Hence one sees two interesting effects. First, from eq. (2.5.2) one observes that instead of a field-independent sticking probability P_e and capture coefficient c_{ng}, TCR gives quantities which *decrease* with field:

$$P_e(\mathscr{E}) = [1 + c_{ne}\, t_n\, n_e\, \chi]^{-1}$$
$$c_{n\,\text{eff}}(\mathscr{E}) = c_{ng} + c_{ne}/(1 + c_{ne}\, t_n\, n_e\, \chi) \tag{2.5.6}$$

Here $n_e \equiv e_{ne}/c_{ne}$ refers to capture and emission from the effective excited state, in analogy with n_g. Secondly, from eq. (2.5.4), instead of a field-dependence $e_n = n_g c_{ng} \chi$, a *less strong* field-dependence

$$e_{n\,\text{eff}}(\mathscr{E}) = [c_{ng} + c_{ne}/(1 + n_e c_{ne} t_n \chi)] n_g \chi \tag{2.5.7}$$

is found, and arises from $n_e \to n_e \chi$ and $n_g \to n_g \chi$. In fact, for low fields and high temperatures $e_{n\,\text{eff}}$ has only a weak dependence. The reason is that for $n_e c_{ne} t_n \gg 1$, which is the appropriate approximation,

$$e_{n\,\text{eff}}(\mathscr{E}) \simeq n_g c_{ng} \chi + n_g/n_e t_n \tag{2.5.8}$$

and the second term is liable to dominate. In eqs. (2.5.4) and (2.5.6) we have recovered by an intuitive steady-state argument (2.4.42) and (2.4.46) which had resulted in section 2.4.4 from a treatment of time-dependent effects.

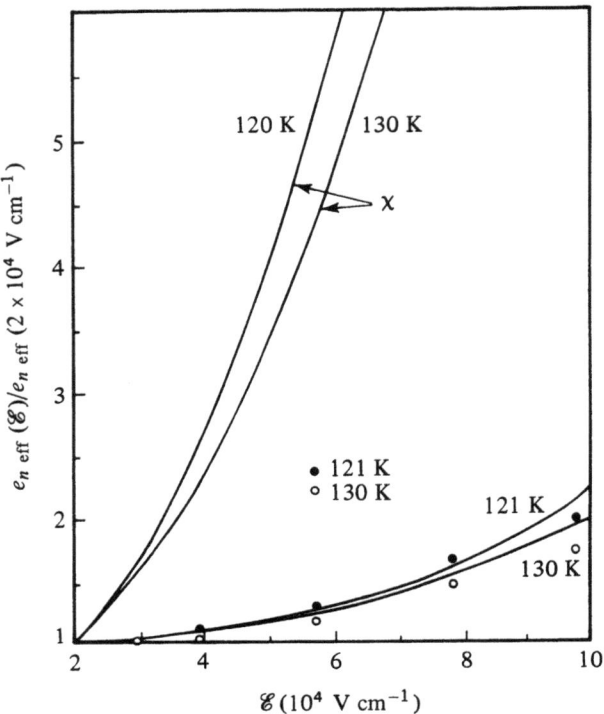

Fig. 2.5.2. The effect of an electric field $\mathscr{E}$ (a) on the Poole–Frenkel factor given by eq. (2.5.5) and (b) on $e_{n\,\mathrm{eff}}$ using eq. (2.5.7) and the data of Table 2.5.1. The experimental points apply to S in Si at 121 K and 130 K [2.5.12]. The value of $e_{n\,\mathrm{eff}}(\mathscr{E})$ is normalized by its value at $\mathscr{E} = 2 \times 10^4$ V cm^{-1}.

In the above the *capture* coefficients c_{ng}, c_{ne} have been treated as independent of electric field. There are at least two corrections to this. First, if the electric field is large enough for the potential maximum to drop below the energy level of the excited state, then $e_{ne}\,\chi = n_e\,c_{ne}\,\chi$ becomes large and one finds eq. (2.5.8) again. Since c_{ne} has now dropped out, our main equations which treat it as field-independent can still be used. Secondly, the trapping coefficients may decrease by reason of hot electron effects [2.5.11], but this matter is not investigated here.

Fig. 2.5.1 shows a comparison of eq. (2.5.6), and Fig. 2.5.2 shows a comparison of eqs. (2.5.7), with experiments on S centers in Si ([2.5.12], figures 10 and 7, respectively). The data used are given in Table 2.5.1. In Fig. 2.5.1 we have added a background capture $c_{n0} = 1.6 \times 10^{-9}$ cm^3 s^{-1}, independent of electric field, which may be due to neutral centers. Such a correction is needed to obtain agreement with the experimental results, and also with the observed capture cross section for S in Si [2.5.13]. At the larger fields capture into excited states becomes insignificant and direct capture into the ground state, usually neglected in cascade models,

Table 2.5.1. *Values of parameters used in Figs.* 2.5.1 *and* 2.5.2

v_{th} (= thermal velocity)	$10^7 (T/300)^{\frac{1}{2}}$ cm s^{-1}	
$\sigma_{ne} \equiv c_{ne}/v_{\text{th}}$	10^{-13} cm^2 (a)	
$\sigma_{ng} \equiv c_{ng}/v_{\text{th}}$	0.25×10^{-16} cm^2 (a)	
$n_e c_{ne} t_n$	$0.0188 T^2 \exp(-0.017/kT)$ (b)	

(a) Chosen to fit the experiments shown in Figs. 2.5.1 and 2.5.2.

(b) Inferred from [2.5.13]; kT is expressed in electron volts.

becomes important. Fig. 2.5.2 for the emission coefficient also shows the pure Poole–Frenkel factor χ to illustrate the drop in the electric field dependence brought about by passing from e_n to $e_{n\,\text{eff}}$ when TCR is envisaged. However, even the present theory will yield an emission coefficient proportional to χ at sufficiently low temperatures. In that case $n_e c_{ne} t_n \ll 1$ and eq. (2.5.7) gives rise to

$$e_{n\,\text{eff}}(\mathscr{E}) = (c_{ng} + c_{ne}) n_g \chi$$

This is consistent with the approximately straight-line law connecting the logarithm of the emission rate and $\mathscr{E}^{\frac{1}{2}}$ found experimentally [2.5.7], and expected from eq. (2.5.5).

Figures 2.5.1 and 2.5.2 show that the model is capable of yielding quantitative agreement with experiment.

Note that tunneling and phonon-assisted tunneling have not been included. They may be important at low temperatures and would increase the value of χ However in Si, for example, these effects can often be neglected [2.5.10].

We now comment on the fact that the hole concentration (p) does not appear explicitly in the simple formulae given in this section. The reason is that the concentrations v_0 and v_g in formula (2.5.3) are regarded as given. A fuller analysis must consider n and p as given by conditions of doping and excitation, while the concentrations v_0, v_g, v_e are derived in terms of n and p. They are subject to $v_0 + v_g + v_e = N_T$, the total concentration of centers. A simple derivation is given in the appendix to this section.

In section 2.5.4 the theory of χ will be improved by taking into account that the emitted electron moves in the conduction band in a region in which the density of states depends on position, and also that this motion is terminated by a scattering event. The comparison with experiment in Figs. 2.5.1 and 2.5.2 which use the older result (2.5.5) is not significantly changed (provided σ_{ng} is lowered to 10^{-17} cm^2, c_{n0} is slightly raised to 1.8×10^{-9} cm^3 s^{-1}, and $n_e c_{ne} t_n = 0.0188 T^2 \exp(-0.028/kT)$). The mean free path is required in this model and was taken as $10^{-3}/T(K)$ cm for Si.

Appendix to section 2.5.2

The results (2.4.42) and (2.4.46) for the effective emission rate $e_{n\,\text{eff}}$ and for the effective capture rate coefficient $c_{n\,\text{eff}}$ have been obtained from an analysis of the transient situation (section 2.4.3) and by an intuitive argument (section 2.5.2). Some readers may feel the need for a firmer foundation of these formulae based on steady-state kinetics. This is supplied in the present appendix.

Using Fig. 2.4.1, let us equate the transition rate R_{gp} from the ground state of the center to the valence band and the transition rate into the ground state $R_{ng} + R_{eg}$. This ensures a steady state for the ground state with transition rate R per unit volume:

$$(v_g p - v_0 p_g)\, c_p = (v_0 n - v_g n_g)\, c_{ng} + \frac{v_e}{t_n} - \frac{v_g}{t'_n} \equiv R$$

This gives

$$v_e = t_n \{R - v_0 n c_{ng} + v_g n_g c_{ng} + v_g/t'_n\}$$

The transition rate out of the conduction band must also be R. Hence

$$R = R_{ng} + R_{ne} = (v_0 n - v_g n_g)\, c_{ng} + (v_0 n - v_e n_e)\, c_{ne}$$

which simply repeats eqs. (2.4.20). Substituting for v_e,

$$R = v_0 n c_{n\,\text{eff}} - v_g e_{n\,\text{eff}}$$

where the effective coefficients are given by eqs. (2.4.42) and (2.4.46).

2.5.3 Schottky effect and Poole–Frenkel effect

If a charge q is in vacuum at a distance x from a plane infinite dielectric of dielectric constant ε then it is attracted to it as if a charge $-[(\varepsilon - 1)/(\varepsilon + 1)]\, q$ were located at a distance x behind the dielectric surface. If ε is large, as in the case of a metal, the image charge is just $-q$ and the image force of attraction is $q^2/(2x)^2$. Between a metal and an insulator the attractive force is $q^2/4\varepsilon x^2$, where ε is the high-frequency dielectric constant of the insulator. In the presence of an electric field $\mathscr{E}$ the electronic potential energy has the form (Fig. 2.5.3a)

$$V(x) = E_{c0} - e\mathscr{E}x - e^2/4\varepsilon x \quad (x > 0)$$

where E_{c0} is the bottom of the conduction band at $x = 0$ in the absence of the image force. The latter can act only above a certain minimum distance of the order of a few Ångstrom. Below such distances, a proper atomic, rather than a macroscopic, treatment is needed.

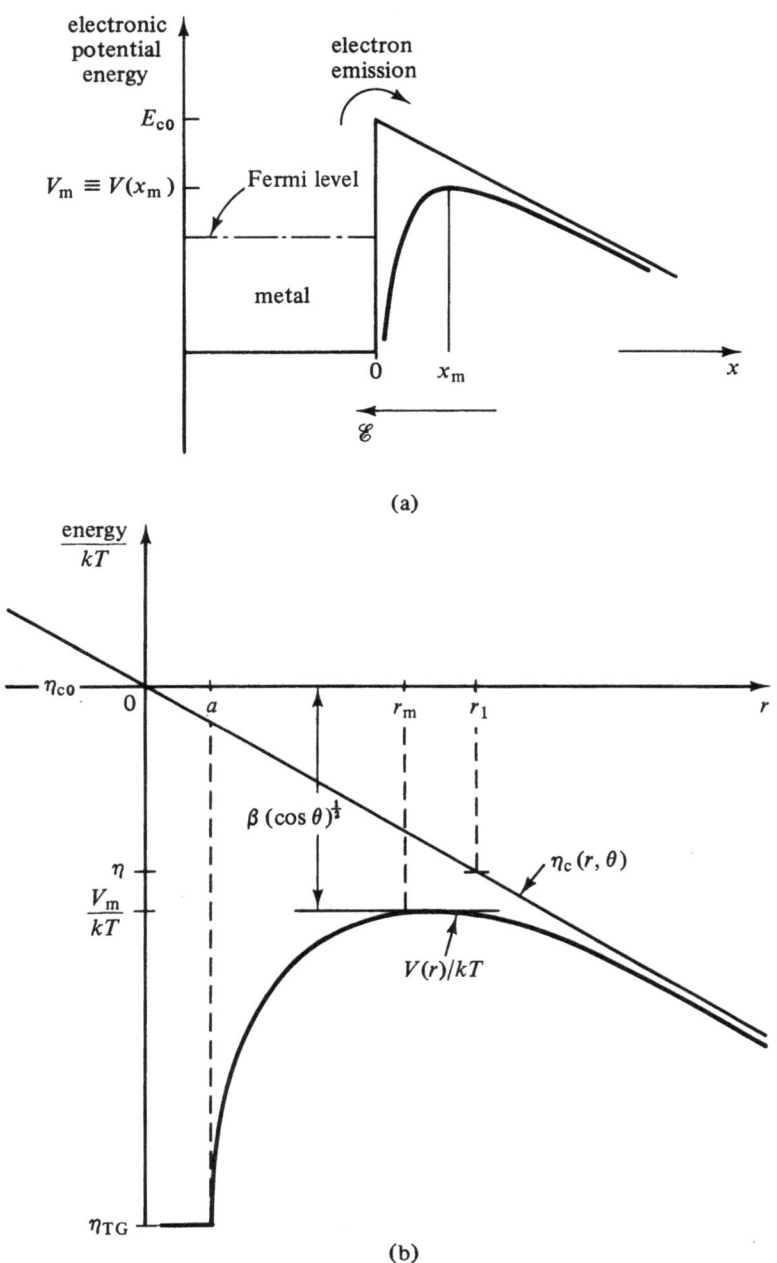

Fig. 2.5.3. (a) Electron emission from a metal into a semiconductor. In the absence of the electric field $\mathscr{E}$, the bottom of the conduction band of the semiconductor is at energy E_{c0}. (b) Electron energies for emission from a positively charged trap. The energy in the final state of an emitted electron is ηkT and it can lie anywhere above V_m which is itself determined by the electric field $\mathscr{E}$ and the value of $\cos\theta$.

On the Sommerfeld model of a metal, electron emission from a metal can take place because electrons can overcome the surface barrier at energy E_{c0}. The electric field reduces the barrier so that the maximum of $V(x)$ occurs at

$$x_m = (e/4\varepsilon\mathcal{E})^{\frac{1}{2}}$$

and has the value

$$V(x_m) = E_{c0} - (e^3\mathcal{E}/\varepsilon)^{\frac{1}{2}}$$

The lowering of the potential energy barrier by the field from E_{c0} to $V(x_m)$ is sometimes called the Schottky effect. Using a Boltzmann distribution, the thermionic emission current I should thus increase exponentially as the square root of the electric field. The $\ln I$ versus $\mathcal{E}^{\frac{1}{2}}$ plots are called Schottky lines, and a straight line relation is often found. Typical values are $\mathcal{E} = 100$ V cm^{-1}, $x_m = 2 \times 10^{-5}$ cm so that a macroscopic treatment is reasonable. For much higher fields the potential energy $V_m(x)$ may be reduced to such an extent that the barrier becomes relatively unimportant.

Such straight lines have been analyzed for impurities in silicon at low temperature [2.5.7]. The agreement with theory is however not always good and depends on the details of the experiment. For Au in Si see figure 10 of [2.5.14]; for S in Si see figure 12 of [2.5.12]. For related studies see [2.5.15]–[2.5.17].

The Schottky paper dates from 1923. In 1938 J. Frenkel pointed out that a positively charged defect which has lost its electrons would have its Coulomb potential energy barrier $e^2/\varepsilon x$ lowered by an electric field 'by a mechanism similar to that of the Schottky effect in the thermoelectronic emission from metals'. In this case, dividing now by kT (Fig. 2.5.3b with $\theta = 0$)

$$\frac{V(x)}{kT} = \eta_{c0} - \frac{e\mathcal{E}x}{kT} - \frac{e^2}{\varepsilon kTx}$$

$$\frac{V_m}{kT} = \eta_{c0} - 2\left(\frac{e^3\mathcal{E}}{\varepsilon}\right)^{\frac{1}{2}}\frac{1}{kT}$$

$$x_m = (e/\varepsilon\mathcal{E})^{\frac{1}{2}} \quad \text{(twice the previous value!)}$$

Assuming a Boltzmann distribution, the emission rate will be enhanced by a factor

$$\chi_F \equiv \exp\left(\eta_{c0} - \frac{V_m}{kT}\right) = \exp\left[2\left(\frac{e^3\mathcal{E}}{\varepsilon}\right)^{\frac{1}{2}}\frac{1}{kT}\right]$$

In the exponent we have twice the quantity which appears in the Schottky effect. But the electric field- and temperature-dependences of the exponent are the same in both cases. Frenkel in this way amended what was at the time called Poole's law.

The Poole–Frenkel effect has to be discussed also by considering the three-dimensional situation, and this was done by Jonscher [2.5.5] and Hartke [2.5.6]. If θ $(-\pi/2 \leqslant \theta \leqslant \pi/2)$ is the angle between the direction of propagation of the electron and the electric field we now have (Fig. 2.5.3b)

$$\frac{V(r)}{kT} = \eta_{c0} - \frac{e\mathscr{E}r\cos\theta}{kT} - \frac{e^2}{\varepsilon kTr}$$

$$\frac{V_m}{kT} = \eta_{c0} - 2\left(\frac{e^3\mathscr{E}\cos\theta}{\varepsilon}\right)^{\frac{1}{2}}\frac{1}{kT}$$

$$r_m = \left(\frac{e}{\varepsilon\mathscr{E}\cos\theta}\right)^{\frac{1}{2}}$$

The lowering of the potential energy maximum is seen to vanish for an electron whose attempt to escape from the defect proceeds at a right angle to the electric field. The electric field has a negligible effect if the electron attempts to escape with a component in the direction of the electric field $(\pi/2 < \theta < \pi)$. If the frequency of the attempt to escape by the electron is $\nu/4\pi$ per unit solid angle then the escape rate is determined by an energy barrier of height E_T for $\pi/2 \leqslant \theta < \pi$. Putting $\eta_T \equiv E_T/kT$ this contributes to the escape rate

$$\frac{\nu}{4\pi}\exp\left(-\eta_T\right)\int_0^{2\pi}d\varphi\int_{\pi/2}^{\pi}\sin\theta\,d\theta = \frac{\nu}{2}\exp\left(-\eta_T\right)$$

It is obtained by an integration over the appropriate solid angle. The second contribution to the solid angle integral arises from $0 \leqslant \theta < \pi/2$ and is with $\beta \equiv (2/kT)(e^3\mathscr{E}/\varepsilon)^{\frac{1}{2}}$

$$\frac{\nu}{4\pi}\exp\left(-\eta_T\right)\int_0^{2\pi}d\varphi\int_0^{\pi/2}\cdot\sin\theta\exp\left[\beta(\cos\theta)^{\frac{1}{2}}\right]d\theta$$

Putting $y^2 \equiv \cos\theta$, this is

$$\frac{\nu}{2}\exp\left(-\eta_T\right)\int_0^1 2y\exp\left(\beta y\right)dy = \nu\beta^{-2}\exp\left(-\eta_T\right)\left[(\beta-1)\exp\beta+1\right]$$

The zero-field release rate is

$$\frac{1}{\tau(0)} = \frac{\nu}{4\pi}\exp\left(-\eta_T\right)\int_0^{2\pi}d\varphi\int_0^{\pi}\sin\theta\,d\theta = \nu\exp\left(-\eta_T\right)$$

Combining these results, one finds [2.5.6]

$$\chi_H \equiv \frac{\tau(0)}{\tau(\mathscr{E})} = \beta^{-2}[1+(\beta-1)\exp\beta]+\tfrac{1}{2}$$

At high fields this is

$$\beta^{-1}(1-\beta^{-1})\exp\beta$$

and is experimentally distinguishable from Frenkel's law

$$\chi_F = \exp\beta$$

2.5.4 The field-dependence of the emission rate [2.5.9]

We shall assume a hydrogen-like donor. The electron ground state is then spherically symmetrical and has no first order Stark effect. The energy, E_{TG} say, to lift the electron from this state to the bottom of the conduction band at the location $r = 0$ of the donor is thus effectively independent of the electric field $\mathscr{E}$ and of the angle θ between a general position r and $\mathscr{E}$. The potential energy $V(r)$ of the electron and the conduction band edge $E_c(r)$ in the presence of the field are given by ($\eta_c \equiv E_c/kT$, etc.)

$$\frac{V(r)}{kT} = \eta_{c0} - \frac{l}{r} - \frac{r}{L}\cos\theta \quad (r > a) \tag{2.5.9}$$

$$\eta_c(r, \theta) = \eta_{c0} - (r/L)\cos\theta \tag{2.5.10}$$

where η_{c0} is the value of $\eta_c(r, \theta)$ at $r = 0$ and

$$l \equiv \frac{ze^2}{\varepsilon kT}, \quad L \equiv \frac{kT}{e\mathscr{E}}$$

Here a is a measure of the radius of the ground state wavefunction and ze is the charge on the donor. The maximum V_m of V occurs at $r = r_m$ where (Fig. 2.5.3b)

$$\eta_m \equiv V_m/kT = \eta_{c0} - \beta(\cos\theta)^{\frac{1}{2}}, \quad \beta^2 \equiv 4l/L \tag{2.5.11}$$

$$r_m = (lL/\cos\theta)^{\frac{1}{2}} \tag{2.5.12}$$

Equation (2.5.10) incorporates the common assumption made in treatments of the Poole–Frenkel effect that the conduction band of a semiconductor is affected by the uniform electric field applied externally, but not by the potential well which represents an ionized donor.

In a one-dimensional argument $\theta = 0$ and the potential energy maximum is lowered by the electric field by an amount given by

$$\eta_{c0} - \eta_m = \beta = \frac{1}{kT}\left(\frac{4ze^3}{\varepsilon}\right)^{\frac{1}{2}}\mathscr{E}^{\frac{1}{2}}$$

Assuming a Boltzmann distribution, the emission rate from the trap is accordingly enhanced by a factor

$$e_{ng}(\mathscr{E})/e_{ng}(0) = \exp \beta$$

This is essentially Frenkel's result (p. 176).

The conduction band electron concentration when the Fermi level is at the donor electron ground state energy will be again denoted by

$$n_g \equiv N_c \exp(\eta_{TG} - \eta_{c0}), \quad \exp(\eta_{TG}) = Z_0/Z_{1g} \tag{2.5.13}$$

where the partition function definition of η_{TG} has also been given (see, for example, formulae (2.4.18)). Here Z_0 refers to the ionized donor and Z_{1g} to the donor with its electron in the ground state. The parabolic density of states per unit volume per unit energy range is [with $N_c \equiv 2(2\pi m_c kT/h^2)^{\frac{3}{2}}$]

$$N[\eta, r, \theta] = \frac{2}{\pi^{\frac{1}{2}}} (kT)^{-1} N_c [\eta - \eta_c(r, \theta)]^{\frac{1}{2}} \tag{2.5.14}$$

Its angle- and position-dependencies arise from the slope of the band edge.

Consider first electrons with a velocity component to the right in Fig. 2.5.3. Then $0 \leqslant \theta \leqslant \pi/2$ and $\cos\theta \geqslant 0$. An electron of energy $E \equiv \eta kT$ enters the conduction band at

$$r_1 = 0 \quad \text{if } \eta > \eta_{c0} \tag{2.5.15}$$

and at

$$r_1 = (\eta_{c0} - \eta) L/\cos\theta \quad \text{if } \eta < \eta_{c0} \tag{2.5.16}$$

This second condition is obtained from $\eta = \eta_c(r, \theta)$. The chance of the electron then surviving in the conduction band without being scattered in the range of distances $(r, r + dr)$ decreases exponentially with r and may be taken to be given by

$$P(r)\, dr = \frac{1}{\lambda} \exp\left(-\frac{r - r_1}{\lambda}\right) dr \tag{2.5.17}$$

Here λ is the mean free path of an electron in the conduction band. The density of states N of eq. (2.5.14) must then be replaced by an r-independent averaged density of states per unit volume per unit energy range

$$\mathscr{N} = \int_{r_1}^{\infty} NP(r)\, dr \tag{2.5.18}$$

The electron is regarded as emitted when it is in the conduction band which has been assumed to be free of the trap potential. Let C_{ng}, C'_{ng} be reaction 'constants' (of dimension $[\mathrm{L^3T^{-1}}]$), respectively, for the capture of an electron from the band at energy E into the ground state of the trap and for the emission from that state into the conduction band. Using the notation of eq. (2.5.13), detailed balance gives again eq. (2.4.6), i.e.

$$C'_{ng}(\eta) = C_{ng}(\eta)\exp\left(\eta_{\mathrm{TG}} - \eta\right) \tag{2.5.19}$$

The emission rate from the ground state of the trap in a given spatial direction (θ, φ) in the presence of the electric field is

$$e_{ng}(\mathscr{E};\theta,\varphi) = kT\int C'_{ng}(\eta,\theta)\,\mathscr{N}(\eta,\theta)[1-f_n(\eta)]\,\mathrm{d}\eta \tag{2.5.20}$$

where $f_n(\eta)$ is the occupation probability of a state of energy $E = \eta kT$. For a nondegenerate semiconductor this gives

$$e_{ng}(\mathscr{E};\theta,\varphi) = kT\int C_{ng}(\eta,\theta)\,\mathscr{N}(\eta,\theta)\exp\left(\eta_{\mathrm{TG}}-\eta\right)\mathrm{d}\eta$$

$$= \frac{n_g kT}{N_c}\iint C_{ng}(\eta,\theta)\,P(r)\,N(\eta,r,\theta)\,\mathrm{e}^{\eta_{co}-\eta}\,\mathrm{d}\eta\,\mathrm{d}r$$

where eqs. (2.5.13) and (2.5.18) have been used. Using eq. (2.5.14)

$$e_{ng}(\mathscr{E};\theta,\varphi) = \frac{2n_g}{\pi^{\frac{1}{2}}}\iint C_{ng}(\eta,\theta)\,P(r)[\eta-\eta_c(r,\theta)]^{\frac{1}{2}}\,\mathrm{e}^{\eta_{co}-\eta}\,\mathrm{d}\eta\,\mathrm{d}r \tag{2.5.21}$$

For given η the r-integration extends from r_1 to ∞. Alternatively for given r the η-integration extends from $\eta_c(r,\theta)$ or $\eta_m(\theta)$, whichever is the larger. If one assumes that the reaction 'constant' is independent of energy then no additional physical assumptions are needed now. One merely has to integrate!

For down-field emission $(\cos\theta \geqslant 0)$ and up-field emission $(\cos\theta \leqslant 0)$ in a given direction one finds with $e_{ng}(0) = n_g C_{ng}$

$$\frac{e_{ng}(\mathscr{E};\theta,\varphi)}{e_{ng}(0)} = \begin{cases} 1 + \left[\dfrac{e\mathscr{E}\lambda}{kT}\cos\theta\right]^{\frac{1}{2}}\exp\left[\dfrac{4ze^3\mathscr{E}\cos\theta}{\varepsilon(kT)^2}\right]^{\frac{1}{2}} & (\cos\theta \geqslant 0) \\[4mm] \left[1 + \dfrac{e\mathscr{E}\lambda}{kT}|\cos\theta|\right]^{-1} & (\cos\theta < 0) \end{cases} \tag{2.5.22}$$

(see the appendix to this section). The down-field emission is approximate and increases exponentially as $(\mathscr{E}\cos\theta)^{\frac{1}{2}}$, while the up-field emission decreases with $\mathscr{E}|\cos\theta|$. For emission in a plane perpendicular to the field the emission rate behaves as if no field were present.

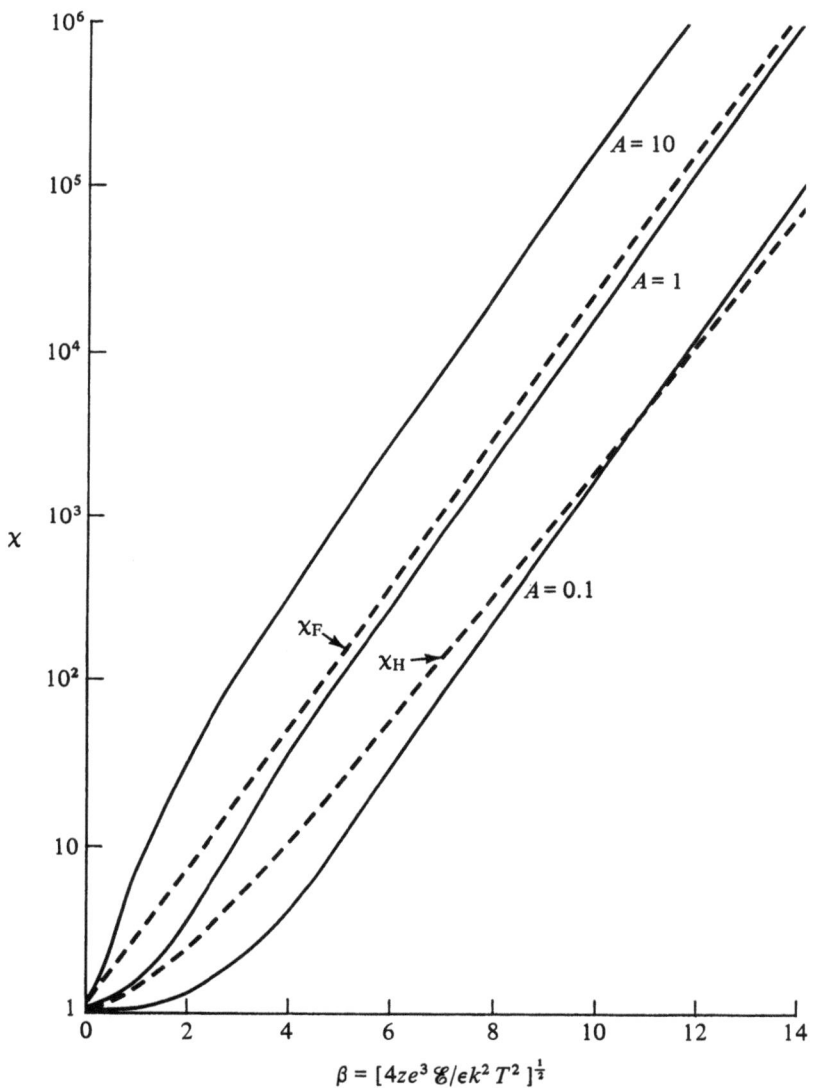

Fig. 2.5.4. Curves of $\chi \equiv e_{ng}(\mathscr{E})/e_{ng}(0)$. Solid curves are based on eq. (2.5.23) for three different values of $A = (\varepsilon k T\lambda/4ze^2)^{\frac{1}{2}}$. Also shown are the original Frenkel form $\chi_F = \exp \beta$ (one dimension) and the corresponding approximate three-dimensional form (eq. (2.5.28)).

The final step is to average over all directions:

$$e_{ng}(\mathscr{E}) = \frac{1}{4\pi} \int e_{ng}(\mathscr{E}; \theta, \varphi) \, d\omega$$

This gives

$$\chi = \frac{e_{ng}(\mathscr{E})}{e_{ng}(0)} = \frac{1}{2} + \frac{1}{2}\left(\frac{\varepsilon kT\lambda}{ze^2}\right)^{\frac{1}{2}} (\exp \beta) \, g(\beta) + \frac{kT}{2e\mathscr{E}\lambda} \ln\left(1 + \frac{e\mathscr{E}\lambda}{kT}\right) \qquad (2.5.23)$$

where

$$g(\beta) \equiv 1 - \frac{2}{\beta} + \frac{2}{\beta^2}[1 - \exp(-\beta)] \qquad (2.5.24)$$

For zero field $g(\beta) \rightarrow g(0) \rightarrow 0$ and $\chi \rightarrow 1$. For low fields and high fields one finds respectively

$$\chi = \begin{cases} 1 + \dfrac{1}{3}\left(\dfrac{e\mathscr{E}\lambda}{kT}\right)^{\frac{1}{2}} & \text{(low } \mathscr{E}) & (2.5.25) \\[3mm] A \exp \beta \left[A \equiv \left(\dfrac{\varepsilon kT\lambda}{4ze^2}\right)^{\frac{1}{2}}\right] & \text{(high } \mathscr{E}) & (2.5.26) \end{cases}$$

Note that eq. (2.5.26) has the same field-dependence as was found by Frenkel's original one-dimensional argument:

$$\chi_F = \exp \beta \qquad (2.5.27)$$

In the Hartke result (2.5.5), derived in section 2.5.3,

$$\chi_H \equiv \tfrac{1}{2} + \frac{1}{\beta^2}[(\beta - 1)\exp \beta + 1] \qquad (2.5.28)$$

the first term arises from the up-field emission and the second term from the down-field emission. The various functions are shown in Figs. 2.5.4 and 2.5.5.

We now make some summarizing remarks.

The discussion of field-enhanced emission furnished the theoretical result (2.5.23), which is additional to the Frenkel formula (2.5.27) and the Jonscher–Hartke formula (2.5.28). The latter two have been in the literature for twenty years. Unfortunately, they are not easy to distinguish experimentally as is clear from the papers cited [2.5.1], [2.5.2], [2.5.7], [2.5.8]. This is true in spite of the fact that the shapes of the two curves are rather different (Fig. 2.5.4). The newer result (2.5.23) adds to these two curves a family of curves labeled by a parameter A which depends on the mean free path. It cannot be expected that at the present stage it

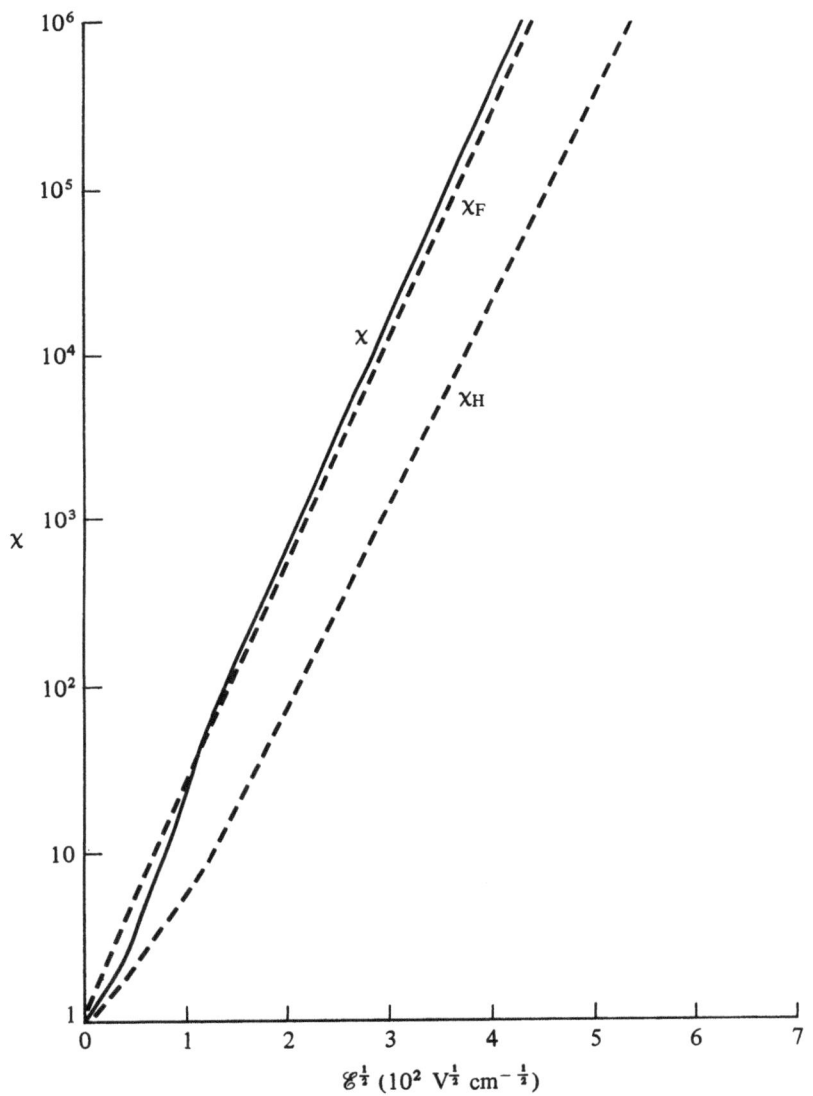

Fig. 2.5.5. As Fig. 2.5.4 but drawn for Si, using $T = 80$ K, $\lambda = [1.4 \times 10^{-3}/T(\text{K})]$ cm, $\varepsilon = 12$, $z = 1$. The resulting value of A is 1.59.

is possible to make an experimental selection from these curves either. However, the new result may suggest some further experimental attacks on the problem. There are two possibilities:

(1) By extrapolating high field data to the zero-field limit ($\beta = 0$) an intercept $\ln A$ is expected which should give one a value of λT which may be a constant or known from other information (see Fig. 2.5.4).

(2) The slope of χ against $\mathscr{E}^{\frac{1}{2}}$ at low fields should behave as $(\lambda/T)^{\frac{1}{2}}$, i.e. increase with λ.

Perhaps one can even combine the observations (1) and (2) in a favorable case and infer the mean free path.

As regards the dependence of capture rates, on a naive theory one might expect them to be field-independent. However, as seen in Fig. 2.5.1, a simple cascade theory using the Hartke formula for χ leads to the expectation that it will decrease with field, in agreement with experiment. The improved expression (2.5.23) for χ can also explain the experiments although the parameters are somewhat different, as noted at the end of section 2.5.2. The idea that capture and emission rates might be equally affected by the electric field (which is sometimes suggested) is not correct.

Appendix: The integrations of section 2.5.4

The integral (2.5.21) is

$$\frac{e_{ng}(\mathscr{E};\theta,\varphi)}{n_g\,C_{ng}} = \frac{2}{\lambda\pi^{\frac{1}{2}}} \int_{\eta_m}^{\infty}\int_{r_1}^{\infty} (\eta-\eta_c)^{\frac{1}{2}} \exp(\eta_{c0}-\eta)\exp-\left(\frac{r-r_1}{\lambda}\right) dr\, d\eta \quad (2.5.29)$$

For $\cos\theta \geqslant 0$ the energy integration can be split at $\eta = \eta_{c0}$. For the first integral (I_1) one can choose the range η_{c0} to ∞ and $r_1 = 0$, and for the second integral (I_2) one covers the range η_m to η_{c0} with r_1 given by eq. (2.5.16). Then I_1 gives an incomplete gamma function

$$\Gamma(\tfrac{3}{2},c) = c^{\frac{1}{2}}\exp(-c) + (\pi^{\frac{1}{2}}/2)\,\mathrm{erfc}(c^{\frac{1}{2}})$$

with $c = r\cos\theta/\lambda$. One finds

$$I_1 = \left(\frac{\lambda\cos\theta}{L}\right)^{\frac{1}{2}} + \frac{1}{\lambda}\int^{\infty}\exp\left(\frac{\cos\theta}{L}-\frac{1}{\lambda}\right) r\,\mathrm{erfc}\left[\left(\frac{r\cos\theta}{\lambda}\right)^{\frac{1}{2}}\right] dr$$

$$I_2 = \left(\frac{\lambda}{L}\cos\theta\right)^{\frac{1}{2}} \{\exp[\beta(\cos\theta)^{\frac{1}{2}}]-1\}$$

Hence

$$\left(\frac{e_{ng}}{n_g\,C_{ng}}\right)_{\mathrm{down}} = 1 + \left(\frac{\lambda}{L}\cos\theta\right)^{\frac{1}{2}}\exp[\beta(\cos\theta)^{\frac{1}{2}}]-\delta \quad (2.5.30)$$

where

$$\delta \equiv \frac{1}{\lambda}\int_0^{\infty}\left[1-\exp\left(\frac{r\cos\theta}{L}\right)\mathrm{erfc}\left(\frac{r\cos\theta}{L}\right)^{\frac{1}{2}}\right]\exp\left(-\frac{r}{\lambda}\right) dr \quad (2.5.31)$$

If $\mathscr{E} \to 0$, $L \to \infty$ and $\delta \to 0$ so that eq. (2.5.30) goes to unity as it should. Neglecting δ, eq. (2.5.30) is the first of eqs. (2.5.22).

The term in square brackets in (2.5.31) is

$$f\left(\frac{r\cos\theta}{L}\right) \equiv 1 - \frac{2}{\pi^{\frac{1}{2}}}\exp\frac{r\cos\theta}{L}\int_{(\frac{r\cos\theta}{L})^{\frac{1}{2}}}^{\infty}\exp(-t^2)\,dt \tag{2.5.32}$$

and is always less than unity [2.5.9]. In the integral for δ the larger values of f are of course weighted by $\exp(-r/\lambda)$ making δ usually negligible.

For $\cos\theta < 0$ the second part of eq. (2.5.23) can be obtained by straightforward integration.

Turning to the average over all directions one has for the down-field contribution

$$\frac{1}{4\pi}\int e_{ng}(\mathscr{E};\theta,\varphi)\,d\omega = \int_0^1 e_{ng}(\mathscr{E};\theta,\varphi)\,d(\cos\theta)$$

$$= \tfrac{1}{2}n_g\,C_{ng}\int_0^1\left[1+\left(\frac{\lambda}{L}x^2\right)^{\frac{1}{2}}\exp\beta x\right]2x\,dx$$

where $x^2 = \cos\theta$. This gives, with $g(\beta)$ defined by eq. (2.5.24),

$$n_g\,C_{ng}[\tfrac{1}{2}+\beta^{-1}(\exp\beta)\,g(\beta)] \tag{2.5.33}$$

For the up-field average one finds with $y \equiv |\cos\theta|$

$$\frac{1}{4\pi}\int e_{ng}(\mathscr{E};\theta,\varphi)\,d\omega = \tfrac{1}{2}n_g\,C_{ng}\int_0^1\frac{dy}{1+(\lambda/L)\,y}$$

$$= \tfrac{1}{2}n_g\,C_{ng}\frac{L}{\lambda}\ln\left(1+\frac{\lambda}{L}\right) \tag{2.5.34}$$

Addition of relation (2.5.33) and eq. (2.5.34) gives the result (2.5.23) of the text.

2.5.5 Thermionic and field emission

Thermionic and field emission represent other mechanisms of generating electrons. Although mature subjects – O.W. Richardson was awarded the 1928 Nobel prize in physics for his work on thermionic phenomena – they have again become important.

In a new field of vacuum microelectronics, old-fashioned vacuum electronics is applied to electron emission on a microscale where one may have 10^7 emitting electron tips per cm² to obtain current densities of up to 1000 A cm⁻². The tunnel effect, which is normally neglected in the theory of the Poole–Frenkel effect, has then to be considered, and it arises when the electron energy E is less than the

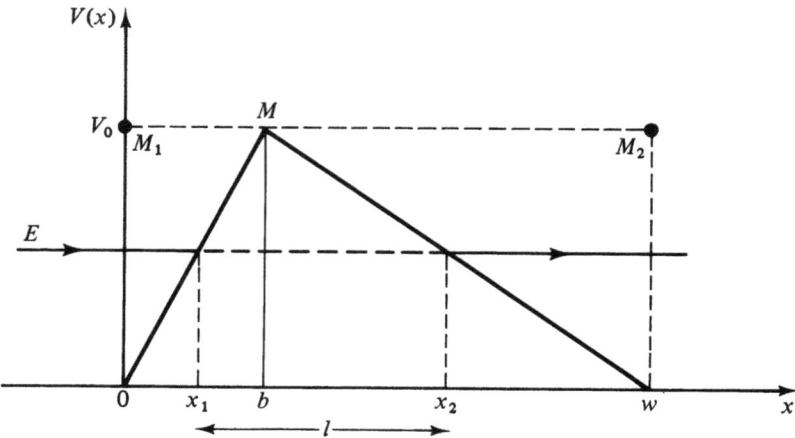

Fig. 2.5.6. A triangular barrier.

potential energy in a region $x_1 \leqslant x \leqslant x_2$ (say) which is classically forbidden. The k-vector is then imaginary in this region and the transmission probability $T(E)$ for the barrier is given approximately by the WKB method (devised by G. Wentzel, H.A. Kramers and L. Brillouin) of quantum mechanics by

$$T(E) \simeq \exp\left[-2\int_{x_1}^{x_2} k \, dx\right] \tag{2.5.35}$$

For the triangular barrier shown in Fig. 2.5.6 the exponent is for an effective mass m^*

$$-\frac{2}{\hbar^2}(2m^*)^{\frac{1}{2}}\left\{\int_{x_1}^{b}\left(\frac{x}{l}V_0-E\right)^{\frac{1}{2}}dx + \int_{b}^{x_2}\left[\frac{w-x}{w-b}V_0-E\right]^{\frac{1}{2}}dx\right\}$$

so that

$$T(E) \simeq \exp\left\{-\frac{4(2m^*)^{\frac{1}{2}}w}{3\hbar V_0}(V_0-E)^{\frac{3}{2}}\right\} \tag{2.5.36}$$

Remarkably, it remains the same for all positions of M in the range M_1 to M_2. In particular for M close to M_1, using for the barrier field $\mathscr{E}$ V_0/we,

$$T(E) \simeq \exp\left\{-\frac{4(2m^*)^{\frac{1}{2}}}{3eh\mathscr{E}}(V_0-E)^{\frac{3}{2}}\right\} \tag{2.5.37}$$

(Strictly speaking the WKB method does not apply for a barrier of the type shown in Fig. 2.5.7, as is clear from Fig. 2.5.8. It does apply for the barrier of Fig.

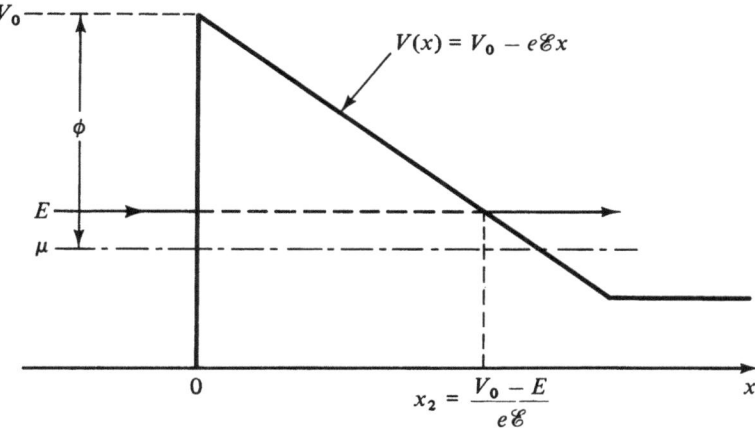

Fig. 2.5.7. Tunneling through a barrier at a metal–semiconductor interface. μ denotes the Fermi level.

2.5.6. So it is best to regard formula (2.5.37) as a limiting case of the result (2.5.36).)

The particle flux impinging on the barrier in the x-direction with kinetic energy in the range $(E, E+\mathrm{d}E)$ is

$$n(E)\,\mathrm{d}E = \frac{2}{h^3}\frac{p_x}{m^*}\,\mathrm{d}p_x \int_{-\infty}^{\infty} \int \frac{\mathrm{d}p_y\,\mathrm{d}p_z}{\exp(\eta_T-\gamma)+1}$$

where the electron energy is $\eta_T kT$ and the electron momentum is (p_x, p_y, p_z). With $u = (p_y^2+p_z^2)/2m^*kT \equiv p^2/2m^*kT$ and $(p_y, p_z) = p(\cos\theta, \sin\theta)$, $\mathrm{d}p_y\,\mathrm{d}p_z = p\,\mathrm{d}p\,\mathrm{d}\theta$ and $\eta_T = \eta + u$

$$n(E)\,\mathrm{d}E = \frac{2}{h^3}\,\mathrm{d}E \int_{p=0}^{\infty} \int_{\theta=0}^{2\pi} \frac{p\,\mathrm{d}p\,\mathrm{d}\theta}{\exp(\eta-\gamma)\exp(u)+1}$$

$$= 4\pi m^* kT h^{-3} \ln[1+\exp(\gamma-\eta)]\,\mathrm{d}E$$

This is the so-called electron supply function. The particle flux impinging on the barrier is for a current density j

$$\frac{j}{e} = \int_A^{\infty} T(E)\,n(E)\,\mathrm{d}E = \frac{m^*kT}{2\pi^2 h^3} \int_A^{\infty} T(E) \ln[1+\exp(\gamma-\eta)]\,\mathrm{d}E \qquad (2.5.38)$$

The lowest energy, A, from which tunneling is possible is the bottom of the conduction band of the metal in the case of metal–vacuum field emission. It can be the bottom of the conduction band of the semiconductor for emission from a metal into a semiconductor (Fig. 2.5.7). The result (2.5.38) will be applied to two cases of electron generation: thermionic and field emission, which are both of interest in vacuum microelectronics.

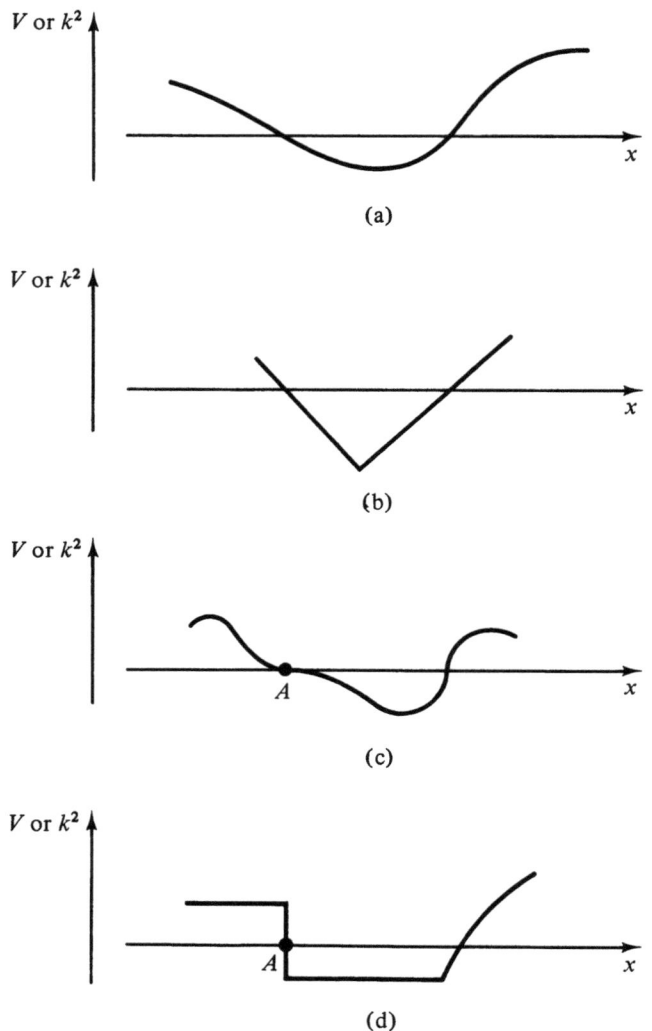

Fig. 2.5.8. The WKB method can be used for cases (a) and (b). It is not applicable to cases (c) and (d) because of the potential at the points A.

For *thermionic emission* without an applied electric field there is a potential step of height V_0 at $x = 0$ ($\mathscr{E} = 0$ in Fig. 2.5.7). Using $A = V_0$, partial integration and

$$T(E) = \begin{cases} 0 & (E < V_0) \\ 1 & (E > V_0) \end{cases}$$

one has

$$\frac{j}{e} = \tfrac{1}{2}\Phi_1 F_1[(\mu - V_0)/kT] \tag{2.5.39}$$

Here $F_1(a) \equiv \Gamma(2)^{-1} \int_0^\infty x[\exp(x-a)+1]^{-1} dx$ is a Fermi integral and

$$\Phi_1 \equiv \frac{m^*(kT)^2}{\pi^2\hbar^3} = \frac{m^*}{2\pi m}\left(\frac{kT}{I_0}\right)^2 \frac{1}{a_0^2 t_0} = 3.74 \times 10^{15} \frac{m^*}{m}\left(\frac{kT}{I_0}\right)^2 \text{Å}^{-2}\,\text{s}^{-1}$$

is a unit of flux. $V_0 - \mu$ is the barrier height above the Fermi level, i.e. the metal work function φ for emission into a vacuum. The Richardson–Dushman equation for the thermionic emission from a metal,

$$j = A^* T^2 \exp(-\varphi/kT), \quad A^* \equiv e\Phi_1/2 \tag{2.5.40}$$

results for the Boltzmann distribution. One should find a straight line in a Richardson plot of

$$\ln(j/T^2) = \ln A^* - \varphi/kT$$

against $1/T$, which yields the work function.

In the semiconductor context thermionic emission between two materials of different effective masses poses interesting problems [2.5.18].

If there is an applied field present one arrives at the Fowler–Nordheim theory (1928), originally devised to explain *field emission* from a (cold) metal into a vacuum and now of crucial importance in vacuum microelectronics. It can also be used for the (external) emission from a degenerate semiconductor into another semiconductor. The degenerate gas (or, equivalently, the low temperature approximation) tells us that for $E > \mu$ the number of electrons available for tunneling is relatively small, while for $E < \mu$ the barrier rapidly becomes thicker. Hence most tunneling electrons come from the neighbourhood of the Fermi level. We need therefore take only two terms in the Taylor expansion of $f(x) \equiv \ln T(E)$ about $E = \mu$ using formula (2.5.37)

$$f(E_z) \simeq f(\mu) + \left(\frac{df(E)}{dE}\right)_\mu (E-\mu) = -C + D(E-\mu)$$

Then

$$C \equiv \frac{4(2m^*)^{\frac{1}{2}}(V_0-\mu)^{\frac{3}{2}}}{3e\hbar\mathscr{E}}$$

$$D \equiv \frac{3}{2}\frac{4}{3}\frac{[2m^*(V_0-\mu)]^{\frac{1}{2}}}{e\hbar\mathscr{E}}$$

Another low temperature approximation is

$$\ln[1+\exp(\gamma-\eta)] = \begin{cases} 0 & (\eta > \gamma) \\ \gamma - \eta & (\eta < \gamma) \end{cases}$$

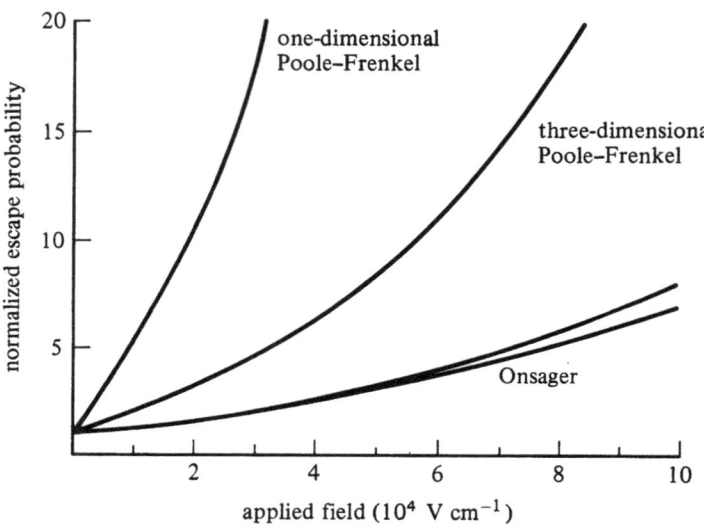

Fig. 2.5.9. Comparison of typical escape probabilities [2.5.26]. The Onsager theory is given for two thermalisation distances.

The particle flux becomes

$$\frac{j}{e} = \frac{m^*(kT)^2}{2\pi^2\hbar^3}\exp\left(-C-D\mu\right)\int^\gamma(\gamma-\eta)\exp\left(kTD\eta\right)d\eta$$

The integral may be taken from $-\infty$ without serious error. One finds the Fowler–Nordheim result

$$\frac{j}{e} = \frac{m^*}{2\pi^2\hbar^3 D^2}\exp\left(-C\right)$$

$$= \frac{e^2\mathscr{E}^2}{16\pi^2\hbar(V_0-\mu)}\exp-\frac{4(2m^*)^{\frac{1}{2}}(V_0-\mu)^{\frac{3}{2}}}{3eh\mathscr{E}}$$

$$= \tfrac{8}{9}\Phi_2\left(\frac{\mathscr{E}}{\mathscr{E}_0}\right)^2\exp\left(-\frac{\mathscr{E}_0}{\mathscr{E}}\right) \tag{2.5.41}$$

We have introduced another unit of flux and a unit of electric field:

$$\Phi_2 \equiv \frac{m^*(V_0-\mu)^2}{\pi^2\hbar^3}, \quad \mathscr{E}_0 \equiv \frac{4}{3}\frac{(2m^*)^{\frac{1}{2}}(V_0-\mu)^{\frac{3}{2}}}{eh} \tag{2.5.42}$$

If the total current is known, the emitting area can be obtained from m^* and $V_0-\mu$. Image force corrections [2.5.19], the effect of nonparabolic densities of states [2.5.20], and emission from a valence band [2.5.21] have also been discussed.

It has been found recently that the pre-breakdown currents between vacuum-insulated high voltage electrodes is higher than expected from eq. (2.5.41) and has been attributed to microstructure defects on or near the emitter surface. This can possibly lead to a thermionic component in the field emission current, and it is an area of current study [2.5.22], [2.5.23].

The Fowler–Nordheim formula (2.5.41) gives numerically

$$j = 1.54 \times 10^{-6} \frac{\mathscr{E}^2}{(V_0-\mu)} 10^{-2.97 \times 10^7 (V_0-\mu)^{\frac{3}{2}}/\mathscr{E}}$$

where j is in A cm^{-2}, $\mathscr{E}$ in V cm^{-1}, $V_0 - \mu$ in eV. Note that there is no explicit temperature-dependence (so prominent in thermionic emission and the Poole–Frenkel effect), though there is a weak implicit dependence through μ. As $\mathscr{E} \to 0$ the field emission current also vanishes. The reason is that the distance between the points for which $E = V(x)$ increases for shallower slopes (and given E), thus making the barrier eventually impenetrably thick.

2.5.6 Geminate recombination

In some materials such as organic crystals and amorphous or disordered substances there is a fair possibility that a carrier when excited is still within a range of its attracting center so that it has a good chance of falling back to its original state. This is called *geminate recombination*; the term is derived from the Latin word *geminatus*, meaning 'arrange in pairs'. The root is familiar from the constellation Gemini (the twins), the third sign of the zodiac. The reduction in the thermal ionization energy needed to separate the charge carriers as a result of an applied electric field then gives rise to field-dependent photogeneration, photo-conductivity, etc. The process is rather like the Poole–Frenkel effect, but the theory also includes the random walk, or Brownian motion, of the current carriers involved. It was developed by Onsager. In one version it leads for high fields to the same exponential dependence as the one-dimensional Poole–Frenkel effect [2.5.24]. Another variant [2.5.25] leads to a much gentler field-dependence; this is shown in Fig. 2.5.9 [2.5.26]. We shall briefly discuss the result of the latter theory which is much used in connection with amorphous semiconductors (although we only touch on their properties in this book).

One considers carriers in thermal equilibrium with an isotropic medium of dielectric constant ε. This theory then leads to a probability of dissociation of a charge pair given by

$$p(r, \theta, \mathscr{E}) = \exp(-A) \exp(-B) \sum_{i=0}^{\infty} \sum_{j=0}^{\infty} \frac{A^i}{i!} \frac{B^{i+j}}{(i+j)!} \tag{2.5.43}$$

which we take from equation (11) of Onsager's paper without proof. Here

$$A \equiv 2q/r \equiv e^2/\varepsilon k Tr, \quad B \equiv (e\mathscr{E}r/2kT)(1+\cos\theta)$$

where r is the initial separation of oppositely charged carriers, one of which is at the origin, and θ is the angle between the vector from the origin to the other carrier and the applied electric field $\mathscr{E}$. It follows that

$$p(r,\theta,0) = p(r,\pi,\mathscr{E}) = \exp(-e^2/\varepsilon k Tr) \tag{2.5.44}$$

One can also write

$$p(r,\theta,0) = \exp(-r_{\mathrm{C}}/r) \tag{2.5.45}$$

where

$$e^2/\varepsilon r_{\mathrm{C}} = kT \tag{2.5.46}$$

defines the Coulombic capture radius: the Coulomb energy at r_{C} is just kT. From eq. (2.5.44) one sees that if one moves exactly against the direction of $\mathscr{E}$ the probability of dissociation increases with r, but is independent of the magnitude of $\mathscr{E}$. One finds for $p(r,\theta,\mathscr{E})$ up to terms in $\mathscr{E}^2$,

$$p(r,\theta,\mathscr{E}) = \left[1 + \frac{e^3\mathscr{E}}{\varepsilon k^2 T^2}(1+\cos\theta) + \left(\frac{e^4}{4\varepsilon^2 k^2 T^2} - \frac{e^2 r}{2\varepsilon k T} + r^2\right)\frac{e^2\mathscr{E}^2}{4k^2 T^2}(1+\cos\theta)^2\right]$$

$$\times \exp\left(-\frac{2e^2}{\varepsilon k Tr}\right) \tag{2.5.47}$$

Thus if one *fixes* $p(r,0,\mathscr{E})$ we see that increasing $\mathscr{E}$ forces one to reduce r. It follows that the contours of constant probability are crowded together in the direction of $\mathscr{E}$, see Fig. 2.5.10 [2.5.27]. (The r^2-term in eq. (2.5.45) appears to be usually omitted [2.5.28] and [2.5.29], p. 485.)

It is often convenient to assume a spherically symmetric probability distribution, $g(r)$, for the initial separation between the ions of each ion pair. Then the carrier quantum yield, or the carrier generation efficiency, is

$$\Phi(\mathscr{E}) = \Phi_0 \int g(r) p(r,\theta,\mathscr{E}) \, d\mathbf{r} \tag{2.5.48}$$

(This relation is formally the same if g depends on θ.) The efficiency Φ_0 of production of thermalized ion pairs per absorbed photon is assumed to be independent of electric field. With $d\mathbf{r} = r^2 \, dr \sin\theta \, d\theta \, d\varphi$, the terms shown in eq. (2.5.47) give

$$\frac{\Phi(\mathscr{E})}{4\pi\Phi_0} = \left(1 + \frac{e^3}{2\varepsilon k^2 T^2}\mathscr{E}\right)\int_0^\infty e^{-r_{\mathrm{C}}/r} g(r) r^2 \, dr$$

$$+ \frac{1}{3}\frac{e^2\mathscr{E}^2}{4k^2 T^2}\int_0^\infty \left(\frac{r_{\mathrm{C}}^2}{4} - \frac{r_{\mathrm{C}} r}{2} + r^2\right)e^{-r_{\mathrm{C}}/r} g(r) r^2 \, dr + \dots \tag{2.5.49}$$

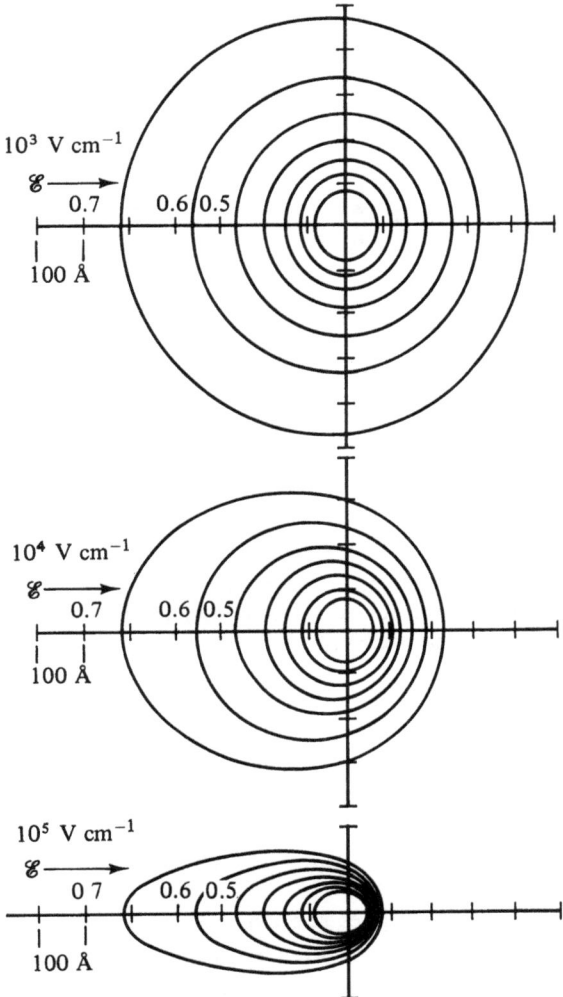

Fig. 2.5.10. Contours of constant probabilities $0.1, 0.2, \ldots, 0.7$ of pair dissociation as a function of separation for three different applied fields [2.5.27]. The left-hand intersections with the r-axis are field-independent. The right-hand intersections are pushed to smaller values by the field. The curves apply to an isotropic dielectric medium of dielectric constant $\varepsilon = 3.02$ at 300 K.

Our stated assumptions make the integrals independent of the field $\mathscr{E}$ so that

$$\frac{\Phi(\mathscr{E})}{\Phi_0} = 1 + C\mathscr{E} + D\mathscr{E}^2 + \ldots \quad (C, D \text{ are constants})$$

For weak fields a plot of $\Phi(\mathscr{E})$ against $\mathscr{E}$ yields a slope/intercept ratio which is independent of adjustable parameters and hence an excellent test for the applicability of the Onsager theory [2.5.27].

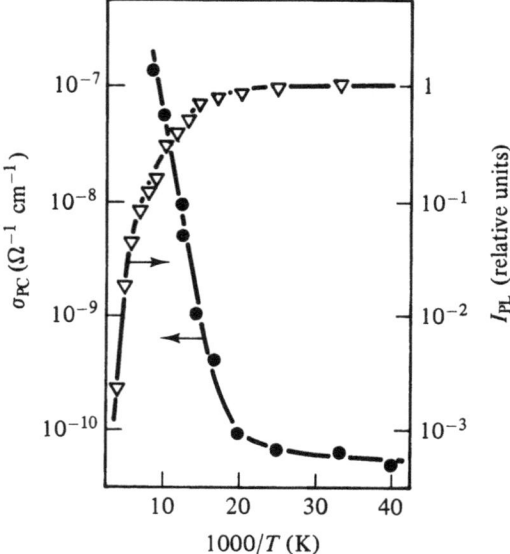

Fig. 2.5.11. The anticorrelation between photoconductivity (σ_{PC}) which rises with temperature (as geminate pairs decompose) and photoluminescence (I_{PL}) which falls with temperature [2.5.33]. The curves are for amorphous hydrogenated Si.

Extensive studies of organic crystals such as anthracene and tetracene have yielded experimental values of C. For anthracene it was found that at 298 K [2.5.30]

$$C = 3.21 \times 10^{-5} \text{ (electrons) and } 3.02 \times 10^{-5} \text{ (holes) cm V}^{-1}$$

and there exist other related studies [2.5.29].

Since geminate recombination is often radiative, an electric field will tend to quench geminate recombination and the associated luminescence. This has also been widely observed, for example, in As chalcogenides [2.5.31], [2.5.32].

The competition between photoconductivity (encouraged by pair dissociation) and photoluminescence (encouraged by geminate recombination) can be studied experimentally, see Fig. 2.5.11, and enables one to estimate for example the quantum efficiency for the generation of mobile carriers. For amorphous hydrogenated Si this efficiency rises steeply from 3×10^{-3} near 50 K to close to unity at room temperatures [2.5.33]. One can understand this by observing that the Coulomb radius (2.5.46) is large (300 Å) at 50 K and reduces to 50 Å at room temperature.

The Onsager theory adds the effect of Brownian motion to the Poole–Frenkel effect or, in recombination terms, to the Langevin theory of recombination of ions in a gas (first proposed in 1903 and mentioned in [2.5.25] and in footnote 16 of

[2.5.34]). Langevin takes the relative drift velocity of the ions of one sign towards the ions of the opposite sign to be in terms of their mobilities v_+, v_-

$$v_D = (v_+ + v_-)(e/r^2)$$

for a distance r between them, as can be seen by using a frame of reference in which the slower ions are instantaneously at rest. If the concentration of the ions is n, their loss due to recombination is

$$-\dot{n} = 4\pi r^2(e/r^2)(v_+ + v_-)n^2 = 4\pi(v_+ + v_-)en^2 = An^2$$

This arises from $4\pi r^2 v_D \, dt$ as the number of negative ions (say) flowing into a sphere of radius r in time dt. The recombination coefficient is then $A = 4\pi e(v_+ + v_-)$. This approach, which now seems somewhat simple, is put into historical perspective by Loeb [2.5.35].

2.6 Cascade capture

2.6.1 Some formulae for cascade capture

In this section we leave the *truncated* cascade *recombination* model of sections 2.4 and 2.5 and consider the *untruncated* cascade *capture*. Thus one passes from two-band considerations to single-band models.

In a simple model of capture the capture cross section is the product of three factors: (i) the probability of collision r_0/λ where r_0 is an effective radius of the sphere of influence of the capturing center and λ is the mean free path for an energy-losing collision of a moving particle; (ii) the geometrical cross section of the sphere of influence; (iii) a factor ($\frac{4}{3}$) arising from averaging over all paths inside the sphere. The result is the cross section

$$\sigma = 4\pi r_0^3/3\lambda \tag{2.6.1}$$

used by Thomson for recombination in gases [2.6.1]. It has been argued [2.6.2], [2.6.3] that the replacement

$$\lambda \to v\tau_E$$

is appropriate where v is the velocity of the moving particle (which is an electron in the solid-state case) and τ_E is the electron energy relaxation time. These authors found

$$\sigma = (4\pi/3v\tau_E)(e^2 z/\varepsilon kT)^3 \tag{2.6.2}$$

where ez is the charge on the center. This arises from a cascade process which includes all the excited states of the center including those with a binding energy less than kT. Lax expected these levels not to make a great contribution as they are

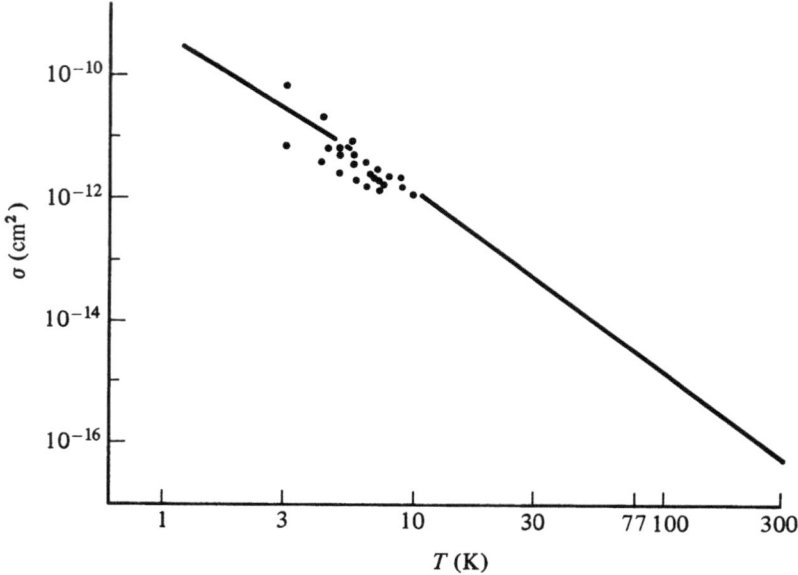

Fig. 2.6.1. Electron-capture cross sections of positively charged centers in Ge for $z = 1$. The continuous curve is based on the interpolation formula (2.6.3).

liable to be easily ionized and hence to have a low sticking probability. The result (2.6.2) applies for $kT \gg m_c s^2$ where m_c is the effective electron mass and s is the velocity of sound. For $m_c s^2 \gg kT$ the T^{-3}-dependence (2.6.2) becomes a T^{-1}-dependence, and the interpolation formula

$$\sigma = \frac{4\pi}{3v\tau_E} \frac{e^2 z}{\varepsilon kT} \left[\frac{e^2 z}{\varepsilon (kT + 2.74 m_c s^2)} \right]^2, \quad [2.74 = (15/2)^{\frac{1}{2}}] \qquad (2.6.3)$$

has been proposed [2.6.3]. For $kT \ll m_c s^2$ this yields the low temperature result

$$\sigma = \frac{8\pi}{45 v\tau_E} \frac{e^2 z}{\varepsilon kT} \left(\frac{e^2 z}{\varepsilon m_c s^2} \right)^2 \qquad (2.6.4)$$

Relation (2.6.3) is compared with data for Sb^+ [2.6.4], [2.6.5] and Fe^+ [2.6.6] in Ge in Fig. 2.6.1 [2.6.3].

For $z = 1$ it is believed that eq. (2.6.3) holds for $T \lesssim 30$ K, 100 K for Ge and Si, respectively, but these limits are raised to 120 K and 400 K for $z = 2$ [2.6.3].

In any case the figure shows that attractive centers act as 'giant traps' which led to the cascade concept as a mechanism [2.6.7]. Neutral centers tend to lie in the 10^{-17}–10^{-15} cm^2 range, while repulsive centers tend to have very small cross sections (10^{-24}–10^{-21} cm^2). The temperature-dependencies can also vary widely as T^{-1} to T^{-4} or even exponentially as $\exp(-E/kT)$, where $E \sim 0.05$ eV. These numbers are rough guides only; for more details see section 5.1 and references cited there.

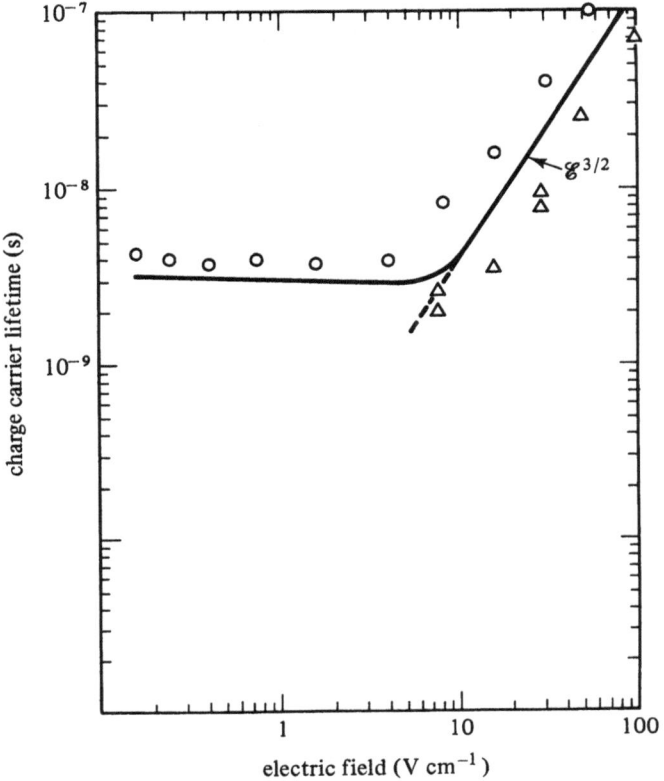

Fig. 2.6.2. Effect of electric field on carrier lifetime at low temperature in comparison with a cascade capture theory. $\bigcirc$, Si: B, $N_B = 9.2 \times 10^{13}$ cm^{-3}, $N_D = 9.2 \times 10^{12}$ cm^{-3}, $T = 4.2$ K. $\triangle$, Si: As, $N_{AS} = 1.4 \times 10^{16}$ cm^{-3}, $N_A = 6.0 \times 10^{12}$ cm^{-3}, $T = 5.5$ K.

It was seen in Figs. 2.5.1–2.5.3 that the effect of an electric field is to decrease capture and increase emission both experimentally and on the basis of the TCR theory. This leads to reduced recombination coefficients and increased lifetimes. One finds this again in the case of cascade *capture* (Fig. 2.6.2). There is broad agreement between the experimental points and the cascade capture theory (solid line) of [2.6.8]. The figure has been added to by Sclar and is reproduced from his article [2.6.9].

The Lax theory has been amended by various authors and has been used in spite of its limited accuracy. In the following subsections 2.6.2–2.6.7 a specific formulation of the theory will be explained [2.6.10].

2.6.2 Implications of the classical conservation laws

(i) Energy

The scattering process will be described by an impact parameter method (Fig. 2.6.3). The scattering center is assumed rigidly fixed at the origin of the coordinates O. The potential energy of an electron in its field is specified by the parameters A, V_m, R, g as follows:

$$V(r) = \begin{cases} -V_m & (0 < r \leqslant R) \\ -Ar^{-g} & (R \leqslant r) \end{cases}$$

(2.6.5)

whence for continuity

$$V_m = AR^{-g}$$

(2.6.6)

Various centers are covered by this definition (see Table 2.6.1).

Suppose now an electron of effective mass m_c and energy

$$E_0 = \tfrac{1}{2}m_c v_0^2$$

(2.6.7)

approaches from infinity, and its position at a general time t is $r(t)$ as measured from O. Then, if its kinetic energy is E, one has at time t

$$E[r(t)] + V[r(t)] = E_0$$

(2.6.8)

Now from eqs. (2.6.5) and (2.6.8) one has for $r(t) > R$, $E = E_0 + Ar^{-g}$, so that

$$r^2 \, dr = -\frac{A^{3/g} \, dE}{g(E - E_0)^{1+3/g}} \quad (r > R)$$

(2.6.9)

(ii) Angular momentum

Suppose the electron's path, if undeviated, would be at a perpendicular distance b from the center of scattering. The angular momentum conservation then implies that

$$r^2(t)\,\dot{\theta}(t) = \text{constant} = bv_0 = b[(2E_0/m_c)]^{\frac{1}{2}}$$

(2.6.10)

Substituting for $\dot{\theta}$ from eq. (2.6.10) in the expression

$$E[r(t)] = \tfrac{1}{2}m_c(\dot{r}^2 + r^2\dot{\theta}^2)$$

(2.6.11)

for the kinetic energy at an arbitrary point along the orbit, one finds

$$\dot{r}^2 = (2/m_c)[E(r) - E_0 b^2/r^2]$$

(2.6.12)

For given b there is therefore a minimum distance of approach, r_m, given by $\dot{r} = 0$, i.e.

$$r^2(t) \geqslant r_m^2 \equiv E_0 b^2/E(r_m)$$

(2.6.13)

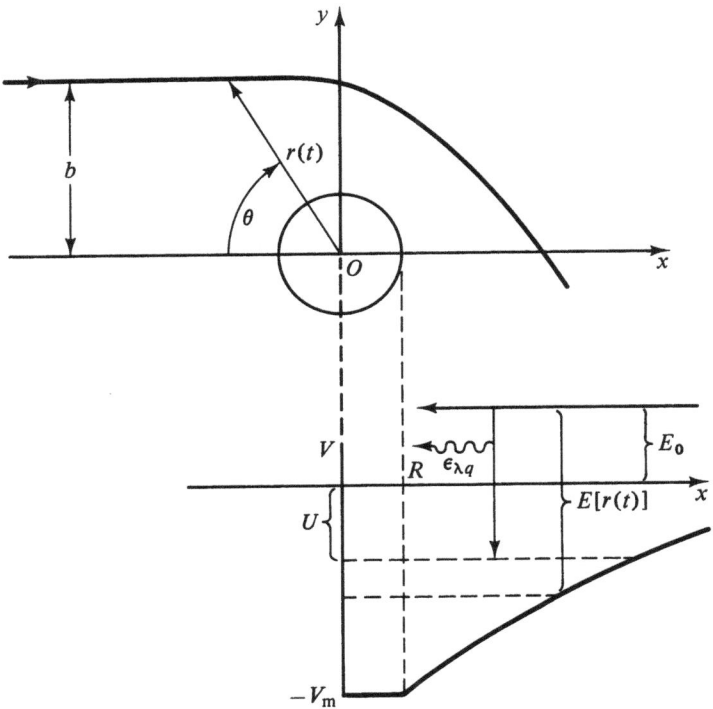

Fig. 2.6.3. The significance of the impact parameters E_0 and b is shown. The electron potential energy in the neighborhood of the center is also shown, and the meaning of various energies is illustrated.

Alternatively, given that some distance r (from O) occurs in the orbit, the impact parameter satisfies

$$b^2 \leqslant b_M^2 \equiv r^2 E(r)/E_0 \qquad (2.6.14)$$

(iii) Boson momentum and energy

These quantities will be denoted by $\mathbf{q}$ and $\varepsilon_{\lambda q}$ respectively, where λ refers to a direction of polarization. Let $\mathbf{p}$ be the momentum of the electron before boson emission. Phonons or photons may be emitted and the theory so far holds for both, though a radiative cascade [2.6.11] is not our concern here.

The nonrelativistic energy–momentum conservation implies

$$Y \equiv \frac{(\mathbf{p}-\mathbf{q})^2}{2m_c} + \varepsilon_{\lambda q} - \frac{p^2}{2m_c} = 0 \qquad (2.6.15)$$

If α is the angle between $\mathbf{p}$ and $\mathbf{q}$, it follows that

$$\frac{q^2}{2m_c} - \frac{pq}{m_c}\cos\alpha + \varepsilon_{\lambda q} = 0 \qquad (2.6.16)$$

The inequality $p \geqslant p \cos \alpha$ implies an inequality for the electron kinetic energy:

$$E[r(t)] \geqslant \frac{q^2}{8m_c}\left[1 + \frac{2m_c}{q^2}\varepsilon_{\lambda q}\right]^2 \equiv E_{min} \qquad (2.6.17)$$

In order that a boson $(\lambda, \mathbf{q})$ be emitted by an electron with impact parameters (E_0, b), it follows from eqs. (2.6.8) and (2.6.17) that V be sufficiently large and negative, and therefore by eqs. (2.6.5) that r be small enough. One finds, assuming $r_0 > R$,

$$r \leqslant r_0, \quad \text{where } r_0^g \equiv \frac{8m_c A}{q^2}\left\{\left[1 + \frac{2m_c \varepsilon_{\lambda q}}{q^2}\right]^2 - \frac{8m_c E_0}{q^2}\right\}^{-1} \qquad (2.6.18)$$

The above equations give the classical background for scattering with emission of bosons, and depends essentially on eq. (2.6.15). This equation can also be obtained quantum mechanically from perturbation theory. The transition probability per unit time for processes involving one *phonon* is in this case governed by an operator of the form

$$C \exp\left(\pm \frac{i}{\hbar}\mathbf{q}\cdot\mathbf{r}\right)$$

where C is independent of the electron coordinate $\mathbf{r}$, and the positive sign is taken for absorption and the negative sign for emission. If the initial and final states are approximated by plane waves of wave momenta $\mathbf{p}$ and $\mathbf{p}'$, respectively, a nonzero transition probability results only if

$$\mathbf{p} = \mathbf{p}' + \mathbf{q} \quad \text{(emission)} \qquad (2.6.19)$$

Energy conservation, eq. (2.6.19), and the assumption of a parabolic band now yields eq. (2.6.15).

(A similar argument holds for photon emission, although C then involves the operator grad_r, which, however, merely introduces a factor $\mathbf{p}$.)

2.6.3 Relations between cross sections and probabilities

The question of whether or not it is classically possible for a given electron to emit a given boson was discussed in section 2.6.2. We consider next the probability of such a process. Let $W_\lambda[E(r(t)), \mathbf{q}] \, d\mathbf{q}$ be the probability per unit time that an electron of kinetic energy E and total energy E_0 emits a boson with polarization λ and with momentum in the range $\mathbf{q}, \mathbf{q} + d\mathbf{q}$. If it does so, the electron energy after the process is $E_0 - \varepsilon_{\lambda q}$ and its binding energy is (see Fig. 2.6.3)

$$U = -(E_0 - \varepsilon_{\lambda q}) = \varepsilon_{\lambda q} - E_0 \qquad (2.6.20)$$

The essential assumption of the Lax cascade theory is that there exists a probability $P(U)$, depending only on U, that an electron which has binding energy U enters the ground state of the atom without first leaving it. This is the *sticking probability* already encountered in eq. (2.4.50) in connection with the TCR model.

Thus the probability that somewhere along its orbit the electron is captured with the creation of one boson with polarization λ, and that it thereafter reaches the ground state without first leaving the atom is

$$
\begin{aligned}
P_\lambda(E_0, b) &= \int_{-\infty}^{\infty} dt \int_{(\varepsilon_{\lambda q} \geq E_0)} W_\lambda(E, \mathbf{q}) \, P(U) \, d\mathbf{q} \\
&= 2 \int_{r_m}^{r_0} \frac{dt}{dr} dr \int_{(\varepsilon_{\lambda q} \geq E_0)} W_\lambda(E, \mathbf{q}) \, P(U) \, d\mathbf{q}
\end{aligned}
\tag{2.6.21}
$$

The r-integration extends, using eqs. (2.6.13) and (2.6.18), from the minimum distance of approach for given b, to the maximum distance for which a collision with energy loss $\varepsilon_{\lambda q}$ can occur. The factor 2 arises from the double transverse of the distance r_m to infinity by the electron.

The transition probability per unit time $W_\lambda[E, \mathbf{q}] \, d\mathbf{q}$ is obtainable from first order time-dependent perturbation theory as

$$
W_\lambda[E, \mathbf{q}] = \frac{V}{(2\pi)^2 \hbar^4} B_\lambda[E, \mu^2, \mathbf{q}] (N_{\lambda q} + 1) \, \delta[E(\mathbf{p}') - E(\mathbf{p}) - \varepsilon_{\lambda q}]
\tag{2.6.22}
$$

where $B_\lambda[E, \mu^2, \mathbf{q}]$ involves quantum mechanical matrix elements, with wavefunctions as yet unspecified, and is dimensionally the square of an energy; $N_{\lambda q}$ is the mean number of bosons in the state λ, $\mathbf{q}$ and μ is the cosine of the angle between $\mathbf{p}$ and $\mathbf{q}$. It will be presumed that $\varepsilon_{\lambda q}$ and hence $N_{\lambda q}$ depend only on the magnitude of $\mathbf{q}$.

In the situation envisaged on p. 197, the parameter in the δ-function (2.6.22) is replaced by Y of eq. (2.6.15) yielding

$$
W_\lambda[E, \mathbf{q}] = \frac{V m_c}{(2\pi)^2 \hbar^4 pq} B_\lambda[E, \mu^2, \mathbf{q}] (N_{\lambda q} + 1) \, \delta\left[\mu - \left(\frac{E_{min}}{E}\right)^{\frac{1}{2}} \right]
\tag{2.6.23}
$$

Integrating over all angular orientations, given the length of $|\mathbf{q}|$, one obtains the probability per unit time that an electron of total energy E_0 and kinetic energy E emits a boson of polarization λ with a momentum of magnitude in the range $(q, q + dq)$:

$$
\int_{\text{(angles)}} W_\lambda[E, \mathbf{q}] \, d\mathbf{q} \equiv W_\lambda[E, q] \, dq = \frac{m_c V}{2\pi \hbar^4 p} q B_\lambda\left[E, \frac{E_{min}}{E}, q \right] (N_{\lambda q} + 1) \, dq
\tag{2.6.24}
$$

Hence eq. (2.6.21) becomes

$$
P_\lambda(E_0, b) = \frac{m_c V}{\pi \hbar^4 p} \int_{r_m}^{r_0} \frac{dt}{dr} dr \int_{(\varepsilon_{\lambda q} \geq E_0)} P(U) \, q B_\lambda\left[E, \frac{E_{min}}{E}, q \right] (N_{\lambda q} + 1) \, dq
\tag{2.6.25}
$$

Table 2.6.1. *The potential energy* (2.6.5) *for various cases*

Type of center	A	g	Unit of cross section, Q, from eq. (2.6.32)
Coulomb	e^2/ε	1	$(m_c^2 VG_\lambda(q)/3\hbar^4)(2e^2/m_c s^2 \varepsilon)^3$
Neutral	$e^2\gamma/2\varepsilon^2$	4	$(m_c^3 VG_\lambda(q)/3\hbar^4)(e^2\gamma/m_c s^2 \varepsilon^2)^{\frac{3}{4}}$
Dipole	$e^2a/2\varepsilon$	2	$(m_c^3 VG_\lambda(q)/3\hbar^4)(e^2a/m_c s^2 \varepsilon)^{\frac{3}{2}}$

To obtain a cross section for electrons of total energy E_0, integrate over all impact parameters b to find

$$\sigma_\lambda(E_0) \equiv 2\pi \int_0^\infty bP_\lambda(E_0, b)\,\mathrm{d}b$$

$$= \frac{2m_c V}{\hbar^4} \int_0^\infty b\,\mathrm{d}b \int_{r_m}^{r_0} \frac{\mathrm{d}t}{\mathrm{d}r}\,\mathrm{d}r \int_{(\varepsilon_{\lambda q} \geqslant E_0)} \frac{q}{p} B_\lambda \left[E, \frac{E_{\min}}{E}, q\right] P(U)(N_{\lambda q}+1)\,\mathrm{d}q$$

$$\equiv \int_{\varepsilon_{\lambda q} \geqslant t_0} \sigma_\lambda(E_0, q)\,\mathrm{d}q \tag{2.6.26}$$

If the magnitude of the momentum of the emitted boson is fixed to a range $(q, q+\mathrm{d}q)$, one obtains the cross section $\sigma_\lambda(E_0, q)$ defined in eq. (2.6.26). By inverting orders of integration and noting eqs. (2.6.13) and (2.6.14)

$$\sigma_\lambda(E_0, q) = \frac{2m_c V}{\hbar^4} q(N_{\lambda q}+1) P(U) \int_0^{r_0} B_\lambda \left[E, \frac{E_{\min}}{E}, q\right] \frac{\mathrm{d}r}{p} \int_0^{b_M} \frac{\mathrm{d}t}{\mathrm{d}r} b\,\mathrm{d}b \tag{2.6.27}$$

This formulation can be applied to photon emission [2.6.11] although this will not be done here.

In the cases of phonon emission, B_λ is a function of q only. It will then be denoted by $G_\lambda(q)$. Hence

$$\sigma_\lambda(E_0, q) = \frac{A^{3/g}m_c VqG_\lambda(q)}{3\hbar^4 E_0}(N_{\lambda q}+1) P(\varepsilon_{\lambda q} - E_0)$$

$$\times \left\{\frac{m_c \varepsilon_{\lambda q}^2}{2q^2}\left[1+\frac{q^2}{2m_c \varepsilon_{\lambda q}}\right]^2 - E_0\right\}^{-3/g} \tag{2.6.28}$$

This expression can be put into the form

$$z\sigma(z, y)\,\mathrm{d}y = B\left\{\frac{x^2}{y^2}\left[1+\frac{y^2}{2x}\right]^2 - z\right\}^{-3/g}\mathrm{d}(y^2) \tag{2.6.29}$$

where

$$z \equiv \frac{E_0}{\frac{1}{2}m_c s^2}, \quad y \equiv \frac{q}{m_c s}, \quad x \equiv \frac{\varepsilon_{\lambda q}}{m_c s^2} \tag{2.6.30}$$

B depends on $\varepsilon_{\lambda q}$ and normally also on q:

$$B \equiv Q(N_{\lambda q}+1) P(\varepsilon_{\lambda q}-E_0) \tag{2.6.31}$$

where

$$Q \equiv \frac{m_c^2 VG_\lambda(q)}{3\hbar^4}\left(\frac{A}{\frac{1}{2}m_c s^2}\right)^{3/g} \tag{2.6.32}$$

is a unit of cross section for a given type of scattering. From eqs. (2.6.24) and (2.6.29)

$$z\sigma_\lambda(z) = \int B\left[\frac{y^2}{x^2+(x-z) y^2+y^4/4}\right]^{3/g} \mathrm{d}(y^2) \tag{2.6.33}$$

Here $\sigma_\lambda(z)$ is the quantity $\sigma_\lambda(E_0)$ of eq. (2.6.26). This completes the general formalism, except for the limits of the integral in eq. (2.6.33) (see p. 202).

2.6.4 Details of special cases covered by the exposition

Table 2.6.1 gives details of the constants A and g which characterize various centers, in accordance with eq. (2.6.6) [2.6.12]. The *Coulombic* center is assumed to have a net charge e, which is positive if the current carriers considered are electrons. It must be negative if the charge carriers are holes. The static dielectric constant is denoted by ε. A *neutral* center is characterized by a polarizability γ, which may be evaluated, for example, from the relation

$$\gamma = \gamma_H \frac{m}{m_c}\left(\frac{I_0}{I}\right)^2 \tag{2.6.34}$$

where m is the free electron mass, and I_0 ($= 13.6$ eV) and I are the ionization energies of a free hydrogen atom and of the impurity, respectively. γ_H ($= 6.66 \times 10^{-25}$ cm^3) is the polarizability of a free hydrogen atom. In certain cases the potential energy due to the defect may be dominated by a *dipolar* contribution. The dipole moment is then denoted by ae in Table 2.6.1, where a is an appropriate distance.

Some important expressions for more important matrix elements $G_\lambda(q)$ are given in Table 2.6.2. Columns 3 and 4 give the longitudinal and transverse phonon contributions to $G_\lambda(q)$ assuming the matrix element

$$M_{mn} \equiv V^{-\frac{1}{2}}\int \varphi_m^* e^{i\mathbf{q}\cdot\mathbf{r}/\hbar}\varphi_n \, \mathrm{d}\tau$$

contains plane waves for the final and initial electron wavefunctions φ_m and φ_n. In Table 2.6.2 Ξ_D and D_0 are phenomenological constants, ε_0 and ε_i are the

Table 2.6.2. *Expressions for $G_\lambda(q)$ (cf. eq. (2.6.28))*

Material	$B_\lambda(E, \mu^2, q)$	$G_l(q)$	$G_t(q)$	Boson
Covalent	$\dfrac{1}{2\rho sq}[\Xi_D(e_\lambda(q)\cdot q)$ $+\Xi_u(a\cdot q)(a\cdot e_\lambda(q))]^2$ $\times |M_{mn}|^2$	$\dfrac{\Xi_D^2 q}{2\rho sV}$	0	Acoustic phonon
Covalent	$\dfrac{\hbar^2}{2\rho\varepsilon_0}[D_0\,a\cdot e_\lambda(q)]^2|M_{mn}|^2$	$\dfrac{\hbar^2 D_0^2\cos^2\chi}{2\rho V\varepsilon_0}$	$\dfrac{\hbar^2 D_0^2\sin^2\chi}{2\rho V\varepsilon_0}$	Optical phonon
Ionic	$\dfrac{2\pi e^2\hbar^2\varepsilon_i}{|q|^2}\left[\dfrac{1}{\varepsilon_\infty}-\dfrac{1}{\varepsilon}\right]|M_{mn}|^2$	$\dfrac{2\pi e^2\hbar^2\varepsilon_i}{Vq^2}\left[\dfrac{1}{\varepsilon_\infty}-\dfrac{1}{\varepsilon}\right]$	0	Optical phonon

characteristic energies ascribed to optical phonons in covalent (Ge, Si, InSb, GaAs, GaP, etc.) and ionic (NaCl, etc.) materials, χ is the angle between the momentum q and the longitudinal axis of the valley in k-space containing the initial and final electron states, $e_\lambda(q)$ is the phonon polarization vector and ε_∞ is the high-frequency dielectric constant.

Although the use of plane waves must be a dubious approximation, it can be supported, to some extent, by the fact that the original calculations by Lax utilized it and the results have, in a number of cases, been in satisfactory agreement with experiment. (That the assumption of plane waves is implicit in Lax's approach is clearly brought out by the present procedure.)

2.6.5 *The total capture cross section*

For *optical* phonons, all phonon wave momenta are assumed to be characterized by a single energy ε_0. Since $\varepsilon_{\lambda q}$ and $G_\lambda(q)$ are independent of q for *optical* phonons in covalent materials, the factor B appearing in eq. (2.6.33) may, in this case, be taken outside the integral. The limits of the integration over q are determined by the conservation of energy and momentum and the nature of the material. From eq. (2.6.15), it may be shown that the upper and lower limits, for one particular electron momentum p, are given by

$$q_u = p+[p^2-2m\varepsilon_0]^{\frac{1}{2}} \simeq 2p \quad [\text{or } q_0 = \hbar(6\pi^2 N)^{\frac{1}{3}}] \tag{2.6.35}$$

$$q_1 = p-[p^2-2m\varepsilon_0]^{\frac{1}{2}} \simeq \frac{m\varepsilon_0}{p} \tag{2.6.36}$$

where only the first term of a binomial expansion has been retained. The alternative in brackets arises as follows. The wavevector $q/\hbar$ must lie within the first Brillouin zone of the reciprocal lattice. Replacing this zone by a sphere of equal volume, whose radius is $q_0/\hbar$, yields an upper limit of q. Hence, the upper limit is taken as q_0 or $2p$ depending on which is the smaller. N is the number of unit cells per unit volume of the crystal.

The wave momentum and energy of *acoustic* phonons are assumed to obey the Debye relations $sq = \varepsilon_q$, the upper limit of q is again q_0 or $2p$, but the lower limit is now zero [2.6.13].

For given ε_q, E_0 and V_m, the maximum kinetic energy that an electron may acquire in the presence of a capturing center is $E_0 + V_m$ by eqs. (2.6.6) and (2.6.8), where $E_0 \leqslant \varepsilon_q$. Hence the maximum value that q_u may assume is, by eq. (2.6.35), $2[2m(E_0 + V_m)]^{\frac{1}{2}}$ or q_0, whichever is the smaller. Similarly, the minimum value q_1 is given by $m\varepsilon_0/[2m(E_0 + V_m)]^{\frac{1}{2}}$ for optical phonons and zero for acoustic modes.

To provide a link with experiment, it is necessary to relate $\sigma_\lambda(E_0)$ to the recombination constant T_1^s (see p. 103), which is an average over all $E_0 > \varepsilon_q$. If electrons infinitely removed from capturing centers obey Maxwell–Boltzmann statistics, then*

$$T_1^s(\text{cm}^3 \text{ s}^{-1}) = \left(\frac{8KT}{\pi m}\right)^{\frac{1}{2}} \sigma \sim \left[\frac{T^{(K)}}{300}\right]^{\frac{1}{2}} 1.5 \times 10^7 \sigma(\text{cm}^2) \tag{2.6.37}$$

where the total cross section is [2.6.7]

$$\sigma = \frac{1}{(KT)^2} \sum_\lambda \int_0^{\varepsilon_m} E_0 \sigma_\lambda(E_0) e^{-E_0/KT} \, dE_0 \tag{2.6.38}$$

and ε_m is the phonon energy ε_0 or ε_i for optical phonons and ε_m is sq_u for acoustic modes.

From eqs. (2.6.38) and (2.6.33), the total capture cross section σ for electron capture with *optical phonon emission* is with $G_\lambda(y) \equiv G_\lambda(q)$

$$\sigma_{opt} = \frac{m_c^2 V}{3\hbar^4} \left(\frac{A}{\frac{1}{2}m_c s^2}\right)^{3/g} \frac{1}{\gamma^2(1 - e^{-\varepsilon_0/KT})} \sum_\lambda \int_0^\beta P[\beta - z] e^{-z/\gamma} \, dz$$

$$\times \int_{y_{min}^2}^{y_{max}^2} G_\lambda(y) \left[\frac{y^2}{x^2 + y^2(x - z) + y^4/4}\right]^{3/g} d(y^2) \tag{2.6.39}$$

where

$$\gamma \equiv \frac{KT}{\frac{1}{2}m_c s^2} \quad \text{and} \quad \beta \equiv \frac{\varepsilon_0}{\frac{1}{2}m_c s^2}$$

* k is a wavevector. We therefore denote Boltzmann's constant by K in this section 2.6.5.

Table 2.6.3. *Constants for the optical phonons*

Material	ε_0 (eV)	sq_0 (eV)	V_m (eV)	$\frac{1}{2}m_c s^2$ (eV)	A (ev Å⁴)
Germanium	0.035	0.032	11	1.13×10^{-5}	11
	[2.6.15]	[2.6.15]		[2.6.7]	
Silicon	0.06	0.049	11	2×10^{-4}	20
	[2.6.7]	[2.6.7]	[2.6.7]	[2.6.7]	[2.6.7]

Values of V_m and A are appropriate for neutral centers.

The corresponding expression for longitudinal *acoustic phonon emission* is discussed in section 2.6.6. (For Coulombic centers it reduces to Lax's expression (2.21) provided the upper limits are replaced by infinity. For acoustic phonon emission and neutral centers one obtains *exactly* Lax's expression (6.23).)

Capture cross sections have been evaluated for a number of cases involving Coulombic and neutral centers with acoustic and optical phonon emission. A summary of constants for optical phonons in Ge and Si is given in Table 2.6.3.

For Coulombic centers, the sticking probabilities at various temperatures and for electron cascade capture with acoustic phonon emission and absorption were taken from earlier work [2.6.7], [2.6.14]. It is believed that these curves may be of the correct shape even for an optical phonon cascade capture at a Coulombic center, since the general form of $P(U)$ would not be expected to change significantly. For example, $P(U)$ must be small when $U < KT$ and approach unity for $U \gtrsim KT$, irrespective of the type of electron–phonon interactions involved.

As regards *neutral* centers, the use of the sticking probabilities just cited leads to temperature-dependencies of cross sections which are not observed in practice. Since the sticking probability for neutral centers is expected to be larger than for Coulombic centers, $P(U)$ for neutral centers was taken to be unity for $U \gtrsim KT$. This may be further justified by noting that the integral (2.6.39) over z is such that the main contributions come from the region $U \gg KT$. In both germanium and silicon the maximum allowed wave momentum is thus q_0, since Table 2.6.3 shows that $2[2m(E_0 + V_m)]^{\frac{1}{2}} > q_0$. In the numerical work presented here on Coulombic centers, q_0 has been also used as upper limit in the integration.

The optical phonon contribution to the capture cross section for electrons into attractive Coulombic centers in silicon resulting from eq. (2.6.39) and Table 2.6.3 is plotted as a function of temperature in Fig. 2.6.4. The 'Herring factor' ω has not been introduced. It is the square of the ratio of the optic to acoustic matrix elements, and is defined by

$$\omega = \frac{s^2 D_0^2 \hbar^2}{\Xi_D^2 \varepsilon_0^2}$$

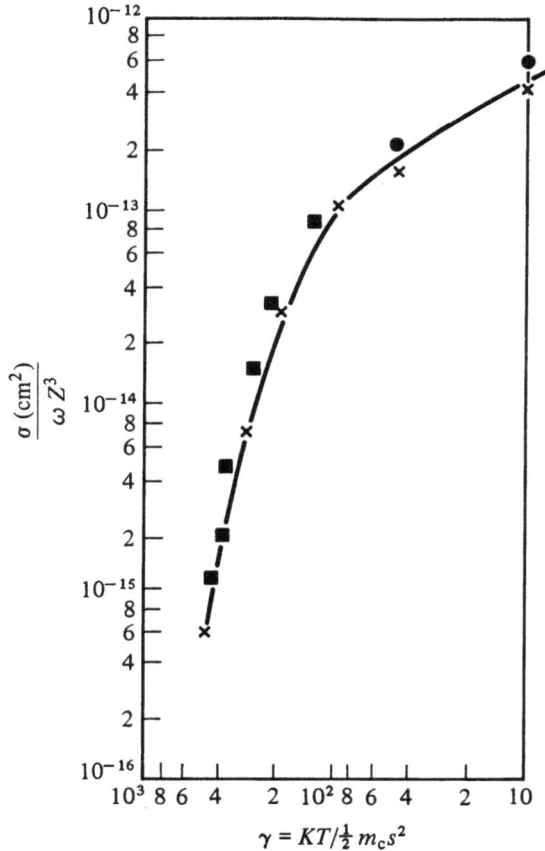

Fig. 2.6.4. Optical phonon contribution to the capture cross section, for electrons, of Coulombic attractive centers in Si according to eq. (2.6.39) and Table 2.6.3. ■, Results from [2.6.7], figure 9. ●, Results obtained using eq. (2.6.39) and the sticking probability of [2.6.7] as given in his figure 4 or equation (B.13). ×, As ●, but using the sticking probability of [2.6.14].

Lax's results are seen to be only slightly altered. From Table 2.6.4 the ratio of the optical to acoustic cross section for germanium is for Coulomb and neutral centers respectively: 479ω and 11ω.

Table 2.6.4 gives some room temperature cross sections obtained from the theory outlined here, but using Lax's sticking probability (as given in his figure 4 or his equation (B.13) when appropriate). They are compared with corresponding results obtained by Lax. The above calculations differ from Lax's in that q_0 was systematically introduced into all cases, for both optical and acoustic phonons. Only in the case of acoustic phonon emission at neutral centers are the approaches equivalent. This is in agreement with Table 2.6.4. The present results are consistently smaller than those of Lax, but the effect is small, except in the case of optical and acoustic phonon contributions to capture by Coulombic centers in Ge, where the restriction imposed by q_0 appears to be important. Since q_0 is larger for Si, one would expect his results to be more accurate in this case. Table 2.6.4 also gives the

Table 2.6.4. *Room temperature cross sections and recombination coefficients*

(i) Material	(ii) Center	(iii) Phonon	(iv) σ summed over the polarization index λ		(v) σ from [2.6.7] (10^{-17} cm²)	(vi) Correction factor for [2.6.7]	(vii) Recombination coefficient T_1^8 equivalent to column (iv) (10^{-8} cm³ s⁻¹)
			value (10^{-17} cm²)	equation used§			
Ge	Coulomb	acoustic	$0.014Z^3$	(2.6.33)	$0.051Z^{3*}$	0.27	$2.131 \times 10^{-4}Z^3$
Ge	Coulomb	optical	$6.7Z^3\omega$	(2.6.39)	$66Z^3\omega$	0.10	$0.102Z^3\omega$
Si	Coulomb	optical	$1200Z^3\omega$	(2.6.39)	$1500Z^3\omega$†	0.80	$15.8Z^3\omega$
Ge	neutral	acoustic	6.0	(2.6.33)	5.1*	0.98	0.091
Ge	neutral	optical	66ω	(2.6.39)	68.2ω*	0.97	1.00ω
Si	neutral	optical	330ω	(2.6.39)	440ω‡	0.75	4.35ω

* Worked out from information given in [2.6.7].
† [2.6.7], p. 1514.
‡ Based on [2.6.7], equation (6.22).
§ In conjunction with data in Table 2.6.2 for covalent materials.

recombination coefficients (2.6.37) and utilizes the result (2.6.41), below, for acoustic phonons.

The reasons for the significance of q_0, in the case of acoustic phonon emission in germanium, may be explained in the following way. One might anticipate that the restrictions imposed by q_0 should be small on the grounds that an electron with kinetic energy of order KT should dominate the cascade process for nondegenerate bands. Thus, cutting off at $sq_0 > KT$, might be thought to be insignificant. In fact, as seen from Fig. 2.6.3, in the neighborhood of an attractive Coulombic or neutral center, the kinetic energy E can be considerably greater than KT, and hence the emitted phonons can also have energy greater than KT. By restricting the emitted phonon wavevectors to the first Brillouin zone, the contribution arising from these energetic phonons is eliminated.

Again, one might suspect that the cut-off may have little effect, since the transition probability (2.6.22) is proportional to $(N_{sq}+1)$, which rapidly decreases with increasing emitted energy. This factor, however, is almost completely canceled by the sticking probability, which has the opposite tendency of favoring transitions involving larger q's, a tendency which is enhanced with increasing temperature. These considerations explain the importance of sq_0 in room temperature calculations.

In the case of optical phonons in covalent materials, the treatment given here cannot be easily compared with Lax's work. (For example, the limits appearing in his equation (4.5) have been given in terms of optical phonon energies, and not in terms of the phonon wave momenta, as here in eqs. (2.6.35) and (2.6.36).)

The temperature-dependencies of cross sections have not been determined in all cases. One would expect them to be approximately the same as those obtained by Lax. For example, for the optical phonon contributions to capture by neutral centers, the temperature in eq. (2.6.39) is contained in

$$\frac{1}{\gamma^2(1-e^{-\varepsilon_0/KT})}\int_0^\beta P[\beta-z]\,e^{-z/\gamma}\,dz$$

Assuming that the dominant part of the integral comes from $z = 0$, and assuming $P[\beta]$ is unity, this becomes

$$\frac{1}{\gamma} \frac{1}{1 - e^{-\varepsilon_0/KT}} = \frac{m_c s^2}{2KT} \frac{1}{1 - e^{-\varepsilon_0/KT}}$$

Thus σ is approximately temperature-independent for $\varepsilon_0 < KT$ and behaves as T^{-1} for $\varepsilon_0 > KT$.

Lastly, we observe a difference between the formalism given here (eq. (2.6.39) and Lax's presentation in the case of capture by neutral centers with optical phonon emission. Lax's cross section (his equation (6.20)) contains the trap radius R reciprocally. However, it does not appear in our results if $2[2m(E_0 + V_m)]^{\frac{1}{2}} > q_0$ (as in Ge and Si). Only if this inequality is the other way round, is R introduced through V_m. Thus the introduction of the limit q_0 leads to an amendment of earlier formulae also in this case.

2.6.6 The longitudinal acoustic phonon cascade capture cross section

The counterpart to eq. (2.6.39) (optical phonon emission) for capture with acoustic phonon emission is readily obtainable from the general theory of section 2.6.3. To this end one need merely use eq. (2.6.28) with $G_1(q) = \Xi_D^2 q/2\rho s V$ from Table 2.6.2 and, using eq. (2.6.20)

$$\varepsilon_{\lambda q} \to sq = E_0 + U \tag{2.6.40}$$

The result is best stated in terms of

$$z \equiv \frac{E_0}{\frac{1}{2}m_c s^2}, \quad \eta \equiv \frac{U}{\frac{1}{2}m_c s^2}, \quad \gamma \equiv \frac{kT}{\frac{1}{2}m_c s^2}, \quad l_c^{-1} \equiv \frac{m^2 \Xi_D^2 kT}{\pi \hbar^4 \rho s^2}$$

$$\sigma_1 \equiv \frac{\pi}{12} \left(\frac{A}{\frac{1}{2}m_c s^2}\right)^{3/g} \frac{1}{\gamma l_c}$$

where l_c is a mean free path and σ a cross section. Then

$$z\sigma_1(z, \eta) = \frac{\sigma_1 P(\eta)}{1 - \exp[-(z+\eta)/\gamma]} \frac{(z+\eta)^2}{\{[1 + \frac{1}{4}(z+\eta)]^2 - z\}^{3/g}}$$

The integration over all $|q|$, introduced in eq. (2.6.26), becomes here an integration over all phonon energies, i.e. all η, by virtue of formula (2.6.40), and yields a cross section for given incident energy E_0. Integration of a Maxwell distribution of incident energies, as in eq. (2.6.38), yields an average cross section

$$\sigma = \frac{1}{\gamma^2} \int_0^{z_m} e^{-z/\gamma} \, dz \int_0^{z_m - z} z\sigma_1(z, \eta) \, d\eta \tag{2.6.41}$$

where

$$z_m \equiv \frac{sq_u}{\frac{1}{2}m_c s^2}$$

Here q_u is given in eq. (2.6.35). Equation (2.6.41) gives the required result and has been used in the numerical work given in Table 2.6.4.

2.6.7 Remarks on the relevant literature

It is desirable to give a physical discussion of the origin of the temperature dependencies of the cross sections which can be very varied as noted in section 2.6.1 and at the end of section 2.6.5. This is however still somewhat controversial and we refer to the review [2.6.3] up to 1977. There is also a review of cascade capture up to 1974 in [2.6.16]. From these reviews it will be noticed that the period of greatest activity in this area was in the 1960s followed of course by applications (for example to piezoelectric effects [2.6.17] and photovoltaics [2.6.18]). It is therefore perhaps not surprising that in a comparatively recent review of recombination in narrow-gap semiconductors [2.6.19] cascade capture received only slight attention. Nonetheless attempts have been made to set up and solve the set of $M+1$ equations in the M donor states (of which $M-1$ are excited states) and the conduction band electron concentration. Numerical results have been obtained for both the steady-state and transient condition [2.6.20], [2.6.21].

In the case of noncrystalline materials the mechanisms of nonradiative recombination such as occur in cascade capture are not yet clear [2.6.22]. The multiphonon transitions (the theory is briefly outlined in chapter 6) are believed to be widely responsible. The energy gap E_G has then to be interpreted as the mobility gap and the more phonons are required to bridge it the smaller is the probability of its occurrence. Deep impurities reduce the number of phonons required for one event and therefore favor multiphonon capture relative to shallower impurities. The temperature-dependence tends to be exponential in this case, as seen from Fig. 6.1.1, below. Recombination mechanisms for amorphous Si have been reviewed in [2.6.23].

2.7 Surface recombination and grain boundary barrier heights

2.7.1 Recombination statistics for surface states

In the history of the transistor the importance of the electrical properties of surfaces played a key role. This is easily realized by perusal of the appropriate literature cited in the Introduction. Surface studies continue to play a crucial part in semiconductor research. Interfaces for the design of heterostructures are of wide interest, the passivation of surfaces so as to reduce surface recombination is important, as are studies of small displacements of surface atoms from the positions they would have in the bulk (this is called surface reconstruction). Surface spectroscopy and scanning tunneling microscopy give in addition accurate information of surface defects and their densities [2.7.1], [2.7.2]. We here pursue some recombination aspects of surfaces.

For our purposes we assume that the number of states per unit area at the surface and falling within the energy gap of the bulk semiconductor is $\mathcal{N}(E_t)\,dE_t$ where dE_t is the energy range considered. Let dn_t, dp_t be the number of full and empty traps, respectively, both per unit area. Then

$$dp_t = \mathcal{N}\,dE_t - f_t\,\mathcal{N}\,dE_t = (1-f_t)\,\mathcal{N}\,dE_t = \mathcal{N}\,dE_t - dn_t \qquad (2.7.1)$$

where f_t is the occupation probability of a trap. The electron and hole trapping rates per unit area into the increment of states $\mathcal{N}\,dE_t$ are (with G, H given by eqs. (2.2.11) and $r = 1$)

$$du_e = G(n\,dp_t - n_1\,dn_t), \quad du_h = H(p\,dn_t - p_1\,dp_t) \qquad (2.7.2)$$

From detailed balance in equilibrium

$$n_1 = \left(n\frac{dp_t}{dn_t}\right)_{eq} = n_{eq}\left(\frac{1-f_t}{f_t}\right)_{eq} = n_{eq}\exp\left(\eta_t - \gamma_{eq}\right) = N_c\exp\left(\eta_t - \eta_c\right)$$

where η_t is the reduced trapping level energy and γ_{eq} is the reduced quasi-Fermi level in equilibrium. Similarly one can verify the usual relation for p_1 (which is easily derived also from eqs. (2.2.13) with $r = 1$)

$$p_1 = N_v\exp\left(\eta_v - \eta_t\right)$$

In the steady state eqs. (2.7.2) give as analogue of eq. (2.3.19)

$$dn_t = f_t\,\mathcal{N}(E_t)\,dE_t = \frac{nG + p_1 H}{(n+n_1)G + (p+p_1)H}\mathcal{N}(E_t)\,dE_t \qquad (2.7.3)$$

Putting eq. (2.7.3) back into eqs. (2.7.2), the steady-state recombination rate per unit area into traps in the energy range dE_t is (cf. eqs. (2.2.17), (2.3.5))

$$du = (du_e)_{\text{steady state}} = (du_h)_{\text{steady state}}$$
$$= \frac{GH(np - n_1 p_1)}{G(n+n_1) + H(p+p_1)}\mathcal{N}(E_t)\,dE_t \qquad (2.7.4)$$

There are the usual results except (a) the recombination rate is per unit area (not per unit volume), (b) the trap concentration per unit volume has been replaced by an incremental trap concentration per unit area

$$N_t \rightarrow \mathcal{N}(E_t)\,dE_t$$

In a p-type semiconductor, putting $n_1 p_1 = n_{eq}p_{eq}$ and $p \sim p_{eq}$,

$$du = (n - n_{eq})\,G(E_t)\,\mathcal{N}(E_t)\,dE_t$$

so that the minority-carrier 'lifetime' involves an integral over the spectrum of surface states:

$$\tau_n^{-1} = \frac{u}{n - n_{eq}} \sim \int G(E_t)\,\mathcal{N}(E_t)\,dE_t$$

Similarly the minority-carrier 'lifetime' in an n-type material satisfies

$$\tau_p^{-1} \sim \int H(E_t)\,\mathcal{N}(E_t)\,dE_t$$

For bulk material $[G] = [H] = [L^3 T^{-1}]$, $[\mathcal{N}(E_t)\,dE_t] = [L^{-3}]$ and we have the usual situation except that we have a spectrum $\mathcal{N}(E_t)\,dE_t$ of traps per unit volume; as already considered in section 2.3.6. We now simply change the dimension for surfaces:

$$[G] = [H] = [L^3 T^{-1}], \quad [\mathcal{N}(E_t)\,dE_t] = [L^{-2}]$$

so that τ_n^{-1}, τ_p^{-1} are surface recombination velocities. This reinterpretation makes the theory outlined in section 2.2.4 applicable to surfaces. (For the effect of extra carriers due to tunneling, etc. see [2.7.3]).

Now f_t can be thought of as a modified Fermi distribution

$$f_t(E_t) = \frac{nG + p_1 H}{(n+n_1)G + (p+p_1)H} = \frac{1}{1 + g\exp(\eta_t - \gamma)} \tag{2.7.5}$$

where, as in eq. (2.3.22)

$$g = \frac{n_1 G + pH}{nG + p_1 H}e^{\gamma - \eta_t} = \frac{N_c\exp(\eta_t - \eta_c)G + N_v\exp(\eta_v - \gamma_h)H}{N_c\exp(\gamma_e - \gamma)\exp(\eta_t - \eta_c)G + N_v\exp(\eta_v - \gamma)H} \tag{2.7.6}$$

γ is an unspecified Fermi level which becomes γ_{eq} in equilibrium; g is seen to become unity in this case.

2.7.2 A schematic grain boundary model [2.7.4], [2.7.5]

Fig. 2.7.1 shows a steady-state energy diagram for a model p-type grain due to acceptors at energy E_A. Surface recombination takes place through N_{DS} donor states and N_{AS} acceptor states at energy E_s at the grain surface $x = 0$. From Fig. 2.7.1 one sees that

$$\eta_A(x) = \eta_A(w) - e\varphi(x)/kT \tag{2.7.7}$$

The assumptions to be made are as follows: (a) The surface of the grain is taken as a flat surface between identical grains. This makes a one-dimensional treatment appropriate. (b) The recombination rates in the bulk and at the surface are regarded as not too different, so that one may assume parallel and flat quasi-Fermi levels of known separation $(\mu_e - \mu_h)$. Thus bulk recombination need not be introduced explicitly; its rate can be worked out from $\gamma_e - \gamma_h$ once appropriate

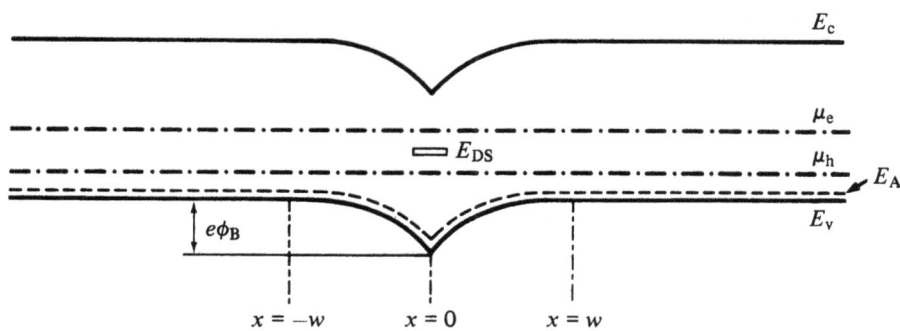

Fig. 2.7.1. Model of a grain boundary at $x = 0$.

recombination center densities and trapping cross sections are settled. (c) The carriers in the grains are nondegenerate. (d) They recombine at the surface and in the bulk by the SRH process. (e) The bulk recombination centers contribute only negligibly to the space charge when compared to the effect of the dopants, either N_D^+ or N_A^-.

The departure from equilibrium in the model is imposed by taking various specific separations between the quasi-Fermi levels as given. This separation can be established by injection or optical excitation which has to overcome recombination in the bulk of the grain. In this way of looking at the problem it is therefore not necessary to introduce bulk recombination explicitly, and the band edges follow also the law (2.7.7).

To obtain the barrier height $e\varphi_B$ an integration of Poisson's equation

$$-\frac{d^2\varphi(x)}{dx^2} = \frac{4\pi e}{\varepsilon}(p-n-N_A^-) \quad [\varphi(0) \equiv \varphi_B, \varphi(w) \equiv 0] \tag{2.7.8}$$

is required. Here $N_A^-(x)$ is the concentration of charged acceptors, p and n are electron and hole concentrations. Using eq. (2.7.7)

$$N_A^-(x) = N_A\{\exp[\eta_A(x)-\gamma_h]+1\}^{-1} = N_A\left\{\exp\left[\eta_A(w)-\frac{e\varphi(x)}{kT}-\gamma_h\right]+1\right\}^{-1} \tag{2.7.9}$$

$$n(x) = N_c\exp[\gamma_e-\eta_c(x)] = n(w)\,\zeta(x) \tag{2.7.10}$$

$$p(x) = N_v\exp[\eta_v(x)-\gamma_h] = p(w)/\zeta(x)$$

A standard notation has been used with $\gamma_e \equiv \mu_e/kT$, $\gamma_h \equiv \mu_h/kT$, and

$$\zeta(x) \equiv \exp[e\varphi(x)/kT]$$

By introducing the readily obtained quantities $n(w)$ and $p(w)$ the two variables $n(x)$, $p(x)$ have now been replaced by the single variable $\zeta(x)$ whose value at $x = 0$ will

be calculated later. Two boundary conditions are needed to integrate Poisson's
equation. At the edge of the space-charge region, $x = w$

$$\left.\frac{d\varphi}{dx}\right|_{x=w} = 0 \tag{2.7.11}$$

At the grain boundary surface, $x = 0$, Gauss's law is used, by taking a short pill
box of unit area at right angles to the grain boundary surface. Then if charges q_i
are contained within it, in a usual notation,

$$\int_A \mathscr{E} \cdot \mathbf{n}\, dA = \sum_i \frac{q_i}{\varepsilon}$$

In the limit of an infinitely short pill box $\sum q_i$ becomes the charge density Q_s per
unit area in the surface states. The field $\mathscr{E}$ emerges from each end of the pill box
so that

$$-2\left.\frac{d\varphi}{dx}\right|_{x=0} = \frac{4\pi Q_s}{\varepsilon} \tag{2.7.12}$$

where the charge per unit area on the surface is

$$Q_s = \int_{E_{vS}}^{E_{cS}} e\{N_{DS}(E_S)[1 - f(E_S, e\varphi_B)] - N_{AS}(E_S)f(E_S, e\varphi_B)\}\, dE_S \tag{2.7.13}$$

N_{DS} and N_{AS} are the numbers per unit area (of the grain boundary surface) of
donor and acceptor traps; $f(E_S, e\varphi_B)$ is the occupation probability of a surface
energy level E_s in the presence of a barrier height $e\varphi_B$.

The steady-state trap occupation at the grain boundary $x = 0$ is given by eq.
(2.7.5) with $\gamma \equiv \gamma_h$ and

$$g \equiv g(E_S, e\varphi_B) = \frac{n_1(E_S)\, G(0) + [p(w)/\zeta(0)]\, H(0)}{n_1(E_S)\, G(0) \exp(\gamma_e - \gamma_h) + [p(w)/\zeta(0)]\, H(0)} \tag{2.7.14}$$

2.7.3 The grain boundary barrier height

The Poisson equation for the potential $\varphi(x)$ is

$$-\nabla^2\varphi = F(\varphi)$$

where

$$F(\varphi) \equiv \frac{4\pi e}{\varepsilon}\left\{ p(w)\,\zeta(x)^{-1} - n(w)\,\zeta(x) - \frac{N_A}{\exp[\eta_A(w) - \gamma_h]\,\zeta(x)^{-1} + 1} \right\}$$

If we multiply by φ' ($\equiv d\varphi/dx$) and integrate from 0 to w we get

$$-\int_0^w [\varphi'(x)\nabla^2\varphi]\,dx = f(0) - f(\varphi_B) \qquad (2.7.15)$$

where

$$f(\varphi) \equiv \int^\varphi F(\varphi)\,d\varphi$$

The result is

$$f(0) - f(\varphi_B) = \frac{4\pi kT}{\varepsilon}\left\{ n(w)[\zeta(0) - 1] + p(w)[\zeta(0)^{-1} - 1]\right.$$

$$\left. + N_A\left[\ln\zeta(0) - \ln\frac{e^{\eta_A(w) - \gamma_h} + 1}{\zeta(0)^{-1}e^{\eta_A(w) - \gamma_h} + 1}\right]\right\}$$

The left-hand side of eq. (2.7.15) is, by eqs. (2.7.11) and (2.7.12)

$$-\tfrac{1}{2}[\varphi'(w)^2 - \varphi'(0)^2] = 2\left(\frac{\pi Q_s}{\varepsilon}\right)^2$$

Thus we obtain our main result, a relation for the barrier height $\varphi_B = (kT/e)\ln\zeta(0)$:

$$n(w)[\zeta(0) - 1] + p(w)[\zeta(0)^{-1} - 1] + N_A\left[\ln\zeta(0) - \ln\left(\frac{e^{\eta_A(w) - \gamma_h} + 1}{\zeta(0)^{-1}e^{\eta_A(w) - \gamma_h} + 1}\right)\right]$$

$$= \frac{\pi e^2}{2\varepsilon kT}\left[\int_{E_{vS}}^{E_{cS}}\left(\frac{N_{DS}(E_S)}{1 + g(E_S, e\varphi_B)^{-1}e^{\gamma_h - \eta_S}} - \frac{N_{AS}(E_S)}{1 + g(E_S, e\varphi_B)e^{\eta_S - \gamma_h}}\right)dE_S\right]^2 \qquad (2.7.16)$$

The form (2.7.5) of the Fermi–Dirac distribution has been used here and comes in through the surface charge Q_S. The equilibrium barrier height is also given by eq. (2.7.16) provided $g(E_S, e\varphi_B)$ is replaced by unity.

In order to locate the quasi-Fermi levels one can assume electrical neutrality in the grain bulk

$$p(w) - n(w) - N_A^- = 0 \qquad (2.7.17)$$

Together with

$$n(w)p(w)\exp(\gamma_h - \gamma_e) = n_1(E_S)p_1(E_S) = n_i^2$$

this gives a quadratic equation for $p(w)$ whose solution is with $N_A^- = N_A$

$$p(w) = N_A/2 \pm [N_A^2/4 + n_i^2\exp(\gamma_e - \gamma_h)]^{\frac{1}{2}} \qquad (2.7.18)$$

Now regard as given: kT, ε, E_G, N_c, N_v, n_i, $G(0)$, $H(0)$, N_A, N_{DS}, $E_{DS} - E_{vs}$ and $\mu_e - \mu_h$. The values of n_1, p_1 can then be obtained and so can $p(w)$ from eq. (2.7.18). From this one can get $\mu_h - E_v(w)$ and so $E_c(w) - \mu_e$ since $\mu_e - \mu_h$ is given. Then eq. (2.7.16), or an equivalent relation, can be used to find $\zeta(0)$ and so the barrier height $\varphi(0) = \varphi_B$.

Note that this procedure does not require the depletion approximation which is frequently used, see for example [2.7.6]–[2.7.8]. It would remove the $n(w)$ and $p(w)$ terms from the left-hand side of eq. (2.7.16).

In order to obtain quantitative results we shall assume

$$N_{AS} = 0, \quad \varepsilon = 11.8, \quad E_G = 1.1 \text{ eV}, \quad N_c = 2.8 \times 10^{19} \text{ cm}^{-3},$$

$$N_v = 1 \times 10^{19} \text{ cm}^{-3}, \quad n_i = 4.67 \times 10^{9} \text{ cm}^{-3}, \quad kT = 0.026 \text{ eV},$$

$$G(0) = 10^{-7} \text{ cm}^3 \text{ s}^{-1}, \quad H = 10^{-9} \text{ cm}^3 \text{ s}^{-1}, \quad E_{DS} - E_{vs} = 0.91 \text{ eV} \quad (2.7.19)$$

We also note for rough orientation that according to the depletion approximation n and p are neglected in the bulk, so that the charge in the surface and the barrier height are by a double integration of Poisson's equation

$$Q_S \propto \begin{cases} -N_A w & (p\text{-type grain}) \\ +N_D w & (n\text{-type grain}) \end{cases} \quad (2.7.20)$$

$$\varphi_B \propto \begin{cases} N_A w^2 \propto Q_S/N_A & (p\text{-type grain}) \\ N_D w^2 \propto Q_S/N_D & (n\text{-type grain}) \end{cases} \quad (2.7.21)$$

Consider a p-type grain in equilibrium. For low bulk acceptor doping N_A the Fermi level μ_{eq} lies above the energy E_{DS} of the surface donors. There are few ionized donors, a small Q_S and hence little band bending (for example, by virtue of (2.7.21)). As N_A is increased the band edges, locked rigidly to the level E_{DS}, may be thought as moving down as if pulled by μ_{eq}, which however moves down more markedly. Thus the ionized donor concentration N_{DS}^+ increases and Q_S and φ_{B0} increase. When $\mu_{eq} < E_{DS}$, $N_{DS} \sim N_{DS}^+$ and Q_S has reached its maximum value so that further increase of N_A is by (2.7.21) expected to lead to a drop in φ_{B0}. These ideas are borne out by Fig. 2.7.2 which shows the maxima of φ_{B0} according to eq. (2.7.16). These occur when $\mu_e \sim E_{DS}$ and so move to higher N_A values as N_{DS} is increased.

Upon illumination the quasi-Fermi levels separate and surface donors previously ionized become neutral. The drop in charge Q_S leads to a drop in the steady-state (and now nonequilibrium) barrier height φ_B. This is shown in Fig. 2.7.3 [2.7.9] which gives good agreement with experiment. It required the conversion of a uniform carrier generation rate into a quasi-Fermi level separation [2.7.10].

Various other comparisons with experiment can be made [2.7.11] and are broadly satisfactory. If the band bending is so strong that in equilibrium the

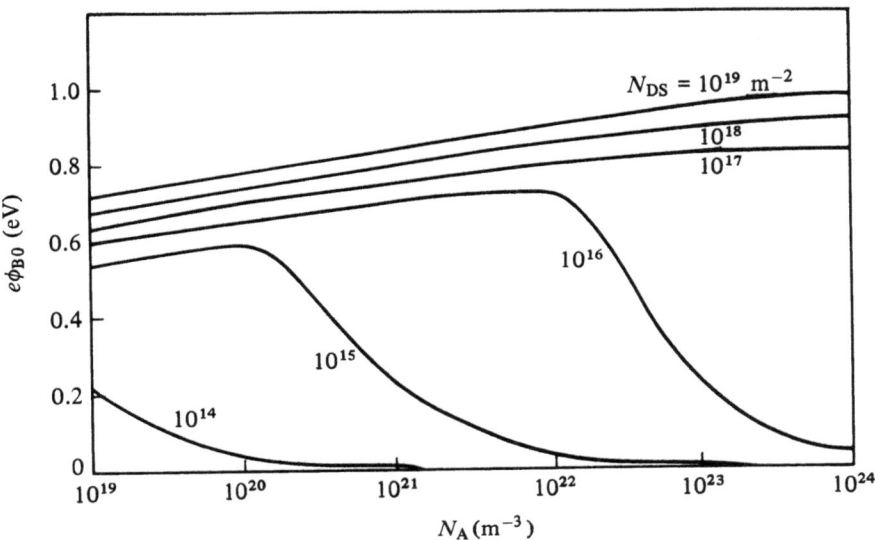

Fig. 2.7.2. The equilibrium barrier height of a p-type grain [2.7.5].

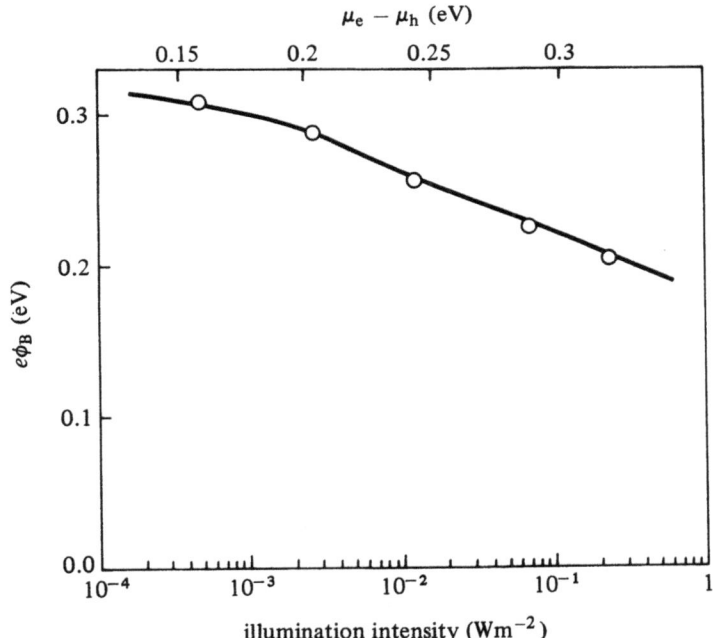

Fig. 2.7.3. Barrier height for an isolated grain boundary in p-type silicon. The points are for sample B-12 of [2.7.8]. The curve is for the theory outlined here with $T = 293$ K, $E_{DS} - E_{vS} = 0.77$ eV, $N_{DS} = 2.25 \times 10^{11}$ cm^{-2}, $N_A = 3 \times 10^{15}$ cm^{-3}.

conduction band reaches the Fermi level at $x = 0$ the electrons in the neighborhood
of $x = 0$ form an inversion layer and this effect has also been discussed for grain
boundary recombination [2.7.12].

For a cylindrical ($\eta = 1$) or spherical grain ($\eta = 2$), $\varphi = \varphi(r)$ and

$$-\nabla^2 \varphi(r) = -\varphi''(r) - (\eta/r)\,\varphi'(r) = F(\varphi)$$

so that (2.7.15) can be written

$$\varphi'(r = 0)^2 - \eta \int_0^{\varphi_B} \frac{\varphi'(r)}{r}\,\mathrm{d}\varphi = -f(\varphi_B) \qquad (2.7.22)$$

This is again just the integrated Poisson equation with the integrated space charge
in the bulk semiconductor on the right-hand side. The first term brings in the
surface space charge Q_s per unit area via an application of Gauss's theorem. This
type of theory shows that the barrier height is lowered by grain curvature. This
lowering increases with curvature [2.7.13]. It leads also to a rather general way of
looking at 'equal-area' rules, of which the Maxwell construction is the most
famous [2.7.14].

For recent experimental work, note for example the reduction in grain boundary
recombination by fast annealing [2.7.15].

2.8 Recombination at dislocations

It was pointed out by Shockley that an edge dislocation in a semiconductor can
give rise to dangling bonds and hence localized (usually acceptor) levels [2.8.1].
This was largely confirmed and dislocations were experimentally introduced into
materials by plastic deformation. Thus an n-doped specimen can become p-type
after deformation. The dislocation number could be estimated from X-ray rocking
curves and from etch pit counts. The flow of impurities to such dislocations was
studied, as were their optical properties.

Here we wish to touch briefly on the recombination properties of dislocations
using the barrier model [2.8.2] reviewed in [2.8.3]. One-dimensional energy bands
can be associated with edge-type dislocations and transitions between them (and
also involving the main bands of a semiconductor) change the lifetimes of carriers.
In the early work the SRH type of statistics was used [2.8.4] for the theoretical
description and Read's model of a dislocation [2.8.5] was tested by the study of
Hall constant, conductivity and magnetoresistance [2.8.6]. For example the
dislocation acceptor level in n-type germanium was thus inferred to lie at 0.50 eV
below the conduction band edge. The general concepts of sections 2.2 and 2.3 can
be applied, using the idea of a common equilibrium Fermi level for the relevant
dislocation band and for the crystal. A rigid shift of all dislocation states due to

their line charge was originally assumed and actually gave reasonable agreement with experiment. The temperature-dependence of the lifetime required, however, the introduction of tunneling through the extremum of the potential distribution [2.8.7], [2.8.8]. Recombination radiation from dislocations in silicon was discovered later [2.8.9].

According to conventional ideas a neutral dislocation will have some states occupied by electrons while others are empty. On equilibration with the rest of the crystal the dislocation may get charged positively by losing electrons to *the p-type material*. Thus Fig. 2.7.1 becomes applicable. Equivalently, negatively charged dislocations can occur in *n*-type material. For this Fig. 2.7.1 has to be inverted. In any case, the charged dislocation line is surrounded by a space charge of opposite sign – negatively charged acceptors in our case. A potential difference then appears between the charged dislocation and the bulk material. This will act as a barrier for holes and as a well for electrons, and conversely for *n*-type material. This is the 'barrier model' of recombination through dislocations.

It has been established more recently that with increasing dislocation density N_d the carrier lifetime τ can either increase or decrease or it can be independent of N_d. According to unmodified SRH statistics increase in N_d should simply cause a decrease in τ. One explanation of why this is not always so is presumably that one has to include the effect of the space charge region as in the barrier model noted above. For *p*-type material this leads to Fig. 2.7.1. The recombination properties of dislocations, stacking faults and other extended defects remain relatively poorly understood and theories are difficult to test. A recent simple theory [2.8.10], which assumes a rectangular potential barrier, will, however, be given next, as the subject is technologically important.

The following notation will be adopted:

v = volume of the space charge region of unit length of a dislocation;

$f \equiv N_d v$ fraction of total volume occupied by space charge regions;

φ_0 = magnitude of the potential barrier near dislocations at equilibrium;

$\Delta\varphi$ = change in φ as the charge carrier injection level is increased;

φ = potential barrier near dislocation = $\varphi_0 - \Delta\varphi$ – a barrier for holes, a well for electrons for *p*-type material;

$\Delta n'$, $\Delta p'$ = concentrations of nonequilibrium carriers in the crystal;

$\Delta n' \exp(\varphi/kT)$, $\Delta p' \exp(-\varphi/kT)$ = concentrations of nonequilibrium carriers near a dislocation.

It follows that the volume-averaged nonequilibrium charge carrier concentrations are

$$\Delta p = (1-f)\Delta p' + f\Delta p' \exp(-\varphi/\mathrm{kT}) \equiv \alpha\Delta p'$$
$$\Delta n = (1-f)\Delta n' + f\Delta n' \exp(\varphi/kT) \equiv \beta\Delta n'$$

(2.8.1)

The total concentrations of carriers in the dislocation space charge regions are

$$n = (n_{eq} + \Delta n') \exp(\varphi/kT) = (n_{eq} + \Delta n/\beta) \exp(\varphi/kT)$$
$$p = (p_{eq} + \Delta p') \exp(-\varphi/kT) = (p_{eq} + \Delta p/\alpha) \exp(-\varphi/kT)$$

(2.8.2)

where n_{eq}, p_{eq} are the equilibrium concentrations in the crystal.

Now as in eq. (2.7.4) the recombination rate per unit volume is

$$u = \frac{(np - n_1 p_1) GHN_t}{G(n + n_1) + H(p + p_1)}$$

$$= \frac{(n_{eq} \Delta p/\alpha + p_{eq} \Delta n/\beta + \Delta n \Delta p/\alpha\beta) GHN_t}{\left[n_{eq} + \dfrac{\Delta n}{\beta} + n_1 \exp\left(-\dfrac{\varphi}{kT}\right) \right] G \exp\left(\dfrac{\varphi}{kT}\right) + \left[p_{eq} + \dfrac{\Delta p}{\alpha} + p_1 \exp\dfrac{\varphi}{kT} \right] H \exp\left(-\dfrac{\varphi}{kT}\right)}$$

where N_t ($\propto N_d$) is the concentration of dislocation recombination centers. The minority-carrier lifetime is for $\Delta n = \Delta p$

$$\tau = \frac{\Delta n}{u} =$$

$$\alpha \frac{\left[n_{eq} + \dfrac{\Delta n}{\beta} + n_1 \exp\left(-\dfrac{\varphi}{kT}\right) \right] G \exp\left(\dfrac{\varphi}{kT}\right) + \left[p_{eq} + \dfrac{\Delta n}{\alpha} + p_1 \exp\left(\dfrac{\varphi}{kT}\right) \right] H \exp\left(-\dfrac{\varphi}{kT}\right)}{(n_{eq} + p_{eq}\, \alpha/\beta + \Delta n/\beta) GHN_t}$$

(2.8.3)

This relation applies also to the minority-carrier lifetime *in n-type material*, provided φ is replaced by $-\varphi$ [2.8.11].

If $\varphi = 0$ one obtains the lifetimes outside the dislocation space charge region.

Consider now *n-type material* with trapping at acceptor levels, then we can put $H \gg G$ and eq. (2.8.3) gives

$$\tau = \frac{p_{eq} + \Delta n/\alpha + p_1 \exp(-\varphi/kT)}{(n_{eq} + p_{eq}\, \alpha/\beta + \Delta n/\beta) GN_t} \alpha \exp\left(\frac{\varphi}{KT}\right)$$

By (2.8.1) with $\Delta p' \to p_{eq}$

$$\alpha p_{eq} = \left[1 + f + f \exp\frac{\varphi}{kT} \right] p_{eq} \sim f \exp\left(\frac{\varphi}{kT}\right) p_{eq} \sim N_d v \exp\left(\frac{\varphi}{kT}\right) p_{eq}$$

Hence

$$\tau \sim \frac{N_d v p_{eq} \exp(\varphi/kT) + \Delta n}{n_{eq} GN_t} \exp\left(\frac{\varphi}{kT}\right)$$

Thus one can see that for small Δn an increase of τ with N_d may be expected due to an increase of φ_0. For large Δn one may expect a decrease in τ as N_d is increased because N_t increases. This is merely *one* example of the application of such a theory. In fact the more detailed analysis of section 2.7 should be applicable to a simple model of a dislocation, but this does not appear to have been attempted yet. A complicating feature arises from the interaction among single sites on a dislocation, due to their close proximity. An occupied site is expected to reduce the occupation probability of neighboring sites. The capture cross section *of a site* is thus expected to decrease as a dislocation becomes more heavily occupied.

The application of the scanning electron microscope ('SEM') to the study of semiconductors by the use of an electron beam induced conductivity (the 'EBIC' method) has led to important progress in the understanding of recombination properties of defects and dislocations, as discussed in [2.8.12]. (Although of wider application to defects, this topic is included here for convenience under the heading of dislocations.) It is then useful to consider the quantity $C \equiv (I_0 - I_d)/I_0$ where I_0 and I_d are currents induced far from a defect and at a defect, respectively. This so-called 'contrast' measures the recombination strength of the defect. In *n*-type material defects tend to be negatively charged and so act as recombination centers for the beam-induced minority carriers. With higher beam currents the centers tend to be neutralized, thus reducing the average recombination efficiency. This reduction of contrast with beam current is well documented in both Si [2.8.13] and CdTe [2.8.14]. For a theory of the beam-induced current at grain boundaries in polycrystalline materials see [2.8.15].

The method has revealed additional contrast at the slip planes swept by moving dislocations during plastic deformation. This is believed to be due to electrically active centers left in the wake of the dislocations. These centers are expected to give rise to a potential barrier (as discussed in section 2.7); but this should decrease with increasing injection [2.8.13], [2.8.16].

The slip planes are a source of anisotropy for both electron mobility (which is decreased in a direction perpendicular to the slip plane) and dislocation speeds (which are increased along the slip plane) as a result of dislocation movement, and these interesting effects are worthy of further study. The Colloque 4 of vol. 44 (1983) of the *Journal de Physique* contains reports of much additional work on recombination at dislocations (see also [2.8.17]–[2.8.19]).

3

Auger effects and impact ionization (mainly for bands)

3.1 Introduction

Auger transitions have already been encountered (for example in section 2.1.1). They have been of interest in solids since 1940 when soft X-ray emission was still the main method of probing the density of electron states in solids and the probability of electronic transitions back into an atomic inner shell. Skinner [3.1.1], working at the University of Bristol, thought the low-energy tail in the soft X-ray spectrum of Na might be due to lifetime broadening of the empty conduction band state left behind by an electron which had made the transition back into the atomic rump. This short lifetime might be due to electron collisions of the type shown later (Fig. 5.2.9). By the uncertainty principle $\Delta E \Delta t \gtrsim \hbar$ this would then lead to the broadening effect. For an almost ideal metal like Na the second electron would have to make a transition across the Fermi level in order to find an empty state with a good probability. A straightforward application of theory in the years 1947–9 to obtain the considerable (1 eV) tail was plagued by a divergence: one obtained an infinitely long tail! This led to the idea that a screened electron–electron interaction has to be used in these problems to take approximate account of correlations in the motion of the electrons [3.1.2]. This screening is thus a many-body effect and it was precisely here, in the discussion of the electron collisions within a band, that it was first introduced, albeit in a semi-empirical manner. It was widely used later as part of the Bohm–Pines theory of plasma oscillations which came in the 1950s [3.1.3]. We shall meet this tailing effect again when we discuss recombination spectra of degenerate semiconductors and of electron–hole drops (see Fig. 5.2.9 and section 5.2.7). In nondegenerate materials there are fewer Auger transitions into the vacant electron state and this lifetime broadening is negligible.

The basic effect was discovered in 1925 by Pierre Auger in gases. It was explained as follows. An atom is ionized in an inner shell. An electron drops into the vacancy from a higher orbit and a second electron takes up the energy which is used to eject it from the atom. This effect was subsequently widely invoked in the analysis of X-ray spectra. In solids the effect is roughly analogous and so we have called it again 'the Auger effect', a name now firmly established. One of its characteristics is that it is not radiative – the energy of a downward transition is often used to emit a photon. Since radiation is what is seen in many experiments, the Auger effect is suspected if and when there is less radiation than expected in the first place, and it is rather harder to investigate than radiative transitions.

In the 1950s estimates of the Auger effect in the solid state were made [3.1.4], but it remained essentially a curiosity well into the 1960s, although the recombination statistics were then worked out (cf. section 2.1.1). The theory was put on a firmer foundation in 1958 [3.1.5] and since that time the importance of Auger transitions in semiconductors has been widely recognized. It has proved to be important as limiting the performance of semiconductor lasers [3.1.6], light-emitting diodes [3.1.7], and solar cells [3.1.8], and it can be crucial in transistors and similar devices whose performance is governed by lifetimes. When heavy doping is required, as it is in the drive towards microminiaturization, its importance tends to increase since the Auger recombination rate behaves roughly as n^2p or p^2n compared with the radiative rate which behaves more like np. Being the inverse process to impact ionization (section 2.1.1) avalanche devices (the photodiode, the impact avalanche transit time (IMPATT) diode, and hot electron devices) also utilize this type of process.

Impact ionization can be regarded as an autocatalytic reaction of order 1:

$$e \rightarrow 2e + h \text{ or } h \rightarrow 2h + e$$

i.e. one extra particle is produced of the type present in the first place. This is a key feature for the impact-induced nonequilibrium phase transitions in semiconductors, discussed more fully elsewhere ([3.1.9] and section 4.6.1). Short reviews of the band–band Auger effect appear in [3.1.10] and [3.1.11] and of the band–trap effect in [3.1.12]. See also [3.1.13].

As regards an actual calculation one needs four wavefunctions (for the four states of two electrons) which are relevant after the many-electron problem has been reduced to a two-electron problem. The four wavevectors imply a twelve-fold integration in **k**-space to obtain a result relevant for all possible states of the two electrons. However, momentum and energy conservation usually reduces this to an eight-fold integration. It is clear that such a calculation requires (a) good wavefunctions and (b) accurate integrations. Ideal conditions as regards these two aspects can never be achieved, but there have been improvements over the years.

Here we shall concentrate on the broad principles and analytical results, leaving the reader to go to the original literature for more accurate or numerical estimates. These tend to go out of date in any case and they can only be quoted, not developed, *in extenso*, in this book. In addition, the many-electron nature of the problem implies that another approximation is inherent in the treatment described here in general terms, apart from the use of perturbation theory. This is clear if one considers that the electron interactions are screened twice: once by the exponential screening factor and a second time by the dielectric constant. So there is some double counting, and the treatment of the effect as uncorrelated electronic transitions mediated by screened Coulomb interactions is another approximation. The collective effects which enter require more sophisticated field theoretic methods which take care of electron–hole correlations, plasmon effects and the effect of free excitons [3.1.14]–[3.1.16] (the so-called excitonic Auger effect. For Auger effects involving bound excitons see section 5.4.1). These calculations confirm broadly the results obtained by the simpler methods used here for energy gaps large compared with the plasmon energies, provided the high frequency dielectric constant is used and doping is not too heavy [3.1.15]. Thus there are significant corrections in the narrow gap lead compounds for example. The full calculation of the band–band Auger effect requires considerable computational effort as discussed further in section 3.6.6.

The radiative transition involves basically only one electron (and possibly a phonon and/or an exciton) and two states and is therefore wave mechanically simpler. Hence one can develop the wave mechanical concepts for the Auger effect in a manner which also includes those for the radiative transitions (but not the other way around). We can therefore achieve some economy in letting the present chapter precede the chapter on radiative recombination.

After these historical remarks we shall explain the essential wave mechanics in sections 3.2–3.4. This will be followed by an analysis of other physically relevant concepts and results. We turn briefly to the Schrödinger wave equation first.

The fact that quantum mechanics takes account of the possibility of interference between different observations means that the mathematics involved cannot be that of classical mechanics. In this brief summary we shall take it as basic that observation on the x-coordinate and the x-component of the linear momentum p_x of a particle interfere. The same applies to the y-coordinate and p_y, and also to the pair (z, p_z). The reason is that a choice of coordinate system cannot affect a physical property, such as the interference between observations. Expressed algebraically, the classical observable p_x is then replaced by an operator P_x and the interference shall correspond to the mathematical fact that xP_x and $P_x x$ differ. The sequence of observations is here represented by reading the product from right to left: P_x is followed by x in the first product and x is followed by P_x in the second product. A simple realization is to 'represent' the initial state of the system by

$\psi(x_1, y_1, z_1, t)$ (in some sense) so that the departure from classical mechanics is illustrated by the fact that

$$(P_x x - x P_x) \qquad (3.1.1)$$

is nonzero. It is found that one can put, if $\hbar$ is a constant,

$$p_x \to P_x = \frac{\hbar}{i} \frac{\partial}{\partial x} \qquad (3.1.2)$$

whence the difference is

$$\frac{\hbar}{i} \left[\frac{\partial}{\partial x} (x\Psi) - x \frac{\partial}{\partial x} \Psi \right] = \frac{\hbar}{i} \Psi$$

. Here Ψ is a wavefunction and represents a quantum state of the whole system. As the above relation holds for all Ψ, one is led to

$$i(P_x x - x P_x) = \hbar I \qquad (3.1.3)$$

where the operator I is the identity which replaces the function on which it acts: $I\Psi = \Psi$ represents the quantum mechanical state of the system, h is Planck's constant and $\hbar = h/2\pi$. The arrow in eq. (3.1.2) gives the operator which is to represent a classical variable.

In this formalism the noncommutation of P_x and x represents the interference between the observations on the classical observables x and p_x. Indeed, in the limit $\hbar \to 0$ one regains classical mechanics. Now in relativity theory the normal vectors $\mathbf{p} \equiv (p_x, p_y, p_z)$ and $\mathbf{r} \equiv (x, y, z)$ are parts of so-called four-vectors

$$(\mathbf{p}, E/c), \quad (\mathbf{r}, ct) \qquad (3.1.4)$$

Thus the relations (3.1.2) and (3.1.4) suggest

$$E \to H = a \frac{\partial}{\partial t}$$

where the Hamiltonian H is the operator representing the classical energy. The constant a has the same dimension as $\hbar$ and it will be taken as $-\hbar/i$.

This suggests that for one particle

$$\left\{ \frac{p^2}{2m} + V(r) = E \right\} \to \left\{ \left[\frac{p^2}{2m} + V(r) \right] \Psi = H\Psi - \frac{\hbar}{i} \frac{\partial \Psi}{\partial t} \right\} \qquad (3.1.5)$$

The Hamiltonian is to be expressed in terms of momenta and coordinates as above. Generalizing for an n-particle system,

$$H(\mathbf{P}_1, \mathbf{P}_2, \ldots, \mathbf{P}_n; \mathbf{r}_1, \ldots, \mathbf{r}_n; t) \Psi = -\frac{\hbar}{i} \frac{\partial \Psi(\mathbf{r}_1, \ldots, \mathbf{r}_n, t)}{\partial t} \qquad (3.1.6)$$

This is the many-particle Schrödinger equation with operators $P_{1x}, \ldots$ representing momentum components.

The kinetic energy operator is taken as

$$\sum_{j=1}^{n} P_j^2/2m_j$$

and the wavefunction can normally be factorized

$$\Psi(r_1, \ldots, r_n, t) = \psi(r_1, \ldots, r_n)\, \varphi(t) \tag{3.1.7}$$

Hence for a time-independent Hamiltonian, e.g. for

$$H(\mathbf{P}_1, \ldots, \mathbf{P}_n; \mathbf{r}_1, \ldots, \mathbf{r}_n) \to -\sum_{1}^{n} \frac{\hbar^2}{2m_j} \nabla_j^2 + V(\mathbf{r}_1, \ldots, \mathbf{r}_n) \tag{3.1.8}$$

one has from eq. (3.1.6) $H\psi\varphi = -\dfrac{\hbar}{i}\psi\dfrac{\partial\varphi}{\partial t}$, i.e.

$$\frac{H\psi}{\psi} = -\frac{\hbar}{i}\frac{\partial\varphi/\partial t}{\varphi} \tag{3.1.9}$$

The left-hand side depends only on the coordinates, the right-hand side depends only on time, so that each side is equal to one and the same constant. We shall denote it by E as it is dimensionally an energy. The time-dependent part of eq. (3.1.9) is

$$\frac{\partial\varphi/\partial t}{\varphi} = -(\mathrm{i}/\hbar)E$$

whence

$$\varphi(t) = A \exp(-\mathrm{i}Et/\hbar) \tag{3.1.10}$$

A being a normalizing constant.

The other equation is the *time-independent Schrödinger equation*

$$H\psi(\mathbf{r}_1, \ldots) = \left[\sum_{j} P_j^2/2m_j + V(\mathbf{r}_1, \ldots)\right]\psi(\mathbf{r}_1, \ldots) = E\psi(\mathbf{r}_1, \ldots) \tag{3.1.11}$$

and leads to an eigenvalue problem. This means that, given the form of V and appropriate boundary conditions, we look for functions ψ_j and corresponding numbers E_j which will solve eq. (3.1.11). The complete solution (3.1.7) is then of a form which uses eq. (3.1.10) and is

$$\Psi(\mathbf{r}_1, \ldots, \mathbf{r}_n; t) = \sum_{j} c_j \exp\left(-\frac{\mathrm{i}}{\hbar}E_j t\right)\psi_j \tag{3.1.12}$$

Differentiating with respect to time, eq. (3.1.12) yields eq. (3.1.6), as required:

$$-\frac{\hbar}{i}\frac{\partial\Psi}{\partial t} = \sum_{j} E_j c_j \exp\left(-\frac{\mathrm{i}}{\hbar}E_j t\right)\psi_j$$

since

$$H\Psi = \sum_{j} c_j \exp\left(-\frac{\mathrm{i}}{\hbar}E_j t\right)H\psi = \sum_{j} c_j E_j \exp\left(-\frac{\mathrm{i}}{\hbar}E_j t\right)\psi_j$$

In writing eq. (3.1.12) as the solution of a general class of problems, it was assumed that the set of functions ψ_n are orthonormal and that they are a complete set. These assumptions are not always fulfilled, but they can be used for the purposes of this brief review.

It is remarkable that such a general equation (3.1.12) can be given for a very large class of many-particle systems. It can be used for investigating many general properties of quantum mechanical systems. In specific cases one is, however, hampered by the difficulty of obtaining the ψ_j's, the E_j's and the c_j's. Approximate methods have been developed to deal with this problem.

3.2 Fermi's golden rule

Considering now only one particle for simplicity, suppose a system satisfies at an initial instant $t = t_0$

$$\Psi(\mathbf{r}, t_0) = \sum_n c_n \exp\left(-\frac{i}{\hbar} E_n^{(0)} t_0\right) \psi_n^{(0)}(\mathbf{r}) \tag{3.2.1}$$

where the $\psi_n^{(0)}$ and $E_n^{(0)}$ come from solutions of Schrödinger's equation

$$H(0)\psi_n^{(0)} = E_n^{(0)} \psi_n^{(0)}(\mathbf{r}) \tag{3.2.2}$$

In the absence of a perturbation, eq. (3.2.1) goes over into

$$\Psi(\mathbf{r}, t) = \sum_n c_n \exp\left[-\frac{i}{\hbar} E_n^{(0)} t\right] \psi_n^{(0)}(\mathbf{r}) \tag{3.2.3}$$

for all $t > t_0$.

We check now that eq. (3.1.6) can be satisfied if the Hamiltonian has been perturbed by an outside influence $H^{(1)}$ to become

$$H = H^{(0)} + H^{(1)} \tag{3.2.4}$$

We then have from eqs. (3.2.3) and (3.2.4)

$$-\frac{\hbar}{i}\dot{\psi} = \sum_n c_n E_n^{(0)} \exp\left(-iE_n^{(0)} t/\hbar\right) \psi_n^{(0)}(\mathbf{r})$$

$$-\frac{\hbar}{i} \sum_n \dot{c}_n \exp\left(-iE_n^{(0)} t/\hbar\right) \psi_n^{(0)}$$

and, as $H^{(1)}$ acts only on the $\psi^{(0)}$'s,

$$[H^{(0)} + H^{(1)}]\psi = \sum_n c_n E_n^{(0)} \exp\left(-iE_n^{(0)} t/\hbar\right) \psi_n^{(0)}$$

$$+ \sum_n c_n \exp\left(-iE_n^{(0)} t/\hbar\right) H^{(1)} \psi_n^{(0)}$$

Equating these expressions, as required by eq. (3.1.6),

$$-\frac{\hbar}{i} \sum_n \dot{c}_n \exp\left(-iE_n^{(0)} t/\hbar\right) \psi_n^{(0)} = \sum_n c_n \exp\left(-iE_n^{(0)} t/\hbar\right) H^{(1)} \psi_n^{(0)} \tag{3.2.5}$$

This relation tells one that if there is no perturbation, $H^{(1)} = 0$, the expansion parameters c_n in eq. (3.2.5) are independent of time, and eq. (3.2.3) remains a valid expansion. If, however, there is a perturbation, eq. (3.1.6) is no longer satisfied unless the 'constants' c_n are allowed to vary with time. This 'method of the variation of constants' then leads to eq. (3.1.6). Regarding the $\psi_n^{(0)}$ as vectors, forming the normal scalar product with $\psi_m^{(0)}$ and using the orthonormality,

$$-\frac{\hbar}{i}\dot{c}_m = \sum_n [H^{(1)}]_{mn} c_n \exp i\omega_{mn} t \tag{3.2.6}$$

where

$$[H^{(1)}]_{mn} \equiv (\psi_m^{(0)}, H^{(1)}\psi_n^{(0)})$$

is a matrix element and

$$\hbar\omega_{mn} \equiv E_m^{(0)} - E_n^{(0)} \tag{3.2.7}$$

Our main problem is to solve this set of homogeneous linear differential equations. In vector notation,

$$-\frac{\hbar}{i}\dot{\mathbf{c}} = A\mathbf{c}, \quad A \equiv \begin{pmatrix} H_{11}^{(1)} & H_{12}^{(1)} e^{i\omega_{12}t} \dots \\ H_{21}^{(1)} e^{-i\omega_{12}t} & H_{22}^{(1)} & \dots \\ \dots & \dots & \dots \end{pmatrix}$$

The result (3.2.6) is so far exact, but we need to make approximations now. Assume the system was in a state i initially, i.e. at $t = 0$. Then, writing $c_n(t)$ for c_n,

$$c_n(0) = \delta_{ni} \tag{3.2.8}$$

Now assume

$$c_n(t) \ll c_i(t) \sim 1 \text{ for } n \neq i \text{ and all } t \text{ considered} \tag{3.2.9}$$

One then has approximately from eq. (3.2.6)

$$-\frac{\hbar}{i}\dot{c}_f = [H^{(1)}]_{fi} e^{i\omega_{fi}t} \quad (f \neq i)$$

It follows that

$$c_f(t) = -\frac{i}{\hbar}\int_0^t H_{fi}^{(1)} e^{i\omega_{fi}t'} dt' \tag{3.2.10}$$

If the perturbation ceases at $t = T$ while the $c_f(t)(f \neq i)$ are still small, relation (3.2.9) is satisfied and the c's again have constant values. They are given by eq. (3.2.10) with $t = T$.

An important case arises if $H^{(1)}$ is independent of time when

$$c_f(t) = \frac{1}{\hbar}H_{fi}^{(1)}(1 - e^{i\omega_{fi}t})/\omega_{fi}$$

and hence

$$|c_f(t)|^2 = 2|H_{fi}^{(1)}|^2 \frac{1 - \cos \omega_{fi} t}{(\hbar\omega_{fi})^2} \tag{3.2.11}$$

Table 3.2.1. *Some values of* $y \equiv (1 - \cos \omega t)/\omega^2$

$\omega^{(1)} = 0$	$\omega^{(2)} = 3\pi/t$	$\omega^{(3)} = 5\pi/t$
$y^{(1)} = 0.500t^2$	$y^{(2)} = \dfrac{2}{9\pi^2}t^2 \sim 0.023t^2$	$y^{(3)} = \dfrac{2}{25\pi^2}t^2 \sim 0.008t^2$
	$\dfrac{y^{(2)}}{y^{(1)}} \sim 0.046$	$\dfrac{y^{(3)}}{y^{(1)}} \sim 0.016$

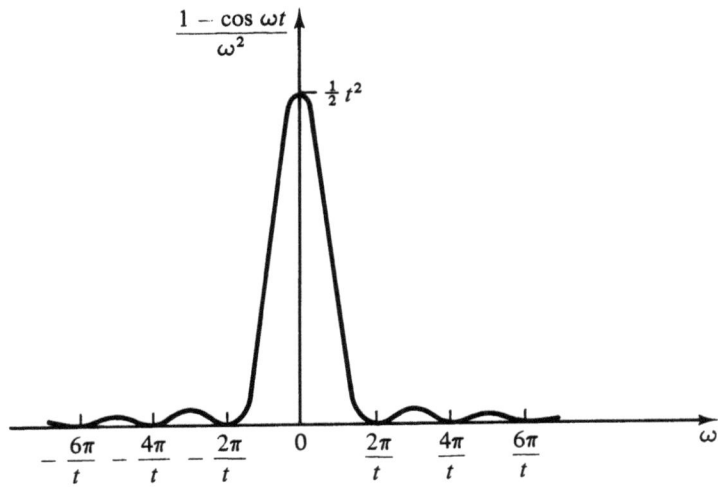

Fig. 3.2.1. This figure illustrates eq. (3.2.11).

The quantity (3.2.11) is in a sum (3.2.3) the probability that the system will be found in state f at time t. This is the content of the probability interpretation of quantum mechanics. This probability expression holds subject to relation (3.2.9), i.e. the time t must not be too long, for even a very small perturbation which is constant in time but acts for a long time will eventually lead to a violation of relation (3.2.9).

The time must also not be too small. Fig. 3.2.1 gives a plot of a factor $y(\omega)$ in the probability $|c_f|^2$. It has a central maximum which becomes narrower and higher, approaching a Dirac δ-function (see Appendix A) as t increases. Thus a reasonable probability of finding the system with a final energy differing from the initial energy by some amount Δ can be obtained only if the time is such as to place the system within the central maximum. This happens if

$$\omega = \Delta/\hbar < 2\pi/t$$

in violation of the uncertainty principle. Hence energy is approximately conserved. The half-way positions between adjacent minima of the curve $y(\omega)$ are given in Table 3.2.1 and are seen to yield much smaller values than the central maximum, indicating again approximate energy conservation. Note that the *total* energy is of course always conserved. For example, if the process considered is photon emission, the electron makes a transition between states of *different* unperturbed energies, but the overall energy is conserved.

For the reasons given one can make the replacement

$$\frac{1-\cos x}{x^2} \to \pi\delta(x) \quad x \equiv \frac{t}{\hbar}(E_f^{(0)}-E_i^{(0)}) \tag{3.2.12}$$

where the multiplier π is due to the fact that (see the appendix to this section)

$$\int_{-\infty}^{\infty} \frac{1-\cos x}{x^2}\,\mathrm{d}x = \pi \int_{-\infty}^{\infty} \delta(x)\,\mathrm{d}x = \pi$$

It follows that eq. (3.2.11) yields for the probability of a transition $i \to f$ in time t

$$|c_f(t)|^2 = 2\pi|H_{fi}^{(1)}|^2\,\delta\!\left[\frac{t}{\hbar}(E_f^{(0)}-E_i^{(0)})\right]\frac{t^2}{\hbar^2}$$

$$= \frac{2\pi t}{\hbar}|H_{fi}^{(1)}|^2\,\delta(E_f^{(0)}-E_i^{(0)}) \tag{3.2.13}$$

It will be denoted by T_{if} when we proceed with this calculation in section 3.7.

One additional observation may be of interest. Let $\mathcal{N}(E_f^{(0)})\,\mathrm{d}E_f^{(0)}$ be the number of final states in the energy range $(E_f^{(0)}, E_f^{(0)}+\mathrm{d}E_f^{(0)})$, then

$$\int|c_f(t)|^2\,\mathcal{N}(E_f^{(0)})\,\mathrm{d}E_f^{(0)} = \frac{\hbar}{t}\int|c_f(t)|^2\,\mathcal{N}(E_f^{(0)})\,\mathrm{d}x$$

$$= \frac{2t\pi}{\hbar}\int|H_{fi}^{(1)}|^2\,\mathcal{N}(E_f^{(0)})\,\delta(x)\,\mathrm{d}x$$

$$= \frac{2t\pi}{\hbar}|H_{fi}^{(1)}|^2\,\mathcal{N}(E_i^{(0)}) \tag{3.2.14}$$

A kind of compromise has occurred here. For transitions to states for which the unperturbed energies are strictly conserved the probability grows as t^2 according to eq. (3.2.11). For transitions to other states it oscillates. If one integrates over final states, one finds a total probability into final states which does neither: it increases as t; the *rate* of change of probability is *constant* in time. Equations (3.2.11) and (3.2.13) are variants of what has become known as Fermi's golden rule of perturbation theory. The transition probabilities per unit time were treated as themselves independent of time in chapter 2. This receives a justification from the present considerations.

It is also worth noting that usually $H^{(1)}$ is a Hermitian operator: $H_{fi} = H_{if}^*$ where the asterisk denotes the complex conjugate. Hence $|H_{fi}|^2 = H_{fi}H_{fi}^* = H_{if}^*H_{if} = |H_{if}|^2$ and the transition probability per unit time is the same for the $i \to f$ transition as for the reverse transition. This *principle of microscopic reversibility* must be carefully distinguished from the *principle of detailed balance*. The latter involves the occupation probabilities furnished by statistical mechanics and always holds in thermal equilibrium. Microscopic reversibility on the other hand involves neither.

There are many other aspects of perturbation theory for which reference should be made to books on quantum mechanics. For example, one can show by going to higher order in the perturbation that the total probability of finding the system in any one of its unperturbed states remains at unity for the period of time envisaged. This means that

$$|c_i(t)|^2 + \sum_{f(\neq i)} |c_f(t)|^2 = 1$$

Furthermore the theory has to be amended if the unperturbed states are degenerate, i.e. if in the series of increasing energies $E_1^{(0)}$, $E_2^{(0)}$, $E_3^{(0)}$,... referring to quantum states, $1, 2, 3, ...$ some of the energy values are equal:

$$E_1^{(0)} \leqslant E_2^{(0)} \leqslant E_3^{(0)} \leqslant ...$$

One must next evaluate the matrix element, and this means that a better understanding of many-electron wavefunctions is required.

Appendix: Proof that $I = \displaystyle\int_{-\infty}^{\infty} \frac{1 - \cos x}{x^2} dx = \pi$ ***(see eq. (3.2.12))***

First apply a partial integration to find

$$I = \int_{-\infty}^{\infty} \frac{\sin x}{x} dx = 2 \int_{0}^{\infty} \frac{\sin x}{x} dx$$

Next replace

$$\frac{1}{x} \text{ by } \int_{0}^{\infty} \exp(-ux) du$$

so that

$$I = 2 \int_{0}^{\infty} du \int_{0}^{\infty} e^{-ux} \sin x \, dx = 2\mathrm{Im} \left\{ \int_{0}^{\infty} du \int_{0}^{\infty} e^{(i-u)x} dx \right\}$$

where 'Im' indicates the 'imaginary part of'. Hence

$$I = 2\,\mathrm{Im} \left\{ \int_{0}^{\infty} du \frac{1}{-u+i} \left| e^{(i-u)x} \right|_{x=0}^{x=\infty} \right\}$$

$$= -2\,\mathrm{Im} \left\{ \int_{0}^{\infty} \frac{-u-i}{u^2+1} du \right\}$$

$$= 2 \int_{0}^{\infty} \frac{du}{u^2+1} = 2 \sin^{-1} \frac{u}{(u^2+1)^{\frac{1}{2}}} \Big|_{0}^{\infty} = 2\frac{\pi}{2}$$

Hence $I = \pi$, as required. Since $1 - \cos x = 2 \sin^2 x/2$ one can make the replacements

$$\frac{1 - \cos x}{x^2} \to \pi \delta(x), \quad \frac{\sin^2 x}{x^2} \to \pi \delta(x), \quad \text{used in } (4.3.14) \tag{3.2.15}$$

inside an integral if $x = 0$ is within the range of x-values.

3.3 Reduction of the many-electron problem

3.3.1 Introduction

It must be appreciated that the matrix elements to be evaluated for the transition probabilities furnished by perturbation theory involve many-electron wavefunctions. Each electron moves under the influence of the remaining $m-1$ electrons in the system, and for large m the theoretical situation is almost intractable. The m-body problem is not solved even in classical mechanics. One must, therefore, look for approximate methods of reducing the problem to a simpler more tractable one. In this section we outline one such method.

Approximate procedures such as the Hartree and Hartree–Fock schemes are discussed in books on quantum mechanics and solid state theory. Other approximate procedures use second quantization, collective variables, etc. As these matters are not the main concern here, only a short but self-contained discussion will be given which shows how matrix elements involving ordinary one-electron wavefunctions arise in an approximate way from the m-electron problem. *The main results, which will be useful in the later work, are eqs. (3.3.17), (3.3.21) and (3.3.26) to (3.3.28). Busy readers can skip six or more pages.*

3.3.2 Determinantal eigenfunctions

Consider a typical electron, labelled by the suffix i. Its spin may be specified by the eigenvalues of its spin angular momentum component for a standard direction, usually labelled as the z-coordinate. The operator

$$\sigma_{iz} \equiv \frac{\hbar}{2} \begin{pmatrix} 1 & 0 \\ 0 & -1 \end{pmatrix}$$

represents this component, which is one of the three Pauli spin operators. Its eigenvalues are $(\pm \hbar/2)$, and they correspond to normalized eigenvalues denoted by η_{i+} and η_{i-}, or more briefly by $\eta_{i\alpha}$ ($\alpha = +$ or $-$). These define a two-dimensional spin space for electron i. The suffix i is a reminder that the spin space for the system as a whole increases its dimensionality by a factor of two for each added electron. The suffixes $+$ and $-$ are explained by

$$\sigma_{iz}\,\eta_{i+}(+\hbar/2) = \frac{\hbar}{2}\eta_{i+}\left(+\frac{\hbar}{2}\right), \quad \sigma_{iz}\,\eta_{i-}(-\hbar/2) = -\frac{\hbar}{2}\eta_{i-}\left(-\frac{\hbar}{2}\right)$$

$$\eta_{i+}\left(-\frac{\hbar}{2}\right) = \eta_{i-}\left(\frac{\hbar}{2}\right) = 0 \tag{3.3.1}$$

The quantity $\pm\hbar/2$ in brackets gives the value of the z-component of spin. In general arguments this *spin variable* will be denoted by ξ_i. One may think of eq. (3.3.1) as stating that only an electron of spin $\hbar/2$ can be accommodated by η_{i+}, and not an electron of spin $-\hbar/2$, and similarly for η_{i-}.

A usual Schrödinger equation for one electron is

$$H(i)\,\varphi(i) = \varepsilon(i)\,\varphi(i) \quad \text{(one-electron approximation)}$$

where i stands for *all* the variables (space r_i *and* spin ξ_i) of the ith electron and $\varepsilon(i)$ is the one-electron energy. It is in the simpler cases possible to separate the equation into space- and spin-dependent equations. Then a typical $\varphi_j(i)$ may be replaced by $\psi_{ia}(\mathbf{r}_i)\,\eta_{ia}(\xi_i)$, provided the spin state α and the orbital or translational state a, when both specified, are labelled 'state j' of the electron i. If it is understood that η depends only on spin and ψ only on the space coordinates, no confusion results from writing

$$H(i)\,\varphi_j(i) = \varepsilon_j(i)\,\varphi_j(i) \tag{3.3.2}$$

$$\varphi_j(i) = \psi_a(i)\,\eta_\alpha(i) \quad (j = 1, 2, \ldots) \tag{3.3.3}$$

and this will be done in the sequel. The suffix j stands for the translational *and* spin quantum number: $\mathbf{j} = (a, \alpha)$.

The product formation exhibited in eq. (3.3.3) is valid only if spin–orbit interaction has been treated in some approximate manner; strictly speaking this interaction links spin and space variable so as to prevent the factorization (3.3.3). The effect of spin–orbit interaction can be treated as a further perturbation.

Passing to a system of m indistinguishable particles, suppose that, similarly, interparticle interactions have been taken into account only approximately, so that the Hamiltonian for the m-particle problem can be written as a sum of terms each of which depends on one particle only. This cannot ever be done exactly because two-body forces depending on $|\mathbf{r}_1 - \mathbf{r}_2|$ cannot in this way be taken into account exactly. Nonetheless, making this assumption, the Schrödinger equation has the form

$$\left[\sum_{i=1}^{m} H(i)\right]\Phi = E\Phi \tag{3.3.4}$$

Product-formation is again a useful device. For, using the notation (3.3.2), if

$$\Phi \equiv \varphi_a(1)\,\varphi_b(2)\ldots\varphi_j(m) \tag{3.3.5}$$

then substitution in eq. (3.3.4) shows that the Schrödinger equation is satisfied with

$$E = \varepsilon_a(1) + \varepsilon_b(2) + \ldots + \varepsilon_j(m) \tag{3.3.6}$$

The solution (3.3.5) remains formally valid if $a = b$, i.e. if two particles are in the same state. Furthermore, it is symmetrical if the particles are permuted among the states. It thus violates the Pauli exclusion principle for fermions, i.e. the requirement that an m-fermion wavefunction must be antisymmetrical under a permutation of any two particles among two occupied states. Both objections are, however, avoided by a determinantal wavefunction

$$\Phi = C \begin{vmatrix} \varphi_1(1)\,\varphi_1(2)\ldots\varphi_1(m) \\ \varphi_2(1)\,\varphi_2(2)\ldots\varphi_2(m) \\ \varphi_m(1)\ldots\varphi_m(m) \end{vmatrix} \tag{3.3.7}$$

Expanding,

$$\Phi = C \sum_{j=1}^{m!} (-1)^{p_j}\, P_j[\varphi_1(1)\,\varphi_2(2)\ldots\varphi_m(m)] \tag{3.3.8}$$

The m states are here labelled by suffixes 1 to m, particle designations are given in parentheses and P_j is one of the $m!$ permutation operators which permutes the assignment of particles to states in the product expansion of the determinant. The 'parity' p_j of the permutation P_j is the number of pair-interchanges from which it can be generated. Thus the permutation

$$\downarrow \begin{matrix} 1\,2\,3\,4 \\ 4\,3\,2\,1 \end{matrix} \text{ arises from } \downarrow \begin{matrix} 1\,2\,3\,4 \\ 1\,3\,2\,4 \end{matrix} \text{ and } \downarrow \begin{matrix} 1\,2\,3\,4 \\ 4\,2\,3\,1 \end{matrix}, \text{ i.e. } p = 2.$$

An odd permutation is one of odd parity. Thus the term $(-1)^{p_j}$ has the effect of weighting odd permutations with a factor -1 and even permutations with a factor $+1$. The wavefunction changes sign if the space and spin coordinates of particles 1 and 2 are interchanged, as this is an interchange of two columns in the determinant. It vanishes if two states are the same, as this makes two rows of the determinant identical. (For bosons the -1 is replaced by $+1$ in eq. (3.3.8).) C is a normalization constant, defined by

$$|C|^2 \int \ldots \int \Phi^*\Phi\, d\tau_1' \ldots d\tau_m' = 1 \tag{3.3.9}$$

Here $d\tau_i'$ denotes an integration $d\mathbf{r}_i$ and a summation over the spin variable ξ_i. Indistinguishability of particles is assured as each particle occurs in each state in eq. (3.3.7).

If the unperturbed Hamiltonian H has the form (3.3.4), products (3.3.5) are eigenfunctions. Since permutations in the assignment of particles to states do not change the eigenvalue (3.3.6), it follows that a superposition of any number of functions $P_j\Phi$ (Φ given by eq. (3.3.5)) with constant coefficients b_j are also eigenfunctions:

$$H[\sum b_j P_j \Phi] = E[\sum b_j P_j \varphi]$$

Thus the functions (3.3.8) are satisfactory eigenfunctions.

3.3.3 Matrix elements between determinantal wavefunctions

We now consider how to evaluate matrix elements of operators between eigenfunctions of an unperturbed Hamiltonian. Suppose now that the initially and finally occupied one-fermion states are specified by wavefunctions

$$\Theta_1, \Theta_2, \ldots, \Theta_m \text{ and } \varphi_1, \ldots, \varphi_m$$

respectively. The matrix element of an operator U between initial and final states of the whole system is from eq. (3.3.8)

$$U_{if} = C_i^* C_f \sum_{j,k=1}^{m!} (-1)^{p_j+p_k} \int \cdots \int P_j[\Theta_1^*(1)\ldots\Theta_m^*(m)]$$
$$\times UP_k[\varphi_1(1)\ldots\varphi_m(m)]\, d\tau_1' \ldots d\tau_m'$$

In order to simplify this expression it will be assumed that

(i) the operator U is symmetrical in the particles. This is essential as we are considering a system of identical particles;

(ii) U is a sum of terms, each of which involves the variables of *at most* two particles. This is a reasonable restriction to two-body forces;

(iii) the Θ_i and φ_j are functions which come from the same orthonormal set of functions, and if a Θ is equal to a φ the attached suffix is the same.

The last phrase is convenient for purposes of exposition as it enables one to assert for example that the sets Θ_i and φ_j are 'identical' when they are, in fact, only 'identical except for possible renumbering'. The operator U will often be a perturbation $H^{(1)}$, but other interpretations of U can occur.

By virtue of assumption (i) the operator P_k can be brought immediately behind the summation sign:

$$U_{if} = C_i^* C_f \sum_{k=1}^{m!} P_k \left[\int \cdots \int \sum_{l=1}^{m!} (-1)^{p_l} P_l[\Theta_1^*(1)\ldots\Theta_m^*(m)] \right.$$
$$\left. \times U\varphi_1(1)\ldots\varphi_m(m)\, d\tau_1' \ldots \varphi\tau_m' \right] \qquad (3.3.10)$$

Here $P_k^{-1} P_j$, which is just one of the $n!$ permutation operators, has been denoted by P_l. Also, since the parities of P_k^{-1} and of P_k are the same, the parity of P_l is $p_l = p_j + p_k$. The k-sum yields $m!$ identical terms since it is applied to an expression from which the coordinates (space and spin) have been integrated out. Permuting the assignment of particles to states, let P_l lead to the result that the state occupied by particle $l1$ initially is occupied by particle 1 after the permutation, etc., i.e.

$$P_l = \begin{pmatrix} l1 & l2 & lm \\ 1 & 2 & m \end{pmatrix}\!\downarrow$$

Equation (3.3.10) now becomes

$$U_{if} = C_i^* \, C_f m! \sum_{l=1}^{m!} (-1)^{p_l} \left[\sum_{j=1}^{m-1} \int \Theta_{lj}^*(j)\, \varphi_j(j)\, \mathrm{d}\tau_j' \right]$$
$$\times \int \Theta_{lm}^*(m)\, U\varphi_m(m)\, \mathrm{d}\tau_m' \qquad\qquad (3.3.11)$$

We now verify that permutations acting on particles can be replaced by permutations acting on states. Suppose P_l acts on the particles and is given by

$$P_l = \begin{pmatrix} 1 & 2 & 3 & 4 & 5 & 6 & 7 & 8 & 9\dots \\ 3 & 2 & 4 & 1 & 5 & 7 & 6 & 8 & 9\dots \end{pmatrix}\!\downarrow$$

i.e. the state which had particle 1, has particle 3 after the permutation, etc.
   In this case

$$P_l[\Theta_1^*(1)\dots\Theta_m^*(m)] = \Theta_1^*(3)\,\Theta_2^*(2)\,\Theta_3^*(4)\,\Theta_4^*(1)\,\Theta_5^*(5)\,\Theta_6^*(7)\,\Theta_7^*(6)\dots$$
$$= \Theta_4^*(1)\,\Theta_2^*(2)\,\Theta_1^*(3)\,\Theta_3^*(4)\,\Theta_5^*(5)\,\Theta_7^*(6)\,\Theta_6^*(7)\dots$$
$$= \sum_{j=1}^{m} \Theta_{lj}(j)$$

so that the permutation can be applied to the states. Furthermore, note that if P_l involves two interchanges, for example

$$P_l = \begin{pmatrix} 1 & 2 & 3 & 4 & 5\dots m \\ 2 & 1 & 4 & 3 & 5\dots m \end{pmatrix}\!\downarrow \quad \text{or} \quad \begin{pmatrix} 1 & 2 & 3 & 4\dots m \\ 2 & 3 & 1 & 4\dots m \end{pmatrix}\!\downarrow$$

at least *three* electrons change their states in going from a product of Θ's to a product of φ's. By assumption (ii) only two coordinates appear in a typical term of U, and by assumption (iii) one orthogonality integral is then left in the term of eq. (3.3.11) under consideration. The contribution to eq. (3.3.11) from such permutations therefore vanishes. Thus one need consider in the l-sum of eq. (3.3.11) only the identity and *single interchanges*.

Equation (3.3.11) simplifies to our main result of this section:

$$U_{if} = C_i^* C_f m! \quad (A+B)$$ (3.3.12)

The term A arises from the identity

$$A = \prod_{j=1}^{m-1} \left[\int \Theta_j^*(j)\, \varphi_j(j)\, d\tau_j' \right] \int \Theta_m^*(m)\, U\varphi_m(m)\, d\tau_m'$$ (3.3.13)

The term B arises from single interchanges. These are denoted now as follows

$$lj \rightarrow [rs]\,j$$

if electrons r and s are interchanged:

$$B = -\sum_{\substack{r,s=1 \\ (r<s)}}^{m} \prod_{j=1}^{m-1} \left[\int \Theta_{[rs]j}^*(j)\, \varphi_j(j)\, d\tau_j' \right] \int \Theta_{[rs]m}^*(m)\, U\varphi_m(m)\, d\tau_m'$$ (3.3.14)

The restriction $r < s$ ensures that an interchange is not counted twice.

3.3.4 Special cases

Three special cases are of importance.

Case I $U = I$, $\Theta_i = \varphi_i$ $(i = 1,\ldots,m)$

In this case the interchanges imply that each term in eq. (3.3.14) contains orthogonality integrals, so that $B = 0$. The determinantal wavefunctions are identical, so that $C_i = C_f$. Since the Θ's are normalized, $A = 1$. Hence $U_{if} = |C|^2 m!$, but by the normalization integral for a determinantal wavefunction, eq. (3.3.9), $U_{if} = 1$ in this case. Hence with assumption (iii) one finds from eq. (3.3.12)

$$|C|^2 = 1/m!$$ (3.3.15)

Case II $U = \sum_{i=1}^{m} U(i)$

As in Case I, one finds $B = 0$. One finds also from eq. (3.3.13)

$$U_{if} = A = \sum_{k=1}^{m} \left[\sum_{\substack{j=1 \\ (j \neq k)}}^{m} \int \Theta_j^*(j)\, \varphi_j(j)\, d\tau_j' \right] \left[\int \Theta_k^*(k)\, U_k(k)\, \varphi_k(k)\, d\tau_k' \right]$$

Three possibilities are of interest.

IIa $\Theta_i = \varphi_i$ (all i): *energy expectation value*
One has

$$U_{if} = \sum_{k=1}^{m} \int \Theta_k^*(1)\, U(1)\, \Theta_k(1)\, d\tau_k' \tag{3.3.16}$$

If U is the Hamiltonian, this is its expectation value in the state given by the Θ's. This type of Hamiltonian can arise if two-body forces are taken into account in some approximate manner as in the Hartree or Hartree–Fock schemes.

IIb $\Theta_i = \varphi_i$ (all i, except $i = k$, $\Theta_k \neq \varphi_k$): *one-electron transition*
The orthogonality integral

$$\int \Theta_k^*(i)\, \varphi_k(i)\, d\tau_i'$$

is present in all terms of U_{if}, except for the term arising from $U(k)$, as follows from assumption (iii). Hence

$$U_{if} = \int \Theta_k^*(k)\, U(k)\, \varphi_k(k)\, d\tau_k'$$

Since a state present in the first m-electron wavefunction is replaced by another which is present in the second wavefunction, one can think of this case as relevant for a transition $\Theta_k \to \varphi_k$. Now U can be thought of as some perturbation operator which induces one-electron transitions of this type in first order. The Θ's and φ's can differ by more than one function if perturbation theory is taken to higher orders. But this will not concern us here. The interaction giving rise to the transition may conserve overall energy via photons or phonons, but this need also not concern us here.

In a semiconductor, a transition, say of an electron into a defect, may change the whole spectrum of relevant states, though possibly only slightly. Assumption (iii), however, implies that the orthonormal set is fixed once and for all. Hence such changes can be taken into account in the present formalization only as an additional perturbation.

On developing the result further,

$$U_{if} = \left[\sum_{\xi_k} \eta_\alpha^*(\xi_k)\, \eta_\beta(\xi_k) \right] \int \psi_a^*(k)\, U(k)\, \psi_b(k)\, d\tau_k$$

Here eq. (3.3.3) has been used and U has been assumed independent of spin. From the properties of the spin functions, a nonzero contribution arises only if $\alpha = \beta$, so that in this case

$$U_{if} = \delta_{\alpha\beta} \int \psi_a^*(k)\, U(k)\, \psi_b(k)\, d\tau_k \tag{3.3.17}$$

and the spin eigenfunction must be unchanged. This completes the reduction of the n-electron problem to a one-electron problem. It is used for radiative transitions, see section 4.3.

Case IIc $\Theta_i = \varphi_i$ is violated for two or more states
In these cases $U_{if} = 0$. This shows that in first-order perturbation theory the operator U of Case II can induce nothing more complicated than electron transitions of the type $\Theta_k \to \varphi_k$, discussed in IIb.

$$\textit{Case III } \ U = \sum_{k<l} U(k,l)$$

This operator arises, for example, if the difference between the Hartree or Hartree–Fock Hamiltonian and the exact Hamiltonian acts as a perturbation U. For the exact Hamiltonian includes the electron–electron interactions

$$\sum_{k<l} e^2/|\mathbf{r}_k - \mathbf{r}_l|$$

and this contributes an operator of the type under consideration here to the perturbation.

If a typical term from U, say $U(k,l)$, is substituted for the general U which occurs in eq. (3.3.13) one finds the following contribution to A:

$$\left[\prod_{\substack{j=1 \\ (j \neq k,\,l)}}^{n} \int \Theta_j^*(j)\,\varphi_j(j)\,d\tau_j' \right]\left[\int\int \Theta_k^*(k)\,\Theta_l^*(l)\,U(k,l) \right.$$
$$\left. \times\, \varphi_k(k)\,\varphi_l(l)\,d\tau_k'\,d\tau_l' \right] \quad (3.3.18)$$

The contribution to B is zero for interchanges which do not involve both k and l since such interchanges generate by assumption (iii) at least one orthogonality integral, and possibly two. Assuming that $k < l$, the only contribution to B in eq. (3.3.14) arises from $r = k$, $s = l$, and one finds the following contribution to B:

$$\left[\prod_{\substack{j=1 \\ (j \neq k,\,l)}}^{n} \int \Theta_j^*(j)\,\varphi_j(j)\,d\tau_j' \right]\left[\int\int \Theta_l^*(k)\,\Theta_k^*(l)\,U(k,l) \right.$$
$$\left. \times\, \varphi_k(k)\,\varphi_l(l)\,d\tau_k'\,d\tau_l' \right] \quad (3.3.19)$$

The expressions for A and B are found by summing formulae (3.3.18) and (3.3.19) over all k and l with $k < l$.

IIIa $\Theta_i = \varphi_i$ *(all i)*
One finds, replacing dummy variables k and l by 1 and 2,

$$U_{if} = \sum_{k<l} \int\int [\Theta_k^*(1)\,\Theta_l^*(2) - \Theta_l^*(1)\,\Theta_k^*(2)]\,U(1,2)\,\Theta_k(1)\,\Theta_l(2)\,d\tau_1'\,d\tau_2'$$

Fig. 3.3.1. A one-electron transition in an energy level scheme.

This is the type of contribution to an energy expectation value (3.3.16) which arises from two-body forces.

IIIb $\Theta_i = \varphi_i$ *(all i, except i = g)*, $\Theta_g \neq \varphi_g$.
One can put $k = g$ or $l = g$ in the expressions (3.3.18) and (3.3.19) (when summed over k and l), but the result will not be required.

IIIc $\Theta_i = \varphi_i$ *is violated for two states i = f, g; f < g*
In the sums over k and l the only nonzero contributions arise from

$$k = f \text{ and } l = g$$

Hence eq. (3.3.12) becomes

$$U_{if} = \int\int \left[\Theta_f^*(1)\,\Theta_g^*(2) - \Theta_g^*(1)\,\Theta_f^*(2) \right] U(1,2)\,\Theta_f(1)\,\Theta_g(2)\,d\tau_1'\,d\tau_2' \quad (3.3.20)$$

As in the treatment of eq. (3.3.16) it is useful to appeal to the product form (3.3.3) with the following notation:

$$\Theta_f \to \psi_a\,\eta_\alpha, \quad \Theta_g \to \psi_b\,\eta_\beta, \quad \varphi_f \to \psi_c\,\eta_\gamma, \quad \varphi_g \to \psi_d\,\eta_\delta$$

Equation (3.3.20) becomes

$$\begin{aligned}
U_{if} = \int\int \Big[& \psi_a^*(1)\,\psi_b^*(2) \sum_{\xi_1\xi_2} \eta_\alpha^*(1)\,\eta_\beta^*(2)\,\eta_\gamma(1)\,\eta_\delta(2) \\
& - \psi_b^*(1)\,\psi_a^*(2) \sum_{\xi_1\xi_2} \eta_\beta^*(1)\,\eta_\alpha^*(2)\,\eta_\gamma(1)\,\eta_\delta(2) \Big] U(1,2)\,\psi_c(1)\,\psi_d(2)\,d\tau_1\,d\tau_2 \\
& = M_D\,\Delta_D - M_E\,\Delta_E
\end{aligned} \quad (3.3.21)$$

The abbreviations used here are

$$\Delta_D \equiv \left[\sum_{\xi_1} \eta_\alpha^*(\xi_1)\,\eta_\gamma(\xi_1) \right]\left[\sum_{\xi_2} \eta_\beta^*(\xi_2)\,\eta_\delta(\xi_2) \right] = \delta_{\alpha\gamma}\,\delta_{\beta\delta} \quad (3.3.22)$$

$$\Delta_E \equiv \left[\sum_{\xi_1} \eta_\beta^*(\xi_1)\,\eta_\gamma(\xi_1) \right]\left[\sum_{\xi_2} \eta_\alpha^*(\xi_2)\,\eta_\delta(\xi_2) \right] = \delta_{\beta\gamma}\,\delta_{\alpha\delta} \quad (3.3.23)$$

$$M_D \equiv \int\int \psi_a^*(1)\,\psi_b^*(2)\,U(1,2)\,\psi_c(1)\,\psi_d(2)\,d\tau_1\,d\tau_2 \quad (3.3.24)$$

$$M_E \equiv \int\int \psi_b^*(1)\,\psi_a^*(2)\,U(1,2)\,\psi_c(1)\,\psi_d(2)\,d\tau_1\,d\tau_2 \quad (3.3.25)$$

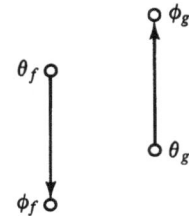

Fig. 3.3.2. A two-electron transition in an energy level scheme.

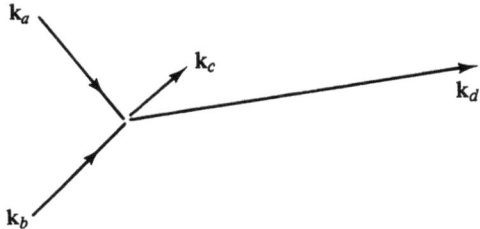

Fig. 3.3.3. A two-electron transition as collision.

The suffixes D and E stand for 'direct' and 'exchange'.

Whereas Case IIb describes a single-electron transition (Fig. 3.3.1), Case IIIc implies a two-electron transition (Fig. 3.3.2). It can also be interpreted as an electron collision (Fig. 3.3.3).

One can visualize eq. (3.3.21) most simply by a table of possible spin assignments, making the convention that α always refers to 'spin up' (see Table 3.3.1). Note from Table 3.3.1 that the quantum mechanical treatment implies a form of spin angular momentum conservation.

Squaring and summing over all spin assignments,

$$|U_{if}|^2 = 2[\overbrace{|M_D|^2 + |M_E|^2}^{\text{unlike spins}} + \overbrace{|M_D - M_E|^2}^{\text{like spins}}]$$

(3.3.26)

no   spin
spin   flip
flip

The factor of two arises because η_α may be chosen in two ways. It must be omitted if the effect of spin is included in the density of states when an integration over all transitions $i \to f$ is performed. The last term in eq. (3.3.26) tends to be troublesome when one sums over all possible states a, b, c, d, because it leads to complicated integrations. It can be avoided for calculations with uncertainty of a factor of two by noting from eq. (3.3.26) that

$$|U_{if}|^2 = 2\beta(|M_D|^2 + |M_E|^2) \quad 1 < \beta < 2$$

(3.3.27)

Table 3.3.1. *Contributions to the matrix element of different spin configurations*

Initial states		Final states		Δ_D	Δ_E	U_{if}	
α	β	γ	δ				
$\uparrow$	$\uparrow$	$\uparrow$	$\uparrow$	1	1	$M_D - M_E$	like spins
$\uparrow$	$\downarrow$	$\uparrow$	$\downarrow$	1	0	M_D ⎱	unlike spins
$\uparrow$	$\downarrow$	$\downarrow$	$\uparrow$	0	1	$-M_E$ ⎰	spins
$\uparrow$ any other assignments				0	0	0	

We have here used

$$|M_D|^2 + |M_E|^2 > |M_D - M_E|^2 \tag{3.3.28}$$

This inequality is clearly connected with the fact that collisions between electrons of unlike spins are more probable than the analogous collisions for like spins. This effect is to be expected since electrons of like spin tend to keep apart, an effect known as the 'exclusion hole'.

Case IIId $\Theta_i = \varphi_i$ *is violated for three or more states*
This corresponds to Case IIc and $U_{if} = 0$.

3.4 Matrix elements of two-body and Coulomb potentials

3.4.1 A general argument

A rather general expression for the 'direct' matrix element M_D of eq. (3.3.24) is obtained by considering a two-body perturbation energy which is specified by its Fourier coefficients U_k:

$$U(1,2) \equiv U(\mathbf{r}_1 - \mathbf{r}_2) = U_0 + \sum_{\substack{\text{all } \mathbf{k} \\ (k \neq 0)}} U_k e^{i\mathbf{k} \cdot (\mathbf{r}_1 - \mathbf{r}_2)} \tag{3.4.1}$$

Writing $\mathbf{r} = \mathbf{r}_1 - \mathbf{r}_2$, one has quite generally

$$V^{-2} \int_V \int U(\mathbf{r}_1 - \mathbf{r}_2) e^{-i\mathbf{k}' \cdot (\mathbf{r}_1 - \mathbf{r}_2)} d\tau_1 d\tau_2 = V^{-1} \int_V U(\mathbf{r}) e^{-i\mathbf{k}' \cdot \mathbf{r}} d\tau = U_{k'} \tag{3.4.2}$$

The last expression arises by assuming the expansion (3.4.1) to be valid and by substituting it for $U(r)$ and using Appendix B, eq. (B.11). Equation (3.4.2) gives the recipe for evaluating the Fourier coefficient. The volume of the material has been denoted by V.

Let wavefunctions $\psi_a(1)$, $\psi_b(2)$, $\psi_c(1)$, $\psi_d(2)$ be also Fourier analyzed according to the scheme (see Appendixes B and C)

$$\psi_a(1) = \frac{1}{V^{\frac{1}{2}}} \sum_{\text{all } \mathbf{k}_a} p_a(\mathbf{k}_a) e^{i\mathbf{k}_a \cdot \mathbf{r}_1} \tag{3.4.3}$$

where the p_a's are complex numbers. The matrix element is

$$M_{\mathrm{D}} = V^{-2} \sum_{\text{all } \mathbf{k}} U_{\mathbf{k}} \sum_{\text{all } \mathbf{k}_a} \sum_{\mathbf{k}_b} \sum_{\mathbf{k}_c} \sum_{\mathbf{k}_d} I\, p_a^*(\mathbf{k}_a)\, p_b^*(\mathbf{k}_b) \times p_c(\mathbf{k}_c)\, p_d(\mathbf{k}_d)$$

where

$$I \equiv \int_V e^{i(\mathbf{k}_c - \mathbf{k}_a + \mathbf{k}) \cdot \mathbf{r}_1}\, d\tau_1 \int_V e^{i(\mathbf{k}_d - \mathbf{k}_b - \mathbf{k}) \cdot \mathbf{r}_2}\, d\tau_2$$

This yields (by eq. (B.11))

$$I = V\delta_{\mathbf{k}, \mathbf{k}_a - \mathbf{k}_c}\, V\delta_{\mathbf{k}, \mathbf{k}_d - \mathbf{k}_b}$$

For $\mathbf{k} \neq 0$ the contribution to M_{D} is

$$M_{\mathrm{D}}^{(1)} \equiv \sum_{\text{all } \mathbf{k}_a} \sum_{\mathbf{k}_b} \sum_{\mathbf{k}_c} \sum_{\mathbf{k}_d} U_{\mathbf{k}_a - \mathbf{k}_c}\, p_a^*(\mathbf{k}_a)\, p_b^*(\mathbf{k}_b)\, p_c(\mathbf{k}_c)\, p_d(\mathbf{k}_d)\, \delta_{\mathbf{k}_a + \mathbf{k}_b, \mathbf{k}_c + \mathbf{k}_d}$$

For $\mathbf{k} = 0$ it is more convenient to write the contribution to M_{D} as

$$M_{\mathrm{D}}^{(0)} \equiv U_0 \int \psi_a^*(\mathbf{r}_1)\, \psi_c(\mathbf{r}_1)\, d\tau_1 \int \psi_b^*(\mathbf{r}_2)\, \psi_d(\mathbf{r}_2)\, d\tau_2$$

By the orthonormality of the ψ's

$$M_{\mathrm{D}}^{(0)} = U_0\, \delta_{a,c}\, \delta_{b,d} \tag{3.4.4}$$

and this vanishes since at least one of the states a, b does not occur among the final occupied states.

This seems a somewhat uninstructive result. If, however, the wavefunctions have dominant Fourier coefficients, one can do a little better. For if $U_{\mathbf{k}}$ varies slowly with $\mathbf{k}$, and if with $\psi_a \neq \psi_c$

$$p_a \text{ peaks at } \mathbf{k}_a = \mathbf{k}_1 \text{ and } p_c \text{ at } \mathbf{k}_c = \mathbf{k}_{1'}$$

where $\mathbf{k}_1$ and $\mathbf{k}_{1'}$ are appropriate wavevectors, then one finds the approximate result

$$M_{\mathrm{D}} \equiv M_{\mathrm{D}}^{(0)} + M_{\mathrm{D}}^{(1)} \simeq M_{\mathrm{D}}^{(1)} \simeq U_{\mathbf{k}_1 - \mathbf{k}_{1'}}\, F \tag{3.4.5}$$

where F is a correction factor given by

$$F = \sum_{\text{all } \mathbf{k}_a} \sum_{\mathbf{k}_b} \sum_{\mathbf{k}_c} \sum_{\mathbf{k}_d} p_a^*(\mathbf{k}_a)\, p_b^*(\mathbf{k}_b)\, p_c(\mathbf{k}_c)\, p_d(\mathbf{k}_d)\, \delta_{\mathbf{k}_a + \mathbf{k}_b, \mathbf{k}_c + \mathbf{k}_d} \tag{3.4.6}$$

Although the rest of this section looks mathematical, it contains merely simple variations on the theme (3.4.5). Note that similar arguments hold for the exchange term (3.3.25).

3.4.2 The use of free-electron functions

The most extreme form of 'peaking' occurs if plane waves are used. In that case we may put, if $\mathbf{k}_2$, $\mathbf{k}_{2'}$ are additional appropriate wavevectors,

$$p_a(\mathbf{k}_a) = \delta_{\mathbf{k}_a, \mathbf{k}_1}, \quad p_b(\mathbf{k}_b) = \delta_{\mathbf{k}_b, \mathbf{k}_2}, \quad p_c(\mathbf{k}_c) = \delta_{\mathbf{k}_c, \mathbf{k}_{1'}}, \quad p_d(\mathbf{k}_d) = \delta_{\mathbf{k}_d, \mathbf{k}_{2'}} \qquad (3.4.7)$$

The function F becomes

$$F^{(F)} = \delta_{\mathbf{k}_1 + \mathbf{k}_2, \, \mathbf{k}_{1'} + \mathbf{k}_{2'}}$$

where the superscript (F) stands for *free*, and eq. (3.4.5) is easily seen to be exact in this case. We write for the free-electron case

$$M_{\mathrm{D}} \rightarrow M_{\mathrm{D}}^{(F)} = U_{k_1 - k_{1'}} \delta_{\mathbf{k}_1 + \mathbf{k}_2, \, \mathbf{k}_{1'} + \mathbf{k}_{2'}} \qquad (3.4.8)$$

The significance of $F^{(F)}$ is that it allows a nonzero contribution only if the sum of the initial wavevectors in a perturbation calculation is equal to the sum of the final wavevectors. This is analogous to the conservation of linear momentum in a classical mechanical collision problem in which momenta $\mathbf{k}_1$, $\mathbf{k}_2$ go over into momenta $\mathbf{k}_{1'}$, $\mathbf{k}_{2'}$ after the collision.

3.4.3 The use of four Bloch functions

For an electron in a perfect lattice the Bloch function is more appropriate than the plane wave. Thus one can put (by eq. (C.13)).

$$\psi_a(1) = \frac{1}{V^{\frac{1}{2}}} u_a(\mathbf{k}_1, \mathbf{r}_1) \, \mathrm{e}^{\mathrm{i}\mathbf{k}_1 \cdot \mathbf{r}_1} = \frac{1}{V^{\frac{1}{2}}} \sum_{\mathbf{L}_1} v_a(\mathbf{k}_1, \mathbf{L}_1) \, \mathrm{e}^{\mathrm{i}(\mathbf{L}_1 + \mathbf{k}_1) \cdot \mathbf{r}_1} \qquad (3.4.9)$$

where u_a is lattice periodic and is expanded in terms of the lattice vectors $\mathbf{L}$ of k-space, so that in eq. (3.4.3)

$$p_a(\mathbf{k}_a) = \sum_{\mathbf{L}_1} v_a(\mathbf{k}_1, \mathbf{L}_1) \, \delta_{\mathbf{L}_1 + \mathbf{k}_1, \mathbf{k}_a} \qquad (3.4.10)$$

Either by integrating as before, or by substituting these expressions for p in the formula for M_{D}, one finds for Bloch functions, a matrix element (distinguished by a superscript (B))

$$M_{\mathrm{D}}^{(B)} = U_0 \delta_{a,c} \delta_{b,d} + \sum_{\mathbf{L}_1, \mathbf{L}_2, \mathbf{L}_{1'}, \mathbf{L}_{2'}} U_{\mathbf{L}_1 - \mathbf{L}_{1'} + \mathbf{k}_1 - \mathbf{k}_{1'}}$$

$$\times v_a^*(\mathbf{k}_1, \mathbf{L}_1) v_b^*(\mathbf{k}_2, \mathbf{L}_2) v_c(\mathbf{k}_{1'}, \mathbf{L}_{1'}) v_d(\mathbf{k}_{2'}, \mathbf{L}_{2'}) \delta_{\mathbf{L}_1 + \mathbf{L}_2 + \mathbf{k}_1 + \mathbf{k}_2, \, \mathbf{L}_{1'} + \mathbf{L}_{2'} + \mathbf{k}_{1'} + \mathbf{k}_{2'}}$$

$$(3.4.11)$$

One sees that momentum is now conserved in the amended sense ('Umklapp-type processes')

$$\mathbf{k}_1 + \mathbf{k}_2 - \mathbf{k}_{1'} - \mathbf{k}_{2'} = \text{lattice vector} \tag{3.4.12}$$

These terms must be organized a little:

$$M_D^{(B)} = U_0 \delta_{a,c} \delta_{b,d} + M_{D1}^{(B)} + M_{D2}^{(B)} + M_{D3}^{(B)} \tag{3.4.13}$$

where

$$\left. \begin{array}{l} \text{for } M_{D1}^{(B)} \ \mathbf{L}_1 = \mathbf{L}_{1'}, \mathbf{L}_2 = \mathbf{L}_{2'} \\[4pt] \text{for } M_{D2}^{(B)} \ \mathbf{L}_1 + \mathbf{L}_2 = \mathbf{L}_{1'} + \mathbf{L}_{2'}, \mathbf{L}_1 \neq \mathbf{L}_{1'} \\[4pt] \text{for } M_{D3}^{(B)} \ \mathbf{L}_1 + \mathbf{L}_2 \neq \mathbf{L}_{1'} + \mathbf{L}_{2'}, \mathbf{L}_1 \neq \mathbf{L}_{1'} \end{array} \right\} \tag{3.4.14}$$

It is convenient to observe next that by eq. (3.4.9)

$$F_{ac} \equiv \frac{1}{V} \int_V u_a^*(\mathbf{k}_1, \mathbf{r}) \, u_c(\mathbf{k}_{1'}, \mathbf{r}) \, \mathrm{d}\tau$$

$$= \sum_{\mathbf{L}_1} v_a^*(\mathbf{k}_1, \mathbf{L}_1) \, v_c(\mathbf{k}_{1'}, \mathbf{L}_1) \tag{3.4.15}$$

$$F_{bd} \equiv \frac{1}{V} \int_V u_b^*(\mathbf{k}_2, \mathbf{r}) \, u_d(\mathbf{k}_{2'}, \mathbf{r}) \, \mathrm{d}\tau$$

$$= \sum_{\mathbf{L}_2} v_b^*(\mathbf{k}_2, \mathbf{L}_2) \, v_d(\mathbf{k}_{2'}, \mathbf{L}_2) \tag{3.4.16}$$

These are just the functions which turn up in $M_{D1}^{(B)}$. Thus, using eq. (3.4.8) and assuming that there is no contribution from U_0.

$$M_{D1}^{(B)} = U_{\mathbf{k}_1 - \mathbf{k}_{1'}} \delta_{\mathbf{k}_1 + \mathbf{k}_2, \mathbf{k}_{1'} + \mathbf{k}_{2'}} F_{ac} F_{bd}$$

$$= M_D^{(F)} F_{ac} F_{bd} \tag{3.4.17}$$

A completely similar argument shows that for the exchange term, to be denoted by $M_E^{(B)}$,

$$M_{E1}^{(B)} = U_{\mathbf{k}_2 - \mathbf{k}_{1'}} \delta_{\mathbf{k}_1 + \mathbf{k}_2, \mathbf{k}_{1'} + \mathbf{k}_{2'}} F_{ad} F_{bc}$$

$$= M_E^{(F)} F_{ad} F_{bc} \tag{3.4.18}$$

Thus, provided one can show that $M_{D2}^{(B)}$, $M_{D3}^{(B)}$ are negligible, one has indeed confirmed a form of eq. (3.4.5) for this case, both for M_D and M_E.

For InSb the smallest nonzero reciprocal lattice vector is of the order of 1 Å^{-1}, while a wavevector corresponding to the energy gap is of order 0.024 Å^{-1}. Thus if a nonzero $\mathbf{L}$ occurs on the right-hand side of eq. (3.4.12), some of the electron momenta would be so great as to be weighted with small probabilities of occurrence in the later integrations over all possible states $\mathbf{k}_1$, $\mathbf{k}_2$, $\mathbf{k}_{1'}$, $\mathbf{k}_{2'}$. This justifies that $M_{\mathrm{D3}}^{(\mathrm{B})}$ can often be neglected.

One must now consider the Fourier coefficient of the potential. For a screened Coulomb potential $(e^2/\varepsilon r)\,e^{-\lambda r}$ (by eq. (C.14)).

$$U_{\mathbf{k}} = \frac{4\pi e^2}{\varepsilon V(\lambda^2 + k^2)} \tag{3.4.19}$$

where λ^{-1} is the screening radius. Thus, the larger wavevector $\mathbf{k}$ in eq. (3.4.19), the smaller the $U_{\mathbf{k}}$, and this means from eqs. (3.4.11) and (3.4.14) that one should often be able to neglect $M_{\mathrm{D2}}^{(\mathrm{B})}$ as a first approximation.

Since the exchange term M_{E} can be treated similarly, we have arrived at an understanding of U_{if} through eq. (3.3.27) with

$$\left.\begin{aligned}
|M_{\mathrm{D}}| &\simeq \frac{4\pi e^2}{\varepsilon V} \frac{|F_{ac} F_{bd}|}{\lambda^2 + |\mathbf{k}_1 - \mathbf{k}_{1'}|^2}\, \delta_{\mathbf{k}_1 + \mathbf{k}_2,\, \mathbf{k}_{1'} + \mathbf{k}_{2'}} \\[2mm]
|M_{\mathrm{E}}| &\simeq \frac{4\pi e^2}{\varepsilon V} \frac{|F_{ad} F_{bc}|}{\lambda^2 + |\mathbf{k}_2 - \mathbf{k}_{1'}|^2}\, \delta_{\mathbf{k}_1 + \mathbf{k}_2,\, \mathbf{k}_{1'} + \mathbf{k}_{2'}}
\end{aligned}\right\} \tag{3.4.20}$$

The neglect of sums of type M_{D2}, M_{D3} *makes these results approximate.* They are again of type (3.4.5). In special cases the neglect of the Umklapp-type terms can be justified by direct calculations [3.4.1].

We now collect our results. The probability of a transition in time t is given by the Fermi golden rule (3.2.13). For the squared matrix element which occurs one must substitute eq. (3.3.27),

$$|H_{fi}^{(1)}|^2 = |U_{if}|^2 = 2\beta[|M_{\mathrm{D}}|^2 + |M_{\mathrm{E}}|^2]\ (1 < \beta < 2) \tag{3.4.20'}$$

The individual terms are given approximately by eqs. (3.4.20).

It will be noted that the use of Bloch functions gives rise to *overlap integrals* (3.4.15) which are characteristic of this procedure. Their properties are known to a limited extent and are discussed in section 3.6.

One may regard the effect of the screening radius λ^{-1} as a generalization of two-body terms which must be in the Hamiltonian due to Coulomb interactions among the electrons. However, the use of determinantal wavefunctions is not satisfactory in that the only correlation among the motion of the particles which they take into account is that which arises from the requirements of antisymmetry. There are other correlations as a result of which the electrons move in a background of the positive charges of the host ions so as to inhibit the interaction effects of the electron charges on each other. As this effect is neglected in the determinantal wavefunction, it may be introduced into the Hamiltonian instead. It appears there, of course as a fairly crude approximation, in the form of a screened interaction.

The possibility of applying perturbation theory to electron interactions in a gas receives a justification from similar considerations. The 'switching on' of the Coulomb interactions between electrons at $t = 0$ is of course impossible. They are 'on' all the time. But because the electrons move against a background of positive charge, the Coulomb interactions have little effect over long distances. One might hope therefore that their effect is small enough for a perturbation treatment. A full many-body theory can get over some of these difficulties, but is not attempted here.

It has already been seen that for given λ the Fourier coefficients decrease rapidly as k increases. Indeed for an unscreened potential, U_0 diverges. This corresponds to the infinite range of this potential in real space. The effect of screening is that the Fourier coefficient (3.4.19) which corresponds to the average energy

$$U_0 = \frac{1}{V^2} \int U(\mathbf{r}_1, \mathbf{r}_2) \, \mathrm{d}\tau_1 \, \mathrm{d}\tau_2 = \frac{4\pi e^2}{\varepsilon V \lambda^2}$$

is finite.

3.4.4 Localized states: the case of two Bloch functions

If one of the electrons which makes a transition has a localized state as initial and/or final state, localized (rather than Bloch) wavefunctions must be used for these states. Let them be associated with electron 1 in the matrix element M_D of eq. (3.3.24). The indistinguishability of electrons 1 and 2 is restored by virtue of the existence of the exchange term. It is now best to change the notation slightly to

$$M_\mathrm{D} \int \int \psi_1^*(1) \, \psi_2^*(2) \, U(1, 2) \, \psi_{1'}(1) \, \psi_{2'}(2) \, \mathrm{d}\tau_1 \, \mathrm{d}\tau_2$$

Introducing Bloch functions (3.4.9) for ψ_2 and $\psi_{2'}$ and leaving ψ_1, $\psi_{1'}$ unspecified,

$$M_\mathrm{D} = \frac{e^2}{\varepsilon V} \int \int \frac{e^{-\lambda|\mathbf{r}_1 - \mathbf{r}_2|}}{|\mathbf{r}_1 - \mathbf{r}_2|} \left[\sum_{\mathbf{L}_2} \sum_{\mathbf{L}_{2'}} v_2^*(\mathbf{k}_2, \mathbf{L}_2) \, v_{2'}(\mathbf{k}_{2'}, \mathbf{L}_{2'}) \right.$$
$$\left. e^{i(\mathbf{k}_{2'} - \mathbf{k}_2 + \mathbf{L}_{2'} - \mathbf{L}_2) \cdot \mathbf{r}_2} \right] \psi_1^*(\mathbf{r}_1) \, \psi_{1'}(\mathbf{r}_1) \, \mathrm{d}\tau_1 \, \mathrm{d}\tau_2$$

Inserting the Fourier expansion (3.4.19) $\sum_k U_k \exp[i\mathbf{k} \cdot (\mathbf{r}_1 - \mathbf{r}_2)]$ of the screened Coulomb potential, the $\mathrm{d}\tau_2$-integral can be carried out first and the $\mathrm{d}\tau_1$-integral can be handled separately. One has a sum over all wavevectors $\mathbf{k}$, and the $\mathrm{d}\tau_2$-integral is simply

$$\int e^{i(\mathbf{\mu} - \mathbf{k}) \cdot \mathbf{r}_2} \, \mathrm{d}\tau_2 = V \delta_{\mathbf{k}, \mathbf{\mu}} \quad (\mathbf{\mu} \equiv \mathbf{k}_{2'} - \mathbf{k}_2 + \mathbf{L}_{2'} - \mathbf{L}_2) \tag{3.4.21}$$

One finds

$$M_\mathrm{D} = \frac{4\pi e^2}{\varepsilon V} \sum_{\mathbf{L}_2} \sum_{\mathbf{L}_{2'}} v_2^*(\mathbf{k}_2, \mathbf{L}_2) \, v_{2'}(\mathbf{k}_{2'}, \mathbf{L}_{2'}) \frac{N(\mathbf{\mu}, 1, 1')}{\lambda^2 + \mu^2} \tag{3.4.22}$$

where

$$N(\mu, 1, 1') \equiv \int_V \psi_1^*(\mathbf{r}_1)\,\psi_{1'}(\mathbf{r}_1)\,e^{i\mu\cdot\mathbf{r}_1}\,d\tau_1 \;[\equiv N,\text{ say}] \tag{3.4.23}$$

Localized state wavefunctions can be expressed as superpositions of Bloch functions. This idea will be applied to ψ_1 and $\psi_{1'}$ using $F_\mathbf{k}$, $G_\mathbf{k}$ as superposition constants. One then finds with an obvious notation, including $\mathbf{M} \equiv \mathbf{L} - \mathbf{L}'$ for a lattice vector in $\mathbf{k}$-space,

$$N = V^{-1}\sum_\mathbf{k}\sum_\mathbf{l}\sum_\mathbf{L}\sum_{\mathbf{L}'} F_\mathbf{k}^* G_\mathbf{l}\, v_1^*(\mathbf{k},\mathbf{L})\, v_{1'}(\mathbf{l},\mathbf{L}') \int_V e^{i(\mathbf{l}+\mathbf{L}'-\mathbf{k}-\mathbf{L}+\mu)\cdot\mathbf{r}}\,d\tau$$

$$= V^{-1}\sum_\mathbf{k}\sum_\mathbf{l}\sum_\mathbf{L}\sum_\mathbf{M} F_\mathbf{k}^* G_\mathbf{l}\, v_1^*(\mathbf{k},\mathbf{L})\, v_{1'}(\mathbf{l},\mathbf{L}-\mathbf{M})\left(\frac{1}{V}\int_V e^{i(\mu+\mathbf{l}-\mathbf{k}-\mathbf{M})\cdot\mathbf{r}}\,d\tau\right)$$

$$= \sum_\mathbf{k}\sum_\mathbf{l}\sum_\mathbf{M} F_\mathbf{k}^* G_\mathbf{l}\left(\frac{1}{V}\int u_1^*(\mathbf{k},\mathbf{r}')\, u_{1'}(\mathbf{l},\mathbf{r}')\, e^{i\mathbf{M}\cdot\mathbf{r}'}\,d\tau'\right)\left(\frac{1}{V}\int e^{i(\mu+\mathbf{l}-\mathbf{k}-\mathbf{M})\cdot\mathbf{r}}\,d\tau\right) \tag{3.4.24}$$

The last step is useful because it shows that N contains a generalization of the overlap integrals (3.4.15), (3.4.16):

$$F_{1,1'}^{(-\mathbf{M})} \equiv V^{-1}\int u_1^*(\mathbf{k},\mathbf{r})\, u_{1'}(\mathbf{l},\mathbf{r})\, e^{i\mathbf{M}\cdot\mathbf{r}}\,d\tau$$

$$= \sum_\mathbf{L}\sum_{\mathbf{L}'} v_1^*(\mathbf{k},\mathbf{L})\, v_{1'}(\mathbf{l},\mathbf{L}')\,\frac{1}{V}\int e^{i(\mathbf{L}'-\mathbf{L}+\mathbf{M})\cdot\mathbf{r}}\,d\tau$$

$$= \sum_\mathbf{L} v_1^*(\mathbf{k},\mathbf{L})\, v_{1'}(\mathbf{l},\mathbf{L}-\mathbf{M}) \tag{3.4.25}$$

Thus $F_{a,c}^{(0)}$ is the overlap integral of type (3.4.15). Provided one can find the functions F, G, u_1, $u_{1'}$, each of which presents difficulties, eq. (3.4.22) with eq. (3.4.24) is formally exact. Indeed, these results apply to the case of four Bloch functions as well. However, just as in the case of four Bloch functions, it is a good approximation to neglect some terms of the Umklapp type. In the present case we shall neglect all terms with $\mathbf{M} \neq 0$. Equation (3.4.24) with $\mathbf{M} \neq 0$ will be denoted by $N_2(\mu, 1, 1')$. The terms with $\mathbf{M} = 0$ will be denoted by $N_1(\mu, 1, 1')$. Hence

$$N(\mu, 1, 1') = N_1(\mu, 1, 1') + N_2(\mu, 1, 1') \simeq N_1(\mu, 1, 1') \tag{3.4.26}$$

To make further progress one needs a more specific idea of the Auger transition to be modeled. Suppose for example that state $1'$ is a shallow donor and state 1 is in the same (conduction) band as states 2 and $2'$. Then $F_\mathbf{k}$ and $G_\mathbf{l}$ might be Fourier transforms of a free Coulomb-type wavefunction and a bound hydrogen-type wavefunction, respectively. Keeping this process in mind, a further approximation may be applied to N_1. Two possibilities may be noted:

(a) Approximate $\mathbf{k}$ (in eq. (3.4.24)) by a suitable value $\mathbf{k}_1$. This might be the value of $\mathbf{k}$ for which the Auger process considered has an optimum probability. This eliminates the k-sum. The vector $\mathbf{l}$ can be found from the second integral in which it will also be assumed that $\mathbf{M} = 0$ and $\boldsymbol{\mu} = \mathbf{k}_{2'} - \mathbf{k}_2$, which is consistent with the assumption that Umklapp-type processes may be neglected. One then has $\mathbf{l} \rightarrow \mathbf{k} + \mathbf{M} - \boldsymbol{\mu} \rightarrow \mathbf{k}_1 + \mathbf{k}_2 - \mathbf{k}_{2'}$ and

$$N(\boldsymbol{\mu}, 1, 1') \sim N_1(\boldsymbol{\mu}, 1, 1')$$

$$\sim F_{\mathbf{k}_1}^* \, G_{\mathbf{k}_1 + \mathbf{k}_2 - \mathbf{k}_{2'}} \frac{1}{V} \int_V u_1^*(\mathbf{k}_1, \mathbf{r}) \, u_{1'}(\mathbf{k}_1 + \mathbf{k}_2 - \mathbf{k}_{2'}, \mathbf{r}) \, d\tau$$

Substituting this result and eq. (3.4.25) into eq. (3.4.22), one finds for approximation (a)

$$M_D \simeq \frac{4\pi e^2}{\varepsilon V} G_{\mathbf{k}_1 + \mathbf{k}_2 - \mathbf{k}_{2'}} \frac{F}{\lambda^2 + |\mathbf{k}_{2'} - \mathbf{k}_2|^2} \tag{3.4.27}$$

where with $\mathbf{L}_2 \sim \mathbf{L}_{2'}$

$$F \equiv F_{\mathbf{k}_1}^* \left\{ \frac{1}{V} \int u_1^*(\mathbf{k}_1, \mathbf{r}) \, u_{1'}(\mathbf{k}_1 + \mathbf{k}_2 - \mathbf{k}_{2'}, \mathbf{r}) \, d\tau \right\} \left\{ \frac{1}{V} \int u_2^*(\mathbf{k}_2, \mathbf{r}) \, u_{2'}(\mathbf{k}_{2'}, \mathbf{r}) \, d\tau \right\}$$

(b) A superior method is to regard the first integral in eq. (3.4.24) as independent of $\mathbf{k}$ and $\mathbf{l}$ and assign to it a value corresponding to suitable choices $\mathbf{k}_0$ and $\mathbf{l}_0$, say. The sums over $\mathbf{k}$ and $\mathbf{l}$ are now separated and suggest the introduction of the following wavefunctions:

$$\psi_f(\mathbf{r}) \equiv V^{-\frac{1}{2}} \sum_{\mathbf{k}} F_{\mathbf{k}} \exp(i\mathbf{k} \cdot \mathbf{r}) \tag{3.4.28}$$

$$\psi_b(\mathbf{r}) \equiv V^{-\frac{1}{2}} \sum_{\mathbf{l}} G_{\mathbf{l}} \exp(i\mathbf{l} \cdot \mathbf{r}) \tag{3.4.29}$$

whence [3.4.2]

$$N(\boldsymbol{\mu}, 1, 1') \sim N_1(\boldsymbol{\mu}, 1, 1')$$

$$\sim \left(\int \psi_f^*(\mathbf{r}) \, \psi_b(\mathbf{r}) \, e^{i\boldsymbol{\mu} \cdot \mathbf{r}} \, d\tau \right) \left(\frac{1}{V} \int u_1^*(\mathbf{k}_0, \mathbf{r}) \, u_{1'}(\mathbf{l}_0, \mathbf{r}) \, d\mathbf{r} \right) \tag{3.4.30}$$

The last (overlap) integral may again be approximated by unity.

The neglect of the Umklapp-type processes also yields a simplification in eq. (3.4.22), if the argument of eq. (3.4.25) is used with $\mathbf{M} = 0$. One finds

$$M_D = \frac{4\pi e^2}{\varepsilon V} \frac{F_{22'}}{\lambda^2 + |\mathbf{k}_{2'} - \mathbf{k}_2|^2} N_1 \tag{3.4.31}$$

where, as in eq. (3.4.16),

$$F_{2,2'} = \frac{1}{V} \int u_2^*(\mathbf{k}_2, \mathbf{r}) \, u_{2'}(\mathbf{k}_{2'}, \mathbf{r}) \, d\tau \tag{3.4.32}$$

This is again a result of the form (3.4.5). To pass to the matrix element of the Hamiltonian one can proceed as in eq. (3.4.20′). Formulae of this type are also useful in theoretical approaches to impact ionization rates, as these require similar matrix elements.

3.4.5 Four Bloch states: inclusion of Umklapp-type terms

The electron collision process when all states are in energy bands gives rise to the matrix element (3.4.11) which was approximated by neglecting $M_{\mathrm{D2}}^{(\mathrm{B})}$ and $M_{\mathrm{D3}}^{(\mathrm{B})}$ of eq. (3.4.13). One can retain a sum over the $\mathbf{k}$-space vectors $\mathbf{L}$ explicitly and so obtain a formally more exact result. In a sense this is more elegant, but it postpones the point at which further approximations have to be made to a later stage. In any case, this alternative way of writing the result is readily explained as follows.

Starting with eq. (3.4.11), and assuming that one electron at least changes its band, the U_0-term can be omitted. Then, using eq. (3.4.19),

$$M_{\mathrm{D}}^{(\mathrm{B})} = \frac{4\pi e^2}{\varepsilon V} \sum_{\mathbf{L}_1, \mathbf{L}_{1'}} \frac{v_{n_1}^*(\mathbf{k}_1, \mathbf{L}_1) \, v_{n_{1'}}(\mathbf{k}_{1'}, \mathbf{L}_{1'})}{|\mathbf{L}_{1'} - \mathbf{L}_1 + \mathbf{k}_{1'} - \mathbf{k}_1|^2 + \lambda^2} Z \tag{3.4.33}$$

where a, b, c, d have been replaced by band numbers n_1, n_2, $n_{1'}$, $n_{2'}$ and

$$Z \equiv Z(n_2, n_{2'}, \mathbf{k}_1 + \mathbf{L}_{1'} - \mathbf{k}_1 - \mathbf{L}_1)$$

$$\equiv \sum_{\mathbf{L}_2, \mathbf{L}_{2'}} v_{n_2}^*(\mathbf{k}_2, \mathbf{L}_2) \, v_{n_{2'}}(\mathbf{k}_{2'}, \mathbf{L}_{2'}) \, \delta_{\mathbf{L}_{1'} - \mathbf{L}_1, \, \mathbf{L}_2 - \mathbf{L}_{2'} + \mathbf{k}_1 + \mathbf{k}_2 - \mathbf{k}_{1'} - \mathbf{k}_{2'}}$$

With

$$\mathbf{L}' \equiv \mathbf{L}_{1'} - \mathbf{L}_1, \quad \mathbf{L} \equiv \mathbf{L}_{1'} + \mathbf{L}_{2'} - \mathbf{L}_1 - \mathbf{L}_2 = \mathbf{L}_{2'} - \mathbf{L}_2 + \mathbf{L}' \tag{3.4.34}$$

this is

$$Z(n_2, n_{2'}, \mathbf{k}_{1'} - \mathbf{k}_1 + \mathbf{L}') = \sum_{\mathbf{L}, \mathbf{L}_2} v_{n_2}^*(\mathbf{k}_2, \mathbf{L}_2) \, v_{n_{2'}}(\mathbf{k}_{2'}, \mathbf{L} - \mathbf{L}' + \mathbf{L}_2) \delta_{\mathbf{k}_1 + \mathbf{k}_2 - \mathbf{k}_{1'} - \mathbf{k}_{2'}, \, \mathbf{L}} \tag{3.4.35}$$

Contributions to Z arise only if the algebraic $\mathbf{k}$-sum yields a $\mathbf{k}$-space lattice vector $\mathbf{L}$. (The value $\mathbf{L} = 0$ yields often the most important contribution.) Thus

$$M_{\mathrm{D}}^{(\mathrm{B})} = \frac{4\pi e^2}{\varepsilon V} \sum_{\mathbf{L}_1} \sum_{\mathbf{L}'} \frac{v_{n_1}^*(\mathbf{k}_1, \mathbf{L}_1) \, v_{n_{1'}}(\mathbf{k}_{1'}, \mathbf{L}_1 + \mathbf{L}')}{|\mathbf{L}' - \mathbf{k}_1 + \mathbf{k}_{1'}|^2 + \lambda^2} Z \tag{3.4.36}$$

Recall next the overlap integral (3.4.25)

$$F_{n_1, n_{1'}}^{(\mathbf{L}')} \equiv \sum_{\mathbf{L}_1} v_{n_1}^*(\mathbf{k}_1, \mathbf{L}_1) \, v_{n_{1'}}(\mathbf{k}_{1'}, \mathbf{L}_1 + \mathbf{L}') \tag{3.4.37}$$

Then we may also write

$$Z \equiv Z(n_2, n_{2'}, \mathbf{k}_{1'} - \mathbf{k}_1 + \mathbf{L'}) = F_{n_2, n_{2'}}^{(\mathbf{L} - \mathbf{L'})} \qquad (3.4.38)$$

provided it is understood that

$$\mathbf{k}_1 + \mathbf{k}_2 - \mathbf{k}_{1'} - \mathbf{k}_{2'} = \mathbf{L} \qquad (3.4.39)$$

is a lattice vector in **k**-space (and possibly zero). One finally has

$$M_D^{(B)} = \frac{4\pi e^2}{\varepsilon V} \sum_{\mathbf{L'}} \frac{F_{n_1, n_{1'}}^{(\mathbf{L'})} F_{n_2, n_{2'}}^{(\mathbf{L} - \mathbf{L'})}}{|\mathbf{L'} + \mathbf{k}_{1'} - \mathbf{k}_1|^2 + \lambda^2} \qquad (3.4.40)$$

For the exchange process one replaces $\mathbf{k}_1$ by $\mathbf{k}_2$, so that

$$M_E^{(B)} = \frac{4\pi e^2}{\varepsilon V} \sum_{\mathbf{L'}} \frac{F_{n_1, n_{1'}}^{(\mathbf{L'})} F_{n_2, n_{2'}}^{(\mathbf{L} - \mathbf{L'})}}{|\mathbf{L'} + \mathbf{k}_{1'} - \mathbf{k}_1|^2 + \lambda^2} \qquad (3.4.41)$$

Note that the overlap integrals $F_{n_1, n_2}^{(\mathbf{L})}$ are generalizations of the quantities (3.4.15) and (3.4.16) in the sense that

$$F_{ac} = F_{ac}^{(0)}, \quad F_{bd} = F_{bd}^{(0)} \qquad (3.4.42)$$

as already pointed out in eq. (3.4.25).

3.5 Threshold energies for impact ionization

3.5.1 Introduction

The importance of impact ionization in devices has been recognized for a long time. It has come to the forefront again in recent years because of the development of optical communication systems using optical fibers. A very promising detector in such a system is the avalanche photodiode. Low noise can be achieved in such systems if the electron and hole impact ionization coefficients have very different values. These avalanches are the result of many impact ionization events, and the rate at which these events occur depends on quantum mechanical probabilities, which are rather complicated to estimate theoretically. However, some guidance can be obtained from the threshold energy of the impact ionizing carrier. This is the least energy the carrier must have in order for the impact ionization process to proceed. Below this threshold value the process is forbidden by the conservation laws of energy and momentum. As no other physical principles are involved, the threshold calculations are quasi-classical and are much easier than the probability calculations referred to above.

Table 3.5.1. *Some threshold kinetic energies*

Line	$(\varepsilon_{2'})_{th}$	$(\varepsilon_1)_{th} = (\varepsilon_2)_{th}$	$(\varepsilon_{1'})_{th}$	Impact ionizing carrier and band	
1	$\dfrac{2m_h + m_c}{2m_h + m_c - m_s}(E_G - \Delta)$	$\dfrac{m_s m_h (E_G - \Delta)}{(2m_h + m_c)(2m_h + m_c - m_s)}$	$\dfrac{m_s m_c (E_G - \Delta)}{(2m_h + m_c)(2m_h + m_c - m_s)}$	h(S)	CHHS
2	$m_s \to m_\ell,\ \Delta \to 0$ in line 1: $\dfrac{(2m_h + m_c)E_G}{2m_h + m_c - m_\ell}$	$\dfrac{m_\ell m_h E_G}{(2m_h + m_c)(2m_h - m_c - m_\ell)}$	$\dfrac{m_\ell m_c E_G}{(2m_h + m_c)(2m_h + m_c - m_\ell)}$	h(L)	CHHL
3	$m_h \to m_\ell$ in line 1: $\dfrac{(2m_\ell + m_c)(E_G - \Delta)}{2m_\ell + m_c - m_s}$	$\dfrac{m_s m_\ell (E_G - \Delta)}{(2m_\ell + m_c)(2m_\ell + m_c - m_s)}$	$\dfrac{m_s m_c (E_G - \Delta)}{(2m_\ell + m_c)(2m_\ell + m_c - m_s)}$	h(S)	CLLS

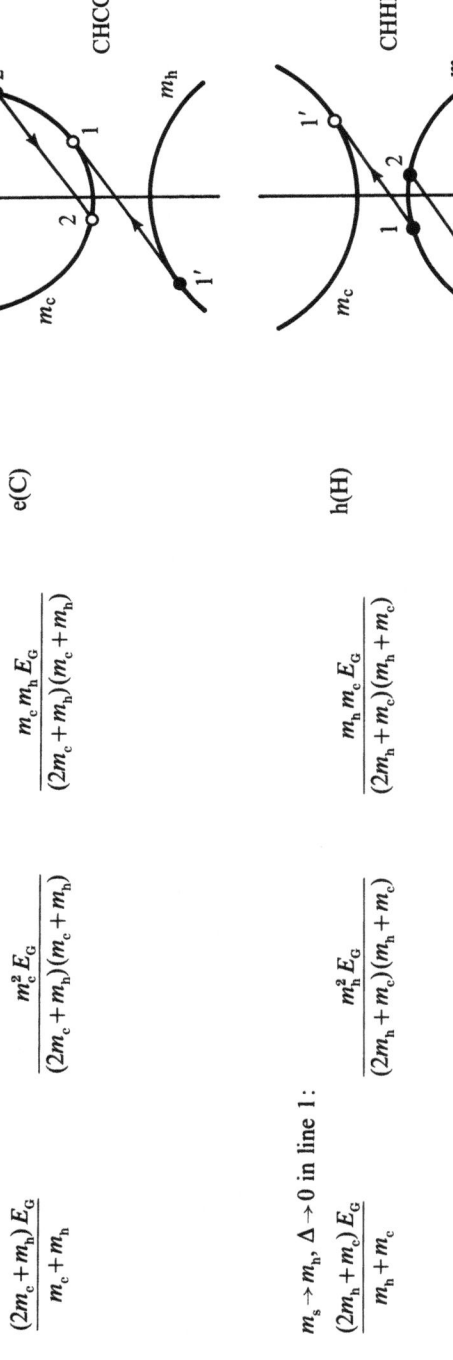

CHCC

CHHH

4   Interchange m_c and m_h in line 5:

$$\frac{(2m_c+m_h)E_G}{m_c+m_h}$$

$$\frac{m_c^2 E_G}{(2m_c+m_h)(m_c+m_h)}$$

$$\frac{m_c m_h E_G}{(2m_c+m_h)(m_c+m_h)}$$

   e(C)

5    $m_s \to m_h$, $\Delta \to 0$ in line 1:

$$\frac{(2m_h+m_c)E_G}{m_h+m_c}$$

$$\frac{m_h^2 E_G}{(2m_h+m_c)(m_h+m_c)}$$

$$\frac{m_h m_c E_G}{(2m_h+m_c)(m_h+m_c)}$$

   h(H)

Arrows indicate transitions by electrons. 2′ is the impact ionizing state. In lines 1 and 3 it is assumed that $E_G > \Delta$. C, S, H, L denote conduction, split-off, heavy hole, light hole bands, respectively. There is just one valence band in line 5, denoted by H. The hole masses m_l, m_h, m_s are taken as positive.

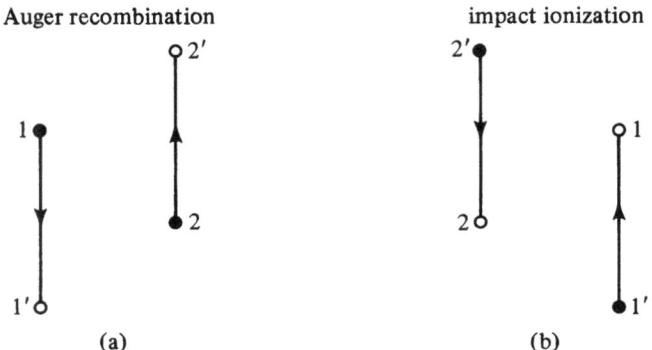

Fig. 3.5.1. Formal energy diagram of an Auger process and matching impact ionization process. For a CHCC process state 1′ is in the valence band; the remaining states are in the conduction band.

3.5.2 *Two isotropic parabolic bands*

Consider the CHCC process of Table 3.5.1 first (see also Fig. 3.5.1). The energetic (impact ionizing) state is labeled 2′. The final states of electrons 1 and 2 are labeled 1′ and 2′ as shown. The conditions of energy and momentum conservation are

$$E_c(\mathbf{k}_{2'}) = E_c(\mathbf{k}_1) + E_c(\mathbf{k}_2) - E_v(\mathbf{k}_{1'}) \tag{3.5.1}$$

$$\mathbf{k}_{2'} = \mathbf{k}_1 + \mathbf{k}_2 - \mathbf{k}_{1'} \tag{3.5.2}$$

where the probability of Umklapp-type processes (3.4.12) is neglected at the outset and total energies in the conduction and valence bands are given. Imagine now arbitrary variations in $\mathbf{k}_1$, $\mathbf{k}_2$, $\mathbf{k}_{1'}$ then these determine $\mathbf{k}_{2'}$ by eq. (3.5.2) and hence $E_c(\mathbf{k}_{2'})$ by eq. (3.5.1). Using the energy conservation condition (3.5.1) for both the varied and the unvaried state gives, on taking the difference,

$$\nabla E_c(\mathbf{k}_{2'}) \cdot d\mathbf{k}_{2'} = \nabla E_c(\mathbf{k}_1) \cdot d\mathbf{k}_1 + \nabla E_c(\mathbf{k}_2) \cdot d\mathbf{k}_2 - \nabla E_v(\mathbf{k}_{1'}) \cdot d\mathbf{k}_{1'} \tag{3.5.3}$$

This vanishes at an extremum of $E_c(\mathbf{k}_{2'})$. For small but arbitrary variations

$$d\mathbf{k}_{2'} = d\mathbf{k}_1 + d\mathbf{k}_2 - d\mathbf{k}_{1'}$$

where one may regard the three variations on the right as independent. Substituting for $d\mathbf{k}_{2'}$ in eq. (3.5.3), one finds an equation involving only the three independently variable wavevector increments:

$$[\nabla E_c(\mathbf{k}_1) - \nabla E_c(\mathbf{k}_{2'})] \cdot d\mathbf{k}_1 + [\nabla E_c(\mathbf{k}_2) - \nabla E_c(\mathbf{k}_{2'})] \cdot d\mathbf{k}_2$$
$$- [\nabla E_v(\mathbf{k}_{1'}) - \nabla E_c(\mathbf{k}_{2'})] \cdot d\mathbf{k}_{1'} = 0$$

It follows that

$$\nabla_{\mathbf{k}_1} E_c(\mathbf{k}_1) = \nabla_{\mathbf{k}_2} E_c(\mathbf{k}_2) = \nabla_{\mathbf{k}_{1'}} E_v(\mathbf{k}_{1'}) \tag{3.5.4}$$

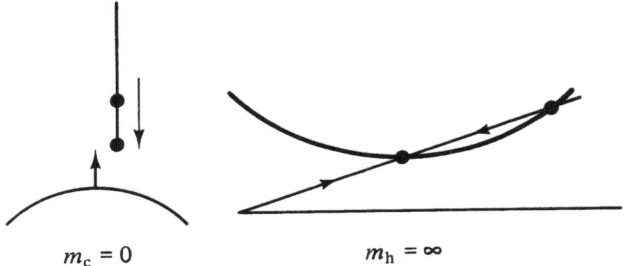

$$m_c = 0 \qquad\qquad m_h = \infty$$

Fig. 3.5.2. Limiting cases for impact ionization.

This is a useful threshold condition for general bands [3.5.1]. It states equality of the three group velocities (see eq. (D.19)) as a threshold condition; parabolicity of the bands has not yet been assumed.

For isotropic parabolic bands with band edge energies E_c, E_v, eq. (3.5.4) is (with $m_v \equiv |m_v|$)

$$\nabla_{\mathbf{k}_1}(E_c + \hbar^2 k_1^2/2m_c) = \nabla_{\mathbf{k}_2}(E_c + \hbar^2 k_2^2/2m_c) = \nabla_{\mathbf{k}_{1'}}(E_v - \hbar^2 k_{1'}^2/2m_v)$$

so that at threshold

$$k_1 = k_2 = -\frac{m_c}{m_v}k_{1'}, \quad k_{2'} = 2k_1 - k_{1'} = \frac{2m_c + m_v}{m_c}k_1 \qquad (3.5.5)$$

We use eq. (3.5.1) in a form which shows the kinetic energies ε for the various states:

$$E_c + \varepsilon_{2'} = E_c + \varepsilon_1 + E_c + \varepsilon_2 - (E_v - \varepsilon_{1'})$$

where $E_G = E_c - E_v$. Hence

$$\varepsilon_{2'} = E_G + \varepsilon_1 + \varepsilon_2 + \varepsilon_{1'} \qquad (3.5.6)$$

At threshold this becomes, using eqs. (3.5.5),

$$\frac{\hbar^2 k_1^2}{2m_c}\left(\frac{2m_c + m_v}{m_c}\right)^2 = E_G + \frac{\hbar^2 k_1^2}{2m_c}\left[2 + \frac{m_c}{m_v}\left(\frac{m_v}{m_c}\right)^2\right]$$

Hence the entries in line 4 of Table 3.5.1 result (if one rewrites m_v as m_h). Note that $(\varepsilon_{2'})_{th}$ always exceeds E_G, and reaches $2E_G$ if $m_c \gg m_h$, $\frac{3}{2}E_G$ if $m_c = m_h$ and E_G if $m_h \gg m_c$. Formulae of this type are important as they give a rough idea of how an Auger recombination rate might depend on temperature. One would expect it to do so via an Arrhenius factor $\exp[-(\varepsilon_{2'})_{th}/kT]$. The smaller the threshold energy the weaker the expected temperature dependence. Similarly, impact ionization is expected to be easier the smaller the threshold energy. One can see from Fig. 3.5.2 that energy and momentum conditions are readily satisfied in the two limiting cases $m_c = 0$ and $m_v = \infty$ of CHCC to yield $(\varepsilon_1)_{th} = (\varepsilon_2)_{th} = (\varepsilon_{1'})_{th} = 0$ and $(\varepsilon_{2'})_{th} = E_G$.

For a hole-induced impact ionization process a simple interchange of m_c and m_h yields the threshold energies relevant in this case (line 5 of Table 3.5.1).

3.5.3 Three isotropic parabolic bands

One need not base one's analysis of threshold on the type of argument that has been used for two bands. As a variant we give here a purely algebraic method for the CHHS process. Of the ten impact ionization processes originally envisaged for InSb it is this process, as well as the CHCC process, which were judged to be the most important ones [3.5.2]. One may say that this process is hole-initiated since the state 2' in the split-off valence band has to be vacant originally, and one is searching for the state of least kinetic energy $\varepsilon_{2'}$ which will allow this process to proceed.

Taking the origin of the energy at the top of the split-off valence band, energy conservation gives

$$\left(\Delta - \frac{\hbar^2 k_1^2}{2m_v}\right) + \left(\Delta - \frac{\hbar^2 k_2^2}{2m_v}\right) = \left(E_G + \Delta + \frac{\hbar^2 k_{1'}^2}{2m_c}\right) - \frac{\hbar^2 k_{2'}^2}{2m_s} \qquad (3.5.7)$$

Multiplying by $2m_c/\hbar^2$, putting $E_G - \Delta = \hbar^2 k_a^2/2m_c$, and replacing k_1 by eq. (3.5.2),

$$\frac{m_c}{m_s} k_{2'}^2 - \frac{m_c}{m_v}(k_1^2 + k_2^2) - (\mathbf{k}_1 + \mathbf{k}_2 - \mathbf{k}_{2'})^2 = k_a^2$$

Therefore, labeling the left-hand terms (1) to (4),

$$\left(\frac{m_c}{m_s} - 1\right) k_{2'}^2 - \frac{m_c}{m_v}(k_1^2 + k_2^2) - (\mathbf{k}_1 + \mathbf{k}_2)^2 + 2\mathbf{k}_{2'} \cdot (\mathbf{k}_1 + \mathbf{k}_2) = k_a^2$$

$$\qquad (1) \qquad\qquad (2) \qquad (3) \qquad\qquad (4)$$

In order to achieve an expression for $k_{2'}^2$ as a sum of squares, write term (2) as

$$-\frac{m_c}{2m_v}(k_1^2 + k_2^2 + 2\mathbf{k}_1 \cdot \mathbf{k}_2) - \frac{m_c}{2m_v}(k_1^2 + k_2^2 - 2\mathbf{k}_1 \cdot \mathbf{k}_2)$$

$$\qquad\qquad (a) \qquad\qquad\qquad\qquad (b)$$

$$+\frac{2m_v}{2m_v + m_c} k_{2'}^2 - \frac{2m_v}{2m_v + m_c} k_{2'}^2$$

$$\qquad\qquad\qquad\qquad (c) \qquad\qquad\qquad (d)$$

Take (c) in with (1); take (a) in with (3) so as to obtain

$$\left(\frac{m_c}{m_s}-1+\frac{2m_v}{2m_v+m_c}\right)k_{2'}^2-\left[\left(1+\frac{m_c}{2m_v}\right)^{\frac{1}{2}}(\mathbf{k}_1+\mathbf{k}_2)\right]^2-\frac{m_c}{2m_v}(\mathbf{k}_2-\mathbf{k}_1)^2$$

$$\quad (1) \qquad\qquad (c) \qquad\qquad\qquad\underset{(3)}{\uparrow}\quad\underset{(4)}{\uparrow} \qquad\qquad\qquad\qquad (b)$$

$$+2\mathbf{k}_{2'}\cdot(\mathbf{k}_1+\mathbf{k}_2)-\frac{2m_v}{2m_v+m_c}k_{2'}^2=k_a^2$$

$$\qquad\qquad (4) \qquad\qquad\qquad (d)$$

Now put

$$\mathbf{k}_0=\left(1+\frac{m_c}{2m_v}\right)^{\frac{1}{2}}(\mathbf{k}_1+\mathbf{k}_2)-\left(1+\frac{m_c}{2m_v}\right)^{-\frac{1}{2}}\mathbf{k}_{2'} \qquad\qquad (3.5.8)$$

Then $-k_0^2$ incorporates the terms (3), (a); (4); (d). One finds

$$\frac{m_c}{m_s}\frac{2m_v+m_c-m_s}{2m_v+m_c}k_{2'}^2=k_0^2+\frac{m_c}{2m_v}(\mathbf{k}_2-\mathbf{k}_1)^2+k_a^2$$

$$\quad (1) \qquad (c) \qquad\qquad\qquad (b)$$

These are general results without, so far, any reference to thresholds. We now look for the smallest kinetic energy of the impact ionizing state, i.e. for the smallest value of $k_{2'}^2$ compatible with energy and momentum conservation. It arises if $\mathbf{k}_0=0$ and $\mathbf{k}_1=\mathbf{k}_2$. Therefore, assuming $E_G>\Delta$, i.e. $k_a^2>0$,

$$k_{2'}^2=\frac{m_s}{m_c}\frac{2m_v+m_c}{2m_v+m_c-m_s}k_a^2 \qquad\qquad (3.5.9a)$$

$$\frac{2m_v+m_c}{2m_v}2\mathbf{k}_1=\mathbf{k}_{2'} \qquad\qquad (3.5.9b)$$

From eqs. (3.5.9a) $(\varepsilon_{2'})_{th}$ is found as given in line 1 of Table 3.5.1. Equation (3.5.9b) is found from eq. (3.5.8), $\mathbf{k}_0=0$, and $\mathbf{k}_1=\mathbf{k}_2$. It leads to $(\varepsilon_1)_{th}$ as given in line 1 of Table 3.5.1. Lastly, $(\varepsilon_{1'})_{th}$ can be found by substituting in eq. (3.5.7). These are helpful results and can be used to recover the conclusions for two bands, as shown in Table 3.5.1.

In addition to the two methods discussed, the common group velocity and the algebraic method, other methods can be used, notably a minimization procedure in which one of the conservation laws is introduced as a subsidiary condition. It is looked after by a Lagrangian multiplier in this method [3.5.3].

A striking property of the three-band impact ionization process is that the threshold kinetic energy of the impact ionizing carrier vanishes if $E_G=\Delta$. Both

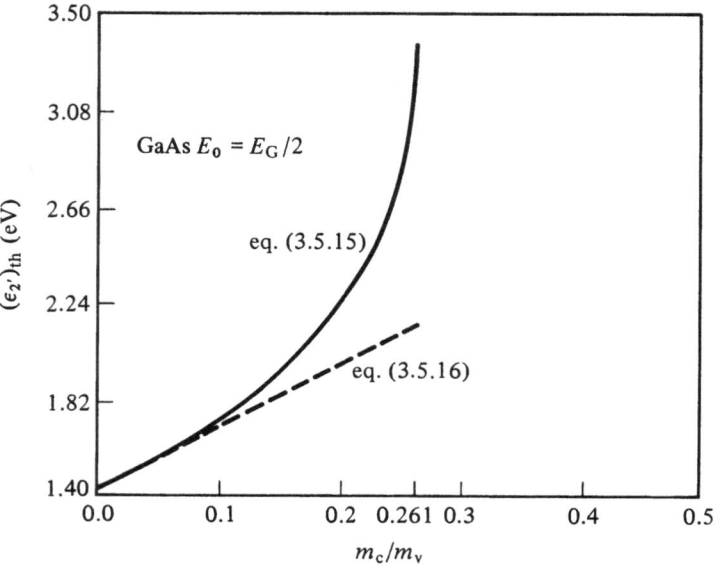

Fig. 3.5.3. Threshold energy for impact ionization for given E_0 as a function of m_c/m_v.

electrons then make vertical transitions between extrema. In a material such as $Ga_{1-x}Al_xSb$ this can be achieved at $x = 0.065$ at 300 K [3.5.4]. The resultant hole-to-electron ionization coefficient becomes very large at this value. This is a desirable property for avalanche photodiodes as it leads to low noise.

Note that the threshold energies of Table 3.5.1 can be summarized by the formula

$$E_T \equiv (\varepsilon_{2'})_{th} = -\frac{m^*\varepsilon}{m_{2'} + m^*} \tag{3.5.10}$$

where ε is $E_G - \Delta$ or E_G and

$$m^* \equiv m_{1'} - m_1 - m_2 \tag{3.5.11}$$

Here E_T is regarded as negative if the ionizing particle is a hole and the m's are regarded as negative for holes. This result will be obtained by a general argument in section 3.5.7.

3.5.4 One isotropic parabolic, one isotropic nonparabolic band

Consider a CHCC process in which both bands are isotropic and the conduction band is nonparabolic. However, the nonparabolicity will be neglected for states 1 and 2 in the conduction band since they lie in general close enough to the minimum for the effect of nonparabolicity to be negligible. Thus eq. (3.5.5) will hold again. We shall take it in the form

$$(k_1)_{th} = (k_2)_{th} = \frac{m_c}{2m_c + m_v}(k_{2'})_{th}, \quad (k_{1'})_{th} = -\frac{m_v}{2m_c + m_v}(k_{2'})_{th}$$

where m_c is now the effective mass *at the bottom* of the conduction band. The energy conservation condition is for an arbitrary $E-\mathbf{k}$ relation at state $2'$,

$$E_c + (\varepsilon_{2'})_{th} = 2[E_c + (\varepsilon_1)_{th}] - [E_v - (\varepsilon_{1'})_{th}]$$

Hence

$$(\varepsilon_{2'})_{th} = E_G + \frac{\hbar^2}{m_c}\left(\frac{m_c}{2m_c + m_v}\right)^2 (k_{2'})_{th}^2 + \frac{\hbar^2}{2m_c}\cdot\frac{m_c}{m_v}\left(\frac{m_v}{2m_c + m_v}\right)^2 (k_{2'})_{th}^2$$

$$= E_G + \frac{m_c}{2m_c + m_v}\frac{\hbar^2(k_{2'})_{th}^2}{2m_c} \tag{3.5.12}$$

The conduction band will be taken in the simple form, due to E.O. Kane, already mentioned in eq. (1.6.16),

$$\frac{\hbar^2 k^2}{2m_c} = \varepsilon\left(1 + \frac{\varepsilon}{2E_0}\right) \tag{3.5.13}$$

which yields

$$\varepsilon_{2'} = \left[\left(1 + \frac{\hbar^2 k_{2'}^2}{m_c E_0}\right)^{\frac{1}{2}} - 1\right]E_0 \tag{3.5.14}$$

One sees that only for large E_0 is the parabolic condition

$$\varepsilon_{2'} = \hbar^2 k_{2'}^2 / 2m_c$$

approached. Combining eqs. (3.5.12) and (3.5.14) and rearranging, one finds [3.5.5]

$$(\varepsilon_{2'})_{th} = \frac{m_c + m_v}{m_c}\left\{1 \pm \left[1 - \frac{2m_c(2m_c + m_v)E_G}{(m_c + m_v)^2}\frac{E_G}{E_0}\right]^{\frac{1}{2}}\right\}E_0 \tag{3.5.15}$$

To examine the case of weak nonparabolicity, expand this result to second order in E_G/E_0 to find

$$(\varepsilon_{2'})_{th} = \frac{2m_c + m_v}{m_c + m_v}E_G\left[1 + \frac{(2m_c + m_v)m_c}{(m_c + m_v)^2}\frac{E_G}{2E_0}\right] \tag{3.5.16}$$

In first approximation this agrees with line 4 of Table 3.5.1. The result (3.5.16) was known [3.5.6] before the full result (3.5.15). They are compared in Fig. 3.5.3. The full result shows something not revealed by eq. (3.5.16), namely that m_c/m_v must be small enough for given E_G/E_0, otherwise the threshold energy becomes infinitely

large, showing that impact ionization cannot occur (Fig. 3.5.5 for GaAs [3.5.5]).
Fig. 3.5.4 shows the dependence of the threshold energy on $E_G/2E_0$ together with
the experimental result for GaAs which is, however, known only approximately
[3.5.7].

The formula (3.5.15) yields a critical or 'starred' value of m_c/m_v

$$\left(\frac{m_c}{m_v}\right)^* = \frac{[(E_G/E_0)^2 + 2(E_G/E_0)]^{\frac{1}{2}} - E_G/E_0 + 1}{4E_G/E_0 - 1} \tag{3.5.17}$$

above which it fails. For $E_0 = E_G/2$, for example,

$$\left(\frac{m_c}{m_v}\right)^* = \frac{2\sqrt{2}-1}{7} = 0.261$$

These starred values at which impact ionization ceases since $(\varepsilon_{2'})_{th} \to \infty$ are,
however, unreliable, since large values of $(\varepsilon_{2'})_{th}$ imply that states 1 and 2 move
higher into the conduction band at threshold and so begin to 'feel' the
nonparabolicity, contrary to the assumptions on which eq. (3.5.12) and hence
subsequent equations are based. In these ranges of higher $(\varepsilon_{2'})_{th}$ eq. (3.5.15) is
however still useful in showing the existence of such a critical value of m_c/m_v.

We now give an improved estimate to replace eq. (3.5.17). Write the Kane band
(3.5.13) as a hyperbola $(y+c)^2/a^2 - x^2/b^2 = 1$:

$$\frac{(E+E_0)^2}{E_0^2} - \frac{k^2}{m_c E_0/\hbar^2} = 1$$

where $E = 0$ at $k = 0$, and note that its asymptotes have the form $y = \pm(a/b)x - c$.
Hence the asymptotes to the Kane band are given by

$$E = \pm\left(\frac{\hbar^2 E_0}{m_c}\right)^{\frac{1}{2}} k - E_0$$

The valence band is given by

$$E = -E_G - \hbar^2 k^2/2m_v$$

At an intersection of the two $(E-k)$ curves the k-values are given by

$$\frac{\hbar^2 k^2}{2m_v} + \left(\frac{\hbar^2 E_0}{m_c}\right)^{\frac{1}{2}} k + E_G - E_0 = 0$$

and cease to be real if

$$\frac{m_c}{m_v} \geqslant \frac{1}{2[(E_G/E_0)-1]} \equiv \left(\frac{m_c}{m_v}\right)_{crit} \tag{3.5.18}$$

This 'critical' value lies above the 'starred' value for given E_G/E_0, indicating a
later switch-off of this impact ionization process as m_c/m_v is increased. For

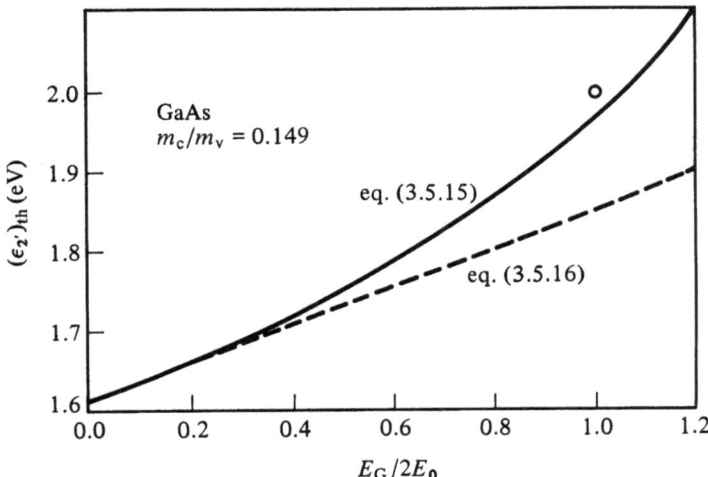

Fig. 3.5.4. Threshold energy for impact ionization for given m_c/m_v as a function of $E_G/2E_0$.

$E_0 = E_G/2$, for example, $(m_c/m_v)_{crit} = 0.5 \geqslant (m_c/m_v)^* = 0.26$. That relation (3.5.18) gives a switch-off may be seen from Fig. 3.5.5. If therefore one has a simple Kane-like conduction band and a parabolic valence band these relations give one a guide on how to manipulate m_c/m_v or E_G/E_0 depending on whether the impact ionization process is desired or not.

A stronger nonparabolicity, specified by a parameter k_0, can be induced by bending the $\varepsilon_{2'}(k_{2'})$-characteristic over to present a maximum, for example by substituting not eq. (3.5.13) into eq. (3.5.12), but, following R.I. Taylor [3.5.5 and Ph.D. thesis, University of Durham, 1987],

$$\varepsilon_{2'}(k_{2'}^2) = \frac{\hbar^2 k_{2'}^2}{2m_c}\left[1 - \left(\frac{k_{2'}}{k_0}\right)^2\right]$$

where k_0 is an appropriate parameter. The band becomes parabolic as $k_0 \to \infty$. By solving eq. (3.5.12) one finds two solutions:

$$(k_{2'\pm})^2 = \frac{m_c + m_v}{2m_c + m_v}\frac{k_0^2}{2}\left\{1 \pm \left[1 - \left(\frac{k_{0m}}{k_0}\right)^2\right]^{\frac{1}{2}}\right\}$$

where

$$\frac{\hbar^2 k_{0m}^2}{2m_c} \equiv 4\left[\frac{2m_c + m_v}{m_c + m_v}\right]^2 E_G$$

defines the value k_{0m} of k_0. For real $(k_{2'\pm})^2$, k_0^2 must not be too small: $k_0^2 \geqslant k_{0m}^2$, and this is a condition for Auger effects to be possible. The appropriate region in an

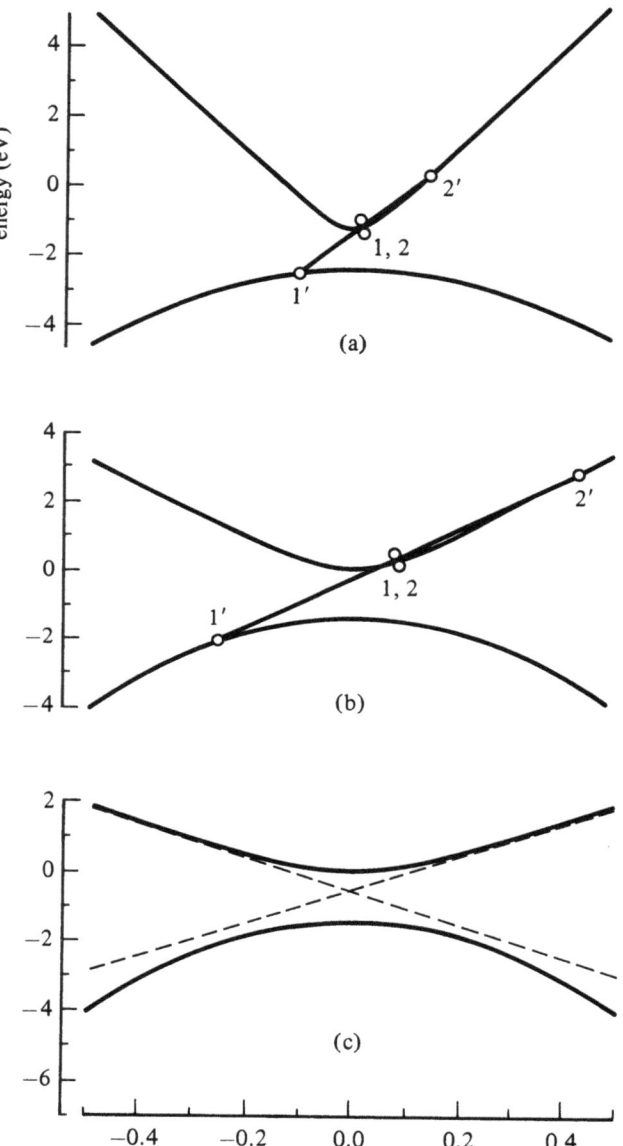

Fig. 3.5.5. Quadruplets of states at the threshold of impact ionization for given E_G/E_0. The horizontal axis gives aK/π where $a = 5.65$ Å (GaAs). Also $m_c/m_v = 0.2(m_c/m_v)_{crit}$, $(m_c/m_v)_{crit}$, $1.2(m_c/m_v)_{crit}$ for (a), (b), (c), respectively. In case (c) impact ionization cannot occur.

energy versus k_2^2-plot is the part of the $\varepsilon_{2'}(k_{2'}^2)$-curve which lies above the $(\varepsilon_{2'})_{\mathrm{th}}$-curve (3.5.12). This is illustrated in Fig. 3.5.6 which also furnishes an example of an anti-threshold.

3.5.5 *The relation between Auger recombination and impact ionization in bands*

The connection between impact ionization and Auger effect recombination depends on the fact that, given the four states, the relevant matrix element in an perturbation calculation is the same in both cases (see Fig. 3.5.1). The state 2′ represents an energetic electron which initiates the impact ionization, or the energetic 'Auger electron' produced by the recombination. On the general grounds explained earlier, the Coulomb interaction between the electrons is the relevant perturbation in both cases.

There is a second connection, which arises from the integration over all states. The 'most probable transitions' occur when the probability product Θ for the occupation of states 1 and 2 and the probability of vacancy of states 1′ and 2′ is maximized over all permitted states 1, 2, 1′, 2′. This maximization of Θ is subject to momentum and energy conservation. For whatever band structure or in whatever bands these states are, the probability of vacancy of the state 2′ of the Auger electron can be taken as unity. Then one has for nondegenerate bands using energy conservation,

$$
\begin{aligned}
\Theta &= \exp\{F_1 - \eta_1 + F_2 - \eta_2 + \eta_{1'} - F_{1'}\} \\
&= \exp(F_1 + F_2 - F_{1'})\exp(\eta_{1'} - \eta_1 - \eta_2) \\
&= \exp(F_1 + F_2 - F_{1'} - F_{2'})\exp(F_{2'} - \eta_{2'})
\end{aligned}
\tag{3.5.19}
$$

where the reduced quasi-Fermi levels have been labeled in a general manner, even though some of these levels will coincide.

In impact ionization the analogous problem is to minimize the energy of the state 2′ of the 'energetic electron' over all states 1, 2, 1′, 2′ compatible with energy and momentum conservation and with the condition that impact ionization actually occurs. This will yield the threshold energy $(E_{2'})_{\mathrm{th}}$ for the process. It is given by the maximization of $\exp(-\eta_{2'})$. This is precisely the maximization of eq. (3.5.19). Hence the following theorem results [3.5.8].

Theorem 1
Of all quadruplets of states in nondegenerate bands, the most probable Auger recombination quadruplets are identical with the quadruplets which lead to the threshold of impact ionization.
(3.5.20)

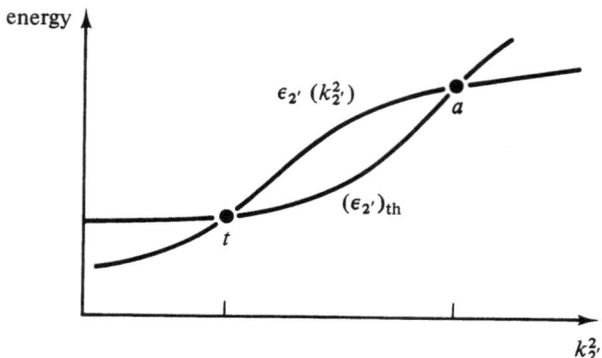

Fig. 3.5.6. Schematic illustration of threshold t and anti-threshold a. Impact ionization can occur for $k_t < k_{2'} < k_a$.

This theorem links the two effects in a very direct manner.

Sometimes an anti-threshold is found (Fig. 3.5.6). In such cases the impact ionizing energy reaches a maximum value compatible with energy and momentum conservation beyond which impact ionization is prevented by these constraints. Hence [3.5.9]:

Theorem 2

Of all quadruplets of states in nondegenerate bands, the least probable Auger recombination quadruplets are identical with the quadruplets which lead to the anti-threshold of impact ionization.   (3.5.21)

For the possibility of the additional saddle thresholds, the reader is referred to the literature [3.5.10].

3.5.6 Some related relevant literature

Graphical identifications of thresholds were used by Franz [3.5.11] and they were used in theoretical work on quantum efficiencies [3.5.12], where the impact ionization threshold is also important, and in connection with the impact ionization probability in Si [3.5.13]. Using more complete band structure determinations, graphical work was also done in [3.5.1] and [3.5.14]. The analytical formula for two direct isotropic parabolic bands goes back to [3.5.15], and it was later extended to three direct isotropic parabolic bands [3.5.3], [3.5.16], [3.5.17] and to indirect materials [3.5.18]–[3.5.21]. A numerical treatment for indirect materials was given in [3.5.22]. Eventually, rather general analytical formulae were obtained for up to three nonisotropic parabolic bands [3.5.9]–[3.5.10]. More recent numerical work on the threshold energies of the GaAs and the GaAs–GaSb system based on their band structures has confirmed an expected dependence on orientation [3.5.23], [3.5.24]. For comprehensive reviews see [3.5.7] and [3.5.25]. A series of lowest thresholds in cubic Si and Ge have, however, been identified more recently [3.5.26].

3.5.7 General theory of threshold energies

In this subsection a generalized theory of threshold energies will be given which allows for Umklapp-type processes, for indirect band extrema and for saddle points in the band structure. One pays for this generality by having to make allowances for some algebraic complexity. By putting this subsection at the end of section 3.5 we can see that the earlier and simpler results stand on their own. This subsection can in fact be omitted by the reader who is interested only in the more straightforward cases.

Put for energy and momentum in the four states $j = 1, 2, 1', 2'$

$$E_j = \varepsilon_j + \varphi_j(\mathbf{X}_j), \quad \mathbf{k}_j = \mathbf{K}_j + \mathbf{X}_j \tag{3.5.22}$$

where ε_j, $\mathbf{K}_j$ represent reference levels for energy and momentum which correspond to a band extremum. Momentum conservation states

$$\mathbf{k}_{1'} + \mathbf{k}_{2'} - \mathbf{k}_1 - \mathbf{k}_2 \pm \mathbf{q} + \mathbf{L} = \mathbf{0} \tag{3.5.23}$$

where $\mathbf{L}$ is a vector joining lattice points in $\mathbf{k}$-space and $\mathbf{q}$ is the wavevector of a phonon which is absorbed (upper sign) or emitted (lower sign) during the process. If $\mathbf{L}$ is nonzero, one has Umklapp-type processes. Let

$$\mathbf{Q} \equiv \mathbf{K}_1 + \mathbf{K}_2 - \mathbf{K}_{1'} - \mathbf{K}_{2'} \mp \mathbf{q} - \mathbf{L} \tag{3.5.24}$$

Then eq. (3.5.23) implies

$$\mathbf{X}_{1'} + \mathbf{X}_{2'} - \mathbf{X}_1 - \mathbf{X}_2 = \mathbf{Q} \tag{3.5.25}$$

In the phononless case $(\mathbf{q} = 0)$ $\mathbf{L}$ is chosen so as to place $\mathbf{Q}$ into the first Brillouin zone. Energy conservation, allowing for a phonon of energy $\hbar\omega_q$, is

$$E_{1'} + E_{2'} - E_1 - E_2 \pm \hbar\omega_q = 0 \tag{3.5.26}$$

Eliminating $\mathbf{k}_{1'}$, it implies, using eq. (3.5.25),

$$\varepsilon + \varphi_{1'}(\mathbf{X}_1 + \mathbf{X}_2 - \mathbf{X}_{2'} + \mathbf{Q}) + \varphi_{2'}(\mathbf{X}_{2'}) - \varphi_1(\mathbf{X}_1) - \varphi_2(\mathbf{X}_2) \pm \hbar\omega_q = 0 \tag{3.5.27}$$

Here ε collects together the energy origins of the bands:

$$\varepsilon \equiv \varepsilon_{1'} + \varepsilon_{2'} - \varepsilon_1 - \varepsilon_2 \tag{3.5.28}$$

The condition (3.5.27) incorporates both energy and momentum conservation for a general four-band situation. Some bands can of course be identical.

Our main assumption is that the states $1'$, 1 and 2 are in parabolic bands:

$$\varphi_j(\mathbf{X}_j) = \sum_{i=1}^{3} \frac{\hbar^2 X_{ji}^2}{2m_{ji}} \quad (j = 1', 1, 2; i = x, y, z, \text{ i.e. } i = 1, 2, 3 \text{ for convenience})$$

The principal effective masses have been denoted by m_{j1}, m_{j2} and m_{j3}. They may be positive or negative. The single constraint (3.5.27) is now

$$\varepsilon + \varphi_{2'}(\mathbf{X}_{2'}) + \sum_{i=1}^{3} \frac{\hbar^2}{2}\left[\frac{(X_{1i} + X_{2i} - X_{2'i} + Q_i)^2}{m_{1'i}} - \frac{X_{1i}^2}{m_{1i}} - \frac{X_{2i}^2}{m_{2i}}\right] \pm \hbar\omega_q = 0 \tag{3.5.29}$$

The impact ionizing state $2'$ may be in a nonparabolic band which can be of practical importance. Its coordinates $X_{2'i}$ are mixed up with the other six coordinates X_{1i}, X_{2i}. The variable coordinates may be separated from each other by a linear transformation to new variables x_{1i}, x_{2i} which yields the equivalent constraint

$$F(\mathbf{X}_{2'}, \mathbf{x}_1, \mathbf{x}_2) \equiv A(\mathbf{X}_{2'}) - \sum_{i=1}^{3} (\sigma_{1i} x_{1i}^2 + \sigma_{2i} x_{2i}^2) = 0 \qquad (3.5.30)$$

where we have in effect diagonalized a quadratic form,

$$A(\mathbf{X}_{2'}) \equiv \varepsilon + \varphi_{2'}(\mathbf{X}_{2'}) + \sum_{i=1}^{3} \frac{\hbar^2}{2} \frac{(X_{2'i} - Q_i)^2}{m_i^*} \qquad (3.5.31)$$

and

$$m_i^* \equiv m_{1'i} - m_{1i} - m_{2i} \qquad (3.5.32)$$

(cf. eq. (3.5.11)). Also the $\{\sigma_{ji}\}$ are each $+1$ or -1 with the signs given by

$$\sigma_{1i} \equiv \text{sign} \left(\frac{1}{m_{1i}} - \frac{1}{m_{1'i}} \right) \qquad (3.5.33a)$$

$$\sigma_{2i} \equiv \text{sign} \left(\frac{1}{m_{2i}} - \frac{1}{m_{1'i}} \right) \qquad (3.5.33b)$$

$$\sigma_{3i} = \text{sign} \left\{ \sigma_{2i} \frac{m_{1'i} m_i^*}{(m_{1i} - m_{1'i})(m_{2i} - m_{1'i})} \right\} \qquad (3.5.33c)$$

The results (3.5.30)–(3.5.33) are established in Appendix E. Thresholds now correspond to turning points of $E_{2'}$ when $\mathbf{x}_1$, $\mathbf{x}_2$ and $\mathbf{X}_{2'}$ are varied, subject to eq. (3.5.30) being fulfilled. The latter condition is taken into account by a Lagrangian multiplier λ. Since $E_{2'}$ depends on $\mathbf{X}_{2'}$, but not on $\mathbf{x}_1$, and $\mathbf{x}_2$, the nine conditions are

$$\frac{\partial}{\partial x_{ji}} (E_{2'} - \lambda F) = 0, \quad \text{i.e. } \lambda x_{ji} = 0 \quad (j = 1, 2; i = 1, 2, 3) \qquad (3.5.34)$$

$$\frac{\partial}{\partial X_{2'i}} (E_{2'} - \lambda F) = 0, \quad \text{i.e. } \frac{\partial E_{2'}(\mathbf{X}_{2'})}{\partial X_{2'i}} + \lambda \frac{\partial A(\mathbf{X}_{2'})}{\partial X_{2'i}} = 0 \quad (i = 1, 2, 3) \qquad (3.5.35)$$

It is clear that if $\lambda \neq 0$ all x_{ji} vanish by conditions (3.5.34) and hence

$$x_{ij} = 0 \qquad (3.5.36a)$$

$$A(\mathbf{X}_2) = 0 \quad (\alpha\text{-threshold}) \qquad (3.5.36b)$$

by eq. (3.5.30). If $\lambda \neq 0$ we call it an α-threshold, and one sees that

eqs. (3.5.36) and (3.5.30) imply each other.

Class β thresholds arise if $\lambda = 0$, so that only conditions (3.5.35) remain to be satisfied in the form

$$\partial E_{2'}(\mathbf{X}_{2'})/\partial X_{2'i} = 0 \quad (\beta\text{-threshold}) \qquad (3.5.37)$$

Thresholds occur in this case only at those points of zero slope of $E_{2'}(\mathbf{X}_{2'})$ which also satisfy eq. (3.5.30) for some $(\mathbf{x}_1, \mathbf{x}_2)$.

The threshold conditions used require the vanishing of $E_{2'}$-derivatives as stated in the condition (3.5.34) and (3.5.35). This includes the possibility that $E_{2'}$ has a maximum (rather than a minimum) value in a conduction band. For a valence band it includes the possibility that $E_{2'}$ has a minimum (rather than a maximum) value. These additional points thus included in the analysis are not thresholds in the normal sense, which is that for higher than threshold energies in the conduction band (and for lower energies in the valence band) impact ionization may proceed. They are instead anti-thresholds. These are such that for higher than anti-threshold energies in a conduction band (and for lower energies in a valence band) impact ionization by a conduction band electron is no longer possible. In addition, there are saddle thresholds. Here the $E_{2'}$-derivatives vanish but the sign of the second derivative depends on the direction in the $(X_{2'}, x_1, x_2)$-space which is considered.

A classification of these various types of threshold may be based on eq. (3.5.30), by saying that a threshold is of type M_s when s of the six σ_{ij}'s in eq. (3.5.30) have value -1, the remaining $6-s$ σ_{ij}'s then have the value $+1$. For the simplest case of a process involving a conduction band minimum and a valence band maximum, eqs. (3.5.33) show that the processes are of type M_0, and eq. (3.5.31) shows that $A(X_{2'}) \geq 0$. Processes involving a conduction band maximum and a valence band minimum are of type M_6 and $A(X_{2'}) \leq 0$. Processes of type $M_1, \ldots, M_5$ are of the saddle type.

In order to advance to specific solutions, assume the band containing state $2'$ to be parabolic. Then eq. (3.5.36b) gives

$$A(\mathbf{X}_{2'}) \equiv \varepsilon + \sum_{i=1}^{3} \frac{\hbar^2}{2}\left[\frac{(X_{2'i}-Q)^2}{m_i^*} + \frac{X_{2'i}^2}{m_{2'i}}\right] = 0 \quad \text{(one condition)} \tag{3.5.38}$$

since $\varphi_{2'}(\mathbf{X}_{2'}) = \sum_{i=1}^{3} \hbar^2 X_{2'i}^2/2m_{2'i}$. Also conditions (3.5.35) require

$$\frac{X_{2'i}}{m_{2'i}}[(1+\lambda)m_i^* + \lambda m_{2'i}] = \lambda Q_i \quad \text{(three conditions)} \tag{3.5.39}$$

for a class α threshold. For largest matrix elements the momentum transfer should be as small as possible, i.e. all the $\mathbf{K}_j$-vectors should be in the same plane. Choosing the z-plane, $\mathbf{K}_j = (0, 0, K_j)$, and assuming $\mathbf{q} = 0$ and $\mathbf{L} = (0, 0, L_3)$, eq. (3.5.24) shows that

$$\mathbf{Q} = (0, 0, Q), \quad \text{where } Q = K_{13} + K_{23} - K_{1'3} - K_{2'3} - L_3 \tag{3.5.40}$$

One therefore has to solve the four conditions (3.5.38) and (3.5.39) subject to eq. (3.5.40). The three possible results for $\mathbf{X}_{2'}$, and the four threshold energies which result, are given in Table 3.5.2. The threshold energies are measured from the extremum of $E_{2'}(\mathbf{X}_{2'})$. In the xz- and yz-planes one has only one threshold each since the components of Q vanish for the x- and y-directions. They are denoted by $E_T(3)$ and $E_T(4)$.

For most semiconductors (Ge, Si, GaAs, GaP, InP, $Ga_xIn_{1-x}As_yP_{1-y}$ etc.) the energy surfaces approximate to ellipsoids of revolution centered on the $\langle 100 \rangle$- or $\langle 111 \rangle$-axes. Then the two effective masses in a direction transverse to the symmetry direction may be represented by a single effective mass (m_t). The effective mass for the longitudinal direction is denoted by m_1. Hence one can put

$$m_{2'1} = m_{2'2} \to m_t, \quad \mathrm{m}_{2'3} \to m_1, \quad m_i^* \equiv m_{1'3} - m_{13} - m_{23}$$

Table 3.5.2. Threshold energies E_T and values of $X_{2'i}$ for threshold conditions (3.5.38)–(3.5.40)

These results appear in a slightly different form in section 4.8 of [3.5.25]

Solution number	$X_{2'1}^2$	$X_{2'2}^2$	$X_{2'3}$	E_T
(i)	0	0	$\dfrac{Qm_{2'3} \pm [-m_3^* m_{2'3}(Q^2+R^2)]^{\frac12}}{\tilde m_3}$	$E_T(1,2) = \dfrac{m_{2'3} - m_3^*}{\tilde m_3^2}\dfrac{\hbar^2 Q^2}{2} - \dfrac{m_3^*}{\tilde m_3}\varepsilon$ $\pm \dfrac{m_{2'3} m_3^*}{\tilde m_3^2}\hbar^2 Q\left[-\dfrac{Q^2}{m_{2'3} m_3^*} - \dfrac{2\varepsilon}{\hbar^2}\left(\dfrac{1}{m_{2'3}} + \dfrac{1}{m_3^*}\right)\right]^{\frac12}$ (3.5.41) (The square root has to be positive)
(ii)	0	$\dfrac{m_2^* m_{2'2}}{\tilde m_2}\left[\dfrac{R^2}{\tilde m_3} + \dfrac{(m_3^* m_{2'2}^2 + m_2^{*2} m_{2'3})Q^2}{(m_3^* m_{2'2} - m_2^* m_{2'3})^2}\right]$ (This quantity has to be positive)	$\dfrac{m_2^* m_{2'3} Q}{m_3^* m_{2'2} - m_2^* m_{2'3}}$	$E_T(3,4) = \dfrac{m_i^* m_{2'i}}{(m_i^* m_{2'3} - m_3^* m_{2'i})\tilde m_i}\dfrac{\hbar^2 Q^2}{2} - \dfrac{m_i^*}{\tilde m_i}\varepsilon \left\{ \begin{matrix} i = 1 \\ i = 2 \end{matrix} \right.$ (3.5.42)
(iii)	$\dfrac{m_1^* m_{2'i}}{\tilde m_1}\left[\dfrac{R^2}{\tilde m_3} + \dfrac{(m_3^* m_{2'1}^2 + m_1^{*2} m_{2'3})Q^2}{(m_3^* m_{2'1} - m_1^* m_{2'3})^2}\right]$ (This quantity has to be positive)	0	$\dfrac{m_1^* m_{2'3} Q}{m_3^* m_{2'1} - m_1^* m_{2'3}}$	

Notation: R is defined by $\hbar^2 R^2/2m_3 = \varepsilon_{1'} + \varepsilon_{2'} - \varepsilon_1 - \varepsilon_2 (\equiv \varepsilon)$.
m_i^* and $\tilde m_i$ are defined by $\tilde m_i \equiv m_{2i} + m_i^* = m_{1'i} + m_{2'i} - m_{1i} - m_{2i}$ ($i = 1, 2, 3$).
Q is defined in eq. (3.5.24).

While this does not greatly simplify eqs. (3.5.41) and (3.5.42), the two expressions (3.5.42) go over into a single one.

A more specialized case arises for Ge in the ⟨111⟩-direction when states 1 and 2 are at the zone center and states 1′ and 2′ at opposite zone edges (Fig. 3.5.7(b)). Thus, by eq. (3.5.24) with

$$\mathbf{q} = 0, \quad \mathbf{K}_{1'} = -\frac{\mathbf{L}_1}{2}, \quad \mathbf{K}_{2'} = \frac{\mathbf{L}_1}{2}$$

$$\mathbf{Q} = 0 + 0 + \frac{\mathbf{L}_1}{2} - \frac{\mathbf{L}_1}{2} - \mathbf{L}$$

Choose $\mathbf{L} = 0$ which puts $\mathbf{Q}$ into the first Brillouin zone at $\mathbf{Q} = 0$. Assume states 1 and 2 can be described by two effective masses $m_\parallel$ and $m_\perp$ and states 1′ and 2′ by a single effective mass:

$$m_{11} = m_{12} = m_{21} = m_{22} \equiv m_\perp, \quad m_{13} = m_{23} \equiv m_\parallel$$
$$m_{1'1} = m_{1'2} = m_{1'3} \equiv -m_v, \quad m_{2'1} = m_{2'2} = m_{2'3} \equiv m_0$$

Then $m_3^* = m_{1'3} - m_{13} - m_{23} = -m_v - 2m_\parallel$

$$\tilde{m}_t = m_0 - m_v - 2m_\perp$$

The fact that $\mathbf{Q} = 0$ leads to considerable simplification and one obtains [3.5.9] a result first given in [3.5.20]:

$$E_{\mathrm{T}}(1) = E_{\mathrm{T}}(2) = \frac{m_v + 2m_\parallel}{m_0 - m_v - 2m_\parallel} \varepsilon$$

$$E_{\mathrm{T}}(3) = \frac{m_v + 2m_\parallel}{m_0 - m_v - 2m_\perp} \varepsilon$$

It applies to Ge.

Results obtained previously in this section can also be re-derived as special cases of eqs. (3.5.41) and (3.5.42). We need to observe only that for $Q = 0$ one recovers formula (3.5.10).

3.6 Bloch overlap integrals

3.6.1 A sum rule for the integrals

It was seen in section 3.4 that when one evaluates the matrix elements of the Coulomb potential using Bloch functions, one comes across overlap integrals of the modulating parts of Bloch wavefunctions. It is therefore desirable to establish some familiarity with the main properties of these integrals. A convenient notation is

$$F_{n\mathbf{k}}^{n'\mathbf{k+q}} \equiv V^{-1} \int_V u_{n'}^*(\mathbf{k+q}, \mathbf{r}) u_n(\mathbf{k}, \mathbf{r}) \, \mathrm{d}\mathbf{r} \qquad (3.6.1)$$

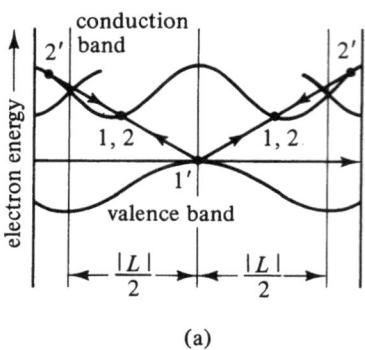
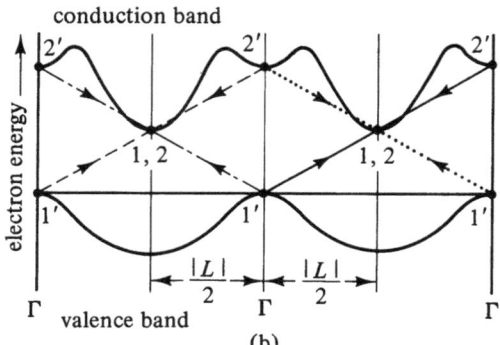

(a)   (b)

Fig. 3.5.7. (a) Impact ionization by an energetic electron in an indirect material with extrema away from the zone edge (e.g. Si) in a repeated zone scheme. Two equivalent processes are shown, arrows show electron transitions. (b) Impact ionization by an electron in an indirect material with band extrema at the zone edge (e.g. GaP, Ge) involving impact ionization from the direct conduction band minimum in a repeated zone scheme. Four equivalent processes are shown, arrows show electron transitions.

For small enough $\mathbf{q}$ the following sum rule will be proved [3.6.1]:

$$\sum_{n'(\neq n)} [E_{n'}(\mathbf{k}) - E_n(\mathbf{k})] |F_{n\mathbf{k}}^{n'\mathbf{k}+\mathbf{q}}|^2 = \frac{\hbar^2}{2m}\left[q^2 - m \sum_{i,j=1}^{3} q_i q_j \left(\frac{1}{m^*}\right)_{ij}^{n\mathbf{k}} \right] \tag{3.6.2}$$

Here m is the normal electron mass, and the effective mass tensor (D.25) has been used.

First step in the proof of eq. (3.6.2)
Obtain an expression for $u_{n'}(\mathbf{k}+\mathbf{q})$ in terms of $u_{n'}(\mathbf{k})$ by perturbation theory. Recall that if $\{u_n\}$ are eigenfunctions of an unperturbed Hamiltonian H_0 and if E_n is the corresponding eigenvalue, then a perturbed Hamiltonian may be written as $H = H_0 + H' = H_0 + \lambda H''$. To the first order in the perturbation (i.e. in λ) the eigenfunctions of H are

$$\psi_n = u_n + \sum_{n'(\neq n)} \frac{H'_{n'n}}{E_n - E_{n'}} u_{n'} \tag{3.6.3}$$

assuming $E_n \neq E_{n'}$ if $n \neq n'$ (nondegenerate case).

    In order to pass from a known situation at a wavevector $\mathbf{k}$ (at a band extremum for example), to an unknown one not far away at wavevector $\mathbf{k}+\mathbf{q}$ (say), we apply eq. (3.6.3) to the Schrödinger equation for an electron in a periodic potential, which may, by eq. (D.8), be written

$$H_{\mathbf{k}} u_n(\mathbf{k}) = E_n(\mathbf{k}) u_n(\mathbf{k}), \quad H_{\mathbf{k}} \equiv e^{-i\mathbf{k}\cdot\mathbf{r}} H e^{i\mathbf{k}\cdot\mathbf{r}} \tag{3.6.4}$$

In the presence of spin–orbit interaction, the unmodified Hamiltonian is

$$H = \frac{p^2}{2m} + a(\boldsymbol{\sigma} \times \nabla U) \cdot \mathbf{p} + U \tag{3.6.5}$$

where U is the periodic potential, σ_i are the dimensionless Pauli spin matrices and $a = \hbar/4m^2c^2$ is a constant. The last form of writing comes from the definition of the velocity operator, eq. (D.16),

$$\mathbf{v} = (i/\hbar)[H, \mathbf{r}] = \nabla_{\mathbf{p}} H = \frac{1}{m}\mathbf{p} + a\boldsymbol{\sigma} \times \nabla U \tag{3.6.6}$$

For any operator $A(\mathbf{p}, \mathbf{r})$ one knows (Appendix D) that

$$e^{-i\mathbf{k} \cdot \mathbf{r}} A(\mathbf{p}, \mathbf{r}) e^{i\mathbf{k} \cdot \mathbf{r}} = A(\mathbf{p} + \hbar\mathbf{k}, \mathbf{r}) \tag{3.6.7}$$

Using eqs. (3.6.5)–(3.6.7)

$$H_{\mathbf{k+q}} - H_{\mathbf{k}} = \frac{1}{2m}[(\mathbf{p} + \hbar\mathbf{k} + \hbar\mathbf{q})^2 - (\mathbf{p} + \hbar\mathbf{k})^2] + a(\boldsymbol{\sigma} \times \nabla U) \cdot \hbar\mathbf{q} \tag{3.6.8}$$

and also

$$\hbar\mathbf{q} \cdot \mathbf{v_k} = \frac{1}{m}\mathbf{p} \cdot \hbar\mathbf{q} + \frac{\hbar^2}{m}\mathbf{k} \cdot \mathbf{q} + a(\boldsymbol{\sigma} \times \nabla U) \cdot \hbar\mathbf{q}$$

Equation (3.6.8) is the Hamiltonian to be used in eq. (3.6.4).

To pass from $\mathbf{k}$ to $\mathbf{k+q}$ in a band, one may say that the exact Hamiltonian is $H_{\mathbf{k}}$ and the approximate one $H_{\mathbf{k+q}}$. According to eq. (3.6.8) the perturbation is therefore

$$H' = H_{\mathbf{k+q}} - H_{\mathbf{k}} = \frac{\hbar^2 q^2}{2m} + \hbar\mathbf{q} \cdot \mathbf{v_k} \tag{3.6.9}$$

Since the $u_n(\mathbf{k})$ form a complete set for periodic functions, one can use eq. (3.6.3). Only nondiagonal elements occur in eq. (3.6.3), hence the first term on the right-hand side of eq. (3.6.9) is irrelevant. One finds, the suffix distinguishing Cartesian components,

$$u_{n'}(\mathbf{k+q}) = u_{n'}(\mathbf{k}) + \sum_{n''(\neq n')} \sum_{i=1}^{3} C_{n''n'}^{\mathbf{k}i} q_i u_{n''}(\mathbf{k}) \tag{3.6.10}$$

where

$$C_{n''n'}^{\mathbf{k}i} \equiv \frac{\hbar}{E_{n'}(\mathbf{k}) - E_{n''}(\mathbf{k})} \int_V u_{n''}^*(\mathbf{k})(\mathbf{v_k})_i u_{n'}(\mathbf{k}) \, d\mathbf{r}$$

This procedure contains the essence of what is called the $\mathbf{k} \cdot \mathbf{p}$ approximation. We shall neglect terms of order q^2 compared with terms of order unity, and $u_{n'}(\mathbf{k}+\mathbf{q})$ of eq. (3.6.10) is then still normalized in the same sense as $u_{n'}(\mathbf{k})$.

Second step
Derivation of a formal expression for the modulus squared of the overlap integral. Inserting eq. (3.6.10) into eq. (3.6.1) and noting that

$$V^{-1} \int_V u_{n'}^*(\mathbf{k}) u_n(\mathbf{k}) \, d\mathbf{r} = \delta_{n'n}$$

one finds that the term in q vanishes if $n = n'$. Thus

$$F_{n\mathbf{k}}^{n\mathbf{k}+\mathbf{q}} = 1 + O(q^2) \sim 1 \tag{3.6.11}$$

If $n \neq n'$, however, the first term in eq. (3.6.10) yields an orthogonality integral, and

$$F_{n\mathbf{k}}^{n'\mathbf{k}+\mathbf{q}} = \sum_{i=1}^{3} q_i C_{nn'}^{ki}$$

In fact, one is interested in $|F|^2$. Noting that $\mathbf{v}$ is a Hermitian operator, one finds

$$|F_{n\mathbf{k}}^{n'\mathbf{k}+\mathbf{q}}|^2 = \frac{\hbar^2}{[E_{n'}(\mathbf{k}) - E_n(\mathbf{k})]^2} \sum_{i,j=1}^{3} q_i q_j \cdot \int u_n^*(\mathbf{k}) v_{ki} u_{n'}(\mathbf{k}) \, d\mathbf{r} \int u_{n'}^*(\mathbf{k}) v_{kj} u_n(\mathbf{k}) \, d\mathbf{r}$$

These are just the terms occurring in the oscillator strength (D.32). Hence

$$[E_{n'}(\mathbf{k}) - E_n(\mathbf{k})] |F_{n\mathbf{k}}^{n'\mathbf{k}+\mathbf{q}}|^2 = \frac{\hbar^2}{2m} \sum_{i,j=1}^{3} q_i q_j f_{kij}^{n'n} \tag{3.6.12}$$

Third step
Summing over all n' ($\neq n$) and using the f-sum rule (D.31), one finally arrives at the result (3.6.2).

A second general result follows from the Cauchy–Schwarz inequality

$$\left| \int v^*(\mathbf{r}) w(\mathbf{r}) \, d\mathbf{r} \right|^2 \leqslant \left(\int |v(\mathbf{r})|^2 \, d\mathbf{r} \right) \left(\int |w(\mathbf{r})|^2 \, d\mathbf{r} \right)$$

Interpreting v and w as $V^{-\frac{1}{2}} u_{n'\mathbf{k}+\mathbf{q}}$ or $V^{-\frac{1}{2}} u_{n\mathbf{k}}$ this gives

$$|F_{n\mathbf{k}}^{n'\mathbf{k}+\mathbf{q}}|^2 \leqslant 1 \tag{3.6.13}$$

so that by eq. (3.6.2)

$$\sum_{n'(\neq n)} [E_{n'}(\mathbf{k}) - E_n(\mathbf{k})] \geqslant \frac{\hbar^2}{2m} \left[q^2 - m \sum_{i,j=1}^{3} q_i q_j \left(\frac{1}{m^*} \right)_{ij}^{n\mathbf{k}} \right] \tag{3.6.14}$$

A third general result is the orthonormality relation

$$F_{nk}^{n'k} = \delta_{nn'} \tag{3.6.15}$$

3.6.2 Overlap integrals for a scalar effective mass

In the case of the InSb-type of band structure near $\mathbf{k} = 0$ one has to consider four bands. Numbering them in order of decreasing energy, one may put $n = 1$ for the conduction band, $n = 2$ for the heavy-hole band, $n = 3$ for the light-hole band and $n = 4$ for the split-off band. The main energy gap is

$$E_1(\mathbf{k}) - E_2(\mathbf{k}) = E_G$$

One can also assume scalar effective masses

$$\left(\frac{1}{m^*}\right)_{ij}^{n\mathbf{k}} = \frac{1}{m_{n\mathbf{k}}}\delta_{ij} \quad (n = 1, 2, 3, 4)$$

The theorem (3.6.2) yields [3.6.1]

$$\sum_{n'}[E_{n'}(\mathbf{k}) - E_n(\mathbf{k})]|F_{n\mathbf{k}}^{n'\mathbf{k}+\mathbf{q}}|^2 = \frac{\hbar^2 q^2}{2m_{n\mathbf{k}}}\left[\frac{m_{n\mathbf{k}}}{m} - 1\right] \tag{3.6.16}$$

It follows that

$$|F_{2\mathbf{k}}^{1\mathbf{k}+\mathbf{q}}|^2 = \frac{\hbar^2 q^2}{2m_{2\mathbf{k}} E_G}\left(\frac{m_{2\mathbf{k}}}{m} - 1\right) + \sum_{n'=3}^{4}\frac{E_n(\mathbf{k}) - E_{n'}(\mathbf{k})}{E_G}|F_{2\mathbf{k}}^{n'\mathbf{k}+\mathbf{q}}|^2$$

The overlap between valence bands is expected to be small [3.6.2] and $m_{2\mathbf{k}}$ to be negative, so that

$$|F_{2\mathbf{k}}^{1\mathbf{k}+\mathbf{q}}|^2 \simeq \left(1 + \frac{|m_{2\mathbf{k}}|}{m}\right)\frac{\hbar^2 q^2}{2|m_{2\mathbf{k}}| E_G} \tag{3.6.17}$$

This is in agreement with the orthogonality condition when $q = 0$. Thereafter the overlap integral rises *initially* as q^2; departures from eq. (3.6.17) occur when q is no longer small.

In order to advance to a numerical estimate of eq. (3.6.17) we shall consider a CHCC impact transition involving the conduction band and the heavy-hole valence band for InSb. Note that for threshold $\mathbf{k}_1$ and $\mathbf{k}_{1'}$ are antiparallel. The momentum transfer per electron at threshold, using eq. (3.5.5), is

$$q = k_1 - k_{1'} = \left(1 + \frac{m_h}{m_c}\right)k_1$$

By line 4 of Table 3.5.1

$$\frac{\hbar^2 q^2}{2m_c} = \left(\frac{m_c + m_h}{m_c}\right)^2 (\varepsilon_1)_{th} = \left(\frac{m_c + m_h}{m_c}\right)^2 \left(\frac{m_c^2}{(2m_c + m_h)(m_c + m_h)}\right) E_G$$

$$= \frac{m_c + m_h}{2m_c + m_h} E_G$$

Hence [3.6.1]

$$|F_{2k}^{1k+q}|^2 = \frac{\hbar^2 q^2}{2m_h E_G}\left(1 + \frac{m_h}{m}\right) = \frac{m_c}{m_h}\frac{m_c + m_h}{2m_c + m_h}\left(\frac{m_h}{m} + 1\right)$$

$$= \frac{m_c + m_h}{2m_c + m_h}\left(\frac{m_c}{m} + \frac{m_c}{m_h}\right) = \frac{1 + \mu}{1 + 2\mu}\left(\frac{m_c}{m} + \mu\right)$$

With $m_c \sim 0.0114m$, $m_h \sim -0.18m$ one has $\mu \sim 0.063$ and $|F|^2 \sim 0.07$. It is remarkable how analytical and numerical (rough) estimates of as complicated a quantity as an overlap function can thus be obtained by a judicious blend of theory and band parameter values.

In arriving at eq. (3.6.17), only one term has, in fact, been retained in the n'-sum of eq. (3.6.2). This is not in general justifiable. If it is, it leads to an additional result as follows. The left-hand side of a resulting equation such as (3.6.17)

$$|F_{nk}^{n'k+q}|^2 = \frac{\hbar^2 q^2}{2m_{nk}[E_{n'}(\mathbf{k}) - E_n(\mathbf{k})]}\left(\frac{m_{nk}}{m} - 1\right) \tag{3.6.18}$$

is unchanged by an interchange of $(n, \mathbf{k})$ and $(n', \mathbf{k}+\mathbf{q})$. This can be seen from

$$V^2|F_{nk}^{n'k+q}|^2 = \left[\int u_{n'}^*(\mathbf{k}+\mathbf{q}, \mathbf{r})\, u_n(\mathbf{k}, \mathbf{r})\, d\mathbf{r}\right]\left[\int u_n^*(\mathbf{k}, \mathbf{r})\, u_{n'}(\mathbf{k}+\mathbf{q}, \mathbf{r})\, d\mathbf{r}\right]$$

$$= V^2|F_{n'k+q}^{nk}|^2$$

Hence the use of eq. (3.16.18) without restriction implies

$$\frac{\hbar^2 q^2}{2[E_{n'}(\mathbf{k}) - E_n(\mathbf{k})]}\left(\frac{1}{m} - \frac{1}{m_{nk}}\right) = \frac{\hbar^2 q^2}{2[E_n(\mathbf{k}) - E_{n'}(\mathbf{k})]}\left(\frac{1}{m} - \frac{1}{m_{n'k+q}}\right)$$

Thus the condition for consistency

$$\frac{2}{m} \simeq \frac{1}{m_{nk}} + \frac{1}{m_{n'k+q}} \tag{3.6.19}$$

has to be satisfied, and this will usually not be the case.

A different method of estimating $|F_{1k}^{2k+q}|$ is to argue [3.6.3] that near $\mathbf{k} = 0$

$$[E_2(\mathbf{k}) - E_1(\mathbf{k})]|F_{1k}^{2k+q}|^2 \sim [E_3(\mathbf{k}) - E_1(\mathbf{k})]|F_{1k}^{3k+q}|^2 \gg [E_4(\mathbf{k}) - E_1(\mathbf{k})]|F_k^{4k+q}|^2$$

since the gap between the conduction band and the split-off valence band is larger than that for any other pair of relevant bands. One then finds from eq. (3.6.16) with $n = 1$ that

$$|F_{1k}^{2k+q}|^2 \sim |F_{1k}^{3k+q}|^2 \sim \frac{\hbar^2 q^2}{4 m_{1k} E_G} \left(1 - \frac{m_{1k}}{m} \right) \qquad (3.6.17')$$

3.6.3 The effect of the overlap integral on the matrix element of the Coulomb potential

The wavevector dependence (3.6.17) of the overlap integral has an important effect on the matrix element M_D of the screened Coulomb potential, when taken with respect to Bloch states. This was given in eqs. (3.4.20) as

$$|M_D| = \frac{4\pi e^2}{\varepsilon V} \frac{|F_{ac} F_{bd}|}{\lambda^2 + |\mathbf{k}_1 - \mathbf{k}_{1'}|^2} \, \delta_{\mathbf{k}_1 + \mathbf{k}_2, \, \mathbf{k}_{1'} + \mathbf{k}_{2'}}$$

where the main contributions in the wavefunctions for the states a, b, c and d are assumed to come from reduced wavevectors $\mathbf{k}_1$, $\mathbf{k}_2$, $\mathbf{k}_{1'}$, $\mathbf{k}_{2'}$, respectively. If the states b and d are in the same band, eq. (3.6.11) shows that for wavevectors which are not too different $|F_{bd}|^2$ is of order unity. If, however, state a corresponds to a wavevector $\mathbf{k}_1$ in one band and state c to a wavevector $\mathbf{k}_{1'}$ in another band, then, using eq. (3.6.17), one finds for small enough $|\mathbf{k}_1 - \mathbf{k}_{1'}|$

$$|F_{ac}|^2 = |F_{vk_{1'}}^{ck_1}|^2 = \left(1 + \frac{|m_{vk_{1'}}|}{,m} \right) \frac{\hbar^2 (\mathbf{k}_1 - \mathbf{k}_{1'})^2}{2|m_{vk_{1'}}| [E_c(\mathbf{k}_1) - E_v(\mathbf{k}_{1'})]} \qquad (3.6.20)$$

If one neglects the screening parameter λ, as is often justifiable for large enough momentum transfer per electron, one finds

$$|M_D|^2 = \left(\frac{4\pi e^2 \hbar}{\varepsilon V} \right)^2 \frac{1}{2E_G} \left(\frac{1}{m} + \frac{1}{|m_{vk_{1'}}|} \right) \frac{\delta_{\mathbf{k}_1 + \mathbf{k}_2, \, \mathbf{k}_{1'} + \mathbf{k}_{2'}}}{|\mathbf{k}_1 - \mathbf{k}_{1'}|^2} \qquad (3.6.21)$$

Thus one has the interesting result [3.6.4] that the inverse *second* power in the momentum transfer, familiar from the usual Coulomb matrix element $|M_D|$, has been replaced by the inverse *first* power.

This effect can be converted back from momentum space to real space, by considering the amended carrier interaction potential energy

$$W(r) = \frac{e^2 r_0}{\varepsilon r^2}, \quad r_0 \equiv \frac{2}{\pi} \left[\frac{1}{m} + \frac{1}{|m_v|} \right]^{\frac{1}{2}} \left[\frac{\hbar^2}{2 E_G} \right]^{\frac{1}{2}} \qquad (3.6.22)$$

instead of the usual Coulomb potential $U(r)$. The factor by which the Coulomb potential has been multiplied is r_0/r. The matrix element of $W(\mathbf{r}_1 - \mathbf{r}_2)$ with respect to *free-electron* functions is then exactly eq. (3.6.21). Thus the equivalent potential (3.6.22) is approximate

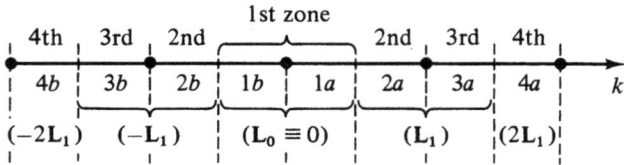

Fig. 3.6.1. *k*-space for a one-dimensional lattice.

but it incorporates some effects of the overlap functions. The periodic potential is seen to change the force $\mathbf{F}(r)$ between two carriers by the factor

$$\frac{F_W(r)}{F_U(r)} = \frac{2r_0}{r} \qquad (3.6.23)$$

Thus the repulsive Coulomb interaction is increased by this effect for $r < 2r_0$, but is weakened for $r > 2r_0$. For InSb, using $|m_v| = 0.3m$ and $E_G = 0.22$ eV at 77 K, r_0 is of the order of 5.5 Å. These effects have not, as yet, found any applications, and hold only in so far as eq. (3.6.20) is valid.

3.6.4 The overlap integrals in the limit of a small periodic potential

We shall now make some remarks about the behavior of these overlap integrals as the two reduced wavevectors which enter into it range over the zone. This can be done if the periodic potential is assumed to be small. Let

$$\mathbf{K} = \mathbf{L} + \mathbf{k} \qquad (3.6.24)$$

be a wavevector in the extended zone scheme and let $\mathbf{k}$ be the corresponding wavevector in the reduced zone scheme. Then the Bloch wavefunction for the state $\mathbf{K}$ may be written in the two schemes

$$\psi(\mathbf{K}, \mathbf{r}) = U(\mathbf{K}, \mathbf{r}) e^{i\mathbf{K} \cdot \mathbf{r}} \qquad (3.6.25)$$

$$\psi_L(\mathbf{k}, \mathbf{r}) = u_L(\mathbf{k}, \mathbf{r}) e^{i\mathbf{k} \cdot \mathbf{r}} \qquad (3.6.26)$$

where instead of a band number the reciprocal lattice vector $\mathbf{L}$ has been used as an auxiliary label. In the limit of a small periodic potential $U(\mathbf{K}, \mathbf{r}) \to 1$ because eq. (3.6.25) becomes a plane wave. Since eqs. (3.6.25) and (3.6.26) are strictly equal,

$$u_L(\mathbf{k}, \mathbf{r}) = U(\mathbf{k} + \mathbf{L}, \mathbf{r}) e^{i\mathbf{L} \cdot \mathbf{r}} \to e^{i\mathbf{L} \cdot \mathbf{r}}$$

In any one zone, different vectors $\mathbf{L}$ will be appropriate to different parts which may be called subzones. It then follows that in the limit of a vanishing periodic potential (the so-called empty lattice case)

$$\frac{1}{V} \int_V u_L^*(\mathbf{k}', \mathbf{r}) u_L(\mathbf{k}, \mathbf{r}) \, d\mathbf{r} = \frac{1}{V} \int_V e^{i(\mathbf{L} - \mathbf{L}') \cdot \mathbf{r}} \, d\mathbf{r} = \delta_{L, L'}$$

Thus, in this limit, one would expect a step function behavior of the integral in the sense that it is unity if the states are in the same subzone of a given zone, and zero otherwise.

This situation is illustrated in the following figures. For a one-dimensional crystal, the reciprocal lattice points are indicated by large dots in Fig. 3.6.1 and the reciprocal lattice vectors are $\mathbf{L}_0 = 0$ or $n\mathbf{L}_1$ $(n = \pm 1, \pm 2, \ldots)$, where $\mathbf{L}_1$ is shown in Fig. 3.6.1. Regions of *k*-

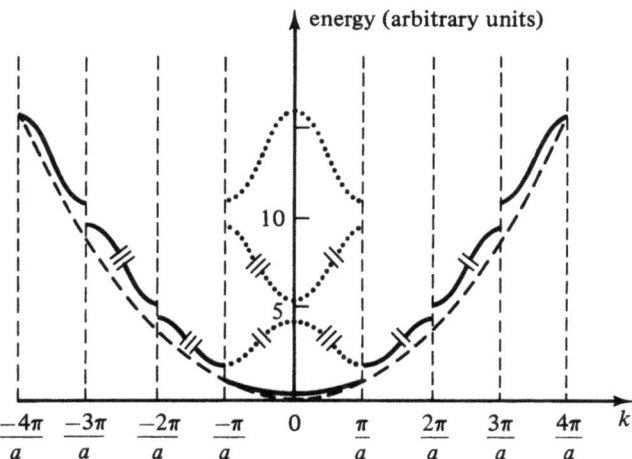

Fig. 3.6.2. The extended and reduced wavevector scheme for one dimension.

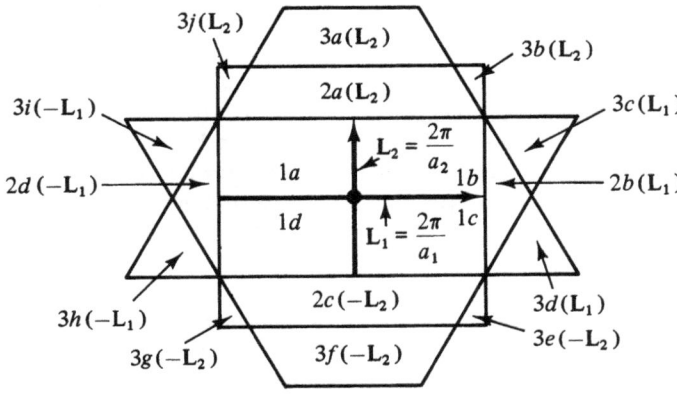

Fig. 3.6.3. Reciprocal lattice and the first three zones for a rectangular lattice.

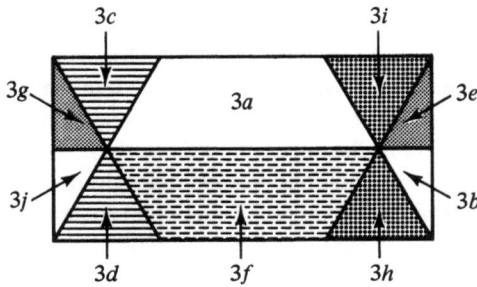

Fig. 3.6.4. Reduction of the third zone of a rectangular lattice into the first zone.

space labeled $2a$, $2b$ refer to two distinct subzones of the second zone. For each subzone in a two-dimensional case the appropriate vector $\mathbf{L}$ from eq. (3.6.24) is given in brackets in Fig. 3.6.3. Fig. 3.6.2 shows the extended (full curves) and the reduced (dotted curves) wavevector scheme. Sections of the graph corresponding to states between which the overlap integral

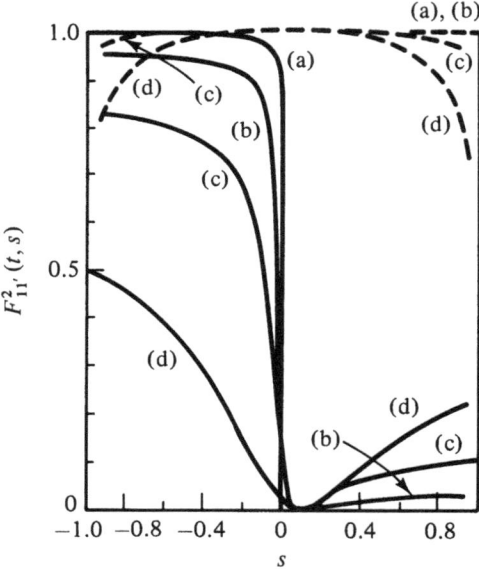

Fig. 3.6.5. $|F|^2$ as a function of a reduced valence band wavevector s for various values of binding constants P; (a) $P = 0.1$; (b) $P = 0.4$; (c) $P = 1.0$; (d) $P = 5.0$. s attains the value unity at the zone boundary. These curves hold for a reduced conduction band wavevector $t = 0.1$. The broken curves give $|F_{11'}(t, s)|^2 + |F_{12}(t, s)|^2$ where $F_{12}(t, s)$ is the corresponding overlap integral when both wavevectors are in the conduction band [3.6.5].

is unity in the free-electron limit are marked by the same number of lines. For small periodic potentials the step function behavior of the free-electron limit is still noticeable and can be verified from Fig. 3.6.5, based on the Kronig–Penney model.

Figs. 3.6.3 and 3.6.4 illustrate the reduction of the third zone for a two-dimensional lattice. One would expect the overlap integrals between any two states in the region which is made up of subzones a, b and j to be unity in the free-electron limit. The same applies to the region made up of subzones f, g and e, to the region made up of subzones c and d and to the region made up of subzones i and h. This splits the third zone into ten subzones or four regions. States in different regions have zero overlap in the free-electron limit.

A mathematical analysis of Brillouin zones in d dimensions, which includes an asymptotic estimate of the number of subzones, has recently been given [3.6.6]. This approach lends itself to a simple proof of the following propositions: (a) All zones have equal volumes, (b) each zone can be translated into the first zone so as to fill it exactly, by translating different subzones by appropriate reciprocal lattice vectors [3.6.7]. The argument hinges on the following definition of the nth Brillouin zone B_n: It is the set of points in a d-dimensional Euclidean space (the reciprocal space) which have the origin O as their (not necessarily unique) nth nearest lattice point. Thus B_1 has O as its center. The arguments involved do not require recourse to matters not essential to the problem, such as the Schrödinger equation, periodic boundary conditions or k-vector quantization, which occur in the more conventional discussions.

3.6.5 Fourier analysis of the overlap integrals

The product of the modulating parts of Bloch functions are lattice periodic, so that the Fourier analysis is

$$u_{n'}^*(\mathbf{k}+\mathbf{q},\mathbf{r})\,u_n(\mathbf{k},\mathbf{r}) = \sum_{\mathbf{L}} (I_{n\mathbf{k}}^{n'\mathbf{k}+\mathbf{q}})_L\, e^{i\mathbf{L}\cdot\mathbf{r}} \tag{3.6.27}$$

the sum extending over lattice vectors in the reciprocal lattice. The coefficients are

$$(I_{n\mathbf{k}}^{n'\mathbf{k}+\mathbf{q}})_L \equiv \frac{1}{V}\int u_{n'}^*(\mathbf{k}+\mathbf{q},\mathbf{r})\,u_n(\mathbf{k},\mathbf{r})\,e^{i\mathbf{L}\cdot\mathbf{r}}\,d\mathbf{r} \tag{3.6.28}$$

so that with the notation (3.6.1)

$$(I_{n\mathbf{k}}^{n'\mathbf{k}+\mathbf{q}})_0 = F_{n\mathbf{k}}^{n'\mathbf{k}+\mathbf{q}} \tag{3.6.29}$$

Note that

$$(I_{n\mathbf{k}}^{n'\mathbf{k}'})_{\mathbf{k}''} = 0 \text{ unless } \mathbf{k}'' \text{ is a lattice vector in the reciprocal lattice} \tag{3.6.30}$$

The Fourier coefficients (3.6.28) are generalizations of the overlap integrals discussed so far in this section. It will be shown that they can be given a simple physical interpretation [3.6.4], [3.6.8] in terms of the velocity operator (or momentum operator) matrix elements which enter into the theory of radiative transitions.

To establish this result the Hamiltonian (3.6.5) will be used. Note that

$$H_{\mathbf{q}} = \frac{(\mathbf{p}+\hbar\mathbf{q})^2}{2m} + a(\sigma\times\nabla U)\cdot(\mathbf{p}+\hbar\mathbf{q}) + U$$

$$= H + \frac{\hbar^2 q^2}{2m} + \hbar\mathbf{q}\cdot\mathbf{v}$$

Premultiplying by $e^{i\mathbf{q}\cdot\mathbf{r}}$,

$$H\,e^{i\mathbf{q}\cdot\mathbf{r}} = e^{i\mathbf{q}\cdot\mathbf{r}}H + e^{i\mathbf{q}\cdot\mathbf{r}}\left[\frac{\hbar^2 q^2}{2m} + \hbar\mathbf{q}\cdot\mathbf{v}\right] \tag{3.6.31}$$

Taking matrix elements with respect to Bloch functions $\psi_{n'}(\mathbf{k}')$, $\psi_n(\mathbf{k})$,

$$\left[E_{n'}(\mathbf{k}') - E_n(\mathbf{k}) - \frac{\hbar^2 q^2}{2m}\right](I_{n\mathbf{k}}^{n'\mathbf{k}'})_{\mathbf{k}'-\mathbf{k}-\mathbf{q}} = \hbar\mathbf{q}\cdot(n'\mathbf{k}'|e^{i\mathbf{q}\cdot\mathbf{r}}\mathbf{v}|n\mathbf{k}) \tag{3.6.32}$$

In the absence of spin–orbit interaction, one has on the right-hand side the operator $\dfrac{1}{m}e^{i\mathbf{q}\cdot\mathbf{r}}\mathbf{p}$

which governs the radiative processes involving a photon of wavevector $\mathbf{q}$. Note that eq. (3.6.32) is exact in the sense that it is independent of perturbation theory.

Multiplying eq. (3.6.32) by its complex conjugate,

$$|(I_{nk}^{n'k'})_{k'-k-q}|^2 = \frac{\hbar^2}{2m}\left[\sum_{i,j=1}^{3} q_i q_j f_{n'n}^{ij}(\mathbf{q})\right] \frac{E_{n'}(\mathbf{k}) - E_n(\mathbf{k})}{[E_{n'}(\mathbf{k}') - E_n(\mathbf{k}) - \hbar^2 q^2/2m]^2} \tag{3.6.33}$$

where a generalized oscillator strength, eq. (D.32), has been introduced:

$$f_{n'n}^{ij}(\mathbf{q}) \equiv 2m \frac{(n\mathbf{k}|e^{-i\mathbf{q}\cdot\mathbf{r}}v_i|n'k')(n'k'|e^{i\mathbf{q}\cdot\mathbf{r}}v_j|n\mathbf{k})}{E_{n'}(\mathbf{k}) - E_n(\mathbf{k})} \tag{3.6.34}$$

This is an exact form of the perturbation result (3.6.12) provided one chooses $\mathbf{k}' = \mathbf{k} + \mathbf{q}$, but it has not been much applied to individual semiconductors.

3.6.6 Numerical estimates

Apart from results based on the Kronig–Penney model [3.6.5] estimates for planar crystals have been given. These suggested that eq. (3.6.11) is valid to within 10% provided q is less than about $\frac{1}{4}$ of the effective radius of the first Brillouin zone [3.6.9]. The use of sums rules [3.6.1], [3.6.3] has already been mentioned. The connection (3.6.32) with the matrix elements for the radiative transitions enables one also to make estimates of overlap integrals based on optical data [3.6.3]. Lifetime data in which Auger recombination is important can also be analyzed theoretically, leaving the overlap integrals as parameters to be identified from experiments. This was the original method [3.6.10]. More recently numerical approaches directly from band structure determinations [3.6.11], $\mathbf{k} \cdot \mathbf{p}$ perturbation theory and the pseudo-potential method of band theory have been used [3.6.12], [3.6.13], and improved computational schemes have been developed [3.6.14].

In spite of its elegance, the identification of overlap integrals from the sum rule (3.6.2) encounters the obvious difficulty that it does not provide enough equations to identify all the F's. Physical intuition (which is not quantitative) has to be used. It is obvious on general grounds that a frontal numerical attack on the problem of Auger recombination using full band structure details, Fermi–Dirac statistics, the wavevector dependence of the dielectric function and including all Umklapp-type terms is bound to succeed and to throw up overlap integrals as a by-product. This has been attempted only in recent work, which has also found these integrals to be strongly direction-dependent. For a novel use of perturbation theory and improved discussion of GaAs and III–V compounds see [3.6.15] This has led to the introduction of an impact ionization threshold surface such that the distance from the origin of a point on the surface gives the value of the threshold energy for the direction involved [3.6.15]. For n-type Si 25 000 terms have been summed [3.6.16] to show that the normal Auger effect will account for the experimental results. For p-type Si the phonon-assisted Auger effect (Fig. 3.6.6) seems to dominate [3.6.12], [3.6.17], which was believed to govern also the n-type material by earlier authors

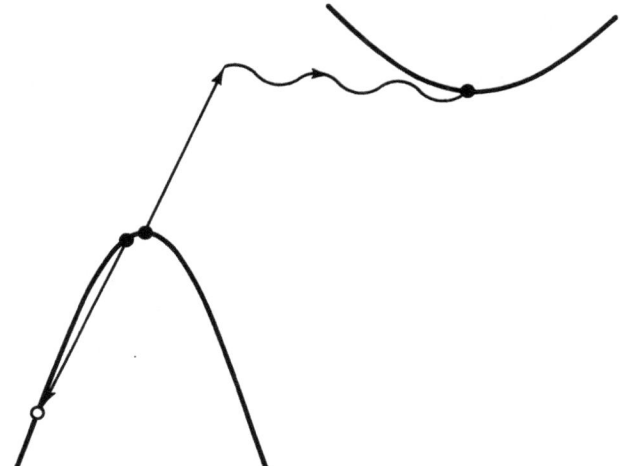

Fig. 3.6.6. Auger process with phonon emission.

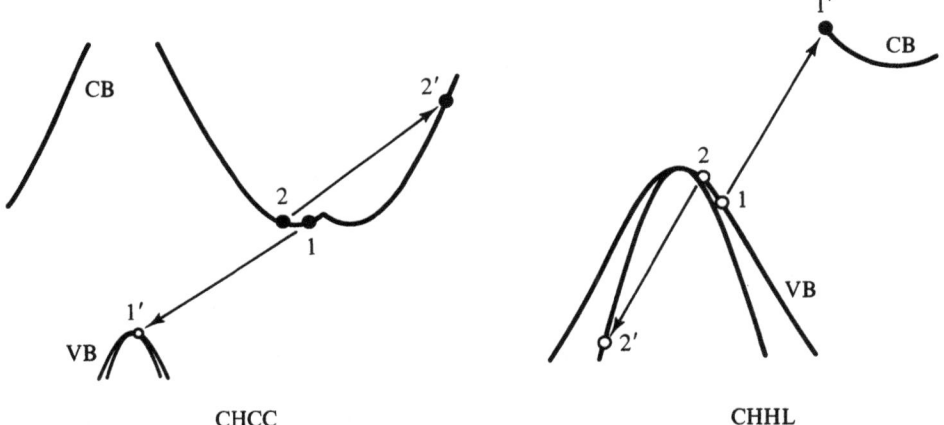

CHCC   CHHL

Fig. 3.6.7. Quadruplets of states for Auger processes in Si [3.6.16]. The small
maximum in the conduction band is caused by a band crossing at the X-point
of the Brillouin zone.

[3.6.18]. A concensus has not yet established itself on this point. However, the
CHHL process, which is liable to be the most important one in *p*-type Si, cannot
involve states which are as near the band edges as in the CHCC process for *n*-type
Si (Fig. 3.6.7), and this could allow the phonon-assisted Auger process to be more
important in this case.

3.7 Auger and impact ionization probabilities

3.7.1 General (approximate) formulae for impact ionization probabilities per unit time and Auger recombination rates

In this section we pull together various conclusions obtained earlier in chapter 3, in order to obtain some physically important results. This is therefore a key section of this chapter.

We ask for the probability T_{if} of a transition in time t from an initial state i to a final state f, where we have in mind an impact ionization process. Let us consider the CHCC process of Table 3.5.1 as being one of the more important ones. For this calculation we use eq. (3.2.13):

$$T_{if} = \frac{2\pi t}{\hbar} |H_{if}|^2 \, \delta(E_f - E_i) \tag{3.7.1}$$

The electrons not shown remain in their states and section 3.3 shows how one can reduce the many-electron problem involved to a two-electron problem. Of course, strictly speaking *all* electrons are affected by the transition considered. But this effect is neglected here. From eq. (3.3.27)

$$|H_{if}|^2 = 2\beta(|M_D|^2 + |M_E|^2) \quad (1 < \beta < 2) \tag{3.7.2}$$

For four Bloch functions this becomes, by virtue of eqs. (3.4.40) and (3.4.41)

$$|H_{if}|^2 = \frac{32\pi^2 e^4}{\epsilon^2 V^2} \beta(|\Sigma_D|^2 + |\Sigma_E|^2) \, \delta_{\mathbf{k}_1 + \mathbf{k}_2 - \mathbf{k}_{1'} - \mathbf{k}_{2'}, \mathbf{L}} \tag{3.7.3}$$

The summations of eqs. (3.4.40) and (3.4.41) are denoted by Σ_D and Σ_E. They have dimension L^2 and are carried out over reciprocal lattice vectors.

The algebraic form of the impact ionization probability P per unit volume per unit time for a band–band or a band–trap process can be obtained from a dimensional analysis. By eqs. (3.7.1) and (3.7.3) it must be proportional to e^4 and a term $|\Sigma|^2$, which can stand for $|\Sigma_D|^2 + |\Sigma_E|^2$. In addition some dependence on an effective mass m_c, an electron energy E and Planck's constant may be expected. So let us write (with b, c, d constants which are to be determined)

$$[P] = [e^4 V^{-1} |\Sigma|^2 \hbar^b E^c m_c^d][L^{-3}T^{-1}]$$

Hence

$$M^2 L^6 T^{-4} \cdot L^{-3} \cdot L^4 \cdot M^b L^{2b} T^{-b} \cdot M^c L^{2c} T^{-2c} \cdot M^d = L^{-3} T^{-1}$$

Using $t_0 = \hbar^3/e^4 m \sim 2.42 \times 10^{-17}$ s, one finds for some number a

$$[P] = \left[\frac{e^4 |\Sigma|^2 E^2 m_c^3}{\hbar^7 V} \right] = \left[\frac{1}{t_0 V} \frac{m_c}{m} \right] \left[\frac{m_c^2 |\Sigma|^2 E^2}{\hbar^4} \right]^a$$

where the last factor is dimensionless. This form of the result with $a = 1$ is in fact found in eq. (3.7.19), below. (One has to use $\hbar^4 = 4m^2r_1^4I_0^2$.) The magnitude will be decided by the (missing) numerical coefficient as well as by the detailed interpretation of $|\Sigma|^2$ and of E^2.

The impact ionization per unit volume per unit time from state $2'$ is

$$p(2')\,P(2') = \frac{p(2')}{Vt}\int\dots\int[1-p(1)]\,p(1')\,[1-p(2)]\,T_{if}\,\mathrm{d}S_1\,\mathrm{d}S_{1'}\,\mathrm{d}S_2$$

where $p(i)$ is the occupation probability of state i. The impact ionization rate per unit volume, given that there is an electron in state $2'$, i.e. if $p(2') = 1$, is $P(2')$. The symbols $\mathrm{d}S_i$ indicate an integration or summation over all translational states, the summation over spin variables having been carried out in the expression for T_{if}. Hence

$$P(2') = \frac{2\pi}{\hbar V}\int\dots\int[1-p(1)]\,p(1')\,[1-p(2)]\,|H_{if}|^2$$
$$\times\,\delta(E_f - E_i)\,\mathrm{d}S_1\,\mathrm{d}S_{1'}\,\mathrm{d}S_2 \tag{3.7.4}$$

which has the required dimension $\mathrm{L}^{-3}\mathrm{T}^{-1}$. We now remove the matrix element from the integral and assign the threshold values to the wavevectors which occur in it, as shown for example in eq. (3.4.20):

$$P(2') = \frac{2\pi}{\hbar V}|H_{if}|_{\mathrm{th}}^2\int\dots\int[1-p(1)]\,p(1')\,[1-p(2)]$$
$$\times\,\delta(E_f - E_i)\,\mathrm{d}S_1\,\mathrm{d}S_{1'}\,\mathrm{d}S_2 \tag{3.7.5}$$

This is of course a serious approximation, but the probability factors $p(i)$ decrease rapidly as one pushes the states 1, 1', 2 away from the band edges, so that the matrix element at threshold is in fact more heavily weighted by those probabilities than is the matrix element away from threshold. One could argue that eq. (3.7.5) gives therefore an upper limit for $P(2')$. Although the use of threshold values in $|H_{if}|_{\mathrm{th}}^2$ is quite specific, one may alternatively average $|H_{if}|^2$ as an approximation and again remove it from the integral.

The CHCC process, introduced to fix ideas, has not so far been used, and the above procedure is in fact applicable in principle even to impact ionization from localized states (see section 5.3.3). In fact each of the three states 1, 1', 2 could be localized, except that $|H_{if}|_{\mathrm{th}}^2$ depends of course on the wavefunctions of the states involved. But if we do not commit ourselves to eq. (3.7.3), one can integrate eq. (3.7.5) over the localized states involved, and each such integration produces the concentration, c_j say, of occupied or empty traps, as appropriate. There can be at most three such integrations as shown in Fig. 3.7.1. In general let there be μ of

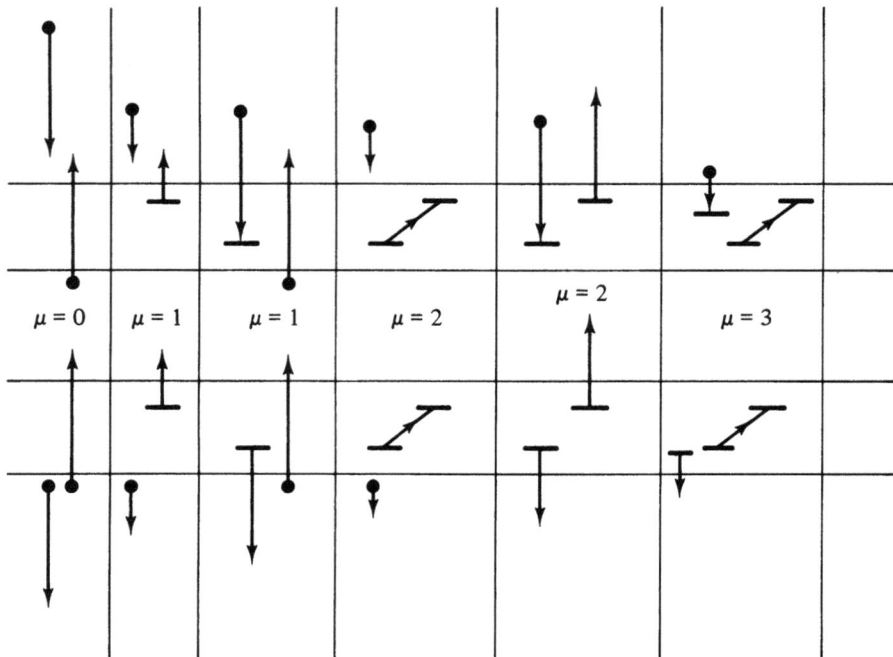

Fig. 3.7.1. Values of μ for various impact ionization processes.

them ($\mu = 0$, 1, 2 or 3). This leaves band integrations only, denoted by dS_j, which are still dimensionless quantities and each involves a wavevector and a three-fold integration. The result is

$$P(2') = \frac{2\pi}{\hbar} V^{\mu-1} |H_{if}|_{th}^2 c_1 \ldots c_\mu \int \delta(E_f - E_i) \, dS_1 \ldots dS_{3-\mu} \tag{3.7.6}$$

where

$$\int p(j) \, dS_j = c_j V \quad (j = 1, \ldots, \mu; \mu = 0, 1, 2 \text{ or } 3) \tag{3.7.7}$$

It has here been assumed that nondegenerate statistics applies to the states in the bands so that

$$0 < p(i) \ll 1 \quad \text{(conduction band)}$$

$$p(i) \sim 1 \quad \text{(valence band)}$$

This has enabled us to remove the remaining probability factors from the integral (3.7.5). We are, of course, still in a *constant matrix element approximation*.

We have obtained a remarkable formula (3.7.6) for the impact ionization rate as essentially a multiple integration of an energy delta function [3.7.1]. It counts the number of states on an energy shell.

If one reverses the arrows in Fig. 3.7.1 one has the corresponding Auger recombination processes. For the CHCC Auger process, let the recombination coefficient be $A(2')$. Then in thermal equilibrium, denoted by '0',

$$A(2') n_0^2 p_0 \, dS_{2'} = p_0(2') \, P(2') \, dS_{2'}$$

Here we have utilized detailed balance between forward and reverse processes involving an incremental range of states $2'$, n_0 and p_0 being equilibrium concentrations of electrons and holes. The dimension of $A(2')$ is $L^6 T^{-1}$. The recombination coefficient for the whole process is B_1 or B_2 for the band–band or T_1, T_2, T_3 or T_4 for the band–trap effect. With a neutral notation (A) it is [3.7.2]

$$A = \frac{1}{a_{10} a_{1'0} a_{20}} \int p_0(2') \, P(2') \, dS_{2'} \tag{3.7.8}$$

One sees that the theoretical or experimental estimation of $P(2')$ is useful also for the Auger effect. Just as optical absorption data can tell one about radiative recombination, so impact ionization data can tell us about Auger recombination. This will be elaborated in section 5.2.5.

3.7.2 The impact ionization rate for a simple case

Return again to our standard CHCC process and the impact ionization rate per unit volume (3.7.6). We need to integrate over all states 1, 1' and 2. However, because of eq. (3.4.39) and since $\mathbf{k}_{2'}$ is given, only two of the three wavevectors are independent, and we take them to refer to states 1 and 2. Using $\mathbf{X}_j$ to denote the wavevector measured from the appropriate origin, as in eqs. (3.5.22), (3.7.6) (with $\mu = 0$) and eq. (3.7.3) yield

$$P(2') = \frac{2\pi}{\hbar V} |H_{if}|_{\text{th}}^2 \int \delta(E_f - E_i) \left(\frac{V}{8\pi^3}\right)^2 dX_1 \, dX_2$$

$$= \frac{e^4}{V \hbar \pi^3 \varepsilon^2} |\Sigma|_{\text{th}}^2 \int \delta(E_f - E_i) \, dX_1 \, dX_2 \tag{3.7.9}$$

Here $|\Sigma|^2$ has dimension $[L^4]$ and is by eqs. (3.4.40) and (3.4.41)

$$|\Sigma|^2 \equiv |\Sigma_D|^2 + |\Sigma_E|^2 = \sum_L \left[\left| \sum_{L'} \frac{F_{n_2, n_{1'}}^{(L')} F_{n_2, n_{2'}}^{(L-L')}}{|\mathbf{L}' + \mathbf{k}_{1'} - \mathbf{k}_1|^2 + \lambda^2} \right|^2 \right.$$

$$\left. + \left| \sum_{L'} \frac{F_{n_1, n_{1'}}^{(L')} F_{n_2, n_{2'}}^{(L-L')}}{|\mathbf{L}' + \mathbf{k}_{1'} - \mathbf{k}_2|^2 + \lambda^2} \right|^2 \right] \delta_{\mathbf{k}_1 + \mathbf{k}_2 - \mathbf{k}_{1'} - \mathbf{k}_{2'}, \, \mathbf{L}} \tag{3.7.10}$$

The main contribution arises from $\mathbf{L} = 0$. One often takes $|\Sigma|^2$ to be simply proportional to the product of the squares of two overlap integrals ($L = L' = 0$). This means that Umklapp processes are treated as negligible.

Now suppose the bands of states 1, 1' and 2 are parabolic with principal effective masses

$$m_{1i}, m_{1'i}, m_{3i} \quad (i = 1, 2, 3)$$

Then, upon eliminating $\mathbf{k}_{1'}$ by means of the momentum conservation condition, the expression for $E_f - E_i$ will contain product terms as already noted in connection with eq. (3.5.29). Once again we resort to the change of variables described in Appendix E which has the Jacobian

$$J = \frac{8}{\hbar^6} \prod_{i=1}^{3} \left[\frac{m_{1i} m_{1'i} m_{2i}}{\sigma_{1i} \sigma_{2i} (m_{1'i} - m_{1i} - m_{2i})} \right]^{\frac{1}{2}} \tag{3.7.11}$$

One obtains

$$P(2') = \frac{\beta e^4}{V \hbar \pi^3 \varepsilon^2} |\Sigma|_{\text{th}}^2 \cdot J \cdot I(A) \tag{3.7.12}$$

where

$$I(A) \equiv \int \delta \left[A(\mathbf{X}_{2'}) - \sum_{i=1}^{3} (\sigma_{1i} x_{1i}^2 + \sigma_{i2} x_{2i}^2) \right] d\mathbf{x}_1 \, d\mathbf{x}_2 \tag{3.7.13}$$

and $A(\mathbf{X}_{2'})$ is given by eq. (3.5.31). The integral in the transformed variables can readily be carried out. It depends on $A(\mathbf{X}_{2'})$ which is known if $\mathbf{k}_{2'}$ is given since by eq. (3.5.22)

$$\mathbf{k}_{2'} = \mathbf{K}_{2'} + \mathbf{X}_{2'}$$

It also depends on the nature of the band structure involved through the number of values $+1$ and -1 among the six σ_{ji}'s. A systematic exposition of the various integrals $I(A)$ which result is given in [3.7.3].

Before we give some general results let us work out the integral for the simplest case when all σ's are equal to unity. This corresponds to a transition involving simple conduction and valence bands. For in this case

$$m_{1i}, m_{2i} > 0, \quad m_{1'i} < 0$$

whence the $\sigma_{1i}, \sigma_{2i} = 1$ by eq. (3.5.33). Introduce

$$y^2 \equiv \sum_{j=1}^{6} y_j^2 \equiv \sum_{i=1}^{3} (x_{1i}^2 + x_{2i}^2)$$

so that

$$y_1 = x_{11}, \quad y_2 = x_{12}, \quad y_3 = x_{13}, \quad y_4 = x_{21}, \quad y_5 = x_{22}, \quad y_6 = x_{23}$$

Then

$$dx_1\, dx_2 = dy_1 \ldots dy_6 = dy = b_6\, y^5\, dy \quad (b_6 \equiv \pi^3)$$

since a volume element in the six-dimensional space of y is denoted by dy. For spherical symmetry it can be written as dy multiplied by the surface area of a five-dimensional sphere. In three dimensions, as another example, we would have

$$dr = b_3\, r^2\, dr \quad (b_3 \equiv 4\pi)$$

For n dimensions one has

$$dy = b_n\, y^{n-1}\, dy \quad (b_n \equiv 2\pi^{n/2}/\Gamma(n/2))$$

In the present case

$$I(A) = \pi^3 \int_{-\infty}^{\infty} y^5 \delta[A(\mathbf{X}_{2'}) - y^2]\, dy = \frac{\pi^3}{2} \int_{\text{all } z} z^2 \delta(A - z)\, dz$$

where we have put

$$z = y^2 \quad dy = dz/2y = dz/2z^{\frac{1}{2}}$$

The integral is zero for negative A and one finds

$$I(A) = \frac{\pi^3}{2}[A(\mathbf{X}_{2'})]^2 \quad \text{if } A(\mathbf{X}_{2'}) > 0 \tag{3.7.14}$$

$$= \frac{\pi^3}{2}\left[\varepsilon + \varphi_{2'}(\mathbf{X}_{2'}) \sum_{i=1}^{3} \frac{\hbar^2}{2} \frac{(X_{2'i} - Q_i)^2}{m_{1'i} - m_{1i} - m_{2i}}\right]^2 \tag{3.7.15}$$

where eq. (3.5.31) has been used.

All essential elements are now to hand to find from eq. (3.7.12)

$$P(2') = \frac{\beta}{\varepsilon^2 V t_0} \frac{|\Sigma|_{\text{th}}^2}{r_1^4} \prod_{i=1}^{3} \left[\frac{m_{1i}\, m_{2i}\, m_{1'i}}{(m_{1'i} - m_{1i} - m_{2i})\, m_0^2}\right]^{\frac{1}{2}} \frac{[A(\mathbf{X}_{2'})]^2}{I_0^2} \tag{3.7.16}$$

Here we have used the Bohr radius, the ionization energy of the H atom and the atomic unit of time:

$$r_1 \equiv \frac{\hbar^2}{e^2 m} = 0.528\ \text{Å}, \quad I_0 \equiv \frac{me^4}{2\hbar^2} = 13.6\ \text{eV}, \quad t_0 \equiv \frac{\hbar^3}{e^4 m} = 2.42 \times 10^{-17}\ \text{s}$$

$$\tag{3.7.17}$$

As a result, only the first factor in eq. (3.7.16) carries a dimension.

To be even more specific, assume a direct semiconductor so that $\mathbf{Q} = 0$ and consider two isotropic parabolic bands with band edge energies E_c, E_v. Then $\varepsilon = E_v + E_c - 2E_c$ by eq. (3.5.28) and with $\mu \equiv m_c/m_v$

$$A(X_{2'}) = -E_G + (E_{2'} - E_c) - (E_{2'} - E_c)\frac{m_c}{2m_c + m_v} = \frac{1+\mu}{1+2\mu}(E_{2'} - E_c) - E_G$$

(3.7.18)

Now the threshold energy for the CHCC process is, from line 4 of Table 3.5.1,

$$E_{th} = \frac{1+2\mu}{1+\mu}E_G$$

so that

$$A(\mathbf{X}_{2'}) = \frac{1+\mu}{1+2\mu}(E_{2'} - E_c - E_{th})$$

The product of masses in eq. (3.7.16) is

$$\prod_{i=1}^{3}\left[\frac{-m_c^2 m_v}{-m_v - 2m_c}\frac{1}{m_0^2}\right]^{\frac{1}{2}} = (1+2\mu)^{-\frac{3}{2}}\left(\frac{m_c}{m_0}\right)^3$$

Hence one can reduce eq. (3.7.16) to

$$P(2') = \frac{\beta}{\varepsilon^2 V t_0}\frac{|\Sigma|_{th}^2}{r_1^4}\frac{(1+\mu)^2}{(1+2\mu)^{\frac{7}{2}}}\left(\frac{m_c}{m_0}\right)^3\left(\frac{E_{2'} - E_c - E_{th}}{I_0}\right)^2$$

(3.7.19)

This is our final result for the impact ionization rate per unit volume involving two simple parabolic bands. For other band structures one can go back to our more general formula (3.7.12). The $|\Sigma|^2$-factor can also be approximated as indicated above. But problems arise from the fact that the conduction and valence band orbitals near the band edges have often s- and p-symmetry, respectively. In that case a careful analysis is needed to ensure that one obtains a nonzero result. For GaAs, using a simplified form of eq. (3.7.19), it has been estimated [3.7.4] that

$$VP(2') \sim 5 \times 10^{12}\left(\frac{E_{2'} - E_c - E_{th}}{E_G}\right)^2 \text{ s}^{-1}$$

The dependence $P(2') \propto (E_{2'} - E_c - E_{th})^2$ goes back to [3.7.5].

The older literature used 'hard' thresholds in the sense that the impact ionization probability was assumed to change abruptly from zero to unity at the threshold. In eq. (3.7.19) we have, on the contrary, a 'soft' threshold. These are

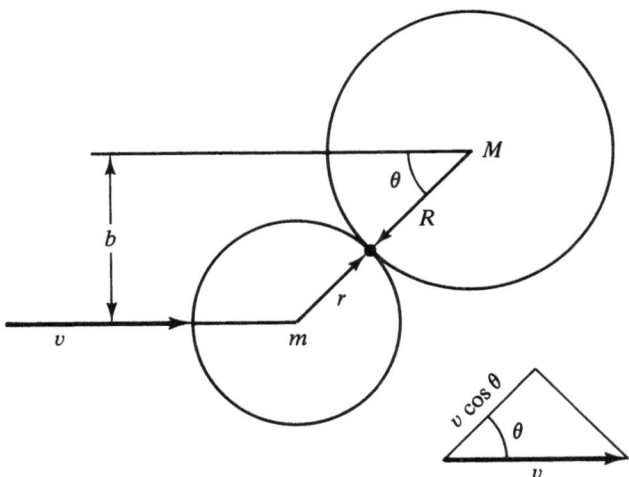

Fig. 3.7.2. Diagram showing the impact parameter b for colliding spheres.

known to be more accurate since the probabilities of occupation or vacancy of states (which enter the analysis) are gently changing functions. If the bands are anisotropic then the threshold energy depends on the direction of the wavevector and one can construct a closed surface in **k**-space such that the square of the distance from the origin in a given direction gives the threshold energy for that direction. This will produce additional softness. This matter has been investigated for ellipsoidal parabolic bands and can yield

$$P(2') = W_{ph}\,s\left(\frac{E_{2'}-E_c-E_{th}}{E_G}\right)^m$$

where $m = 2$, $\frac{5}{2}$ or 3 [3.7.6]. We have also introduced a phonon scattering rate W_{ph} and a factor s. Values believed necessary for Si and GaAs have decreased from $s \sim 100$ in the late 1970s to $s \sim 10^{-3}$–10^{-1} [3.7.7]. This also softens the threshold.

A simple classical treatment yielding a *first* power in $E_{2'}/E_G - 1$, and so an even softer threshold, is here instructive [3.7.8]. For simplicity we now replace E_G by a threshold energy $E_T \equiv \frac{1}{2}Mv_T^2$ (not related necessarily to semiconductors) and $E_{2'}$ by $E \equiv \frac{1}{2}mv^2$. These considerations assume that one regards the impact ionizing particle as a sphere of radius r and mass m, the atom to be ionized as a sphere of radius R and mass M and supposes finally that the velocity component of the ionizing particle along the line of centers has to exceed a critical (threshold) value v_T for ionization. The velocity at right angles to the line of centers plays then no part. If b is the impact parameter for the collision (see Fig. 3.7.2) one needs (if m, M, r, R, E_T, E are given) $v\cos\theta > v_T$, or alternatively $b < b_T$ for ionization. We shall calculate b_T below.

We have

$$\sin \theta = \frac{b}{r+R}, \quad \cos^2 \theta = 1 - \frac{b^2}{(r+R)^2} > \left(\frac{v_T}{v}\right)^2 = \frac{m}{M}\frac{E_T}{E}$$

Hence the equation for b_T is

$$1 - \left(\frac{b_T}{r+R}\right)^2 = \frac{mE_T}{ME}$$

If N targets exist per unit volume in a sample of cross-sectional area A and length l, the total target area (assuming no obscuration) is

$$A_T = N \cdot Al \cdot S_T, \quad S_T \equiv \pi b_T^2 = \pi(r+R)^2 \left(1 - \frac{mE_T}{ME}\right)$$

where S_T is the cross section for one successful ionization attempt. If l is the mean free path, $A_T = A$, i.e. $l = 1/NS_T$. Hence the ionization rate is

$$\frac{1}{\tau(E)} = \frac{v}{l} = vNS_T = \frac{2\pi(r+R)^2 NE_T}{(2mE)^{\frac{1}{2}}}\frac{m}{M}\left(\frac{ME}{mE_T} - 1\right)$$

since $v/E = 2/(2Em)^{\frac{1}{2}}$. If $r = R$ and $m = M$

$$\frac{1}{\tau(E)} = \frac{8\pi r^2 NE_T}{(2mE)^{\frac{1}{2}}}\left(\frac{E}{E_T} - 1\right)$$

This is a simple model of a soft threshold.

3.7.3 Classification and general properties of results using the constant matrix element approximation

The constant matrix element approximation (p. 281) has led us to rather general formulae, which it may be worth interpreting a little more. In Table 3.7.1 ten Auger processes are shown on band diagrams, together with the formal kinetic expressions for the Auger and impact ionization rates. Appropriate reaction coefficients have been invented to do this. The first four processes do not involve localized states ($\mu = 0$, $c_1 \ldots c_\mu = 1$ in eq. (3.7.6)). Processes 5–8 involve one localized state. Processes 9 and 10 are donor–acceptor transitions ($\mu = 2$). In processes 3 and 4 momentum is conserved by phonons so that one has to integrate over nine independent wavevector components arising from $\mathbf{k}_1$, $\mathbf{k}_{1'}$, $\mathbf{k}_2$. The number of independent wavevector components of electrons in bands is denoted by r in the table. Because of the elimination of a $\mathbf{k}$ by momentum conservation, $r = 6$ for

Table 3.7.1. *This shows how the 'first 70' semiconductor Auger effects are obtained, viz. as $\sum(r+1)$ for all ten types of processes. M_0 and M_r are the most important effects, M_6 being replaced by M_4 in a two-dimensional theory. The notation M_s is also used for van Hove singularities. Auger effects with three localized states have also been discussed but are not shown*

Number of process	1	2	3	4	5	6	7	8	9	10
Energy diagram of direct Auger process	(diagram)	(diagram)	As 1 and 2, but phonon-assisted		(diagram)	(diagram)	(diagram)	(diagram)	(diagram)	(diagram)
Auger recombination rate per unit volume	$B_1 n^2 p$	$B_2 np^2$	$B_1^{\mathrm{ph}} n^2 p$	$B_2^{\mathrm{ph}} np^2$	$T_1 n^2 p_{t1'}$	$T_2 npp_{t1'}$	$T_3 pnn_{t1'}$	$T_4 p^2 n_{t1'}$	$T_1^{\mathrm{DA}} nn_{t1} p_{t1'}$	$T_2^{\mathrm{DA}} pn_{t1} p_{t1'}$
Impact ionization rate per unit volume	$Y_1 n$	$Y_2 p$	$Y_1^{\mathrm{ph}} n$	$Y_2^{\mathrm{ph}} p$	$X_1 nn_{t1}$	$X_2 pn_{t1'}$	$X_3 np_{t1'}$	$X_4 pp_{t1'}$	$X_1^{\mathrm{DA}} np_{t1} n_{t1'}$	$X_2^{\mathrm{DA}} pp_{t1} n_{t1'}$
μ	0	0	0	0	1	1	1	1	2	2
$c_1 \ldots c_\mu$					n_{t1}	$n_{t1'}$	$p_{t1'}$	$p_{t1'}$	$p_{t1} n_{t1'}$	$p_{t1} n_{t1'}$
r	6	6	9	9	6	6	6	6	3	3
$r+1$	7	7	10	10	7	7	7	7	4	4
For 'normal' bands the process shown is of type	M_0	M_6	M_0	M_9	M_0	M_6	M_0	M_6	M_0	M_3

Table 3.7.2. *Auger recombination reaction constants.* E_i, E_f *are energies of the initial and final states, respectively. The concentration of occupied and empty traps has been denoted by* n_t *and* p_t, *respectively;* $I \equiv p_0(2') P_0(2') dS_{2'}$. *The last column of Table 3.7.3 should also be consulted*

Number of process	$\dfrac{P(2')}{(2\pi/\hbar)\|U_{if}\|^2}$	Auger recombination coefficient $[L^6 T^{-1}]$
1 2	$\left.\begin{array}{c}\\\\\end{array}\right\}\dfrac{1}{V}\displaystyle\int \delta(E_f - E_i)\,dS_1\,dS_2$	$B_1 = I/n_0^2 p_0$ $B_2 = I/n_0^2 p_0$
3 4	$\left.\begin{array}{c}\\\\\end{array}\right\}\dfrac{1}{V}\displaystyle\int \delta(E_f - E_i)\,dS_1\,dS_2\,dS$	$\left.\begin{array}{c}B_1^{ph}\\B_2^{ph}\end{array}\right\}$ formally as for $\left\{\begin{array}{c}B_1\\B_2\end{array}\right.$
5 6	$\left.\begin{array}{c}\\\\\end{array}\right\}n_t\displaystyle\int \delta(E_f - E_i)\,dS_1\,dS_2$	$T_1 = I/n_0^2 p_{t0}$ $T_2 = I/n_0 p_0 p_{t0}$
7 8	$\left.\begin{array}{c}\\\\\end{array}\right\}p_t\displaystyle\int \delta(E_f - E_i)\,dS_1\,dS_2$	$T_3 = I/n_0 p_0 p_{t0}$ $T_4 = I/p_0^2 n_{t0}$
9 10	$\left.\begin{array}{c}\\\\\end{array}\right\}n_t p_t V\displaystyle\int \delta(E_f - E_i)\,dS_2$	$T_1^{DA} = I/n_0 n_{t0} P_{t0}$ $T_2^{DA} = I/p_0 n_{t0} p_{t0}$

processes 1 and 2. Assume that states 1, 1' 2 are in parabolic bands along the lines of eqs. (3.5.22) with

$$\varphi_j(\mathbf{X}_j) = \sum_{i=1}^{3} \hbar^2 X_{ji}^2 / 2m_{ji} \qquad (3.7.20)$$

Then one can make an appropriate change of variables and collect the new independent variables together as a vector $\mathbf{y}$ of r components. This corresponds to the earlier procedure which led to eq. (3.5.30). One finds from eq. (3.7.6)

$$P(2') = B \int \delta\left[A(2') - \sum_{i=1}^{r} \sigma_i y_i^2\right] dy_1 \dots dy_r \qquad (3.7.21)$$

where B and A are independent of $\mathbf{y}$ and are given in terms of the quantities introduced in eqs. (3.7.6), (3.5.22) and (3.7.20).

For each of our ten basic processes, each of the r principal effective masses can be positive or negative yielding 2^r different band structures, where $r = 3$, 6 or 9. However, since all of the σ's enter into eq. (3.7.21) in the same way, the integrals (3.7.21) depend only upon the number, s say, of negative σ's. This reduces the *essentially different* expressions for the Auger reaction constant of Table 3.7.2 to the situations $s = 0, 1, \dots, r$. Thus each process may be of $r+1$ different types M_s. The total number of such Auger processes is 70, where we have summed the $(r+1)$ rates for the ten basic processes. The number of distinct impact ionization processes is of course also 70. The choice of σ's has been made such that for normal processes between conduction band minima and valence band maxima one has a type M_0-process when the energetic carrier in state $2'$ is an electron, and type M_r when it is a hole. These classifications are shown in the last row of Table 3.7.1. The description of processes of type M_s has been explained in section 3.5.7.

The types M_0 to M_r are also a classification of the (van Hove) critical points of the density of states. Type M_0 corresponds to a minimum of $(D - \sum_{i=1}^{r} \sigma_i y_i^2)$ in y-space, M_r to a maximum, and M_1 to M_{r-1} to saddle points which, just as in the radiative case, may arise for conduction band maxima, valence band minima or saddle points in the band structure. An example of such a saddle type for process 4 is afforded by placing the $1'$ state in a conduction band maximum such as that separating direct and indirect minima in III–V materials like GaAs, GaP, etc. For such a band maximum, with the 1 and 2 states in valence band maxima, a process of type M_6 results. If, as appears to be the case in GaP, this band extremum is in fact a saddle point when viewed in three dimensions, there results a process of type M_7 or M_8.

There are of course other processes which have not been given in the table. These include Auger and impact ionization processes involving excitons (see, for example, [3.7.9] and [3.7.10]), pairs of particles bound to the same center (see, for example, [3.7.11]), transitions involving several localized states (see, for example, [3.7.12] and [3.7.13]), etc.

Some special instances of eqs. (3.7.6) and (3.7.8) are given in Table 3.7.2. Examples of eq. (3.7.21) are given in Table 3.7.3. Note that in the latter table σ_{r+3} ($r = 1, 2, 3$) refers to the set of σ's associated with the effective masses of the second electron, and σ_{r+6} refers to the set of σ's associated with the effective masses of the third electron.

In order to illustrate further the step from eq. (3.7.6) to eq. (3.7.21) we consider here the specific case of processes 3 and 4. All states are band states and Table 3.7.2 with eq. (3.5.22) yields

$$P(2') = \frac{1}{V}\frac{2\pi}{\hbar}\left(\frac{V}{8\pi^3}\right)^3 |H_{if}|^2_{th} \int d\mathbf{X}_1 \, d\mathbf{X}_2 \, d\mathbf{X}_{1'} \, \delta[\varepsilon + \varphi_{1'}(\mathbf{X}_{1'}) + \varphi_{2'} - \varphi_1(\mathbf{X}_1) - \varphi_2(\mathbf{X}_2)] \qquad (3.7.22)$$

where $(E_f - E_i)$ is now given explicitly, the phonon energy being neglected in comparison with the electron energies, and $\varepsilon \equiv \varepsilon_{1'} + \varepsilon_{2'} - \varepsilon_1 - \varepsilon_2$. This expression may be applied to all direct, indirect and Umklapp processes, provided a reasonable approximation to $|H_{if}|^2_{th}$ may be obtained. Transformations

$$y_{ji} = \left(\frac{2|m_{ji}|}{\hbar^2}\right)^{\frac{1}{2}} X_{ji}$$

are now introduced, casting eq. (3.7.22) into the form (3.7.21), so that

$$P(2') = \frac{1}{V}\frac{2\pi}{\hbar}\left(\frac{2^{\frac{3}{2}}}{\hbar^3}\frac{V}{8\pi^3}\right)^3 |H_{if}|^2_{th} \prod_{i=1}^{3} |m_{1i} \, m_{2i} \, m_{1'i}|^{\frac{1}{2}} \int \delta\left(\varepsilon + \varphi_{2'} - \sum_{i=1}^{9} \sigma_i y_i^2\right) d^9 y \qquad (3.7.23)$$

which yields the entry in row 2 of Table 3.7.3. The σ_i are here again each associated with one effective mass:

$$m_{1i} = \sigma_i |m_{1i}|, \quad m_{2i} = \sigma_{i+3}|m_{2i}|, \quad -m_{1'i} = \sigma_{i+6}|m_{1'i}|$$

Table 3.7.3. *Identification of terms in the impact ionization integral* (3.7.21) $\mathbf{Q} \equiv \mathbf{k}_{1'} + \mathbf{k}_{2'} - \mathbf{k}_1 - \mathbf{k}_2$; $\varepsilon \equiv \varepsilon_{1'} + \varepsilon_{2'} - \varepsilon_1 - \varepsilon_2$.
All bands involved other than the band of state 2' are assumed parabolic.

$p_r \equiv \mathrm{sgn}\,(1/m_{2r} - 1/m_{1'r})$, $q_r^2 \equiv (4\sigma_r p_r m_{1r} m_{2r}/((m_{1r} - m_{1'r})(m_{2r} - m_{1r}))$, where $r = 1, 2$ and 3

Number of process	Reaction constant	Number of distinct Auger rates $= (r+1)$	$\dfrac{B}{(2\pi/\hbar)\,\lvert H_{\mathrm{f}f}\rvert_{\mathrm{th}}^2}$	Values of σ_i	$A(2')$	$P(2')$ depends on
1	B_1	7	$\dfrac{1}{V}\left(\dfrac{2^{\frac{3}{2}}\,V}{\hbar^3\,8\pi^3}\right)^2 \displaystyle\prod_{i=1}^{3}\left\lvert \dfrac{m_{1i}\,m_{2i}\,m_{1'i}}{m_{1'i}-m_{1i}-m_{2i}}\right\rvert^{\frac{1}{2}}$	$\sigma_r = \mathrm{sgn}\left(\dfrac{1}{m_{1r}}-\dfrac{1}{m_{1'r}}\right)$ $\sigma_{r+3} = -\mathrm{sgn}\,(p_r - \sigma_r q_r)$	$\varepsilon + \varphi_{2'}\ +$ $\displaystyle\sum_{i=1}^{3}\dfrac{\hbar^2}{2}\dfrac{(X_{2'i}-Q_i)^2}{m_{1'i}-m_{1i}-m_{2i}}$	$\mathbf{k}_1, \mathbf{k}_2, \mathbf{k}_2$ ($\mathbf{k}_{1'}$ has been removed by momentum conservation)
2	B_2	7	$\dfrac{1}{V}\left(\dfrac{2^{\frac{3}{2}}\,V}{\hbar^3\,8\pi^3}\right)^2 \displaystyle\prod_{i=1}^{3}\lvert m_{1i}\,m_{2i}\,m_{1'i}\rvert^{\frac{1}{2}}$			
3	B_1^{ph}	10	$\dfrac{1}{V}\left(\dfrac{2^{\frac{3}{2}}\,V}{\hbar^3\,8\pi^3}\right)^3 \displaystyle\prod_{i=1}^{3}\lvert m_{1i}\,m_{2i}\,m_{1'i}\rvert^{\frac{1}{2}}$	$\sigma_r = \mathrm{sgn}\,(m_{1r})$ $\sigma_{r+3} = \mathrm{sgn}\,(m_{2r})$ $\sigma_{r+6} = \mathrm{sgn}\,(-m_{1'r})$	$\varepsilon + \varphi_{2'}$	$\mathbf{k}_1, \mathbf{k}_2, \mathbf{k}_{1'}, \mathbf{k}_2'$
4	B_2^{ph}	10				
5	T_1	7	$\left(\dfrac{2^{\frac{3}{2}}\,V}{\hbar^3\,8\pi^3}\right)^2 \displaystyle\prod_{i=1}^{3}\lvert m_{1i}\,m_{2i}\rvert^{\frac{1}{2}} \times \begin{cases} n_{t1'} \\ p_{t1'} \end{cases}$	$\sigma_r = \mathrm{sgn}\,(m_{1r})$ $\sigma_{r+3} = \mathrm{sgn}\,(m_{2r})$	$\varepsilon + \varphi_{2'}$	$\mathbf{k}_1, \mathbf{k}_2, \mathbf{k}_2'$. Momentum $\mathbf{k}_{1'}$ of trap state is not involved
8	T_4	7				
6	T_2	7	$\left(\dfrac{2^{\frac{3}{2}}\,V}{\hbar^3\,8\pi^3}\right)^2 \displaystyle\prod_{i=1}^{3}\lvert m_{1i}\,m_{2i}\rvert^{\frac{1}{2}} \times \begin{cases} n_{t1'} \\ p_{t1'} \end{cases}$	$\sigma_r = \mathrm{sgn}\,(-m_{1r})$ $\sigma_{r+3} = \mathrm{sgn}\,(m_{2r})$	$\varepsilon_{1'} - \varepsilon_{2'} + \varepsilon_2 - \varepsilon_1 + \varphi_{2'}$	$\mathbf{k}_1, \mathbf{k}_2, \mathbf{k}_2'$. Momentum $\mathbf{k}_{1'}$ of trap state is not involved
7	T_3	7				
9	T_2^{DA}	4	$V\left(\dfrac{2^{\frac{3}{2}}\,V}{\hbar^3\,8\pi^3}\right)^3 \displaystyle\prod_{i=1}^{3}\lvert m_{2i}\rvert^{\frac{1}{2}}\,p_{t1}\,n_{t1'}$	$\sigma_r = \mathrm{sgn}\,(m_{2r})$	$\varepsilon + \varphi_{2'}$	$\mathbf{k}_2, \mathbf{k}_2'$. Momentum $\mathbf{k}_1, \mathbf{k}_{1'}$ of trap state is not involved
10	T_2^{DA}	4				

We now turn to the δ-function integrals. Suppose that in eq. (3.7.21) one has p σ's with value -1, and $r-p$ with value $+1$. One is then interested in integrals of the type

$$I_{r,p}(A) \equiv \int_{-\infty}^{\infty} \delta(A - x_1^2 + x_2^2)\, dy_1 \ldots dy_r$$

where

$$x_1^2 \equiv \sum_{i=1}^{p} y_i^2 \quad x_2^2 \equiv \sum_{i=p+1}^{r} y_i^2$$

Two limiting cases are particularly important: $p = 0$, corresponding to a process of type M_0, and $p = r$, corresponding to a process M_r. The integrals are straightforward and yield

$$I_{r,0}(A) \quad (A \geqslant 0) = I_{r,r}(A) \quad (A < 0) = \frac{\pi^{r/2}}{\Gamma\left(\dfrac{r}{2}\right)} |A|^{(r-2)/2}$$

For the opposite signs of $A(\mathbf{X}_{2'})$ the integrals vanish. This is a simple generalization of the result derived in eq. (3.7.14) which corresponds to $r = 6$. For more details see [3.7.14] and [3.7.3].

3.8 Auger lifetimes

3.8.1 Basic results to be used

In order to estimate the Auger effect lifetime one integrates over all possible positions in $\mathbf{k}$-space of each of the four band states involved. Collecting results from which to start, we have eq. (3.7.1) for the probability T_{if} of a transition $i \to f$ in terms of the matrix element $|H_{if}|^2$ which for Coulomb interactions is given by eq. (3.3.26) together with eq. (3.4.20). Thus we have again eq. (3.7.4), i.e. using

$$|H_{if}|^2 = \frac{32\pi^2 e^4}{\varepsilon^2 V^2} \left\{ \frac{|F_{ac} F_{bd}|^2}{[\lambda^2 + (k_1 - k_{1'})^2]^2} + \frac{|F_{ad} F_{bc}|^2}{[\lambda^2 + (k_2 - k_{1'})^2]^2} \right.$$

$$\left. + \left| \frac{F_{ac} F_{bd}}{\lambda^2 + (k_1 - k_{1'})^2} - \frac{F_{ad} F_{bc}}{\lambda^2 + (k_2 - k_{1'})^2} \right|^2 \right\} \delta_{\mathbf{k}_1 + \mathbf{k}_2, \mathbf{k}_{1'} + \mathbf{k}_{2'}} \tag{3.8.1}$$

$$P = \frac{2t}{\hbar^2 V} \int \ldots \int \Phi |H_{if}|^2 \frac{1 - \cos x}{x^2} \, dS_1 \, dS_{1'} \, dS_2 \, dS_{2'}$$

where P is the recombination probability per unit volume per unit time and $x \equiv (t/\hbar)(E_f - E_i)$. The probability factor Φ in the integrand is with the assumption of quasi-Fermi levels and a CHCC process (see Table 3.5.1 or Fig. 2.1.1(b))

$$\Phi = p(1)\,p(2)[1 - p(1')][1 - p(2')]$$

$$\left[1 - \frac{1 - p(1)}{p(1)} \frac{1 - p(2)}{p(2)} \frac{p(1')}{1 - p(1')} \frac{p(2')}{1 - p(2')} \right]$$

Of course in eq. (3.7.4) we had only impact ionization. Here we have included impact ionization as the reverse of Auger recombination and we have also integrated over all four band states (one will be removed by momentum conservation). The twelve-fold integral has thus, with energy conservation, only eight independent variables of integration. Now using energy conservation $E_1 + E_2 = E_{1'} + E_{2'}$

$$\Phi = p(1)p(2)[1-p(1')][1-p(2')]$$
$$[1 - \exp(\eta_1 - F_e + \eta_2 - F_e + F_h - \eta_{1'} + F_e - \eta_{2'})]$$
$$= p(1)p(2)[1-p(1')][1-p(2')][1 - \exp\{-(F_e - F_h)\}]$$

For small departures from equilibrium $\delta F_e \equiv F_e - F_{eq}$, $\delta F_h = F_h - F_{eq}$ this is

$$\Phi = \{p(1)p(2)[1-p(1')][1-p(2')]\}_{eq}(\delta F_e - \delta F_h)$$

Also the excess electron concentration is

$$\delta n = n - n_{eq} = N_c[F_{\frac{1}{2}}(F_{eq} + \delta F_e - \eta_c) - F_{\frac{1}{2}}(F_{eq} - \eta_c)] = n_{eq}\alpha_c\,\delta F_e$$

where

$$\alpha_c \equiv I_{-\frac{1}{2}}(F_{eq} - \eta_c)/I_{\frac{1}{2}}(F_{eq} - \eta_c)$$

Similarly we shall use

$$\alpha_v \equiv I_{-\frac{1}{2}}(\eta_v - F_{eq})/I_{\frac{1}{2}}(\eta_v - F_{eq})$$

Finally we have

$$P = \frac{1}{2}\frac{64\pi^2 e^4 t}{V^3\hbar^2\varepsilon^2}\left[\frac{\delta n}{\alpha_c n_{eq}} + \frac{\delta p}{\alpha_v p_{eq}}\right]\left(\frac{V}{8\pi^3}\right)^3\int\cdots\int(p(1)p(2)[1-p(1')][1-p(2')])_{eq}$$
$$\{\ldots\}\cdot\frac{1-\cos x}{x^2}\,\mathrm{d}\mathbf{k}_1\,\mathrm{d}\mathbf{k}_{1'}\,\mathrm{d}\mathbf{k}_{2'} \tag{3.8.2}$$

where the details of the brace in eq. (3.8.1) have been omitted. The factor $V/8\pi^3$ is the number of wavevectors per unit volume of $\mathbf{k}$-space. For nondegenerate material $\alpha_c = \alpha_v = 1$ and the equation then simplifies somewhat. The factor of $\frac{1}{2}$ arises because integration over all 'direct' transitions generates all exchange transitions. The factor $\frac{1}{2}$ must be applied if the integrations extend over all states in order to avoid counting transitions twice.

In the rest of this section 3.8, a method of obtaining the transition probability P will be outlined. It has the merit of comparative simplicity, but it gives only a qualitative guide. This may be traced back to the treatment of the overlap integrals and the use of parabolic bands. Improved methods involve much more numerical calculation, but one can find the equivalent to overlap integrals as a by-product of a band structure calculation (see section 3.6.6). Our simple argument will illustrate the main principles.

3.8.2 Two approximations: treatment of the exchange terms, use of most probable transitions

(a) The quantity (3.8.2) can be written schematically as a sum of three terms. With

$$\mathbf{g} \equiv \mathbf{k}_1 - \mathbf{k}_{1'}, \quad \mathbf{h} \equiv \mathbf{k}_2 - \mathbf{k}_{1'}, \quad d\mathbf{k} \equiv d\mathbf{k}_1 \, d\mathbf{k}_{1'} \, d\mathbf{k}_{2'}$$

these are

$$P_1 \equiv \int \frac{|a_1|^2}{(\lambda^2 + g^2)^2} f(\mathbf{k}) \, d\mathbf{k} \quad (\text{'direct' transition})$$

$$P_2 \equiv \int \frac{|a_2|^2}{(\lambda^2 + h^2)^2} f(\mathbf{k}) \, d\mathbf{k} \quad \begin{array}{l}(\text{'exchange' transition in which } \mathbf{k}_1, \mathbf{k}_2 \text{ are} \\ \text{interchanged relative to } P_1)\end{array}$$

$$P_3 \equiv \int \left| \frac{a_1}{\lambda^2 + g^2} - \frac{a_2}{\lambda^2 + h^2} \right|^2 f(\mathbf{k}) \, d\mathbf{k}$$

where $a_1 = F_{ac} F_{bd}$, $a_2 = F_{ad} F_{bc}$, and

$$f(\mathbf{k}) = \frac{32\pi^2 e^4 t}{V^3 \hbar^2 \varepsilon^2} \left[\frac{\delta n}{\alpha_c n_{eq}} + \frac{\delta p}{\alpha_v p_{eq}} \right] \left(\frac{V}{8\pi^3} \right)$$

$$\times (p(1)\,p(2)\,[1 - p(1')]\,[1 - p(2')])_{eq} \left(\frac{1 - \cos x}{x^2} \right)$$

Thus, in an obvious notation,

$$P_3 = \left(\int_{g>h} + \int_{h>g} \right) \left| \frac{a_1}{\lambda^2 + g^2} - \frac{a_2}{\lambda^2 + h^2} \right|^2 f(\mathbf{k}) \, d\mathbf{k}$$

$$< \int_{g>h} \frac{|a_2|^2}{(\lambda^2 + h^2)^2} f(\mathbf{k}) \, d\mathbf{k} + \int_{h>g} \frac{|a_1|^2}{(\lambda^2 + g^2)^2} f(\mathbf{k}) \, d\mathbf{k}$$

Hence

$$P_3 < P_2 + P_1 = 2P_1 \tag{3.8.3}$$

The last relation can be proved by relabelling the vector $\mathbf{k}_2$ in P_2 as $\mathbf{k}_1$ or the vector $\mathbf{k}_1$ in P_1 as $\mathbf{k}_2$. This is possible since both vectors are integrated out. It follows that

$$2P_1 < P < 4P_1 \tag{3.8.4}$$

It is thus possible to obtain a rough understanding of P from P_1, which deals with the 'direct' transition.

The uncertainty by a factor of two suggested in relation (3.8.3) is a maximum uncertainty as seen from the derivation of the result. Normally P_3 is smaller than

$2P_1$, since P_3 corresponds to collisions between electrons of like spin, and because of the exclusion hole these are less likely than collisions of electrons with opposite spin. This result is a slight variant of eq. (3.3.27).

(b) For a nondegenerate semiconductor $p(2')$ is negligible so that the probability factor is ($\eta_G \equiv \eta_c - \eta_v$)

$$\exp(F_e - \eta_1 + F_e - \eta_2 + \eta_{1'} - F_h)_{eq} = \exp(F_{eq} - \eta_c - \eta_G) \exp Y \qquad (3.8.5)$$

where Y is the wavevector-dependent part of the probability factor. In fact, with $\mu = m_c / m_v$,

$$Y = -(\eta_1 - \eta_c) - (\eta_2 - \eta_c) - (\eta_v - \eta_{1'}) = -\frac{\hbar^2}{2m_c kT}(k_1^2 + k_2^2 + \mu k_{1'}^2)$$

Its most probable ('m.p.') value is obtainable from the threshold values for impact ionization (theorem 1 of section 3.5.5), i.e. from line 4 of Table 3.5.1. Hence

$$Y_{\text{m.p.}} = -\frac{\hbar^2 k_G^2}{2m_c kT}\left[\frac{2\mu^2}{(2\mu+1)(\mu+1)} + \frac{\mu}{(2\mu+1)(\mu+1)}\right] = -\frac{\mu}{\mu+1}\eta_G$$

For the most probable transitions the probability factor is

$$\exp\left(F_{eq} - \eta_c - \frac{1+2\mu}{1+\mu}\eta_G\right) \qquad (3.8.6)$$

This indicates an activation energy $[(1+2\mu)/(1+\mu)]E_G$ [3.8.1] to which we return in eq. (3.8.12). In this situation $\mathbf{k}_1$ and $\mathbf{k}_{1'}$ are counter-directed so that the momentum transfer is

$$(k_1 + k_{1'})_{\text{m.p.}} = (1+\mu)(k_{1'})_{\text{m.p.}} = (1+\mu)\frac{k_G}{(1+2\mu)^{\frac{1}{2}}(1+\mu)^{\frac{1}{2}}}$$

$$= \left(\frac{1+\mu}{1+2\mu}\right)^{\frac{1}{2}} k_G \qquad (3.8.7)$$

The evaluations are useful also in connection with assumption (ii) (given below). One observes that only in the limit $\mu = 0$ can an electron make a transition between the band edges. On the other hand, if the effective masses in the conduction and the valence bands are the same, the most probable triplet of states $\mathbf{k}_1$, $\mathbf{k}_2$ and $\mathbf{k}_{1'}$ lies at energies of $\frac{1}{6}E_G$ from the nearest band edge, and the energy transfer for each electron is $\frac{4}{3}E_G$. Normally one will have $0 < \mu < 1$.

Since the most probable value of the momentum transfer g is of order k_G and the screening parameter λ occurs only in combination with g, it is clearly reasonable to neglect λ^2 compared with g^2 in the integral, and this will be done below.

Note that the term 'most probable transitions' in this subsection has the meaning given to it in connection with the factor Φ. The most probable transitions in the strict sense would require one to maximize in some way the *whole* integrand in eq. (3.8.2). This would, however, yield a negligible gain considering the complication involved, since the exponential term must clearly dominate in the integral.

In order to evaluate eq. (3.8.2) the following assumptions will be made:

(i)   Electron and hole gases will be assumed to be nondegenerate. This reduces the Fermi distributions to Boltzmann distributions and $\alpha_c \sim \alpha_v \sim 1$.

(ii) The overlap integrals will be evaluated for the wavevectors which occur for the most probable transitions and then be removed from the main integration. They will then be denoted by

$$F_1 \equiv (F_{ac})_{\text{m.p.}}, \quad F_2 \equiv (F_{bd})_{\text{m.p.}}, \quad F_3 \equiv (F_{ad})_{\text{m.p.}}, \quad F_4 \equiv (F_{bc})_{\text{m.p.}}$$

(iii) Parabolic bands with extrema at $\mathbf{k} = 0$ will be considered.

(iv) In the first place only the term involving $|M_D|^2$ in the brace of eq. (3.8.2) will be treated. Because of eq. (3.8.3) this is adequate within a factor of two.

(v)   The screening length will be neglected. This was justified above.

3.8.3 The integration [3.8.2]

If we now put

$$\mathbf{j} \equiv \mathbf{k}_1 + \mu \mathbf{k}_{1'}$$

it follows that we can substitute in $(\hbar^2 k_G^2 / 2m_c \equiv E_c - E_v \equiv E_G)$

$$\frac{2m_c}{t\hbar} x = k_{2'}^2 - \mu k_{1'}^2 - k_2^2 - k_1^2 - k_G^2$$

the quantities

$$\mathbf{k}_{1'} = \frac{\mathbf{j} - \mathbf{g}}{1 + \mu}, \quad \mathbf{k}_2 = \mathbf{k}_{2'} - \mathbf{g}, \quad \mathbf{k}_1 = \frac{\mathbf{j} + \mu \mathbf{g}}{1 + \mu}$$

to find

$$\frac{2m_c}{t\hbar} x = 2\mathbf{k}_{2'} \cdot \mathbf{g} - \frac{j^2}{1 + \mu} - \frac{1 + 2\mu}{1 + \mu} g^2 - k_G^2 \tag{3.8.8}$$

The probability product (3.8.5) is therefore

$$\exp\left\{ -\frac{\hbar^2}{2m_c kT}\left(k_{2'}^2 - k_G^2 - \frac{2m_c}{t\hbar} x \right) \right\} \exp\left(F_{eq} - \eta_c - \eta_G \right)$$

Replacing the variables

$$\mathbf{k}_{1'} \text{ by } \mathbf{j}, \quad \text{and } \mathbf{k}_1 \text{ by } \mathbf{g}$$

which yields a Jacobian for the transformation of $(1+\mu)^{-3}$, one can now rewrite the integral for P_1 as

$$P_1 = \frac{32\pi^2 e^4 t |F_1 F_2|^2}{V^2 \hbar^2 \varepsilon^2 (1+\mu)^3} \left(\frac{\delta n}{n_{eq}} + \frac{\delta p}{p_{eq}} \right) \left(\frac{V}{8\pi^3} \right)^3 \int \cdots \int$$

$$\times p(1)\, p(2)[1 - p(1')] \frac{1 - \cos x}{g^4 x^2}\, d\mathbf{k}_{2'}\, d\mathbf{g}\, d\mathbf{j}$$

$$= A \int \exp\left(-\frac{\hbar^2 k_{2'}^2}{2m_c kT} \right) d\mathbf{k}_{2'} \int \frac{1}{g^4}\, d\mathbf{g} \int_{-\infty}^{x_1} \exp\left(\frac{\hbar x}{ikT} \right) \frac{1 - \cos x}{x^2} j\, dx \qquad (3.8.9)$$

where

$$A = \frac{32\pi^2 e^4 t |F_1 F_2|^2}{V^2 \hbar^2 \varepsilon^2 (1+\mu)^3} \left[\frac{\delta n}{n_{eq}} + \frac{\delta p}{p_{eq}} \right] \left(\frac{V}{8\pi^3} \right)^3 \cdot \frac{4\pi m_c (1+\mu)}{t\hbar} \exp\left[F_{eq} - \eta_c \right]$$

We have also used

$$dx = \frac{-t\hbar}{2m_c} \frac{2j\,dj}{1+\mu}, \quad d\mathbf{j} = 4\pi j^2\, dj = \frac{-4\pi m_c (1+\mu)}{t\hbar} j\, dx$$

and x_1 is the maximum value of x. It occurs in eq. (3.8.8) when $j = 0$.

The x-integration yields energy conservation as it is of the form $\pi \int f(x)\, \delta(x)\, dx = \pi f(0)$. Using the j-value from eq. (3.8.8) and putting y for the cosine of the angle between $\mathbf{k}_{2'}$ and $\mathbf{g}$,

$$P_1 = 2A\pi^2 (1+\mu)^{\frac{1}{2}} \int e^{-\eta_{2'}} d\mathbf{k}_{2'} \int_0^{\infty} \frac{dg}{g^2} \int_{y_1}^1 \left[2k_{2'} gy - \frac{1+2\mu}{1+\mu} g^2 - k_G^2 \right]^{\frac{1}{2}} dy \quad (3.8.10)$$

The lower limit y_1 arises from the need for x_1 in eq. (3.8.9) to be positive for energy conservation, i.e.

$$\frac{2m_c}{t\hbar} x_1 \equiv 2k_{2'} gy - \frac{1+2\mu}{1+\mu} g^2 - k_G^2 \geqslant 0$$

Hence the lower limit in the y-integral does not contribute and

$$P_1 = \frac{2\pi^2}{3} A(1+\mu)^{\frac{1}{2}} \int \frac{\exp(-\eta_{2'})}{k_{2'}} d\mathbf{k}_{2'}$$

$$\times \int \frac{\{2k_{2'} g - [(1+2\mu)/(1+\mu)] g^2 - k_G^2\}^{\frac{3}{2}}}{g^3} dg$$

One has to ensure in the g-integration that $y_1 < 1$, i.e.

$$2k_{2'}gy_1 \equiv \frac{1+2\mu}{1+\mu}g^2 + k_G^2 \leqslant 2k_{2'}g$$

With

$$u \equiv \frac{1+\mu}{1+2\mu}k_{2'}, \quad v \equiv \left(\frac{1+\mu}{1+2\mu}\right)^{\frac{1}{2}}k_G$$

this means that the limits for g, g_1 and g_2 must satisfy $g_{1,2}^2 - 2ug_{1,2} + v^2 = 0$, i.e.

$$-z_1 \leqslant z \equiv g - u \leqslant z_1 \equiv (u^2 - v^2)^{\frac{1}{2}}$$

Hence

$$P_1 = \frac{2\pi^2}{3}A(1+\mu)^{\frac{1}{2}}\left(\frac{1+2\mu}{1+\mu}\right)^{\frac{1}{2}}\int e^{-\eta_{2'}}\frac{dk_{2'}}{k_{2'}}\int_{-z_1}^{z_1}\frac{(z_1^2-z^2)^{\frac{3}{2}}}{(z+u)^3}dz$$

The z-integral has the value

$$\frac{3\pi}{2v}(u-v)^2 \tag{3.8.11}$$

which is obtained in the appendix to this section. It follows that

$$P_1 = \frac{\pi^3}{v}A(1+\mu)^{\frac{1}{2}}\left(\frac{1+2\mu}{1+\mu}\right)^{\frac{3}{2}}\int e^{-\eta_{2'}}(u-v)^2(k_{2'})^{-1}\,dk_{2'}$$

The $k_{2'}$-integral will be converted into an $E_{2'}$-integral, and an Auger effect 'activation' or 'threshold energy'

$$E_{th} \equiv \frac{1+2\mu}{1+\mu}E_G \tag{3.8.12}$$

suggested by eqs. (3.7.19) and (3.8.5), will be introduced. Then

$$(u, v) = \frac{1+\mu}{1+2\mu}\frac{2m_c}{\hbar^2}(E_{2'}^{\frac{1}{2}}, E_{th}^{\frac{1}{2}})$$

The last integral is

$$\int_{\eta_{th}}^{\infty} e^{-\eta_{2'}}(\eta_{2'}^{\frac{1}{2}} - \eta_{th}^{\frac{1}{2}})^2\,d\eta_{2'} = B[(2\eta_{th}+1)e^{-\eta_{th}} - 2\eta_{th}^{\frac{1}{2}}\Gamma(\tfrac{3}{2}, \eta_{th})]$$

where the incomplete gamma function has been used:

$$\Gamma(a, \eta) \equiv \int_{\eta}^{\infty} e^{-x}x^{a-1}\,dx$$

Hence eq. (3.8.10) becomes

$$P_1 = \frac{\sqrt{2}m_c^{\frac{5}{2}}e^4(kT)^2 |F_1 F_2|^2}{\pi^2(1+\mu)^{\frac{3}{2}}\hbar^6\epsilon^2 E_G^{\frac{1}{2}}}\left[\frac{\delta n}{n_{eq}} + \frac{\delta p}{p_{eq}}\right]\exp{(F_{eq} - \eta_c - \eta_{th})}f(\eta_{th}) \qquad (3.8.13)$$

where

$$f(\eta) \equiv 1 + 2\eta - 2\eta^{\frac{1}{2}}e^{\eta}\Gamma(\tfrac{3}{2}, \eta)$$

and this function tends to unity for large η since

$$\Gamma(\tfrac{3}{2}, \eta) = \eta^{\frac{1}{2}}e^{-\eta}\left[1 + \frac{1}{2\eta} - \frac{1}{4\eta^2} + \cdots\right]$$

3.8.4 The Auger lifetimes

The relevant lifetime which corresponds to the limit $2P_1$ of the probability per unit time is

$$\tau_1 = \frac{\delta n}{2P_1} = \frac{(1+\mu)^{\frac{3}{2}}\epsilon^2\eta_G^{\frac{1}{2}}\pi^{\frac{1}{2}}}{4(m_c/m)|F_1 F_2|^2}\cdot\frac{\hbar^3}{me^4}\cdot\left[1 + \frac{\delta p\,n_{eq}}{\delta n\,p_{eq}}\right]^{-1}\cdot\frac{e^{\eta_{th}}}{f(\eta_{th})} \qquad (3.8.14)$$

The second factor carries the dimension and is the universal constant t_0 whose value is 2.4×10^{-17} s. The third factor allows for a nonequilibrium concentration of holes. The last factor is for large η_{th} simply $\exp{(\eta_{th})}$ and yields a kind of activation energy which is of order of magnitude of the band gap. Using relation (3.8.4) the electron lifetime τ, allowing for the exchange term, satisfies

$$\tfrac{1}{2}\tau_1 < \tau < \tau_1$$

where a value near the upper limit is most likely. Substituting some of the numerical values,

$$\tau_1 = (10.7 \times 10^{-18})\frac{(1+\mu)^{\frac{3}{2}}\epsilon^2\eta_G^{\frac{1}{2}}}{(m_c/m)|F_1 F_2|^2}\left[1 + \frac{\delta p\,n_0}{\delta n\,p_0}\right]^{-1}\frac{e^{\eta_{th}}}{f(\eta_{th})} \quad \text{(seconds)} \qquad (3.8.15)$$

This is exact within the model and differs from the later modification [3.8.3] of the original result [3.8.1] only by the factor $\eta_{th}/f(\eta_{th})$. This difference arises from the more accurate integrations performed here.

In a more complete expression for the lifetime one has to include in P_1 the effect of hole–hole collisions (transition B_2 in Fig. 2.1.1). The probability is given by eq. (3.8.13), except that the bands are interchanged. Instead of η_{th} there then occurs the quantity

$$\eta'_{th} = \frac{1+2/\mu}{1+1/\mu}\eta_G = \left(\frac{1+2\mu}{1+\mu} + \frac{1-\mu}{1+\mu}\right)\eta_G \equiv \eta_{th} + \eta' \qquad (3.8.16)$$

The total transition probability per unit time is

$$P_{1,\,\text{tot}} = P_1\left\{1 + \frac{1}{\mu}\frac{f(\eta'_{\text{th}})}{f(\eta_{\text{th}})}\left|\frac{F_3 F_4}{F_1 F_2}\right|^2 e^{-\eta'}\right\} \tag{3.8.17}$$

The expressions still contain the overlap integrals

$$F_1 \equiv (F_{ac})_{\text{m.p.}} = \frac{1}{V}\left\{\int u_c^*(\mathbf{k}_1, \mathbf{r})\, u_v(\mathbf{k}_{1'}, \mathbf{r})\, d\mathbf{r}\right\}_{\text{m.p.}} = \{I_{v\mathbf{k}_1}^{c\mathbf{k}_1}\}_{\text{m.p.}}$$

$$F_2 \equiv (F_{bd})_{\text{m.p.}} = \frac{1}{V}\left\{\int u_c^*(\mathbf{k}_2, \mathbf{r})\, u_c(\mathbf{k}_{2'}, \mathbf{r})\, d\mathbf{r}\right\}_{\text{m.p.}} = \{I_{c\mathbf{k}_2}^{c\mathbf{k}_2}\}_{\text{m.p.}}$$

where the notation (3.6.1) has been used. For the most probable transitions, the first integral can be estimated from formula (3.6.17), and the second integral is from eq. (3.6.11) of order unity. For the first integral we need the value for $q_{\text{m.p.}} = (|\mathbf{k}_1 - \mathbf{k}_{1'}|)_{\text{m.p.}} = (k_1 + k_{1'})_{\text{m.p.}}$ of eq. (3.8.7). Hence eq. (3.6.17) becomes

$$\left|I_{2\,\mathbf{k}_{1'}}^{1\,\mathbf{k}_1 + q}\right|^2 = \left(1 + \frac{|m_v|}{m}\right)\frac{\hbar^2}{2|m_v|}\cdot\frac{1+\mu}{1+2\mu}\,k_G^2\cdot\frac{2m_c}{\hbar^2 k_G^2}$$

$$= \frac{1+\mu}{1+2\mu}\left(\frac{m_c}{m} + \mu\right)$$

For InSb one may choose

$$m_c = 0.0114m, \quad |m_v| = 0.18m$$

whence

$$|F|^2 \doteqdot \left|I_{2\,\mathbf{k}_{1'}}^{1\,\mathbf{k}_1 + q}\right|^2 \doteqdot 0.07$$

This is a typical value. However, as has been emphasized at the end of section 3.6.6, while the sum rule is of course correct, the contributions of different terms is difficult to estimate. Thus, in spite of its elegance, greater accuracy can be obtained by numerical procedures which utilize details of the band structure involved.

Some numerical values are given in Table 3.8.1 which, however, does not exhaust what can be found in the literature. It gives values of the reaction coefficients B_1, B_2 rather than the lifetimes τ_1, whose dominant term exceeds in any case 1 μs at room temperature already for energy gaps in excess of about 0.30 eV. The reaction coefficients can however be used to compute shorter Auger lifetimes at the higher carrier concentrations.

The effect of degeneracy on the mass action laws is brought out clearly in Fig. 3.8.1 ([3.8.10]). The mass action laws $B_1 n^2 p$ or $B_2 np^2$ are valid for a range of Fermi levels where the curves are flat. For a narrow gap semiconductor it is clearly often

Table 3.8.1. *Phononless Auger coefficients at 300 K (unless stated otherwise) in 10^{-31} cm^6 s^{-1} using conduction band and heavy-hole band concentrations to define the coefficients (see Table 3.5.1)*

Material (energy gap, eV)	Values	Type of identification	Comment	Reference
Si (1.12)	2.8, 0.99	Experimental CHCC, CHHH	Little dependence on doping or temperature	[3.8.4]
	≤0.5	Theoretical and experimental CHCC	Uses hot electron generation	[3.8.5]
	2.7	Theoretical CHCC	Major calculation involving more than 25000 terms	[3.8.6]
	1	Experimental CHHH	Minority-carrier lifetime. Little dependence on temperature. Highly doped phonon-assistance conjectured	[3.8.7]
Ge (0.66)	0.2	Experimental $B_1 + B_2$	Decay of recombination radiation	[3.8.8]
	0.07	Theoretical sum of four transition types	Uses sum rule for overlap integrals	[3.8.9]
InSb (0.17)	1.4×10^5, 9×10^4 Variable	Theoretical CHCC, CHHL. Agrees with experiment	Intrinsic case. Full four-band $\mathbf{k} \cdot \mathbf{p}$ calculation	[3.8.10], Figure 5 [3.8.11], [3.8.12]
	8	Experimental CHHL		[3.8.13]
GaP (2.26)	0.0165	Theoretical CHHL	Monte Carlo calculation based on $\mathbf{k} \cdot \mathbf{p}$ theory	[3.8.13]
	2.81	Theoretical CHHL	Averaged overlap functions	[3.8.14]

Material (band gap)	Value	Type	Description	Ref.
GaAs (1.42)	160, 4.64 / 47.2, 6.4	Theoretical CHCC, CHHS	Averaged overlap integral. Parabolic bands and improved bands and including phonon assistance	[3.8.15]
	1000	Experimental CHHS	From luminescence, 77 K, but weakly temperature dependent	[3.8.16]
GaSb (0.72)	20000	Theoretical CHCC	Averaged overlap integral parabolic bands	[3.8.15]
	520, 53700	Theoretical CHCC, CHHS	Improved bands and including phonon assistance	
	1	Experimental CHHS	From luminescence, 77 K, but weakly temperature dependent	[3.8.16]
	13, 270	Theoretical CHCC, CHHH	Averaged overlap integral parabolic band	[3.8.17]
	6000	Theoretical CHCC		[3.8.15]
	0.6, 210	Theoretical CHCC, CHHS	Improved band and including phonon assistance	[3.8.16]
$In_{0.72}Ga_{0.28}As_{0.6}P_{0.4}$ (1.08)	230	Experimental CHHH	Photodecay	[3.8.18]
	500	Experimental CHHS	From emission band	[3.8.19]
	900	Experimental CHHS	From emission of p-type diode laser, Zn-doping $\sim 10^{18}$ cm^{-3}	[3.8.20]

Table 3.8.1. (*Contd.*)

Material (energy gap, eV)	Values	Type of identification	Comment	Reference
$Cd_{0.2}\,Hg_{0.8}\,Te$	2.5×10^6, 0.9×10^6	Theoretical CHCC, CHHL	166 K; lower temperatures also given. Four-band **k·p** approximation	[3.8.21]
	Variable	Experimental		[3.8.22]
$In\,As_{1-x}\,Sb_x$				
InAs, InSb (0.37) (0.17)	22000, 4.2×10^5		$\Delta = 0.39,\ 0.79$ eV	
$x = 0.3$ (0.18)	120000		x for largest Auger effect $\Delta = 0.25$ eV	
$In_{1-x}\,Ga_x\,As$		Theoretical CHHS		[3.8.23]
InAs, GaAs (0.37) (1.4)	22000, 65		$\Delta = 0.39,\ 0.34$ eV	
$x = 0.5$ (0.74)	3800	Four-band Kane model The maximum is reached for $\left\lvert \dfrac{E_G - \Delta}{kT} \right\rvert \sim \dfrac{2m_v}{m_s}$	$\Delta = 0.33$ eV	
$GaSb_{1-x}\,As_x$				
GaSb, GaAs (0.7) (1.4)	1600, 65		$\Delta = 0.8,\ 0.34$ eV	
$x = 0.3$ (0.7)	7900		x for largest Auger effect $\Delta = 0.61$ eV	

Material		Values	Description	Ref.
$Pb_{1-x} Sn_x$ Te				
PbS		4500, 3800	Experimental, theoretical	
(0.32)			(using theory due to	[3.8.24]
$x = 0.17$ (0.207)		43000, 45000	[3.8.25])	
$Ga_{1-x} Al_x$ As				
GaAs		1.08, 10	Theoretical CHCC, CHHS	
(1.4)				
			Calculation	
			of laser	
			threshold	
$x = 0.1$		1.01, 8	current	[3.8.26]
$x = 0.2$		0.8, 5		

Additional information can be found as follows: InGaAsP [3.8.27]; lead compounds [3.8.28]; $Hg_{1-x} Cd_x$ Te [3.8.29]; GaSb [3.8.30]; $InAs_{1-x} Sb_x$ [3.8.31]; $Ga_x In_{1-x}$ Sb [3.8.32]; HgTe [3.8.33].

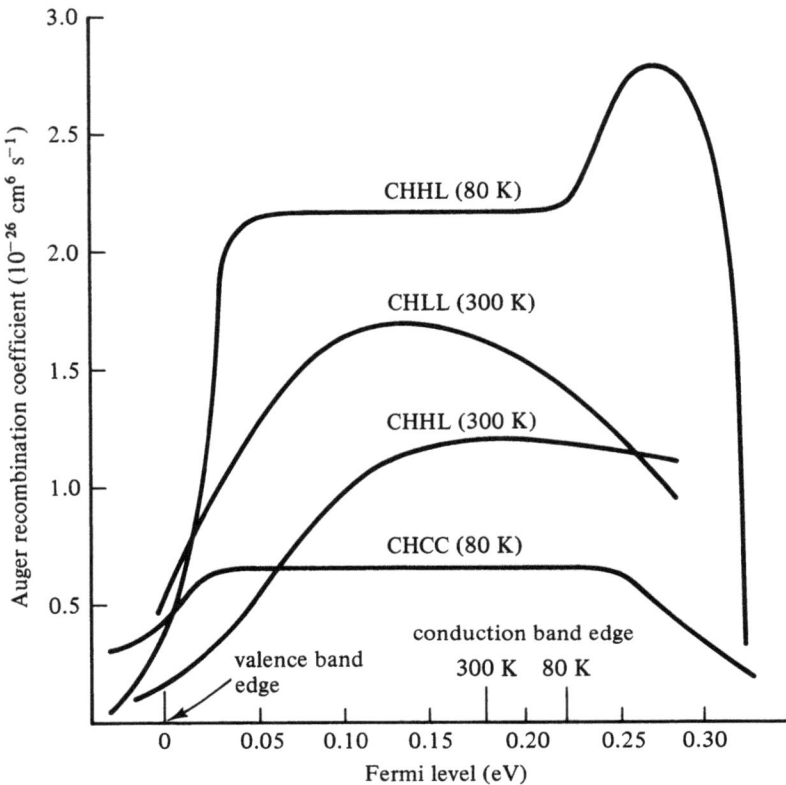

Fig. 3.8.1. The quantities recombination rate/$n_{eq}^2 p_{eq}$ for the CHCC process and recombination rate/$n_{eq} p_{H\,eq}^2$ for the CHLH process in InSb where $p_{H\,eq}$ is the equilibrium concentration of holes in the heavy-hole band.

not valid. If one neglects momentum conservation and adopts the constant matrix element assumption, then the **k**-vector integrations are independent and yield the simple mass action laws. In fact, the matrix element often vanishes for the threshold of the Auger transitions if **k·p** wavefunctions are used and the constant matrix element assumption is then very misleading. Even an $n^x p^y$ representation for the Auger effect is under these circumstances valid only for a limited range of Fermi levels [3.8.28], [3.8.34].

As expected from the formulae of Table 3.5.1 for the threshold or activation energies, one sees from Table 3.8.1 a rapid increase of Auger coefficients as the relevant band gap decreases. For indirect semiconductors the considerable change of momentum per electron reduces the band–band effect relative to direct materials also by virtue of the factor $[\lambda^2 + |k_1 - k_{1'}|^2]^{-2}$ in the matrix element. Thus one would expect a reduction of the Auger mass action coefficients B_1, B_2 as one adds Al to GaAs in view of the fact that GaAs has a smaller (and direct) energy gap than AlAs

(1.52 eV as against 2.25 eV at 77 K) which is in addition an indirect material. This is confirmed by detailed calculations on the $Ga_{1-x}Al_xAs$ system [3.8.26]. These show that the CHHS process is more important than the CHCC process. This may be understood from the threshold energies which involve $E_G - \Delta$ for the former and E_G for the latter.

The reason why the theoretical treatments of band–band Auger effects have moved from analytical to numerical methods resides in: (a) the complication of the band structures and their important effect on the estimates. One can say equivalently that the treatment of the overlap integrals is hard; (b) the multidimensional nature of the integrations; (c) the need to include Umklapp-type processes for the larger band-gap materials; (d) the need to include the dielectric function and appropriate screening functions which are not constants.

3.8.5 Notes on experimental procedures

Early experimental indications of the band–band Auger effect were given by Moss for PbS [3.8.35], suggesting $B_{1,2} \sim 2 \times 10^{-28} \ cm^6 \ s^{-1}$ [3.8.36]. He used the photoelectromagnetic ('PEM') effect and photoconductivity. These were also used for InSb in 1959 [3.8.37]. An early experimental value of a coefficient was given for Tl_2S by Ostrowski ([3.8.38], $B_{1,2} \sim 10^{-31}$–$10^{-30} \ cm^6 \ s^{-1}$). Te was also studied in 1960 [3.8.39]. At higher temperatures the lifetime against temperature curves for different samples of the same material all converge towards one curve, $\tau \propto n^2$, indicating an unavoidable intrinsic mechanism for this material (viz. the band–band Auger effect).

The kinetics (see Fig. 2.1.1) also gives Auger coefficients by virtue of a rate dependence proportional to n^2p or p^2n. But the simple power law dependence can be hidden. For if P is the recombination probability per unit volume per unit time and Δn is the concentration of added electrons,

$$n = n_{eq} + \Delta n, \quad p = p_{eq} + \Delta n$$

then

$$\frac{1}{\tau_n} = \frac{P}{n - n_{eq}} = B_1 \frac{n(np - n_{eq}p_{eq})}{n - n_{eq}} = B_1(n_{eq} + \Delta n)(n_{eq} + p_{eq} + \Delta n)$$

$$\frac{1}{\tau_p} = B_2(p_{eq} + \Delta n)(n_{eq} + p_{eq} + \Delta n)$$

so that the total Auger lifetime is given by

$$\frac{1}{\tau} = \frac{1}{\tau_n} + \frac{1}{\tau_p} = [B_1(n_{eq} + \Delta n) + B_2(p_{eq} + \Delta n)](n_{eq} + p_{eq} + \Delta n) \qquad (3.8.18)$$

Table 3.8.2. *Conditions for the reduction of eq. (3.8.18) to eq. (3.8.19)*

	C	n_1
$n_{eq} \gg p_{eq},\ \Delta n$	B_1	n_{eq}
$p_{eq} \gg n_{eq},\ \Delta n$	B_2	p_{eq}
$\Delta n \gg n_{eq},\ p_{eq}$	$B_1 + B_2$	Δn
$\Delta n \ll n_{eq} = p_{eq} = n_i$	$B_1 + B_2$	n_i

Thus one obtains the form

$$\frac{1}{\tau} = Cn_1^2 \tag{3.8.19}$$

only in the special cases given in Table 3.8.2 [3.8.40].

If there is an activation energy $E_{th} = \theta E_G$ and $n_1 \sim n_i$ one finds

$$\tau \propto e^{(\theta+1)\eta_G} \left[= \exp \frac{2\mu+1}{\mu+1} \eta_G \text{ for the CHCC process} \right] \tag{3.8.20}$$

which is a result we have seen already in Table 3.5.1, line 4, and in eq. (3.7.18).

Although the law (3.8.19) may be hidden, a super-linear rise of the reciprocal lifetime at least *suggests* an Auger effect. An exponential temperature dependence according to relation (3.8.20) similarly no more than hints at an Auger effect. Note that the phonon-assisted effect can reduce the activation energy to zero since the phonon can take up (or supply) energy, so that Auger effects can occur even in the absence of a law (3.8.20).

Further, for intrinsic material in the steady state with generation rate g per unit volume

$$\frac{dn}{dt} = g - B_1\,n(g)^3 = 0, \quad \text{i.e. } n(g) = (g/B_1)^{\frac{1}{3}}$$

This was found via $n \propto g^{\frac{1}{3}}$, $\tau \propto g^{-\frac{2}{3}}$ for the high excitation rates $g \sim 10^{28}$ cm^{-3} s^{-1} already in early work in Si [3.8.41] leading to $B_1 \sim 5 \times 10^{-30}$ cm^6 s^{-1}.

Great interest in Auger effects in heterojunction lasers (and also in light-emitting diodes) arose in the 1980s because of the temperature dependence $I_{th} = I \exp(T/T_0)$ of the threshold current. The deleterious effect of a finite T_0 value can be explained by current leakage over the heterobarrier, optical losses due to parasitic (inter-valence band) absorption, nonradiative recombination at defects (in the bulk or surface) or band–band Auger recombination. The latter effect has proved to supply an important part of the explanation (see Table 3.8.1 and [3.8.18]). A technique using pumping pulses of only ~ 100 ps length and optical transmission

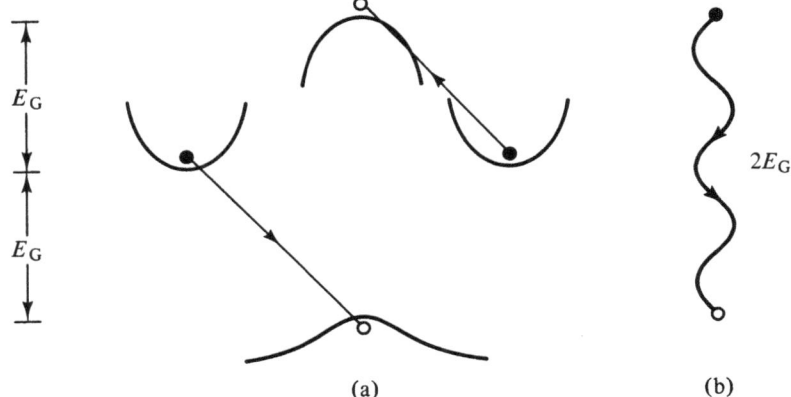

Fig. 3.8.2. A two-electron radiative transition: (a) an Auger process; (b) the radiative step. Full circles are initial states, open circles are final states. The arrows indicate electron transitions.

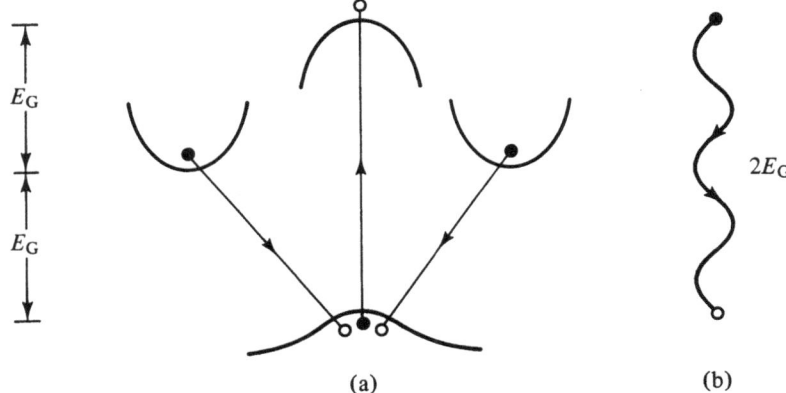

Fig. 3.8.3. A three-electron radiative transition: (a) a three-electron process; (b) the radiative step. Full circles are initial states, open circles are final states. Compare with Fig. 3.8.2. The arrows indicate electron transitions.

as a function of the delay time between the effect of this pulse and a weaker probe pulse has been used in [3.8.18] and elsewhere [3.8.42]. Another ingenious method utilizes the change with bias current of both the spontaneous emission from a light-emitting diode and the differential carrier lifetime. It led to $B_2 \sim 1-2 \times 10^{-29}$ cm^6 s^{-1} for the CHSH process in InGaAsP [3.8.43].

More spectacular and more convincing is the detection of the energetic Auger electron or hole. For this purpose one may look for the weaker luminescence emitted when this carrier recombines radiatively. This has been done for the band–band process in Si [3.8.44] and for the band–impurity process in GaAs [3.8.45]. It leads to the so-called $2E_G$-emission (Fig. 3.8.2). Its rate limiting step is the rate of the Auger process per unit volume, $B_1 n^2 p$, which populates the high energy level. The radiative process then proceeds at a rate $B'_1 n^2 p^2$ per unit volume with $B'_1 \sim 10^{-61}$ cm^9 s^{-1} [3.8.44]. The Auger electron has also been identified from the velocity distribution in GaAs [3.8.46], an experimental approach which does not seem to have been developed or applied further. These effects are rather weak and so the experiments are hard.

A $2E_G$-emission can also be obtained from a three-electron (or second-order) Auger process suggested in [3.8.47] and shown in Fig. 3.8.3. In this case the rate limiting process proceeds itself at a rate per unit volume of Cn^2p^2 with C possibly as large as 2×10^{-48} cm^9 s^{-1} [3.8.48]–[3.8.49] in PbSe with carrier concentration of 10^{18} cm^{-3}. Although no experimental data are to hand, it should be looked for in narrow gap semiconductors.

Appendix: Proof of formula (3.8.11)

We require

$$I = \int_{-A}^{A} \frac{(A^2 - x^2)^{\frac{3}{2}}}{(x+B)^3} dx \quad (B > A)$$

Writing $y = x/B$, $D \equiv A/B < 1$, and then $y = D\cos\theta$,

$$I = B \int_{-D}^{D} \frac{(D^2 - y^2)^{\frac{3}{2}}}{(1+y)^3} dy$$

$$= BD^4 \int_{0}^{\pi} \frac{\sin^4\theta \, d\theta}{(1 + D\cos\theta)^3}$$

$$= BD^4(1+p^2)^3 \int_{0}^{\pi} \frac{\sin^4\theta \, d\theta}{(1+p^2 - 2p\cos\theta)^3}$$

where

$$p \equiv -\frac{1}{D} + (D^{-2} - 1)^{\frac{1}{2}}, \quad \text{i.e. } D = -\frac{2p}{1+p^2} \quad \text{and } p^2 < 1$$

By using the expansion

$$8\sin^4\theta = \cos 4\theta - 4\cos 2\theta + 3$$

the integral reduces to a sum of integrals of the form

$$I(n, m, p) \equiv \int_{0}^{\pi} \frac{\cos nx \, dx}{(1 - 2p\cos x + p^2)^m} = \frac{\pi p^{2m+n-2}}{(1-p^2)^{2m-1}}$$

$$\times \sum_{k=0}^{m-1} \binom{m+n-1}{k} \binom{2m-k-2}{m-1} \left(\frac{1-p^2}{p^2}\right)^k$$

which are, for $p^2 < 1$, given in tables [3.8.50]. In fact,

$$I = \frac{BD^4}{8}(1+p^2)^3 [I(4,3,p) - 4I(2,3,p) + 3I(0,3,p)]$$

$$= \frac{3\pi}{8} BD^4 \frac{(1+p^2)^3}{1-p^2}$$

$$= \frac{3\pi}{2(B^2-A^2)^{\frac{1}{2}}} [B - (B^2-A^2)^{\frac{1}{2}}]^2$$

With $A \to z_1$, $B \to u$, $(B^2-A^2)^{\frac{1}{2}} \to (u^2-z_1^2)^{\frac{1}{2}} \to v$. This is formula (3.8.11).

4

Radiative recombination (mainly for bands)

4.1 Introduction

After a brief review of the statistics of radiative recombination which introduces the transition probability per unit time per unit volume (B_{IJ}) and stimulated and spontaneous emission in section 4.2, these concepts are derived quantum mechanically by second quantization in section 4.3. However, a reader willing to accept eq. (4.4.1) can skip that section. Next, emission rates as well as optical band–band absorption phenomena are discussed in sections 4.4 and 4.5. The simple relationships which exist by virtue of detailed balance between emission and absorption make theory and experiment on absorption relevant to emission problems. The emphasis on absorption depends on the fact that in a semiconductor under normal conditions the conduction band states are occupied with a probability less than unity, typically following a Boltzmann distribution exp $(-E/kT)$. There are therefore comparatively few initial states for radiative emission from a band, and their number drops rapidly as E increases. In absorption these states are all available and one can also sample higher lying states where band structure effects, e.g. nonparabolicity, may play a part. The importance of absorption is due to the fact that it is often experimentally more accessible than emission studies.

Both emission and absorption are affected by the crucial difference between direct and indirect semiconductors. If one writes the transition rate per unit volume as $R = Bnp$, then for the case of Fig. 4.1.1 R is dominated by the electrons in the direct minimum, while in Fig. 4.1.2 R is dominated by the electrons in the indirect minimum. Yet for the same concentrations n and p, the direct material offers more radiative recombination because B is greater. For example in the case of Ge it was

312

$$E_c(0) < E_c(k_1)$$

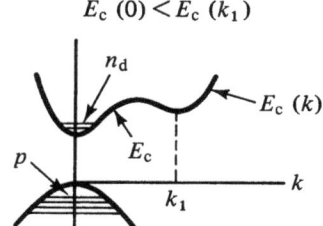

Fig. 4.1.1. A direct semiconductor.

$$E_c(0) > E_c(k_1)$$

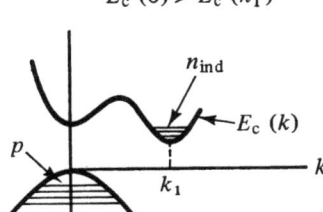

Fig. 4.1.2. An indirect semiconductor.

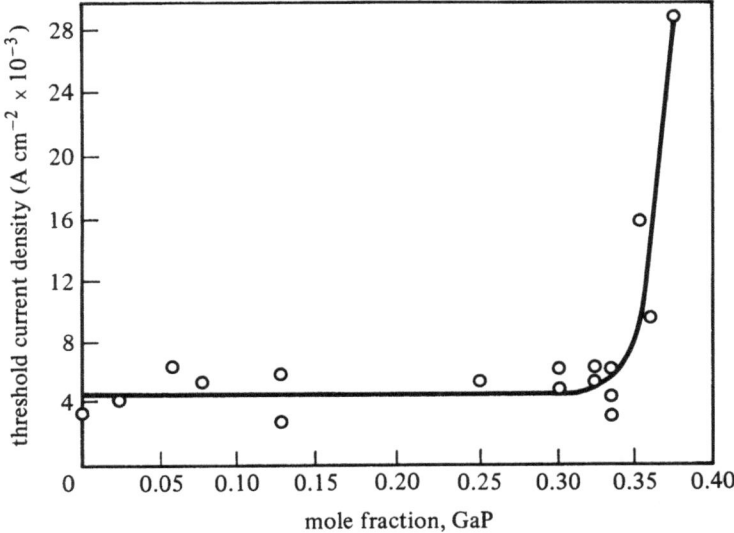

Fig. 4.1.3. Lowest values of laser threshold current density at 77 K as a function of mole fraction of GaP for $Ga(As_{1-x}P_x)$ junction lasers [4.1.2].

found [4.1.1] by Haynes and Nilsson (shortly before Haynes's premature death) that

$$B_{direct} \sim 10^{-8} \ cm^3 \ s^{-1}, \quad B_{indirect} \sim 10^{-15} \ cm^3 \ s^{-1}$$

Thus the direct radiative transitions are more probable by the tremendous factor of 10^7 than the indirect ones; it is only the low population of the higher direct

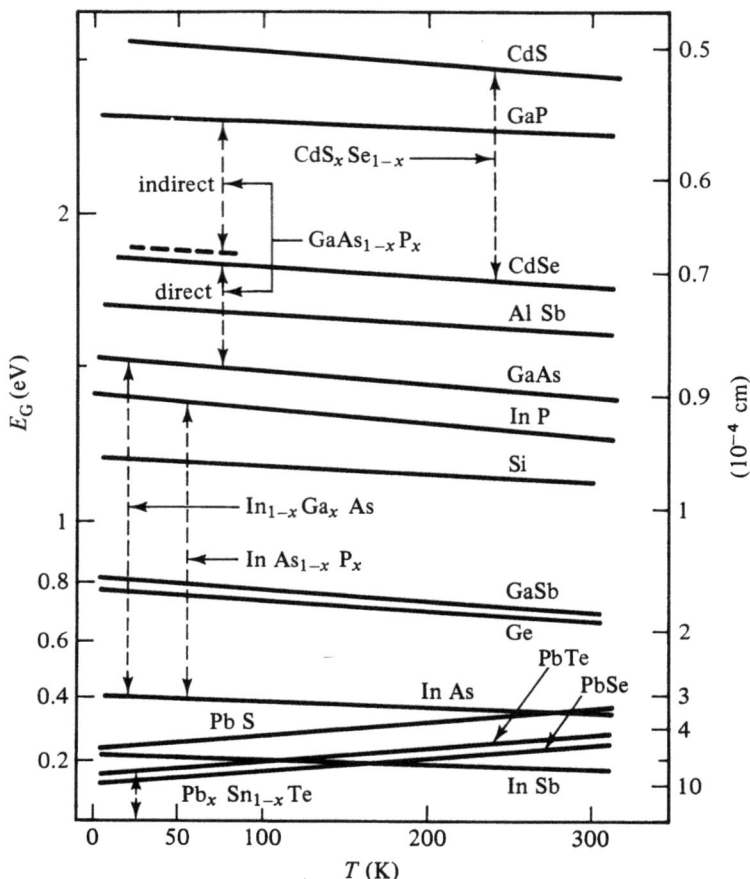

Fig. 4.1.4. Energy gaps and corresponding wavelengths as a function of temperature for various materials [4.1.3].

minimum in Ge which makes them unimportant. The principle of dialectics which bids you look for the unity of opposites pays handsome rewards in the form of certain compounds such as $GaAs_{1-x}P_x$ which is direct for $x = 0$ and indirect with a larger gap for $x = 1$. Forward biased p–n junctions based on GaP emit at about 5500 Å (in the green). If P is added to GaAs single crystals of increasing x-values are formed, and one can study the laser threshold current densities for these specimens (Fig. 4.1.3). It increases rapidly above $x \sim 0.35$ when the material becomes indirect, which is detrimental to laser action. A number of energy gaps are shown for general orientation in Fig. 4.1.4.

Not only for lasers but also for light-emitting diodes, photosensors and the like, good radiative transition rates are desirable. Some numerical values are given in Table 4.1.1 together with estimated radiative lifetimes τ_r. They are much larger

Table 4.1.1. *Recombination coefficients, and lifetimes derived from them, for radiative band–band transitions* [4.1.4]

Direct materials

	GaAs	GaSb	InSb
B (cm^3 s^{-1})	7.2×10^{-10}	2.4×10^{-10}	4.6×10^{-11}
τ_r (s) (for carrier concentration of 10^{18} cm^{-3})	1.3×10^{-9}	4.2×10^{-9}	2.2×10^{-8}
E_G (eV) at 300 K	1.435	0.72	0.18

Indirect materials

	GaP	Si	Ge
B (cm^3 s^{-1})	5.4×10^{-14}	1.8×10^{-15}	5.3×10^{-14}
τ_r (s) (for carrier concentration of 10^{18} cm^{-3})	1.9×10^{-5}	5.6×10^{-4}	1.9×10^{-5}
E_G (eV) at 300 K	2.26	1.12	0.66

than the actually measured lifetimes since other recombination mechanisms tend to dominate.

Holonyak and his co-workers at Urbana also showed that $GaAs_{1-x}P_x$ exhibits interesting temperature effects near the direct–indirect transition. An initially low resistance material (300 K) first gains and later loses resistance as the crystal is cooled. The idea of the explanation is that an indirect donor level at energy E_{Di} (Fig. 4.1.5) plays a part [4.1.5]. The indirect minimum rises above the Fermi level on cooling and discharges its electrons into the indirect donor level so that the resistance increases (150 K). Further cooling raises this level itself above the Fermi level so that its electrons pass to the direct minimum, thus lowering the resistance and providing good radiative recombination. This discussion relies on the indirect gap increasing more rapidly than the direct gap as the temperature is lowered.

The proper treatment of radiative transitions requires the use of determinantal wavefunctions and the matrix element between such functions must then be reduced to matrix elements between one-electron functions. This is done explicitly for example by Johnson [4.1.6]. In our case we can simply appeal to eq. (3.3.17) and proceed on that basis.

4.2 A review of the statistical formalism for single-mode radiative transitions

Consider first a two-level system in single-mode operation. In this picture one has an initial quantum state of an electronic system and a final state. The system can, of course, be an atomic one, but this possibility will be neglected as our main

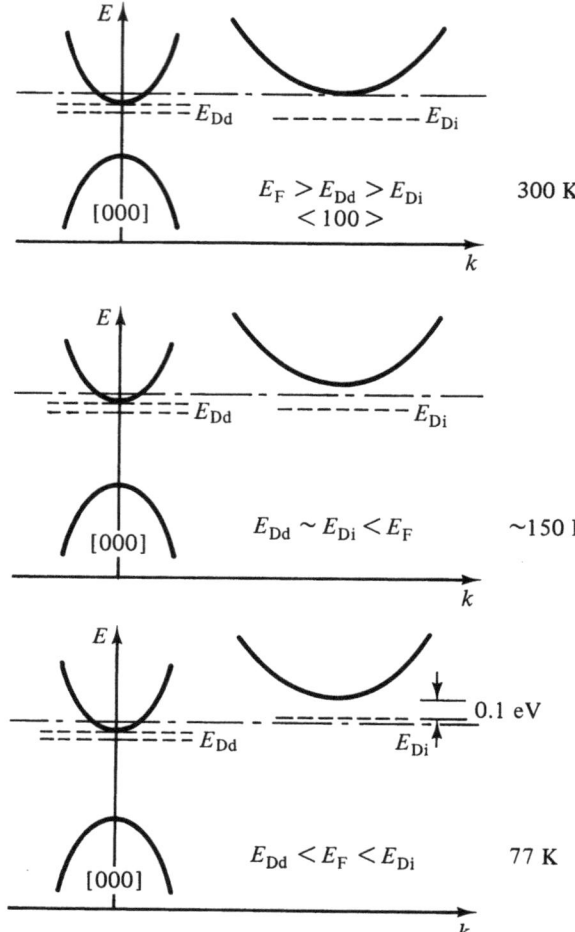

Fig. 4.1.5. Schematic E–k diagram of $GaAs_{1-x}P_x$ showing the relative positions of the direct and indirect conduction band minima as a function of temperature. The dot–dash lines indicate Fermi level positions [4.1.4].

interest is in semiconductors. The energy difference $E_I - E_J > 0$ is assumed to be matched exactly to the frequency of the single radiation mode. We shall later relax this assumption. Then quantum mechanical perturbation theory shows that the presence of the radiation leads to transition probabilities per unit time per unit volume between the electronic states of the form

$$B_{JI}^{SM}(E_k)\, N(E_k),\quad B_{IJ}^{SM}(E_k)\, N(E_k),\quad B_{IJ}^{SM}(E_k) \qquad (4.2.1)$$

They represent in turn: absorption, stimulated emission and spontaneous emission. $N(E_k)$ is the mean number of photons in the mode considered, and the coefficient B has a superfix SM to note that it applies to single-mode operation. The initial state is represented by the first suffix, so that the first term refers to a gain in the energy of the electronic system. In fact, as will be seen below,

$$B_{IJ}^{SM} = B_{JI}^{SM} \tag{4.2.2}$$

The transition probabilities given above presume that the initial electronic state is occupied and that the final state is empty. In fact, there normally exists a probability p_I of the electronic state I being occupied, and therefore a probability $1-p_I$ of the state being empty. With this notation the three transition rates per unit volume which correspond to eq. (4.2.2) are respectively

$$\{u_{IJ}^{abs}, u_{IJ}^{st}, u_{IJ}^{sp}\}^{SM}$$
$$= B_{IJ}^{SM}(E_k)\{p_J(1-p_I)N(E_k), p_I(1-p_J)N(E_k), p_I(1-p_J)\} \tag{4.2.3}$$

Let us write the condition for the steady state of the photon number N, which is of course excitation-dependent in the form

$$aN = bN + 1 \tag{4.2.4}$$

where $b = 1$, but we shall also allow $b = 0$ for purposes of exposition. From eq. (4.2.3)

$$a = (B_{JI}/B_{IJ})[p_J/(1-p_J)][(1-p_I)/p_I] \tag{4.2.5}$$

and hence

$$N = 1/(a-b) \tag{4.2.6}$$

This harmless looking equation can be made to yield exciting results.

(i) *Classical, Boltzmann era.* Put $\eta_I - \eta_J = h\nu/kT$ and we have in equilibrium, denoted by suffix 0,

$$b = 0, \quad p_{I0} = \exp(\gamma_0 - \eta_I), \quad 1 - p_{I0} \simeq 1 \quad \text{(for all } I) \tag{4.2.7}$$

Hence

$$a = (B_{JI}/B_{IJ})\exp(\eta_I - \eta_J) \tag{4.2.8}$$

This pre-Einstein classical position leads via eq. (4.2.6) to

$$N_0 = (B_{IJ}/B_{JI})\exp(-h\nu/kT) \tag{4.2.9}$$

As $T \to \infty$ we do not have the expected $N_0 \to \infty$, but

$$N_0 \to B_{IJ}/B_{JI} = \text{finite} \tag{4.2.10}$$

which was regarded by Einstein as unphysical and led to his introduction of stimulated emission.

(ii) *Classical, Einstein era.* We adopt now eqs. (4.2.7) with $b = 1$. This gives via eq. (4.2.6)

$$N_0 = \{(B_{JI}/B_{IJ})\exp(h\nu/kT) - 1\}^{-1} \qquad (4.2.11)$$

One finds now that $N_0 \to \infty$ as $T \to \infty$, as one would expect, only if $B_{IJ} = B_{JI}$ so that

$$N_0 = \{\exp(h\nu/kT) - 1\}^{-1}, \; B_{IJ} = B_{JI} \qquad (4.2.12)$$

The Bose–Einstein ' -1 ' has thus its origin in $b = 1$, eq. (4.2.6), i.e. in the stimulated emission. One learns from it that a radiation field enhances radiative emission, an effect of crucial importance in lasers and masers.

(iii) *Quantum, Einstein era.* The above holds for electron populations in sets of levels I and J and they are now known to be subject to Fermi–Dirac probabilities. How is it that the Einstein result $b = 1$, or eq. (4.2.12), of 1916 is correct even though the Fermi–Dirac distribution came only in 1926? This is due to the 'play of probabilities' which assigns to

$$(1 - p_{I0})/p_{I0} \text{ the expression } \exp(\eta_I - \gamma_0) \qquad (4.2.13)$$

in the Maxwell–Boltzmann cases (i) and (ii), *as well as* in the case of the Fermi–Dirac distribution. It is a kind of mathematical accident.

(iv) *Post-Einstein era.* One can use eq. (4.2.6) to cover recent work. Suppose the system receives energy at a constant rate which causes increased absorption and also increased emission. Then the situation can be arranged to be a nonequilibrium steady state, in which the occupation probabilities p_I, p_J remain constant in time. One can again use eq. (4.2.7), but without the suffices '0' as one is away from equilibrium. Thus detailed balance cannot be used. Replace the equilibrium Fermi level by two quasi-Fermi levels $(\mu_I, \mu_J) \equiv kT(\gamma_I, \gamma_J)$ for states I and J, respectively. Then eq. (4.2.11) becomes by a kind of 'play of probabilities'

$$N_{\text{st.st.}} = \left[\frac{B_{JI}}{B_{IJ}}\frac{p_J}{1-p_J}\frac{1-p_I}{p_I} - 1\right]^{-1} = \left[\frac{B_{JI}}{B_{IJ}}\exp\left(\frac{h\nu}{kT} - (\gamma_I - \gamma_J)\right) - 1\right]^{-1} \qquad (4.2.14)$$

One thus discovers that the quasi-Fermi levels of the electronic system impose their difference, as a kind of photon electrochemical potential, on the steady-state photon distribution. We need again $B_{IJ} = B_{JI}$. In equilibrium

$$\gamma_I = \gamma_J = \gamma_0 \qquad (4.2.15)$$

and eq. (4.2.14) reduces to the Planck formula (4.2.12). The result (4.2.14) depends on two assumptions: the existence of a steady state and of two quasi-Fermi levels for the electronic system [4.2.1].

In a *p–n* junction the two bulk materials several diffusion lengths away from the junction are approximately in equilibrium even if a modest current is flowing. On one side one has then just one quasi-Fermi level, say μ_I, and on the other side one has just one quasi-Fermi level, say μ_J. Then the difference, as already noted in section 2.2.2,

$$\mu_I - \mu_J = q\varphi \qquad (4.2.16)$$

corresponds to the applied voltage φ. This is in agreement with eq. (4.2.15), in that $\varphi = 0$ implies no current and hence thermal equilibrium is possible.

Statistical mechanics teaches one the rule that, *in equilibrium*, occupation probabilities of individual quantum states are always less for states of higher energies. This ensures that the total energy of a quantum system converges, the occupation probability p being a kind of convergence factor. Assuming, on the contrary, that

$$p = \text{constant for all energies } E_I$$

then the total energy of the system is

$$\sum_I E_I p = p \sum_I E_I \to \infty$$

It diverges if energies $E_I \to \infty$ as $I \to \infty$. However, if one is away from equilibrium, the above rule can be suspended and one can have population inversion. In the present notation the condition for it is from eq. (4.2.14) [4.2.2]

$$h\nu = E_I - E_J < \mu_I - \mu_J = q\varphi \qquad (4.2.17)$$

If $h\nu$ is the energy gap, eq. (4.2.17) implies that at least one band must be degenerate. Laser action in indirect materials was also considered early in the 1960s [4.2.3]–[4.2.4], including direct gap laser action in indirect materials [4.2.5] and related general questions [4.2.6]. In particular, if several groups of states $l = 1, 3, 5, \ldots$ (say) can lose electrons and several groups of states can by gaining electrons be converted into states $g = 2, 4, \ldots$ (say), then the condition (4.2.17) for negative absorption generalizes [4.2.7] to

$$h\nu < \sum_l \mu_l - \sum_g \mu_g$$

This result holds if phonon or impurity levels participate. For reviews of the optoelectronic aspects of this work see [4.2.8]–[4.2.11].

Using eqs. (4.2.14) and (4.2.2), one finds

$$N_{\text{st.st.}} = \left[\exp \frac{h\nu - q\varphi}{kT} - 1 \right]^{-1} \qquad (4.2.18)$$

which becomes negative (and therefore meaningless) for population inversion. Even for $h\nu = q\varphi$ there is trouble because the steady-state photon occupation diverges. This does not correspond to a 'death ray', but is the result of imperfect modeling; for example, the leakage of photons from the cavity has been neglected. This problem was analyzed in [4.2.12]. The formula (4.2.18) arises also in connection with solar cells and has been discussed in [4.2.13]–[4.2.15].

4.3 The quantum mechanics of radiative transitions

4.3.1 Introduction

The purpose of this section is to derive the result (4.3.59) (or (4.4.1)) which is basic to the quantum theory of radiative transitions but which requires a certain amount of formalism for its derivation. This section can therefore be omitted by readers willing to assume the validity of the result (4.4.1).

It is convenient to start the discussion of the interaction of matter and radiation by a semiclassical procedure which uses quantum mechanical perturbation theory for the electronic transition associated with the emission or absorption of radiation. It uses, however, a classical description of the radiation field. This merely enters through the electromagnetic vector potential $\mathbf{A}$ in the Hamiltonian

$$
\begin{aligned}
H_{\text{s.c.}} &= \frac{1}{2m}\left[\mathbf{P} - \frac{e}{c}\mathbf{A}(\mathbf{r},t)\right]^2 + e\varphi(\mathbf{r},t) + U(\mathbf{r},t) \\
&= \frac{1}{2m}P^2 - \frac{e}{2mc}(\mathbf{A}\cdot\mathbf{P} + \mathbf{P}\cdot\mathbf{A}) + e\varphi + U
\end{aligned}
\tag{4.3.1}
$$

where m is the mass of the electron, P is its momentum operator, φ is the scalar electrostatic potential and U is the electron potential energy. In eq. (4.3.1) the term in c^{-2} has been omitted. This procedure yields the correct form of the absorption and stimulated emission probabilities. It does not, however, explain the occurrence of spontaneous emission. To understand this phenomenon in this theory, one has to rely on a variant of Einstein's original statistical argument which leads to eq. (4.2.12) [4.3.1], [4.3.2].

In the correct treatment [4.3.3], the electromagnetic field is quantized and it occurs, as H_t say, in second-quantized form in the Hamiltonian

$$
H = H_{\text{s.c.}} + H_t
\tag{4.3.2}
$$

of the system. This enables one to describe the field as a superposition of normal modes, or equivalent oscillators. A quantum mechanical oscillator can emit radiation in the absence of an external field, and hence spontaneous emission occurs on the same footing as absorption or stimulated emission in this theory.

If one goes to higher order than the first in perturbation theory one reaches a domain lying beyond the reach of the Fermi golden rule. In that case one can see the possibility of processes in which any number of photons (and also phonons) can take part. The square of the vector potential in the electron–photon interaction (4.3.1) (which also allows for higher order processes) must then not be neglected. Ten photons or more can, under favorable conditions, be observed to participate in a transition in atomic systems.

4.3.2 First-order time-dependent perturbation theory

Consider a quantum mechanical system with Hamiltonian

$$H = H_0 + \lambda V(r, t) \tag{4.3.3}$$

where H_0 is the unperturbed Hamiltonian, and λ is a constant which measures the strength of the perturbation. The (unperturbed) eigenfunctions of H_0 have the form

$$\psi_I(\mathbf{r}) e^{-iE_I^0 t/\hbar}, \quad H_0 \psi_I = E_I^0 \psi_I, \tag{4.3.4}$$

where the E_I^0 are the eigenvalues, and the ψ_I depends on all coordinates (briefly denoted by $\mathbf{r}$) specifying the system but not on the time t. Since the functions (4.3.4) form a complete orthonormal set, an eigenfunction φ of H can be expanded

$$\varphi = \sum_R C_R(t) \psi_R e^{-iE_R^0 t/\hbar} \tag{4.3.5}$$

Since φ must also satisfy the Schrödinger equation

$$i\hbar \frac{\partial \varphi}{\partial t} = H\varphi \tag{4.3.6}$$

eq. (4.3.5) must be inserted in eq. (4.3.6) to yield

$$\sum_R (i\hbar \dot{C}_R + E_R^0 C_R) \psi_R e^{-iE_R^0 t/\hbar} = \sum_R E_R^0 e^{-iE_R^0 t/\hbar} \psi_R C_R + \lambda \sum_R V C_R \psi_R e^{-iE_R^0 t/\hbar}$$

Multiplying by $\psi_I^* \exp(iE_I^0 t/\hbar)$, integrating over all $\mathbf{r}$ and using the orthonormality of the ψ's,

$$\dot{C}_I = -\frac{i}{\hbar} \lambda \sum_R e^{i(E_I^0 - E_R^0)t/\hbar} V_{IR} C_R \tag{4.3.7}$$

where

$$V_{IR} \equiv \int \psi_I^* V \psi_R \, dr \tag{4.3.8}$$

can depend only on t.

If the unperturbed state is $R = J$ (say), and occurs for $t \leqslant t_0$ before the perturbation is 'switched on', then for short times later only the term $R = J$ will contribute in eq. (4.3.7) when $I \neq J$. All other terms will involve higher corrections. Hence

$$C_I = -\frac{i}{\hbar} \lambda \int_{t_0}^t e^{i(E_I^0 - E_J^0)t/\hbar} V_{IJ} \, dt \quad (I \neq J) \tag{4.3.9}$$

Now use eq. (4.3.7) with $I = J$, and insert eq. (4.3.9). To first order in λ,

$$\dot{C}_J = -\frac{i}{\hbar} \lambda V_{JJ} C_J, \quad \text{i.e. } C_J = \exp\left[-\frac{i}{\hbar} \lambda \int_{t_0}^t V_{JJ} \, dt \right] \tag{4.3.10}$$

Thus one has the expansion coefficients in terms of V.

Suppose now we have a periodic disturbance of angular frequency ω, as would happen for an electromagnetic wave. A Hermitian perturbation has then the form (the dagger denotes the Hermitian conjugate)

$$\lambda V = \sum_{\mathbf{k}} [G^{(\mathbf{k})} e^{-i\omega_{\mathbf{k}} t} + G^{(\mathbf{k})\dagger} e^{i\omega_{\mathbf{k}} t}] \tag{4.3.11}$$

where G is an operator which can involve the coordinates $\mathbf{r}$ as well as the momenta, but not the time. G and ω can depend on the wavevectors $\mathbf{k}$. The time integration can then be carried out in eq. (4.3.9), and one finds

$$-C_I = \sum_{\mathbf{k}} [G^{(\mathbf{k})}_{IJ} F_{IJ}(\omega_{\mathbf{k}}) + (G^{(\mathbf{k})\dagger})_{IJ} F_{IJ}(-\omega_{\mathbf{k}})] \quad (I \neq J)$$

where the matrix element of G is defined as in eq. (4.3.8) and

$$F_{IJ}(\omega_{\mathbf{k}}) \equiv \frac{1 - \exp\left[\dfrac{i}{\hbar}(t_0 - t)(E_I^0 - E_J^0 - \hbar\omega_{\mathbf{k}})\right]}{E_I^0 - E_J^0 - \hbar\omega_{\mathbf{k}}} \exp\left[\dfrac{it}{\hbar}(E_I^0 - E_J^0 - \hbar\omega_{\mathbf{k}})\right] \tag{4.3.12}$$

The probability that a state $I \neq J$ will be found at time $t > t_0$ is given by

$$|C_I|^2 = \sum_{\mathbf{k}} \{|G^{(\mathbf{k})}_{IJ} F_{IJ}(\omega_{\mathbf{k}})|^2 + |(G^{(\mathbf{k})\dagger})_{IJ} F_{IJ}(-\omega_{\mathbf{k}})|^2 + \ldots\} \tag{4.3.13}$$

The dots denote terms for which the exponential factors in eq. (4.3.12) do not cancel. These rapidly fluctuating terms do not contribute significantly. The two terms shown contribute only if $E_I^0 - E_J^0 \pm \hbar\omega_{\mathbf{k}}$ can be small. After purely algebraic manipulations which involve the replacement of

$$\frac{\sin^2 y}{y^2} \text{ by } \pi\delta(y), \tag{4.3.14}$$

see eq. (3.2.15), where $\delta(y)$ is the Dirac delta function, one finds for the transition probability per unit time per unit volume

$$P_{IJ} = \frac{1}{V(t - t_0)}|C_I(t)|^2 = \frac{2\pi}{\hbar V} \sum_{\mathbf{k}} [|G^{(\mathbf{k})}_{IJ}|^2 \delta(E_I^0 - E_J^0 - \hbar\omega_{\mathbf{k}})$$
$$+ |G^{(\mathbf{k})\dagger}_{IJ}|^2 \delta(E_I^0 - E_J^0 - \hbar\omega_{\mathbf{k}})] \quad (I \neq J) \tag{4.3.15}$$

This result is basic to the later work. The first delta function shows that a state of higher energy can be reached by the part of the system which does not include the radiation if a quantum of the appropriate frequency is available for absorption. The sum must be retained in eq. (4.3.15) since a number of wavevectors $\mathbf{k}$ may satisfy the delta function condition.

4.3.3 Fourier analysis of the pure radiation field

In order to obtain the Hamiltonian H_t of a free electromagnetic field, it is desirable to avoid the continuous infinity of generalized coordinates and momenta needed to describe it. This set is made enumerable by assuming the field to be enclosed in a cube of side L, and to require that the values of the vector potential $\mathbf{A}$ and its derivatives have the same values at pairs of opposite points on the surface. This is the familiar periodic boundary condition. The permitted wavevectors $\mathbf{k} = (k_1, k_2, k_3)$ for $\mathbf{A}$ form then a discrete set:

$$k_j = \frac{2\pi n_j}{L} \quad (j = 1, 2, 3; n_j = 0, \pm 1, \pm 2, \ldots) \tag{4.3.16}$$

One can now make a spatial Fourier expansion of $\mathbf{A}$ for any time t, in the form

$$\mathbf{A}(\mathbf{r}, t) = \sum_{\mathbf{k}, \lambda} q_{\mathbf{k}\lambda}(t) \mathbf{A}_{\mathbf{k}\lambda}(\mathbf{r}) \tag{4.3.17}$$

Here $\lambda = 1, 2, 3$ for the three mutually perpendicular directions of polarization. Let $\mathbf{e}_{\mathbf{k}\lambda}$ be a triad of orthonormal real unit vectors for each $\mathbf{k}$ (associated with the directions of polarization). One can then, in accordance with usual Fourier expansions, interpret the $\mathbf{A}_{\mathbf{k}\lambda}$ as plane waves (in both the classical and the quantum treatments):

$$\mathbf{A}_{\mathbf{k}\lambda} = \frac{a_{\mathbf{k}}}{V^{\frac{1}{2}}} e^{i\mathbf{k}\cdot\mathbf{r}} \mathbf{e}_{\mathbf{k}\lambda} \tag{4.3.18}$$

so that $\mathbf{A}_{\mathbf{k}\lambda}^* = \mathbf{A}_{-\mathbf{k}\lambda}$, $a_{\mathbf{k}} = a_{-\mathbf{k}}$, $\mathbf{e}_{\mathbf{k}\lambda} = \mathbf{e}_{-\mathbf{k}\lambda}$, where $a_{\mathbf{k}}$ is a real normalization constant and $V = L^3$. The k-sum in eq. (4.3.17) goes over the whole of k-space. Since $\mathbf{A}$ is real (classically) or a Hermitian operator (quantum mechanically) one has from eqs. (4.3.17) and (4.3.18)

$$\mathbf{A} - \mathbf{A}^{\ddagger} = 0 = \sum_{\mathbf{k}, \lambda} (q_{\mathbf{k}\lambda} \mathbf{A}_{\mathbf{k}\lambda} - q_{\mathbf{k}\lambda}^{\ddagger} \mathbf{A}_{\mathbf{k}\lambda}^*) = \sum_{\mathbf{k}, \lambda} (q_{\mathbf{k}\lambda} - q_{-\mathbf{k}\lambda}^{\ddagger}) \mathbf{A}_{\mathbf{k}\lambda}$$

While the asterisk indicates the complex conjugate of an ordinary function, the dagger can here serve the double purpose of indicating the complex conjugate in the classical theory and a Hermitian conjugate operator in its quantum mechanical analogue. The reality condition therefore is

$$q_{\mathbf{k}\lambda}^{\ddagger} = q_{-\mathbf{k}\lambda}$$

so that

$$\dot{q}_{\mathbf{k}\lambda}^{\ddagger} = \dot{q}_{-\mathbf{k}\lambda} \tag{4.3.17'}$$

By (4.3.18) the $\mathbf{A}_{\mathbf{k}\lambda}$ are orthogonal:

$$\int_V \mathbf{A}_{\mathbf{k}\lambda}^* \cdot \mathbf{A}_{\mathbf{k}'\lambda'} \, d\mathbf{r} = a_{\mathbf{k}}^2 \delta_{\mathbf{k}, \mathbf{k}'} \delta_{\lambda, \lambda'} \text{ or } \int_V \mathbf{A}_{\mathbf{k}\lambda} \cdot \mathbf{A}_{\mathbf{k}'\lambda'} \, d\mathbf{r} = a_{\mathbf{k}}^2 \delta_{\mathbf{k}', -\mathbf{k}} \delta_{\lambda, \lambda'} \tag{4.3.19}$$

The magnetic field is

$$\mathbf{H} = \text{curl } \mathbf{A} = i \sum_{\mathbf{k}, \lambda} q_{\mathbf{k}\lambda} (\mathbf{k} \times \mathbf{A}_{\mathbf{k}\lambda}) \tag{4.3.20}$$

For longitudinally polarized waves let us choose $\lambda = 3$, so that

$$\mathbf{e}_{\mathbf{k}3} = \mathbf{k}/|\mathbf{k}| \tag{4.3.21}$$

Then eq. (4.3.20) shows that these waves do not contribute to the magnetic field, and one can omit the value $\lambda = 3$ in eq. (4.3.20). By eq. (4.3.18) this convention also means that div $\mathbf{A}_{\mathbf{k}\lambda}$ is proportional to $\mathbf{k} \cdot \mathbf{e}_{\mathbf{k}\lambda}$ and so vanishes for $\lambda = 1, 2$. Hence for the transverse part of the field

$$\mathbf{A}_{\text{tr}} \equiv \sum_{\substack{\mathbf{k} \\ \lambda=1,2}} q_{\mathbf{k}\lambda} \mathbf{A}_{\mathbf{k}\lambda}, \quad \text{div } \mathbf{A}_{\text{tr}} = 0 \tag{4.3.22}$$

It is convenient to split the electric field into transverse and longitudinal components. These satisfy

$$\mathbf{E}_{tr} = -\frac{1}{c}\frac{\partial \mathbf{A}_{tr}}{\partial t}, \quad \mathbf{E}_{lg} = -\frac{1}{c}\frac{\partial \mathbf{A}_{lg}}{\partial t} - \nabla \varphi$$

φ being the scalar potential, and together they yield Maxwell's equation

$$\mathbf{E} = \mathbf{E}_{tr} + \mathbf{E}_{lg} = -\frac{1}{c}\frac{\partial \mathbf{A}}{\partial t} - \nabla \varphi \tag{4.3.23}$$

From eqs. (4.3.17) and (4.3.22)

$$\mathbf{E}_{tr} = -\frac{1}{c}\sum_{\substack{\mathbf{k}\\ \lambda=1,2}} \dot{q}_{\mathbf{k}\lambda}\mathbf{A}_{\mathbf{k}\lambda} \tag{4.3.24}$$

A free field, sometimes called a pure radiation field, such as occurs for light waves, does not involve electric charges and longitudinal components are absent. For such a field the energy is

$$W = \frac{1}{8\pi}\int_V (H^2 + E_{tr}^2)\,\mathrm{d}\mathbf{r} = \frac{1}{8\pi}\int_V (\mathbf{H}^{\ddagger}\cdot\mathbf{H} + \mathbf{E}^{\ddagger}\cdot\mathbf{E})\,\mathrm{d}\mathbf{r} \tag{4.3.25}$$

The second form is more convenient for the quantum mechanical case. This quantity will be evaluated in order to determine restrictions on $a_{\mathbf{k}}$.

The first integral is, if $\mathbf{k}\times\mathbf{e}_{\mathbf{k}1} = |\mathbf{k}|\,\mathbf{e}_{\mathbf{k}2}$ and $\mathbf{k}\times\mathbf{e}_{\mathbf{k}2} = -|\mathbf{k}|\,\mathbf{e}_{\mathbf{k}1}$,

$$\frac{1}{8\pi}\sum_{\substack{\mathbf{k}\mathbf{k}'\\ \lambda\lambda'=1,2}}\int_V q_{\mathbf{k}'\lambda'}^{\dagger}\,q_{\mathbf{k}\lambda}(\mathbf{k}'\times\mathbf{A}_{\mathbf{k}'\lambda'}^{*})\cdot(\mathbf{k}\times\mathbf{A}_{\mathbf{k}\lambda})\,\mathrm{d}\mathbf{r}$$

$$= \frac{1}{8\pi}\sum_{\mathbf{k}\mathbf{k}'\lambda} q_{\mathbf{k}'\lambda}^{\dagger}\,q_{\mathbf{k}\lambda}|\mathbf{k}||\mathbf{k}'|\int_V \mathbf{A}_{\mathbf{k}'\lambda}^{*}\cdot\mathbf{A}_{\mathbf{k}\lambda}\,\mathrm{d}\mathbf{r} \tag{4.3.26}$$

$$= \frac{1}{8\pi}\sum_{\mathbf{k},\lambda=1,2} a_{\mathbf{k}}^2|\mathbf{k}|^2 q_{\mathbf{k}\lambda}^{\dagger}\,q_{\mathbf{k}\lambda}$$

The second integral is

$$\frac{1}{8\pi c^2}\sum_{\mathbf{k},\lambda=1,2} a_{\mathbf{k}}^2 \dot{q}_{\mathbf{k}\lambda}^{\dagger}\,\dot{q}_{\mathbf{k}\lambda} \tag{4.3.27}$$

This shows that the integrals are easily evaluated.

4.3.4 The free field as a collection of oscillators

Classically eq. (4.3.25) yields

$$W = \frac{1}{8\pi c^2}\sum_{\substack{\mathbf{k}\\ \lambda=1,2}} a_{\mathbf{k}}^2(\dot{q}_{\mathbf{k}\lambda}^{\dagger}\,\dot{q}_{\mathbf{k}\lambda} + |kc|^2 q_{\mathbf{k}\lambda}^{\dagger}\,q_{\mathbf{k}\lambda}) \tag{4.3.28}$$

The energy of a classical harmonic oscillator of mass m and angular frequency ω is

$$\frac{m}{2}(\dot{q}^2 + \omega^2 q^2) = \frac{m}{2}\left[\left(\frac{p}{m}\right)^2 + \omega^2 q^2\right] \tag{4.3.29}$$

This suggests that the energy of the field can be thought of as residing in an enumerable infinity of noninteracting one-dimensional harmonic oscillators with angular frequency

$$\omega_{\mathbf{k}} = \omega_{-\mathbf{k}} = |\mathbf{k}|c \tag{4.3.30}$$

To test this idea we try a Hamiltonian $H_t(p_1, p_2, \ldots, q_1, q_2, \ldots)$ for the transverse part of the field suggested by eqs. (4.3.28) and (4.3.29):

$$H_t = \tfrac{1}{2} \sum_{\substack{\mathbf{k} \\ \lambda-1,2}} A_{\mathbf{k}}[B_{\mathbf{k}}^2 p_{\mathbf{k}\lambda}^{\dagger} p_{\mathbf{k}\lambda} + \omega_{\mathbf{k}}^2 q_{\mathbf{k}\lambda}^{\dagger} q_{\mathbf{k}\lambda}] \tag{4.3.31}$$

$$q_{\mathbf{k}\lambda}^{\dagger} = q_{-\mathbf{k}\lambda}, \quad p_{\mathbf{k}\lambda}^{\dagger} = p_{-\mathbf{k}\lambda}$$

where $A_{\mathbf{k}} = A_{-\mathbf{k}}$ is real and has the dimension of mass, $B = B_{-\mathbf{k}}$ and $A_{\mathbf{k}} B_{\mathbf{k}}$ is a number. In fact, from eq. (4.3.28)

$$A_{\mathbf{k}} \equiv \frac{\mu^2 a_{\mathbf{k}}^2}{4\pi c^2}, \quad \dot{q}_{\mathbf{k}\lambda} \equiv B_{\mathbf{k}} p_{\mathbf{k}\lambda}^{\dagger} \tag{4.3.32}$$

where we have allowed for a refractive index μ of the (nonmagnetic) medium containing the radiation.

Treating eq. (4.3.31) classically, one can use the Hamiltonian equations of motion

$$\dot{q}_{\mathbf{k}\lambda} = \frac{\partial H_t}{\partial p_{\mathbf{k}\lambda}} = A_{\mathbf{k}} B_{\mathbf{k}}^2 p_{\mathbf{k}\lambda}^{\dagger}, \quad \dot{p}_{\mathbf{k}\lambda} = -\frac{\partial H_t}{\partial q_{\mathbf{k}\lambda}} = -A_{\mathbf{k}} \omega_{\mathbf{k}}^2 q_{\mathbf{k}\lambda}^{\dagger} \tag{4.3.33}$$

By eqs. (4.3.32) and (4.3.33) one can eliminate $B_{\mathbf{k}}$ since it must satisfy

$$A_{\mathbf{k}} B_{\mathbf{k}} = 1$$

for consistency. Relations (4.3.33) also imply

$$\ddot{q}_{\mathbf{k}\lambda} = B_{\mathbf{k}} \dot{p}_{\mathbf{k}\lambda}^{\dagger} = -\omega_{\mathbf{k}}^2 q_{\mathbf{k}\lambda} \tag{4.3.34}$$

confirming that eq. (4.3.31) is an oscillator Hamiltonian.

Quantum mechanically, one treats the p's and q's in eq. (4.3.31) as operators subject to the commutation rules

$$\left.\begin{array}{l} \mathrm{i}[p_{\mathbf{k}\lambda}, q_{\mathbf{k}'\lambda'}] = \hbar\delta_{\mathbf{k},\mathbf{k}'}\,\delta_{\lambda,\lambda'} \\ \mathrm{i}[p_{\mathbf{k}\lambda}, p_{\mathbf{k}'\lambda'}] = \mathrm{i}[q_{\mathbf{k}\lambda}, q_{\mathbf{k}'\lambda'}] = 0 \end{array}\right\} \tag{4.3.35}$$

By the Heisenberg equations of motion (see eq. (D.15))

$$\left.\begin{array}{l} \dot{q}_{\mathbf{k}\lambda} = \dfrac{\mathrm{i}}{\hbar}[H_t, q_{\mathbf{k}\lambda}] = p_{\mathbf{k}\lambda}^{\dagger}/A_{\mathbf{k}} \\[3mm] \dot{p}_{\mathbf{k}\lambda} = \dfrac{\mathrm{i}}{\hbar}[H_t, p_{\mathbf{k}\lambda}] = -A_{\mathbf{k}}\omega_{\mathbf{k}}^2 q_{\mathbf{k}\lambda}^{\dagger} \end{array}\right\} \tag{4.3.36}$$

Also, using eqs. (4.3.36)

$$\ddot{q}_{\mathbf{k}\lambda} = \frac{i}{\hbar}[H_{\mathfrak{f}}, \dot{q}_{\mathbf{k}\lambda}] = \frac{i}{A_{\mathbf{k}}\hbar}[H_{\mathfrak{f}}, p_{\mathbf{k}\lambda}^{\dagger}] = \frac{1}{A_{\mathbf{k}}}\dot{p}_{\mathbf{k}\lambda}^{\dagger} = -\omega_{\mathbf{k}}^{2}q_{\mathbf{k}\lambda} \qquad (4.3.37)$$

This is the exact analogue of the classical result embodied in eqs. (4.3.33) and (4.3.34). In fact eqs. (4.3.34) and (4.3.37) suggest that dimensionless quantities $b_{\mathbf{k}\lambda}$, $b_{\mathbf{k}\lambda}'$ exist such that

$$q_{\mathbf{k}\lambda} = [b_{\mathbf{k}\lambda}\,\mathrm{e}^{-\mathrm{i}\omega_{\mathbf{k}}t} + b_{\mathbf{k}\lambda}'^{\dagger}\,\mathrm{e}^{\mathrm{i}\omega_{\mathbf{k}}t}]\,Q_{\mathbf{k}}, \quad Q_{\mathbf{k}} = Q_{-\mathbf{k}} = \mathrm{real} \qquad (4.3.38)$$

Equation (4.3.17') enables one to eliminate the b' by virtue of

$$b_{\mathbf{k}\lambda}' = b_{-\mathbf{k}\lambda} \qquad (4.3.39)$$

$Q_{\mathbf{k}}$, like $A_{\mathbf{k}}$ and $B_{\mathbf{k}}$ above, carries the appropriate dimension. Classically the b's are numbers, quantum mechanically they are operators.

If one applies eq. (4.3.17') to eqs. (4.3.33) or (4.3.36), one also finds an analogous result for the p's:

$$p_{\mathbf{k}\lambda}^{\dagger} = p_{-\mathbf{k}\lambda} \qquad (4.3.40)$$

We now interpret the operators b in terms of the generalized coordinates and momenta and hence determine their commutation relations. By eqs. (4.3.36) and (4.3.38),

$$p_{\mathbf{k}\lambda}^{\dagger} = -\mathrm{i}\omega_{\mathbf{k}}A_{\mathbf{k}}Q_{\mathbf{k}}[b_{\mathbf{k}\lambda}\,\mathrm{e}^{-\mathrm{i}\omega_{\mathbf{k}}t} - b_{-\mathbf{k}\lambda}^{\dagger}\,\mathrm{e}^{\mathrm{i}\omega_{\mathbf{k}}t}] \qquad (4.3.41)$$

Eliminating $b_{-\mathbf{k}\lambda}$ from eqs. (4.3.28) and (4.3.41),

$$b_{\mathbf{k}\lambda} = \frac{1}{2Q_{\mathbf{k}}}\left[q_{\mathbf{k}\lambda} + \frac{i}{\omega_{\mathbf{k}}A_{\mathbf{k}}}p_{\mathbf{k}\lambda}^{\dagger}\right]\mathrm{e}^{\mathrm{i}\omega_{\mathbf{k}}t} \qquad (4.3.42)$$

An analogous relation for $b_{\mathbf{k}\lambda}^{\dagger}$ is obtainable similarly, but is in fact already implied by eq. (4.3.42). From eqs. (4.3.35) and (4.3.42) one now finds

$$[b_{\mathbf{k}\lambda}, b_{\mathbf{k}'\lambda'}^{\dagger}] = \frac{\hbar}{2\omega_{\mathbf{k}}A_{\mathbf{k}}Q_{\mathbf{k}}^{2}}\delta_{\mathbf{k},\mathbf{k}'}\delta_{\lambda,\lambda'} \qquad (4.3.43)$$

$$[b_{\mathbf{k}\lambda}, b_{\mathbf{k}'\lambda'}] = [b_{\mathbf{k}\lambda}^{\dagger}, b_{\mathbf{k}'\lambda'}^{\dagger}] = 0$$

By virtue of eqs. (4.3.38) and (4.3.41) the Hamiltonian $H_{\mathfrak{f}}$ of eq. (4.3.31) can be expressed entirely in terms of the operators b. Utilizing eq. (4.3.43) one finds

$$H_{\mathfrak{f}} = \sum_{\substack{\mathbf{k}\\ \lambda=1,2}} \{(2\omega_{\mathbf{k}}A_{\mathbf{k}}Q_{\mathbf{k}}^{2}/\hbar)\,b_{\mathbf{k}\lambda}^{\dagger}b_{\mathbf{k}\lambda} + \tfrac{1}{2}\}\hbar\omega_{\mathbf{k}} \qquad (4.3.44)$$

It is convenient to define the $Q_{\mathbf{k}}$ now by

$$2\omega_{\mathbf{k}}A_{\mathbf{k}}Q_{\mathbf{k}}^{2} = \hbar, \quad \text{i.e. } a_{\mathbf{k}}Q_{\mathbf{k}} = (2\pi\hbar/\omega_{\mathbf{k}})^{\frac{1}{2}}(c/\mu) \qquad (4.3.45)$$

so that finally

$$H_{\mathfrak{f}} = \sum_{\substack{\mathbf{k}\\ \lambda=1,2}} (b_{\mathbf{k}\lambda}^{\dagger}b_{\mathbf{k}\lambda} + \tfrac{1}{2})\hbar\omega_{\mathbf{k}} \qquad (4.3.46)$$

where

$$[b_{\mathbf{k}\lambda}, b_{\mathbf{k}'\lambda'}^{\dagger}] = \delta_{\mathbf{k},\mathbf{k}'}\delta_{\lambda,\lambda'} \qquad (4.3.47)$$

4.3.5 The number operator

The operator $N_{k\lambda} \equiv b_{k\lambda}^\dagger b_{k\lambda}$ satisfies by virtue of eqs. (4.3.43) and (4.3.46)

$$N_{k\lambda} b_{k\lambda} = (b_{k\lambda} b_{k\lambda}^\dagger - 1) b_{k\lambda} = b_{k\lambda}(N_{k\lambda} - 1) \tag{4.3.48}$$

Taking Hermitian conjugates

$$N_{k\lambda} b_{k\lambda}^\dagger = b_{k\lambda}^\dagger (N_{k\lambda} + 1) \tag{4.3.49}$$

it follows that: (a) if $|N'_{k\lambda}\rangle$ is the eigenfunction of $N_{k\lambda}$ with eigenvalue $N'_{k\lambda}$, then $b_{k\lambda}|N'_{k\lambda}\rangle$ is the eigenfunction of $N_{k\lambda}$ with eigenvalue $N'_{k\lambda} - 1$ and $b_{k\lambda}^\dagger|N'_{k\lambda}\rangle$ is the eigenfunction of $N_{k\lambda}$ with eigenvalue $N'_{k\lambda} + 1$; (b) $N'_{k\lambda} \geq 0$ because $N'_{k\lambda} = \langle N'_{k\lambda}| N_{k\lambda}|N'_{k\lambda}\rangle = b_{k\lambda} N'_{k\lambda}|b_{k\lambda} N'_{k\lambda}\rangle$.

The last expression is the length of a vector and hence non-negative. By repeated application of $b_{k\lambda}$ as in (a) one reaches eventually negative eigenvalues, contrary to (b), unless the eigenvalue zero stops the process in accordance with eq. (4.3.48). Hence the eigenvalue zero of $N_{k\lambda}$ occurs, and the other eigenvalues are the positive integers. The operator $N_{k\lambda}$ is therefore called the number operator. Its eigenvalue gives by eq. (4.3.46) the occupation number of the mode (k, λ).

All the $|N'_{k\lambda}\rangle$ eigenfunctions are assumed normalized. Hence

$$\langle b_{k\lambda}^\dagger N'_{k\lambda}| b_{k\lambda}^\dagger N'_{k\lambda}\rangle = \langle N'_{k\lambda}| b_{k\lambda} b_{k\lambda}^\dagger|N'_{k\lambda}\rangle = \langle N'_{k\lambda}| N_{k\lambda} + 1|N'_{k\lambda}\rangle = N'_{k\lambda} + 1$$
$$\langle b_{k\lambda} N'_{k\lambda}| b_{k\lambda} N'_{k\lambda}\rangle \equiv N'_{k\lambda}$$

It follows that

$$|b_{k\lambda}^\dagger N'_{k\lambda}\rangle \equiv b_{k\lambda}^\dagger|N'_{k\lambda}\rangle = (N'_{k\lambda} + 1)^{\frac{1}{2}}|N'_{k\lambda} + 1\rangle$$
$$|b_{k\lambda} N'_{k\lambda}\rangle \equiv b_{k\lambda}|N'_{k\lambda}\rangle = (N'_{k\lambda})^{\frac{1}{2}}|N'_{k\lambda} - 1\rangle \tag{4.3.50}$$

Finally, one sees from eqs. (4.3.50) that

$$|N'_{k\lambda}\rangle = [N'_{k\lambda}!]^{-\frac{1}{2}}[b_{k\lambda}^\dagger]^{N'_{k\lambda}}|0\rangle \tag{4.3.51}$$

$$\langle N''_{k\lambda}|b_{k\lambda}^\dagger|N'_{k\lambda}\rangle = (N'_{k\lambda} + 1)^{\frac{1}{2}}\delta_{N''_{k\lambda}, N'_{k\lambda} + 1}\,\delta_{\lambda, \lambda'}$$
$$\langle N''_{k\lambda}|b_{k\lambda}|N'_{k\lambda}\rangle = (N'_{k\lambda})^{\frac{1}{2}}\delta_{N''_{k\lambda}, N'_{k\lambda} - 1}\,\delta_{\lambda, \lambda'} \tag{4.3.52}$$

The $b_{k\lambda}^\dagger$ is called the creation operator for mode (k, λ), while $b_{k\lambda}$ is called the annihilation operator.

4.3.6 The perturbation

In order to find the form of the perturbation, the following observations are needed. The vector potential is by eqs. (4.3.17), (4.3.38) and (4.3.39)

$$\mathbf{A} = \sum_{\substack{k \\ \lambda-1, 2}} [b_{k\lambda} \mathbf{A}_{k\lambda} e^{-i\omega_k t} + b_{-k\lambda}^\dagger \mathbf{A}_{k\lambda} e^{i\omega_k t}] Q_k \quad (\mathbf{A}_{k\lambda} \equiv V^{-\frac{1}{2}} a_k e^{i\mathbf{k}\cdot\mathbf{r}} \mathbf{e}_{k\lambda})$$

If in the second sum one replaces $-k$ by k' and uses eqs. (4.3.18), (4.3.30) and (4.3.38), one finds

$$\mathbf{A} = \sum_{\substack{k \\ \lambda-1, 2}} [b_{k\lambda} \mathbf{A}_{k\lambda} e^{-i\omega_k t} + b_{-k\lambda}^\dagger \mathbf{A}_{k\lambda}^* e^{i\omega_k t}] Q_k \tag{4.3.53}$$

Next note that for $j = 1, 2$ or 3

$$i(P_j \mathbf{A} - \mathbf{A} P_j) = \hbar \frac{\partial \mathbf{A}}{\partial x_j}$$

in analogy with $i(f(P_j) X_j - X_j f(P_j)) = \hbar \, \partial f / \partial P_j$ found in eq. (D.16). Multiplying by a unit vector parallel to x_j and summing,

$$i(\mathbf{P} \cdot \mathbf{A} - \mathbf{A} \cdot \mathbf{P}) = \hbar \, \mathrm{div} \, \mathbf{A} = 0 \qquad (4.3.54)$$

The last form holds because of eq. (4.3.22).

The eigenfunctions of the unperturbed electron system are denoted by $|I\rangle$, corresponding to the Hamiltonian $H_0 \equiv P^2/2m + e\varphi + U$. The wavefunctions $|N'_{k\lambda}\rangle$ correspond to the free radiation field, specified by the Hamiltonian H_t of eq. (4.3.46). The combined system has the Hamiltonian (4.3.2), so that, with e denoting the charge on the electron,

$$H = H_0 + H_t + H_1, \quad H_1 \equiv -\frac{e}{mc} \mathbf{A} \cdot \mathbf{P} \qquad (4.3.55)$$

where eq. (4.3.54) has been used. Using eqs. (4.3.53) and (4.3.18) one finds

$$H_1 = \sum_{\mathbf{k}} [G^{(\mathbf{k})} \mathrm{e}^{-\mathrm{i}\omega_k t} + G^{(\mathbf{k})\dagger} \mathrm{e}^{\mathrm{i}\omega_k t}] \qquad (4.3.56)$$

where

$$G^{(\mathbf{k})} \equiv -\frac{e a_{\mathbf{k}}}{mc V^{\frac{1}{2}}} Q_{\mathbf{k}} \sum_{\lambda=1,2} b_{\mathbf{k}\lambda} \, \mathrm{e}^{\mathrm{i}\mathbf{k}\cdot\mathbf{r}} \mathbf{e}_{\mathbf{k}\lambda} \cdot \mathbf{P}$$

Thus the perturbation has the form (4.3.11) and the transition probability is given by eq. (4.3.15). Using eq. (4.3.45) it requires the calculation of a matrix element of

$$G^{(\mathbf{k})} = -\left(\frac{2\pi\hbar}{V\omega_k}\right)^{\frac{1}{2}} \frac{e}{\mu m} \sum_{\lambda=1,2} b_{\mathbf{k}\lambda} \, \mathrm{e}^{\mathrm{i}\mathbf{k}\cdot\mathbf{r}} \mathbf{e}_{\mathbf{k}\lambda} \cdot \mathbf{P} \qquad (4.3.57)$$

4.3.7 The transition probability

The unperturbed problem corresponds to product wavefunctions

$$\prod_{\mathbf{k}} \prod_{\lambda=1}^{2} \{|N'_{\mathbf{k}\lambda}\rangle\}$$

for the field and $|I\rangle$ for the electron system. As a result of the transition to be considered only one radiation mode gains or loses a photon. By eq. (4.3.52) the modes which are left unchanged do not contribute to the matrix element. Let the mode for which there is a change be the $(\mathbf{k}, \lambda)$-mode: the remaining modes can then be neglected. Let the initial state be $|I\rangle |N'_{\mathbf{k}\lambda}\rangle$ and let the electron system make a transition to the state $|J\rangle$. Let the final state of the radiation mode $(\mathbf{k}, \lambda)$ be $|N''_{\mathbf{k}\lambda}\rangle$. Then by eq. (4.3.52) a nonzero matrix element is obtained only if $N''_{\mathbf{k}\lambda} = N'_{\mathbf{k}\lambda} + 1$, when the contribution comes from $G^{\dagger}$, and $E^0_J < E^0_I$

(emission). Alternatively $N''_{k\lambda} = N'_{k\lambda} - 1$, when the contribution comes from G and $E_J^0 > E_I^0$ (absorption). Hence the matrix element is

$$
\left.\begin{array}{c} |G_{IJ}^{(k)\dagger}| \\ |G_{IJ}^{(k)}| \end{array}\right\} = -\left(\frac{2\pi\hbar}{V\omega_k}\right)^{\frac{1}{2}} \frac{e}{\mu m} \begin{cases} \sum\limits_{\lambda=1,2} M_{IJ}^-(N'_{k\lambda}+1)^{\frac{1}{2}} & \text{(emission)} \\ \sum\limits_{\lambda=1,2} M_{IJ}^+(N'_{k\lambda})^{\frac{1}{2}} & \text{(absorption)} \end{cases}
$$

where

$$
M_{IJ}^\pm \equiv M_{IJ}^\pm(\mathbf{k},\lambda) \equiv |\langle I|\, e^{\pm i\mathbf{k}\cdot\mathbf{r}}\mathbf{e}_{k\lambda}\cdot\mathbf{P}|J\rangle| \tag{4.3.58}
$$

Substitution in eq. (4.3.15) yields for the transition probability per unit time per unit volume

$$
P_{IJ} = \frac{4\pi^2 e^2}{\mu^2 V^2 m^2} \sum_k \frac{1}{\omega_k} \sum_{\lambda=1,2} \begin{cases} (M_{IJ}^-)^2 (N'_{k\lambda}+1)\,\delta(E_I^0 - E_J^0 - \hbar\omega_k) \\ (M_{IJ}^+)^2 N'_{k\lambda}\,\delta(E_I^0 - E_J^0 + \hbar\omega_k) \end{cases} \tag{4.3.59}
$$

4.4 Interband emission rates

4.4.1 The rate of single-mode allowed radiative transitions

It is our object in this section to attain some physical results without having to develop a great deal of mathematical formalism. For this reason it is desirable to adopt the formula (4.3.59) from quantum mechanics whose derivation can be omitted by readers not interested in the details. The formula gives the radiative transition probability per unit time per unit volume between quantum states I and J as

$$
P_{IJ}(E_k) = \frac{4\pi^2 e^2 \hbar}{\mu^2 m^2 V^2 E_k} \sum_{k\,\lambda=1,2} \begin{cases} (M_{IJ}^-)^2 [N(E_k)+1]\,\delta(E_I - E_J - E_k) & \text{(emission)} \\ (M_{IJ}^+)^2 N(E_k)\,\delta(E_I - E_J + E_k) & \text{(absorption)} \end{cases} \tag{4.4.1}
$$

Here I refers to the initial and J to the final state, thus explaining the difference between the delta functions. The photon occupation number occurs in the manner expected from, for example, eq. (4.2.3). M_{IJ} is the matrix element of the electron momentum $\mathbf{P}$:

$$
M_{IJ}^\pm \equiv M_{IJ}^\pm(\mathbf{k},\lambda) \equiv |\langle I|\, e^{\pm i\mathbf{k}\cdot\mathbf{r}}\mathbf{e}_{k\lambda}\cdot\mathbf{P}|J\rangle| \tag{4.4.2}
$$

Here E_k is the energy of a photon mode of wavevector $\mathbf{k}$ and either polarization, and $\mathbf{e}_{k\lambda}$ is the appropriate unit vector giving the direction of polarization. Also, m is the electron rest mass, V is the volume of the cavity, and μ is the refractive index of the material.

These formulae assume (a) *one* photon mode of the required energy E_k, (b) one electronic state I and one electronic state J, and (c) probability unity for the initial

state to be occupied by an electron and the final state to be empty. These
assumptions will be removed in sections 4.4.2 and 4.4.3. These results hold only for
direct (i.e. phononless) transitions since phonon contributions are not allowed for
in formula (4.4.1); see section 4.5.2.

To gain familiarity with this formula we shall first check that it is dimensionally
correct, noting from Coulomb's law that

$$[e^2] = [\text{energy} \times \text{distance}],$$

from the choice of centimeter–gram–second units that μ is dimensionless, from
$\int \delta(E)\,dE = 1$ that the energy delta function has the dimension $[\text{energy}^{-1}]$, and
from eq. (4.4.2) that the matrix element has the dimension of momentum. One
then finds from formula (4.4.1)

$$[P_{IJ}] = \left[\frac{\text{energy}\,L \cdot \text{energy}\,T}{M^2 \cdot L^6 \cdot \text{energy}}(MLT^{-1})^2 \frac{1}{\text{energy}}\right] = [L^{-3}T^{-1}]$$

as required.

In order to gain further familiarity with formula (4.4.1) we shall average $(M^\pm)^2$
over polarizations and over the directions of the photon wavevector $\mathbf{k}$. For this
purpose it will be assumed that the radiation field is isotropic and unpolarized, so
that the photon occupation numbers depend only on the magnitude of $\mathbf{k}$, not on
its direction, and that they are also independent of λ. One finds

$$P_{IJ}(E_k) = \frac{4\pi^2 e^2 \hbar}{\mu^2 m^2 V^2 E_k}|M_{IJ}|_{\text{av}}^2 \begin{cases} [N(E_k)+1]\,\delta(E_I - E_J - E_k) \\ N(E_k)\,\delta(E_I - E_J + E_k) \end{cases} \equiv B_{IJ}^{\text{SM}} \begin{cases} [N(E_k)+1] \\ N(E_k) \end{cases}$$

$$\text{(4.4.3)}$$

This gives a quantum mechanical identification of the B's. The details of the
averaging of the matrix element are given in the remainder of this section, but may
be omitted on first reading.

Let $d\Omega_k$ be an element of solid angle for the integration over the angles made by $\mathbf{k}$ for
given k with the real and orthonormal set $\{\mathbf{e}_{k\lambda}\}$. One normally takes

$$\mathbf{e}_{k3} = \mathbf{k}/|\mathbf{k}| \tag{4.4.4}$$

so that $\lambda = 3$ characterizes longitudinal polarization. It can then be shown that only
$\lambda = 1, 2$ contribute to the matrix element. The averaged matrix element in eq. (4.3.3) is

$$|M_{IJ}^\pm|_{\text{av}}^2 \equiv \frac{1}{8\pi}\sum_{\lambda-1,2}\int (M_{IJ}^\pm)^2\,d\Omega_k = \frac{1}{8\pi}\sum_{\lambda-1,2}\int |\langle I|(\mathbf{e}_{k\lambda}\cdot\mathbf{P})e^{\pm i\mathbf{k}\cdot\mathbf{r}}|J\rangle|^2\,d\Omega_k$$

$$\equiv \frac{1}{8\pi}\sum_{\lambda-1,2}\int |\mathbf{e}_{k\lambda}\cdot\mathbf{Q}|^2\,d\Omega_k \tag{4.4.5}$$

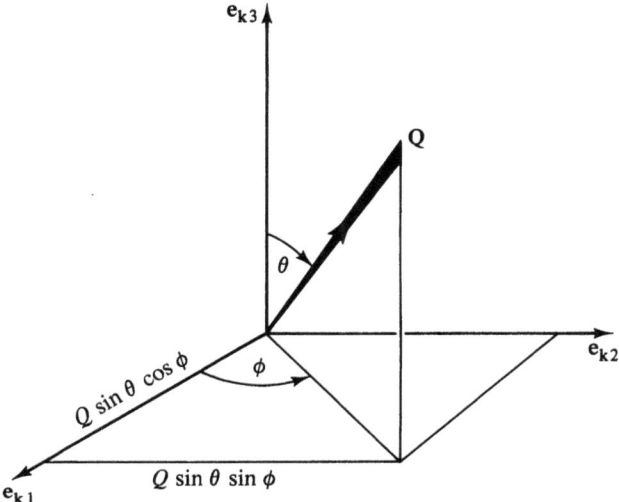

Fig. 4.4.1. Diagram illustrating the averaging of the matrix element.

The average involves normalizing factors $1/2$, $1/4\pi$ for the sum over λ and the integral over $d\Omega_{\mathbf{k}}$ respectively. Neglecting the usually small photon wavevector in $\exp(i\mathbf{k}\cdot\mathbf{r})$, $\mathbf{Q}$ is independent of $\mathbf{k}$:

$$\mathbf{Q} \equiv \langle I|\mathbf{P}|J\rangle \tag{4.4.6}$$

The averaging can therefore be done by arranging the triad in all possible orientations relative to $\mathbf{Q}$ or, equivalently, by regarding the triad as fixed and letting $\mathbf{Q}$ range over all orientations. The latter concept will be adopted here. This procedure is not justifiable for forbidden transitions since by eq. (4.4.37), below, the angles enter differently in that case.

One now finds (see Fig. 4.4.1)

$$|M_{IJ}|^2_{\mathrm{av}} = \frac{1}{8\pi}\int (|e_{\mathbf{k}1}\cdot\mathbf{Q}|^2 + |e_{\mathbf{k}2}\cdot\mathbf{Q}|^2)\,d\Omega_{\mathbf{k}}$$

$$= \frac{1}{8\pi}\int |\mathbf{Q}|^2 \sin^2\theta(\cos^2\varphi + \sin^2\varphi)\,d\Omega_{\mathbf{k}}$$

$$= \frac{1}{8\pi}\int |\mathbf{Q}|^2 \sin^2\theta\cdot\sin\theta\,d\theta\,d\varphi$$

$$= \frac{2\pi}{8\pi}\int_0^\pi |\mathbf{Q}|^2 \sin^3\theta\,d\theta$$

Now

$$\int_0^\pi \sin^3\theta\,d\theta = \left[\tfrac{1}{3}\cos^3\theta - \cos\theta\right]_0^\pi = \tfrac{1}{3}(-1-1)-(-1-1) = \tfrac{4}{3}$$

It follows that

$$|M_{IJ}|^2_{\mathrm{av}} = \tfrac{1}{3}|\mathbf{Q}|^2 = \tfrac{1}{3}\sum_{b-x,y,z}|Q_b|^2 = \tfrac{1}{3}\sum|\langle I|P_b|J\rangle|^2 \tag{4.4.7}$$

The averaged matrix element requires, therefore, merely an estimate of the matrix elements of the Cartesian components of the electron momentum. A great deal can, however, be accomplished without actually estimating the matrix element (4.4.7), but by regarding it merely as a parameter.

It remains to verify that a photon wavevector is normally much shorter than an electron wavevector. We shall consider them in turn.

(1) *Electron*

The energy E corresponding to a wavevector of length k in a band of effective mass m^* is

$$E = \frac{\hbar^2 k^2}{2m} \frac{m}{m^*} = \frac{(1.04 \times 10^{-27})^2 (10^8 k)}{2 \times 9.1 \times 10^{-28}} 0.624 \times 10^{12}$$

The centimeter–gram–second system has been used, with k in cm^{-1} converted to reciprocal angstroms, $\hbar$ in erg–seconds and m in grams. The last factor converts E from ergs to electron volts. Thus

$$E = 3.71[k^2(\text{Å}^{-2})] \frac{m}{m^*} \, eV \tag{4.4.8}$$

Typically, the electron energy is $k_B T$ (for the moment we denote Boltzmann's constant by k_B) which is $1/40$ eV at room temperature, whence

$$k = 0.0871(m^*/m) \, \text{Å}^{-1} \tag{4.4.9}$$

(2) *Photon*

In this case we take $\hbar$ in electron volt–seconds and use

$$E = cp = c\hbar k = 3 \times 10^{10} \times 0.662 \times 10^{-15} \times 10^8 k \, (ev)$$

so that

$$E = 1986[k(\text{Å}^{-1})] \, eV \tag{4.4.10}$$

The photon energy is of order of the band gap, i.e. $E \sim 1$ eV, so that

$$k = 0.000\,50[E(eV)] \, \text{Å}^{-1} \tag{4.4.11}$$

Thus eq. (4.4.11) can indeed be neglected compared with eq. (4.4.9). This is used in the next subsection to arrive at the k-selection rule.

4.4.2 Multi-mode operation, spectral emission functions and k-selection rule for allowed transitions

In this section the constraints (a), (b) and (c) on the theory, noted at the beginning of section 5.4.1, will be removed. The extension from single-mode to multi-mode operation requires an expression, $\mathcal{N}(E)$, for the number of photon modes per unit energy. This enables one to remove constraint (a). It can be achieved by replacing the energy delta function in eq. (4.4.3) as follows:

$$\delta(E_I - E_J \pm E_k) \to \mathcal{N}(E_k)\delta_{E_I - E_J, E_k} \qquad (4.4.12)$$

it being now assumed that $E_I > E_J$. This effectively multiplies the transition probability by the number of modes in the energy range, since each mode can contribute equally. Use

$$\mathcal{N}(E_k) = \frac{l}{2}\frac{\mu^3 V E_k^2}{\pi^2 c^3 \hbar^3} \qquad (4.4.13)$$

for l directions of polarization. Thus $l = 1$ for polarized radiation and $l = 2$ for unpolarized radiation. Hence eq. (4.4.3) becomes

$$P_{IJ}(E_k) = \frac{4le^2\mu E_k}{2m^2 V c^3 \hbar^2}|M_{IJ}|^2_{av}\begin{cases} N(E_k)+1 \\ N(E_k) \end{cases} \equiv B_{IJ}(E_k)\begin{cases} [N(E_k)+1] \\ N(E_k) \end{cases} \qquad (4.4.14)$$

the energy-conserving delta function being understood. This expression differs from the single-mode expression, and therefore the superfix SM is now omitted. Both B's can, however, be used in the statistical theory of section 4.2. Rough identifications of $|M_{IJ}|^2_{av}$ in terms of energy gaps and effective masses, if known, can be achieved by the f-sum rule, as explained in eqs. (D.36) and (D.37).

The spontaneous emission term which is the last term in eq. (4.2.3), was in the older literature written as $A_{IJ}p_{I0}$, A being the so-called Einstein A-coefficient. The coefficient governing induced emission and absorption was called the Einstein B-coefficient. In eq. (4.2.12) we considered the equilibrium of photons when $A_{IJ}/B_{JI} = 1$. (If one treats the energy *flux* one would get a different ratio.) One can obtain an alternative expression for B_{IJ} as follows. Apply the Heisenberg equation of motion (D.15) to the momentum matrix element

$$\frac{1}{m}\mathbf{p}_{IJ} = \frac{d}{dt}\mathbf{r}_{IJ} = \frac{i}{\hbar}(H\mathbf{r} - \mathbf{r}H)_{IJ} = \frac{i}{\hbar}(E_I - E_J)\mathbf{r}_{IJ}$$

Using this result in eq. (4.4.14), one obtains a fairly standard [4.4.1] expression for the A-coefficient:

$$B_{IJ}(E_k) = \frac{l}{2}\frac{4e^2\mu E_k^3}{\hbar^4 c^3 V}|r_{IJ}|^2_{av} \qquad (4.4.14')$$

In solids one has to add a further factor $(\mathscr{E}_{eff}/\mathscr{E})^2$, where $\mathscr{E}_{eff}$ is the field effective in inducing the transitions and $\mathscr{E}$ is the average electric field.

In order to remove constraint (b) on the theory one has to sum over the number of single-electron states, $g_{IJ}(E_k)\,dE_k$, say, which can contribute to the photon emission in the energy range $(E_k, E_k + dE_k)$. For this purpose the electron quantum state I is allowed to range over the appropriate group, i say, of quantum states, and the state J is similarly allowed to range over the appropriate group, j say. To remove constraint (c) at the same time, the probabilities p_I, $1-p_I$, etc., used in section 4.2 will also be introduced. One can then write for multi-mode operation that the total rate of spontaneous and net stimulated emission per unit volume, respectively, is

$$R_{ij}^{sp} = \int r_{ij}^{sp}(E)\,dE \quad R_{ij}^{st} = \int r_{ij}^{st}(E)\,N(E)\,dE \tag{4.4.15}$$

where the r's are spectral functions. They are rates per unit volume per unit energy range. Here we have used the abbreviations

$$r_{ij}^{sp}(E_k) = \sum_{I \in i}\sum_{J \in j} u_{IJ}^{sp}(E_k)\,g_{IJ}(E_k)\,\delta_{E_I - E_J, E_k}$$

$$= \sum_{I,J} B_{IJ}(E_k)\,p_I(1-p_J)\,g_{IJ}(E_k)\,\delta_{E_I - E_J, E_k} \tag{4.4.16}$$

$$r_{ij}^{st}(E_k) = \sum_{I,J} [u_{IJ}^{\prime st}(E_k)/N(E_k)]\,g_{IJ}(E_k)\,\delta_{E_I - E_J, E_k}$$

$$= \sum_{I,J} B_{IJ}(E_k)\,(p_I - p_J)\,g_{IJ}(E_k)\,\delta_{E_I - E_J, E_k} \tag{4.4.17}$$

The prime on u^{st} means that it is the *net* stimulated emission

$$u_{IJ}^{\prime st} = u_{IJ}^{st} - u_{IJ}^{abs}$$

The two terms on the right-hand side are identical, except for the probability factors which on subtraction yield

$$p_I(1-p_J) - p_J(1-p_I) = p_I - p_J$$

as shown in eq. (4.4.17). For single-mode operation, the superfixes SM have to be attached to r_{ij} and B_{IJ} and eq. (4.4.3) must be used for the latter. Otherwise the theory goes through as before. Our main concern now is to obtain closed expressions for the quantities (4.4.16) and (4.4.17). We shall discuss in section 4.4 mainly two cases: direct semiconductors with parabolic bands either with **k**-selection or without **k**-selection.

In order to see how **k**-selection comes about, note that it is *not* actually what one would expect classically, see Fig. 4.4.2. If a particle of initial momentum $\mathbf{k}_i$ receives a kick due to a particle (a photon in this case) of momentum $\mathbf{k}$, the resulting final momentum is expected to be

$$\mathbf{k}_f = \mathbf{k}_i + \mathbf{k} \tag{4.4.18}$$

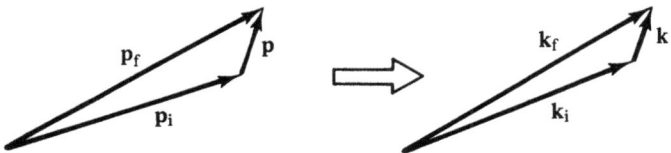

Fig. 4.4.2. Momentum conservation leads to wavevector conservation and violates **k**-selection $\mathbf{k}_i = \mathbf{k}_f$.

In fact this is, strictly speaking, what results from the use of quantum mechanics, as will now be shown.

Consider the matrix element (4.4.2) for emission

$$\langle I|\,e^{-i\mathbf{k}\cdot\mathbf{r}}|\mathbf{e}_{\mathbf{k}\lambda}\cdot\mathbf{P}|J\rangle = \mathbf{e}_{\mathbf{k}\lambda}\cdot\langle I|\,e^{-i\mathbf{k}\cdot\mathbf{r}}\mathbf{P}|J\rangle \qquad (4.4.19)$$

before the averaging process (4.4.5) has been carried out. In the simplest case one can use normalized free-electron functions, or plane waves

$$\langle I| = V^{-\frac{1}{2}}\langle e^{-i\mathbf{k}_i\cdot\mathbf{r}}|, \qquad |J\rangle = V^{-\frac{1}{2}}|e^{i\mathbf{k}_f\cdot\mathbf{r}}\rangle \qquad (4.4.20)$$

where V is the volume of the crystal. The required matrix element is

$$V^{-1}\int_V e^{-i(\mathbf{k}_i+\mathbf{k})\cdot\mathbf{r}}\frac{\hbar}{i}\nabla\,e^{i\mathbf{k}_f\cdot\mathbf{r}}\,d\mathbf{r} = \hbar\mathbf{k}_f\frac{1}{V}\int_V e^{i(\mathbf{k}_f-\mathbf{k}_i-\mathbf{k})\cdot\mathbf{r}}\,d\mathbf{r} = \hbar\mathbf{k}_f\,\delta_{\mathbf{k}_f,\mathbf{k}_i+\mathbf{k}} \quad (4.4.21)$$

which implies (4.4.18).

We have here replaced **P** by $(\hbar/i)\nabla$ as usual. This is, however, only an intuitive argument, since the wavefunctions $\langle I|$ and $|J\rangle$ ought to be orthogonal, so that one finds a zero matrix element when the operator used is the identity. Plane waves do not have this property; for $\mathbf{k}_i = \mathbf{k}_f$ they are just identical functions. Nonetheless, the argument suggests that the matrix element for emission is zero when momentum conservation is violated. An analogous argument holds for absorption. The **k**-selection rule now results if the photon wavevector **k** is neglected in accordance with eqs. (4.4.9)–(4.4.11).

A better argument is obtained by using Bloch functions

$$\langle I| = V^{-\frac{1}{2}}u^*_{n\mathbf{k}_i}(\mathbf{r})\,e^{-i\mathbf{k}_i\cdot\mathbf{r}}, \qquad |J\rangle = V^{-\frac{1}{2}}u_{m\mathbf{k}_f}(\mathbf{r})\,e^{i\mathbf{k}_f\cdot\mathbf{r}}$$

where n and m are the band indices, so that for emission n refers to the conduction band and m refers to the valence band. Then from eq. (D.11)

$$\mathbf{P}|J\rangle = V^{-\frac{1}{2}}\mathbf{P}u_{m\mathbf{k}_f}\,e^{i\mathbf{k}_f\cdot\mathbf{r}} = V^{-\frac{1}{2}}e^{i\mathbf{k}_f\cdot\mathbf{r}}(\mathbf{P}+\hbar\mathbf{k}_f)\,u_{m\mathbf{k}_f} \qquad (4.4.22)$$

The matrix element in eq. (4.4.19) becomes

$$\frac{1}{V}\int_V e^{i(\mathbf{k}_f-\mathbf{k}_i-\mathbf{k})\cdot\mathbf{r}}u^*_{n\mathbf{k}_i}(\mathbf{P}+\hbar\mathbf{k}_f)\,u_{m\mathbf{k}_f}\,d\mathbf{r}$$

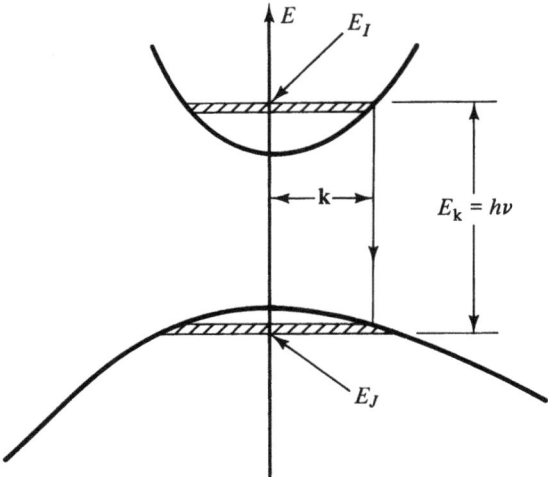

Fig. 4.4.3. Radiative transition with **k**-selection. Multi-mode operation occurs as E_k can be varied.

Since the last three factors in the integrand form a lattice-periodic function, eq. (B.10) may be applied to yield

$$\langle I|\mathbf{P}|J\rangle = \frac{\delta_{\mathbf{k}_f,\mathbf{k}_i+\mathbf{k}}}{\Omega}\int_\Omega u_{n\mathbf{k}_i}^*(\mathbf{P}+\hbar\mathbf{k}_f)\,u_{m\mathbf{k}_f}\,d\mathbf{r}$$

$$+\sum_{\substack{\mathbf{m}\\(\mathbf{K}_\mathbf{m}\neq0)}}\frac{\delta_{\mathbf{k}_f,\mathbf{k}_i+\mathbf{k}+\mathbf{K}_\mathbf{m}}}{\Omega}\int_\Omega u_{n\mathbf{k}_i}^*\,e^{i\mathbf{K}_\mathbf{m}\cdot\mathbf{r}}(\mathbf{P}+\hbar\mathbf{k}_f)\,u_{m\mathbf{k}_f}\,d\mathbf{r}\qquad(4.4.23)$$

The sum is over all reciprocal lattice vectors and Ω is the volume of a unit cell. The sum has been split into a term with $\mathbf{m}=0$ (so that $\mathbf{K}_\mathbf{m}=0$) and terms with $\mathbf{m}\neq0$ which will be neglected since they involve rather larger wavevectors $\mathbf{k}_i$, $\mathbf{k}_f$. We have then again a **k**-selection rule.

If one neglects the photon wavevector **k**, as explained at the end of section 4.4.1, $\mathbf{k}_i=\mathbf{k}_f$ and the $\hbar\mathbf{k}_f$-matrix element in the first term of eq. (4.4.23) vanishes since Bloch functions of different bands with the same wavevector are orthogonal. Even the **P**-matrix element may vanish at the **k**-vector, $\mathbf{k}_0$ say, of interest. One then has 'forbidden' transitions, and to discuss these a Taylor expansion to first order in $\mathbf{k}-\mathbf{k}_0$ is desirable:

$$M_{IJ}^\pm \to \mathbf{e}_{\mathbf{k}_{ph}\lambda}\cdot\mathbf{P}_{vc}(\mathbf{k}) = \mathbf{e}_{\mathbf{k}_{ph}\lambda}\cdot\mathbf{P}_{vc}(\mathbf{k}_0) + \{\nabla_\mathbf{k}[\mathbf{e}_{\mathbf{k}_{ph}\lambda}\cdot\mathbf{P}_{vc}(\mathbf{k})]\}_{\mathbf{k}-\mathbf{k}_0}\cdot(\mathbf{k}-\mathbf{k}_0)$$

The photon wavevector is here no longer denoted by **k** but by $\mathbf{k}_{ph}$ to reserve **k** for electrons. Hence

$$M_{IJ}^\pm \to \begin{cases} \mathbf{e}_{\mathbf{k}_{ph}\lambda}\cdot\mathbf{P}_{vc}(\mathbf{k}_0) & \text{(allowed)} \\ \{\nabla_\mathbf{k}[\mathbf{e}_{\mathbf{k}_{ph}\lambda}\cdot\mathbf{P}_{vc}(\mathbf{k})]\}_{\mathbf{k}-\mathbf{k}_0}\cdot(\mathbf{k}-\mathbf{k}_0) & \text{(forbidden)} \end{cases}\qquad(4.4.24)$$

The forbidden case can arise in a direct semiconductor if the periodic part of the Bloch functions have the same symmetry, say s-type symmetry, at say $\mathbf{k}_0 = 0$. Away from this value allowed transitions are again possible.

Note incidentally that the result (4.4.22) shows that

$$\mathbf{P}\psi_{n\mathbf{k}}(r) \equiv \mathbf{P}|n\mathbf{k}\rangle = \hbar\mathbf{k}\psi_{n\mathbf{k}}(r) + \frac{\hbar}{i}[\nabla u_{n\mathbf{k}}(r)]\,\mathrm{e}^{\mathrm{i}\mathbf{k}\cdot\mathbf{r}}$$

The second term ensures that $\mathbf{P}$ and the electron Hamiltonian H do not both have Bloch functions as eigenfunctions. If they did, they would commute and the periodic potential $U(\mathbf{r})$ in $H = p^2/2m + U(\mathbf{r})$ prevents this. Thus $\hbar\mathbf{k}$ is not an eigenvalue of the electron momentum, and to call it that would be a misnomer. The term 'crystal momentum of the electron' is commonly used instead. (See also Appendix D.)

4.4.3 Spectral emission functions with and without k-selection for multi-mode operation and direct transitions

For multi-mode operation, k-selection and parabolic spherical bands one can put for direct semiconductors (see Fig. 4.4.3)

$$E_I = E_c + \frac{\hbar^2 k^2}{2m_c}, \quad E_J = E_v - \frac{\hbar^2 k^2}{2|m_v|} \tag{4.4.25}$$

where E_c, E_v are the energies at the band edges, m_c, and $m_v = -|m_v|$ are the effective masses and the occurrence of the same k in both relations is an expression of the k-selection principle. Incorporating the Kroenecker delta of eq. (4.4.17) it follows that

$$E_k = E_I - E_J = E_G + \frac{\hbar^2 k^2}{2m^*} \tag{4.4.26}$$

where the energy gap is E_G and the joint density-of-states effective mass (1.5.16) is given by

$$\frac{1}{m^*} = \frac{1}{m_c} + \frac{1}{|m_v|}, \quad \text{i.e. } m^* = \frac{m_c|m_v|}{m_c + |m_v|} \tag{4.4.27}$$

This is dominated by the smaller of the two masses m_c, $|m_v|$. Note that on p. 29 we gave a rather more sophisticated approach to this density of states. However it is in small print for possible omission on first encounter. Using the density of states including the factor two for spin

$$\sum_{I \in i}\sum_{J \in j} \delta_{E_I - E_J, E_k}\, g_{IJ}(E_k)\,\mathrm{d}E_k = (2m^*)^{\frac{3}{2}}\frac{V}{2\pi^2\hbar^3}(E_k - E_G)^{\frac{1}{2}}\,\mathrm{d}E_k \tag{4.4.28}$$

This result is obtainable from eq. (4.4.26) in the usual way by equating the left-hand side of eq. (4.4.28) to

$$\frac{V}{4\pi^3}\,dk = \frac{V}{\pi^2}\cdot k\cdot k\,dk = \frac{V}{\pi^2}\cdot\left[(E_k-E_G)\frac{2m^*}{\hbar^2}\right]^{\frac{1}{2}}\cdot\frac{m^*}{\hbar^2}\,dE_k \qquad (4.4.29)$$

At $E_k = E_G$ we then have a sudden rise in $\mathcal{N}(E_k)$ from zero and this point is of type M_0 (Fig. 1.5.2). Since E_I, E_J are fixed for the sum over I and J, $p_I - p_J = p_i - p_j$ can be taken out of the sum in eq. (4.4.17) and one has finally for $l = 2$

$$r_{ij}^{st}(E_k) = \frac{4e^2 E_k\mu}{Vm^2c^3\hbar^2}|M_{IJ}|_{av}^2(p_i-p_j)(2m^*)^{\frac{3}{2}}\frac{V}{2\pi^2\hbar^3}(E_k-E_G)^{\frac{1}{2}}$$

This holds only for $E_k = h\nu \geqslant E_G$ and r_{ij}^{st} vanishes otherwise. (If the bands are not spherical one might find at $E_k = E_G$ for example a point of type M_1 or M_2 (Fig. 1.5.4)).
    Thus

$$r_{ij}^{st}(E_k) = A(E_k-E_G)^{\frac{1}{2}}(p_i-p_j) \qquad (4.4.30)$$

where

$$A \equiv \frac{2\mu e^2 E_k(2m^*)^{\frac{3}{2}}}{\pi^2 m^2 c^3\hbar^5}|M_{IJ}|_{av}^2 \sim 1.89\times10^{39}\ \mathrm{erg}^{-\frac{3}{2}}\ \mathrm{cm}^{-3}\ \mathrm{s}^{-1}\ \text{for GaAs}$$

Since

$$E_i = E_c + \frac{|m_v|}{m_c+|m_v|}(E_k-E_G)$$

and

$$E_j = E_v - \frac{m_c}{m_c+|m_v|}(E_k-E_G)$$

therefore

$$p_i^{-1} = 1+\exp\left[\frac{|m_v|}{m_c+|m_v|}(\eta_k-\eta_G)-(\gamma_e-\eta_c)\right]$$

$$p_j^{-1} = 1+\exp\left[\frac{-m_c}{m_c+|m_v|}(\eta_k-\eta_G)-(\gamma_h-\eta_v)\right]$$

Note that it has been assumed that the electronic spin is preserved in the transition. If it were not, a second factor of two would be appropriate in the density of states.
    As regards the form of the spectral function r_{ij}^{st}, it vanishes if $E_k - E_G = 0$ and if $p_i = p_j$. The first zero expresses the fact that the model does not permit radiative

emission at $E_k < E_G$. The second zero occurs if the exponents in p_i^{-1} and p_j^{-1} are the same. This happens if

$$\left(\frac{h\nu}{kT} \equiv\right)\eta_k = \gamma_e - \gamma_h \tag{4.4.31}$$

Between these two zeroes r_{ij}^{st} must have a maximum, as confirmed by Fig. 4.4.4. As this is plotted for $\mu_e - E_c = 5$ meV and $E_v - \mu_h = 11.8$ meV the second zero is expected at a value of the horizontal coordinate given by

$$E_k - E_G = \mu_e - \mu_h - E_c + E_v = 16.8 \text{ meV}$$

in agreement with the figure.

Turning to r_{ij}^{sp}, it is given by

$$r_{ij}^{sp}(E_k) = A(E_k - E_G)^{\frac{1}{2}}p_i(1 - p_j) \tag{4.4.32}$$

It differs from eq. (4.4.30) only in the replacement of p_J by $p_I p_J$. Since this is smaller, one sees at once that for a given system

$$r_{ij}^{sp}(E_k) > r_{ij}^{st}(E_k) \tag{4.4.33}$$

We next deal with electronic transitions without k-conservation or spin conservation. Then we get a product of terms (4.4.28), i.e.

$$\sum_{I,J} g_{IJ}(E_I, E_J)\delta_{E_I - E_J, E_k} \, dE_I \, dE_J$$

$$= 2\left(\frac{V}{\pi^2\hbar^3}\right)^2 (m_c m_v)^{\frac{3}{2}}(E_I - E_c)^{\frac{1}{2}}(E_v - E_J)^{\frac{1}{2}}\delta_{E_I - E_J, E_k} \, dE_I \, dE_J$$

Inserting this expression into eq. (4.4.17), one now finds after the integration over E_J

$$r_{ij}^{st}(E_k) = C \int_{E_c}^{E_c + E_k - E_G} (E_I - E_c)^{\frac{1}{2}}(E_v - E_I + E_k)^{\frac{1}{2}}$$

$$\times \left[\frac{1}{e^{\eta_I - \gamma_e} + 1} - \frac{1}{e^{\eta_I - \eta_k - \gamma_h} + 1}\right] dE_I \tag{4.4.34}$$

Here

$$C \equiv \frac{8\mu e^2 E_k (m_c m_v)^{\frac{3}{2}} V}{\pi^4 m^2 \hbar^8 c^3}|M_{IJ}|^2_{av} \tag{4.4.35}$$

and it is assumed that the matrix element depends only on E_k. The dimension of C is readily verified to be

$$[C] = \frac{ML^3T^{-2} \cdot ML^2T^{-2} \cdot M^3 \cdot L^3}{M^2 \cdot M^8 L^{16}T^{-8} \cdot L^3T^{-3}} M^2 L^2 T^{-2} = \frac{1}{(\text{energy})^3}L^{-3}T^{-1}$$

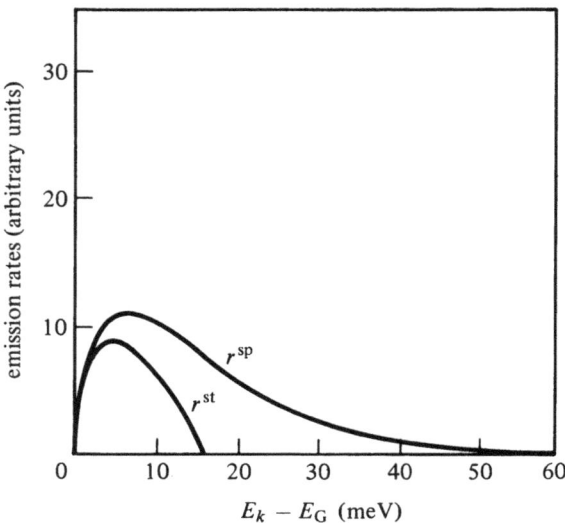

Fig. 4.4.4. Emission rates with **k**-selection rule calculated from eqs. (4.4.30) and (4.4.31).

In eq. (4.4.34) one finds eq. (4.4.35) multiplied by (energy)^2 leaving r^{st} with the correct dimension (energy)$^{-1}$L^{-3}T^{-1}.

The spectral function for spontaneous emission is again identical with r^{st} except for the probability factors, and one finds for $l = 2$, using eq. (4.4.14),

$$r_{ij}^{sp}(E_k) = C \int_{E_c}^{E_c+E_k-E_G} (E_l - E_c)^{\frac{1}{2}}(E_v - E_l + E_k)^{\frac{1}{2}}$$

$$\times \frac{\exp(\eta_l - \eta_k - \gamma_h)\,dE_l}{[\exp(\eta_l - \gamma_e) + 1][\exp(\eta_l - \eta_k - \gamma_h) + 1]} \qquad (4.4.36)$$

The relaxation of **k**-conservation increases the emission rates since more transitions are then allowed and this is illustrated for the case of GaAs in Figs. 4.4.4 and 4.4.5 [4.4.2]. The group theoretical treatment of selection rules, of which **k**-conservation is the simplest, is beyond the present scope, but see for reviews [4.4.3] and [4.4.4]. The question of **k**-selection arises again in quantum well lasers, experiments being sometimes more easily explained without them [4.4.5], and sometimes with them [4.4.6]. The theoretical arguments are not entirely in favor of **k**-selection. For if impurities participate in the radiative process they can take up momentum so that the momentum of the electronic system by itself need no longer be conserved.

The theory leading to eq. (4.4.32) has been broadly verified and we give the example of InSb at 77 K [4.4.7]; see also [4.4.8] and Figs. 4.4.6 and 4.4.7. Using the

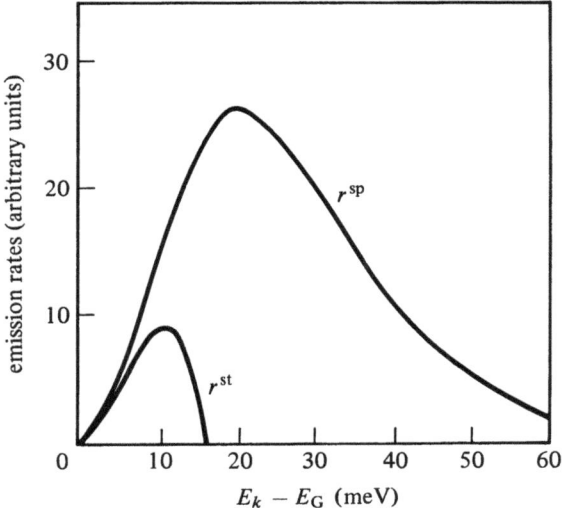

Fig. 4.4.5. Emission rates without **k**-selection rule calculated from eqs. (4.4.34) and (4.4.36).

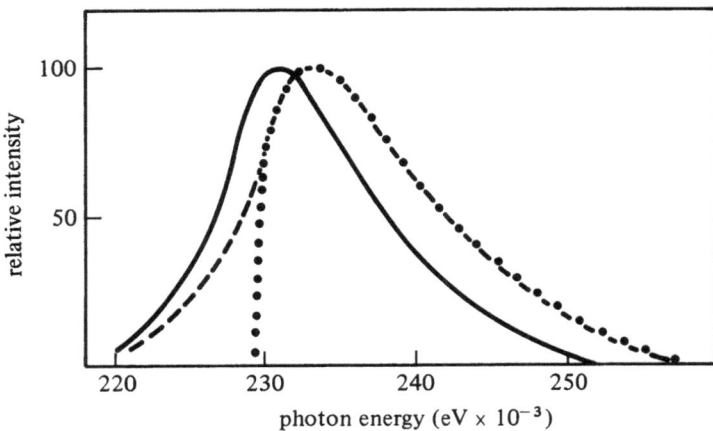

Fig. 4.4.6. Normalized emission spectrum of a p-type sample ($p = 5 \times 10^{15}$ cm^{-3}) of InSb at 77 K. The dashed curve is the observed curve (solid line) after correction for self-absorption. The points are calculated for direct transitions [4.4.7].

various equations leading to (4.4.31), together with $A \propto \nu^2$, the emission intensity for a degenerate electron gas and nondegenerate hole gas is

$$I(\nu) \propto \nu^2 (h\nu - E_G)^{\frac{1}{2}} \exp\left[-\frac{m_c}{m_c - |m_v|} \frac{h\nu - E_G}{kT} \right]$$

$$\times \left\{ 1 + \exp\left[\frac{|m_c|}{m_c + |m_v|} \frac{(h\nu - E_G)}{kT} - (\gamma_e - \eta_c) \right] \right\}^{-1}$$

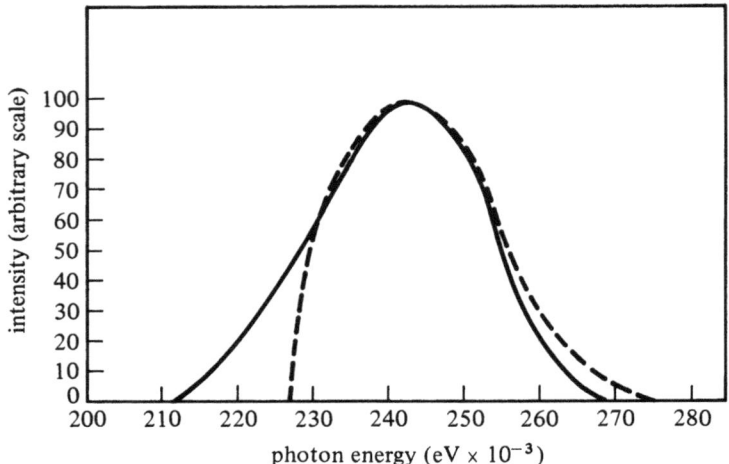

Fig. 4.4.7. Normalized emission spectrum for an *n*-type sample of InSb ($n = 5 \times 10^{16}$ cm^{-2}) at 77 K. The solid curve is measured. The dashed curve is calculated for direct transitions [4.4.7].

The factor $\exp[-(\gamma_h - \eta_v)]$ has been taken into the proportionality constant. This result has been used in Figs. 4.4.6 and 4.4.7 for the theoretical curves with $m_c = 0.015m$, $|m_v| = 0.45m$, $E_G = 0.229$ eV, $(\gamma_e - \eta_c)kT = 30$ *meV*. The experimental long-wavelength tail was attributed to emission from impurities.

From eq. (4.4.17), using B_{IJ} as given by eq. (4.4.14), we can obtain transition rates which have not been averaged over polarization vectors to find a result analogous to eq. (4.4.30) but different in two respects. The average sign is omitted from the matrix element and the assumption that there is no dependence on angles is dropped. This can be done by multiplying by $\cos\theta\,d\theta\,d\varphi/4\pi$ and supplying the angular integral sign. This allows for the introduction of the $\cos\theta$ term due to the angle θ between the vectors

$$\{\nabla_k[e_{k_{ph}\lambda} \cdot P_{vc}(k)]\}_{k-k_0} \quad \text{and} \quad (k-k_0)$$

which arises by eq. (4.4.24) if the forbidden transitions are considered. One finds

$$r_{ij}^{st} = A_1\left\{\int |e_{k_{ph}\lambda} \cdot P_{vc}(k)|^2 (E_k - E_G)^{\frac{1}{2}}\frac{\cos\theta\,d\theta\,d\varphi}{4\pi}\right\}(p_i - p_j) \qquad (4.4.37)$$

where

$$A_1 \equiv \frac{l}{2}\frac{2\mu e^2 E_k (2m^*)^{\frac{3}{2}}}{\pi^2 m^2 c^3 \hbar^5}$$

If the matrix element can be taken out of the integral as a constant, one finds again eq. (4.4.30) for the allowed transitions. For the forbidden transitions a factor $\cos^2 \theta$ arises from the scalar product with $(\mathbf{k} - \mathbf{k}_0)$ leading to an angular term

$$\frac{1}{4\pi} \int_{-1}^{1} \cos^2 \theta \, d(\cos \theta) \int_0^{2\pi} d\varphi = \tfrac{1}{3}$$

The factor $|\mathbf{k} - \mathbf{k}_0|^2$ leads to a term $(2m^*/\hbar^2)(E_k - E_G)$ by energy consideration so that in this case

$$r_{ij}^{\text{st}} = \frac{l}{2} \frac{2\mu e^2 E_k (2m^*)^{\frac{5}{2}}}{3\pi^2 m^2 c^3 \hbar^7} |\{\boldsymbol{\nabla}_k [\mathbf{e}_{\mathbf{k}_{\text{ph}}\lambda} \cdot \mathbf{P}_{\text{vc}}(\mathbf{k})]\}_{\mathbf{k} = \mathbf{k}_0}|^2 (E_k - E_G)^{\frac{3}{2}} \qquad (4.4.38)$$

Note that $E_k = h\nu$ and that $\mathbf{k}_{\text{ph}}$ is the photon wavevector.

4.5 Interband absorption, emission and their relationship

4.5.1 Absorption–emission relationships and radiative recombination in terms of optical data

The absorption coefficient for single-mode operation, $\alpha_{IJ}^{\text{SM}}(\nu)$, can be defined in terms of the net absorption rate per unit volume, $u_{IJ}^{\prime \text{abs SM}}$, and the photon flux $F(N_\nu)$ by

$$u_{IJ}^{\prime \text{abs SM}} = F(N_\nu) \alpha_{IJ}^{\text{SM}}(\nu) (\equiv -u_{IJ}^{\prime \text{st SM}}) \qquad (4.5.1)$$

The prime shows as in eq. (4.4.17) that the stimulated emission rate has been subtracted out, so that

$$u_{IJ}^{\prime \text{abs SM}} = u_{IJ}^{\text{abs SM}} - u_{IJ}^{\text{st SM}} = B_{IJ}^{\text{SM}} p_I (1 - p_J) N(E_k) \left[\frac{(1 - p_I) p_J}{p_I (1 - p_J)} - 1 \right]$$

By the play of probabilities (4.2.13), and using eq. (4.4.31), this is

$$u_{IJ}^{\prime \text{abs SM}} = u_{IJ}^{\text{sp SM}} N(E_k) \left(\exp \frac{h\nu - q\varphi}{kT} - 1 \right) \qquad (4.5.2)$$

This gives the relation between spontaneous emission and absorption for single-mode operation. It also holds normally for multi-mode operation, see eq. (4.5.7).

The flux is obtained by noting that the number of photons crossing area A in time t at an angle Θ to the normal is equal to the number of photons in a right circular cylinder of base area $A \cos \Theta$ and length ct. Hence

$$F(N_\nu) A t = (N_\nu / V) \cdot A \cos \Theta \cdot c' t$$

where $c' = c/\mu$, μ being the refractive index. Also N_ν / V is the volume concentration of photons in a mode of frequency ν. (In this section it is convenient to write N_ν, E_ν instead of $N(E_k)$, E_k for the occupation number and energy of a mode.) Hence

$$F(N_\nu) = N_\nu c \cos \Theta / \mu V \qquad (4.5.3)$$

Thus for $\Theta \sim \pi/2$

$$\alpha_{IJ}^{SM}(v) = (\mu V/c)[B_{IJ}^{SM} p_I(1-p_J)]\left[\exp\frac{hv-q\varphi}{kT}-1\right]$$

Here B_{IJ}^{SM} is given by eq. (4.4.3). The photon mode occupation number has cancelled out since both $u_{IJ}'^{st}$ and $F(N_v)$ are linear in it. The quantity in the first square bracket is just u_{IJ}^{spSM}. The result is

$$\alpha_{IJ}^{SM}(v) = \frac{4\pi^2 e^2 \hbar}{\mu V m^2 c E_v}|M_{IJ}|_{av}^2 p_I(1-p_J)\left(\exp\frac{hv-q\varphi}{kT}-1\right)\delta(E_I-E_J-E_v)$$

(4.5.4)

where I is the level of higher energy. The exponential term represents the contribution of actual absorption. The -1 represents the contribution of stimulated emission.

There is an alternative way of writing this result. Let us consider the ratio

$$\frac{u_{IJ}'^{stSM}}{u_{IJ}^{spSM}} = \frac{B_{IJ}^{SM} N_v[p_I(1-p_J)-p_J(1-p_I)]}{B_{IJ}^{SM} p_I(1-p_J)} = N_v\left(1-\exp\frac{hv-q\varphi}{kT}\right)$$

(4.5.5)

Hence eq. (4.5.4) may also be written

$$\alpha_{IJ}^{SM}(v) = \frac{\mu V}{c}u_{IJ}^{spSM}\left(\exp\frac{hv-q\varphi}{kT}-1\right) = -\frac{\mu V}{cN_v}u_{IJ}'^{stSM}$$

(4.5.6)

Using the number of photon modes in an energy increment dE_v, $\mathcal{N}(E_v)dE_v$, the photon flux for *multi-mode* operation is

$$F(N_v)\mathcal{N}(E_v)dE_v$$

Noting also that the net absorption rate is the net emission rate with its sign reversed, eq. (4.5.1) can be written as

$$\alpha_{IJ}(v) = -\frac{r_{IJ}^{st}(E_v)N_v}{F(N_v)\mathcal{N}(E_v)}$$

Summing over the electronic states and using eq. (4.4.13) one obtains finally

$$\alpha_{ij}(v) = -\frac{r_{ij}^{st}(E_v)N_v}{F(N_v)\mathcal{N}(E_v)} = -r_{ij}^{st}\cdot\frac{\mu V}{c}\cdot\frac{2\pi^2 c^3 \hbar^3}{lV\mu^3 E_v^2}$$

(4.5.7)

Normally $l=2$ (unpolarized radiation) may be chosen.

Can one introduce the spontaneous emission rate into the multi-mode absorption coefficient just as it was possible to do into the single-mode absorption coefficient? That the answer is 'yes' depends on eqs. (4.4.16) and (4.4.17) and on the fact that if H_{IJ} is an appropriate function of I and J

$$\alpha_{ij}(\nu) = \underbrace{\sum_{I \in i}\sum_{J \in j} H_{IJ} p_I (1-p_J)}_{\text{spontaneous}} = \underbrace{\sum_{I \in i}\sum_{J \in j} H_{IJ}(p_I - p_J)}_{\text{stimulated}} \left[1 - \exp\frac{h\nu - q\varphi}{kT}\right]^{-1}$$

where we have made use of the play of probabilities (4.2.13). This is, however, justifiable only if (1) the sums go over states I, J of fixed energy difference, and (2) the distributions are of the quasi-Fermi type. Here (1) is justified because of the Kroenecker delta in eqs. (4.4.16) and (4.4.17). It follows that there are two useful alternatives:

$$\alpha_{ij}(\nu) = -\frac{2\pi^2 c^2 \hbar^3}{l\ \mu^2 E_\nu^2} \times \begin{cases} r_{ij}^{\text{st}}(E_\nu) \\[2ex] \left(1 - \exp\dfrac{h\nu - q\varphi}{kT}\right) r_{ij}^{\text{sp}}(E_\nu) \end{cases} \tag{4.5.8}$$

where the superfix SM has again been omitted for multi-mode operation. Inverting eq. (4.5.8) and using eq. (4.4.15)

$$r_{ij}^{\text{sp}}(E_\nu) = -\frac{l\mu^2 E_\nu^2 \alpha_{ij}(\nu)}{\pi^2 c^2 \hbar^3 \{1 - \exp[(h\nu - q\varphi)/kT]\}} \tag{4.5.9}$$

$$R^{\text{sp}} = -\frac{l(kT)^3}{2\pi^2 c^2 \hbar^3} \int_0^\infty \frac{\mu^2 \alpha_{ij}(\nu) x^2 \, dx}{1 - \exp(x - q\varphi/kT)} \tag{4.5.10}$$

This gives one a way of estimating the total spontaneous radiation rate per unit volume from a knowledge of $q\varphi$, which is negligible near equilibrium, and the optical data

$$\mu(\nu), \quad \alpha(\nu)$$

For $l = 2$ the coefficient in front of the integral is

$$6.83(T/300)^3 \times 10^{18} \ (\alpha_{ij} \text{ is in cm}^{-1}, T \text{ in K})$$

From the method of derivation the results (4.5.7) to (4.5.10) hold for all absorption mechanisms: allowed or forbidden, direct or indirect.

If one puts $q\varphi = 0$ and replaces α by

$$(4\pi\mu\nu/c) \times (\text{absorption index}) \tag{4.5.11}$$

then eq. (4.5.10) gives a result due to van Roosbroeck and Shockley [5.5.1] and applied extensively to the estimation of radiative lifetimes [5.5.2].

On inserting eq. (4.4.30) into eq. (4.5.8) one finds for direct transitions with $hv > E_G$

$$\alpha_{ij}(v) = A'(hv - E_G)^{\frac{1}{2}}(p_i - p_j) \qquad (4.5.12)$$

where

$$A' \equiv \frac{e^2(2m^*)^{\frac{3}{2}}}{\mu cm\hbar^2}f_{ij}, \quad f_{ij} \equiv \frac{4}{lmE_v}|M_{IJ}|^2_{av} \qquad (4.5.13)$$

Normally one assumes $p_i - p_j = 1$. If $hv < E_G$ then $\alpha(v) = 0$. Here f is the oscillator strength of the transition. The origin of this concept is as follows. Suppose an atom is regarded as a rigid ion with a single outer electron. Then the behavior of this electron in the presence of light can often be simulated by a number of oscillators, each of charge e and mass m/f_{ij}. There is one oscillator for each admissible frequency $(E_i - E_j)/h$. The result (4.5.12) was obtained in essence already in 1956 [4.5.3]. Indeed, the basic theory of radiative transitions in semiconductors which has been outlined here dates from this period. If one wishes to give a standard reference, one might turn to the 1963 paper by Lasher and Stern [4.5.4]; see also [4.5.5]. We have followed in part [4.5.6], but see also [4.5.7] to [4.5.9]. The absorption coefficient for forbidden direct transitions in multi-mode operation is found by inserting eq. (4.4.38) into eq. (4.5.8) when one finds

$$\alpha_{ij}(v) = \frac{2e^2(2m^*)^{\frac{5}{2}}}{3\mu m^2 c\hbar^4 E_v}|\{\nabla_k[e_{k_{ph}\lambda} \cdot P_{vc}(k)]\}_{k=k_0}|^2(hv - E_G)^{\frac{3}{2}} \qquad (4.5.14)$$

4.5.2 Interband indirect transitions and excitonic effects

Indirect radiative transitions are important for indirect semiconductors such as Ge and Si. A phonon of energy hq is absorbed (top sign) or emitted (bottom sign) and energy conservation yields (Fig. 4.5.1)

$$hv \pm hq = E_G + \varepsilon_c + \varepsilon_v \qquad (4.5.15)$$

where ε_c is the kinetic energy of the conduction band electron and ε_v the kinetic energy of the valence band hole. For the most likely transitions ε_c, ε_v are small so that the phonon wavevector, q^* say, corresponding to the phonon energy hq (and determined by a phonon dispersion relation) is $q^* = k_c - k_v \sim k_0$, where k_0 is the displacement in k-space between the extrema. One needs to use second-order perturbation theory which involves not only the radiative matrix element but also the phonon-interaction matrix element. As an approximation we shall treat them as independent of v, k_c and k_v so that the transition rate is simply proportional to

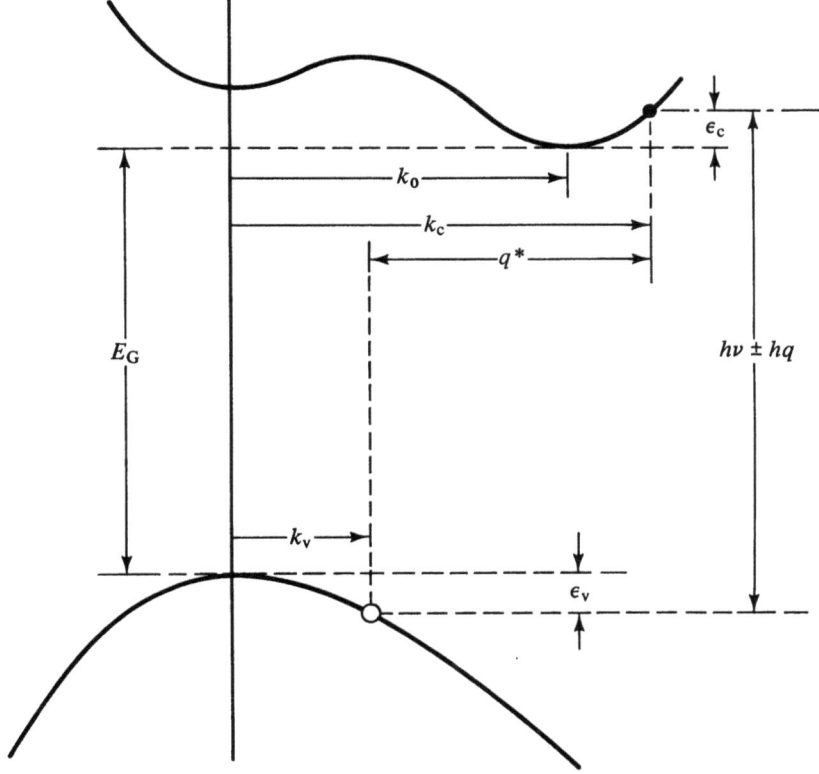

Fig. 4.5.1. Indirect radiative absorption in which an electron at k_v is promoted to a state k_c. The top sign applies to the case when a phonon is also absorbed by the electron system. The bottom sign holds if a phonon is emitted by the electron system.

the product of the densities of states at k_c and k_v subject to eq. (4.5.15) in which v and q are constants. So the v-dependence of the absorption coefficient is expected to depend on

$$\int (\varepsilon_c \varepsilon_v)^{\frac{1}{2}} \delta(h\nu \pm hq - E_G - \varepsilon_c - \varepsilon_v)\, d\varepsilon_c\, d\varepsilon_v$$

$$= \int_0^a (a - \varepsilon_v)^{\frac{1}{2}} \varepsilon_v^{\frac{1}{2}}\, d\varepsilon_v = \frac{\pi a^2}{8} \quad (a \equiv h\nu \pm hq - E_G)$$

Hence if the coefficient A denotes a slowly varying function of v, one finds proportionality to $A(h\nu - E_G \pm hq)^2$, i.e. a higher power than for direction transitions. The coefficient is complicated and for the details specialized discussions [4.5.3], [4.5.7], [4.5.8], [4.5.10] and [4.5.11] must be consulted. They lead, however, to the form of result obtained here and summarized in Table 4.5.1 which is

Table 4.5.1. *Frequency dependence of absorption coefficients between parabolic bands*

Exciton effects neglected	Exciton effects included $(t = 1, 2, 3, \ldots)$
Direct, allowed	
$hv \leqslant E_G : 0$	Lines at $\hbar\omega = E_G - \dfrac{E_{ex}(1)}{t^2}$
	$\left[E_{ex}(1) \equiv \dfrac{m^*}{m\varepsilon^2} 13.6 \text{ eV} \right]$
	with oscillator strength $f_t \propto 1/t^3$
$hv \geqslant E_G : \alpha_1 = \dfrac{A_1}{\omega}(hv - E_G)^{\frac{1}{2}}$	$\alpha_1' = \alpha_1 \dfrac{\pi z \exp(\pi z)}{\sinh(\pi z)}$ for the continuum
$A_1 \equiv \dfrac{2e^2}{\mu m^2 c} \left(\dfrac{2m^*}{\hbar^2}\right)^{\frac{3}{2}} \left[\mathbf{e}_{\mathbf{k}_{ph}\lambda} \cdot \mathbf{P}_{vc}(0) \right]^2$	$\left[z \equiv \left(\dfrac{E_{ex}(1)}{hv - E_G} \right)^{\frac{1}{2}} \right]$
Direct, forbidden	
$hv \leqslant E_G : 0$	Lines at $hv = E_G - \dfrac{E_{ex}(1)}{t^2}$
	with oscillator strength $f_t = (t^2 - 1)/t^5$
$hv \geqslant E_G : \alpha_2 = \dfrac{A_2}{\omega}(hv - E_G)^{\frac{3}{2}}$	$\alpha_2' = \alpha_2 \left(1 + \dfrac{1}{z^2}\right) \dfrac{\exp(\pi z)}{\sinh(\pi z)}$ for the
$A_2 \equiv \dfrac{2e^2}{3\mu m^2 c} \left(\dfrac{2m^*}{\hbar^2}\right)^{\frac{5}{2}} |\nabla_\mathbf{k} \mathbf{e}_{\mathbf{k}_{ph}\lambda} \cdot \mathbf{P}_{vc}(\mathbf{k})|^2_{\mathbf{k}=0}$	continuum
Indirect, allowed	
Phonon absorption	*Bound exciton, continuous absorption*
$hv \leqslant E_G - hq : \alpha_3 = 0$	$\alpha_3' \propto \dfrac{1}{t^3}(hv - E_G + hq)^{\frac{1}{2}}$
$hv \geqslant E_G - hq : \alpha_3 = A_3 \dfrac{(hv - E_G + hq)^2}{\exp(hq/kT) - 1} \left(A_3 \propto \dfrac{1}{hv} \right)$	$\alpha_3' \propto (hv - E_G + hq)^{\frac{3}{2}}$
Phonon emission	*Bound exciton, continuous absorption*
$hv \leqslant E_G + hq : \alpha_4 = 0$	$\alpha_4' \propto \dfrac{1}{t^3}(hv - E_G + hq)^{\frac{1}{2}}$
$hv \geqslant E_G + hq : \alpha_4 = A_3 \dfrac{(hv - E_G - hq)^2}{1 - \exp(-hq/kT)}$	$\alpha_4' \propto (hv - E_G - hq)^{\frac{3}{2}}$
Indirect, forbidden	
Phonon absorption	*Bound exciton, continuous absorption*
$hv \leqslant E_G - hq : \alpha_5 = 0$	$\alpha_5' \propto (hv - E_G + hq)^{\frac{3}{2}} \dfrac{t^2 - 1}{t^5}$

Table 4.5.1. (*Cont.*)

$hv \geqslant E_{\mathrm{G}} - hq : \alpha_5 = A_5 \dfrac{(hv - E_{\mathrm{G}} + hq)^3}{\exp(hq/kT) - 1}$	$\alpha_5' \propto (hv - E_{\mathrm{G}} + hq)^{\frac{5}{2}}$
Phonon emission	*Bound exciton, continuous absorption*
$hv \leqslant E_{\mathrm{G}} + hq : \alpha_6 = 0$	$\alpha_6' \propto (hv - E_{\mathrm{G}} - hq)^{\frac{3}{2}} \dfrac{t^2 - 1}{t^5}$
$hv \geqslant E_{\mathrm{G}} + hq : \alpha_6 = A_5 \dfrac{(hv - E_{\mathrm{G}} - hq)^3}{1 - \exp(-hq/kT)}$	$\alpha_6' \propto (hv - E_{\mathrm{G}} - hq)^{\frac{5}{2}}$

developed from [4.5.12]. The absorption coefficient is in the case of phonon absorption also proportional to the phonon occupation number, $N_{\mathbf{q}}$ say, of a mode of wavevector $\mathbf{q}^*$; in the case of emission it is proportional (as in the photon case (4.2.10)) to $N_{\mathbf{q}} + 1$. Since phonons are bosons one finds from eq. (1.2.8) with $a = -1$

$$N_{\mathbf{q}} = \left[\exp\frac{hq}{kT} - 1\right]^{-1}, \quad N_{\mathbf{q}} + 1 = \left[1 - \exp\left(-\frac{hq}{kT}\right)\right]^{-1}$$

These factors are included in Table 4.5.1. Combining these results, the indirect absorption coefficient is

$$\alpha = \begin{cases} A\left\{\dfrac{(hv - E_{\mathrm{G}} + hq)^2}{\exp(hq/kT) - 1} + \dfrac{(hv - E_{\mathrm{G}} - hq)^2}{1 - \exp(-hq/kT)}\right\} & (E_{\mathrm{G}} + hq \leqslant hv) \quad (4.5.16) \\[3mm] A\dfrac{(hv - E_{\mathrm{G}} + hq)^2}{\exp(hq/kT) - 1} & (E_{\mathrm{G}} - hq \leqslant hv \leqslant E_{\mathrm{G}} + hq) \\[2mm] & \hspace{3cm} (4.5.17) \\[2mm] 0 & (hv \leqslant E_{\mathrm{G}} - hq) \end{cases}$$

To complete the picture one has to write down such expressions for longitudinal and transverse acoustic and optical modes. In each case one can also obtain the corresponding expressions for $r^{\mathrm{st}}(E_{\mathrm{v}})$ and $r^{\mathrm{sp}}(E_{\mathrm{v}})$ by use of eq. (4.5.8). In the case of Si all four cases are needed to fit experimental results, the phonon energies being (in meV):

long. ac: 57.7, trans. ac: 18.5, long. op: 91, trans. op: 120

The expression (4.5.16) was used for comparison with experiment in the important original letters from the (then) Radar Research Establishment on Ge [4.5.13] and Si [4.5.14] (Figs. 4.5.2 and 4.5.3), showing that direct transitions do not

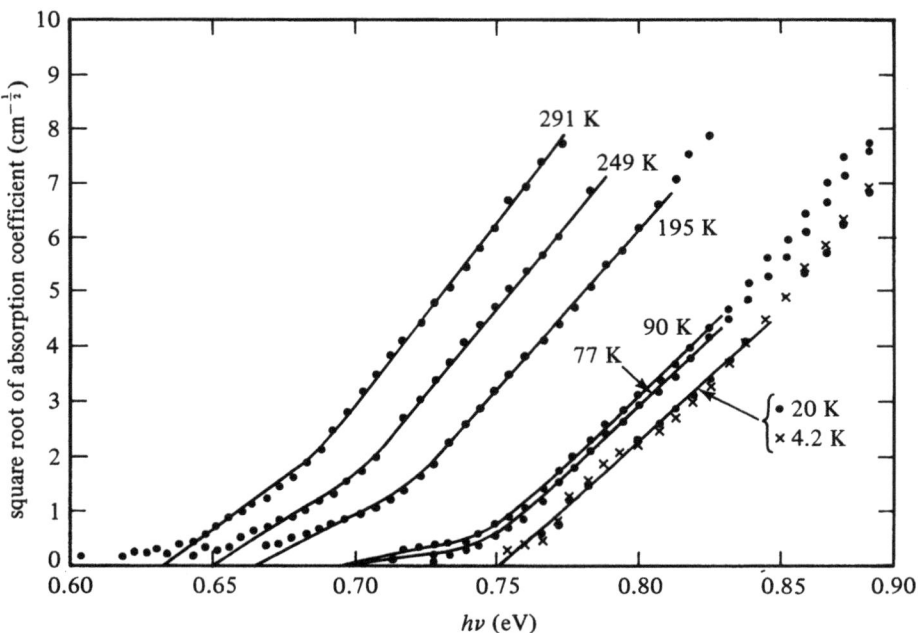

Fig. 4.5.2. Dependence of absorption constant on photon energy for 60 Ω-cm Ge [4.5.13].

play an important part. More detailed discussions were given in their full papers [4.5.15]. The work is well reviewed by R.A. Smith [4.5.10], who was head of the RRE Physics Department when these researches were pursued. Other work of a similar nature followed, and Fig. 4.5.4 gives (as a further example) results for alpha SiC where eq. (4.5.16) was again used ($hq \sim 90$ meV) [4.5.16].

The Coulomb interaction between the current carriers which played so important a part in chapter 3, has so far been neglected in this chapter. One of its consequences is that it leads to electron–hole bound states. These excitons are somewhat analogous to hydrogen atoms. The reduced mass (4.4.27) requires re-interpretation for hydrogen of course, e.g. $m_c \rightarrow$ electron mass, $|m_v| \rightarrow$ proton mass. For the hydrogen atom $m^* \sim m_c$, whereas m^* lies more evenly between m_c and $|m_v|$ for excitons. Since the system is in a medium of dielectric constant ε, the results (1.8.2) and (1.8.4) apply to yield exciton radii

$$\tilde{r}_t \equiv (\varepsilon m/m^*)\, t \times 0.528 \text{ Å} \quad (r_1 \equiv \hbar^2/me^2 = 0.528 \text{ Å}) \tag{4.5.18}$$

and exciton binding energies

$$E_{xt} \equiv \frac{m^*}{m\varepsilon^2}\frac{1}{t^2} \times 13.6 \text{ eV} \quad (I_0 \equiv me^4/2\hbar^2 = 13.6 \text{ eV}) \tag{4.5.19}$$

for principal quantum numbers $t = 1, 2, \ldots$. here $m^{*-1} \equiv m_c^{-1} + m_h^{-1}$. This intuitive approach to eqs. (4.5.18) and (4.5.19) must be replaced by a properly founded

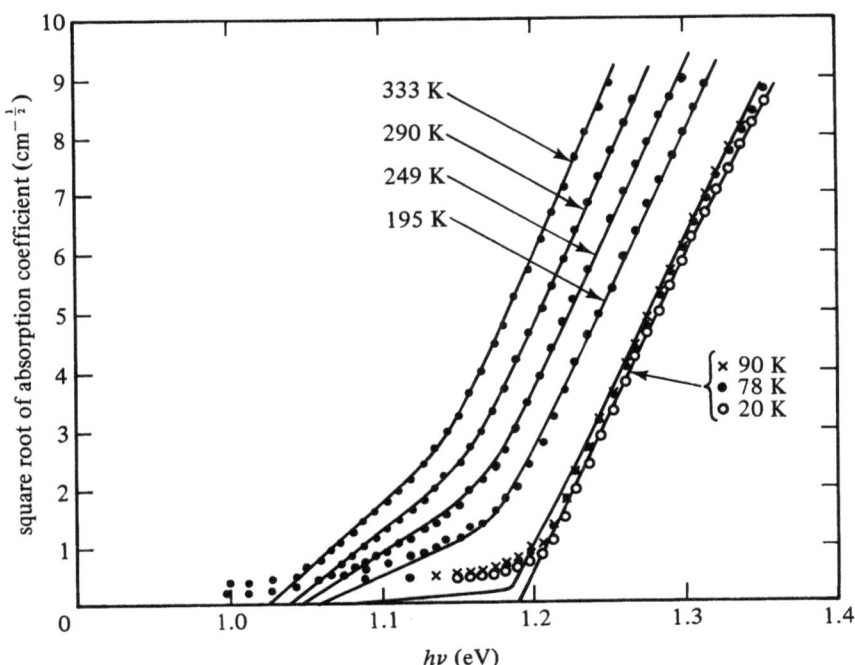

Fig. 4.5.3. Dependence of absorption constant on photon energy for Si [4.5.14].

effective mass approximation for excitons [4.5.17], [4.5.18]. This has been achieved in close analogy with the theory for shallow donors [4.5.19] and acceptors [4.5.20]. The effect of excitons rapidly decreases as the temperature is raised since even E_{x1}, the binding energy of the ground state which exceeds that of the excited states $t = 2, 3, \ldots$, is of the order of only a few meV. In fact, with $\varepsilon = 16$ and $m_c/m \sim 0.1$ (the case of Ge) one has $\bar{r}_1 \sim 80$ Å, so that an exciton responds to *average* semiconductor properties, and $E_{x1} \sim 5$ meV. This corresponds to kT at T = 60 K. Above this temperature thermal excitation is expected to break up the excitons. The situation is even more delicate since screening of the Coulomb attraction (see section 3.4.3) also weakens the binding. This is discussed on p. 437, below, and for example in [4.5.21]. One finds that the observation of exciton effects is favored by low temperatures and pure samples. In very pure cuprous oxide exciton peaks have been observed up to $t = 9$ at 4.2 K. They are part of the so-called yellow series. For a review see [4.5.22].

In Table 4.5.1 modifications of the hv-dependence of the absorption coefficient as a result of exciton participation are given. They are largely due to Elliott [4.5.23], reviewed in [4.5.24]. Owing to their complication, some of the A-coefficients have not been identified in Table 4.5.1. The exciton enhancement of absorption by the so-called Sommerfeld factor $\pi z \exp \pi z / \sin(\pi z)$ has been observed

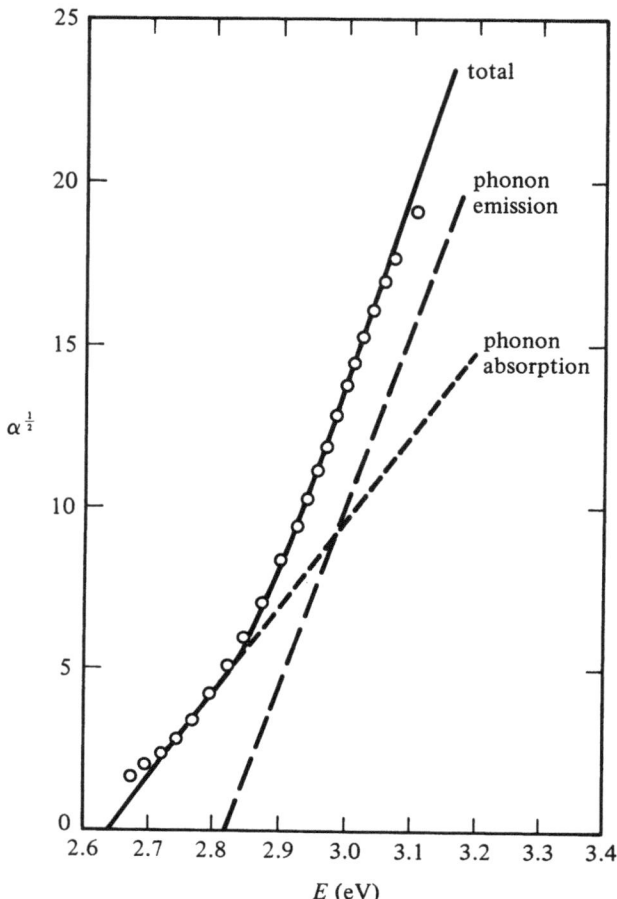

Fig. 4.5.4. Comparison of observed and calculated absorption in alpha SiC at 700 K [4.5.16].

very clearly, for example in semi-insulating GaAs [4.5.25], showing the expected decrease at the higher temperatures (Fig. 4.5.5). For less pure samples the α_1-expression of Table 4.5.1 is expected to apply, but the absorption edge is blurred by the 'Urbach rule' effect [4.5.26]

$$\alpha \propto \exp(-h\nu/kT), \tag{4.5.20}$$

giving a much gentler rise (Fig. 4.5.6). Experiment shows a shift of the edge to higher energies as doping is increased (Fig. 4.5.7) [4.5.27]. This is partly due to the filling up of the conduction band so that an electron has to bridge a greater energy in order to absorb a photon; this is the Burstein–Moss effect [4.5.28], [4.5.29]. It must however be emphasized that in the case of the heavy doping envisaged in Fig. 4.5.7 the fluctuations in the carrier density and of the impurity concentration as

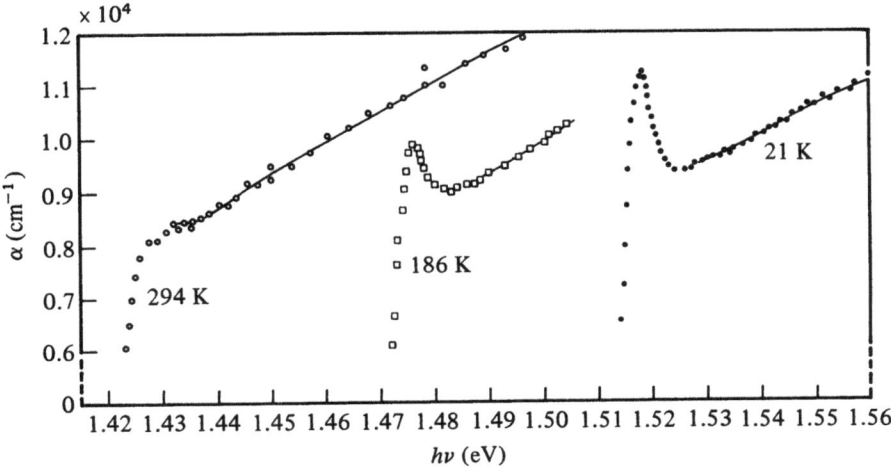

Fig. 4.5.5. Intrinsic absorption in GaAs of less than 10^{10} carriers per cm³. The bound exciton peaks disappear as the temperature is raised [4.5.25].

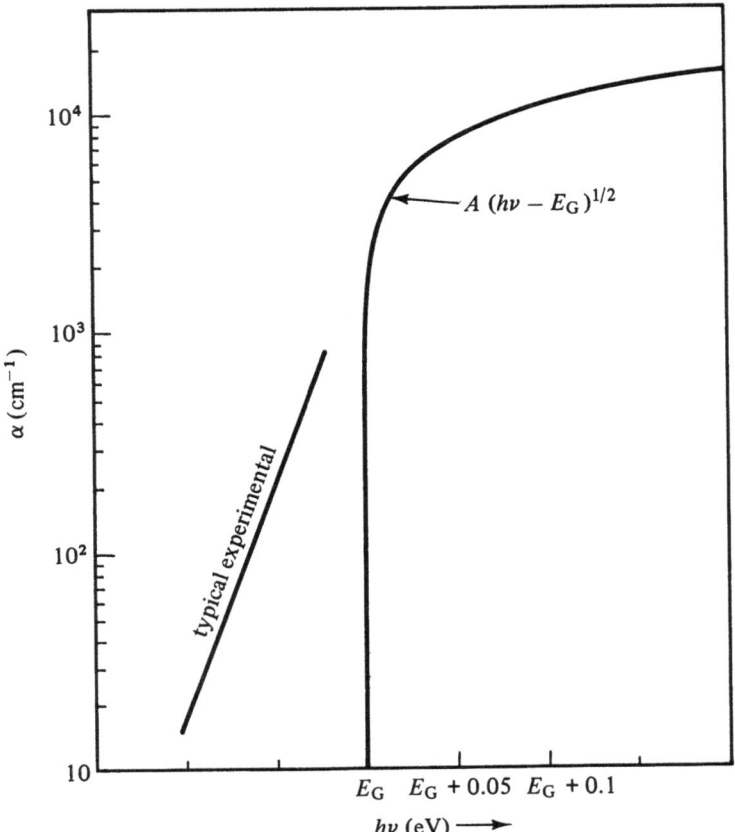

Fig. 4.5.6. Qualitative behavior of an absorption edge [4.5.27].

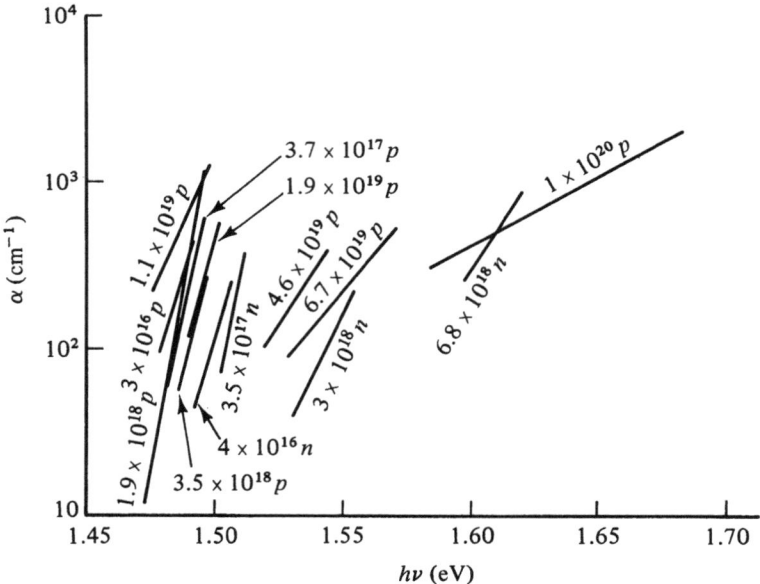

Fig. 4.5.7. Absorption edge of *n*- and *p*-type GaAs at 5 K [4.5.27]. Whether sample is *n*-type or *p*-type is also indicated.

well as additional scattering effects will make the simple theory outlined here highly approximate. The Urbach tail is believed to be due to thermal disorder (atomic vibrational displacements) in crystalline materials, and to structural disorder in the case of amorphous materials. In addition band gap estimates from absorption data are blurred by free carrier excitations which do not arise in the case of photoluminescence.

4.5.3 Analytical interband emission rates via absorption rates

By eq. (4.5.10) each absorption coefficient implies radiative recombination in the form of spontaneous emission. Neglecting the frequency dependence of the refractive index we shall consider an absorption coefficient whose frequency dependence has the form

$$\alpha(v) = B(hv - AkT)^b = B(kT)^b(\eta - A)^b \quad (\eta = hv/kT)$$

where B, b and A are constants. Table 4.5.1 shows that this covers many cases of interest in so far as they can be handled analytically. Absorption rises at $\eta = A$ and emission is made convergent by the Bose statistics factor

$$\frac{1}{e^\eta - 1} = \frac{e^{-\eta}}{1 - e^{-\eta}} = e^{-\eta}(1 + e^{-\eta} + e^{-2\eta} + \ldots)$$

Now A is of the order E_G/kT and exceeds unity so that the first few terms in the expansion are adequate. Taking only the first term, the integral for R^{sp} involves

$$J = \int_A^\infty (\eta - A)^b \eta^2 e^{-\eta} \, d\eta = e^{-A} \int_0^\infty x^b (x+A)^2 e^{-x} \, dx$$
$$= e^{-A} \Gamma(b+1) \{ A^2 + 2(b+1)A + (b+2)(b+1) \} \tag{4.5.21}$$

Special cases are given in Table 4.5.2.

For the *direct* allowed spontaneous emission rate one finds for example using eq. (4.5.12) in eq. (4.5.10) as well as eq. (4.5.21)

$$R^{sp} = \frac{l}{2} \frac{16(\pi kT)^{\frac{7}{2}} e^2 (2m^*)^{\frac{3}{2}} \mu f_{cv}}{mc^3 h^5} e^{-\eta_G} (\eta_G^2 + 3\eta_G + \tfrac{15}{4}) \tag{4.5.22}$$

Formulae of this type go back to R.N. Hall [4.5.30]; see also [4.5.2]. The first term usually dominates. For a nondegenerate semiconductor $R^{sp} = Bnp = Bn_i^2$ so that the recombination coefficient for direct allowed spontaneous radiation is, using eq. (1.12.10),

$$B \equiv \frac{R^{sp}}{n_i^2} = \frac{l}{2} \frac{(2\pi)^{\frac{1}{2}} \mu e^2 h f_{cv} E_G^2}{mc^3 (kT)^{\frac{3}{2}} (m_c + m_h)^{\frac{3}{2}}} \tag{4.5.23}$$

A convenient lifetime is the 'intrinsic' radiative lifetime

$$\tau_i \equiv \frac{n_i}{2R} = \frac{2mc^3 h (m_c m_v)^{\frac{3}{4}}}{l \, 8\pi^2 e^2 \mu f_{cv} E_G^2} \left(\frac{m_c + m_v}{m_c m_v} \right)^{\frac{3}{2}} \exp \frac{\eta_G}{2} \tag{4.5.24}$$

from which the radiative lifetime

$$\tau \equiv 2n_i \tau_i / (n+p)_{eq}$$

is easily derived.

A similar argument for the spontaneous emission by interband allowed *indirect* transitions depends on substituting eqs. (4.5.16) and (4.5.17) into eq. (4.5.10). One finds

$$R^{sp} = \frac{l}{2} A \frac{8\pi (kT)^5 \mu^2 (kT)^2}{c^2 h^3 [\exp(hq/kT) - 1]} X \tag{4.5.25}$$

where

$$X \equiv 2\exp\left(-\eta_G + \frac{hq}{kT}\right) \left[\left(\eta_G - \frac{hq}{kT}\right)^2 + 6\left(\eta_G - \frac{hq}{kT}\right) + 12 \right]$$
$$+ 2\exp(-\eta_G) \left[\left(\eta_G + \frac{hq}{kT}\right)^2 + 6\left(\eta_G + \frac{hq}{kT}\right) + 12 \right]$$

Tale 4.5.2. *The integral J*

b	A		
$\frac{1}{2}$	E	allowed direct	$e^{-A}\dfrac{\pi^{\frac{1}{2}}}{2}\left[A^2+3A+\dfrac{15}{4}\right]$
$\frac{3}{2}$ $\Big\{$	E_G	forbidden direct	$e^{-A}\dfrac{3\pi^{\frac{1}{2}}}{4}\left[A^2+3A+\dfrac{35}{4}\right]$
	$E_G\pm hq$	indirect allowed with exciton	
2	$E_G\pm hq$	allowed indirect	$e^{-A}2[A^2+6A+12]$
$\frac{5}{2}$	$E_G\pm hq$	forbidden indirect with excitons	$e^{-A}\dfrac{15\pi^{\frac{1}{2}}}{8}\left[A^2+7A+\dfrac{63}{4}\right]$
3	$E_G\pm hq$	forbidden indirect	$e^{-A}6[A^2+8A+20]$

If one neglects hq compared with E_G and assumes that the η_G-terms dominate then

$$X \simeq 2\eta_G^2[1+\exp(hq/kT)]\exp(-\eta_G)$$

so that

$$R^{sp} = \frac{l}{2}A\frac{16\pi\mu^2(kT)^3}{c^3h^3}E_G^2\exp(-\eta_G)\coth\left(\frac{hq}{2kT}\right) \tag{4.5.26}$$

This leads to

$$B = \frac{l}{2}A\frac{\mu^2h^3E_G^2}{2\pi^2c^2(m_c m_v)^{\frac{3}{2}}}\coth\left(\frac{hq}{2kT}\right) \tag{4.5.27}$$

and

$$\tau_1 = \frac{2}{lA}\frac{\pi^{\frac{1}{2}}c^2(m_c m_v)^{\frac{3}{4}}}{2^{\frac{5}{2}}\mu^2(kT)^{\frac{3}{2}}E_G^2}\exp\left(\frac{\eta_G}{2}\right)\coth\left(\frac{hq}{2kT}\right) \tag{4.5.28}$$

In fact, A is inversely proportional to $h\nu$, $A = A_1/h\nu$ say [4.5.7]. One then finds by analogous arguments

$$R^{sp} \sim \frac{l}{2}A_1\frac{16\pi\mu^2(kT)^3E_G}{c^2h^3}e^{-\eta_G}\coth\left(\frac{hq}{2kT}\right) \tag{4.5.29}$$

Thus in this approximation AE_G^2 in eq. (4.5.26) is replaced by $A_1 E_G$ in eq. (4.5.29). A similar change would have to be applied to eq. (4.5.27) and (4.5.28) to yield the improved expressions for the recombination coefficient B and the radiative lifetime τ_1.

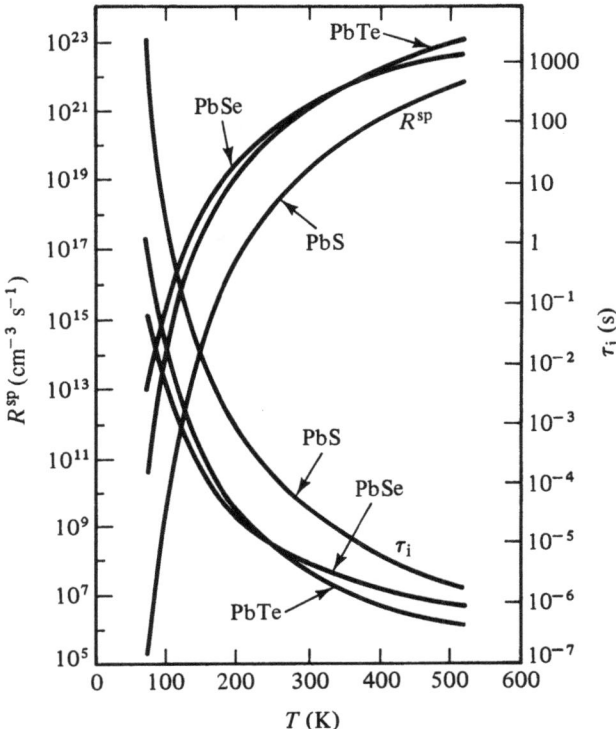

Fig. 4.5.8. Inferred radiative recombination and intrinsic lifetimes for some lead compounds as a function of temperature.

The temperature dependence of the effective recombination cross section

$$\sigma_r \equiv R^{sp}/n_i^2 v = B/v$$

(v = average thermal velocity of the carriers, and proportional to $T^{\frac{1}{2}}$) is obtainable as

$$\sigma_r \propto T^{-2} \text{ (direct)}, \quad T^{-\frac{1}{2}} \coth{(hq/2kT)} \text{ (indirect)}$$

This agrees with a closely related analytical discussion [4.5.31] via the absorption index (and an integration in the limit of large enough E_G/kT if hq/kT is neglected). An application to some lead compounds from this reference is shown in Fig. 4.5.8.

4.6 Recombination–generation induced phase transitions and chaos

4.6.1 Nonequilibrium phase transitions due to generation–recombination

Now that Auger and radiative transitions have been studied, a discussion of some more general aspects of recombination is in order: fluctuations, phase transitions and chaos in semiconductors. Although by no means appropriate only to radiative recombination, it is convenient to include these topics in the present chapter. The 'route to chaos', to use a currently fashionable concept, which will be traced here depends on recurrence relations. Such relations will therefore be used first to obtain some simpler but important results in the fluctuation theory for semiconductors. Section 4.6.2 is more advanced and not needed for section 4.6.3.

Assume that there are n electrons in a conduction band. (We revert to *numbers* instead of *concentrations* in this section.) Let $g(n)\,dt$, $r(n)\,dt$ be the probabilities that one electron enters or leaves the band in time dt. Then the probability distribution function $P(n)$, assumed to depend on one variable only, satisfies

$$\dot{P}(n) = r(n+1)\,P(n+1) + g(n-1)\,P(n-1) - P(n)[g(n)+r(n)] \tag{4.6.1}$$

In the steady state the solution of the above difference equation is (by induction or otherwise)

$$P(n) = \left[\prod_{i=0}^{n-1} g(i) \Big/ \prod_{j=1}^{n} r(j)\right] P(0) \tag{4.6.2}$$

The most probable value, n_0, of n arises from $d\ln P(n)/dn = 0$, i.e. for large n_0

$$g(n_0) = r(n_0) \quad (n_0 \gg 1) \tag{4.6.3}$$

Next, expanding generation and recombination rates about n_0,

$$g(n) - r(n) = -\frac{n-n_0}{\tau} + \dots, \quad \tau \equiv \frac{1}{r'(n_0) - g'(n_0)} \tag{4.6.4}$$

Thus the processes considered lead in general (for $\tau > 0$) back to the steady state with a relaxation time τ, provided the disturbance was small.

To obtain the distribution near the most probable value, put

$$\ln P(n) = \int_0^{n-1} [\ln g(i)]\,di - \int_1^n [\ln r(j)]\,dj \tag{4.6.5}$$

One then has

$$\frac{d\ln P(n)}{dn} = \ln g(n) - \ln r(n)$$

$$\frac{d^2 \ln P(n)}{dn^2} = \frac{g'(n)}{g(n)} - \frac{r'(n)}{r(n)}$$

At $n = n_0$ the first derivative vanishes and one recovers eq. (4.6.3). Hence

$$\ln P(n) = \ln P(n_0) - \tfrac{1}{2}(n-n_0)^2 (r'/r - g'/g)_{n-n_0} + \dots$$

If one confines oneself to the neighborhood of n_0, one has

$$P(n) = P(n_0)\exp[-\tfrac{1}{2}(n-n_0)^2/\tau g(n_0)] \tag{4.6.6}$$

The variance is then found to be

$$\sigma^2 \equiv \langle (n-n_0)^2 \rangle = (2\pi)^{\frac{1}{2}} P(n_0) [\tau g(n_0)]^{\frac{3}{2}} \tag{4.6.7}$$

A normal distribution is found on normalizing eq. (4.6.5) to unity, which yields

$$P(n_0) = [2\pi\tau g(n_0)]^{-\frac{1}{2}}$$

A large variance is liable to result from slow relaxation (large τ) and large steady-state rates (4.6.3). The results (4.6.4) to (4.6.7) were given early on by the late R.E. Burgess [4.6.1], [4.6.2] and are sometimes referred to as the recombination–generation theorem.

As an example consider an intrinsic semiconductor with some band–band processes and an additional process. The additional process will not be specified, except by the lifetime L^{-1} which would be found if it were to act by itself. The band–band processes to be used are (see p. 103) Y^s, Y_1, B^s. Thus

$$g(n) = Y^s + Y_1 n, \quad r(n) = Ln + B^s n^2 \tag{4.6.8}$$

It follows that eq. (4.6.3) yields

$$n_0 = \begin{cases} n_1 \equiv 0 & \tag{4.6.9} \\ (n_2 \equiv (Y_1 - L)/B^s & \end{cases} (Y^s = 0) \qquad \tag{4.6.10}$$

$$n_0 = \begin{cases} n_3 \equiv n_1^* \equiv \dfrac{L-Y_1}{2B^s}\left\{\left[1 + \dfrac{4Y^s B^s}{(L-Y_1)^2}\right]^{\frac{1}{2}} - 1\right\} & (Y^s \neq 0, Y_1 < L) \\[4mm] n_4 \equiv n_2^* \equiv \dfrac{Y_1 - L}{2B^s}\left\{1 + \left[1 + \dfrac{4B^s Y^s}{(Y_1 - L)^2}\right]^{\frac{1}{2}}\right\} & (Y^s \neq 0, Y_1 > L) \end{cases} \tag{4.6.11}$$

Thus there are two solutions if $Y^s = 0$ and two solutions if $Y^s \neq 0$; n_1^* converges to n_1 and n_2^* converges to n_2 if $Y^s \to 0$. In all, one has to consider four cases. The relaxation times are given by eq. (4.6.4) and are

$$\{\tau_1, \tau_2, \tau_3 = \tau_4\} = \left\{\dfrac{1}{L - Y_1}, \dfrac{1}{Y_1 - L}, \dfrac{1}{[(Y_1 - L)^2 + 4B^s Y^s]^{\frac{1}{2}}}\right\} \tag{4.6.12}$$

This result is instructive as it shows that the first solution requires $L > Y_1$ for stability. For $Y_1 > L$ it becomes unstable and (if $Y^s = 0$) the second solution takes over. One can thus infer the existence of a second-order 'phase transition' from the state $n = n_1 = 0$ to the state $n = n_2 > 0$ as the impact ionization coefficient is increased, for example by increasing the electric field. One sees from eq. (4.6.7) that (as usual in statistical mechanics) the fluctuations become very large near the transition point $L \sim Y_1$. The solution $n = 0$ is available only if $Y^s = 0$. This is analogous to the condition $H \to 0$ in the ferromagnetic phase transition, H being the externally applied magnetic field (Fig. 4.6.1).

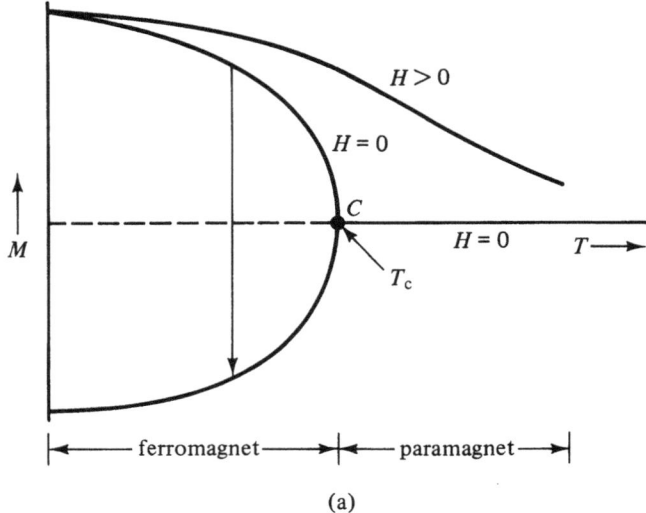

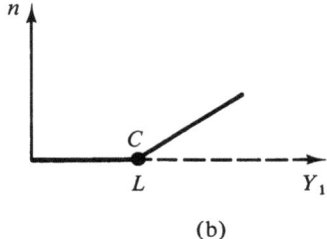

Fig. 4.6.1. Second-order phase transitions at points C. (a) On raising the temperature the magnetization M vanishes at $T = T_c$ provided there is no applied magnetic field H. Because of fluctuations there can be domains with oppositely oriented magnetization as $H \to 0$ provided $T < T_c$. (b) On increasing impact ionization (by increasing the electric field) the stable electron concentration n in an intrinsic semiconductor can suddenly increase from zero. This assumes $Y^s = 0$.

In the simple example given above one need not argue via the recombination–generation theorem; it suffices to write down the reaction rate law

$$\dot{n} \equiv Y^s + (Y_1 - L)n - B^s n^2 = 0 \qquad (4.6.13)$$

for the steady state. On solving the quadratic equation one recovers eqs. (4.6.11).

The vanishing of a polynomial of order two gives two solutions, and this simple phenomenon is responsible for the phase transition. A simple stability analysis of the solutions can be used to confirm the conclusions obtained from eq. (4.6.12). This field is reviewed in [4.6.3]. This and more complicated examples, in which the quadratic is replaced by higher-order polynomials, show that it is possible to illustrate in the field of semiconductors the general idea that a nonequilibrium system can have a remarkable reservoir of structure and that this can lead to new 'organization' of the system, i.e. to new phases. Models for switching which exhibit hysteresis have also been based on these ideas [4.6.4].

4.6.2 Recurrence relations for M-level systems

Section 4.6.1 can be generalized from one variable n to $M-1$ independent occupation numbers denoted by a vector

$$\mathbf{n} = (n_1, n_2, \ldots, n_{M-1})$$

The Mth number (n_M) is assumed to be present but eliminated by electron conservation so as to leave variables which are independent. We also introduce: the maximum electron numbers N_i of level i; the normalized probability $P_t(\mathbf{n})$ of finding the system in state $\mathbf{n}$ at time t:

$$\sum_{\mathbf{n}=0}^{\mathbf{N}} P_t(\mathbf{n}) = 1 \tag{4.6.14}$$

the average of any function $f(\mathbf{n})$ at time t:

$$\langle f(\mathbf{n}) \rangle_t \equiv \sum_{\mathbf{n}=0}^{\mathbf{N}} f(\mathbf{n}) \, P_t(\mathbf{n}) \tag{4.6.15}$$

and the departure from the average at time t

$$\boldsymbol{\alpha} \equiv \mathbf{n} - \langle \mathbf{n} \rangle_t \equiv \mathbf{n} - \mathbf{n}_{0t} \tag{4.6.16}$$

The various transitions which can occur are labeled by a vector $\boldsymbol{\varepsilon}$ such that (if it is not the zero vector) the component ε_i $(i = 1, \ldots, M-1)$ gives the number of electrons which are added to the level i as a result of the transition $\boldsymbol{\varepsilon}$. The components can be negative, positive or zero. Given that the set of occupation numbers is $\mathbf{n}$, we then denote the appropriate transition rate by $p_\varepsilon(\mathbf{n})$. This notation is illustrated in Fig. 4.6.2 with the definition

$$p_0(\mathbf{n}) \equiv - \sum_{\varepsilon \neq 0} p_\varepsilon(\mathbf{n})$$

It follows that

$$\sum_\varepsilon p_\varepsilon(\mathbf{n}) = 0 \tag{4.6.17}$$

The following boundary conditions rule out unphysical values:

$$p_\varepsilon(\mathbf{n}) = P_t(\mathbf{n}) = 0 \quad \text{if for some } i \quad n_i > N_i \quad \text{or} \quad n_i < 0 \tag{4.6.18}$$

$$p_\varepsilon(\mathbf{n}) = 0 \qquad\quad \text{if for some } i \quad n_i > N_i - \varepsilon_i \quad \text{or} \quad n_i < -\varepsilon_i$$

The difference equations (4.6.1) for the probabilities is now generalized to

$$\frac{\partial}{\partial t} P_t(\mathbf{n}) = \sum_{\varepsilon \neq 0} P_t(\mathbf{n}-\varepsilon) \, p_\varepsilon(\mathbf{n}-\varepsilon) - P_t(\mathbf{n}) \sum_{\varepsilon \neq 0} p_\varepsilon(\mathbf{n})$$

$$= \sum_\varepsilon P_t(\mathbf{n}-\varepsilon) \, p_\varepsilon(\mathbf{n}-\varepsilon) \tag{4.6.19}$$

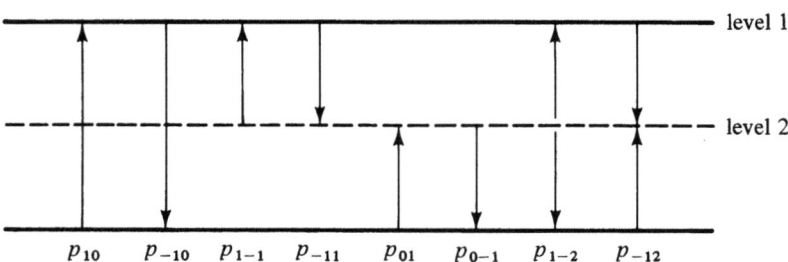

Fig. 4.6.2. The transition rates $p_\varepsilon(\mathbf{n})$ are shown formally for various transitions ε. It is assumed that there are $M = 3$ levels.

In particular, in a steady state the suffixes t may be dropped throughout, and

$$\sum_\varepsilon P(\mathbf{n}-\varepsilon)\, p_\varepsilon(\mathbf{n}-\varepsilon) = 0 \qquad (4.6.20)$$

This equation, together with the boundary conditions (4.6.18) determines the unknown probability $P(\mathbf{n})$ in terms of the transition rates $p_\varepsilon(\mathbf{n})$, which are assumed given. It is sufficient for the present purpose, however, to identify the first and second moments of the n_i. This represents an easier problem. There are $\frac{1}{2}(M-1)(M+2)$ unknowns, namely $n_{01}, n_{02}, \dots,$ n_{0M-1}; $\alpha_1^2, \alpha_1 \alpha_2, \dots, \alpha_{M-1}^2$.

Using the boundary conditions it may be shown that on multiplying eq. (4.6.19) by a function $f(\mathbf{n})$ and summing over all $\mathbf{n}$,

$$\frac{\mathrm{d}}{\mathrm{d}t}\langle f(\mathbf{n})\rangle_t = \sum_\varepsilon \langle f(\mathbf{n}+\varepsilon)\, p_\varepsilon(\mathbf{n})\rangle_t \qquad (4.6.21)$$

In the steady state

$$\sum_\varepsilon \langle f(\mathbf{n}+\varepsilon)\, p_\varepsilon(\mathbf{n})\rangle = 0 \qquad (4.6.22)$$

Choosing in turn $f(\mathbf{n}) = n_i$ and $\alpha_i \alpha_j$ respectively, eqs. (4.6.22) and (4.6.17) yield

$$\sum_\varepsilon \varepsilon_i \langle p_\varepsilon(\mathbf{n})\rangle = 0 \quad (i = 1, \dots, M-1) \qquad (4.6.23)$$

$$\sum_\varepsilon \{\varepsilon_i \langle \alpha_j p_\varepsilon(\mathbf{n})\rangle + \varepsilon_j \langle \alpha_i p_\varepsilon(\mathbf{n})\rangle + \varepsilon_i \varepsilon_j \langle p_\varepsilon(\mathbf{n})\rangle\} = 0 \quad (i,j = 1, \dots, M-1) \qquad (4.6.24)$$

The equations may be expanded to second order in α, using

$$\left.\begin{aligned}
D_\varepsilon^{lm}(\mathbf{n}_0) &\equiv \frac{1}{2}\left[\frac{\partial^2}{\partial n_i \partial n_m} p_\varepsilon(\mathbf{n})\right]_{\mathbf{n}=\mathbf{n}_0} \\[2mm]
E_\varepsilon^l(\mathbf{n}_0) &\equiv \left[\frac{\partial}{\partial n_l} p_\varepsilon(\mathbf{n})\right]_{\mathbf{n}=\mathbf{n}_0}
\end{aligned}\right\} \qquad (4.6.25)$$

to find

$$\sum_\varepsilon \varepsilon_i \left[p_\varepsilon(\mathbf{n}_0) + \sum_{l,\, m=1}^{M-1} D_\varepsilon^{lm}(\mathbf{n}_0)\langle \alpha_l \alpha_m\rangle\right] = 0 \qquad (4.6.26)$$

$$\sum_\varepsilon \sum_{l=1}^{M-1} E_\varepsilon^l(\mathbf{n}_0)[\varepsilon_i \langle \alpha_j \alpha_l\rangle + \varepsilon_j \langle \alpha_i \alpha_l\rangle] = -\sum_\varepsilon \varepsilon_i \varepsilon_j \left[p_\varepsilon(\mathbf{n}_0) + \sum_{l,\, m=1}^{M-1} D_\varepsilon^m(\mathbf{n}_0)\langle \alpha_l \alpha_m\rangle\right] \qquad (4.6.27)$$

There are $M-1$ equations (4.6.26) and $\frac{1}{2}M(M-1)$ equations (4.6.27) which is just the correct number to identify the unknowns.

This procedure depends on the neglect of terms of third and higher orders in α, and is therefore valid only for small fluctuations from the steady state. This is consistent with the further approximation to be made,

$$|p_\varepsilon(\mathbf{n_0})| \gg \left| \sum_{l,m=1}^{M-1} D_\varepsilon^{lm}(\mathbf{n_0}) \langle \alpha_l \alpha_m \rangle \right| \qquad (4.6.28)$$

which considerably simplifies the algebra involved in solving eqs. (4.6.26) and (4.6.27). Using eqs. (4.6.28), eqs. (4.6.26) and (4.6.27) become for $i,j = 1, \ldots, M-1$,

$$\sum_\varepsilon \varepsilon_i p_\varepsilon(\mathbf{n_0}) = 0 \qquad (4.6.29)$$

$$\sum_\varepsilon \sum_{l=1}^{M-1} E_\varepsilon^l(\mathbf{n_0})[\varepsilon_i \langle \alpha_j \alpha_l \rangle + \varepsilon_j \langle \alpha_i \alpha_l \rangle] = - \sum_\varepsilon \varepsilon_i \varepsilon_j p_\varepsilon(\mathbf{n_0}) \qquad (4.6.30)$$

From the $M-1$ equations (4.6.29) the $M-1$ unknowns $\mathbf{n_0}$ can be found. From the $\frac{1}{2}M(M-1)$ linear equations (4.6.30) the $\langle \alpha_i \alpha_j \rangle$ can also be found. The first and second moments of the fluctuating variables $\mathbf{n_0}$ can now be obtained from this theory. Thus eq. (4.6.30) is an M-level generalization of the recombination–generation theorem while eq. (4.6.29) generalizes eq. (4.6.3).

The statistics of charge carrier fluctuations in an M-level semiconductor thus yields the variances $\langle \alpha_i \alpha_j \rangle$ when the above equations are solved for them explicitly. Typically this requires identifications in terms of mass action laws such as $p_{-11} = T_1^s n(N-n_t)$, and this has been carried out for $M = 2$ and 3 [4.6.5], [4.6.6]. It leads one to recombination–generation noise theory for semiconductors as follows. In noise theory one is interested in the spectral density $G_{ij}(\omega)$ of the fluctuations for constant temperature and volume, and this is defined by

$$G_{ij}(\omega) = 2\mathcal{R} \int_{-\infty}^{\infty} \langle \alpha_i(0) \alpha_j(t) \rangle \exp(-\omega t) \, dt$$

(see for example [4.6.5]), where $\mathcal{R}$ denotes 'the real part of'. The autocorrelation function $\langle \alpha_i(0) \alpha_j(t) \rangle$ tells one how long a fluctuation in level j persists on average, given a fluctuation in level i. Then one shows that $G_{ij}(\omega)$ is a superposition of the form

$$G_{mn}(\omega) = \sum_{k=1}^{M-1} A_{mn}(k) \frac{\tau_k^2}{1+\omega^2 \tau_k^2} \quad (m,n = 1, \ldots, M-1)$$

where the τ_k's are the $M-1$ lifetimes involved in the problem, and the A's are appropriate coefficients. In this way one arrives at the frequency dependence of the spectral fluctuation density. The experimental recombination–generation spectra do tend to exhibit a falling curve as the angular frequency increases with smoothed out steps in the neighborhood of $\omega \sim \tau_k^{-1}$. Among other sources of noise which are added to recombination–generation noise are those due to diffusion and drift. By assuming distributions of relaxation times τ_k one can generate theoretically a great variety of frequency dependences. Among these is the well known, but not fully understood, $(1/\omega)$-noise. This is usually referred to as $(1/f)$-noise.

4.6.3 A recombination–generation model for chaos

We shall transcribe the essential steps of the original work [4.6.7], which was expounded for holes (numbers, not concentrations, denoted by p) into the language appropriate to electrons of which there are again assumed to be n in the system.

Consider the system at discrete times $t = k\tau$ where k is an integer and τ is a basic time interval. We assume that the time evolution, corresponding to the Markoff property [4.6.8], is given by a Chapman–Kolmogoroff equation

$$P(n, k+1) - P(n, k) = \sum_{n'}' \{W_k(n, n') P(n', k) - W_k(n', n) P(n, k)\} \qquad (4.6.31)$$

where $\sum'$ excludes the term $n' = n$ in the sum. Here $P(n, k)$ is the probability of finding the system at the beginning of the kth time interval with the electron number n, and $W_k(n', n)$ is the probability that during the kth time interval a transition occurs from a number n to n'.

We denote the probabilities that during the kth time interval s generation or recombination events occur by $G_k(n+s, n)$ or $R_k(n-s, n)$, respectively. We can then write

$$W_k(n', n) = \sum_{s=1}^{\infty} [G_k(n+s, n) \delta_{n', n+s} + R_k(n-s, n) \delta_{n', n-s}] \qquad (4.6.32)$$

In the following we assume that the generation–recombination probabilities are homogeneous in time, and accordingly drop the subscript k. Assuming that the single-step transition probabilities are proportional to the basic time interval τ, we set

$$G(n+1, n) = \tau g(n) \quad \text{and} \quad R(n-1, n) = \tau r(n) \qquad (4.6.33)$$

where $g(n)$ and $r(n)$ are the normal single-step generation and recombination probabilities per unit time such as those of eq. (4.6.8). In any case we assume that n is the only independent dynamic variable, others having been eliminated for example by charge neutrality. This holds for spatially homogeneous processes involving a single donor level. The results can be generalized to the case of $N > 1$ dynamic variables, e.g. to multiple impurity levels. The probability for two or more generation or recombination steps in a time τ is $O(\tau)$ where $\lim_{\tau \to 0} [O(\tau)/\tau] = 0$ [4.6.9], therefore

$$W_k(n', n) = \tau[g(n) \delta_{n', n+1} + r(n) \delta_{n', n-1}] + O(\tau) \qquad (4.6.34)$$

The conventional theory now assumes that the elementary generation–

recombination processes occur so frequently that they can be approximated by a continuum in time. One therefore lets $\tau \to 0$ in eq. (4.6.3) using eq. (4.6.34). This gives the master equation

$$\frac{\partial}{\partial t} P(n) = \sum_{n'}{}' \{w(n, n') P(n') - w(n', n) P(n)\} \tag{4.6.35}$$

where $w(n', n) \equiv g(n) \delta_{n', n+1} + r(n) \delta_{n', n-1}$ contains only single-step transitions. We shall, however, keep a finite time interval $\tau > 0$, and must therefore retain all *multistep* transitions $s > 1$ in eq. (4.6.32). Mixed multiple recombination and generation events are also possible, but are omitted here.

The further evaluation of eq. (4.6.31) is possible if the individual generation–recombination events are uncorrelated and $s \ll n$ gives the essential contribution; in that case

$$G(n+s, n) = [\tau g(n)]^s, \quad R(n-s, n) = [\tau r(n)]^s \tag{4.6.36}$$

Inserting eq. (4.6.32) into eq. (4.6.31) and using eq. (4.6.36)

$$P(n, k+1) - P(n, k) = \sum_{s=1}^{\infty} \tau^s \{g(n-s)^s P(n-s, k) + r(n+s)^s P(n+s, k)$$
$$- [g(n)^s + r(n)^s] P(n, k)]\}$$

Multiplying by n and summing over n gives

$$\langle n \rangle_{k+1} - \langle n \rangle_k = \sum_{s=1}^{\infty} s\tau^s [\langle g(n)^s \rangle_k - \langle r(n)^s \rangle_k]$$
$$= \left\langle \frac{\tau g(n)}{[1 - \tau g(n)]^2} \right\rangle_k - \left\langle \frac{\tau r(n)}{[1 - \tau r(n)]^2} \right\rangle_k \tag{4.6.37}$$

where $\langle f(n) \rangle_k \equiv \sum_{n=1}^{\infty} f(n) P(n, k)$ for any function $f(n)$. Equation (4.6.37) can be simplified by the 'mean-field approximation' $\langle f(n) \rangle = f(\langle n \rangle)$, i.e. by neglecting fluctuations. This is a good approximation for sharply peaked distributions. One finds our main result

$$\langle n \rangle_{k+1} = \langle n \rangle_k + \frac{\tau g(\langle n \rangle_k)}{[1 - \tau g(\langle n \rangle_k)]^2} - \frac{\tau r(\langle n \rangle_k)}{[1 - \tau r(\langle n \rangle_k)]^2} \tag{4.6.38}$$

For $\tau \to 0$ this gives the usual deterministic semiconductor rate equation

$$\frac{d}{dt} \langle n \rangle \doteq g(\langle n \rangle) - r(\langle n \rangle) \tag{4.6.39}$$

which follows also directly from the master equation (4.6.35).

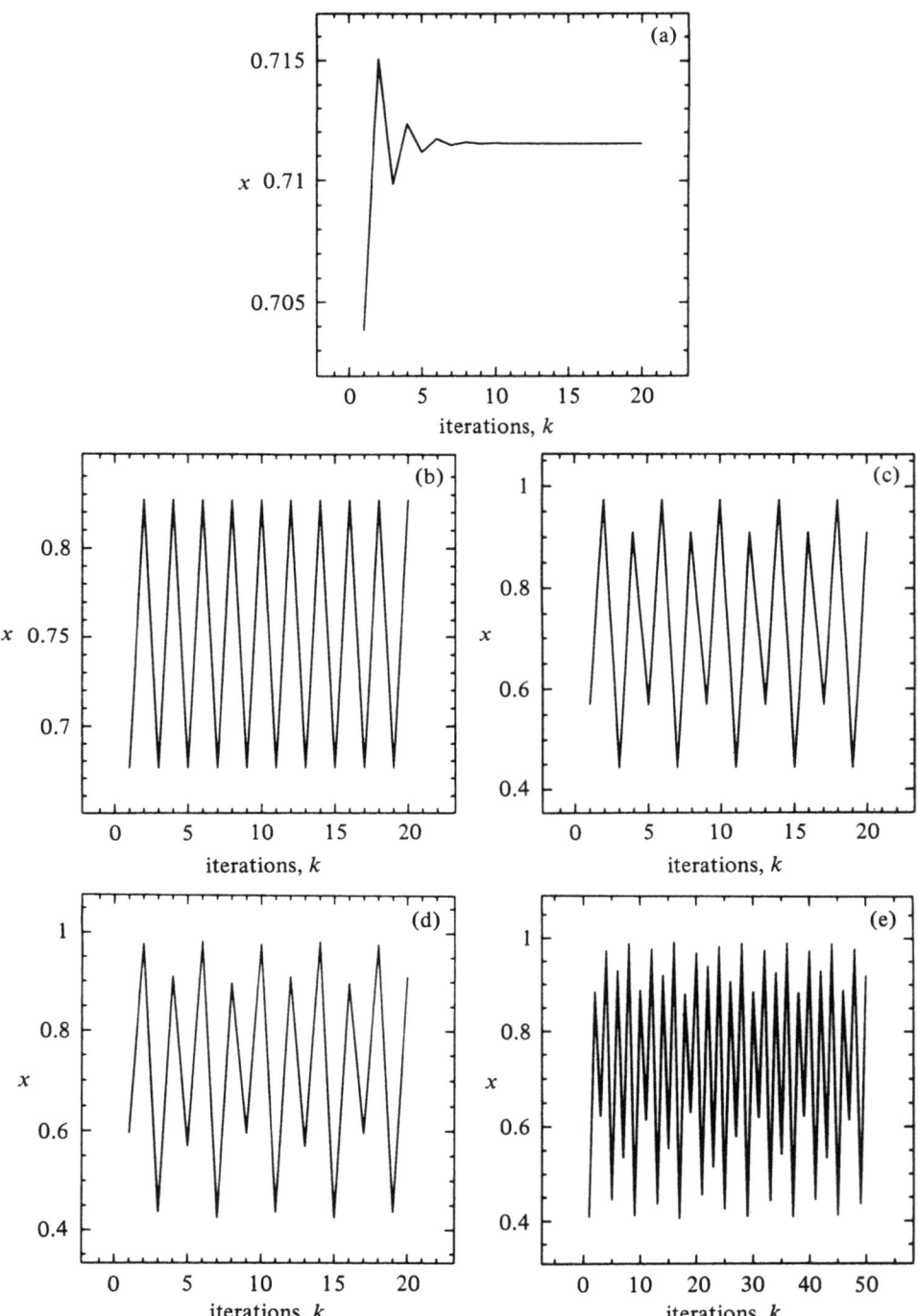

Fig. 4.6.3. Values of x_k versus the number of iterations k according to eq. (4.6.42), starting with $x_k = 0.5$ and discarding the first 50, 100, 500 and 550 iterations in cases (a) $A = 0.80$, (b) $A = 1.00$, (c) $A = 1.15$, (d) $A = 1.16$ and (e) $A = 1.17$, respectively.

Note that the argument leading to eq. (4.6.38) is essentially different from a naive procedure which would start with the generation–recombination rate equation (4.6.39) and then replace $d\langle n \rangle/dt$ by $(\langle n \rangle_{k+1} - \langle n \rangle_k)/\tau$ and $\langle n \rangle$ by $\langle n \rangle_k$. Then the different result

$$\langle n \rangle_{k+1} - \langle n \rangle_k = \tau[g(\langle n \rangle_k) - r(\langle n \rangle_k)] \tag{4.6.40}$$

would be obtained. But this argument is flawed by the need to assume that all transitions take place at the discrete times $k\tau$. In the argument leading to eq. (4.6.38), on the other hand, the transitions are distributed throughout the time interval τ, and multiple transitions are explicitly allowed for; they lead to the denominators in eq. (4.6.38).

For the simplest reaction kinetics we use the capture of a conduction band electron into an impurity (T_1^S) and the impact ionization (X_1) of an impurity by a conduction band electron. Introduce a reduced mean electron number

$$x_k \equiv \langle n \rangle_k/(N_D - N_A) \quad (0 \leqslant x_k \leqslant 1) \tag{4.6.41}$$

Then eq. (4.6.38) gives the recurrence relation

$$x_{k+1} = f(x_k) \tag{4.6.42}$$

where $\quad f(x) \equiv x\left\{1 + \dfrac{A(1-x)}{[1 - A(N_D - N_A)x(1-x)]^2} - \dfrac{B(C+x)}{[1 - B(N_D - N_A)x(C+x)]^2}\right\}$

Here

$$A \equiv X_1(N_D - N_A)\tau, \quad B \equiv T_1^S(N_D - N_A)\tau, \quad C \equiv N_A/(N_D - N_A)$$

where (because of our change from concentrations to numbers) X_1 and T_1^S have the dimension $(\text{time})^{-1}$ without the usual volume dimension. For p-type material and assumptions analogous to the above one finds precisely the same equation (4.6.42). The impact ionization coefficient, essentially A, is chosen to be the control parameter in obtaining $\langle n \rangle_k$, essentially x_k, as a function of the number of iterations k. This is illustrated in Fig. 4.6.3, using $B = 0.24$, $C = 0.25$ [4.6.7]. From one final value (~ 0.7114) of x_k for $A = 0.8$, we pass to 2, 4 and 8 values as A is increased. In Fig. 4.6.3(e) one finally has 'chaos'. This means broadly that initial conditions which are only slight variants of each other will lead normally, for the given constants, to quite different temporal developments of the system. Very similar curves are found for the famous logistic equation. In this $f(x) = rx(1-x) \ (0 \leqslant r \leqslant 4)$ and r takes the place of A; chaos is first encountered for $r = 3.57$.

One has thus gained an attractive way of looking at the origin of chaos in semiconductors. One has to attribute it to the occurrence of autocatalysis in the form of impact ionization, and to the discrete nature of the generation–recombination events. Both these elements are essential for the model. There are of course other mechanisms for chaos in semiconductors. These have been surveyed recently [4.6.10].

5

Defects

5.1 Introduction

The word 'flaw' was introduced by Shockley and Last [5.1.1] 'for the purpose of distinguishing between imperfections with multiple possibilities for charge condition and ordinary donors and acceptors' following the early work on the statistics of these centers described in [5.1.2].

In fact it is a convenient term which encompasses both chemical and physical defects and it is sometimes used in this generalized sense [5.1.3]. We shall here use the term *defect* and use an amended form of Blakemore's classification scheme (Fig. 5.1.1). This is not done to give a systematic discussion of these defect classes (this requires a major book), but to indicate the variety of possibilities. (An As atom sitting on a Ga site (As_{Ga}) in GaAs is a typical *anti-site defect*.) There is a further classification [5.1.4] which depends on the value of

$$\Delta z \equiv \text{valency of the impurity} - \text{valency of the host}$$

On this basis what is often called an isoelectronic defect, defined by $\Delta z = 0$, is more logically named isovalent, since the *total* number of electrons on the atom involved to which the term 'isoelectronic' refers is often of minor relevance. Also on this basis one could *define* donor states as 'positively charged states of a defect', but this causes difficulty with a simple donor ionization reaction

$$D^+ + e \rightleftharpoons D^\times$$

since the right-hand side would not then be a 'donor'. However, we will not attempt to solve this problem of definition and classification here.

The isoelectronic impurities can be used in indirect materials since they have sometimes wavefunctions with Fourier components in k-space which extend from

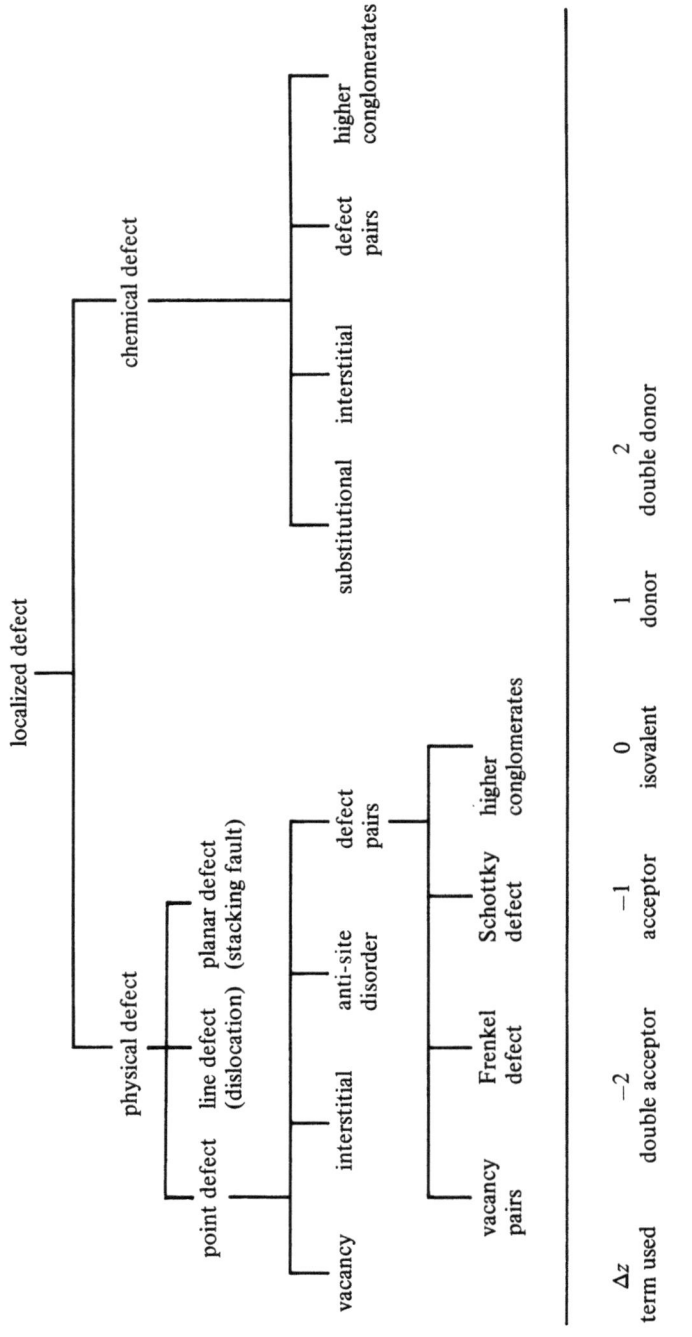

Fig. 5.1.1. Defect types.

one extremum (near which the level lies) to the other extremum thus facilitating a radiative transition. This is useful if the host material has an energy gap which is appropriate for the desired frequency in a light-emitting diode. In the absence of the isoelectronic defect the radiative transition (or the oscillator strength) may not be strong enough in an indirect material, see section 5.4.1.

In addition defects whose levels are shallow (i.e. lie near the band edge) control the conductivity via the number of current carriers in the bands. They are normally ionized at room temperatures, so that donors have given up electrons, acceptors have accepted electrons, as was seen in section 1.12.1. Nonuniformity of doping can further add to the usefulness of shallow defects. Doping can be tailored by ion implantation to provide desirable impurity distributions, the abrupt p–n junction being the simplest example.

The experimental approaches for studying defects by luminescence and optical absorption have been successfully employed for shallow defects and also for some deeper levels.

The deeper defects tend to govern the recombination lifetime of materials as seen in section 2.3.6, p. 137. Fast switches require short lifetimes, solar cells and lasers require long lifetimes, and so on with other devices.

Much work has been devoted to deep lying defect levels, i.e. those for which the hydrogenic model is no longer applicable. We give some account of the hydrogenic model and the radiative properties which it implies in sections 5.2.1 and 5.2.2. For deep levels it is inadequate, and other theoretical approaches have been developed for simulating the localized states. The problem is to model the inner part of the potential or the charge density up to a critical radius, r_0 say, and then to deal with the potential for $r > r_0$. Of course, beyond the problem of the radial dependence there are also questions of angular dependences. The ways of arriving from theory at reasonable estimates of localized energy levels of deep centers is in fact complicated and specialized, using tight-binding or self-consistent or semi-empirical Hamiltonians, Green's functions, numerical methods, etc. Apart from multiphonon transitions, discussed in section 5.4.4 and in detail in chapter 6, these matters are here left to specialized monographs [5.1.5]–[5.1.8] (after all, [5.1.8] alone covers 770 pages!). Note, however, that the deep levels tend to couple to acoustic modes while the shallower levels are more delocalized with weaker vibronic coupling which is largely to the optical modes.

As regards the experimental identification of deep levels a popular procedure [5.1.9], [5.1.10] is to arrange for capture and emission of carriers from defects by electrical fields or by photoionization, and to monitor the return to a steady state by capacitance or current measurements. These and other variants of the deep-level transient spectroscopy ("DLTS") method can yield the parameters of the defects and they are based on the ideas developed on p. 152ff. An excellent summary of the parameters of defects in the most important semiconductors has recently been issued [5.1.11]. The subject of deep centers is discussed at the international biennial 'Lund conferences' which started in Lund, Sweden, in 1977 (they are not now confined to Sweden).

As an important preliminary (but simplified) view of the recombination properties of deep centers, and as an introduction to chapter 6, we here note two characteristics.

The first is the energy gap law, which states that the multiphonon nonradiative transition rate when an energy of transition E is involved is approximately

$$T = A \exp(-\alpha E) \tag{5.1.1}$$

where A and α are approximate constants. Improved theories give more complicated generalizations of this formula [5.1.12]. A rough idea of the origin of such an exponential decay with energy gap can be obtained as follows [5.1.13]. The transition probabilities, T_s, for this process with the participation of s phonons are assumed to satisfy

$$T_s = \theta T_{s-1} \quad (0 < \theta \ll 1) \tag{5.1.2}$$

with constant θ. It follows that $T_s = \theta^s T_0$. Now suppose that the process is dominated by a participating phonon number s_0, given by

$$E = s_0 \hbar \omega \tag{5.1.3}$$

where ω is the highest possible angular frequency, then

$$T \simeq T_{s_0} = T_0 e^{-\beta E} \quad \beta = \frac{\ln(1/\theta)}{\hbar \omega} \tag{5.1.4}$$

This is of the form (5.1.1).

The second characteristic is the Dexter–Klick–Russell rule [5.1.14] about the quenching of radiative (fluorescent, luminescent, etc.) transitions. For its discussion we shall use potential energy parabolae as a function of a configuration coordinate Q (Fig. 6.2.1). The latter was introduced in the 1930s, and it is of crucial importance in chapter 6. It is arrived at as a result of a sequence of approximations: (i) The one-electron approximation (section 3.3.1), which neglects correlations among electronic motions. (ii) The adiabatic or Born–Oppenheimer approximation (section 6.3.2), which treats the nuclei as static as far as the electronic motions are concerned. This is a good approximation if the lattice frequencies are smaller than the frequencies of the electrons. This approximation enables one to regard the nuclear wavefunction as depending on the state (but not on the position) of the electrons. Conversely the electronic wavefunction still depends on the frozen positions of the nuclei. (iii) The neglect of this dependency is called the Condon approximation. This is expected to be valid if the electron–lattice interaction is not too large. One now has to consider the defect with its electrons and a small number (n) of neighboring atoms. In this picture an excitation or de-excitation is followed by a relaxation of the reduced system to a new equilibrium configuration. The oscillations about it can be described in terms of the electronic plus vibrational potential energy surfaces regarded as a function of the $3n$ coordinates of the neighboring atoms. (iv) This is reduced to n coordinates by considering only radial motions. (v) One can often regard the mode in which the n atoms move in phase as the most important. One is then left with the one 'configuration' coordinate, Q. The other modes suffer in effect an approximate cancellation.

Direct optical transitions must occur vertically on this diagram since absorption and emission of radiation is very rapid compared with the atomic vibrations. Further, the parabola for the excited states is expected to be broader than that for the ground state since it tends to have a less firmly bound outer electron. The two parabolae should therefore intersect. Perturbations modify this (see Fig. 6.3.2, later) but the parabolae still have points of close proximity which are relevant for transitions between them.

Using Fig. 6.2.1(b), an excitation from equilibrium at Q_1 leads to an energy $E_2(Q_1)$ in the excited states which lie above the new equilibrium energy $E_2(Q_2)$. If the parabolae intersect at energy E_c, the rule says that the system will decay by nonradiative multiphonon process to $E_1(Q_1)$ if $E_2(Q_1) > E_c$. This rule is a good guide for F-centers in ionic crystals. However, it is rather unrealistically precise, as one would expect some radiative transitions even if $E_2(Q_1) > E_c$. An improved theory suggests that the ratio of radiative to multiphonon transition probabilities is

$$\left[1 - 2\left(\frac{E_2(Q_1) - E_2(Q_2)}{E_2(Q_1) - E_1(Q_1)}\right)\right]^2 \quad (E_2(Q_2) > E_1(Q_1))$$

For a fuller discussion of the assumptions involved, see [5.1.15].

5.2 Radiative recombination involving traps

5.2.1 General theory; the hydrogenic ground state

Most important in semiconductor lasers, light-emitting diodes and other devices are the radiative transitions between a band and an impurity. They are the so-called free–bound transitions of the astrophysicists, and can be studied theoretically in the solid state by using an appropriate effective mass theory. This uses an unperturbed Hamiltonian H and a perturbed Hamiltonian H_T:

$$H = P^2/2m + U(\mathbf{r}), \quad H_T = H + F(\mathbf{r})$$

where $U(\mathbf{r})$ is the periodic potential and the perturbing energy is $F(\mathbf{r})$. The wavefunction $\psi(\mathbf{r})$ of the perturbed problem is expanded in terms of Bloch functions $\psi_{n\mathbf{k}}(\mathbf{r})$. Thus

$$H\psi_{n\mathbf{k}}(\mathbf{r}) = E_n(\mathbf{k})\,\psi_{n\mathbf{k}}(\mathbf{r}), \quad \psi_{n\mathbf{k}}(\mathbf{r}) = V^{-\frac{1}{2}}u_{n\mathbf{k}}(\mathbf{r})\,e^{i\mathbf{k}\cdot\mathbf{r}}$$

and

$$[H+F(\mathbf{r})]\,\psi(\mathbf{r}) = E\psi(\mathbf{r}), \quad \psi(\mathbf{r}) = \sum_{n'}\sum_{\mathbf{k}'} G_{n'}(\mathbf{k}')\,\psi_{n'\mathbf{k}'}(\mathbf{r}) \tag{5.2.1}$$

and therefore

$$\sum_{n'}\sum_{\mathbf{k}'}[E_{n'}(\mathbf{k}')-E_{n'}(0)]\,G_{n'}(\mathbf{k})\,\psi_{n'\mathbf{k}'}(\mathbf{r}) + \sum_{n'}\sum_{\mathbf{k}'}F(\mathbf{r})\,G_{n'}(\mathbf{k}')\,\psi_{n'\mathbf{k}'}(\mathbf{r})$$

$$= [E-E_{n'}(0)]\,\psi(\mathbf{r})$$

Multiplying by $\psi_{n\mathbf{k}}^*(\mathbf{r})$, integrating and using the orthogonality of the Bloch function

$$[E_n(\mathbf{k})-E_n(0)]\,G_n(\mathbf{k}) + \sum_{n'}\sum_{\mathbf{k}'}(n\mathbf{k}|F|n'\mathbf{k}')\,G_{n'}(\mathbf{k}') = [E-E_n(0)]\,G_n(\mathbf{k})$$

Multiplying by $e^{i\mathbf{k}\cdot\mathbf{r}}$ and summing over the Brillouin zone

$$\sum_{\substack{\mathbf{k}\\(\text{B.Z.})}}[E_n(\mathbf{k})-E_n(0)]\,G_n(\mathbf{k})\,e^{i\mathbf{k}\cdot\mathbf{r}} + \sum_{\substack{\mathbf{k}\\(\text{B.Z.})}}\sum_{n'}\sum_{\mathbf{k}'}(n\mathbf{k}|F|n'\mathbf{k})\,G_{n'\mathbf{k}'}\,e^{i\mathbf{k}\cdot\mathbf{r}}$$

$$= [E-E_n(0)]\sum_{\substack{\mathbf{k}\\(\text{B.Z.})}}G_n(\mathbf{k})\,e^{i\mathbf{k}\cdot\mathbf{r}} \tag{5.2.2}$$

Using the 'envelope' function

$$\varphi_n(\mathbf{r}) \equiv V^{-\frac{1}{2}}\sum_{\mathbf{k}}G_n(\mathbf{k})\,e^{i\mathbf{k}\cdot\mathbf{r}} \tag{5.2.3}$$

and coordinates so that $\mathbf{k}=0$ lies near the conduction band minimum (for spherical bands)

$$E_n(\mathbf{k})-E_n(0) = \hbar^2 k^2/2m_c$$

one finds

$$-\frac{\hbar^2}{2m_c}\nabla^2\varphi_n(\mathbf{r}) = \sum_{\mathbf{k}}\frac{\hbar^2 k^2}{2m_c}G_n(\mathbf{k})\,e^{i\mathbf{k}\cdot\mathbf{r}} = \sum_{\mathbf{k}}[E_n(\mathbf{k})-E_n(0)]\,G_n(\mathbf{k})\,e^{i\mathbf{k}\cdot\mathbf{r}}$$

Substituting into eq. (5.2.2), one finds the effective mass equation

$$-\frac{\hbar^2}{2m_c}\nabla^2\varphi_n(\mathbf{r})+C_n(\mathbf{r}) = [E-E_n(0)]\,\varphi_n(\mathbf{r})\tag{5.2.4}$$

where

$$C_n(\mathbf{r}) \equiv V^{-\frac{1}{2}}\sum_{n'}\sum_{\mathbf{k'}} G_{n'}(\mathbf{k'})\,e^{i\mathbf{k}\cdot\mathbf{r}}\sum_{\mathbf{k}}(n\mathbf{k}|F|n'\mathbf{k'})\,e^{i(\mathbf{k}-\mathbf{k'})\cdot\mathbf{r}}$$

With often reasonable approximations one can argue [5.2.1] that

$$C_n(\mathbf{r}) \simeq F(\mathbf{r})\,\varphi_n(\mathbf{r})\tag{5.2.5}$$

Then eq. (5.2.4) is the Schrödinger equation for $\varphi_n(\mathbf{r})$ still using $F(\mathbf{r})$ as a perturbation, but the periodic potential $U(\mathbf{r})$ has been removed. The cost is that an effective mass m_c has replaced the normal electron mass. This argument still goes through for nonspherical bands [5.2.1], [5.2.2].

The beauty of this result is that if the Schrödinger equation has been solved for a perturbation $F(\mathbf{r})$ in the atomic case, the solution can be readily adapted to the case of a condensed matter situation. In particular, for a Coulomb potential $-e^2/\varepsilon r$, $\varphi_n(\mathbf{r})$ is a H atomic function modified by the insertion of the dielectric constant and an effective mass. The $G_n(\mathbf{k})$ are, by eq. (5.2.3), Fourier coefficients of the known hydrogen functions. For large r, when the perturbation is small, this function must approach that of the nearest band. It is therefore reasonable to take only the contribution from the conduction band, $n = c$ say, to find from eq. (5.2.1) a donor electron wavefunction

$$\psi(\mathbf{r})\to\psi_D = V^{-\frac{1}{2}}\sum_{\mathbf{k}} G_c(\mathbf{k})\,u_{c\mathbf{k}}(\mathbf{r})\,e^{i\mathbf{k}\cdot\mathbf{r}} \approx V^{-\frac{1}{2}}u_{c0}(\mathbf{r})\,\varphi_c(\mathbf{r})\tag{5.2.6}$$

In the last form we have used only the band-edge Bloch function. For an acceptor we have similarly

$$\psi_A = V^{-\frac{1}{2}}\sum_{\mathbf{k}} G_v(\mathbf{k})\,u_{v\mathbf{k}}(\mathbf{r})\,e^{i\mathbf{k}\cdot\mathbf{r}} \approx V^{-\frac{1}{2}}u_{v0}(\mathbf{r})\,\varphi_v(\mathbf{r})\tag{5.2.7}$$

More sophisticated theories have been constructed to arrive at these types of results. For example one can reduce the interband coupling terms, which are completely suppressed in eq. (5.2.6), by a series of canonical transformations [5.2.3], [5.2.4]. One also has to consider the many-electron aspect of the theory [5.2.5], [5.2.6]. While all bands are involved in eq. (5.2.4), the reduction to one band by the use of eq. (5.2.5) does violence to the problem if deep traps are involved. It is almost equivalent to the use of a hydrogen-like model. However, we are not entering a discussion of the unsolved problem of deep levels here. Throughout this chapter we shall proceed semi-empirically to see how far a simple theory will take us.

We shall consider a transition from a conduction band state specified by the wavefunction

$$\psi_{ck}(\tau) = V^{-\frac{1}{2}} \sum_L v_c(\mathbf{k}, \mathbf{L}) \, e^{i(\mathbf{k}+\mathbf{L})\cdot\mathbf{r}} \tag{5.2.8}$$

where an expansion in terms of momentum eigenfunctions v has been made, as already encountered in eq. (3.4.9), for example. The sum is over all lattice vectors in $\mathbf{k}$-space. If the final state of the electron is in an acceptor, one can rewrite eq. (5.2.7) as

$$\psi_A = V^{-\frac{1}{2}} \sum_L \sum_k v_v(\mathbf{k}, \mathbf{L}) \, G_v(\mathbf{k}) \, e^{i(\mathbf{k}+\mathbf{L})\cdot\mathbf{r}} \tag{5.2.9}$$

The matrix element of the momentum operator is $M_{ij} \to M_{CA}$, i.e.

$$\langle A|\mathbf{P}|ck\rangle = V^{-1} \int \sum_{k'} \sum_{L'} v_v^*(\mathbf{k}', \mathbf{L}') \, G_v^*(\mathbf{k}') \, e^{-i(\mathbf{k}'+\mathbf{L}')\cdot\mathbf{r}}$$

$$\times \frac{\hbar}{i} \nabla[\sum_L v_c(\mathbf{k}, \mathbf{L}) \, e^{i(\mathbf{k}'+\mathbf{L}')\cdot\mathbf{r}}] \, d\mathbf{r}$$

$$= \hbar V^{-\frac{1}{2}} \int_V [\sum_{k'} \sum_{L'} v_v^*(\mathbf{k}' \cdot \mathbf{L}') \, G_v^*(\mathbf{k}') \, e^{-i(\mathbf{k}+\mathbf{L})\cdot\mathbf{r}}]$$

$$\times [\sum_L (\mathbf{k}+\mathbf{L}) \, v_c(\mathbf{k}, \mathbf{L}) \, e^{i(\mathbf{k}+\mathbf{L})\cdot\mathbf{r}}] \, d\mathbf{r}$$

$$= \hbar \sum_{L, L'} v_v^*(\mathbf{k}+\mathbf{L}-\mathbf{L}', \mathbf{L}') \, G_v^*(\mathbf{k}+\mathbf{L}-\mathbf{L}') \, v_c(\mathbf{k}, \mathbf{L})(\mathbf{k}+\mathbf{L})$$

By eq. (B.11) the $\mathbf{r}$-integration forces $\mathbf{k}' = \mathbf{k}+\mathbf{L}-\mathbf{L}'$. Assume as before (e.g. in eq. (3.4.20)) that the main contribution arises from $\mathbf{L} = \mathbf{L}'$, and one finds

$$\langle A|\mathbf{P}|ck\rangle = \hbar \sum_L (\mathbf{k}+\mathbf{L}) \, v_v^*(\mathbf{k}, \mathbf{L}) \, v_c(\mathbf{k}, \mathbf{L}) \, G_v^*(\mathbf{k})$$

$$= \langle v\mathbf{k}|\mathbf{P}|ck\rangle \, G_v^*(\mathbf{k}) \tag{5.2.10}$$

[The last form of writing depends on

$$\langle v\mathbf{k}|\mathbf{P}|ck\rangle = \hbar V^{-1} \sum_{L, L'} \int_V v_v^*(\mathbf{k}, \mathbf{L}') \, v_c(\mathbf{k}, \mathbf{L})(\mathbf{k}+\mathbf{L}) \, e^{i(\mathbf{L}-\mathbf{L}')\cdot\mathbf{r}} \, d\mathbf{r}$$

$$= \hbar \sum_L v_v^*(\mathbf{k}, \mathbf{L}) \, v_c(\mathbf{k}, \mathbf{L})(\mathbf{k}+\mathbf{L})] \tag{5.2.11}$$

The matrix element for the radiative transition "conduction band → acceptor" has thus been reduced to a product of the kth Fourier coefficient of the atomic function $\varphi(\mathbf{r})$ appropriate to the perturbation $F(\mathbf{r})$ and a matrix element between the bands. Note that all wavefunctions $\psi(\mathbf{r})$, $\psi_{nk}(\mathbf{r})$, $\varphi_n(\mathbf{r})$ have dimension $L^{-\frac{3}{2}}$; $u_{nk}(\mathbf{r})$, $v_c(\mathbf{k}, \mathbf{L})$, $G_v(\mathbf{k})$ are dimensionless.

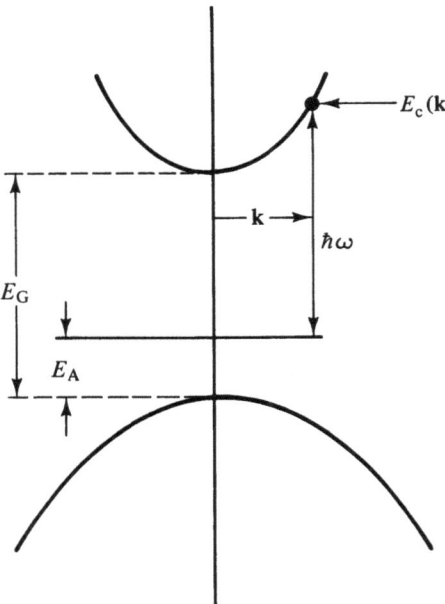

Fig. 5.2.1. A radiative capture.

Multiplying by a polarization vector and averaging over the polarization directions as in section 4.4.1 we find

$$|M_{IJ}|^2_{av} \rightarrow |M_{CA}|^2_{av} = |\langle v\mathbf{k}|\mathbf{P}|c\mathbf{k}\rangle G_v(\mathbf{k})|^2$$

The single-mode transition rate (4.4.3) per unit volume from a state $|c\mathbf{k}\rangle$ into one acceptor atom is

$$P_{CA} = \frac{4\pi^2 e^2}{\mu^2 m^2 V^2 \omega}|\langle v\mathbf{k}|\mathbf{P}|c\mathbf{k}\rangle G_v(\mathbf{k})|^2 \begin{cases} [N(\omega)+1]\,\delta & \text{(emission)} \\ N(\omega)\,\delta & \text{(absorption)} \end{cases} \quad (5.2.12)$$

Note that the photon energy $E_\mathbf{k}$ has been replaced by $\hbar\omega$ and $\mathbf{k}$ now refers to the electron. The energy conserving Dirac δ-functions require (Fig. 5.2.1) that

$$\hbar\omega = E_G - E_A + E_c(\mathbf{k}) - E_c(0) \quad (5.2.13)$$

The absorption coefficient is given by eq. (4.5.4) with the probability factors replaced by the number of conduction band states per unit energy in a volume V multiplied by the number $N_A^\times V$ of neutral acceptors able to accept electrons [5.2.7], [5.2.8]:

$$\alpha_{AC} = \frac{4\pi^2 e^2}{\mu V m^2 c\omega}\left|\langle v\mathbf{k}|\mathbf{P}|c\mathbf{k}\rangle G_v(\mathbf{k})\right|^2 \frac{2^{\frac{1}{2}} m_c^{\frac{3}{2}}[E_c(k)-E_c(0)]^{\frac{1}{2}} V}{\pi^2 \hbar^3}(N_A^\times V)$$

$$= \frac{2^{\frac{5}{2}} e^2 m_c^{\frac{3}{2}}}{\mu m^2 c\hbar^3 \omega}[E_c(\mathbf{k})-E_c(0)]^{\frac{1}{2}} N_A^\times V|\langle v\mathbf{k}|\mathbf{P}|c\mathbf{k}\rangle G_v(\mathbf{k})|^2 \quad (5.2.14)$$

The recombination rate per unit volume from a state $|c\mathbf{k}\rangle$ is eq. (5.2.12) multiplied by the number of electromagnetic modes per unit energy in the volume V for two directions of polarizations ($\mu^3\omega^2 V/\pi^2\hbar c^3$) and the number of relevant acceptors, and integrated over $\hbar\omega$ (which leads to eq. (5.2.13)):

$$U_{CA}(\mathbf{k}) = \frac{4e^2\mu\omega}{m^2 c^3 \hbar V}(N_A^\times V)|\langle v\mathbf{k}|\mathbf{P}|c\mathbf{k}\rangle G_v(\mathbf{k})|^2 [1+N(\omega)] \qquad (5.2.15)$$

This yields an estimate of a radiative lifetime of the electrons in a state $|c\mathbf{k}\rangle$:

$$\tau_{CA} = 1/U_{CA}(\mathbf{k})\,V$$

We shall, however, want a more general result.

Assuming Maxwell–Boltzmann statistics, the probability $P(E_c)\,\mathrm{d}E_c$ of finding an electron in the energy range $(E_c, E_c+\mathrm{d}E_c)$ is for a density of states behaving as E_c^s [5.2.9]

$$P(E_c) = \frac{E_c^s \exp(-E_c/kT)}{\Gamma(s+1)(kT)^{s+1}} \qquad (0 \leqslant E_c < \infty,\ s \text{ is a constant}) \qquad (5.2.16)$$

In our case $s = \frac{1}{2}$ is most common, and only $G_v(\mathbf{k})$ is liable to depend significantly on E_c, so that integration gives, with $E_c \equiv E_c(k) - E_c(0)$

$$U_{CA} = \frac{U_{CA}(\mathbf{k})}{|G_v(\mathbf{k})|^2} \frac{2}{\pi^{\frac{1}{2}}(kT)^{\frac{3}{2}}} \int_0^\infty \exp\left(-\frac{E_c}{kT}\right)|G_v(\mathbf{k})|^2 E_c^{\frac{1}{2}} \,\mathrm{d}E_c \qquad (5.2.17)$$

The above treatment in terms of Fourier coefficients is useful as it furnishes a rather clear idea of the structure of the argument. However, in the absence of an expression for $G_v(\mathbf{k})$, it is still somewhat too general. Let us therefore apply these results to the ground state of a hydrogen-like atom so that

$$\varphi(r) = (\pi \tilde{r}_{A1}^3)^{-\frac{1}{2}}\exp\left(-\frac{r}{\tilde{r}_{A1}}\right) \qquad (5.2.18)$$

where the radius $\tilde{r}_{At}$ of the tth orbit may be defined by the experimentally accessible energy E_{At} of an acceptor level of principal quantum number t above the valence band:

$$E_{At} = \hbar^2/2m_A\,\tilde{r}_{At}^2 \qquad (5.2.19)$$

(Substitute D for A in the case of donors.) This replaces the hydrogenic result

$$E_t = me^4/2\hbar^2 t^2 = \hbar^2 t^2/2mr_t^2 \qquad (5.2.20)$$

where the Bohr orbit is given by eq. (1.8.2): $r_t = \hbar^2 t^2/me^2$, a relation we do not use here. We use t instead of n for the principal quantum number, as the symbol n is used in a number of other contexts. The mass m_A may be taken to be the mass at the top of the valence band. To leave it general here, as often in the literature, is to give more flexible semi-empirical formulae. But they are of dubious theoretical foundations. One finds for the dimensionless Fourier coefficient of the ground state wavefunction

$$G_v(\mathbf{k}) = V^{-\frac{1}{2}} \int_0^\infty \varphi(r)\, e^{-i\mathbf{k}\cdot\mathbf{r}}\, d\mathbf{r} = \frac{8\pi^{\frac{1}{2}} \tilde{r}_{A1}^{\frac{3}{2}}}{V^{\frac{1}{2}}[1 + (r_{A1}k)^2]^2} \tag{5.2.21}$$

It has a sharp maximum as a function of $\tilde{r}_{A1}$ when $(\tilde{r}_{A1}k)^2 = \frac{3}{5}$, whence

$$[G_v(k)]_{max} = 5.54(\tilde{r}_{A1})^{\frac{3}{2}}_{max} / V^{\frac{1}{2}}$$

Substituting in the above results one finds from eqs. (5.2.12), (5.2.14) and (5.2.16) equations (6), (9), and (14) of Dumke's paper [5.2.8]. The V^{-3} factor which then occurs in eq. (5.2.12) is needed for dimensional reasons, but it is not explicit in Dumke's equation (6). Also the factor of four included in his equation (9) for the degeneracy of the localized level is here regarded as incorporated in $N_A^\times$.

Because of their importance we give below two results arising from eq. (5.2.21)

$$\alpha_{AC} = \frac{2^7 \pi e^2 m_c^{\frac{3}{2}} [\hbar\omega - E_G + E_A]^{\frac{1}{2}}}{\mu c m^2 \omega (m_A E_A)^{\frac{3}{2}}} \frac{|\langle v\mathbf{k}|\mathbf{P}|c\mathbf{k}\rangle|^2 N_A^\times}{[1 + m_c(\hbar\omega - E_G + E_A)/m_A E_A]^4} \tag{5.2.22}$$

$$U_{CA} = \frac{64\sqrt{2}\pi\mu e^2 \hbar^2 \omega}{m^2 c^3 (m_A E_A)^{\frac{3}{2}} V} \frac{|\langle v\mathbf{k}|\mathbf{P}|c\mathbf{k}\rangle|^2 N_A^*}{[1 + m_c E_c/m_A E_A]^4} [1 + N(\omega)] \tag{5.2.23}$$

Here $E_A \equiv E_{A1}$, and k^2 has been eliminated by means of

$$\hbar^2 k^2/2m_c \equiv E_c(\mathbf{k}) - E_c(0) \equiv E_c(= \hbar\omega - E_G - E_A) \tag{5.2.24}$$

so that $(\tilde{r}_1 k)^2$ is $(\hbar^2/2m_A E_A)(2m_c E_c/\hbar^2)$.

In fact eq. (5.2.22) can be written in a form first given in [5.2.7]:

$$\alpha_{AC} = \frac{2^7 \pi e^2}{\mu c m^2 \omega} \frac{m_c}{m_A E_A} |\langle v\mathbf{k}|\mathbf{P}|c\mathbf{k}\rangle|^2 N_A^\times \cdot \frac{x^{\frac{1}{2}}}{(1+x)^4} \tag{5.2.25}$$

where

$$x \equiv (\tilde{r}_1 k)^2 = \frac{m_c E_c}{m_A E_A} = \frac{m_c}{m_A} \frac{\hbar\omega - E_G + E_A}{E_A} \tag{5.2.26}$$

The last factor in eq. (5.2.25) captures the frequency dependence of the absorption coefficient. Functional forms for some excited states appear in Table 5.2.1.

For the analogous transitions from a donor level to the valence band the above theory has to be changed by the substitutions

$$\alpha_{AC}, U_{CA}, G_v \rightarrow \alpha_{VD}, U_{DV}, G_c$$

$$m_c, E_c(k) - E_c(0), m_A E_A \rightarrow m_v, E_v(0) - E_v(k), m_D E_D$$

$$N_A^\times \rightarrow N_D^\times$$

Table 5.2.1. *Some functions $G(t, l)$ of eq. (5.2.29) in terms of x given by eq.*
(5.2.26)

(t, l)	$G(t, l)$
$(1, 0)$	1
$(2, 0)$	$2^2(1-x)^2(1+x^2)^{-2}$
$(2, 1)$	$2^4 x(1+x)^{-2}$
$(3, 0)$	$(3x-1)^2(x-3)^2(1+x)^{-4}$
$(3, 1)$	$2^5 \times 3 \times x(1-x)^2(1+x)^{-4}$
$(3, 2)$	$2^7 x^2(1+x)^{-4}$
$(4, 0)$	$2^4(1-x)^2(1-6x+x^2)^2(1+x)^{-6}$
$(4, 1)$	$2^6 x(5x^2-14x+5)^2/5(1+x)^6$
$(4, 2)$	$2^{10}x^2(1-x)^2(1+x)^{-6}$
$(4, 3)$	$2^{12}x^3/5(1+x)^6$

The valence band is often degenerate and a summation over the valence bands
involved is needed. This is done for example in the reviews [5.2.10] and [5.2.11].

If m_c/m_A (or m_v/m_D) is small enough, the denominator of α_{AC} (or α_{VD} for the
valence band – donor absorption) is broadly independent of frequency. The
absorption is then similar in shape, but reduced in magnitude, to the corresponding
band–band absorption coefficient (4.5.12). It is also displaced by the energy E_A (or
E_D).

The recombination rate per unit volume (5.2.17) becomes for the ground state,
using eq. (5.2.21)

$$U_{CA} = \frac{64\sqrt{2\pi}\mu e^2\hbar^2\omega}{m^2 c^3(m_A E_A)^{\frac{3}{2}} V}|\langle v\mathbf{k}|\mathbf{P}|c\mathbf{k}\rangle|^2 N_A^\times \Gamma(\beta) \qquad (5.2.27)$$

where

$$\Gamma(\beta) \equiv \frac{2}{\sqrt{\pi}}\beta^{\frac{3}{2}}\int_0^\infty x^{\frac{1}{2}}e^{-\beta x}(1+x)^{-4}\,dx \qquad (5.2.28)$$

and $\beta \equiv m_A E_A/m_c kT$. The function $\Gamma(\beta)$ reflects the spread of electrons over the
energy levels and it has the value 0.12 for $\beta = 1$ [5.2.8]. The important range of β
values lies above this, whence when $\Gamma(\beta)$ approaches unity. (It decreases with β.)
One sees that the recombination rate increases as the temperature is decreased. The
rate is also proportional to the defect concentration $N_A^\times$ (or $N_D^\times$) so that an electron
can be received by the defect (or provided by it). These remarks transfer into
analogous ones for the radiative *lifetime*.

The theory given can of course not be applied as it stands to heavy doping situations when overlap between wavefunctions, band gap narrowing, Coulomb interactions between carriers, effective mass changes etc., have to be taken into account, as discussed, for example, in [5.2.12]–[5.2.14]. Detailed experimental verification of the formulae is hampered by the fact that for low doping the impurity radiation may not be strong enough and for heavier doping one runs into band tailing and the allied problems mentioned above. Nevertheless, satisfactory comparisons of theoretical and experimental curves have occasionally been made. In the case of radiative lifetimes for GaAs as a function of temperature this was done in [5.2.15] and [5.2.16]. Improvements in the theory have been made and will be outlined in sections 5.2.3 and 5.2.4.

5.2.2 Hydrogenic excited states

For transitions into excited states of a trap it is convenient to have available the Fourier coefficients G_k (cf. eq. (3.4.27)) of the wavefunctions of these states. As in eq. (5.2.20) t will denote the principal quantum number, l the angular momentum quantum number, and hydrogenic atoms will be considered. The main result needed [5.2.17]–[5.2.20] is that the square of the Fourier coefficient is

$$|G_k(t,l)|^2 = \frac{64\pi}{V} \frac{\tilde{r}_t^3}{[1+(k\tilde{r}_t)^2]^4} G(t,l) \tag{5.2.29}$$

Here $G(t,l)$ are the simple functions given in Table 5.2.1. In generalization of eq. (5.2.24) we shall put

$$E_{2'} - E_c = \hbar^2 k_2^2/2m_c, \quad E_v - E_{2'} = \hbar^2 k_{2'}^2/2m_v \tag{5.2.30}$$

if the energetic particle is in the conduction and valence band, respectively. Also

$$E_{2'} - E_c = \frac{\hbar^2}{2m_D \tilde{r}_{Dt}^2} \quad \text{or} \quad \frac{\hbar^2}{2m_A \tilde{r}_{At}^2} \tag{5.2.31}$$

so that

$$(k_{2'}\tilde{r}_{Dt})^2 = \begin{cases} m_c/m_D \\ m_v/m_D \end{cases} \quad \text{and} \quad (k_{2'}\tilde{r}_{At})^2 = \begin{cases} m_c/m_A \\ m_v/m_A \end{cases}$$

for donors and acceptor, respectively. This may be summarized by

$$(k\tilde{r}_t) = m^*/m_{D,A} \tag{5.2.32}$$

where $m^* = m_c$ or m_v.

The frequency dependence of the absorption coefficient for an excited hydrogenic state (t, l) is by a simple extension of eqs. (5.2.14) and (5.2.21)

$$\alpha_{AC}(t, l) \propto G(t, l) \frac{x^{\frac{1}{2}}}{(1+x)^4} \tag{5.2.33}$$

Some of these curves are illustrated in Fig. 5.2.2. Curve (a) was already given by Eagles [5.2.7]. The other curves show the reduction in the matrix element for excited states [5.2.21]. Indeed $G(\mathbf{k})$ vanishes if $t - l$ is even in the special case when $x = 1$, as shown in curve (b). No experimental verification seems to be known of the phenomenon that $G(\mathbf{k})$ can oscillate between nonzero values and zero values as $t - l$ changes from an even to an odd number in the $x = 1$ case.

We add the analytical expression for $G(t, l)$:

$$G(t, l) \equiv \frac{2^{4l} t (l!)^2 (2l+1)(t-l+1)! \, c^{2l}}{(t+l)! (1+c^2)^{2(l+2)}} \left[\frac{(2l+2)_{t-l-1}}{(t-l-1)!} \right]^2$$

$$_3F_2\left(-t+l+1, t+l+1, l+1; 2l+2, l+\tfrac{3}{2}; \frac{4c^2}{(1+c^2)^2} \right) \tag{5.2.33'}$$

Here $c \equiv k\tilde{r}_t$ has been used, $_3F_2$ is a generalized hypergeometric function and

$$(a)_p \equiv a(a+1)(a+2)\ldots(a+p-1) \quad \text{with } (a)_0 = 1, (a)_1 = a$$

An enthusiast wanting to check Table 5.2.1 can do so most easily for the cases $t = l+1$ since then $_3F_2 = 1$.

5.2.3 Tail states due to random potential fluctuations

The above theory is limited not only by unsolved problems of the defect energy levels which lie near mid-gap, but also by the assumption of light doping. This enables one to use hydrogenic models and the density of states (1.4.10). Heavier doping broadens the defect levels and an impurity or defect band of levels is formed with its maximum density near the energy of the original level. With more doping the Coulomb interaction is increasingly screened and the defect band moves towards its adjacent band and a merging occurs. The semiconductor is then said to be heavily doped, and there is a band tail. A reversible increase in the mid-gap density of states (in fact in the neutral dangling bond defects) as a result of strong illumination of amorphous hydrogenated Si has also been found and is currently still under study. This is the so-called Staebler–Wronski effect which is detrimental for the relatively cheap but relatively inefficient amorphous solar cells. Much interest in studying these problems has come from the applications of amorphous materials [5.2.14], [5.2.22], [5.2.23].

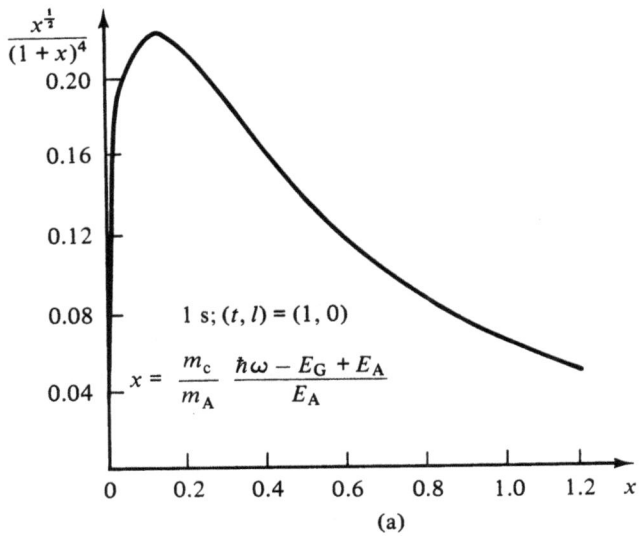

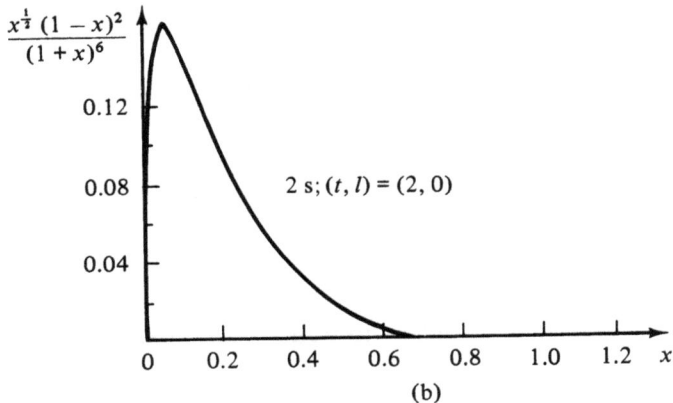

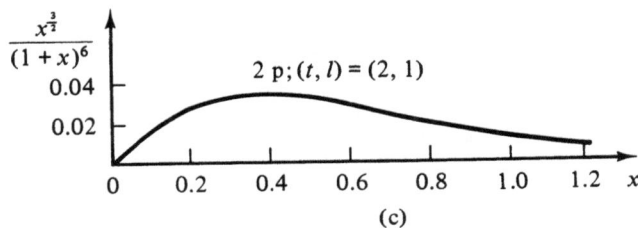

Fig. 5.2.2. Approximate frequency dependence (in terms of x of eq. (5.2.26)) of optical absorption from an hydrogenic acceptor level (t, l) to a conduction band.

Here we consider the question of the density of states due to randomly distributed defects by a semiclassical procedure due to Kane [5.2.24]. Its quantum mechanical counterpart [5.2.25], the Feynman path integral method [5.2.26], as well as other quite sophisticated procedures have also been used, but we shall content ourselves here with the first method, following [5.2.26], even though it is widely believed to overestimate the density-of-states function. Useful reviews exist, see for example [5.2.27] and [5.2.28].

Assume N randomly distributed impurities located at R_i ($i = 1, 2, \ldots, N$). Let the electron-impurity ion potential be $v(\mathbf{r} - \mathbf{R}_i)$ so that $\varphi(x) = \sum_{i=1}^{N} v(\mathbf{r} - \mathbf{R}_i)$ is the fluctuating one-electron potential. Neglecting the Coulomb interaction between the current carriers (which should be largely screened out) we consider effectively a single electron. The probability of finding the potential φ in the range $(\varphi, \varphi + d\varphi)$ is $P(\varphi) \, d\varphi$, where

$$P(\varphi) = \int \cdots \int \frac{d\mathbf{R}_1}{V} \cdots \frac{d\mathbf{R}_N}{V} \delta \left[\varphi - \sum_{i=1}^{N} v(\mathbf{r} - \mathbf{R}_i) \right]$$

$$= \frac{1}{2\pi} \int \cdots \int \frac{d\mathbf{R}_1}{V} \cdots \frac{d\mathbf{R}_N}{V} \int_{-\infty}^{\infty} dt \exp it \left[\varphi - \sum_{i=1}^{N} v(\mathbf{r} - \mathbf{R}_i) \right]$$

$$= \frac{1}{2\pi} \int_{-\infty}^{\infty} dt \exp(it\varphi) \left\{ \int \frac{d\mathbf{R}}{V} \exp[-itv(\mathbf{r} - \mathbf{R})] \right\}^{N}$$

where the integral representation of the δ-function has been used. Writing $x^N = [1 + (x - 1)]^N$, one can put in the limit of $N \to \infty$, $x^N \to \exp[(x - 1)N]$, so that

$$P(\varphi) = \frac{1}{2\pi} \int_{-\infty}^{\infty} dt \exp \left\{ it\varphi + \frac{N}{V} \int d\mathbf{R} \left[e^{-itv(\mathbf{r} - \mathbf{R})} - 1 \right] \right\}$$

If the interaction is weak enough it suffices to truncate an expansion

$$\exp(-itv) \simeq 1 - itv - \tfrac{1}{2} t^2 v^2$$

so that

$$P(\varphi) = \frac{1}{2\pi} \int_{-\infty}^{\infty} dt \exp \left\{ it\varphi - it \frac{N}{V} \int d\mathbf{R} v(\mathbf{r} - \mathbf{R}) - \frac{t^2 N}{2V} \int d\mathbf{R} v^2(\mathbf{r} - \mathbf{R}) \right\}$$

$$= \frac{1}{2\pi} \int_{-\infty}^{\infty} dt \exp \left[it(\varphi - \varphi_0) - \frac{t^2}{2} \xi \right] \tag{5.2.34}$$

In the last step we have used the mean value of $\varphi(\mathbf{r})$

$$\varphi_0 \equiv \langle \varphi(\mathbf{r}) \rangle \equiv \int \cdots \int \frac{d\mathbf{R}_1}{V} \cdots \frac{d\mathbf{R}_N}{V} \sum_{i=1}^{N} v(\mathbf{r} - \mathbf{R}_i) = \sum_{i=1}^{N} \int \frac{d\mathbf{R}_i}{V} v(\mathbf{r} - \mathbf{R}_i)$$

$$\equiv N \int \frac{d\mathbf{R}}{V} v(\mathbf{r} - \mathbf{R}) \tag{5.2.35}$$

We have also used the mean square value of φ:

$$\xi \equiv \langle \varphi(\mathbf{r})^2 \rangle = \sum_{i=1}^{N} \int \cdots \int \frac{d\mathbf{R}_1}{V} \cdots \frac{d\mathbf{R}_N}{V} v^2(\mathbf{r} - \mathbf{R}_i) = N \int \frac{d\mathbf{R}}{V} v^2(\mathbf{r} - \mathbf{R}) \quad (5.2.36)$$

The electron 'sees' the mean φ_0 and the fluctuation ξ.

Using the formula $\int_{-\infty}^{\infty} dx \exp(-ax^2 + bx) = (\pi/a)^{\frac{1}{2}} \exp(b^2/4a)$ with $a = \frac{1}{2}\xi^2$, $b = i(\varphi - \varphi_0)$ the integration in eq. (5.2.34) gives

$$P(\varphi) = \frac{\exp[-\frac{1}{2}(\varphi - \varphi_0)^2/\xi]}{(2\pi\xi)^{\frac{1}{2}}} = \frac{\exp[-\frac{1}{2}(\varphi - \varphi_0)^2/\xi]}{\displaystyle\int_{-\infty}^{\infty} d\varphi \exp[-\frac{1}{2}(\varphi - \varphi_0)^2/\xi]} \quad (5.2.37)$$

Such a potential is called a Gaussian random potential since $P(\varphi)$ is a Gaussian distribution about the mean φ_0.

To calculate the resulting electron density of states it seems reasonable, in first approximation, to use

$$\begin{aligned}
\mathcal{N}_{av}(E) &\equiv \int \mathcal{N}[E - \varphi(\mathbf{r})] P(\varphi) \, d\varphi \\
&= \frac{\sqrt{2}m_c^{\frac{3}{2}}}{(2\pi\xi)^{\frac{1}{2}}\pi^2\hbar^3} \int_{-\infty}^{E} [E - \varphi(x)]^{\frac{1}{2}} \exp\left[-\frac{(\varphi - \varphi_0)^2}{2\xi}\right] d\varphi \\
&= \frac{m_c^{\frac{3}{2}}\xi^{\frac{1}{4}}}{2\pi^2\hbar^3} \exp\left[-\frac{(E - \varphi_0)^2}{4\xi}\right] D_{-\frac{3}{2}}\left(-\frac{E - \varphi_0}{\xi^{\frac{1}{2}}}\right)
\end{aligned}$$

The parabolic cylinder function

$$D_{-p}(z) \equiv \frac{\exp(-z^2/4)}{\Gamma(p)} \int_0^{\infty} x^{p-1} \exp(-zx - \tfrac{1}{2}x^2) \, dx \quad (5.2.38)$$

has here been used. The key assumption is that in each volume element one may take the free electron density of states shifted by the deviation of the local potential energy $\varphi(\mathbf{r})$ from the mean energy φ_0. (The latter can be omitted as it is cancelled by the mean potential energy for the free carriers.) This leaves out of account the kinetic energy of localization which may be considerable for the deeper wells. One can also write the results as

$$\mathcal{N}_{av}(E) = \frac{m_c^{\frac{3}{2}}\xi^{\frac{1}{4}}}{2\pi^2\hbar^3} \exp(-\tfrac{1}{2}z^2) D_{-\frac{3}{2}}(-\sqrt{2}z) \quad \left(z \equiv \frac{E - \varphi_0}{(2\xi)^{\frac{1}{2}}}\right) \quad (5.2.39)$$

Fig. 5.2.3 shows that $\mathcal{N}_{av}(E)$ has developed a low-energy tail, which is always associated with disordered systems.

Fig. 5.2.4 (figure 11 of [5.2.29]) shows how the density of states affects the absorption coefficient in the case of GaAs. The Kane theory (5.2.39) is compared

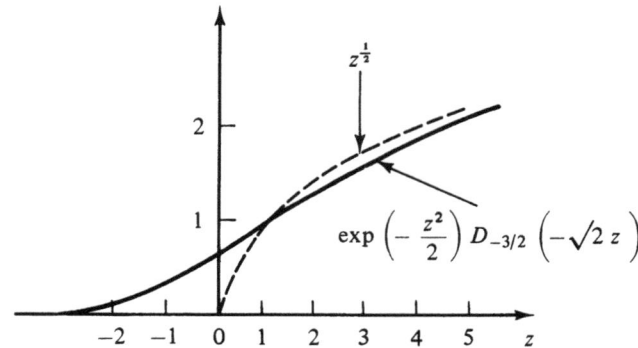

Fig. 5.2.3. A low energy tail in the density of states of a disordered system.

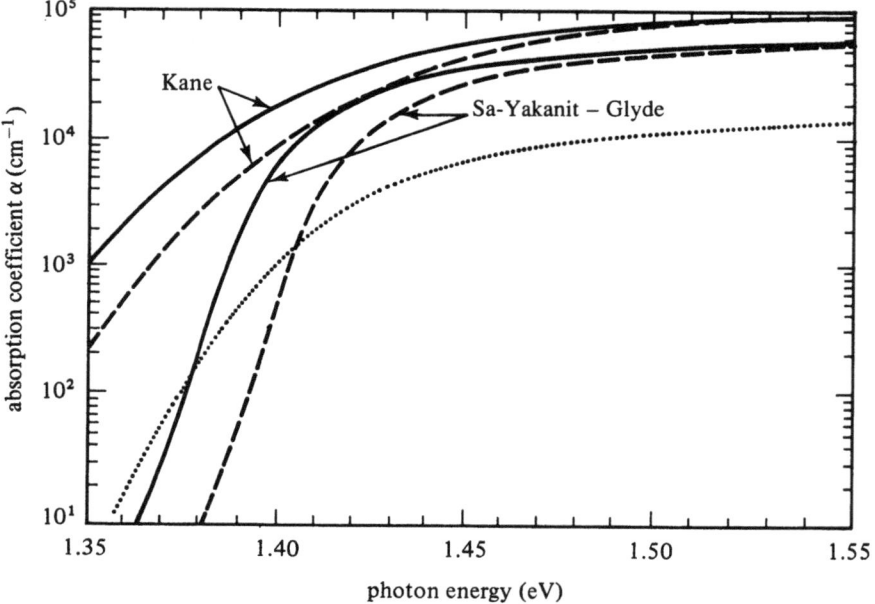

Fig. 5.2.4. Effect of the calculated density of states on the calculated absorption coefficient for p-type GaAs ($p = 1.2 \times 10^{18}$ cm^{-3}). Dashed lines neglect band shrinkage. Full lines use Inkson's band gap shrinkage [5.2.30]. The dotted line is experimental [5.2.31].

with the improved theory due to [5.2.32]. The matrix element (5.2.21) is used in all cases. The details of the theory in [5.2.32] had to be omitted here, but a graphical display of the resulting density of states is given in Fig. 5.2.5.

A Schottky junction tunneling experiment [5.2.33] has yielded a fairly direct determination of the density of states in heavily doped p-type GaAs. The results are compared with a parabolic density of states (aligned to agree with experiment

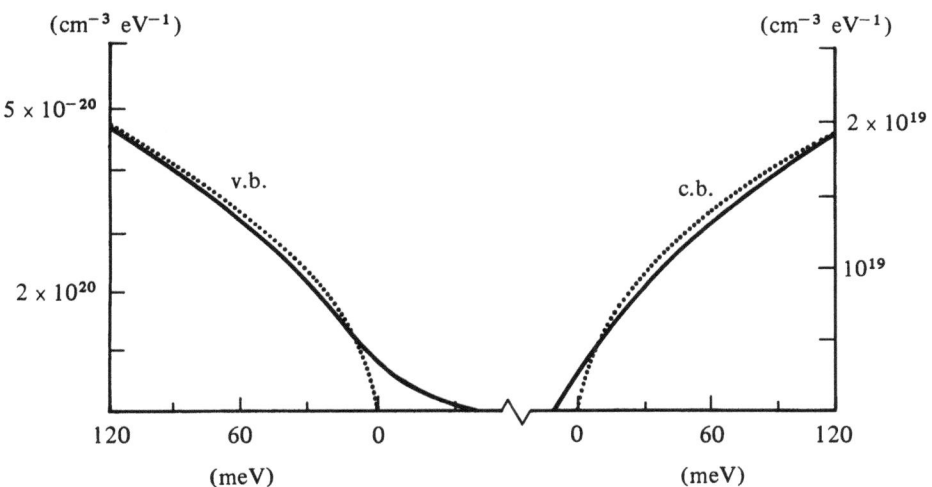

Fig. 5.2.5. Parabolic (dotted) and improved densities of states for GaAs measured from the band edges marked by zero ($N_A = 1.5 \times 10^{18}$ cm^{-3}, $N_D = 3 \times 10^{17}$ cm^{-3}, i.e. $p = 1.2 \times 10^{18}$ cm^{-3}). Electron–hole interactions are neglected.

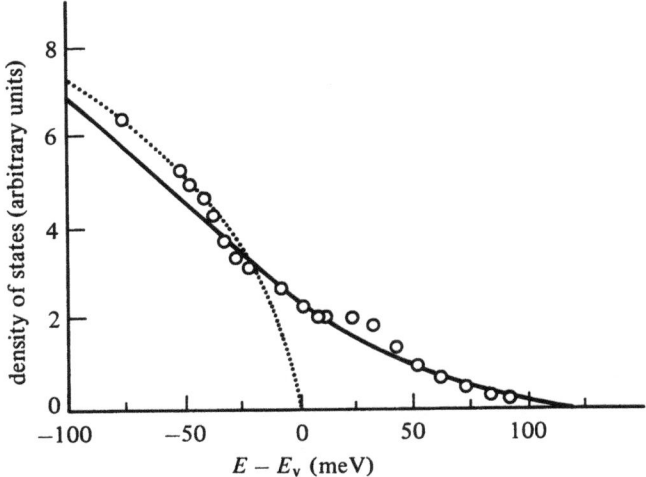

Fig. 5.2.6. The parabolic and improved densities of states [5.2.29] compared with experiments [5.2.33] on p-type GaAs ($T = 4.2$ K, $p = 9.9 \times 10^{18}$ cm^{-3}, screening length 5.03×10^{-7} cm).

at one end) and with the density of states due to [5.2.32] in Fig. 5.2.6 (figure 12 in [5.2.32]).

Turning to the emission of radiation, the existence of low energy tails in heavily doped materials had been established already in the 1960s. Fig. 5.2.7 gives an example for Ge [5.2.34] and similar results had been known already for GaAs [5.2.35] (Fig. 5.2.8); see also [5.2.36] and [5.2.37].

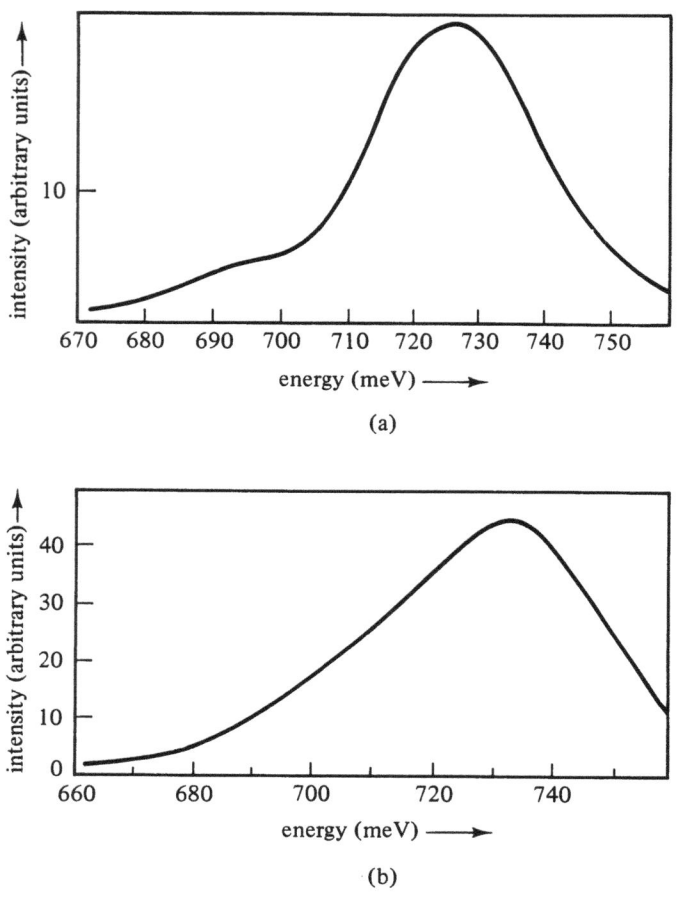

Fig. 5.2.7. Increased low energy tailing in n-type Ge as a result of increased doping. $T = 77$ K. (a) $N_D = 7 \times 10^{18}$ cm^{-3}; (b) $N_D = 3 \times 10^9$ cm^{-3}.

The tails just discussed should disappear if emission can be induced from a degenerate conduction band by electron bombardment, or optical or injection excitation, not requiring a large concentration of donors. Complete disappearance is not expected [5.2.38] owing to the existence of strong lifetime broadening in a degenerate electron gas. (This should be present whether or not the concentration is large.) The electron which has recombined radiatively leaves a hole in the conduction band. If the hole is well below the Fermi level it has a short lifetime since it can be filled by Auger processes in the band, and so broadens each electronic level by the energy–time uncertainty principle. This effect leads to the replacement of each state of the band by a broadened level, and this decreases as the state approaches the Fermi surface. This situation is almost identical with that leading to a broadening of the order of 1 eV in the soft X-ray emission spectra of metals [5.2.39] and already noted in section 3.1. The mechanism is illustrated in

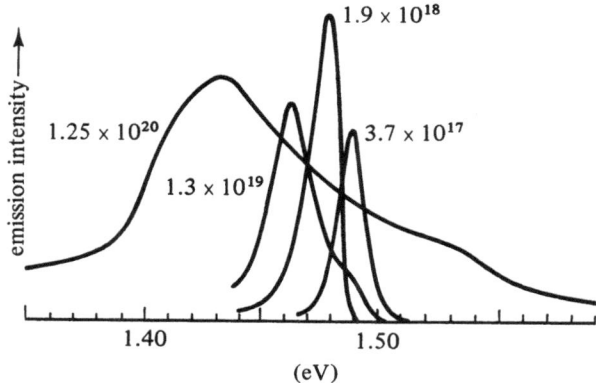

Fig. 5.2.8. Emission spectra at 4.2 K of GaAs doped with the indicated concentrations of Zn acceptors [5.2.35].

Fig. 5.2.9. As the electron vacancy at **k** moves up the band the lifetime broadening decreases since there are fewer possible Auger transitions. The residual density of tail states in a pure material due to this effect is not negligible, but it is small [5.2.39], [5.2.40].

5.2.4 Photoionization and photoneutralization

A rich source of experimental data is provided by photoionization, typically

$$D^{\times} + h\nu \rightarrow D^{+}$$

or photoneutralization, typically

$$A^{-} + h\nu \rightarrow A^{\times}$$

across a substantial part of the forbidden energy gap. The latter condition makes it likely that nonradiative competition is weak, thus facilitating the experiments. The theoretical cross sections for various models are:

$$
\sigma \propto
\begin{cases}
\dfrac{(h\nu + E_A - E_G)^{\frac{1}{2}}}{(h\nu + E_A - E_G + m_A E_A/m_c)^4} & \text{photoneutralization} & (5.2.40) \\[3ex]
\left[\dfrac{4(h\nu - E_D)E_D}{(h\nu)^2}\right]^{\frac{3}{2}} & & \\[3ex]
\dfrac{(h\nu - E_D)^{\frac{3}{2}}}{h\nu[h\nu + E_D(m/m_c - 1)]^2} &
\begin{array}{l}\text{photoionization}\\{}\end{array} &
\begin{array}{l}(5.2.41)\\(5.2.42)\end{array}
\end{cases}
$$

Here m is the normal electron rest mass. Only the first of these expressions is proved here and is obtained at once from eq. (5.2.22) by setting $\sigma \propto \alpha_{CA}$. It holds

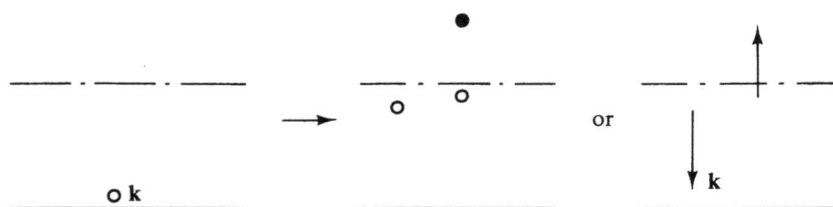

Fig. 5.2.9. Two representations of the filling of an electron vacancy in a Fermi-degenerate band by Auger processes.

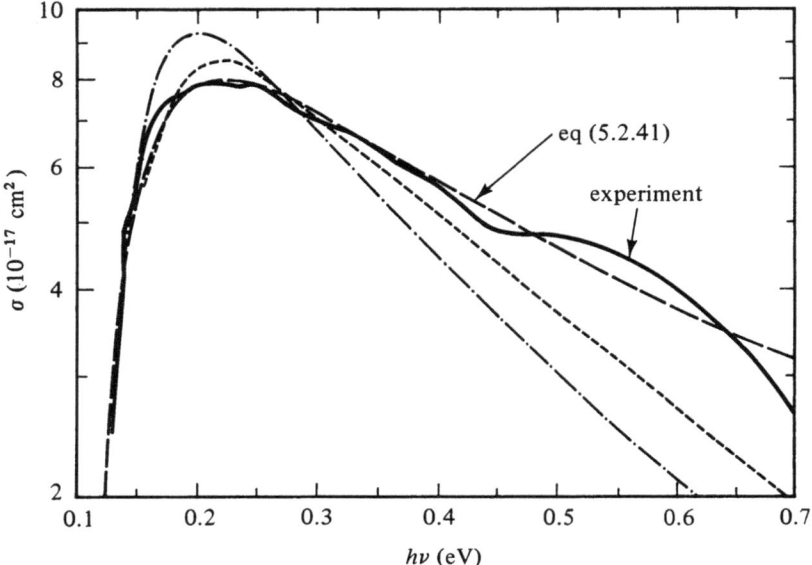

Fig. 5.2.10. Photoionization spectrum for Mn acceptors in GaAs at 20 K (solid line) from [5.2.46], figure 8. Other curves: —— eq. (5.2.41); ―――― quantum-defect model, Coulombic final state; —·—·—·— quantum-defect model, free-particle final state.

for a hydrogenic model. The second expression arises from the δ-function potential energy model [5.2.41]. In this the electron is treated as a plane wave in the band. The third expression was derived in [5.2.42, eqn. (9)] by a modification of the argument leading to eq. (5.2.41). In all cases there is a threshold at $E_G - E_A$ or at E_D. The cross section then rises to a maximum at a value of $h\nu$, respectively, equal to

$$E_G - E_A(1 - m_A/7m_c) \quad \text{or } 2E_D \quad \text{or } 2E_D \text{ (if } m = m_c)$$

At higher photon energies it decays respectively as

$$(h\nu)^{-\frac{7}{2}}, \quad (h\nu)^{-\frac{3}{2}}, \quad (h\nu)^{-\frac{3}{2}}$$

Grimmeiss and Ledebo [5.2.42] make out a good case for their version of the δ-function potential model. Other models have also been championed, notably a model which treats the defect as a hard sphere, the so-called billiard ball model [5.2.43], other quantum defect models, etc. Excellent reviews are given in [5.2.44]–[5.2.47] and no attempt is therefore made here to add another one. However, a typical photoionization spectrum is given in Fig. 5.2.10 showing that, in spite of its approximate nature, the Lucovski model [5.2.41] can sometimes give a reasonably good fit.

5.2.5 Some uses of detailed balance

It is possible to look at the important result (4.5.10) from the point of view of the principle of detailed balance. For this purpose introduce the probability of a photon of vacuum wavelength λ_0 being absorbed per unit time per unit volume

$$P(\lambda_0) = c\alpha(\lambda_0)/V\mu(\lambda_0) \tag{5.2.43}$$

Here μ is the refractive index; this formula can be obtained from eqs. (4.5.1) and (4.5.3) which give for the net normal absorption rate per unit volume

$$\alpha N_v c/\mu V$$

If one considers a single photon one obtains eq. (5.2.43). To find the volume rate of excitation in the solid by photons in the vacuum wavelength range $d\lambda_0$, $P(\lambda_0)$ has to be multiplied by the number of relevant photon modes

$$8\pi\mu^3\lambda_0^{-4} V d\lambda_0$$

and their equilibrium occupation probability

$$[\exp(ch/\lambda_0 kT) - 1]^{-1}$$

A factor $\alpha'(\lambda_0)/\alpha(\lambda_0)$ is also needed since not all photons of wavelength λ_0 will, when absorbed, produce one electron–hole pair. Thus $\alpha'(\lambda_0)/\alpha(\lambda_0) \leqslant 1$ is the probability of this happening per absorbed photon. Hence the equilibrium absorption rate per unit volume of photons in the range $d\lambda_0$ with production of an electron–hole pair is

$$\frac{\alpha'}{\alpha} \frac{8\pi\mu^2\alpha c\lambda^{-4}}{\exp(ch/\lambda_0 kT) - 1} d\lambda_0$$

In terms of frequencies this is with $x \equiv h\nu_0/kT$

$$\frac{8\pi\alpha'}{h^3} \left(\frac{\mu}{c}\right)^2 (kT)^3 \frac{x^2 dx}{\exp x - 1}$$

Detailed balance can be applied for equilibrium conditions and says that these expressions also give the rate per unit volume of spontaneous radiative recombination of electron–hole pairs with the emission of photons in the range $d\lambda_0$ or dv_0. Now the electron–hole recombination rate per unit volume has the form $C(\lambda_0) v_0 p_0 d\lambda_0$, where $C(\lambda_0) d\lambda_0$ is a reaction rate constant of dimension $L^3 T^{-1}$. Hence

$$
\left.
\begin{aligned}
C(\lambda_0) n_0 p_0 &= \frac{8\pi\mu^2\alpha' c\lambda_0^{-4}}{\exp(ch/\lambda_0 kT)-1} \\[2em]
\text{or} \\[1em]
C(v_0) n_0 p_0 &= \frac{8\pi\alpha'}{h^3}\left(\frac{\mu}{c}\right)^2 (kT)^3 \frac{x^2\,dx}{\exp x-1}
\end{aligned}
\right\} \qquad (5.2.44)
$$

This is essentially the result (4.5.10), now obtained from detailed balance [5.2.48].

Departing from equilibrium, the nett recombination per unit volume is for the range $d\lambda_0$

$$
U(\lambda)_0\,d\lambda_0 = C(\lambda_0)(np-n_0 p_0)\,d\lambda_0 = C(\lambda_0)(n_0 p_0)\left(\frac{np}{n_0 p_0}-1\right)d\lambda_0 \qquad (5.2.45)
$$

Putting $n = n_0+\Delta n$, $p = p_0+\Delta n$

$$
\frac{np}{n_0 p_0}-1 = \frac{n_0+p_0+\Delta n}{n_0 p_0}\Delta n \simeq \left(\frac{1}{n_0}+\frac{1}{p_0}\right)\Delta n
$$

Hence the radiative lifetime τ_r is given by

$$
\frac{1}{\tau_r} \equiv \frac{\int U(\lambda_0)\,d\lambda_0}{\Delta n} = \left(\frac{1}{n_0}+\frac{1}{p_0}+\frac{\Delta n}{n_0 p_0}\right)\int_0^\infty \frac{8\pi^2\alpha' c\lambda_0^{-4}}{\exp(ch/\lambda_0 kT)-1}\,d\lambda_0 \qquad (5.2.46)
$$

where eqs. (5.2.44) and (5.2.45) have been used.

Just as absorption and radiative recombination can be related by detailed balance, so impact ionization and Auger recombination are also connected. The 'high energy' particle, which is an electron excited by a photon in the radiative case, is now replaced by a high energy electron in what has been denoted by state 2' in section 3.5. Let $\beta(2')$ ($\leqslant 1$) be the fraction of electrons in states close to 2' which are available for impact ionization. This allows for transitions out of state 2' by processes other than impact ionization. Let $P(2')$ be the impact ionization probability per unit time per unit volume due to an electron which occupies state 2' with certainty and let $\theta(2')$ be the occupation probability of state 2', denoted by $\theta_0(2')$ in equilibrium. Then, defining a reaction rate constant $D(2')\,dS_{2'}$ for Auger recombination with the certain production of an Auger electron in states in the range $dS_{2'}$, detailed balance is expressed by

$$
D(2') n_{10} n_{20} p_{1'0}\,dS_{2'} = \beta(2')\theta_0(2') P(2')\,dS_{2'} \qquad (5.2.47)
$$

Here the concentration of empty states of type 1′ has been denoted by $p_{1'}$. With 2′ fixed, various states 1, 2, 1′ are possible and they are integrated over in eq. (5.2.46).

Departing from equilibrium, the net Auger recombination rate, each transition leading to an Auger electron in a state in the range $dS_{2'}$ is (in analogy with eq. (5.2.45) and using eq. (5.2.47))

$$u(2')\,dS_{2'} = D(2')(n_1 n_2 p_{1'} - n_{10} n_{20} p_{1'0})\,dS_{2'}$$

$$= \frac{n_1 n_2 p_{1'} - n_{10} n_{20} p_{1'0}}{n_{10} n_{20} p_{1'0}} \beta(2')\,\theta_0(2')\,P(2')\,dS_{2'} \qquad (5.2.48)$$

In a simple band–band process $n_1 = n_2 = n$, $p_{1'} = p$. Assume again $n = n_0 + \Delta n$, $p = p_0 + \Delta n$. The first factor in eq. (5.2.48) is

$$\frac{n^2 p - n_0^2 p_0}{n_0^2 p_0} = (2n_0 p_0 + n_0^2 + 2n_0\,\Delta n + p_0\,\Delta n + (\Delta n)^2)\frac{\Delta n}{n_0^2 p_0} \simeq \left(\frac{2}{n_0} + \frac{1}{p_0}\right)\Delta n$$

The recombination lifetime as limited by the Auger effect is

$$\frac{1}{\tau_A} \equiv \frac{\int u(2')\,dS_{2'}}{\Delta n} \simeq \left(\frac{2}{n_0} + \frac{1}{p_0}\right)\int \beta(2')\,\theta_0(2')\,P(2')\,dS_{2'} \qquad (5.2.49)$$

in analogy with eq. (5.2.46) [5.2.49]. Unfortunately the experimental impact ionization data available is not normally adequate to evaluate τ_A from eq. (5.2.49). The optical data for the calculation of τ_r from eq. (5.2.46) is much more abundant. The analogy between the radiative and the Auger case is brought out in Table 5.2.2.

Returning to the problems of radiative recombination, suppose next that a defect with $j-1$ electrons captures a jth electron from the conduction band with spontaneous emission of radiation. Then the spontaneous emission rate per unit volume due to electrons in an energy range dE is

$$\sigma_r(E)\,v(E)\,\mathcal{N}_c(E)\,p_c(E)\,p_{j-1}\,N_D\,dE \quad [v(E) = (2E/m_c)^{\frac{1}{2}}], \qquad (5.2.50)$$

where σ_r is the electron radiative capture cross section of an ionized donor, N_D is the donor concentration, $v(E)$ is the free electron speed, $\mathcal{N}_c$ is the conduction band density of states (per unit volume per unit energy range), $p_c(E)$ is the conduction band occupation probability, and p_{j-1} is the fraction of donors in this charge state. This is in a steady state equal to the absorption rate less the stimulated emission rate, both per unit volume and for a conduction band energy range dE, i.e. to

$$[p_j[1 - p_c(E)] - p_c(E)\,p_{j-1}]\,\sigma_i(h\nu)\frac{c}{\mu}\,N(h\nu)\,N_D\,dE \qquad (5.2.51)$$

Table 5.2.2. *Schematic detailed balance arguments*

		Radiative	Auger

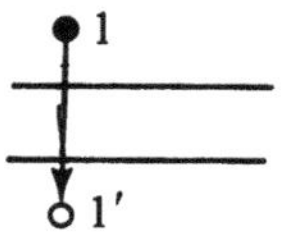

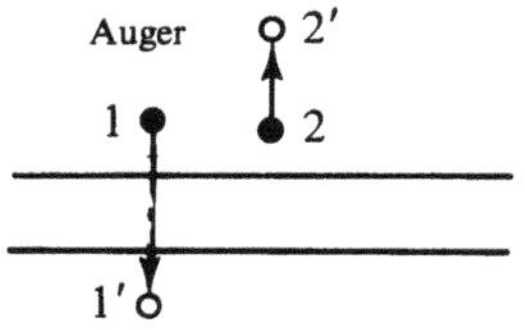

(Radiative diagram: filled circle 1 with arrow down to open circle 1′. Auger diagram: filled circle 1, open circle 2′ above with arrow up, filled circle 2, open circle 1′.)

(1)	Initial object	photon of vacuum wavelength λ_0	Energetic electron in state 2′
(2)	Process of disappearance of initial object	absorption of photon	impact ionization by the energetic electron
(3)	Probability per unit volume per unit time of this process	$P(\lambda_0) = \dfrac{\alpha(\lambda_0)}{V}\dfrac{c}{\mu(\lambda_0)}$	$P(2') = \dfrac{1}{Vt}\displaystyle\int T_{if}\,dS_1\,dS_{1'}\,dS_2$
(4)	Number of states for initial object (in an increment)	$8\pi\mu^3\lambda_0^{-4}\,V\,d\lambda_0$	$dS_{2'}\left(\text{e.g. } \dfrac{V}{4\pi^3}dk_{2'}\right)$
(5)	Number of initial objects (in an increment)	$\dfrac{8\pi\mu^3\lambda_0^{-4}\,V\,d\lambda_0}{\exp\left(\dfrac{ch}{\lambda_0 kT}\right)-1}$	$\theta(2')\,dS_{2'}$
(6)	Rate per unit volume of disappearance of initial objects $= (3)\times(5)$	$\dfrac{8\pi\mu^2\alpha'c\lambda_0^{-4}\,d\lambda_0}{\exp\left(\dfrac{ch}{\lambda_0 kT}\right)-1}$	$\beta(2')\,\theta(2')\,P(2')\,dS_{2'}$
(7)	Symbol for the reaction constant for the production of initial objects in an increment	$C(\lambda_0)\,d\lambda_0$ $[\mathrm{L^3T^{-1}}]$	$D(2')\,dS_{2'}$ $[\mathrm{L^6T^{-1}}]$
(8)	Rate of production per unit volume of initial objects in an increment by detailed balance in thermal equilibrium	$C(\lambda_0)\,n_0 p_0\,d\lambda_0$ $= (6)$ This is the photon emission rate	$D(2')\,n_{10}n_{20}p_{1'0}\,dS_{2'}$ $= (6)$ This is the Auger recombination rate
(9)	Net production rate per unit volume of initial objects in an increment	$C(\lambda_0)(np - n_0 p_0)\,d\lambda_0 =$ $\dfrac{np - n_0 p_0}{n_0 p_0}\times(6)$	$D(2)(n_1 n_2 p_{1'} - n_{10}n_{20}p_{10}p_2)\,dS =$ $\dfrac{n_1 n_2 p_{1'} - n_{10}n_{20}p_{1'0}}{n_{10}n_{20}p_{1'0}}\times(6)$
(10)	Total production rate of initial objects per unit volume	$\int (9) =$ $\dfrac{np - n_0 p_0}{n_0 p_0}\displaystyle\int \dfrac{8\pi\mu^2\alpha'c\lambda_0^{-4}\,d\lambda_0}{\left(\exp\dfrac{c}{\lambda_0}\dfrac{h}{kT}\right)-1}$	$\int (9) =$ $\dfrac{n_1 n_2 p_1 - n_{10}n_{20}p_{10}}{n_{10}n_{20}p_{1'0}}\displaystyle\int \beta(2')\,\theta(2')\,P(2)\,dS_{2'}$

In writing the product of concentrations as a factor in the reaction rate, the absence of Fermi degeneracy is assumed. The occupation probability of state 2′ has been denoted by $\theta(2')$; it is $\theta_0(2')$ in equilibrium.

Here σ_i is the photoionization cross section of a neutral donor, c/μ is the speed of the radiation in the material, and $hv = E_D + E$, where E_D is the energy gap between the conduction band and the donor level. Also $N(hv)\,dE$ is the number of photons per unit volume which is involved. It follows that

$$\frac{\sigma_r(E)}{\sigma_i(hv)} = \frac{cN(hv)}{\mu v(E)\,\mathcal{N}_c(E)} \cdot \left[\frac{p_j}{p_{j-1}} \frac{1-p_c(E)}{p_c(E)} - 1 \right] \qquad (5.2.52)$$

In thermal equilibrium this steady-state condition becomes the detailed balance condition with the equilibrium expression

$$N(hv) = \frac{8\pi\mu^3(hv)^2}{(hc)^3\,[\exp(hv/kT)-1]} \qquad (5.2.53)$$

Hence

$$\frac{\sigma_r(E)}{\sigma_i(hv)} = \frac{\mu^2(hv)^2}{2c^2 m_c E} \cdot \frac{\exp[\eta - \eta(j-\frac{1}{2})]-1}{\exp(hv/kT)-1} \qquad (5.2.54)$$

In eq. (5.2.53) we used the middle expression in eq. (1.3.9) to evaluate (with $p = \mu hv/c$ and $l = 2$) the number of modes per unit volume in a range $d(hv)$:

$$l \cdot \frac{4\pi p^2\,dp}{h^3} = \frac{8\pi(hv)^2\,\mu^2}{c^2 h^3}\,d\left(\frac{hv\mu}{c}\right) = \frac{8\pi\mu^3(hv)^2}{c^3 h^3}\,d(hv)$$

In eq. (5.2.54) we used eq. (1.4.10) with $g = 2$ so that

$$v(E)\,\mathcal{N}_c(E) = \left(\frac{2E}{m_c}\right)^{\frac{1}{2}} \cdot 4\pi \left(\frac{2m_c}{h^2}\right)^{\frac{3}{2}} E^{\frac{1}{2}} = \frac{16\pi E m_c}{h^3} \qquad (5.2.55)$$

Hence

$$\frac{N(hv)}{v(E)\,\mathcal{N}(E)} = \frac{\mu^3(hv)^2}{2m_c c^3 E\,[\exp(hv/kT)-1]}$$

and

$$\left(\frac{p_j}{p_{j-1}}\right)_{\mathrm{eq.}} = \frac{Z_j}{Z_{j-1}} e^\gamma = \exp[\gamma - \eta(j-\tfrac{1}{2})]$$

where $kT\eta(j-\frac{1}{2})$ is the effective energy level of the donor (incorporating degeneracies), and also

$$\left[\frac{1-p_c(E)}{p_c(E)}\right]_{\mathrm{eq.}} = \exp(\eta - \gamma) \qquad \left(\eta \equiv \frac{E}{kT}\right)$$

The second factor in eq. (5.2.54) is of order unity. Hence the exact detailed balance relation (5.2.54) goes over into the approximate relation

$$\frac{\sigma_r(E)}{\sigma_i(h\nu)} = \frac{1}{2m_c E}\left(\frac{\mu h\nu}{c}\right)^2 \tag{5.2.56}$$

between the radiative capture cross section and the photoionization cross section. It is a slight generalization from one-electron centers to j-electron centers of a formula which was considered by Milne for astrophysics [5.2.50], [5.2.51], was shown to be relevant for semiconductors [5.2.52], and was discussed in detail later [5.2.53] (see also [5.2.54]). If phonons participate the argument becomes slightly more complicated as shown in [5.2.55] for the case of band–band recombination.

For hole capture one has similarly

$$\frac{\sigma_r(E)}{\sigma_c(h\nu)} = \frac{1}{2m_v E}\left(\frac{\mu h\nu}{c}\right)^2 \tag{5.2.57}$$

where E is the energy of the hole (measured from the top of the valence band) which is later captured.

In the literature the partition function approach which goes to define the effective energy level $\eta(j-\frac{1}{2})$ is not usually used and one then replaces Z_j/Z_{j-1} by $(g_j/g_{j-1})\exp(-\eta_D)$, where the g's are degeneracy factors and η_D is *presumed* to be a true energy level detectable in optical experiments. Our slightly more general approach allows for the effect of excited states. While here the effect of the degeneracy factor is absorbed in the last factors of eqs. (5.2.52) and (5.2.54) which are of order unity, the degeneracy ratio appears as g_1/g_0 (denoted by w) in [5.2.53], since the approximation there is slightly different. In the case of hole capture the transition is from a j-electron to a $(j-1)$-electron center and a factor g_{j-1}/g_j appears as g_0/g_1 (denoted by β^{-1}) in [5.2.56]. The importance of these factors is however doubtful in view of the neglect of the excited states.

The average of eq. (5.2.50) over the conduction band yields an average capture cross section

$$\bar{\sigma}_r \equiv \frac{\int_0^\infty \sigma_r(E)\,v(E)\,\mathcal{N}_c(E)\,p_c(E)\,dE}{\int_0^\infty v(E)\,\mathcal{N}_c(E)\,p_c(E)\,dE}$$

The denominator is $n\bar{v}$ where n is the electron concentration in the conduction band and $\bar{v}$ is the average speed of electrons in this band. Using eqs. (5.2.55) and (5.2.56) one finds [5.2.53]

$$\bar{\sigma}_r = \frac{8\pi\mu^2}{n\bar{v}h^3c^2}\int_0^\infty (E+E_D)^2\,\sigma_i(h\nu)\,p_c(E)\,dE \tag{5.2.58}$$

where we have replaced $h\nu$ by $E+E_D$. There is a divergence in (5.2.56) as $E\to 0$, i.e. for capture from the bottom of the conduction band. This is due to the fact that eq. (5.2.50) vanishes for $E \to 0$ while eq. (5.2.51) is still finite; the vanishing of eq. (5.2.50) would be removed in a higher order theory. It is also removed by the averaging process which led to eq. (5.2.58).

If one assumes that the electrons are nondegenerate in the conduction band (to which we restrict attention for simplicity) the result for $\bar{\sigma}_r$ can readily be developed further. Using eq. (1.6.11) one can then put

$$p_c(E) = [nh^3/2(2\pi m_c kT)^{\frac{3}{2}}]\exp(-\eta)$$

where the energy zero is chosen at the bottom of the band. The radiative capture coefficient is then ($\eta \equiv E/kT$, $\eta_D \equiv E_D/kT$)

$$c_r \equiv \bar{v\sigma_r} = \frac{2\mu^2}{(2\pi m_c kT)^{\frac{1}{2}} c^2 m_c kT} \int_0^\infty (E+E_D)^2 \sigma_i(h\nu)\exp(-\eta)\,dE \qquad (5.2.59)$$

Let us consider a spectral form

$$\sigma_i(h\nu) = A\frac{(h\nu-E_D)^\lambda E_D^{\theta-\lambda}}{(h\nu)^\theta} \qquad (5.2.60)$$

where A and λ are independent of frequency. This covers the Lucovski form (5.2.41) ($\lambda = \frac{3}{2}$, $\theta = 0$) and a form considered in equation (87) of [5.2.46] ($\lambda = \frac{3}{2}$, $\theta = 1$, $A = 2\sigma$). One finds with $x \equiv E/E_D$

$$c_r = \left(\frac{2}{\pi}\right)^{\frac{1}{2}} \frac{A\mu^2 E_D^3}{c^2 m_c^{\frac{3}{2}}} \int_0^\infty (1+x)^{2-\theta} x^\lambda \exp(-\eta_D x)\,dx \qquad (5.2.61)$$

$$= \left(\frac{2}{\pi}\right)^{\frac{1}{2}} \Gamma(\lambda+1)\frac{A\mu^2(kT)^{\frac{3}{2}}}{c^2 m_c^{\frac{3}{2}}} \eta_D^{(2-\lambda+\theta)/2}\exp\left(\frac{\eta_D}{2}\right)$$

$$\times W_{\frac{2-\lambda+\theta}{2},\,\frac{3+\lambda-\theta}{2}}(\eta_D) \qquad (5.2.62)$$

by using a Laplace transform formula and the Whittaker function. We can avoid this slightly complicated result if, as will often be the case, $\eta_D \gg 1$, when eq. (5.2.61) gives directly

$$c_r \simeq \left(\frac{2}{\pi}\right)^{\frac{1}{2}} \frac{A\mu^2}{c^2 m_c^{\frac{3}{2}}} \Gamma(\lambda+1)(kT)^{\frac{3}{2}}\left(\frac{E_D}{kT}\right)^{2-\lambda} \qquad (5.2.63)$$

which has the form

$$c_r \simeq \frac{3 \times 2^{\frac{1}{2}}}{4}\frac{A\mu^2 kT}{c^2 m_c^{\frac{3}{2}}} E_D^{\frac{1}{2}}$$

for both $\lambda = \frac{3}{2}$ situations noted above. (One can obtain these results of course also by using the asymptotic form of the Whittaker function for large η_D.) One sees that if A is independent of temperature the radiative capture coefficient has a temperature dependence of $T^{\lambda-\frac{1}{2}}$ which contrasts with the usual dependence as $\exp(-E_0/kT)$ (for some activation energy E_0) of the radiationless multiphonon recombination.

The above relations are useful in the analysis of experimental emission lines in terms of photoneutralization or photoionization. One example of such an application is found in Dishman's discussion of recombination involving neutral oxygen in p-type GaP [5.2.56]. He used eq. (5.2.56) to obtain the shape of the emission line (without phonon participation) and to find good agreement with experiment. A related analysis was used [5.2.57] to show that when *negatively* charged gold and platinum impurities in Ge capture thermal electrons into levels lying 0.2 eV below the conduction band, the probability of photon emission varies from 10^{-3} to 10^{-5} (Au) or from 10^{-5} to 10^{-7} (Pt) as the lattice temperature is changed from 35 K to 85 K. These low probabilities are not unexpected.

5.2.6 Donor–acceptor pair radiation

If semiconductors are lightly or moderately doped (so as to avoid impurity bands) donor–acceptor pair radiation can be studied. In his review of this topic [5.2.58], the late Ferd Williams, whose name was in the early days closely associated with these phenomena, traced the history of this study in semiconductors back to 1956 (Reiss, Fuller, and Morin and Prener and Williams). In those days the subject was regarded as belonging to luminescence and in this context acceptors were called activators and donors were called co-activators [5.2.59]. This type of recombination attracted a wider audience at the Paris Symposium on Radiative Recombination in Semiconductors in 1964 by a paper with the almost astrophysical title 'Light from distant pairs' [5.2.60]. Outstandingly beautiful low temperature spectra, giving peaks going beyond next-nearest neighbours ($i = 2$) to high-order nearest neighbours ($i \gtrsim 40$) were produced a few years later. Fig. 5.2.11 [5.2.61] is an example. It gives spectra for the pairs $Zn_{Ga}S_P$ and $Cd_{Ga}S_P$; but other pairs such as $Zn_{Ga}Te_P$, $Zn_{Ga}Se_P$, $Cd_{Ga}Te_P$, etc. have also been obtained in GaP. Fig. 5.2.11 was obtained with a forward biased p–n junction, and with removal of the background radiation and radiation previously identified as due to other mechanisms. The spectra are fitted by lines due to the more distant pairs since higher order interactions distort the result (5.2.64), below, for close pairs. This distortion is represented by the energy E in the equation.

The peaks occur at frequencies v_i ($i = 1, 2, \ldots$) given by

$$h v_i = E_G - E_A - E_D + e^2/\varepsilon R_i[-E] \qquad (5.2.64)$$

where R_i is the distance between the ith order nearest neighbors. This formula can be obtained by means of a five-step cycle as follows.

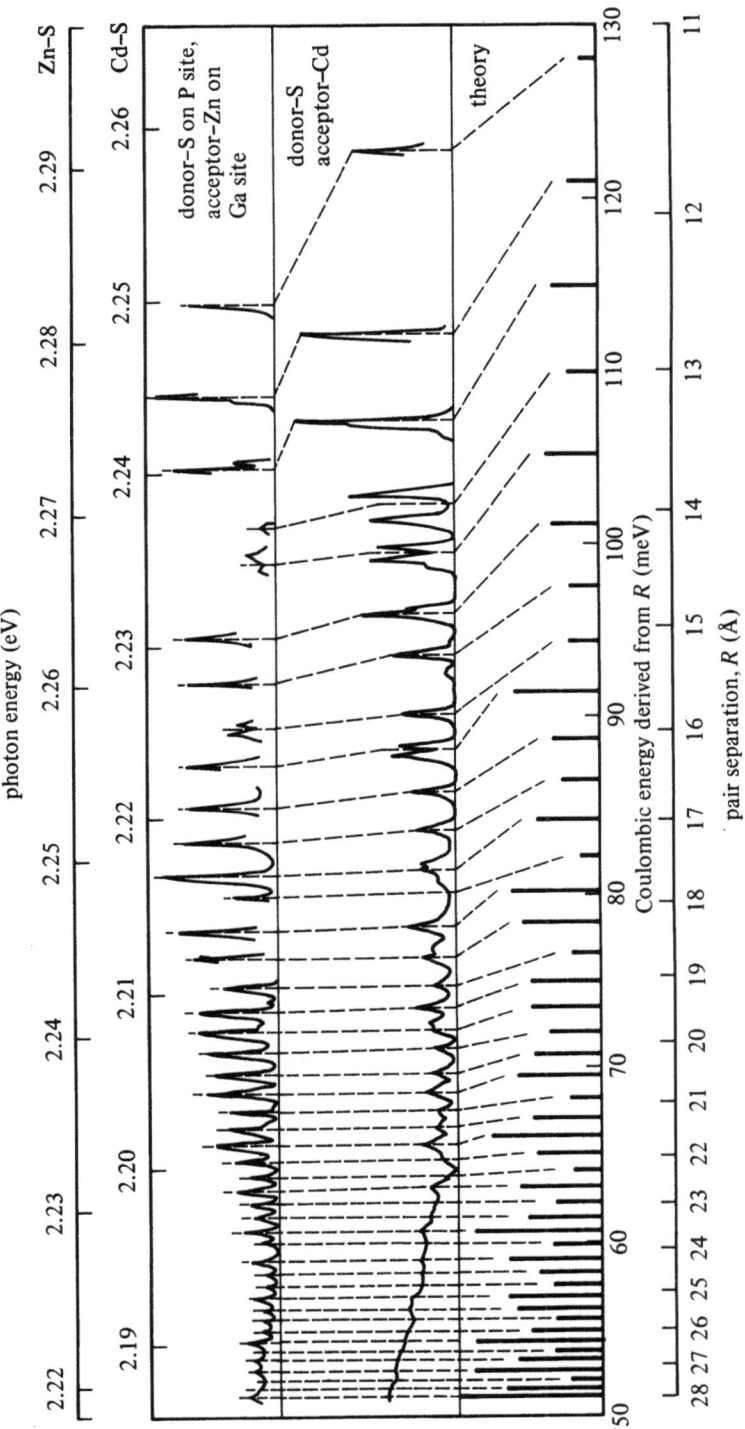

Fig. 5.2.11. Comparison of the positions and intensities of the sharp line pair spectra at 1.6 K corresponding to both Zn–S and Cd–S acceptor–donor pairs with the predicted pair distribution. The lower scales show both the pair separation R and the Coulombic energy ε derived from R. The emission energy scales for the two measured spectra are shown above the figure.

(i) $E_G + D^\times A^\times \rightarrow E_G + D^- A^+ + h\nu$

(ii) $\rightarrow e + p + D^- A^+ + h\nu$

(iii) $\rightarrow e + p + D^- + A^+ + h\nu - \dfrac{e^2}{\varepsilon R}$

(iv) $\rightarrow D^\times + A^\times + h\nu - \dfrac{e^2}{\varepsilon R} + E_D + E_A$

(v) $\rightarrow D^\times A^\times + h\nu - \dfrac{e^2}{\varepsilon R}$

$\qquad + E_D + E_A + E$

These steps can be explained as follows:

(i) A neutral donor–acceptor pair plus energy E_G goes over into a pair in which both centers are ionized, because an electron has jumped from donor to acceptor, emitting a photon of energy $h\nu$. This is the unknown which is to be determined. The energy E_G is still available.

(ii) The energy E_G is used to produce an electron–hole pair.

(iii) The ionized pair is separated from distance R to infinity. This uses up energy $e^2/\varepsilon R$ where ε is an effective dielectric constant.

(iv) The electron recombines with the donor and the hole with the acceptor. This releases energy $E_D + E_A$.

(v) Only the two neutral centers are left, and they are brought together to distance R. Polarizing forces operate and a small amount of energy E is released. The system is in its original condition. Hence energy balance yields an expression for $h\nu$ which is basic in this type of work. E behaves as R^{-6} and can sometimes be neglected.

Comparison of (i) and (v) now yields eq. (5.2.64).

The discrete nature of the peaks is due to the fact that the impurities involved settle in general on lattice sites so that only definite separations R_i are possible. The value of $E_A + E_D$ can sometimes be inferred from the experimental lines by extrapolation to $R_i \rightarrow \infty$. The energy term E is sometimes neglected, or it is identified with a dipole–dipole (or van der Waals) polarization interaction of the form $e^2 b^5 / \varepsilon R_i^6$ where b is a constant estimated as $(6.5)^{\frac{1}{6}} \tilde{r}_1$, where $\tilde{r}_1$ is the Bohr-type radius of the ground state orbit of the more tightly bound member of the pair.

In fact ZnS phosphors were the first materials in which donor–acceptor radiation was hypothesized. However, it is hard to control its stoichiometry and its impurity content. Further it is rather ionic with strong electron–phonon coupling, has relatively large carrier mass and hence relatively large impurity activation energies $E_t \propto m^*$. If eq. (5.2.64) is positive only for relatively close pairs, as in this case, the spectrum is expected to be dominated by the decay between close pairs and the phonon broadening makes it relatively broad and featureless. In such cases therefore spectra such as shown in Fig. 5.2.11 are hard or impossible to obtain. This applies in general to many II–VI compounds.

Interesting experiments can also be made in tracking the decay of luminescence with time following pulsed excitation. The more distant pairs decay later than the nearer pairs. In fact, as one might expect from the overlap of donor and acceptor exponential wavefunctions of the type (5.2.18), the transition probability per unit time for a pair a distance R apart behaves roughly as

$$S(R) = S(0)\,e^{-R/a} \tag{5.2.65}$$

where a is related to the Bohr radii involved, but is of the order of one half the larger radius, i.e. the radius appropriate to the shallower center. Typical values for GaP at He temperatures (extracted from a detailed study of eq. (5.2.65)) are

$$\tilde{r}_{D1} \sim 10.4\ \mathring{A}, \quad \tilde{r}_{A1} \sim 15.8\ \mathring{A}, \quad a \sim 7.7\ \mathring{A} \quad S(0) = 6 \times 10^5\ s^{-1}$$

This applies to the S_P–Zn_{Ga} system [5.2.62]. The notation (5.2.19) has here been used.

One can obtain the donor–acceptor transition matrix element by using eqs. (5.2.6) and (5.2.7)

$$\mathbf{P}_{DA} = \frac{1}{V} \int \left[\sum_{L,k} v_v^*(\mathbf{k}, \mathbf{L})\, G_v^*(\mathbf{k})\, e^{-i(\mathbf{k}+\mathbf{K})\cdot\mathbf{r}} \right] \mathbf{P}$$
$$\times \left[\sum_{L',k'} v_c(\mathbf{k}', \mathbf{L}')\, G_c(\mathbf{k}')\, e^{i(\mathbf{k}'+\mathbf{L}')\cdot(\mathbf{r}-\mathbf{R})} \right] d\mathbf{r}$$

The $\mathbf{r}$-integration yields by eq. (B.11) that $\mathbf{k}' - \mathbf{k} = \mathbf{L} - \mathbf{L}'$ so that with $\mathbf{P} = (\hbar/i)\nabla$

$$\mathbf{P}_{DA} = \hbar \sum_{L,L',k} (\mathbf{k}+\mathbf{L})\, v_v^*(\mathbf{k}, \mathbf{L})\, G_v^*(\mathbf{k})\, v_c(\mathbf{k}+\mathbf{L}-\mathbf{L}', \mathbf{L}')$$
$$\times\, G_c(\mathbf{k}+\mathbf{L}-\mathbf{L}')\, e^{-i(\mathbf{k}+\mathbf{L})\cdot\mathbf{R}}$$

The main contribution arises from $\mathbf{L} = \mathbf{L}'$ so that

$$\mathbf{P}_{DA} \simeq \sum_{L,k} \hbar(\mathbf{k}+\mathbf{L})\, v_v^*(\mathbf{k}, \mathbf{L})\, v_c(\mathbf{k}, \mathbf{L})\, G_v^*(\mathbf{k})\, G_c(\mathbf{k})\, e^{-i(\mathbf{k}+\mathbf{L})\cdot\mathbf{R}}$$
$$\simeq \sum_k \mathbf{P}_{vc}(\mathbf{k})\, G_v^*(\mathbf{k})\, G_c(\mathbf{k})\, e^{-i\mathbf{k}\cdot\mathbf{R}} \tag{5.2.66}$$

Here

$$\mathbf{P}_{vc}(\mathbf{k}) \equiv V^{-1} \int u_v^*(\mathbf{k}, \mathbf{r})\, e^{-i\mathbf{k}\cdot\mathbf{r}}\, \mathbf{P} u_c(\mathbf{k}', \mathbf{r})\, e^{i\mathbf{k}'\cdot\mathbf{r}}\, d\mathbf{r}$$
$$\simeq \sum_L \hbar(\mathbf{k}+\mathbf{L})\, v_v^*(\mathbf{k}, \mathbf{L})\, v_c(\mathbf{k}, \mathbf{L})$$

We have obtained again a relation rather like eq. (5.2.10) but the replacement of the conduction band by a donor as the initial state brings a second Fourier coefficient G into eq. (5.2.66). If $\mathbf{P}_{v,c}(\mathbf{k})$ has its main contribution at some value of $\mathbf{k}$, $\mathbf{k}_0$ (say), then one has from eq. (5.2.66)

$$\mathbf{P}_{DA} \simeq \mathbf{P}_{vc}(\mathbf{k}_0)\, I(\alpha,\rho), \quad \alpha \equiv \frac{\alpha_D}{\alpha_A} \equiv \frac{\tilde{r}_{A1}}{\tilde{r}_{D1}} = \left[\frac{m_D(E_c-E_D)}{m_A(E_A-E_v)}\right]^{\frac{1}{2}}, \quad \rho \equiv \frac{R}{\tilde{r}_{A1}} \quad (5.2.67)$$

where, using eq. (5.2.3),

$$I(\alpha,\rho) \equiv \sum_{\mathbf{k}} G_v^*(\mathbf{k})\, G_c(\mathbf{k})\, e^{-i\mathbf{k}\cdot\mathbf{R}} = \pi(\tilde{r}_{A1}\,\tilde{r}_{D1})^{-\frac{3}{2}} \int_V \exp\left[-\frac{r}{\tilde{r}_{D1}} - \frac{|\mathbf{r}-\mathbf{R}|}{\tilde{r}_{A1}}\right]d\mathbf{r}$$

We give a few steps in the rather tedious evaluation of this integral (which was first given in [5.2.63], equation (34)).

The angle integral is with $d\mathbf{r} = r^2\, dr \sin\theta\, d\theta\, d\varphi$

$$2\pi \int_{-1}^{+1} \exp[-(B-C\cos\theta)^{\frac{1}{2}}]\, d(\cos\theta) \quad \left(B \equiv \frac{r^2+R^2}{\tilde{r}_{A1}^2}, C \equiv \frac{2rR}{\tilde{r}_{A1}^2}\right)$$

$$= \frac{2\pi\tilde{r}_{A1}}{rR}\left\{(|r-R|+\tilde{r}_{A1})\exp\left(-\frac{r-R}{\tilde{r}_{A1}}\right) - (r+R+\tilde{r}_{A1})\exp\left(-\frac{r+R}{\tilde{r}_{A1}}\right)\right\}$$

Hence

$$I(\alpha,\rho) = \frac{2\tilde{r}_{A1}}{(\tilde{r}_{D1}\tilde{r}_{A1})^{\frac{3}{2}}R}\left\{-\int_0^\infty r(R+r+\tilde{r}_{A1})\exp\left[-\left(\frac{R+r}{\tilde{r}_{A1}}+\frac{r}{\tilde{r}_{D1}}\right)\right]dr\right.$$

$$+ \int_0^R r e^{-r/\tilde{r}_{D1}}(R-r+\tilde{r}_{A1})\exp\left(\frac{r-R}{\tilde{r}_{A1}}\right)dr$$

$$\left. + \int_R^\infty r\, e^{-r/\tilde{r}_{D1}}(r-R+\tilde{r}_{A1})\exp\left(\frac{R-r}{\tilde{r}_{A1}}\right)dr\right\}$$

$$= \frac{8\alpha^{\frac{3}{2}}}{(\alpha-1)^3\rho}\{[\rho(\alpha^2-1)+4\alpha]\,e^{-\alpha\rho} + [\rho\alpha(\alpha^2-1)-4\alpha]\,e^{-\rho}\} \quad (5.2.68)$$

The resulting function (5.2.67) behaves *approximately* as eq. (5.2.65), i.e. as

$$I(\alpha,\rho) = C\,e^{-R/a} \quad (5.2.69)$$

For small R, $I(\alpha,\rho)$ rises more slowly than suggested by eq. (5.2.69). In fact,

$$I(\alpha,0) = 8\alpha^{\frac{3}{2}}(1+\alpha)^{-3} < C \quad (5.2.70)$$

See Table 5.2.3 and Fig. 5.2.12. We have assumed $\alpha < 1$. For large R $I(\alpha,\rho)$ goes to zero as $8\alpha^{\frac{3}{2}}(1-\alpha^2)^{-2}\,e^{-\alpha_D R}$.

Table 5.2.3. *Numerical example for radiative donor–acceptor recombination* [5.2.64]

	GaAs (80 K)	CdS (1.6 K)
E_G (eV)	1.5	2.56
$E_c - E_D$ (meV)	6	30
$E_A - E_v$ (meV)	40	160
ε	12.5	7.3
α_D (10^7 cm^{-1}), $\dfrac{m_c}{m}$	0.107, 0.072	0.40, 0.20
α_A (10^7 cm^{-1}), $\dfrac{m_v}{m}$	0.727, 0.050	1.64, 0.63
$I(\alpha, 0)$ of eq. (5.2.70)	0.30	0.50
C of eq. (5.2.69)	0.45	1.1
a (nm or 10 Å) in eq. (5.2.69)	9.4	2.5
T_{DA}^s (10^{-11} cm^3 s^{-1} according to eq. (5.2.73))	88	4.1

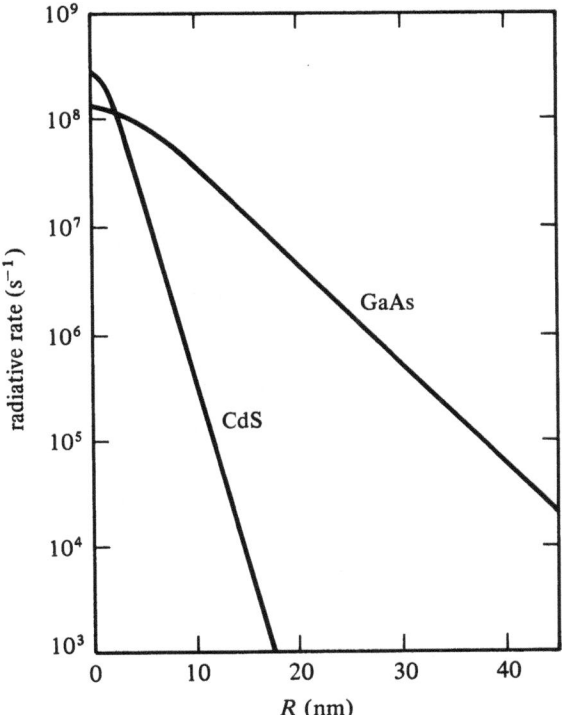

Fig. 5.2.12. The radiative recombination rate per donor–acceptor pair at distance R apart according to eq. (5.2.72).

The total emission rate per unit volume is, from eq. (4.4.14),

$$\sum_{I,J} \frac{4e^2\mu(E_D - E_A)}{m^2\hbar^2 c^3} |M_{IJ}|^2_{av} (N_v + 1) p_I q_J$$

where $E = h\nu = E_D - E_A$ is the energy of the transition, M_{IJ} is the element (5.2.67) of the electron momentum averaged over polarization and the directions of the photon wavevector, N_v is the Bose occupation number of the mode of energy $h\nu$ and will be neglected compared with unity and $p_I q_J$ is the probability of state I being occupied and state J being empty. The matrix element is a product of the interband matrix element which can be estimated as in eq. (D.36) from a four-band f-sum rule appropriate to III–V semiconductors to be

$$\frac{mE_G}{6}\left(\frac{m}{m_c} - 1\right)\frac{E_G + \Delta}{E_G + \frac{2}{3}\Delta}$$

where Δ is the spin–orbit splitting. The second factor may be worked out either by including R-dependence and taking it as $|I(\alpha, \rho)|^2$ as given in eq. (5.2.68), or by averaging over R. We shall use

$$\alpha_0 \equiv \frac{e^2}{\hbar c} = \frac{1}{137}, \quad I_0 \equiv \frac{me^4}{2\hbar^2} = 13.6 \text{ eV}, \quad t_0 \equiv \frac{\hbar^3}{e^4 m} = 2.42 \times 10^{-17} \text{ s} \quad (5.2.71)$$

The *first* method gives the radiative recombination rate per donor–acceptor pair at distance R apart if $p_I = q_J = 1$ is used:

$$\frac{\mu}{6}\left(\frac{m}{m_c} - 1\right)\frac{E_G(E_D - E_A)}{I_0^2}\frac{E_D + \Delta}{E_G + \frac{2}{3}\Delta}|I(\alpha, \rho)|^2 \frac{\alpha_0^3}{t_0} \qquad (5.2.72)$$

Fig. 5.2.12 illustrates this result [5.2.64] which also implies

$$1.48 \times 10^9 |I(\alpha, \rho)|^2 \text{ s}^{-1} \sim 0.13 \times 10^9 \text{ s}^{-1} \quad \text{(GaAs)}$$

$$1.07 \times 10^9 |I(\alpha, \rho)|^2 \text{ s}^{-1} \sim 0.27 \times 10^9 \text{ s}^{-1} \quad \text{(CdS)}$$

The last results apply to the limit of zero separation and agree broadly with the experimental estimates $(0.3 \pm 0.2) \times 10^9 \text{ s}^{-1}$ [5.2.65] and $(0.4 \pm 0.2) \times 10^9 \text{ s}^{-1}$ [5.2.66].

Turning to the *second* method, note that eq. (5.2.70) holds for *one* donor–acceptor pair. Assuming a random distribution of impurities, the average is

$$|\mathbf{P}_{DA}|^2_{av} = \int |\mathbf{P}_{DA}|^2 \frac{d\mathbf{R}}{V} = 4\pi \tilde{r}_{A1}^3 \int_0^\infty |\mathbf{P}_{DA}|^2 \frac{\rho^2 d\rho}{V}$$

$$= |\mathbf{P}_{vc}(\mathbf{k}_0)|^2 \frac{64\pi \tilde{r}_{A1}^3}{V} g(\alpha)$$

where

$$g(\alpha) \equiv (1 + 7\alpha + 17\alpha^2 + 7\alpha^3 + \alpha^4)(1 + \alpha)^{-7}$$

This is again the result of a lengthy integration [5.2.63]. Using the f-sum rule for the first factor, the summation over I and J applies now only to $p_I q_J$ and yields $n_D p_A$, the product of concentrations of occupied donors and empty acceptors. The radiative recombination rate per unit volume can be written as

$$T_{DA}^S \, n_D p_A$$

where T_{Da}^S is the (single-electron) radiative D–A recombination coefficient. Hence

$$T_{DA}^S = \tfrac{32}{3}\pi \mu g(\alpha) \left(\frac{m}{m_c} - 1\right) \frac{E_G(E_D - E_A)}{I_0^2} \frac{E_G + \Delta}{E_G + \frac{2}{3}\Delta} \left(\frac{1}{r_1 \alpha_A}\right)^3 r_1^3 \frac{4e^2 I_0^2}{m\hbar^2 c^3}$$

The last factor is

$$\frac{4e^2}{m\hbar^2 c^3}\left(\frac{me^4}{2\hbar^2}\right)^2 = \left(\frac{e^2}{\hbar c}\right)^3 \frac{me^4}{\hbar^3} = \frac{\alpha_0^3}{t_0}$$

where eq. (5.2.71) has been used. This gives [5.2.64]

$$T_{DA}^S = \frac{32\pi}{3}\left(\frac{m}{m_c} - 1\right)\left(\frac{m}{m_A}\right)^{\frac{3}{2}} \frac{E_G(E_D - E_A)}{I_0^2} \frac{E_G + \Delta}{E_G + \frac{2}{3}\Delta}$$

$$\times \left(\frac{I_0}{E_A - E_v}\right)^{\frac{3}{2}} g(\alpha)\,\mu \frac{(\alpha_0 r_1)^3}{t_0} \tag{5.2.73}$$

where the last factor carries the dimension. Some values of this quantity are given in Table 5.2.3.

The dependence of the peak energy of the broad band pair recombination emission on excitation energy lends itself also to analytical discussion [5.2.67]. Variational methods for optimal wavefunctions have also been used for both allowed and forbidden transitions and for small distances R between the centers [5.2.68]. The use of eq. (5.2.64) for the estimation of the dielectric constant by a regression analysis is also possible [5.2.69].

5.2.7 Notes on excitons

Although exciton effects are, broadly speaking, outside the scope of this book, we have already noted some exciton properties on p. 350, and we review here some additional points relevant to recombination.

At low temperatures ($kT < E_{xt}$) most carriers are bound as excitons, and at even lower temperatures one expects to see excitonic molecules ('biexcitons') and higher conglomerates. Since excitons are Bose particles, Bose–Einstein condensation may be expected at the lowest temperatures. Owing to the mutual attraction of excitons, leading to biexcitons, however, Bose condensation of biexcitons occurs in preference to that of excitons, provided their concentration is high enough. In thin layers, i.e. in quasi-two-dimensional systems, confinement in the third dimension tends to increase the exciton ionization energy, thus leading to a better chance of good exciton concentrations at room temperatures (see chapter 7).

In fact, condensation does occur, but into an electron–hole liquid (EHL). Its binding energy may be expected to be of order E_{ex} and its density of order a_{ex}^{-3}. The phase transition is of first order; it occurs when the exciton gas is supersaturated, and the density is fairly uniform, so that it is reasonable to apply the term 'liquid' to the system. For the present purpose, our main interest resides in the recombination radiation which it emits. In fact the EHL (metallic) phase was first recognized by its characteristic line [5.2.70]. A more recent curve is shown as Fig. 5.2.13 for ultra pure p-type Ge [5.2.71]. The line can be regarded as almost due to a Sommerfeld metal. Its Fermi level lies at energy μ above the bottom of the band and has a low energy tail. The condensation energy φ_s separates it from the free exciton recombination line which lies itself by an amount of order E_{ex} below the band gap energy E_G. The reality of these electron–hole drops was brought home by the photograph [5.2.72] shown in Fig. 5.2.14 using the intrinsic recombination radiation from a nonuniformly compressed sample of Ge. The drop radius is of order 0.3 mm and increases with laser power. The normal theory of radiative transitions (see section 5.5) can be applied to discuss the absorption coefficient.

Among spectra extensively discussed are those of EHL in Si where the low energy tail required a broadening function of the type discussed at the end of p. 388; see [5.2.73], figure 5. The phase diagram and critical points are also discussed in this paper. The field of exciton condensation is reviewed in [5.2.74]; this phenomenon was first proposed in [5.2.75].

A brief remark about the binding of excitons to defects will be made next. Consider the binding of an exciton with effective mass ratio $\sigma \equiv m_e/m_h$ to an ionized donor. The heavier particle will have the tighter orbit. If this were the electron, the hole might see a quasi-neutral object which can no longer bind it. Hence for binding one would expect the hole to be heavier and hence $\sigma < 1$. For binding to an ionized acceptor one would expect $\sigma > 1$. There is thus a critical range of σ for binding in each case. At one time it was suggested that a second region of σ-values for binding might exist. This can be disproved rather elegantly by taking the exciton Hamiltonian in the simple form (using an obvious notation)

$$H = \frac{\mathbf{p}_e^2}{2m_e} + \sigma\frac{\mathbf{p}_h^2}{2m_e} + \frac{e^2}{\varepsilon r_e} + \frac{e^2}{\varepsilon r_h} - \frac{e^2}{\varepsilon|\mathbf{r}_h - \mathbf{r}_e|}$$

The Hellman–Feynman theorem (D.13) (with k_i replaced by σ) now shows

$$\frac{dE_n(\sigma)}{d\sigma} = \left\langle nl \left| \frac{\mathbf{p}_h^2}{2m_e} \right| nl \right\rangle$$

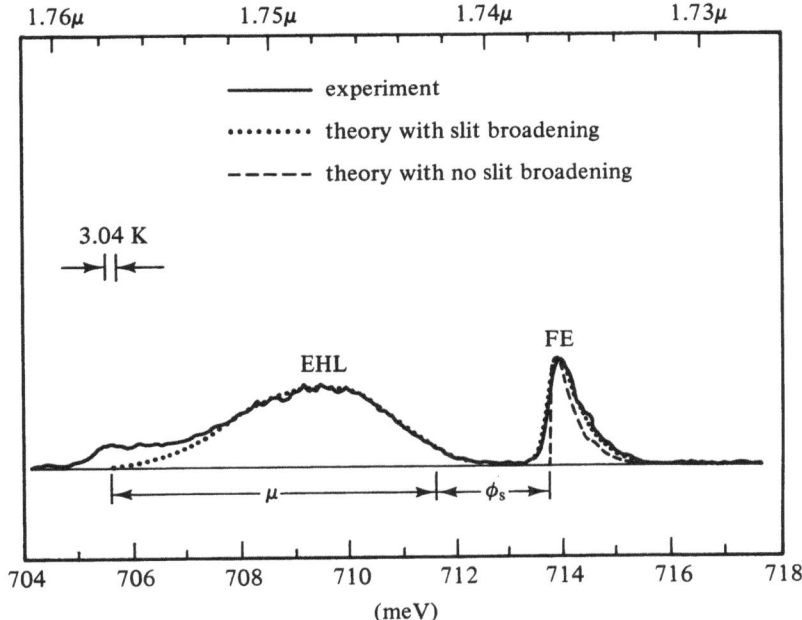

Fig. 5.2.13. The electron–hole drop luminescence in p-type Ge. The free exciton peak is denoted by FE.

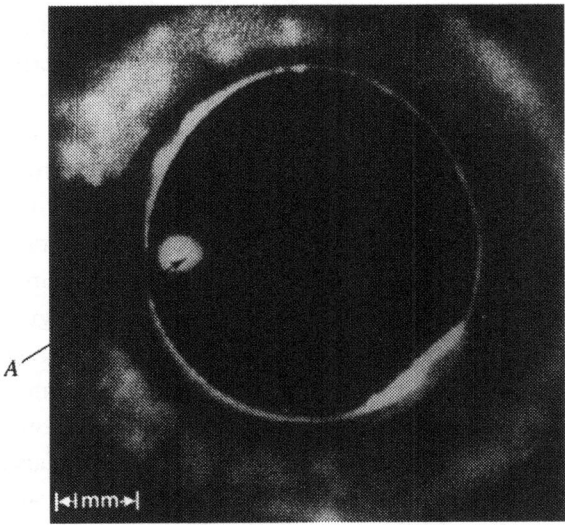

Fig. 5.2.14. Photograph of the intense luminescence due to a long lived electron–hole drop (A) in a 4 mm disk of pure Ge. It is confined by a potential well due to strain fields.

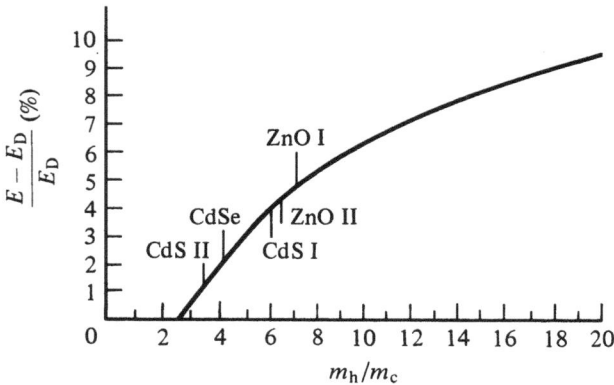

Fig. 5.2.15. Binding energy of an exciton to a singly ionized donor as a function of the exciton mass ratio. The labels I and II refer to different samples.

As this is always positive, there is no second region of binding [5.2.76] (the energy never decreases if σ is increased). A similar argument goes through for biexcitons [5.2.77].

Fig. 5.2.15 gives the binding energy $E-E_D$ of a hole to a neutral donor divided by the hydrogen-like neutral donor binding energy E_D as a function of m_h/m_c. The quantity $E-E_D$ is effectively the binding energy of an exciton to a singly ionized donor. The graph is theoretical and is based on a variational approach. It is derived from [5.2.78], and shows an increase in the binding energy with m_h, as one would expect. No binding seems possible for $m_h/m_e < 2.5$ or $m_e/m_h > 0.4$. The binding energy, i.e. the energy required to remove an exciton from a complex, is found experimentally from the energy by which the free exciton line exceeds the new (bound exciton) line which one finds upon doping a relatively pure material.

Lastly we draw attention to the surprising fact that several excitons can be bound to neutral donors and to neutral acceptors as again revealed by the luminescent intensity (Fig. 5.2.16 taken from [5.2.79]). The figure also shows the free and bound exciton lines for the transverse optical phonon region. The kinetics for the aquisition or loss of additional excitons is of intrinsic interest and has been worked out in [5.2.79]. Recent work concentrates on the effects on the luminescence of symmetries, magnetic field, local potentials, and on the analysis in terms of a shell model, analogous to that used for atoms (and nuclei). In addition the use of a highly sensitive radiation detector has shown the exciting result of *unbound* associations of up to five excitons ('polyexcitons') in ultra-pure silicon. The purity is needed to inhibit excitons being trapped at impurities. The temperature for this effect to show up must be such that there is no excessive thermal decomposition and yet no electron–hole liquid formation, yielding a narrow window of 20–50 K [5.2.80].

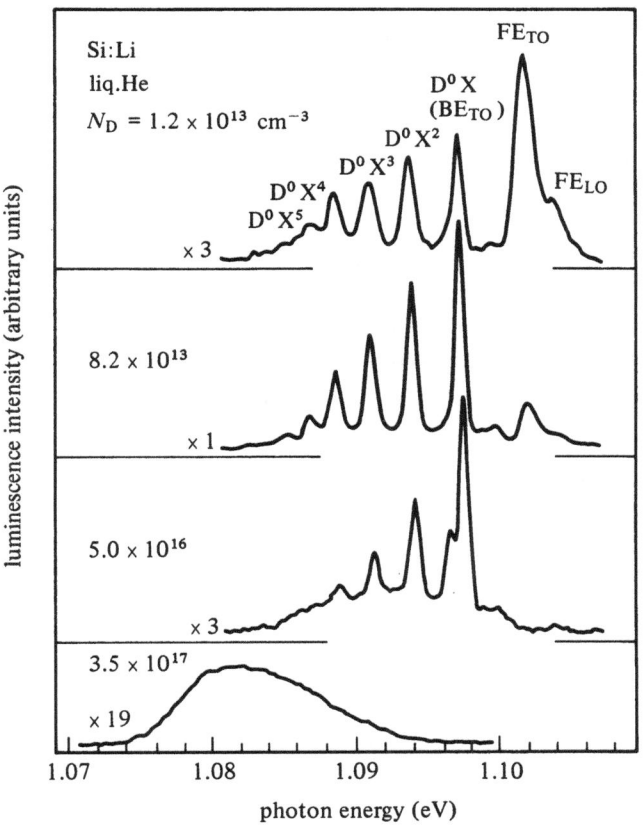

Fig. 5.2.16. Low temperature luminescence spectra in the TO-phonon region of Si doped with Li in the concentration range of 10^{13}–10^{17} cm^{-3}. D^0X^n denotes a complex with n bound excitons.

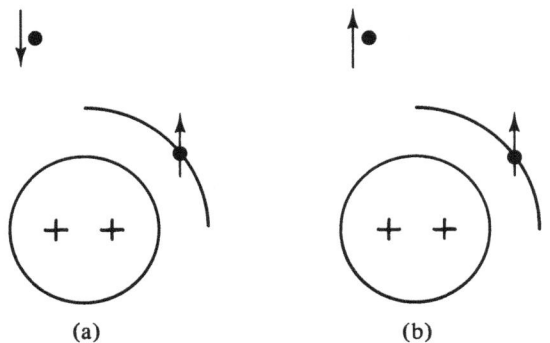

Fig. 5.2.17. The singlet (a) and triplet (b) recombination of an electron with an ionized He atom. The arrows denote spins.

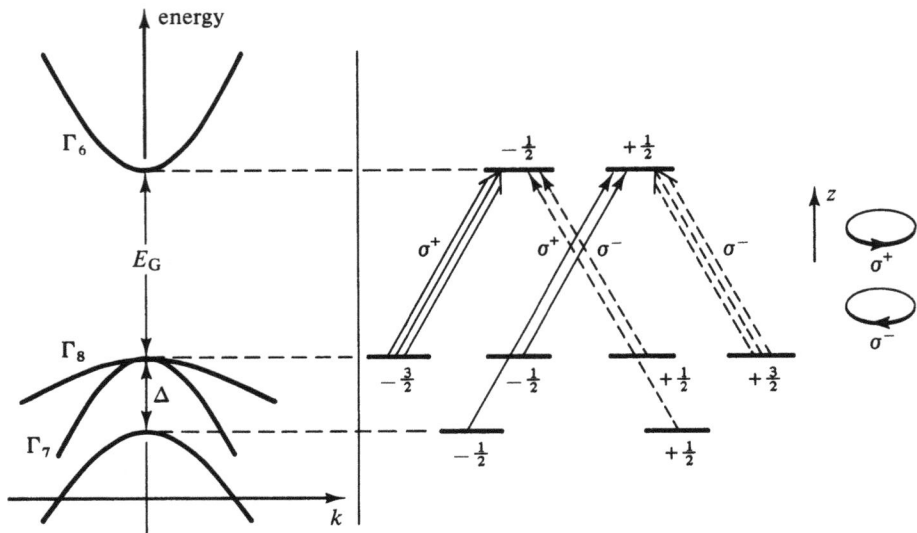

Fig. 5.2.18. The effect of circularly polarized light in diamond type semiconductors (after [5.2.81]).

5.2.8 Points from the literature

Some pointers to the work on spin-dependent recombination and the use of fractals are added here. They are topics that could not be adequately covered in the rest of this book.

(a) It is seen from Fig. 5.2.17 that the triplet recombination of an electron with an ionized helium atom in the ground state is forbidden by the Pauli exclusion principle (triplet recombination). Singlet recombination is allowed. Thus the recombination of a center with an unpaired electron is much stronger for an electron with antiparallel spin than it is for one with a parallel spin, for which it is possible only by virtue of higher order perturbations. Spin oriented electrons and centers can be created by circularly polarized light, say, σ^+ in Fig. 5.2.18. Since the valence band spin state of $-\frac{3}{2}$ is three times as responsive than the valence band spin state of $-\frac{1}{2}$, a preferred matching of photo-excited conduction band electrons and valence band holes results. The consequent photoluminescence is thus enhanced over what one expects for linearly polarized exciting light (which creates no spin polarization). Hence the two types of illumination produce different photoluminescence intensities. This was confirmed first for donor–acceptor photoluminescence in Ge-doped p-type $Ga_{0.6}Al_{0.4}As$ at 77 K [5.2.82].

The excitation of conduction band electrons of concentrations n_+ and n_- for the two spins yields a polarization which can be defined by $P = (n_+ - n_-)/(n_+ + n_-)$. Its initial value for the case envisaged ($n_-/n_+ = 3$) is $P = 0.5$. (In II–VI Wurtzite-type compounds like CdS it would be -1.) This holds only if the exciting light is in the range $E_G < h\nu < E_G + \Delta$. For higher excitation energies contributions from the split-off and other bands tend to lower the initial value of P. Spin relaxation takes place thereafter by interaction with the lattice and has been monitored, the effect of a magnetic field has been studied, and spin-dependent transport has also been investigated. Spin-dependent recombination is thus a very useful tool in solid state physics.

The subject was initially developed by groups of workers in Paris and Leningrad and has been reviewed in [5.2.81], [5.2.83] and [5.2.84], notably the optically detected magnetic resonance ('ODMR') aspect of it. Although activity has quietened, interesting developments still occur. For example, by detecting and analyzing the spin-dependent thermal emission, it has been found possible to correlate the Si dangling bond defect at the Si/SiO_2 interface with a deep hole trap lying 0.3 eV above the valence band [5.2.85], and there have been other developments [5.2.86].

(b) In disordered materials nonexponential decay laws have been found and they can take various forms:

$$\exp\{-C[\ln t/\tau]^\beta\} \quad (\beta \geqslant 1, t > \tau)$$

$$\exp\{-(t/\tau)^\alpha\} \quad (0 < \alpha < 1, t > \tau)$$

$$(t/\tau)^{-\gamma} \quad (\gamma > 0, t > \tau)$$

Some of these occur after electron–hole pair creation in hydrogenated amorphous Si. Such observations can be modeled by means of continuous time random walk models. Both spatial and temporal disorder are found and fractal dimensions are useful in the theory [5.2.87], [5.2.88]. The details are beyond the present scope. It may be noted, however, that even in crystalline semiconductors nonexponential decay laws of the type

$$(1 + t/\tau)^{-\delta}, \quad \delta > 0$$

have been found [5.2.89].

5.3 Auger recombination and impact ionization involving traps without excitons

5.3.1 Construction of universal curves for the T_1 process

Since the wavefunction of a trap is more widely spread in momentum space than the wavefunction of a band state, momentum conservation, which holds for band transitions, is relaxed if traps are involved, with the consequent removal of a threshold energy. As in a phonon-assisted band–band process, the temperature dependence is therefore weak. This furnishes one way of making an experimentally based *a priori* case for the occurrence of either the phonon-assisted Auger effect or the band–trap Auger effect. The other important nonradiative process, the multiphonon emission (or cascade) process has a strong temperature dependence which behaves as $\exp(-E/kT)$ where E is an appropriate barrier energy.

We shall now discuss impact ionization from traps and Auger recombination through traps for the process T_1 (i.e. process number 5 of Table 3.7.1). The theoretical procedure is in principle straightforward and we have already met it in section 3.7.1: One starts with the Fermi golden rule for the probability T_{if} of a transition $i \to f$ in time t; one assumes an electron is with certainty in state $k_{2'}$, multiplies by the appropriate probabilities of occupation or vacancy for states 1, 1' and 2 denoted by $p(1)$, $p(1')$, $p(2)$, and integrates over all translational states indicated by $dS_1 \, dS_{1'} \, dS_2$. One finally divides by the volume V and by t to obtain the impact ionization rate per unit volume. This gives eq. (3.7.4), i.e.

$$P(2') = \frac{2\pi}{\hbar V} \int p(1)\,p(1')\,p(2)\,|H_{if}|^2 \,\delta(E_f - E_i)\,dS_1\,dS_{1'}\,dS_2$$

However, we would like to avoid the approximation of removing the matrix element from the integral as was done in eq. (3.7.5). Instead let there be $n_t V$ occupied traps at state 1'. For the impact ionization process they have to be occupied so that $p(1') = 1$ which gives the integration over $dS_{1'}$. The traps are assumed independent so that, putting also $dS_1 = (V/8\pi^3)\,d\mathbf{k}_1$,

$$P(2') = \frac{n_t V^2}{32\pi^5 \hbar} \int p(1)\,p(2)\,|H_{if}|^2\,\delta(E_f - E_i)\,d\mathbf{k}_1\,d\mathbf{k}_2 \tag{5.3.1}$$

Writing for the ionization energy of the traps with the aid of eq. (5.2.19)

$$E_t = \frac{\hbar^2}{2m_{D,A}\,\tilde{r}_t^2} \equiv \frac{\hbar^2 \alpha^2}{2m_c}$$

one finds that

$$E_f - E_i = (\hbar^2/2m_c)(k_1^2 + k_2^2 + \alpha^2 - k_{2'}^2)$$

and

$$P(2') = \frac{m_c n_t V^2}{16\pi^5 \hbar^3} \int p(1)\,p(2)\,|H_{if}|^2\,\delta(k_1^2 + k_2^2 + \alpha^2 - k_{2'}^2)\,d\mathbf{k}_1\,d\mathbf{k}_2$$

$$= \frac{4m_c e^4 n_t}{\pi^3 \varepsilon^2 \hbar^3} \int p(1)\,p(2)\,\left| \frac{F_{22'}\,N_1}{\lambda^2 + |\mathbf{k}_{2'} - \mathbf{k}_2|^2} \right|^2$$

$$\times \,\delta(k_1^2 + k_2^2 + \alpha^2 - k_{2'}^2)\,d\mathbf{k}_1\,d\mathbf{k}_2 \tag{5.3.2}$$

Table 5.3.1. *Comparison of notations*

McDowell and Coleman [5.3.1]	k	b	$\mathbf{p}$	T	$B(k)$
Here	k_1	α	$\mathbf{p} = \mathbf{k}_{2'} - \mathbf{k}_2$	Q	$B(k_1)$

Note: McDowell and Coleman use a Coulomb gauge in which distances are expressed in terms of the Bohr radius and masses in terms of the electron rest mass.

In the last step we have used the matrix element (3.4.31), which omits the exchange effect.

One has to estimate next the quantity (3.4.30):

$$N_1(\boldsymbol{\mu}, 1, 1') = \int \psi_f^*(\mathbf{r})\, \psi_b(\mathbf{r})\, e^{i\boldsymbol{\mu}\cdot\mathbf{r}}\, d\mathbf{r}$$

$$= V^{-\frac{1}{2}} \int [B(k_1)\, {}_1F_1\, e^{i\mathbf{k}_1\cdot\mathbf{r}_1}]^* \left[\frac{\alpha^{\frac{3}{2}}}{\pi^{\frac{1}{2}}} e^{-\alpha r} \right] e^{i(\mathbf{k}_{2'}-\mathbf{k}_2)\cdot\mathbf{r}}$$

The Coulomb function for the continuum has been used for ψ_f and the ground state function of the hydrogen atom has been used for ψ_b. Here

$$B(k_1) \equiv \Gamma\left(1 + \frac{i\alpha}{k_1}\right) e^{\alpha\pi/2k_1} \tag{5.3.3}$$

and

$${}_1F_1 \equiv {}_1F_1[-i\alpha/k_1, 1, -i(k_1 r + \mathbf{k}_1 \cdot \mathbf{r})]$$

is the confluent hypergeometric function. These functions are given, for example, in [5.3.1] with the change in notation explained in Table 5.3.1. There one also finds the result of the integration. It is

$$|N(\mathbf{k}_{2'}-\mathbf{k}_2, \mathbf{k}_1, \alpha)|^2 = \frac{2^8 \pi \alpha^5}{V} |F(\mathbf{k}_0, \mathbf{l}_0)|^2 \frac{|B(k_1)|^2}{Q^4} \tag{5.3.4}$$

$$\times p^2 \frac{(p - k_1 \cos\delta)^2 + \alpha^2 \cos^2\delta}{(\alpha^2 + p^2 - k_1^2)^2 + 4\alpha^2 k_1^2} \exp\left\{ -\frac{2\alpha}{k_1} \tan^{-1} \frac{2\alpha k_1}{\alpha^2 + p^2 - k_1^2} \right\}$$

where δ is the angle between $\mathbf{p} \equiv \mathbf{k}_{2'} - \mathbf{k}_2$ and $\mathbf{k}_1$,

$$|B(k_1)|^2 \equiv \frac{2\pi\alpha}{k_1[1 - \exp(-2\pi\alpha/k_1)]} \tag{5.3.5}$$

and

$$Q \equiv \alpha^2 + p^2 + k_1^2 - 2pk_1 \cos\delta \tag{5.3.6}$$

Equation (5.3.5) is often referred to as the Sommerfeld factor.

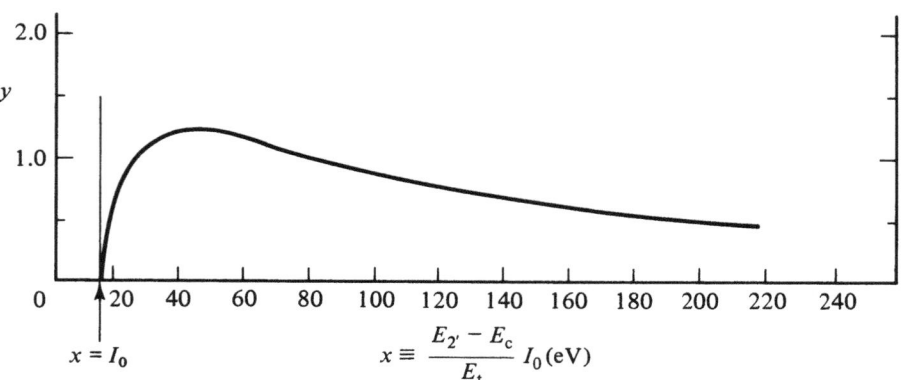

Fig. 5.3.1. Impact ionization cross section for an electron of energy $E_{2'}$ according to eqs. (5.3.2), (5.3.8) and (5.3.9).

The result (5.3.4) has to be substituted into eq. (5.3.2) to yield an expression for $P(k_{2'})$. This can be evaluated numerically; the value of λ is small compared with $|\mathbf{k}_{2'} - \mathbf{k}_2|$ and will be neglected. The result is given in Fig. 5.3.1 as a universal curve [5.3.2], which will now be explained.

Write eq. (5.3.2) as

$$P(2') = \frac{4m_c e^4 n_t}{\pi^3 \varepsilon^2 \hbar^3} I = \frac{4}{\pi^3 \varepsilon^2} \frac{m_c}{m} \frac{n_t}{t_0} I \tag{5.3.7}$$

so that, in accordance to the dimensional analysis of section 3.7.1, I is a dimensionless integral. Divide $P(2')$ by the 'velocity' $\hbar k_{2'}/m_c$ of the carriers of wavevector $k_{2'}$ and multiply by the volume V of the system. One then obtains a scattering cross section $n\sigma(k_{2'})$ referred to the total concentration, n, of free electrons. Hence we put

$$\sigma(k_{2'}) = (m_c V/n\hbar k_{2'}) P(2') \tag{5.3.8}$$

This is rendered dimensionless by dividing by the area, $4\pi \tilde{r}_t^2$ of an effective Bohr sphere, where $\tilde{r}_t$ is the effective Bohr radius of an excited or ground state having principal quantum number t. If E_t is the effective ionization energy of this state, then

$$\tilde{r}_t E_t = \frac{\varepsilon \hbar^2 t^2}{m_c e^2} \cdot \frac{e^4 m_c}{2\varepsilon^2 \hbar^2 t^2} = \frac{e^2}{2\varepsilon}$$

Hence

$$4\pi \tilde{r}_t^2 \frac{n_t}{n} = \frac{\pi e^4}{\varepsilon^2 E_t^2} \frac{n_t}{n}$$

The last factor has been introduced to yield a simple reduced quantity for $P(2')$, namely

$$y \equiv \frac{n}{n_t} \frac{\sigma(k_{2'})}{4\pi\tilde{r}_t^2} = \frac{\alpha^4 V}{\pi^4 k_{2'}} I \quad \left(\alpha^2 \equiv \frac{2m_c E_t}{\hbar^2} \right) \tag{5.3.9}$$

It is shown in Fig. (5.3.1) as a function of ($I_0 \equiv me^4/2\hbar^2 = 13.6$ eV)

$$x \equiv \frac{E_{2'} - E_c}{E_t} I_0 = \left(\frac{k_{2'}}{\alpha} \right)^2 13.6 \text{ eV} \tag{5.3.10}$$

The plot of y against x holds for all E_t and gives the cross section for the solid state case and one spin assignment in analogy with the atomic case (when $E_t = I_0$). For $\varepsilon \sim 16$ and $E_t = 0.1$ eV one has

$$4\pi\tilde{r}_t^2 = 2.53 \; 10^{-11} \text{ cm}^2 \quad \text{and} \quad I_0/E_t = 136$$

The maximum in Fig. 5.3.1 can be thought of as arising from the fact that while the impact ionization probability increases initially with the electron energy $E_{2'}$, beyond a certain energy the electron moves so rapidly that it does not spend enough time near the trap to ionize it with certainty, and $P(2')$ falls again.

Using the result (3.7.8) one can compute the Auger coefficient T_1 from Fig. 5.3.1, and it is given in Fig. 5.3.2 where the following values have been used:

$$T = 300 \text{ K}, m_c = 0.2m, \varepsilon = 16$$

For other host materials at 300 K, the curve can still be used but the numerical values shown for T_1 must be multiplied by $(16 \times 0.2/\varepsilon m_c)^2$.

In the integration of $P(2')$ to obtain the Auger coefficient, the energy range just above the threshold dominates. It is here that good accuracy of $P(2')$ is needed. At higher energies $P(2')$ is weighted by a smaller probability $p(2')$ and there is a smaller contribution to the Auger coefficient. This can be seen from eq. (3.7.8) and has been demonstrated in detail in [5.3.2].

In addition the impact ionization rate per unit volume, specified by the reaction constant X_1, is given by two alternative expressions

$$X_1 nn_t \quad \text{and} \quad V \int_{E_c+E_t}^{\infty} p(2') P(2') \mathcal{N}(E_{2'}) dE_{2'} \tag{5.3.11}$$

where $\mathcal{N}(E_{2'}) = [(2m_c)^{\frac{3}{2}}/2\pi^2\hbar^3](E-E_c)^{\frac{1}{2}}$ is the density of states per unit volume. We shall equate these two expressions, replace $P(2')$ by y according to eqs. (5.3.7) and (5.3.9), and replace $E_{2'} - E_c = \hbar^2 k_{2'}^2/2m_c$ by x according to eq. (5.3.10). Hence one finds that X_1 can also be obtained by an integration from Fig. 5.3.1

$$X_1 = \frac{2}{\pi\varepsilon^2} \frac{m_c}{m} \frac{1}{nt_0} \int_1^{\infty} p(2') \frac{x}{E_0} y \, d\left(\frac{x}{E_0} \right) \quad (t_0 = 2.4.2 \times 10^{-17} \text{ s}) \tag{5.3.12}$$

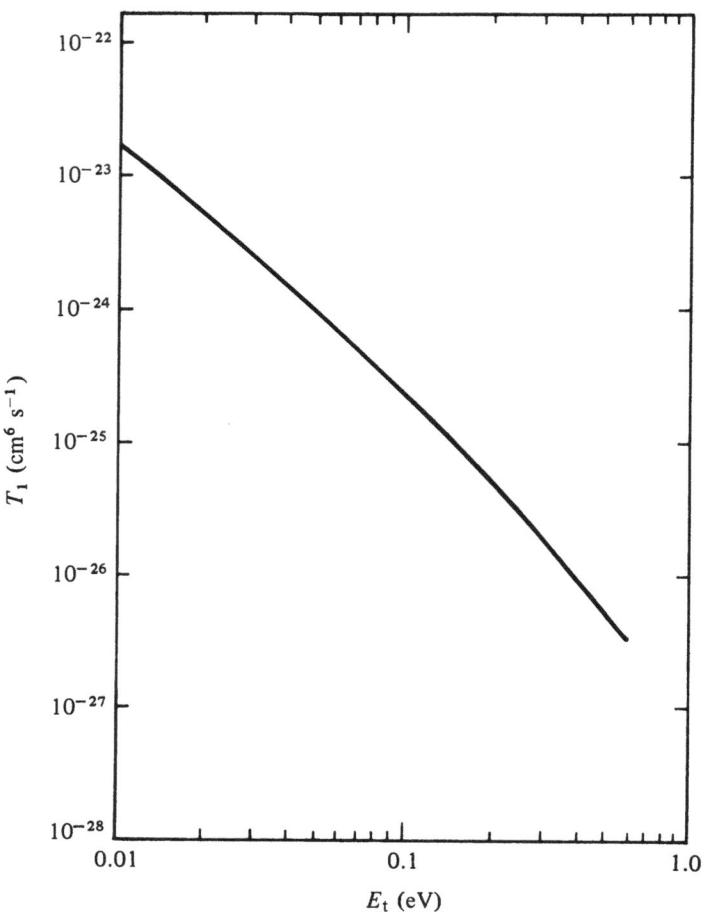

Fig. 5.3.2. Auger coefficient T_1 as deduced from Fig. 6.3.1 using $T = 300$ K, $m_c = 0.2m$ and $\varepsilon = 16$ (Si). Results for other host materials can be obtained by multiplying the value of T_1 shown by $(16 \times 0.2/\varepsilon m_c)^2$.

This result does not appear to be in the literature and has not yet been applied to experimental findings. It can clearly be generalized to a nonparabolic band.

The exchange effect can, after all integrations have been made, be of the same order as the direct effect. However, for indirect materials there is a difference between the effects T_1 and T_4 (see, for example, Table 3.7.1) in which the three band states are in the same band and the effects T_2 and T_3 for which the Auger particle is in a different band. The matrix elements involve

$$|\mathbf{k}_{2'} - \mathbf{k}_2|^2 \quad \text{and} \quad |\mathbf{k}_{2'} - \mathbf{k}_1|^2$$

in the denominator of M_D and M_E, respectively. Since $|\mathbf{k}_{2'} - \mathbf{k}_1|$ is normally of the order of the displacement between the band extrema in k-space, this term is much

Table 5.3.2. *The labeling of four models for estimating a trap Auger effect T_1*

		State 1	
		Bloch function	Continuum hydrogen function
State 1'	Bound hydrogen function	a_1	b_1
	δ-function potential function	a_2	b_2

larger for indirect materials than for direct materials. Therefore M_E can often be neglected in the calculation of T_2 and T_3 for an indirect material.

5.3.2 Introduction of four models

It is desirable to obtain an alternative to the largely numerical information contained in Figs. 5.3.1 and 5.3.2. To achieve this one needs to broaden the discussion. Recall that the matrix element in eq. (5.3.1) involves four wavefunctions $\psi_1(\mathbf{r}_1)$, $\psi_2(\mathbf{r}_2)$, $\psi_{1'}(\mathbf{r}_1)$, $\psi_{2'}(\mathbf{r}_2)$. In the case of the effect T_1, discussed in section 5.3.1, the state 1' is the localized state while the other states are in the conduction band. Let us adopt Bloch functions for states 2 and 2', and for 1' one can adopt in simple cases either bound hydrogenic effective mass functions or wavefunctions based on a δ-function potential. Table 5.3.2 shows straight away how four distinct theories can be constructed.

The labeling of the models is not random, but is based on the two distinct approximations (a) and (b) (section 3.4.4) for the matrix element.

Model b_1 is involved in section 5.3.1. It has been used in plasma work (when one does not need Bloch functions for electron 2) [5.3.3]. Experimental work on plasma is in reasonable agreement with this type of model [5.3.4], [5.3.5]. For recent experimental work using polarized atoms and beams of polarized electrons see [5.3.6]. Curves of a similar shape are also found for impact ionization in solids, for example in ZnS doped with E_r^{+++} [5.3.7]. In fact experimental results for $P(2')$ can in principle be used to obtain a quasi-experimental Auger coefficient by integration as in eq. (3.7.8). This was done for the experimental results on atomic hydrogen [5.3.4] and the resulting curve lies close to the curves of Figs. 5.3.1 and 5.3.2 (see curves in [5.3.2]).

Model a_1 is poor for $P(2')$. However, all four models give somewhat similar results for $P(2')$ near threshold. These values represent the main contribution to the integrations (such as eq. (5.3.11)) for the Auger coefficient. Hence the comparative insensitivity of this coefficient to the model used, as is clear from Fig. 5.3.3 [5.3.2]. If the matrix element is taken out of the integral (5.3.1) and evaluated at threshold one has the 'zero-order approximation' labeled a_1', for model a_1. This is also shown. It is thus not unreasonable to use model a_1' to derive some analytical results in section 5.3.3 for the trap Auger coefficients, provided it is borne in mind that they are only qualitative. But it is of some merit to have them available in addition to the purely numerical curves.

The wavefunctions for electron 1 are properly orthogonal for model b_1. They are not so for the other three models; for example a δ-function type of wavefunction overlaps the Coulomb function. Hence use of these models must be justified, and we do so in Fig. 5.3.3 by comparison with model b_1. In fact model a_2 has yielded reasonably good optical cross sections for moderately deep levels [5.3.8], [5.3.9]. It was also used for calculations of Auger de-excitation [5.3.10] in which an electron makes a transition from an excited Coulomb level of a deep impurity center to its ground state. One can, as was done in [5.3.10], remove the nonorthogonality artificially by defining new orthogonal functions for the free state in terms of the nonorthogonal functions. The resulting wavefunction may however not be a better solution of the exact Schrödinger equation.

Model a_1 has been used for a variety of trap–Auger effect calculations [5.3.11]–[5.3.13], and it will be developed further in section 5.3.3.

Model b_2 does not seem to have had significant use.

5.3.3 Zero-order approximation analytical trap–Auger coefficients in nondegenerate materials

Let us now develop a semiquantitative formula for the trap–Auger effect. For a nondegenerate semiconductor the recombination rate per unit volume can be written in two ways which will be equated:

$$T_1 n^2 p_t = \frac{1}{Vt} \int \Phi \frac{2\pi t}{\hbar} |U_{if}|^2 \delta(E_f - E_i) \, dS_1 \, dS_{1'} \, dS_2 \, dS_{2'} \qquad (5.3.13)$$

To take account of the reverse rate each expression should be multiplied by $1 - \exp(F_t - F_e)$, and this factor then cancels. It comes from

$$T_1 n^2 p_t - X_1 n n_t = T_1 n^2 p_t [1 - (X_1/T_1)(n_t/np_t)]$$

where $X_1/T_1 = (np_t/n_t)_{eq}$ and the play of probabilities (as in eq. (2.2.12)) does the rest. Here Φ is the product of the probability factors for the four states. The Fermi golden rule has been used.

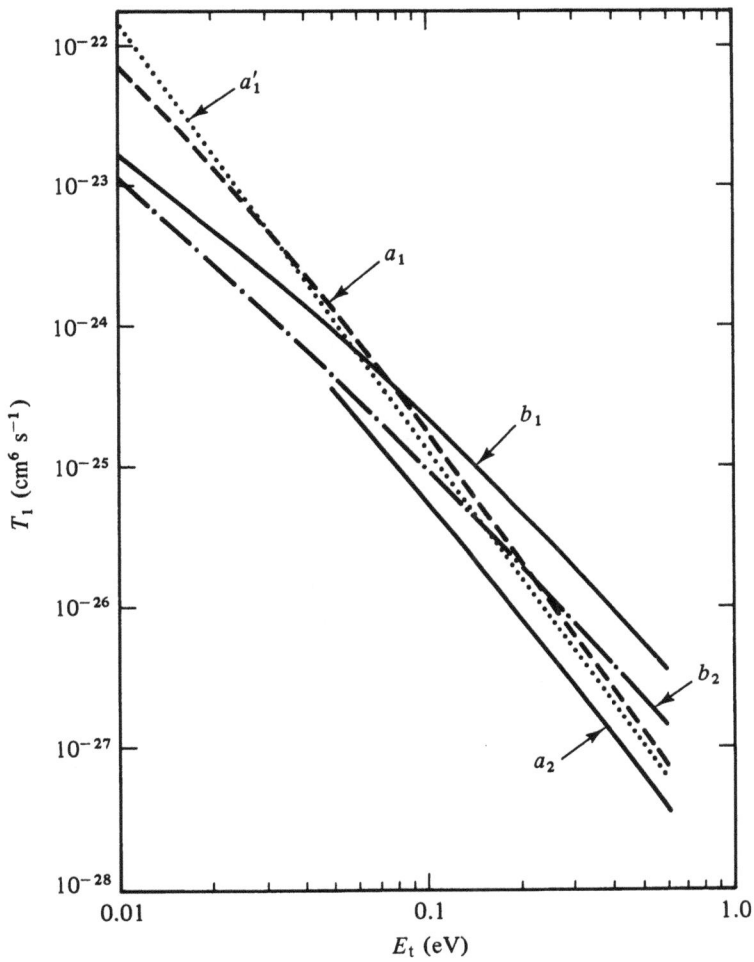

Fig. 5.3.3. Auger recombination coefficients T_1 deduced from curves of the type shown in Fig. 5.3.1 by integration using $T = 300$ K, $m^* = 0.2m$, $\varepsilon = 16$. Results for other host materials at 300 K can be obtained by multiplying the numerical values shown for T_1 by $(16 \times 0.2/\varepsilon m^*)^2$.

Now the number

$$n^2 p_t V^3 = 2^3 \int \Phi \, dS_1 \, dS_{1'} \, dS_2 \qquad (5.3.14)$$

In eq. (5.3.13) the spin is taken into account in U_{if} on the right-hand side so that we here use dS_1 as indicating a sum over translational states only. On the left one therefore needs the factor 2^3 as shown in eq. (5.3.14). From eqs. (5.3.13) and (5.3.14) one has approximately

$$T_1 = \frac{\pi V^2}{4\hbar} \langle |U_{if}|^2 \rangle \int \delta(E_f - E_i) \, dS_{2'}$$

where an averaged matrix element

$$\langle|U_{if}|^2\rangle \equiv \int \Phi |U_{if}|^2 \, dS_1 \, dS_{1'} \, dS_2 \bigg/ \int \Phi \, dS_1 \, dS_{1'} \, dS_2$$

has been introduced. This result holds for all four T_j's. Reverting to T_1, note that

$$E_f - E_i = E_{1'} + E_{2'} - E_1 - E_2 = E_c - E_t + E_{2'} - E_1 - E_2$$

As a further approximation we set $E_1 \sim E_2 \sim E_c$ so that the δ-function ensures that $E_{2'} = E_c + E_t$. One finally has the approximate formula ($1 < \beta < 2$)

$$T_j \simeq \frac{\pi V^2}{4\hbar} \{\langle |U_{if}|^2\rangle \, \mathcal{N}(E_{2'} - E_c)\}_{\text{e.c.}} \quad (j = 1, 2, 3, 4)$$

$$\simeq \frac{\pi V^2}{2\hbar} \beta\{[\langle |M_D|^2\rangle + \langle |M_E|^2\rangle] \, \mathcal{N}(E_{2'} - E_c)\}_{\text{e.c.}} \tag{5.3.15}$$

where eq. (3.4.20') has been used and e.c. indicates that $E_{2'}$ (or $k_{2'}$) has to be evaluated from energy conservation. The density-of-states function $\mathcal{N}$ to be used refers to the band in which state $2'$ lies. For a parabolic band with effective mass m_c and $E_{2'} - E_c = E_{Dt} = \hbar^2/2m_D \tilde{r}_{Dt}^2$ one can replace $\mathcal{N}$ by

$$\frac{(2m_c)^{\frac{3}{2}} (E_{2'} - E_c)^{\frac{1}{2}}}{2\pi^2\hbar^3} V = \frac{m^{*\frac{3}{2}}}{m_{D,A}^{\frac{1}{2}}} \frac{V}{\pi^2\hbar^2\tilde{r}_t}$$

so that

$$T_j = \frac{m^{*\frac{3}{2}}}{m_{D,A}^{\frac{1}{2}}} \frac{V^3}{4\pi\hbar^3\tilde{r}_t} \{\langle |U_{if}|^2\rangle\}_{\text{e.c.}} \tag{5.3.16}$$

Here $m^* = m_c$ or m_v depending on the band of the final state $2'$, and an Auger transition into a hydrogenic state of principal quantum number t is considered. For $\tilde{r}_t$ one must use $\tilde{r}_{Dt}$ or $\tilde{r}_{At}$ depending on whether a shallow donor or acceptor is involved.

To obtain an analytical estimate we shall use eq. (5.3.16) with $\langle|U_{if}|^2\rangle \simeq 2\beta\langle|M_D|^2\rangle$ ($1 < \beta < 2$), neglecting the exchange matrix element M_E and using eq. (3.4.27) for M_D. In M_D there occurs $G_{k_1+k_2-k_{2'}}$. We shall use eq. (5.2.29) with $k_1 \sim k_2 \sim 0$ for this quantity to find in the zero-order approximations and for nondegeneracy and for $j = 1, 2$,

$$T_j = \frac{16 \times 32\pi^2\beta\langle|F|^2\rangle \, G(t, l)}{\varepsilon^2 \left[1 + \dfrac{m^*}{m_{D,A}}\right]^4} \frac{m^2}{m^{*\frac{1}{2}} m_{D,A}^{\frac{3}{2}}} \left(\frac{I_0}{E_t}\right)^3 \frac{r_1^6}{t_0} \tag{5.3.17}$$

The main uncertainties reside (in order of importance) in the choice of numbers for $\langle|F|^2\rangle$, E_t, and β. For the process labeled T_1, $m^* \to m_c$ if donors are involved. For

Table 5.3.3. *Zero-order approximations for Auger band–shallow impurity recombination constants T_1 to T_4 for nondegeneracy.*

All formulae have a numerical factor $(32 \times 16\pi^2\beta/\varepsilon^2)|F|^2 G(t,l)$ and a dimensional factor r_1^6/t_0. β is explained in eq. (3.4.20), I_0 in eq. (5.2.71), and $|F|^2$ are overlap integrals involving modulating parts of Bloch functions and are given in Table 5.3.4. For $G(t,l)$ see Table 5.2.1. If $\beta = 2$ it is useful to note $64 \times 16\pi^2 I_0^3 r_1^6/t_0 = 2.3 \times 10^{-26}$ (eV)^3 cm^6 s^{-1}.

	The trap is a shallow donor at energy E_t below the conduction band	The trap is a shallow acceptor at energy E_t above the valence band
T_1	$\left(\dfrac{m_D}{m_c}\right)^{\frac{1}{2}}\dfrac{m_D^2 m^2}{(m_c+m_D)^4}\left(\dfrac{I_0}{E_t}\right)^3 \propto \tilde{r}_{Dt}^6$	$\left(\dfrac{m}{m_c}\right)^2\left(\dfrac{m_A}{m_c}\right)^{\frac{5}{2}}\left(\dfrac{E_t}{E_G}\right)^{\frac{5}{2}}\dfrac{I_0^3}{E_G^{\frac{3}{2}}(E_G-E_t)^{\frac{3}{2}}}\left[1+\dfrac{m_A-m_c}{m_c}\dfrac{E_t}{E_G}\right]^{-4}$
T_2	$\left(\dfrac{m_D}{m_v}\right)^{\frac{1}{2}}\dfrac{m_D^2 m^2}{(m_v+m_D)^4}\left(\dfrac{I_0}{E_t}\right)^3 \propto \tilde{r}_{Dt}^6$	$\left(\dfrac{m}{m_v}\right)^2\left(\dfrac{m_A}{m_v}\right)^{\frac{5}{2}}\left(\dfrac{E_t}{E_G}\right)^{\frac{5}{2}}\dfrac{I_0^3}{E_G^{\frac{3}{2}}(E_G-E_t)^{\frac{3}{2}}}\left[1+\dfrac{m_A-m_v}{m_v}\dfrac{E_t}{E_G}\right]^{-4}$
T_3	$\left(\dfrac{m}{m_c}\right)^2\left(\dfrac{m_D}{m_c}\right)^{\frac{5}{2}}\left(\dfrac{E_t}{E_G}\right)^{\frac{5}{2}}\dfrac{I_0^3}{E_G^{\frac{3}{2}}(E_G-E_t)^{\frac{3}{2}}}\left[1+\dfrac{m_D-m_c}{m_c}\dfrac{E_t}{E_G}\right]^{-4}$	$\left(\dfrac{m_A}{m_c}\right)^{\frac{1}{2}}\dfrac{m_A^2 m^2}{(m_c+m_A)^4}\left(\dfrac{I_0}{E_t}\right)^3 \propto \tilde{r}_{At}^6$
T_4	$\left(\dfrac{m}{m_v}\right)^2\left(\dfrac{m_D}{m_v}\right)^{\frac{5}{2}}\left(\dfrac{E_t}{E_G}\right)^{\frac{5}{2}}\dfrac{I_0^3}{E_G^{\frac{3}{2}}(E_G-E_t)^{\frac{3}{2}}}\left[1+\dfrac{m_D-m_v}{m_v}\dfrac{E_t}{E_G}\right]^{-4}$	$\left(\dfrac{m_A}{m_v}\right)^{\frac{1}{2}}\dfrac{m_A^2 m^2}{(m_v+m_A)^4}\left(\dfrac{I_0}{E_t}\right)^3 \propto \tilde{r}_{At}^6$

Definition of m_D and m_A, see p. 378; t is principal quantum number

$$E_t \equiv E_c - E_D = \frac{e^4 m_{Dt}}{2\varepsilon^2\hbar^2} \qquad\qquad E_t \equiv E_A - E_v = \frac{e^4 m_{At}}{2\varepsilon^2\hbar^2}$$

Effective masses. m_c, m_v are average effective masses for conduction and valence bands, respectively. The Bohr radius of donors or acceptors is $\tilde{r}_{Dt} = \varepsilon\hbar^2/e^2 m_D$, $\tilde{r}_{At} = \varepsilon\hbar^2/e^2 m_A$ where m_D, m_A are donor and acceptor effective masses. They are obtained from the impurity ionization energy by $E_t = \hbar^2/2m_D\tilde{r}_{Dt}^2$ or $E_t = \hbar^2/2m_A\tilde{r}_{At}^2$, and this leads to the last line of the table.

Table 5.3.4. *Expressions for the overlap integrals F occurring in Table 5.3.3*

$$T_1: V^{-2}\int v_c^*(\mathbf{k},\mathbf{r})\,u_c(0,\mathbf{r})\,\mathrm{d}\tau \int u_c^*(\mathbf{k},\mathbf{r})\,u_c(0,r)\,\mathrm{d}\tau$$
$$T_2: V^{-2}\int v_c^*(\mathbf{k},\mathbf{r})\,u_c(0,\mathbf{r})\,\mathrm{d}\tau \int u_v^*(0,\mathbf{r})\,u_v(\mathbf{k},\mathbf{r})\,\mathrm{d}\tau$$
$$T_3: V^{-2}\int v_c^*(\mathbf{k},\mathbf{r})\,u_v(0,\mathbf{r})\,\mathrm{d}\tau \int u_c^*(0,\mathbf{r})\,u_c(\mathbf{k},\mathbf{r})\,\mathrm{d}\tau$$
$$T_4: V^{-2}\int v_c^*(\mathbf{k},\mathbf{r})\,u_v(0,\mathbf{r})\,\mathrm{d}\tau \int u_v^*(\mathbf{k},\mathbf{r})\,u_v(0,\mathbf{r})\,\mathrm{d}\tau$$

The relations apply to donors. For acceptors replace v_c by v_v. They hold for an impurity wavefunction of the form $\sum_k v_{c,v}(\mathbf{k},\mathbf{r})\,G_k\,\mathrm{e}^{i\mathbf{k}\cdot\mathbf{r}}$ where $v_{c,v}(\mathbf{k},\mathbf{r})\mathrm{e}^{i\mathbf{k}\cdot\mathbf{r}}$ is a Bloch function drawn from the nearest band and $\hbar^2 k^2/2m_D$ is the energy bridged by the recombining electron [5.3.11]. For small k the second factors are of order unity or less. The first factors depend on the impurity wavefunction and are hard to estimate in general.

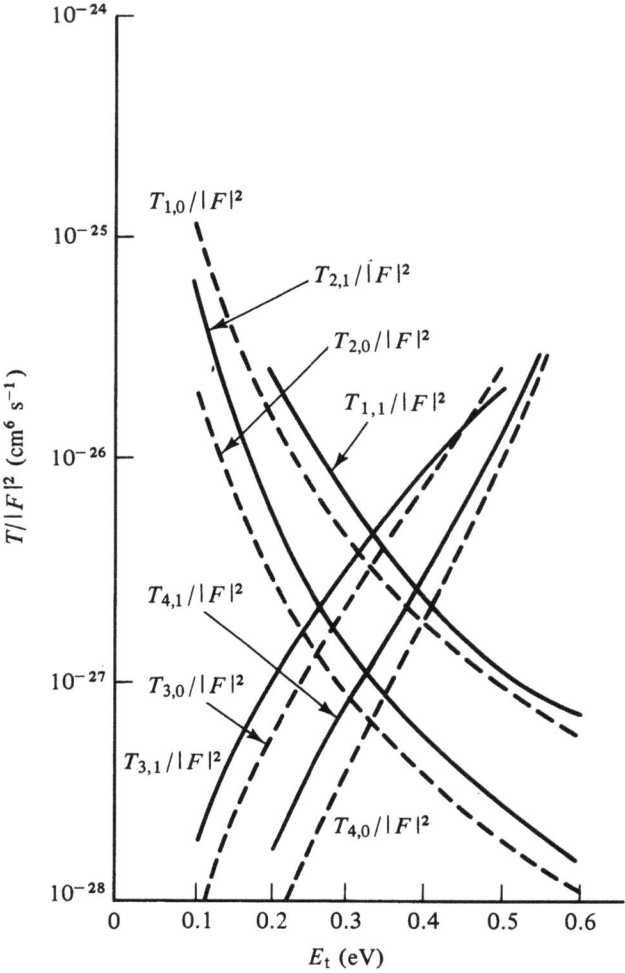

Fig. 5.3.4. The recombination coefficient in the first-order approximation (denoted by $T_{j,1}$) for the ground state ($t = 1$) as a function of trap depth below the conduction band. The zero-order approximations, now denoted by $T_{j,0}$, are compared with the first-order approximations. Parameters: $m_c/m = 0.22$, $m_v/m = 0.39$, $E_G = 0.665$ eV, $\varepsilon = 16$, $T = 300$ K. The approximations for $T_{3,1}/|F|^2$ begin to fail for $E_t > 0.5$ eV. E_t denotes in all cases the energy of the localized level below the conduction band.

T_2, $m^* \to m_v$ if donors are involved. In this way one obtains the first two entries for donors in Table 5.3.3. The other entries are obtained similarly, and all entries can be simplified by assuming $m_D = m_c$ and $m_A = m_v$. The resulting expressions are shown in Figs. 5.3.4–5.3.7 using values for Ge. The temperature dependence of the T_j's is not shown, but it is only slight in the zero- or first-order approximation [5.3.11].

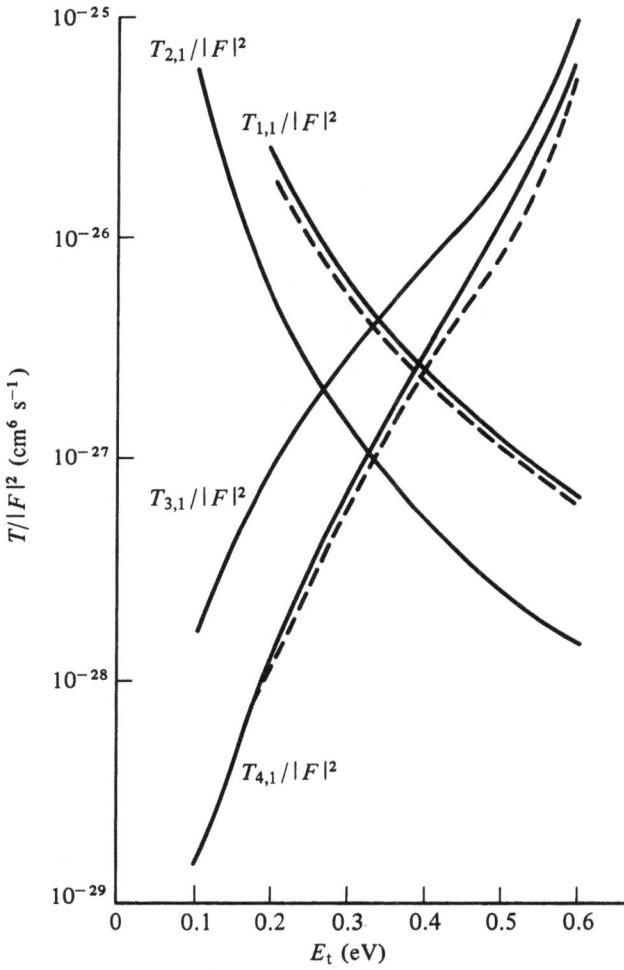

Fig. 5.3.5. The solid lines show $T_{j,1}/|F|^2$ as in Fig. 5.3.4. The dotted lines show $T_{1,1}/|F|^2$ and $T_{4,1}/|F|^2$ in the same approximation, but with the effect of exchange neglected. With the approximations adopted here the exchange terms in $T_{2,1}/|F|^2$ and $T_{3,1}/|F|^2$ are small.

A slightly more sophisticated treatment yields correction terms which, together with the zero-order terms, give the so-called first-order approximation, which will not be discussed here, but its results are shown in Figs. 5.3.4 and 5.3.5 [5.3.11]. It includes the effect of the exchange matrix element M_E. The curves show the expected decrease in T_1 and T_2, and increase in T_3 and T_4, as the impurity level moves down the band gap. The oscillating nature of T_1 and T_2 with l is expected to be smoothed out by improved treatments.

Fig. 5.3.8 shows a calculation of some T_j's for Si [5.3.14]. The larger gap has reduced all the values relative to the case of Ge.

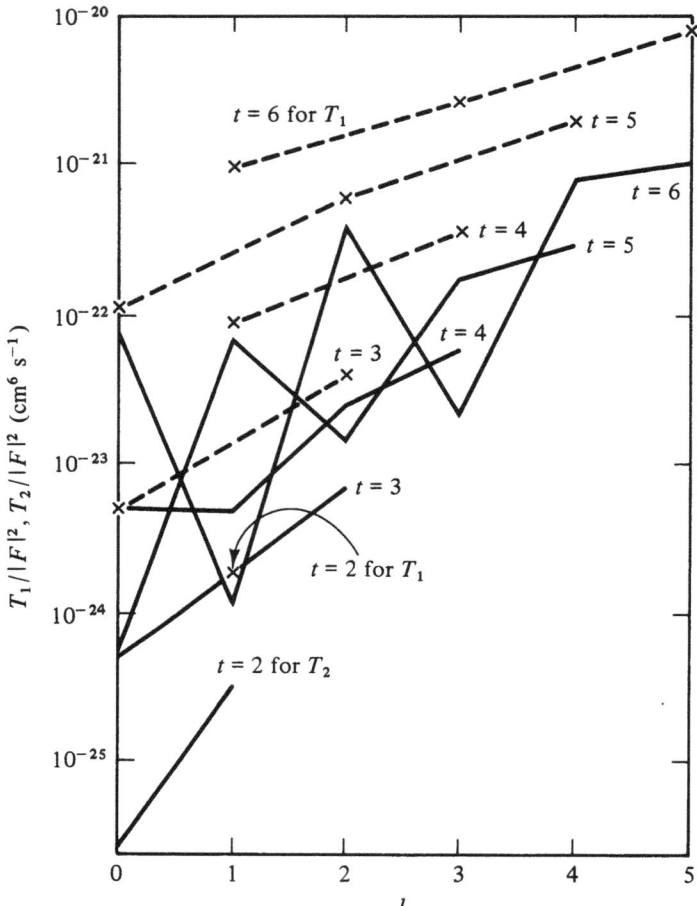

Fig. 5.3.6. $T_1/|F|^2$ and $T_2/|F|^2$ in the zero-order approximation are shown for various values of t and l. The donor states (t,l) have been taken as $(0.26/t^2)$ eV below the bottom of the conduction band. Other parameters are as in Fig. 5.3.4. $T_2/|F|^2$ is represented by a solid line and $T_1/|F|^2$ by a broken line. Note that in this approximation $T_1 = 0$ if $(t+l)$ is even. These values are not shown but interupt the broken curves.

The formulae enable us to compare the pairs of trap–Auger effects illustrated in Fig. 5.3.9 for the case when $m_D = m_c$, $m_A = m_v$ and when the ionization energies of donor and acceptor are assumed equal. One finds for the four cases (a), (b), (c), (d):

$$\text{(a)} \quad \frac{T_3(D)}{T_2(A)} = \left(\frac{m_v}{m_c}\right)^2 \quad \text{(b)} \quad \frac{T_2(D)}{T_3(A)} = \left(\frac{m_c}{m_v}\right)^3$$

$$\text{(c)} \quad \frac{T_1(D)}{T_4(A)} = \left(\frac{m_v}{m_c}\right)^2 \quad \text{(d)} \quad \frac{T_4(D)}{T_1(A)} \simeq \left(\frac{m_c}{m_v}\right)^7 \tag{5.3.18}$$

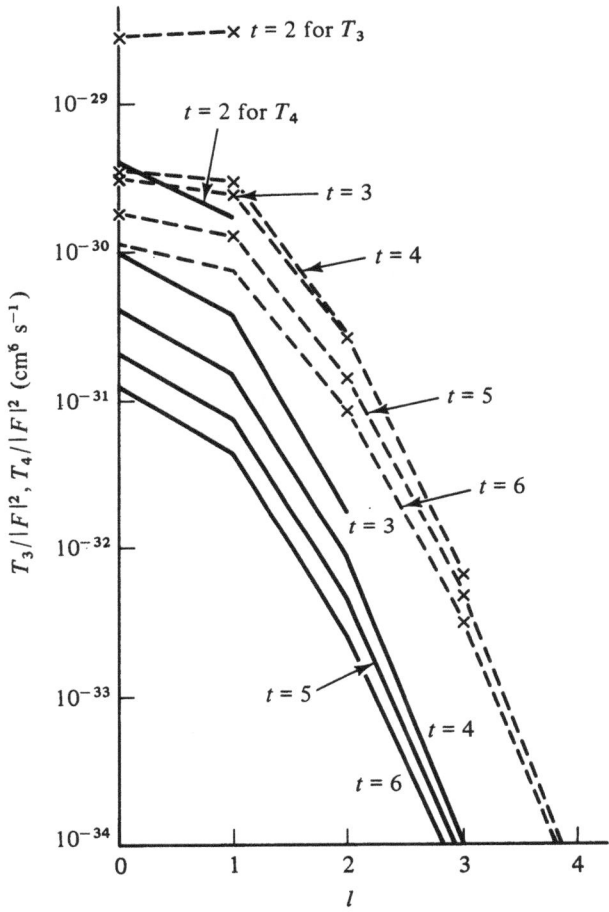

Fig. 5.3.7. Values of $T_3/|F|^2$ and $T_4/|F|^2$ on the same basis as in Fig. 5.3.6. $T_4/|F|^2$ is represented by solid lines and $T_3/|F|^2$ by broken lines.

An additional approximation was needed in (d). Assuming $m_c < m_v$ one sees that, other things being equal, the trap–Auger effect is more important for *that* member of the pair for which the Auger particle which gains energy is an electron rather than a hole. More generally (allowing for $m_v < m_c$) *that* trap–Auger effect 'wins' in which the *lighter* particle gains energy. For shallow donors or acceptors in Si $T_3(D)$ and $T_2(A)$ lie in the region $(10^{-32}, 10^{-30})|F|$ cm^6 s^{-1}, as seen from Fig. 5.3.8.

We now note some early papers. A possible Auger effect involving traps in impure PbS [5.3.15] and in *n*-type Si [5.3.16] was conjectured on experimental grounds by Moss and by Haynes and Hornbeck, respectively. Theoretical estimates were made by Pincherle, Bess and Nagae [5.3.17]. In 1960 a pioneer theoretical trap–Auger coefficient was eventually given by the late Bonch-Bruevich and Gulyaev [5.3.18] and (after some manipulation) it agrees exactly with the donor version of T_3 in Table 5.3.3, provided one takes $\beta = 2$, $m_D = m_c$ and $|F|^2 G(t,l) = 1$. The last condition is now known to be unjustified, as has been explained.

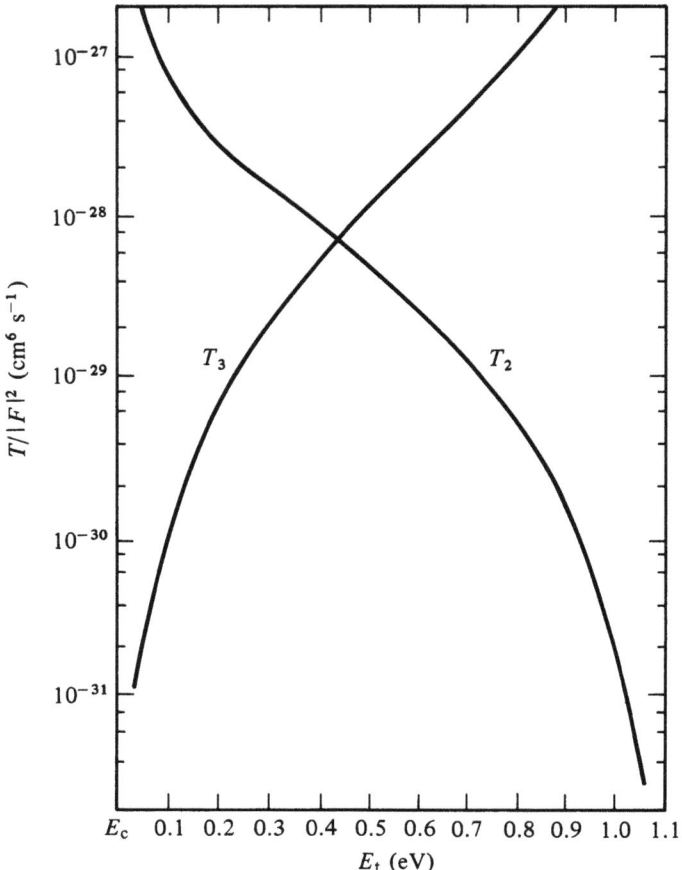

Fig. 5.3.8. Calculated zero-order trap–Auger coefficients for Si ($m_c/m = 1.1$, $m_v/m = 0.59$, $E_G = 1.12$ eV, $\varepsilon = 12$, $T = 300$ K). E_t denotes the energy of the localized level below the conduction band.

(*Misprint warning*: The interesting papers [5.3.14] used T_3 but gave this formula with an incorrect power of $E_G - E_t$ in the first paper and an incorrect power of ε in the second paper.) Some numerical values are given in Table 5.3.5.

   Chosing $T \sim 10^{-28}$ cm^6 s^{-1} as a typical value, a representative cross section σ is obtained by multiplying by a carrier concentration ($\sim 10^{16}$ cm^{-3}) and dividing by a thermal velocity ($\sim 10^7$ cm s^{-1}) to find $\sigma \sim 10^{-19}$ cm^2 which depends of course on the carrier concentrations.

5.3.4 Auger quenching

The transitions discussed here are also relevant to phosphors and electro-luminescent devices in which an electric field can impact-excite an impurity. This then emits radiation on returning to its ground state. However, if the energy is given up to a conduction band electron instead of a photon, the radiative efficiency

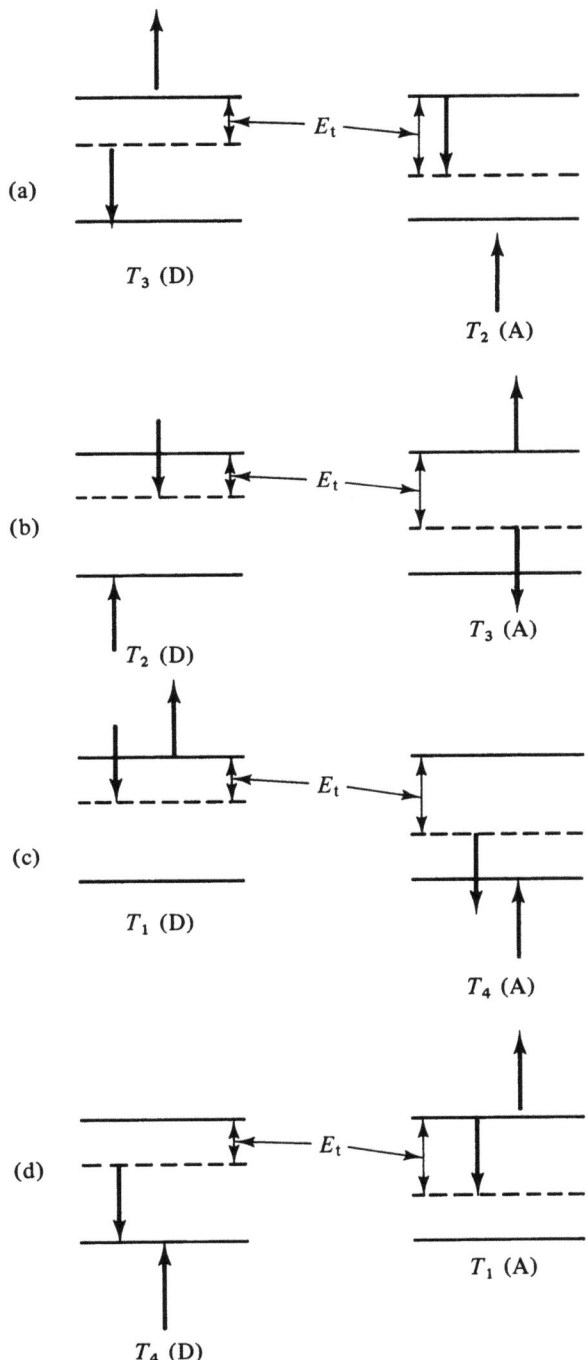

Fig. 5.3.9. Comparison of pairs of trap–Auger effects. (In Table 5.3.3 E_t for acceptors has exceptionally the different meaning, explained there.)

Table 5.3.5. *Some estimates of Auger–trap coefficients (ground state, nondegenerate bands)*

Source		Value $(cm^6 s^{-1})$	Temperature (K)	Reference
			T_1	
(1)	Semi-empirical value for As and P in Ge, $E_t \sim 0.01$ eV	1.5×10^{-23}	6	[5.3.19]
(2)	Theoretical value for Ge, $E_t \sim 0.26$ eV (Cu level), model a_1	$10^{-26}|F|^2$	300	[5.3.11]
(3)	Theoretical value for Ge,			
	$E_t \sim 0.2$ eV, model a_1	$10^{-26}|F|^2$	300	[5.3.2]
	model a_2	$6 \times 10^{-27}|F|^2$	300	
	model b_1	$2 \times 10^{-26}|F|^2$	300	
	model b_2	$10^{-26}|F|^2$	300	
(4)	Theoretical value for GaAs, $E_t \sim 0.005$ eV, model a_1	2×10^{-18}	4	[5.3.20]
(5)	Theoretical value for GaAs, $E_t \sim 0.5$ eV, model a_2	2×10^{-25}	300	[5.3.20]
			T_2	
(6)	As (2)	$2 \times 10^{-27}|F|^2$	300	[5.3.11]
(7)	p-type Si, donor $E_t = 0.78$ eV, Au-doped. Theory, amended model a_1	2.07×10^{-27}	300	[5.3.21]
(8)	p-type Si, Au-doped, minority-carrier lifetime, experimental	$< 10^{-27}$ (if $N_D \sim 10^{14}$ cm^{-3})	300	[5.3.22]
(9)	GaP (Zn, O) temperature dependence of luminescence, experimental	5.2×10^{-26}	300	[5.3.23]
			T_3	
(10)	n-type Si, acceptor, $E_t = 0.55$ eV, Au-doped. Theory, amended model a_1	4.78×10^{-27}	300	[5.3.21]
(11)	n-type Si, Au-doped, minority-carrier lifetime, experimental	10^{-26} (if $N_A \sim 10^{14}$ cm^{-3})	300	[5.3.22]
(12)	Semi-empirical value for Cu in Ge, $E_t \sim 0.4$ eV	10^{-26}	300	[5.3.24], [5.3.25]
(13)	As (2)	$2 \times 10^{-27}|F|^2$	300	[5.3.11]

Table 5.3.5. (*cont.*)

Source	Value (cm^6 s^{-1})	Temperature (K) T_4	Reference
(14) As (2)	$5 \times 10^{-28}\|F\|^2$	300	[5.3.11]
(15) Semi-empirical value from luminescent decay times in *p*-type GaP	2.5×10^{-30}	300	[5.3.26]
	Surface Auger recombination at unspecified trap		
(16) Semi-empirical, Ge	$2 \times (10^{-24} - 10^{-25})$	300	[5.3.27]
(17) Semi-empirical, Si	$2.5 \times (10^{-26} - 10^{-27})$	300	[5.3.27]

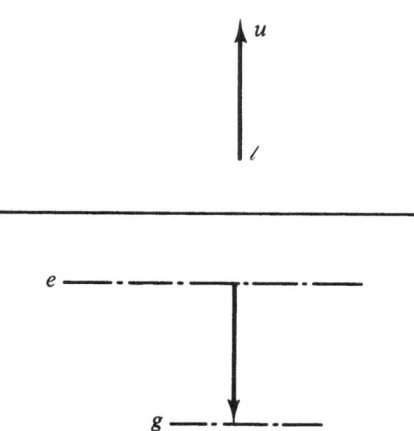

Fig. 5.3.10. A case of Auger quenching.

is impaired and one has an example of Auger quenching or Auger de-excitation (Fig. 5.3.10). A theory of this effect can be produced along the lines of the preceding sections [5.3.28]. It is sometimes expected to be important because the wavefunctions for the ground and excited states of a center enter the matrix element and have a good overlap. However, it must be borne in mind that they are orthogonal, if given correctly, which reduces the overlap to zero.

A slightly different theoretical approach (outlined below) has been used for the promising electroluminescent materials ZnS and ZnSe doped with rare earth ions [5.3.7], [5.3.29], [5.3.30]. (For high fields one must average over a hot electron distribution and this is expected to change the calculated values, but we do not deal with this point here.)

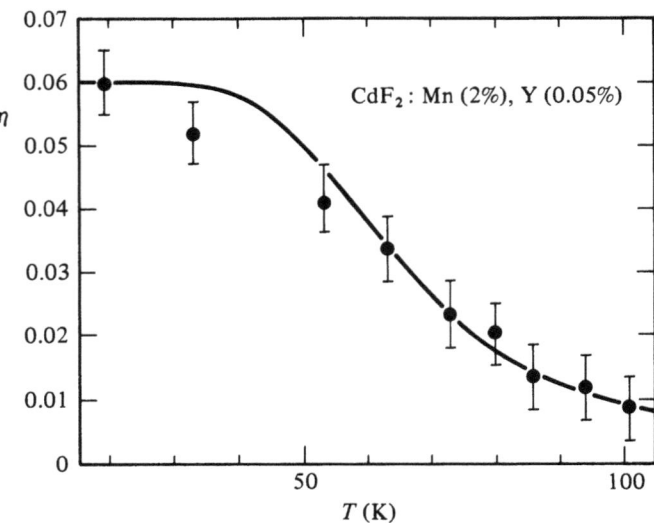

Fig. 5.3.11. Temperature dependence of the Mn^{++} quantum efficiency in highly conducting CdF_2:Mn crystals. The solid line represents a fit to eq. (5.3.19) with the $n(T)$ dependence taken from resistivity measurements.

If the concentration of centers in the excited state is n_e, then the quenching process proceeds at a mass action rate per unit volume of $u = T_q n_e n$, giving the Auger coefficient a unit of $cm^3\ s^{-1}$. The usual unit of $cm^6\ s^{-1}$ is reduced by cm^3 since the knowledge that a center is in an excited state implies that its ground state is available. One may also say that this Auger effect contributes an amount $T_q n$ to the reciprocal lifetime $1/\tau = u/n_e$. The *low* temperature radiative efficiency η_{LT} is not affected by Auger quenching. If τ_{nr} is the lifetime for *other* nonradiative processes, one finds

$$\eta_{LT} \equiv \frac{1/\tau_r}{1/\tau_r + 1/\tau_{nr}} = \frac{1}{1 + \tau_r/\tau_{nr}}$$

$$\eta \equiv \frac{1/\tau_r}{1/\tau_r + 1/\tau_{nr} + T_q n(T)} = \frac{\eta_{LT}}{1 + T_q \eta_{LT} \tau_r n(T)} \tag{5.3.19}$$

The result that the quantum efficiency drops with temperature from η_{LT} to very small values as $n(T)$ increases (for example with temperature) in accordance with eq. (5.3.19) has been confirmed for CdF_2:Mn and ZnS:Mn [5.3.31], see Fig. 5.3.11. (The notation indicates that Mn is the dopant.)

We now calculate the Auger rate, based on [5.3.29]. By eq. (3.2.14) the transition rate from a lower to an upper level of an atom by impact with an electron ($E_\ell \rightarrow E_u$, say) is from Fermi's golden rule $(2\pi/\hbar)|M_D|^2 \mathcal{N}(E_\ell)$, where eq. (3.3.27) has been used, exchange is neglected and the spin degeneracy has not been incorporated in the matrix element. Now

$$M_D \simeq \frac{1}{V^2} \int \frac{e^2}{\varepsilon |\mathbf{r}_1 - \mathbf{r}_2|} e^{i\mathbf{k}_u \cdot \mathbf{r}_1} e^{-i\mathbf{k}_\ell \cdot \mathbf{r}_1} \psi_g(\mathbf{r}_2) \psi_e^*(\mathbf{r}_2) \, d\mathbf{r}_1 \, d\mathbf{r}_2$$

where screening is neglected. The $\mathbf{r}_2$-integral concerns the excitation process within the atom. The electron 1 is not free but is more nearly a Bloch electron under the influence of the atom. This leads to a correction factor, to be denoted by $S_{\ell u}$. The first integral is the Kth ·Fourier component $(\mathbf{K} \equiv \mathbf{k}_u - \mathbf{k}_\ell)$ of the Coulomb interaction and yields $(4\pi e^2 / V\varepsilon K^2) \exp(i\mathbf{K} \cdot \mathbf{r}_2)$ by eq. (C.14). The exponential for small enough $\mathbf{K} \cdot \mathbf{r}_2$ is

$$1 + i\mathbf{K} \cdot \mathbf{r}_2 \tag{5.3.20}$$

This assumes that the center which is being quenched (e.g. a transition metal or rare earth ion) is highly localized. The orthogonality of the $\mathbf{r}_2$-wavefunctions ensures that the contribution from the first term vanishes. Hence

$$|M_D| \simeq \frac{4\pi e^2}{\varepsilon K V} |\mathbf{r}_{eg}| \tag{5.3.21}$$

where $e\mathbf{r}_{eg}$ is the electric dipole matrix element. This holds for given $\mathbf{k}_u$ and $\mathbf{k}_\ell$. Keeping their magnitudes, and hence their energies, fixed, we integrate over all angles between them using

$$K^2 = k_u^2 + k_\ell^2 - 2k_u k_\ell \cos \theta$$
$$d\omega = 2\pi \sin \theta \, d\theta = (2\pi K / k_u k_\ell) \, dK$$

Multiplying $P(\theta, \varphi)$ by $d\omega/4\pi$ and integrating, one finds

$$P \equiv \int P(\theta, \varphi) \, d\omega/4\pi = \frac{32\beta\pi^3 S_{\ell u}^2 e^4 \mathcal{N}(\varepsilon_\ell)}{\varepsilon^2 \hbar V^2 k_u k_\ell} |r_{eg}| \ln \left| \frac{k_u + k_\ell}{k_u - k_\ell} \right| \tag{5.3.22}$$

as the recombination rate where the limits $\theta = \pi$ $(K = k_u + k_\ell)$ and $\theta = 0$ $(K = k_u - k_\ell)$ have been used. Near threshold $k_\ell \ll k_u$, the last term is $2k_\ell/k_u$ and one has

$$P_{th} = \frac{64\beta\pi^3 S_{\ell u}^2 e^4 \mathcal{N}(E_\ell)}{\varepsilon^2 \hbar V^2 k_u^2} |r_{eg}|^2$$

If v_u, v_ℓ are the electron speeds in states $\mathbf{k}_u$, $\mathbf{k}_\ell$ then one can put for a parabolic band (including spin)

$$\mathcal{N}(E_\ell) = 2k_\ell^2 V / \pi h v_\ell$$

Also the Einstein A-coefficient gives from eq. (4.4.14′)

$$VB_{eg}(E_{eg}) \equiv \frac{1}{\tau_r} = \frac{4e^2 \mu E_{eg}^3}{\hbar^4 c^3} |r_{eg}|^2 \left(\frac{\mathscr{E}_{eff}}{\mathscr{E}} \right)^2 \tag{5.3.23}$$

where the ratio of the effective field to the average field has been inserted. The Lorenz form of this local field correction, valid for extreme localization, is $(\mu^2+2)^2/9$ and this is ~ 20 for a typical III–V semiconductor. One then has, introducing also a cross section σ for the process,

$$P = \frac{4\beta\pi S_{\ell u}^2 \hbar^2 c^3 k_\ell}{\varepsilon^2 V \tau_r \mu E_{eg}^3 v_\ell k_u} \left(\frac{\mathscr{E}}{\mathscr{E}_{eff}}\right)^2 \ln\left|\frac{k_u+k_\ell}{k_u-k_\ell}\right| \equiv \frac{v_u}{V}\sigma \qquad (5.3.24)$$

More general results for nonparabolic bands can also be obtained from eq. (5.3.22). In fact [5.3.30]

$$\frac{1}{\varepsilon^2}\left(\frac{\mathscr{E}}{\mathscr{E}_{eff}}\right)^2 = \mu^{-4}$$

giving a μ^{-5}-dependence in eq. (5.3.24).

Experiment agrees broadly with theory [5.3.30], [5.3.32], except for the case of ZnS:Mn. One finds $\sigma \sim 10^{-16}$ cm^2, which is two orders of magnitude greater than what would be obtained from eq. (5.3.24) [5.3.31]. A more sophisticated theory which distinguishes different impurities and can deal with excited states is also available, though somewhat complicated [5.3.7]. Typical values inferred from experiments for T_q are (in cm^3 s^{-1}):

5×10^{-10} (ZnS:Mn; ZnSe:Mn) [5.3.31]
8×10^{-15} (CdF$_2$:Mn^{++}) [5.3.30]
5.3×10^{-15} to 5.4×10^{-14} (CdF$_2$:Gd^{+++}) [5.3.32]
$(2\text{–}6) \times 10^{-11}$ (GaP:Zn,O) [5.3.33]

For a more detailed account of this field see [5.3.30]. Dividing T_q by the thermal velocity one arrives at a (concentration *independent*) cross section which is of order 10^{-16} cm^2 or less, i.e. of the same order of magnitude as was obtained in section 5.3.3.

It may seem puzzling that eq. (5.3.21) is rather different from eq. (3.4.27), corresponding to a replacement

$$\frac{FG_{\mathbf{k}_1+\mathbf{k}_2-\mathbf{k}_{2'}}}{|\mathbf{k}_{2'}-\mathbf{k}_2|^2} \rightarrow \frac{|r_{eg}|}{|\mathbf{k}_{2'}-\mathbf{k}_2|}$$

The reason resides in different approximations for the two cases. To show this, neglect the $\mathbf{L}_2$, $\mathbf{L}_{2'}$ dependence in eq. (3.4.22) of $N(\mu, 1, 1')$, where $\mu \sim \mathbf{k}_{2'}-\mathbf{k}_2$, and expand eq. (3.4.23) as in eq. (5.3.20). One finds

$$N(\mathbf{k}_{2'}-\mathbf{k}_2, 1, 1') = i(\mathbf{k}_{2'}-\mathbf{k}_2) \cdot |\mathbf{r}_{11'}|$$

where $|\mathbf{r}_{11'}|$ corresponds to $|\mathbf{r}_{eg}|$. Hence one obtains eq. (5.3.21) with the aid of eq. (3.4.25) with $\mathbf{M} = 0$. It is thus the expansion (5.3.20) which gives rise to the difference.

5.3.5 More involved arguments and effects

One can develop the semiquantitative results given in section 5.3.3 by taking into account the deformation of the lattice induced by the change of state of the impurity. The theory required involves the electron–phonon interaction which contributes even in first-order perturbation theory. The result is that the theoretical estimate of T_1 given in Table 5.3.3 is enhanced in zero order by a factor $[E_t/(E_t - 2E_1)]^m$ where $2E_1$ is the Stokes shift (and related to the Huang–Rhys factor S of eq. (6.2.3), below). Also $m = 3$ or $\frac{5}{2}$ depending, respectively, on whether models a_1, a_2 or models b_1, b_2 are used [5.3.34]. The temperature dependence remains weak: T_1 drops by a factor of about two as one passes from 100 K to 500 K.

Similarly, one can consider impurity assistance for the *band–band* Auger effect. This relaxes the k-conservation rule and enhances the total Auger transition rate. This calculation requires second-order perturbation theory and has been carried out for the CHHS process in p-type GaAs and p-type GaSb [5.3.35]. It appears that the impurity-assisted process dominates in GaSb and for $p > 10^{19}$ cm^{-3} in GaAs. At these concentrations, however, heavy doping effects may require the theory to be amended.

The multiphonon capture mechanism also enhances Auger capture, and for a defect in GaAs (the so-called B-center) the two mechanisms together can account for the experimental cross sections [5.3.36] as shown in Fig. 5.3.12 [5.3.37]. (Rebsch obtained a typical Auger enhancement factor of four [5.3.38].) With quadratic electron-phonon interaction included, multiphonon theory alone may be able to explain the data [5.3.37]. But really large capture cross sections $\sim 10^{-16}$ cm^2 are not reached.

The Sommerfeld factor (5.3.5) to allow for the Coulomb disturbance of the free state was used in [5.3.39] and [5.3.40]. It was shown graphically that this, too, enhances the Auger rate as expected. The system studied was the Ge acceptor in GaAs.

We now turn to the He model for a two-electron donor. It can 'capture a hole' in an Auger process or, put differently, one electron enters the valence band while the other is promoted to the conduction band. The charge state of the defect changes as follows:

$0 \rightarrow e$, $e \rightarrow 2e$ by capture of a conduction band electron

$2e \rightarrow e$, $e \rightarrow 0$ by capture of a valence band hole

and by the reverse processes. [The two-electron center is in a different energy state (E_2) from the one-electron center (E_1), as already seen in Fig. 2.1.2]. The Auger effect in which the defect loses both electrons, one to the valence band and the

Table 5.3.6. *Experimental cross sections at double donors in Si* [5.3.45]

	T(K)	σ(cm²)	Likely mechanism
Se^x	170	6×10^{-17} ⎫	Auger transition
S^x	170	8×10^{-17} ⎭	
Se$^+$	142	10^{-23} ⎫	Multiphonon or
S$^+$	142	10^{-21} ⎭	radiative transition

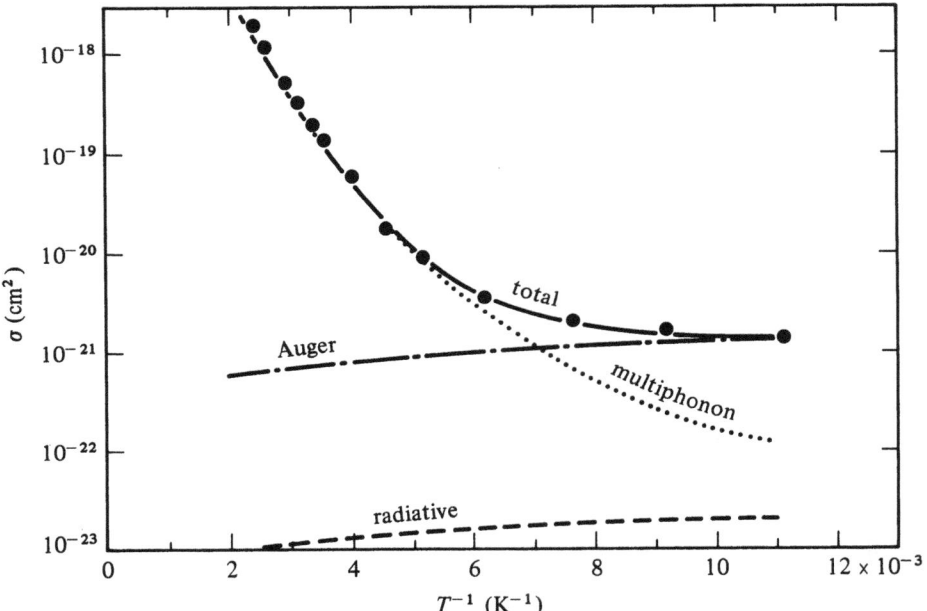

Fig. 5.3.12. A numerical fitting of capture cross section data according to [5.3.37]. (●) Experimental points [5.3.36] for GaAs.

other to the conduction band, was first studied by Jones and Beattie [5.3.41], [5.3.42]. They showed that the Auger coefficient has a maximum when regarded as a function of $E_2 - E_1$. The transition is tantamount to converting a hole into an electron, thus keeping the total number of current carriers constant. It has been discussed theoretically, [5.3.41]–[5.3.44], and some experimental results [5.3.45] are shown in Table 5.3.6 and are in broad agreement with theory [5.3.44]. Double acceptors should behave analogously and Zn in Si has been studied, but it is not clear whether an Auger capture process dominates in this case [5.3.44, 5.3.46].

Table 5.4.1. *Collapse of bound excitons*

('~') denotes here 'analogous to'

C	For e-process $C^+ + e$	For h-process $C^- + h$	Examples of interpretation and references — e-process	Examples of interpretation and references — h-process
Negative total initial charge				
(1) $\otimes^{-\pm}$ $[T_1(D):D^+ + 2e]$	$\otimes^- + e$ $D^\times + e]$	$\otimes^{--} + h$ –		
(2) $\ominus^{-\pm}$ $[T_1(A):A^\times + 2e]$	$\ominus + e$ $A^- + e]$	$\ominus^- + h$ –		
(3) $\ominus^{\pm}$ $[T_3(A):A^- + e + h]$	$\ominus^+ + e$ $A^\times + e]$	$\ominus^- + h$ –	Nonradiative decay of exciton on A^-; Cu$_2$O [5.4.2]	D^- captures second e
Zero total initial charge				
(4) $\ominus^{+\pm}$ $[T_2(A):A^\times + e + h]$	$\ominus^{++} + e$ –	$\ominus + h$ $A^- + h]$		Nonradiative decay of exciton on $A^\times$: CdS [5.4.3]; Cu$_2$O [5.4.2]; Si [5.4.4], [5.4.5]; Ge [5.4.4]
(5) $\otimes^{\pm}(\sim H_2)$ $[T_3(D):D^\times + e + h]$	$\otimes^+ + e$ $D^+ + e]$	$\otimes^- + h$ –	$D^\times$ ionized by collapse: CdS [5.4.3]; Si [5.4.5]	
Positive total initial charge				
(6) $\oplus^{\pm}[\sim H_2^+]$ $[T_2(D):D^+ + e + h]$	$\oplus^{++} + e$ –	$\oplus^- + h$ $D^\times + h]$		
(7) $\oplus^{++}$ $[T_4(D):D^\times + 2h]$	$\oplus^{+++} + e$ –	$\oplus + h$ $D^+ + h]$		
(8) $\otimes^{++}[\sim H_2^+]$ $[T_4(A):A^- + 2h]$	$\otimes^{+++} + e$ –	$\otimes + h$ $A^\times + h]$		D^+ captures e of exciton; nonradiative decay of exciton on D^+:CdS [5.4.2]; GaAs [5.4.6]; Si and Ge [5.4.7]

5.4 More complex effects

5.4.1 Auger effects involving bound excitons

The Auger effect at a double donor can involve two localized particles and one free particle. It corresponds to an Auger effect involving a He atom. If an exciton is bound to a neutral donor one has again a three-particle problem, but the third particle is now also localized and an Auger effect is expected to be encouraged by this close proximity; large cross sections $\sigma \sim 10^{-15}$ cm^2 are possible. In fact, excitons bound to defects can undergo a variety of Auger effects [5.4.1]. The exciton of a complex C can collapse radiatively (giving rise to C', say) and it can also collapse by the Auger effect giving rise to an Auger particle. This can be an electron ('e-process') or a hole ('h-process'):

$$C \rightarrow C' + h\nu \quad \text{or} \quad C^+ + e \quad \text{or} \quad C^- + h \tag{5.4.1}$$

Examples are given in Table 5.4.1. For each case an analogous trap–Auger transition with coefficient T_j is given in square brackets. A He-like analogue of some complexes is also noted. Additional possibilities exist [5.4.8], [5.4.9]. In particular, the defect may be excited rather than ionized.

It is found that the 4.2 K lifetimes of bound excitons are rather short in Si at $\sim (10^{-9}$–$10^{-6})$ s, and behave as

$$\tau \sim \quad 0.845 E_{\mathrm{D}}^{-4.6} \text{ s} \quad \text{for donors} \tag{5.4.2}$$
$$23.8 E_{\mathrm{A}}^{-3.9} \text{ s} \quad \text{for acceptors}$$

where the defect ionization energies E_{D}, E_{A} are expressed in meV [5.4.5]. They can be accounted for as Auger effects. When $\ln \tau$ is plotted against $\ln E_{\mathrm{A, D}}$ one finds a line of negative slope, which was originally found to be -4 for acceptors in GaP [5.4.10]. Furthermore the exciton binding or localization energy is proportional to the ionization energy of the defect; this is the so-called 'Haynes rule'. It seems to hold for donors in direct semiconductors and for donors and acceptors in indirect semiconductors, though there are deviations for acceptors in direct semiconductors. With the aid of this rule the $\tau \propto E_{\mathrm{A}}^{-4}$ law can receive a simple explanation [5.4.10] which is, however, of doubtful reliability.

A striking difference between Si and Ge was found with regard to bound excitons [5.4.4]. On adding 10^{15} cm^{-3} shallow acceptors (In for example) to Si at about 10 K, the lifetime drops from a few microseconds by a factor of 10^3. This is attributed to the capture of free excitons (which limits the lifetime in the pure material) by the acceptors. They then collapse rapidly by Auger effects. In Ge these Auger processes seem to be much slower than the free exciton decay and so doping has little effect. (We assume conditions under which electron–hole drops are not formed.) The explanation is that the holes are more strongly bound to the acceptors in Si than they are in Ge. This leads to a greater spread of the wavefunction in k-space and a larger Auger rate. This in turn can be understood in terms of the Bohr radii which are smaller in Si than they are in Ge. Similarly, the bound exciton lifetimes decrease with the effective Bohr radius of the impurity involved.

There is, however, a limit to the number of free excitons which can be formed, as noted at the end of section 4.5.2. Briefly, exciton formation is impeded by screening which weakens the Coulomb attraction of electron and hole according to the law (1.8.12). For exciton formation one needs the Debye length to exceed the exciton radius

$$l_{\text{Deb}}^2 \equiv \frac{\varepsilon kT}{4\pi n e^2} > r^2 \tag{5.4.3}$$

so that with T in K, E_{x1} in eV and n in cm^{-3}

$$n < \frac{\varepsilon kT}{4\pi e^2}\left(\frac{2\varepsilon E_{x1}}{e^2}\right)^2, \quad \text{i.e. } n\tilde{r}_1^3 < \frac{T}{300 E_{x1}} 10^{-6} \tag{5.4.4}$$

where $E_{x1} \equiv e^2/2\varepsilon r$ is the exciton binding energy ~ 3–4 meV.

For the ratio of radiative to Auger recombination rates, a '($\frac{7}{2}$)-law' arose in connection with excitons bound to neutral donors [5.4.11]. The original argument used an adaptation of the internal conversion coefficient of nuclear physics which yields for the ratio internal conversion probability to radiative probability the quantity $4(\tilde{r}_i \kappa)^{-3}(\tilde{r}_f k)^{-1}$, where $\tilde{r}_i = \hbar^2\varepsilon/e^2 m_i$ is the effective Bohr radius for the initial state of effective mass m_i, κ is the wave number $\omega\mu(\omega)/c = E\varepsilon^{\frac{1}{2}}/\hbar c$ of the photon, $\tilde{r}_f = \hbar^2\varepsilon/e^2 m_f$ is the final state Bohr radius and $k = (2m_f E/\hbar^2)^{\frac{1}{2}}$ is the wavevector of the Auger electron. A factor $\frac{1}{2}$ has been introduced since energy is received by one electron and not by the two K-shell electrons of the internal conversion process.

One finds the law ($\alpha_0 \equiv e^2/\hbar c = \frac{1}{137}$)

$$\frac{u_{\text{rad}}}{u_{\text{Auger}}} = \left(\frac{\varepsilon^{11}}{2^3}\right)^{\frac{1}{2}} \frac{1}{\alpha_0^4}\left(\frac{E}{m^* c^2}\right)^{\frac{7}{2}} \tag{5.4.5}$$

where $m^* \equiv m_i^{\frac{6}{7}} m_f^{\frac{1}{7}}$. If E is in eV

$$\frac{u_{\text{rad}}}{u_{\text{Auger}}} = \left(\frac{\varepsilon}{10}\right)^{\frac{11}{2}}\left(\frac{m}{m^*}\right)^{\frac{7}{2}} 4.129 \times 10^{-7} E^{\frac{7}{2}}$$

Hence for $m^* = 0.305m$, $\varepsilon = 10.2$, $E = 2.31$ eV (GaP) the ratio is 5.5×10^{-4}, indicating the preponderance of Auger transitions. If the formula were reliable (which can hardly be expected) it would give $u_{\text{rad}} \sim u_{\text{Auger}}$ only at $E \sim 6.7$ eV (given $\varepsilon \sim 10$, $m^* = 0.1m$).

A more general approach uses the ratio of the square of the matrix elements for radiative and Auger transitions. The former behaves as E, the energy bridged by an electron, from eq. (5.2.72) or eq. (D36); the latter behaves as E^{-2} from $(e^2/\varepsilon r)^2 = [\sum_{\mathbf{k}}(4\pi e^2/\varepsilon V k^2)\,e^{i\mathbf{k}\cdot\mathbf{r}}]^2 \simeq E^{-2}$, giving a factor E^3. The remaining terms add a factor $E^{\frac{1}{2}}$, giving [5.4.12]

$$\frac{u_{\text{rad}}}{u_{\text{Auger}}} \simeq \left(\frac{E}{I_0}\right)^{\frac{7}{2}} \frac{\alpha_0^3}{v r_1^3} \tag{5.4.6}$$

where I_0 is given by eq. (5.2.71), r_1 is the Bohr radius and v is the carrier concentration.

Note that from a theory of Auger recombination of excitons bound to neutral donors the rough formula

$$\tau^{-1} \sim \frac{\beta^r E_G}{\hbar} \quad (r = 11 \text{ for shallow}, r = 4 \text{ for deep donors})$$

has been suggested [5.4.13]. Here

$$\beta \equiv \left(\frac{m_c}{m\varepsilon^2}\frac{1}{\pi^2}\right)^{\frac{1}{2}} \left(\frac{I_0}{E_G}\right)^{\frac{1}{2}}$$

Excitons as intermediaries for Auger recombination via deep impurities can also be envisaged. If the temperature and the majority-carrier density are low enough, the exciton can on collision with an impurity decompose by an Auger process which may often be of type T_2 or T_3. Whether the exciton is bound briefly to the impurity or not is not clear. But if one has a donor impurity, processes (5) and (6) of Table 5.4.1 are relevant. Such a mechanism would suggest an increase in the lifetime (after initially fairly constant values) with both temperature and majority-carrier concentration. The former would be due largely to the thermal decomposition of excitons. The latter would be due to reduced exciton concentration by increased screening, as discussed in connection with eq. (5.4.3). Such a model has recently been developed and has led to good agreement with experiments on Si:Fe and Si:Cr, though there are some problems if the majority-carrier density exceeds 10^{17} cm^{-3} [5.4.14]. In at least some of these systems (e.g. Si:Pt [5.4.15]) multiphonon mechanisms are ruled out by a small Huang–Rhys factor (6.2.15), see below, of order 0.4 or less. However, cascade processes are reasonable because of the existence of excited states and of phonon replicas (roughly zero-phonon lines displaced by a phonon energy) which show the effectiveness of the electron–phonon interaction. It is expected for deep rather than shallow impurities because their wavefunctions are more localized in space and so more widely spread in k-space. As expected, shallow impurities tend not to show phonon replicas. It is of course possible to combine theoretically several of the main processes – cascade recombination, Auger recombination, multiphonon effects – to obtain a higher-order process. Nature ensures of course that they *all* occur, but does not so readily reveal which dominates.

For the usual donors and acceptors one goes to a *different* column of the periodic table: thus N and P (group V) act as donors in Ge and Si, while Ga and B (group III) act as acceptors. S (group VI) acts as donor in GaP. One can also substitute from the *same* group in the periodic table and replace an atom by another one of the *same* valency:

P can be replaced by N or Bi   (group V) in GaP

Te can be replaced by O   (group VI) in ZnTe

S can be replaced by Te   (group VI) in CdS

These *isoelectronic* substitutes (reviewed in [5.4.16]) have a short-range (non-Coulombic) potential and hence a diffuse electronic wavefunction in momentum space. This provides an overlap with wavefunctions of states at a band extremum, thus effectively converting an indirect into a direct semiconductor. In the indirect material GaP this doping thus increases the radiative efficiency and one has a material from which light-emitting diodes ('LEDs') of various colours have been made.

There is an approximate rule which states that if the difference (electron affinity of impurity) – (electron affinity of the replaced host atom) is positive, an electron can be trapped and in the Coulomb field thus created a hole can also be trapped; conversely a hole can be trapped first and then an electron, if the difference is negative. Thus isoelectronic impurities can also trap excitons. Indeed two neighboring isoelectronic impurities will also bind an exciton and will do so the more strongly the closer they are. In fact this binding energy E_{Bx} decreases with the distance R between the impurities so that the photon energy $h\nu$ emitted also increases. This can be visualized from Fig. 5.4.1 in which the *internal* binding energy E_x of the exciton has been neglected. It illustrates that

$$h\nu + E_{Bx} + E_x = E_G \tag{5.4.7}$$

so that

$$h\nu \text{ increases with } R \tag{5.4.8}$$

Note that an isoelectronic impurity may not have a bound state.

Radiative decay of an exciton at an isoelectronic impurity can be almost 100% efficient, an example is the green-emitting GaP:N diode. The isoelectronic binding

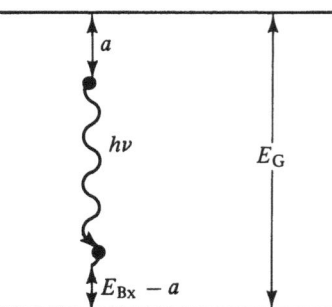

Fig. 5.4.1. Schematic illustration of eq. (5.4.7). The energy a is formal and need not be specified.

is more efficient than exciton binding by a neutral donor, for the close proximity of *three* particles in the latter case provides too great a probability for a competing nonradiative Auger transition.

On increasing the isoelectronic doping the luminescence intensity does not, however, increase indefinitely. Lifetime 'killing' or (what is almost the same thing) radiation quenching soon occurs, see for example [5.4.17]–[5.4.18]. In [5.4.17] the promotion of an impurity band electron into the conduction band was first considered as a quenching mechanism by Auger effect.

5.4.2 Donor–acceptor Auger effect

The donor–acceptor Auger processes can be analyzed on roughly the same basis as the radiative donor–acceptor process. For GaAs at 80 K, for example, theory suggests that the process in which an electron takes up the energy has a coefficient of order $10^{-29}|F|^2$ cm^6 s^{-1} (in which the overlap integral F is still left as a parameter) and $10^{-33}|F|^2$ cm^6 s^{-1} when hole collisions dominate. Furthermore, the ratio of the Auger to the radiative recombination rate decreases as the pair distance increases [5.4.19]. If $E_D - E_A$ is put roughly equal to the band gap the ratio involves again $E_G^{-\frac{7}{2}}$, i.e. the $\frac{7}{2}$-power law noted in eq. (5.4.6). The case when a double donor is involved has also been discussed [5.4.20]. Furthermore, the transfer of the donor–acceptor recombination energy to a third center (donor or acceptor) is possible, leading to its excitation or ionization. This process has recently been identified by an electron-spin-resonance technique in ZnS with a recombination coefficient of order 10^{-30} cm^6 s^{-1} when the energy is transferred to a donor [5.4.21]. This three-center process had already been envisaged by Dishman [5.4.22] for GaP:O and by Sheinkman and co-workers generally [5.4.23] and, for CdS [5.4.24], [5.4.25] and CdI$_2$ [5.4.26].

From the point of view of classifying processes (see Table 3.7.1) it will be noted that in the above discussion we have passed from *Auger processes* involving one localized state (section 5.3.3) to Auger processes involving two localized states. This was envisaged qualitatively already in the 1960s [5.4.27], also for multiply charged impurities [5.4.28]. Energy transfer to a neighboring center means that only one band state is involved. All four states of the Auger process can be localized if the neighboring defect is excited, but not ionized. The roughly analogous band–band process in which *four* distinct bands are involved has still not been observed.

If one takes the many-valley structure of the conduction bands involved into account, a number of possibilities arise (enumerated and analyzed in [5.4.29]) which suggest that in general these cross sections depend weakly on temperature ($\sigma \sim T^{-r}$ with $0 \lesssim r \lesssim 2$) and on electric field if the carrier heating is small. There is a strong dependence on the typical distance R between the complexes. Because of the decay of the wavefunctions involved one would expect an exponential decay with R, familiar from donor–acceptor transitions [5.4.19]. (A dipole–dipole interaction term going as R^{-6} has also been suggested.) There is also a strong dependence on the energy parameters. As has been seen in section 5.3.3, a power law for the decrease in cross section with the energy interval bridged by an electron is expected (in a rough approximation).

5.4.3 Shake effect

A 'shake effect' in X-ray emission from metals has been known for many years and its application to the solid state was suggested *ca.* 1985 [5.4.30], [5.4.31]. In the X-ray case one finds after fast excitation, and therefore high excitation energy, that an electron has been knocked out of its orbit. Assuming that the remaining particles retain their original wavefunction, the system is then in an unrelaxed state Φ whose many-electron wavefunction is not an eigenfunction of the Hamiltonian. It can be regarded as a superposition of states which may include multiply ionized atoms. The decay of the system to equilibrium states Ψ_n with approximate probability $|\langle \Phi, \Psi_n \rangle|^2$ can give rise to a better understanding of satellite bands.

It has been suggested [5.4.30] that this shake effect may reduce the luminescent intensity of the low energy companion line seen upon the decay of an exciton bound to a neutral (As) donor in Si. This line is due to the 2s → 1s decay, the population of the 2s level having been caused by a normal Auger effect leading to defect excitation. In addition polarization mixing observed in luminescence spectra from *n*-doped quantum wells has been discussed in terms of the many-electron shake-up effects [5.4.32].

5.4.4 Recombination-enhanced reactions

As a last class of complex processes, the recombination enhancement of solid state reactions should be mentioned. The energy of the order of a band gap, which must be dissipated in a recombination act, can be channeled into Auger or radiative processes, as already discussed, or into multiphonon emission (chapter 6). Recombination enhancement is linked to the latter mechanism, since potential energy barriers in the solid are then more likely to be overcome with the aid of the additional phonon momenta and energies. Thus one would expect recombination enhanced diffusion of defects and possibly the consequent dissociation of complex defects or creation of new ones [5.4.33]. Defects with near mid-gap levels are liable to have a particularly strong electron–lattice interaction and the recombination enhanced defect migration for such centers has been observed in Si [5.4.33], GaAs and GaP (see the review [5.4.34]) and InP [5.4.35]. So certain is one of these effects, that their absence in certain cases has helped one to infer that the defect in question must be part of an extended complex or defect aggregate which is stable and hard to move through the lattice. This inference was made with respect to the so-called EL2 center in GaAs which is of technological importance in producing semi-insulating material [5.4.36].

Another peculiarity of this center is that incident light will change its electron occupancy only if it is in its normal state, but not if it is in its higher lying metastable state. Thus optical absorption after cooling the sample in the dark to 4 K is shown in Fig. 5.4.2, curve (a). The 4 K absorption drops to curves (b) and (c) after using 40 W tungsten lamp white light for 1 and 10 minutes, respectively. Presumably this is due to the promotion of the level to its metastable state by the incident light. Following this photoquenching of absorption, the initial absorption can be recovered by annealing at 140 K for a few minutes [5.4.37].

We also note the existence of recombination-enhanced dislocation climb and slide [5.4.38].

The movements of defects (and similar effects) by virtue of recombination are important since such phenomena can cause degradation in injection lasers, light-emitting diodes, tunnel diodes, solar cells, etc. (They can also cause the *disappearance* of harmful defects.) The importance of degradation studies, for example of solar cells in a space environment, is obvious – for a review see [5.4.34] and [5.4.39]. Degradation takes place throughout the life of a solar cell reducing its conversion efficiency in due course to a minimum acceptable value, the so-called 'end-of-life efficiency'.

A frequently used and experimentally supported relation for the kinetics of such phenomena depends on the activation energy $E_t \equiv \eta_t kT$ for movement at temperature T in thermal equilibrium, and the energy $E_L \equiv \eta_L kT$ made available

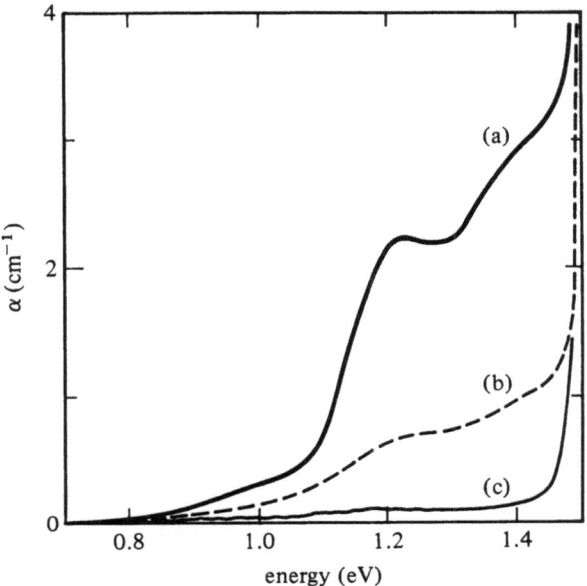

Fig. 5.4.2. Optical absorption spectra recorded at 10 K in the same undoped semi-insulating GaAs-material. (a) After cooling in the dark. (b), (c) After white-light illumination for 1 and 10 minutes, respectively [5.4.37].

to the lattice in a (nonradiative) recombination act. The recombination-enhanced reaction rate per unit volume can then be written as (see e.g. [5.4.33], eq. (7); [5.4.34], eq. (6); [5.4.40], eq. (24))

$$\theta u \exp\left[-(\eta_t - \eta_L)\right]$$

where u is the recombination rate per unit volume and θ is an efficiency factor allowing for the fact that only a fraction of recombination acts can contribute to the motion. Reactions of this type can be classified in various ways. If the energy goes into phonons one has a local heating model which has also been referred to broadly as the phonon-kick or thermal spike model. First introduced to discuss damage to the alkali halide lattice due to bombardment by energetic particles [5.4.41], it was used later to explain the degradation of GaAs tunnel diodes [5.4.42], now no longer of key technological interest. A simple model [5.4.40] using unimolecular reaction theory and subsequently improved [5.4.43], also led (among other results) to the reaction rate $\theta u \exp\left[-(\eta_t - \eta_L)\right]$ noted above. The defect may alternatively be promoted into a state in which the electron is excited, thus in general reducing the effective activation energy for motion of the defect as a whole. This is the local excitation model. Thirdly, the barrier for motion must be expected to depend on the charge state of the defect. The Bourgoin–Corbett model [5.4.44], [5.4.45] stipulates that the defect potential energy for a zero charge state (say) has minima at sites of type A and maxima near sites of type B, whereas for a charged state the situation is reversed. Thus a defect can move through the lattice using the potential energy minima by alternately capturing an electron and a hole. Sites A and B were originally thought of as hexagonal and tetrahedral sites in Ge. A good supply of carriers is needed for this mechanism, but it can be 'athermal', i.e. little thermal energy is needed, so that this can be a low-temperature diffusion process. All three mechanisms appear to have been found, as is clear from various surveys [5.4.34], [5.4.39], [5.4.46]–[5.4.48].

Recombination-enhanced migration of positively (but not of negatively) charged vacancies in Si and its temperature dependence is an additional current puzzle [5.4.49]. In fact, the many-particle aspect of this field makes it hard to develop [5.4.50]. One simply does not know yet in sufficient detail how the defect motion is influenced by its surroundings or what particles are involved in the motion, even though a great deal of experimental detail is available. One has relevant mathematical tools (chapter 6) and one knows that strong defect–lattice coupling must be involved. More physical insight is still needed.

A simple model for recombination enhancement of electrical activity is to suppose that a passive defect P is split into v identical electrically active defects A by the energy liberated by a recombination act: $P \rightleftharpoons vA$. This will serve as a simple illustration. The volume density of P-defects will be denoted by $P(t)$ at time t.

The energy liberated at the defect can be due to the capture of either electrons or holes. Two cases will therefore be distinguished. It will be assumed that the P-defect has a single deep energy level, and that we are dealing with p-type material. The electron energy level on the defect is assumed to be above the equilibrium Fermi level.

(a) *Electron capture*

The capture rate of electrons per unit volume is given by

$$u_p^{(n)} = c_n(1-f)nP(t) \tag{5.4.9}$$

where n is the concentration of electrons in the conduction band, f is the probability that the P-defect level is occupied by an electron, and

$$c_n = \langle v_n \sigma_n^{(P)} \rangle \tag{5.4.10}$$

where v_n is the electron velocity, $\sigma_n^{(P)}$ is the electron capture cross section of a P-defect and the average is with respect to the thermal distribution of electron velocities in the conduction band. The occupation probability f is by eq. (2.3.19)

$$f = \frac{c_n n + c_p p_1}{c_n(n+n_1) + c_p(p+p_1)} \tag{5.4.11}$$

where p is the concentration of holes in the valence band, and n_1, p_1 are given on p. 118.

Substitute eq. (5.4.11) into eq. (5.4.9) and use the assumptions that the defect level is sufficiently far above the equilibrium Fermi level, and that the hole concentration (holes being the majority carriers) is approximately equal to $p_0 + n$ $(p \sim p_0 + n \gg p_1)$. Hence,

$$u_p^{(n)} = c \frac{n + \left(1 + \dfrac{c_n}{c_p}\right) n_2}{n + n_2} n P(t) \tag{5.4.12}$$

where

$$c \equiv \frac{c_n c_p}{c_n + c_p}$$

$$n_2 \equiv \frac{c_n}{c_n + c_p} n_1 + \frac{c_p}{c_n + c_p} p_0 \tag{5.4.13}$$

(Without the above approximation for p, p_0 would be replaced by $p - n + p_1$.)

(b) Hole capture

The capture rate of holes per unit volume is

$$u_p^{(p)} = c_p f p P(t) \tag{5.4.14}$$

where p is the density of holes in the valence band, and c_p is the equivalent of the quantity (5.4.10) for holes. Using the expression (5.4.11)

$$u_p^{(p)} = c \frac{n + \dfrac{c_p}{c_n} p_1}{n + n_2} (p_0 + n) P(t) \tag{5.4.15}$$

(c) Rate equation for the recombination-enhanced defect reaction

Let Π be the probability that a P-defect splits up into v electrically active defects A in one recombination act. A theory of Π is quite complicated. Here, however, the bare existence of the quantity Π is sufficient.

The rate of A-defect creation per unit volume is now

$$\dot{A} = v \Pi u_p(t) \tag{5.4.16}$$

where u_p stands for $u_p^{(n)}$ or $u_p^{(p)}$, whichever is appropriate. Conservation of defects implies

$$v^{-1} A(t) + P(t) = P(0) \tag{5.4.17}$$

and hence, using eqs. (5.4.12) or (5.4.15),

$$\dot{A}(t) = v\Pi\alpha[P(0) - v^{-1}A(t)] \tag{5.4.18}$$

where

$$\alpha \equiv \begin{cases} c\dfrac{n + \left(1 + \dfrac{c_n}{c_p}\right)n_2}{n + n_2}\,n & \text{(electron capture)} \\[4ex] c\dfrac{n + \dfrac{c_p}{c_n}p_1}{n + n_2}(p_0 + n) & \text{(hole capture)} \end{cases}$$

The quantity n depends on time since the creation of A-defects affects the recombination lifetime. The variations of α due to this effect, however, will probably be small in comparison with the variation of A or P, and we shall assume that α is a constant independent of time. The solution is

$$A(t) = vP(0)[1 - e^{-t/\tau}]$$

where

$$\tau = \frac{1}{\Pi\alpha}$$

This indicates a simple exponential rise of $A(t)$ from zero to a saturation value of $vP(0)$ at $t \gg \tau$.

For the use of configuration coordinate diagrams in the discussion of these processes see chapter 6, and recent work by H. Sumi [5.4.50].

6

---

*Multiphonon recombination**

6.1 Introduction

Multiphonon processes in semiconductors belong to a family of electron transfer reactions which spreads over physics, chemistry and biology [6.1.1]. A theory of these processes describes how and in what circumstances electronic energy (ranging from few tenths of an electronvolt to an electronvolt or more) can be converted into the energy of nuclear vibrations. Since a phonon carries much less energy, many phonons must be absorbed or emitted during the transition – hence the name multiphonon transitions. In this chapter, we shall be concerned solely with multiphonon ('MP') transitions at defects which occur with the participation of the defect levels in the band gap, and a brief review of these defect states forms therefore an appropriate starting point for a discussion of this subject.

   The best known examples of impurity levels in the band gap of a semiconductor are the discrete energy levels of charged donors or acceptors, modeled most simply as in eqs. (1.8.2)–(1.8.4). In most *covalent* semiconductors the resulting shallow levels lie close to the respective band edge (within 0.1 eV or so) and the wavefunction spreads over a region extending over many lattice constants. The closely spaced excited states provide a suitable medium for cascade transitions (section 2.6). In *ionic* solids, the Coulomb potential of charged defects is stronger and it can bind electrons in states with energy levels deeper in the band gap. The large energy differences between the bound states and a strong interaction of the bound electron with the lattice create an environment where MP transitions are likely to dominate over other processes. Indeed, the first examples of such multiphonon transitions were encountered in alkali halides [6.1.2]. The interaction

---

* This chapter has been written by T. Markvart, Dept. of Engineering Materials, University of
  Southampton.

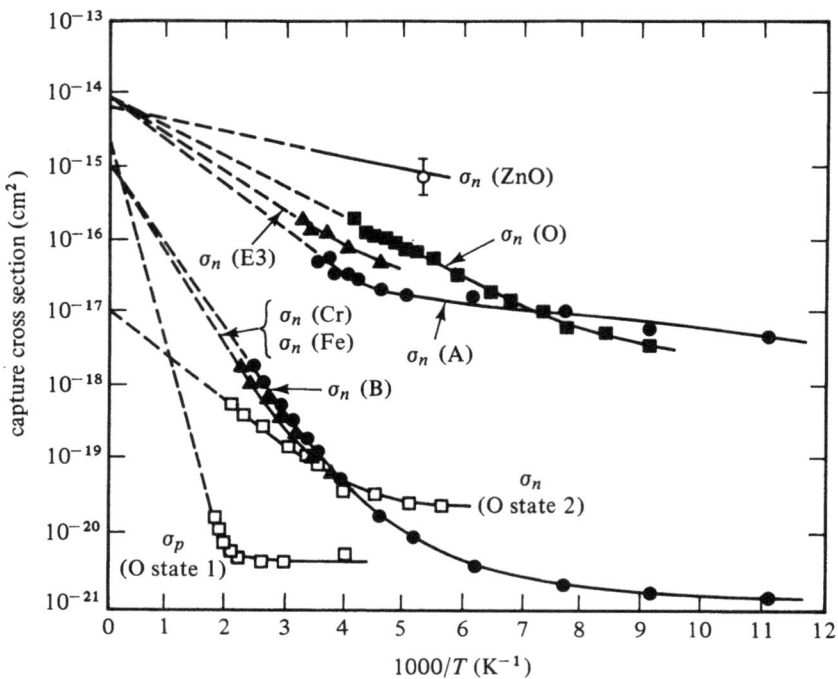

Fig. 6.1.1. The temperature dependence of the capture cross section for various defects in GaAs (filled-in symbols) and GaP (open symbols) [6.1.6].

with the lattice may here be strong enough to trap a carrier, particularly a hole, in a potential well of its own creation. The large amount of energy released into the lattice when the hole recombines with an electron can then result in the displacement of atoms from their lattice position [6.1.3].

Deep levels which lie close to the center of the band gap in covalent semiconductors are of a different origin. They are the bound states of electrons in a short-range potential which is produced by many defects or impurities whose atomic structure differs substantially from the structure of atoms which comprise the host crystal. The theory of deep-level defects is complicated and the reader is referred to the excellent monographs [6.1.4], [6.1.5]. These deep-level defects provide a favorable site for multiphonon recombination in covalent semiconductors. Beside the fact that a deep level forms an effective stepping stone for the transitions across the band gap, these transitions are here assisted by the strong interaction of the bound electron with the lattice motion. The deep electron states represent tightly bound orbitals with linear dimension of a few lattice constants. Since the high electron density associated with the bound electron exerts a large influence on the nearby lattice atoms, the capture or emission of an electron by the defect may substantially modify the vibrational environment near the defect.

Experimental evidence identifying MP transitions has accumulated principally since the mid-1970s by virtue of modern experimental techniques, such as deep level transient spectroscopy [6.1.6]. Multiphonon processes in ionic crystals, however, were observed much earlier, and the foundation for theoretical framework is usually attributed to Huang and Rhys [6.1.2]. MP transitions display a strong dependence on electronic energy which must be dissipated to the lattice, and a characteristic temperature dependence (Fig. 6.1.1). The explanation of these points will be the central theme of this chapter which deals with theory as used for ionic and covalent semiconductors.

6.2 Electron–lattice interaction in semiconductors

6.2.1 Electron states in the presence of electron–lattice interaction

In this section we shall introduce the configuration coordinate (CC) diagram, where a single mode of angular frequency ω is assumed relevant. For initial orientation two potential energy curves are adequate:

$$E_i(Q) = E_i(Q_i) + \tfrac{1}{2}\omega^2(Q - Q_i)^2$$

The ground state $i = 1$ and excited state $i = 2$ are represented by parabolas of equal frequencies but displaced by

$$Q_0 \equiv Q_2 - Q_1 \tag{6.2.1}$$

as they correspond to different equilibrium positions Q_1 and Q_2. The thermal excitation or de-excitation energy is then

$$\Delta E \equiv E_2(Q_2) - E_1(Q_1) \equiv p\hbar\omega \tag{6.2.2}$$

where p denotes the *number of phonons* emitted or absorbed in the multiphonon thermal transition between equilibrium states. The optical absorption transition corresponds to an energy

$$E_2(Q_1) - E_1(Q_1) \equiv p\hbar\omega + \tfrac{1}{2}\omega^2 Q_0^2 = (p + S)\hbar\omega$$

It exceeds the thermal energy and implies a *Stokes shift*

$$\tfrac{1}{2}\omega^2 Q_0^2 \equiv S\hbar\omega \tag{6.2.3}$$

The *Huang–Rhys* factor S characterizes the strength of the electron–phonon interaction which is represented here by the relative displacement of the parabolae (Fig. 6.2.1). For $S \gg 1$, the coupling is strong and for $S \ll 1$, it is weak.

In actual systems the one-dimensional parabolae are replaced by multidimensional (and more complicated) energy surfaces, and one may require more than

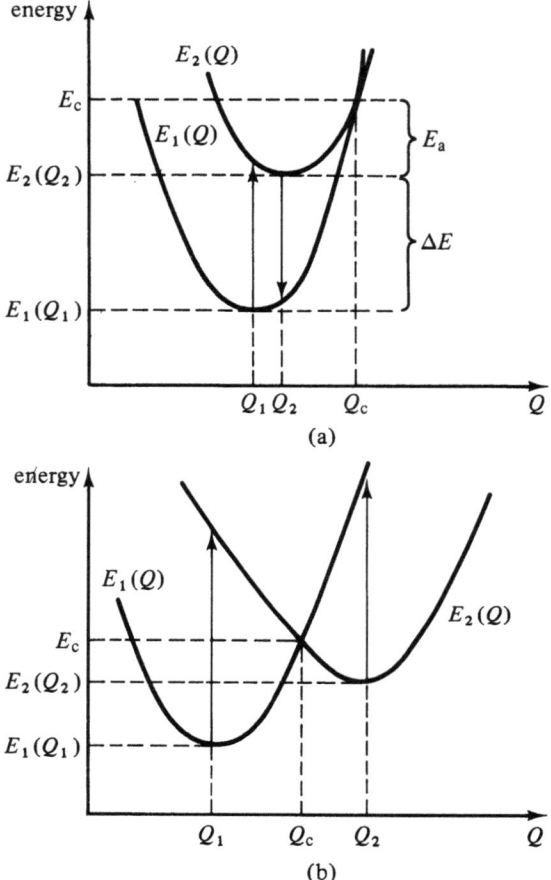

Fig. 6.2.1. Configuration coordinate diagrams for (a) weak and (b) strong interaction. Arrows indicate energy absorption or emission processes. An upward arrow can represent optical absorption and/or impact excitation or ionization. A downward arrow can represent the opposite process: emission or Auger de-excitation or recombination.

two of these. For the first generalization see [6.2.1] and for the second generalization see [6.2.2]. For a general survey see [6.2.3].

Turning to a more general theory, we shall use the following notation: $\mathbf{r}$ is the electron coordinate, the vector $\mathbf{q}$ denotes the set of normal coordinates q_k (k labels here the wavevector and polarization of the mode), $T_{el}(\mathbf{r})$ is the electron kinetic energy operator,

$$T_{el}(\mathbf{r}) = -\frac{\hbar^2}{2m}\nabla_r^2 \tag{6.2.4}$$

$U_{el}(\mathbf{r}, \mathbf{q})$ is the electron potential energy, $U_1(\mathbf{q})$ is the lattice potential energy which, in the harmonic approximation, can be written as

$$U_1(\mathbf{q}) = \tfrac{1}{2} \sum_k \omega(k)^2 q_k^2 \qquad (6.2.5)$$

where $\omega(k)$ are the frequencies of the normal modes. Note that, in accord with the usual definition of the normal modes, the mass factor is absorbed into the q's which therefore have the dimension $\text{mass}^{\frac{1}{2}} \times \text{length}$. $T_1(\mathbf{q})$ is the lattice kinetic energy

$$T_1(\mathbf{q}) = \tfrac{1}{2} \sum_k p_k^2 \qquad (6.2.6)$$

where p_k are conjugate momenta to the normal coordinates q_k. We also assume that the electron Hamiltonian $H_{el}(\mathbf{r}, \mathbf{q})$ varies slowly with the lattice coordinates $\mathbf{q}$ so that the first term in its Taylor series suffices.

The Hamiltonian of the electron–lattice system near a defect can be written in the form

$$
\begin{aligned}
H(\mathbf{r}, \mathbf{q}) &\equiv H_{el}(\mathbf{r}, \mathbf{q}) + H_{vib}(\mathbf{q}) \qquad (6.2.7) \\
&= T_{el}(\mathbf{r}) + U_{el}(\mathbf{r}, \mathbf{q}) + T_i(\mathbf{q}) + U_1(\mathbf{q}) \\
&\approx T_{el}(\mathbf{r}) + U_{el}(\mathbf{r}, 0) + \sum_k \frac{\partial U_{el}(\mathbf{r}, \mathbf{q} = 0)}{\partial q_k} q_k + T_i(\mathbf{q}) + U_1(\mathbf{q}) \\
&= H_{el}(\mathbf{r}, \mathbf{q} = 0) + H_{int}(\mathbf{r}, \mathbf{q}) + T_i(\mathbf{q}) + U_1(\mathbf{q}) \qquad (6.2.8)
\end{aligned}
$$

Let us suppose that an electron is captured in a bound state at the defect with wavefunction $\varphi_n(\mathbf{r})$ which is taken as an eigenfunction of $H_{el}(\mathbf{r}, \mathbf{q} = 0)$ and is therefore independent of the lattice coordinates. For a stationary lattice which reduces $T_1(\mathbf{q})$ to zero, the total energy as a function of the lattice coordinates is then obtained as

$$
\begin{aligned}
E_n(\mathbf{q}) &\equiv \int \varphi_n^*(\mathbf{r}) [H(\mathbf{r}, \mathbf{q}) - T_1(\mathbf{q})] \varphi_n(\mathbf{r}) \, d\mathbf{r} \\
&= \int \varphi_n^*(\mathbf{r}) H_{int}(\mathbf{r}, \mathbf{q}) \varphi_n(\mathbf{r}) \, d\mathbf{r} + U_1(\mathbf{q}) - E_{n0} \qquad (6.2.9)
\end{aligned}
$$

The sum of the first two terms is the lattice potential energy with allowance for the electron–lattice interaction when the state φ_n is occupied by an electron, and

$$E_{n0} = - \int \varphi_n^*(\mathbf{r}) H_{el}(\mathbf{r}, \mathbf{q} = 0) \varphi_n(\mathbf{r}) \, d\mathbf{r} \qquad (6.2.10)$$

represents the electron binding energy in the absence of interaction with the lattice. We may now write

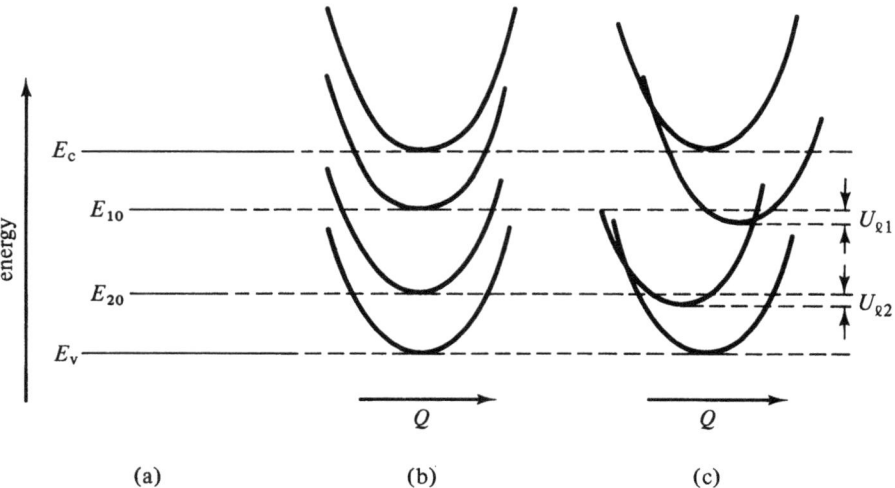

Fig. 6.2.2. The construction of a CC diagram for a defect with two levels in the band gap. The electron energy levels are shown schematically in (a). (b) includes also the vibrational potential energy which is added to the electron energies (a), but does not include electron–lattice interaction. This interaction is shown in (c): the parabolae which correspond to the localized states are shifted along the Q-axis, and their equilibrium energy is lowered by the energy (6.2.14).

$$E_n(\mathbf{q}) = \sum_k [\tfrac{1}{2}\omega(k)^2 q_k^2 + A_n(k)\, q_k] - E_{n0}$$

$$= \sum_k \{\tfrac{1}{2}\omega(k)^2 (q_k - q_{nk})^2\} - U_{1n} - E_{n0} \qquad (6.2.11)$$

where

$$A_n(k) \equiv \int \varphi_n^*(\mathbf{r}) \frac{\partial U_{el}(\mathbf{r}, \mathbf{q} = 0)}{\partial q_k} \varphi_n(\mathbf{r})\, d\mathbf{r} \qquad (6.2.12)$$

$$q_{nk} \equiv -A_n(k)/\omega(k)^2 \qquad (6.2.13)$$

$$U_{1n} \equiv \tfrac{1}{2} \sum \omega(k)^2 q_{nk}^2 = \tfrac{1}{2} \sum A_n(k)^2/\omega(k)^2 \qquad (6.2.14)$$

It is seen that the lattice motion remains a superposition of simple harmonic oscillators but the equilibrium positions are shifted to q_{nk}, and the total energy in lattice equilibrium is lowered by U_{1n} which represents the contribution to the total binding energy from the electron–lattice interaction.

For a defect with two levels in the band gap, say, the electron–phonon interaction can thus be depicted as follows. The electron energy levels are usually represented by the diagram shown in Fig. 6.2.2(a). Including the lattice potential energy gives rise to the CC diagram in Fig. 6.2.2(b, c); only the two defect states and the band edges are shown. In Fig. 6.2.2(b), there is no interaction between the electron and the lattice, but this interaction is present in Fig. 6.2.2(c). It follows from eq. (6.2.8) that if the electron–lattice interaction is linear in the lattice coordinate Q and the vibrations are harmonic, the potential curves will be parabolae with identical frequencies but different equilibrium positions.

The analysis of electron transitions between defect states is facilitated by the vast difference of the electron and nuclear masses. It is usually fair to say that the lattice cannot readjust to sudden changes in electron distribution during an electron transition, and the nuclear coordinates remain practically unchanged. This is a statement of the Franck–Condon principle [6.2.4]–[6.2.6]. Optical and Auger transitions at a defect can thus be shown by vertical arrows joining two points on the energy curves of the CC diagram in the initial and final electron state, and the length of the arrow on the energy scale measures the energy absorbed or emitted in the transition. For a defect in thermal equilibrium, these transitions usually originate near the vibrational equilibrium position Q_1 or Q_2 of the initial electron state (Fig. 6.2.1).

To picture the MP transition we must note that the lattice-independent electron states $\varphi_1(\mathbf{r})$ and $\varphi_2(\mathbf{r})$ provide only an approximate basis. The neglected terms in the Hamiltonian act as a perturbation which induce transitions between these states. These terms are the off-diagonal matrix element of the interaction Hamiltonian H_{int} in eq. (6.2.7) which can be written as

$$V_{nm}(\mathbf{q}) = \int \varphi_n^*(\mathbf{r}) \, H_{\text{int}}(\mathbf{r}, \mathbf{q}) \, \varphi_m(\mathbf{r}) \, d\mathbf{r} = \sum_k B_{nm}(k) \, q_k \qquad (6.2.15)$$

where

$$B_{nm}(k) = \int \varphi_n^*(\mathbf{r}) \frac{\partial U_{\text{el}}(\mathbf{r}, \mathbf{q} = 0)}{\partial q_k} \varphi_m(\mathbf{r}) \, d\mathbf{r} = B_{mn}^*(k) \qquad (6.2.16)$$

In the CC description, eq. (6.2.15) will be written as

$$V_{nm}(Q) = \int \varphi_n^*(r) \, H_{\text{int}}(\mathbf{r}, Q) \, \varphi_m(r) \, d\mathbf{r} \qquad (6.2.17)$$

It is the perturbation (6.2.15) or (6.2.17) that induces the MP transitions. In contrast with the optical and Auger transitions, this perturbation does not couple the defect to any external energy source and, by the Franck–Condon principle, the MP transitions therefore can take place only at the crossing point Q_c of the potential curves, as this corresponds to an arrow of zero length. The energy E_c of the crossing point Q_c is

$$E_c = E_1(Q_c) = E_2(Q_c) = E_2(Q_2) + E_a = E_1(Q_1) + \Delta E + E_a \qquad (6.2.18)$$

where

$$E_a = \frac{(\Delta E - \frac{1}{2}\omega^2 Q_0^2)^2}{4\frac{1}{2}\omega^2 Q_0^2} = \hbar\omega \frac{(p-S)^2}{4S} \tag{6.2.18'}$$

At the moment of the transition from state φ_2 to φ_1 at configuration coordinate Q_c, the lattice potential energy changes from

$$E_2(Q_c) - E_2(Q_2) = E_a$$

to

$$E_1(Q_c) - E_1(Q_1) = \Delta E + E_a$$

while the kinetic energy remains the same on account of momentum conservation. Thus, the energy ΔE is transferred to the lattice. For this reason, the vibrational mode depicted by the two parabolae of Fig. 6.2.1 is called the *accepting mode*. In more general terms, accepting modes are those where the equilibrium lattice positions change with electron state.

6.2.2 Electron–lattice interaction in polar semiconductors

Before turning to ionic, or polar, semiconductors a brief word is in order regarding covalent semiconductors where the electron–lattice interaction H_{int} is dominated by short-range forces. This interaction is often modeled in the deformation-potential approximation [6.2.7], and the electron–lattice interaction energy is described in terms of the bandshift produced by external pressure. The approximations involved in deriving the deformation potentials, however, are suitable mainly for states which extend over a relatively large region of the crystal, and make the deformation potential interaction an unlikely candidate for a mechanism of multiphonon recombination. Electrons in states deep in the band gap interact with the lattice through the lattice dependence of the potential which binds the electron to the defect. A detailed understanding of this mechanism can be gained only by large-scale calculations [6.2.8] which are outside the scope of this book, but a model for this interaction will be described in section 6.5.

In ionic semiconductors, the electron–lattice interaction is due to the Coulomb interaction between the lattice polarization and the electric field of the charge carriers, and this interaction will now be described in some detail. The optical vibrations induce dielectric polarization (dipole moment per unit volume) $\mathbf{P}$ which corresponds to the charge density

$$\rho(\mathbf{r}) = -\operatorname{div}\mathbf{P}$$

This charge density interacts with the Coulomb potential of the charge carrier $\varphi(\mathbf{r})$. From electrostatics, the interaction energy can be written in the equivalent forms

$$H_{\text{int}} = \int \rho(\mathbf{r})\,\varphi(\mathbf{r})\,d\mathbf{r}$$

$$= -\int \varphi(\mathbf{r})\,\text{div}\,\mathbf{P}(\mathbf{r})\,d\mathbf{r} \tag{6.2.19}$$

$$= -\int \mathscr{E}(\mathbf{r})\cdot\mathbf{P}(\mathbf{r})\,d\mathbf{r} \tag{6.2.20}$$

where

$$\mathscr{E} = -\text{grad}\,\varphi$$

is the electric field produced by the charge carrier. Note that the potential φ and electric field $\mathscr{E}$ do not include lattice polarization, but include the electronic polarization of the ions which can be described by the high-frequency dielectric constant ε_{∞}, say. It applies to frequencies high enough so that the lattice cannot follow the external perturbation.

We shall consider, for simplicity, only crystals with two ions in the unit cell where the polarization in the nth unit cell with position vector $\mathbf{R}_n$ is proportional to the relative displacement $\mathbf{u}(\mathbf{R}_n)$ of the ions

$$\mathbf{u}(\mathbf{R}_n) = \mathbf{u}_+(\mathbf{R}_n) - \mathbf{u}_-(\mathbf{R}_n)$$

where $\mathbf{u}_+(\mathbf{R}_n)$ and $\mathbf{u}_-(\mathbf{R}_n)$ are the displacements of the positive and negative ions, respectively. Taking $\mathbf{u}$ as a function of a continuous variable $\mathbf{r}$, we can therefore write

$$\mathbf{P}(\mathbf{r}) = \varepsilon_{\infty}\,\gamma\omega_{1}(M/\Omega)^{\frac{1}{2}}\mathbf{u}(\mathbf{r}) \tag{6.2.21}$$

where the proportionality constant is well known from the theory of vibrations in ionic crystals [6.2.9]. Here, ω_1 is the frequency of longitudinal vibrations, M is the reduced mass of the two ions, Ω is the volume of the unit cell and

$$\gamma = \left\{\frac{1}{4\pi}\left[\frac{1}{\varepsilon_{\infty}} - \frac{1}{\varepsilon_0}\right]\right\}^{\frac{1}{2}} \tag{6.2.22}$$

where ε_0 is the static dielectric constant of the crystal.

By neglecting dispersion, the longitudinal frequency in eq. (6.2.21) was assumed independent of the wavevector k and, in this approximation, the vibrational potential energy summed over all oscillators, one per unit cell, has the form

$$U_1 = \tfrac{1}{2}M\int\frac{d\mathbf{r}}{\Omega}\{\omega_1^2\,u_1^2(\mathbf{r}) + \omega_t^2\,u_t^2(\mathbf{r})\} \tag{6.2.23}$$

The lattice field $\mathbf{u}(\mathbf{r})$ is here decomposed into the longitudinal and transverse components,

$$\mathbf{u}(\mathbf{r}) = \mathbf{u}_1(\mathbf{r}) + \mathbf{u}_t(\mathbf{r}) \tag{6.2.24}$$

where $\mathbf{u}_1(\mathbf{r})$ and $\mathbf{u}_t(\mathbf{r})$ satisfy

$$\text{rot } \mathbf{u}_1(\mathbf{r}) = 0 \tag{6.2.25}$$

$$\text{div } \mathbf{u}_t(\mathbf{r}) = 0 \tag{6.2.26}$$

By virtue of eqs. (6.2.26), (6.2.21) and (6.2.19), only the longitudinal component interacts with the charge carrier. Moreover, from eq. (6.2.20), only the component of the lattice displacement $\mathbf{u}_1$ along $\mathscr{E}(\mathbf{r})$ contributes to the interaction. This component can be written in the form

$$\mathbf{u}_1(\mathbf{r}) = Q\mathscr{E}(\mathbf{r})(\varepsilon_\infty/\alpha^{\frac{1}{2}})(\Omega/M)^{\frac{1}{2}} \tag{6.2.27}$$

where

$$\alpha \equiv \varepsilon_\infty^2 \int \mathscr{E}^2(\mathbf{r}) \, \mathrm{d}\mathbf{r} \tag{6.2.28}$$

Equation (6.2.27), where the factor $(1/\alpha^{\frac{1}{2}})(\Omega/M)^{\frac{1}{2}}$ was introduced for the convenience of normalization, defines the configuration coordinate Q. An electron with wavefunction $\varphi(\mathbf{r})$ represents charge density

$$\rho(\mathbf{r}) = q|\varphi(\mathbf{r})|^2 \tag{6.2.29}$$

The quantity $\alpha/(8\pi\varepsilon_\infty)$ represents the energy of the electrostatic field due to eq. (6.2.29) which is equal to the self-energy of this charge density in its own field:

$$\frac{1}{2}\int \mathrm{d}\mathbf{r} \frac{\rho(\mathbf{r})}{\varepsilon_\infty} \int \frac{\rho(\mathbf{r}')}{|\mathbf{r}-\mathbf{r}'|} \, \mathrm{d}\mathbf{r}'$$

Hence,

$$\alpha = 4\pi q^2 \iint \mathrm{d}\mathbf{r} \, \mathrm{d}\mathbf{r}' \frac{|\varphi(\mathbf{r})|^2 |\varphi(\mathbf{r}')|^2}{|\mathbf{r}-\mathbf{r}'|} \tag{6.2.30}$$

By considering specific wavefunctions, the integral is found to be inversely proportional to the linear dimension of the region occupied by φ [6.2.10]. Hence it follows that the integral (6.2.30), and thus also the shift Q_0 and the lattice relaxation energy, vanish for free electron states. However, they can be appreciable for bound electrons, and the tighter the binding, the larger the shift.

Substituting into eq. (6.2.23) and adding the interaction energy (6.2.20), the vibrational energy with an electron in the bound state is obtained in the form

$$E_1(Q) = \tfrac{1}{2}\omega_1^2 Q^2 - \gamma\omega_1\,\alpha^{\frac{1}{2}}Q = \tfrac{1}{2}\omega_1^2(Q-Q_0)^2 - U_{10}$$

where

$$Q_0 = \gamma\alpha^{\frac{1}{2}}/\omega_1 \tag{6.2.31}$$

$$U_{10} = \tfrac{1}{2}\omega_1^2 Q_0^2 = \tfrac{1}{2}\gamma^2\alpha \tag{6.2.32}$$

The potential energy $E_1(Q)$ is reckoned here from the binding energy $-E_{10}$ (6.2.10) which neglects the interaction with the lattice, and the quantities Q_0 and U_{10} correspond to eqs. (6.2.1) and (6.2.3). The parabola $E_2(Q)$ represents a free electron state for which we take $Q_2 = 0$.

For transitions between two localized states one defines the configuration coordinate Q by

$$\mathbf{u}_1(\mathbf{r}) = Q\{\mathscr{E}_1(\mathbf{r}) - \mathscr{E}_2(\mathbf{r})\}\,(\varepsilon_\infty/\alpha^{\frac{1}{2}})(\Omega/M)^{\frac{1}{2}}$$

where $\mathscr{E}_1$ and $\mathscr{E}_2$ are the electric fields produced by the charge carrier in the two electron states φ_1 and φ_2, and α is now given by

$$\alpha = \varepsilon_\infty^2 \int \{\mathscr{E}_1(\mathbf{r}) - \mathscr{E}_2(\mathbf{r})\}^2\, d\mathbf{r}$$

$$= 4\pi q^2 \iint d\mathbf{r}\, d\mathbf{r}'\, \frac{\{|\varphi_1(\mathbf{r})|^2 - |\varphi_2(\mathbf{r})|^2\}\{|\varphi_1(\mathbf{r}')|^2 - |\varphi_2(\mathbf{r}')|^2\}}{|\mathbf{r}-\mathbf{r}'|}$$

With this value of α, the relative displacement of the equilibrium position is again given by eq. (6.2.31), and the Huang–Rhys parameter can be obtained from eq. (6.2.3).

Taking now H_{int} in the form

$$H_{\mathrm{int}}(\mathbf{R}) = -\varepsilon_\infty\,\gamma\omega_1(M/\Omega)^{\frac{1}{2}} \int \frac{\mathrm{div}\,\mathbf{u}(\mathbf{r})}{|\mathbf{R}-\mathbf{r}|}\,d\mathbf{r} \tag{6.2.33}$$

where the electron coordinate is now denoted by $\mathbf{R}$ to avoid confusion and using the result

$$\mathrm{div}\,\mathscr{E}_i = (4\pi e/\varepsilon_\infty)|\varphi_i|^2 \tag{6.2.34}$$

the off-diagonal matrix element (6.2.15) can be obtained by straightforward algebra:

$$V(Q) = Q\,\frac{4\pi e^2\gamma\omega_1}{\alpha^{\frac{1}{2}}} \int \frac{\{|\varphi_1(\mathbf{r})|^2 - |\varphi_2(\mathbf{r})|^2\}\,\varphi_1^*(\mathbf{r}')\,\varphi_2(\mathbf{r}')}{|r-r'|}\,dr\,dr' \tag{6.2.35}$$

6.3 Multiphonon transition rate at high temperatures

6.3.1 Landau–Zener formula

The probability of transition at the crossing point Q_c of the potential curves which belong to discrete electron states will now be considered in detail. Fig. 6.3.1 shows a part of the CC diagram near the point Q_c. To describe the electron transition between the two electron states $\varphi_1(r)$ and $\varphi_2(r)$ as the system passes through the crossing point at some time t_c, the time evolution of the electron state will be represented by the superposition

$$\varphi(r, t) = a_1(t)\,\varphi_1(r) + a_2(t)\,\varphi_2(r) \tag{6.3.1}$$

where $a_1(t)$ and $a_2(t)$ are time-dependent coefficients. The probability P that the transition occurs in a single passage though the crossing point is then equal to

$$P = |a_2(t_t)|^2 \tag{6.3.2}$$

at some final time $t_f \gg t_c$, subject to the initial condition that, at some initial moment of time $t_{in} \ll t_c$, the system starts in the electron state φ_1,

$$a_1(t_{in}) = 1; \quad a_2(t_{in}) = 0 \tag{6.3.3}$$

Eventually, we shall take the limit $t_{in} \to -\infty$, $t_t \to +\infty$.

The coefficients $a_i(t)$ will be determined from the solution of the time-dependent Schrödinger equation for the wavefunction $\varphi(r, t)$

$$i\hbar \frac{\partial}{\partial t} \varphi(r, t) = H(r, t)\,\varphi(r, t) \tag{6.3.4}$$

Substituting from eq. (6.3.1), multiplying in turn by $\varphi_1^*(r)$ and $\varphi_2^*(r)$ and integrating over r gives, with the use of eq. (6.2.6) as applied to a single vibrational mode,

$$\begin{aligned} i\hbar \dot{a}_1(t) &= E_1(t)\,a_1(t) + V(t)\,a_2(t) \\ i\hbar \dot{a}_2(t) &= E_2(t)\,a_2(t) + V^*(t)\,a_1(t) \end{aligned} \tag{6.3.5}$$

where $V(t) \equiv V_{12}[Q(t)]$ is the matrix element (6.2.8), and the electron states are assumed to be orthogonal. It will be more convenient to define new coefficients $c_1(t)$ and $c_2(t)$ by

$$a_i(t) = c_i(t)\exp\left[-\frac{i}{\hbar}\int_{t_c}^{t} E_i(t')\,dt'\right] \quad (i = 1, 2) \tag{6.3.6}$$

In terms of c_1 and c_2, eqs. (6.3.5) become

$$\dot{c}_1(t) = -\frac{i}{\hbar}V(t)\exp\left[+\frac{i}{\hbar}\int_{t_c}^{t}\delta E(t')\,dt'\right]c_2(t) \tag{6.3.7}$$

$$\dot{c}_2(t) = -\frac{i}{\hbar}V^*(t)\exp\left[-\frac{i}{\hbar}\int_{t_c}^{t}\delta E(t')\,dt'\right]c_1(t) \tag{6.3.8}$$

where

$$\delta E(t) = E_2(t) - E_1(t) \tag{6.3.9}$$

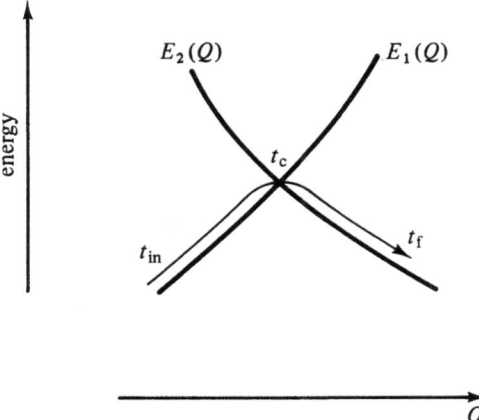

Fig. 6.3.1. The potential curves constructed in the static coupling scheme near Q_c, showing also the relevant times t_{in}, t_c and t_f.

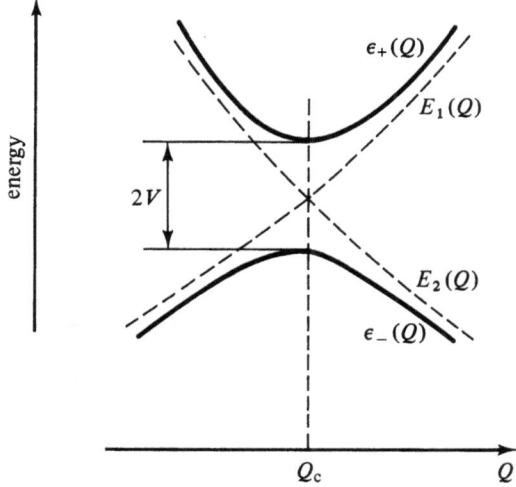

Fig. 6.3.2. Adiabatic potential curves $\varepsilon_\pm(Q)$ near Q_c (full lines). The dashed lines show the potential curves $E_1(Q)$ and $E_2(Q)$ constructed in the static coupling scheme.

We shall now restrict ourselves to the case when the matrix element V (and hence also the transfer probability P) are small, and solve eq (6.3.8) to first order in V. Replacing the right-hand sides by the constant initial values $c_1 = 1$ which, apart from an unimportant phase factor, follows from eq. (6.3.3), and integrating gives

$$c_2(\infty) = -\frac{i}{\hbar} \int_{-\infty}^{+\infty} V^*(t) \exp\left[-\frac{i}{\hbar} \int_{t_c}^{t} \delta E(t')\, dt' \right] dt \qquad (6.3.10)$$

where we took the limit $t_{in} \to -\infty$, $t_f \to +\infty$ as already indicated.

The integrand in eq. (6.3.10) oscillates rapidly except near $t = t_c$ where the phase of the exponent is stationary, and the contribution from the immediate vicinity of t_c will therefore be dominant. This proves that the transition occurs in the neighborhood of the crossing point Q_c, as we have already anticipated from the Franck–Condon principle. The integral in eq. (6.3.10) can be evaluated by the method of stationary phase (see Appendix F), and a standard application of this method gives for the probability that a transition $\varphi_1 \to \varphi_2$ occurs

$$P(v) = |a_2(\infty)|^2 = |c_2(\infty)|^2 = \frac{|V(Q_c)|^2}{\hbar} \frac{2\pi}{v|F_1 - F_2|} \qquad (6.3.11)$$

where, for future convenience, Q_c, rather than t_c, is written for the argument of the matrix element V,

$$F_i = -E_i'(Q_c) \qquad (6.3.12)$$

are the forces and

$$v = dQ/dt \qquad (6.3.13)$$

is the velocity at the crossing point Q_c. Equation (6.3.11) is the celebrated Landau–Zener result [6.3.1], [6.3.2] which is valid as long as $P \ll 1$; in other words, when the crossing point is traversed with large velocity v or when $V(Q_c)$ is small.

6.3.2 Relationship to the Born–Oppenheimer approximation

The standard procedure for separating the electron and lattice variables is based on the Born–Oppenheimer approximation [6.3.3], [6.3.4]. We shall now show how this approach relates to the procedure used in section 6.2 which is usually called the *static coupling scheme* [6.3.5], [6.3.6].

The Born–Oppenheimer method is applicable to situations where the lattice velocities, and hence also the lattice kinetic energy, are small. As $v \to 0$, the Born–Oppenheimer method becomes exact. The appropriate starting point is therefore the Hamiltonian $H(\mathbf{r}, Q)$ (6.2.7) where the lattice kinetic energy is omitted. This Hamiltonian will be denoted by $H_{ad}(\mathbf{r}, Q)$:

$$H_{ad}(\mathbf{r}, Q) = H(\mathbf{r}, \mathbf{q}) - T_l(\mathbf{q}) = H_{el}(\mathbf{r}, Q) + U_l(Q)$$

The electron states $\psi_n(r, Q)$, which are taken as the eigenstates of $H_{ad}(\mathbf{r}, Q)$, depend on the lattice coordinates. The corresponding eigenvalues $\varepsilon_n(Q)$ are then used as the potential energies for the vibrational motion. In the present setting where only two electron levels enter the analysis, these states and energies are easily determined. In the basis formed by the wavefunctions $\varphi_1(r)$ and $\varphi_2(r)$, the Hamiltonian $H_{ad}(\mathbf{r}, Q)$ becomes the 2×2 matrix

$$H_{ad}(\mathbf{r}, Q) = \begin{bmatrix} E_1(Q) & V(Q) \\ V^*(Q) & E_2(Q) \end{bmatrix} \qquad (6.3.14)$$

which can easily be diagonalized. Assuming for simplicity that V is real and constant we obtain the following eigenvalues:

$$\varepsilon_{\pm}(Q) = \tfrac{1}{2}(E_1(Q) + E_2(Q)) \pm \tfrac{1}{2}\{(E_1(Q) - E_2(Q))^2 + 4V^2\}^{\frac{1}{2}} \qquad (6.3.15)$$

The new potential curves $\varepsilon_{\pm}(Q)$ have the form shown in Fig. 6.3.2. It is seen that the intersection of the curves $E_1(Q)$ and $E_2(Q)$ is removed. The minimum separation between the Born–Oppenheimer energy curves $\varepsilon_+(Q)$ and $\varepsilon_-(Q)$ (which is reached at Q_c) is equal to $2V$.

To obtain the eigenstates, it is convenient to define

$$\tan[2\theta(Q)] \equiv 2V/(E_1(Q) - E_2(Q)) \qquad (6.3.16)$$

which gives the eigenstates of matrix (6.3.14) in the form

$$\psi_+(r, Q) = \varphi_1(r) \sin \theta(Q) + \varphi_2(r) \cos \theta(Q) \qquad (6.3.17)$$

$$\psi_-(r, Q) = \varphi_1(r) \cos \theta(Q) - \varphi_2(r) \sin \theta(Q) \qquad (6.3.18)$$

If V is small, $\theta(Q)$ varies from $\theta \approx 0$ for $Q \ll Q_c$ to $\theta \approx \pi/2$ for $Q \gg Q_c$. The electron state $\psi_+(r, Q)$ therefore changes smoothly from φ_2 to φ_1 and ψ_- changes from φ_1 to $-\varphi_2$ as Q passes through the point Q_c. If the passage through the crossing point Q_c is infinitesimally slow, the states $\psi_{\pm}$ are exact and the defect passes from one state φ_1 or φ_2 to the other with probability $P = 1$. Thus, the adiabatic states $\psi_{\pm}$ and the states φ_i $(i = 1, 2)$ which are constructed in the static coupling scheme represent accurate approximations in the limits of small and large velocities v, respectively. This is reflected in the full Zener result [6.3.1] for the probability $P(v)$ which is reproduced here without derivation:

$$P(v) = 1 - \exp\left[-\frac{V^2}{\hbar}\frac{2\pi}{v|F_1 - F_2|}\right] \qquad (6.3.19)$$

Expression (6.3.19) shows that the probability of transition between the adiabatic states, equal to $1 - P(v)$, is exponentially small for small v. However, when the velocity v is sufficiently large,

$$v \gg \frac{V^2}{\hbar}\frac{2\pi}{|F_1 - F_2|} \qquad (6.3.20)$$

$P \ll 1$, and in this limit it is more appropriate to construct the electron states in the static coupling scheme. It is in this limit that most multiphonon transitions take place.

6.3.3 Multiphonon transitions at high temperatures

Using expression (6.3.11), the transition rate per unit time from a state with vibrational energy E is given by

$$W(E) = \frac{\omega}{\pi} P(E) = \frac{\omega}{\pi} \frac{|V(Q_c)|^2}{\hbar} \frac{2\pi}{|F_1 - F_2|[2(E - E_c)]^{\frac{1}{2}}}$$

$$= \frac{|V(Q_c)|^2}{\hbar[S\hbar\omega(E - E_c)]^{\frac{1}{2}}} \qquad (6.3.21)$$

where we noted that the crossing point is passed through with frequency ω/π (twice in a period of vibration), $[2(E - E_c)]^{\frac{1}{2}}$ was substituted for the velocity v, and we used the result

$$|F_1 - F_2| = \omega^2 Q_0 = \omega(2S\hbar\omega)^{\frac{1}{2}} \qquad (6.3.22)$$

which follows from eq. (6.3.12) and is valid for linear electron–lattice interaction. The transition rate at temperature T is then given by the thermal average

$$W(T) = \sum_n p(E_n) W(E_n) \qquad (6.3.23)$$

with the probability distribution

$$p(E) = (1/Z) \exp[-\{E - E_i(Q_i)\}/kT] \qquad (6.3.24)$$

for transitions from state i ($i = 1, 2$). In eq. (6.3.24),

$$Z = \sum \exp[-(n + \tfrac{1}{2})\hbar\omega/kT] = [2\sinh(\tfrac{1}{2}\hbar\omega/kT)]^{-1} \approx kT/\hbar\omega \qquad (6.3.25)$$

The approximate equality holds at high temperature where the present result finds application. Replacing the sum in eq. (6.3.23) by an integral according to the recipe

$$\sum_n \to \int \frac{dE}{\hbar\omega} \qquad (6.3.26)$$

we obtain

$$W(T) = \frac{|V|^2}{\hbar(S\hbar\omega)^{\frac{1}{2}}} \frac{\hbar\omega}{kT} \int_{E_c}^{\infty} \frac{dE}{\hbar\omega} \frac{1}{(E - E_c)^{\frac{1}{2}}} e^{-\{E - E_i(Q_i)\}/kT}$$

$$= \frac{|V(Q_c)|^2}{\hbar} \left[\frac{\pi}{S\hbar\omega kT} \right]^{\frac{1}{2}} \exp\left[-\frac{E_a}{kT} \right] \quad \text{for transitions } \varphi_2 \to \varphi_1 \quad (6.3.27a)$$

$$= \frac{|V(Q_c)|^2}{\hbar} \left[\frac{\pi}{S\hbar\omega kT} \right]^{\frac{1}{2}} \exp\left[-\frac{\Delta E + E_a}{kT} \right] \quad \text{for transitions } \varphi_1 \to \varphi_2$$

$$(6.3.27b)$$

where E_a is given by eq. (6.2.15) and we used the result [6.3.7]

$$\int_0^\infty \frac{e^{-x}}{x^{\frac{1}{2}}} dx = \pi^{\frac{1}{2}} \tag{6.3.28}$$

The integration in eq. (6.3.27) is performed from E_c to ∞ since only the vibrational states with energy in excess of E_c extend through the crossing point Q_c. These states are appreciably populated only at high temperatures. Another mechanism prevails at low temperature, and will be discussed in section 6.4. Equation (6.3.27) therefore gives the high-temperature multiphonon transition rate in the limit of small V. Note that the two transition rates $\varphi_2 \rightarrow \varphi_1$ and $\varphi_1 \rightarrow \varphi_2$ satisfy the detailed balance

$$W(1 \rightarrow 2)/W(2 \rightarrow 1) = \exp(-\Delta E/kT) \tag{6.3.29}$$

6.4 Multiphonon transitions by tunneling

At low temperatures, the multiphonon transition rate $W(T)$ (6.3.27) is small. The reason is simple to see: the crossing point Q_c can be reached only in vibrational states with energy $E > E_c$ which, at low temperatures, have a small probability of thermal occupation. Although, within the realm of classical mechanics, the crossing point is inaccessible for vibrational states with $E < E_c$, their quantum-mechanical wavefunction has a small but finite amplitude at Q_c where the electron transition can take place, in accord with the Franck–Condon principle. We shall say that the transitions then occur by tunneling. In the WKB approximation [6.4.1], the vibrational wavefunction χ_i for a state with $E < E_c$ decays into the classically inaccessible segment between Q_{t1} and Q_{t2} (Fig. 6.4.1b), where it is given by

$$\chi_i(Q) = \left[\frac{\omega}{2\pi\hbar\kappa_i(Q)}\right]^{\frac{1}{2}} \exp\left[-\left|\int_{Q_{ti}}^Q \kappa_i(Q') dQ'\right|\right] \quad (i = 1, 2) \tag{6.4.1}$$

where

$$\kappa_i(Q) = (\sqrt{2}/\hbar)[E_i(Q) - E]^{\frac{1}{2}} \tag{6.4.2}$$

The absolute value in the exponent of eq. (6.4.1) ensures that the wavefunction decays into the classically inaccessible segment regardless whether this is to the right or to the left of the turning point Q_{ti}.

If the matrix element V_{12} is small, the multiphonon transition rate can be calculated by the Golden Rule (3.2.14)

$$W = (2\pi/\hbar)|M|^2 \mathcal{N}_f \tag{6.4.3}$$

where $\mathcal{N}_f$ is the density of final vibrational states. The matrix element M in eq. (6.4.3) is calculated using the total wavefunctions which are equal to the products

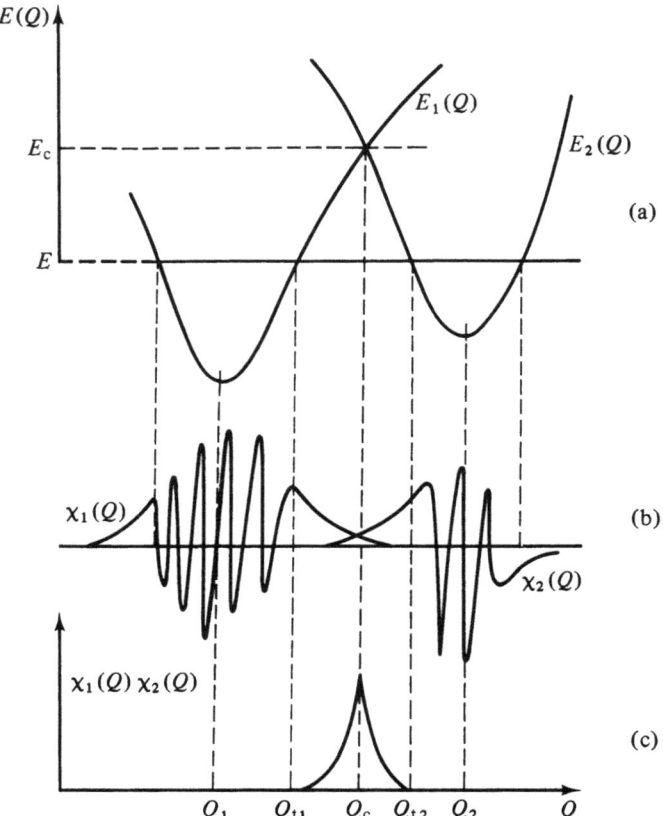

Fig. 6.4.1. (a) The potential energies $E_1(Q)$ and $E_2(Q)$ for the vibrational motion. (b) The vibrational wavefunctions χ_1 and χ_2, and (c) the overlap $\chi_1\chi_2$. This figure is only schematic and χ_1 and χ_2 are assumed real.

of the electron wavefunction $\varphi_i(r)$ $(i = 1, 2)$, with the appropriate vibrational wavefunction $\chi_i(Q)$:

$$M = \int_{-\infty}^{\infty} \mathrm{d}Q \int \mathrm{d}\mathbf{r}\, \chi_1^*(Q)\, \varphi_1^*(\mathbf{r})\, H_{\mathrm{int}}(\mathbf{r}, Q)\, \chi_2(Q)\, \varphi_2(\mathbf{r})$$

$$= \int_{-\infty}^{\infty} \chi_1^*(Q)\, V(Q)\, \chi_2(Q)\, \mathrm{d}Q \qquad\qquad (6.4.4)$$

where $V(Q) \equiv V_{12}(Q)$ is the matrix element (6.2.17).

We consider here, for clarity, the strong coupling case $p < S$ where the classically inaccessible region (which extends from Q_{t1} to Q_{t2} and includes Q_c) separates the equilibrium points Q_1 and Q_2. The obtained result, however, is valid also for weak coupling when $p > S$. Fig. 6.4.1(c) shows the vibrational overlap $\chi_1^*(Q)\chi_2(Q)$

which enters eq. (6.4.4). It is seen that the overlap peaks sharply at the point Q_c, as will presently be verified. The vibrational wavefunction can therefore be approximated by the form (6.4.1) in the classically inaccessible region:

$$M = \frac{\omega}{2\pi\hbar} e^{-\Phi} \int_{-\infty}^{\infty} \frac{V(Q)}{[\kappa_1(Q)\,\kappa_2(Q)]^{\frac{1}{2}}} \exp\left[\int_{Q_c}^{Q} [\kappa_2(Q') - \kappa_1(Q')]\,dQ'\right] dQ \qquad (6.4.5)$$

where

$$\Phi(E) = \int_{Q_{t1}}^{Q_c} \kappa_1(Q)\,dQ + \int_{Q_c}^{Q_{t2}} \kappa_2(Q)\,dQ \qquad (6.4.6)$$

It is seen that, indeed, the exponent in the integral of eq. (6.4.5) has a maximum at the crossing point Q_c, and a small neighborhood of Q_c will therefore give the dominant contribution to the integral. This verifies that the Franck–Condon principle applies also to the tunneling transitions. Including only the contribution from the maximum at Q_c, the integral can be evaluated by the saddle-point method (see Appendix F). The result is

$$M = \omega \frac{V(Q_c)}{[2\pi|F_1 - F_2|]^{\frac{1}{2}}} \left[\frac{\hbar^2}{2(E_c - E)}\right]^{\frac{1}{4}} \exp[-\Phi(E)] \qquad (6.4.7)$$

The transition rate from state φ_2 to state φ_1, say, is now obtained from eq. (6.4.3) by using the density of vibrational states in the form

$$\mathcal{N}_f = 1/\hbar\omega \qquad (6.4.8)$$

Thus,

$$W(E) = (\omega/2\pi)\,P(E)\,e^{-2\Phi(E)} \qquad (6.4.9)$$

where

$$P(E) = \frac{2\pi|V(Q_c)|^2}{\hbar|F_1 - F_2|} \left[\frac{1}{2(E_c - E)}\right]^{\frac{1}{2}} \qquad (6.4.10)$$

Equation (6.4.9) is easy to understand. The first factor is the incidence frequency. The factor P can be interpreted as the probability (6.3.11) that the electron makes a transition at the crossing point Q_c of the potential curves. Since here, however, the energy E lies below the top of the barrier, the velocity is imaginary and was replaced by its absolute value $[2(E_c - E)]^{\frac{1}{2}}$. The last factor in eq. (6.4.9) will be recognized as a typical tunneling factor of the WKB theory.

The tunneling rate at a defect in thermal equilibrium at temperature T is an average of eq. (6.4.9) over the thermal population of vibrational states. Let us assume, for simplicity, that the distribution of vibrational states is continuous. The density of these states is equal to $\hbar\omega$, and the transition rate from state φ_2, say, is then

$$
\begin{aligned}
W(T) &= \int_0^{E_c} W(E)\, p(E)\, \frac{\mathrm{d}E}{\hbar\omega} \\
&= \frac{1}{2\pi\hbar Z} \int_0^{E_c} P(E)\, \mathrm{e}^{-2\Phi(E)}\, \mathrm{e}^{-\{E - E_2(Q_2)\}/kT}\, \mathrm{d}E
\end{aligned}
\tag{6.4.11}
$$

where $p(E)$ is the thermal occupancy factor (6.3.24), and $P(E)$ is given by eq. (6.4.10).

The behavior of the integrand in eq. (6.4.11) is determined principally by the two exponential factors. As the energy increases, the tunneling probability $\mathrm{e}^{-2\Phi(E)}$ increases but the Boltzmann factor decreases. The product of the two terms has a maximum at energy $E^*(T)$ where the total exponent $-2\Phi(E) - \{E - E_2(Q_2)\}/kT$ as a function of E has a maximum, i.e.

$$
\frac{\mathrm{d}}{\mathrm{d}E}[-2\Phi(E)] = \frac{1}{kT}
\tag{6.4.12}
$$

The sharp peak of the integrand indicates that, at a given temperature T, tunneling occurs almost exclusively in a small energy range near $E^*(T)$. The behavior of E^* measured from $E_2(Q_2)$ as a function of temperature can be understood qualitatively by examining the function $\Phi(E)$. At $E = E(Q_2)$, the derivative $-\Phi'(E)$ diverges to $+\infty$, and therefore $E^*(0) = 0$. As the temperature increases, $E^*(T)$ rises, which indicates that tunneling is assisted by thermal activation. At some temperature T_c, the peak of the integrand reaches the energy E_c, and, above this temperature, thermal activation over the barrier dominates over tunneling. The transition rate is then given by the high-temperature formula (6.3.27).

The integral (6.4.11) can be calculated by the saddle-point method (see Appendix F) which gives the general formula for the tunneling rate:

$$
\begin{aligned}
W(T) &= \frac{2\sinh(\hbar\omega/2kT)}{\hbar\{2\pi\Phi''(E^*)\}^{\frac{1}{2}}} P(E^*)\exp\{-2\Phi(E^*) - E^*/kT\} \\
&= \frac{\sinh(\hbar\omega/2kT)\,|V(Q_c)|^2}{\hbar^2|F_1 - F_2|} \left[\frac{2\pi}{(E_c - E^*)\,\Phi''(E^*)}\right]^{\frac{1}{2}} \exp\{-2\Phi(E^*) - E^*/kT\} \\
&= v(E^*)\, p(E^*)\, W(E^*)
\end{aligned}
\tag{6.4.13}
$$

where

$$
v = \frac{1}{\hbar\omega}\left[\frac{2\pi}{\Phi''(E^*)}\right]^{\frac{1}{2}}
$$

is the number of vibrational states which participate effectively in the transition.

For linear electron–lattice interaction the function $\Phi(E)$ can be obtained analytically:

$$\begin{aligned}\Phi(E)\hbar\omega = &-\{(E_a+\Delta E)(E_a-E)\}^{\frac{1}{2}}-(\Delta E+E)\\ &\times \text{arcosh}\,[\{(E_a+\Delta E)/(E+\Delta E)\}^{\frac{1}{2}}]\\ &-\{E_a(E_a-E)\}^{\frac{1}{2}}+E\,\text{arcosh}\,\{(E_a/E)^{\frac{1}{2}}\}\end{aligned} \tag{6.4.14}$$

where E is measured from $E_2(Q_2)$. This gives for the derivative

$$\Phi'(E)\hbar\omega = -\,\text{arcosh}\,\frac{\{E_a(E_a+\Delta E)\}^{\frac{1}{2}}-(E_a-E)}{\{E(E+\Delta E)\}^{\frac{1}{2}}} \tag{6.4.15}$$

and the solution of eq. (6.4.12) for the tunneling energy is then

$$E^* = \frac{\Delta E}{2}\left[\frac{y\cosh\theta-x}{\sinh\theta}-1\right] \tag{6.4.16}$$

where

$$\begin{aligned}\theta &= \hbar\omega/2kT\\ x &= S/(p\sinh\theta) \quad \text{(for } S<p)\\ x &= p/(S\sinh\theta) \quad \text{(for } p<S)\\ y &= (1+x^2)^{\frac{1}{2}}\end{aligned} \tag{6.4.17}$$

which gives for the transition rate at temperature T

$$W(T) = \frac{V(Q_c)^2}{\hbar^2\omega}\left[\frac{2\pi}{py}\right]^{\frac{1}{2}}\exp\left[p\left(\theta+\mu-x\cosh\theta-\log\frac{1+y}{x}\right)\right] \tag{6.4.18}$$

Although the transition rate (6.4.18) was derived for tunneling transitions, it reproduced correctly also the behavior predicted by eq. (6.3.27) in the high-temperature limit, as is shown in [6.4.2].

Another important limiting case corresponds to $T = 0$. Although the present derivation is, strictly speaking, invalid for energies near the vibrational ground state it can be shown [6.4.2] that formulae (6.4.13) and (6.4.18) still hold, giving

$$W(0) = \frac{|V(Q)|^2}{\hbar^2\omega}\left[\frac{2\pi}{p}\right]^{\frac{1}{2}}\exp\{-|p-S|-p|\log(S/p)|\} \tag{6.4.19}$$

The result (6.4.19) is known as the energy-gap law [6.4.3] (the 'gap' refers to $\Delta E = p\hbar\omega$) and, as written, is valid for strong or weak coupling.

For future reference, we shall also write down the results for the vibrational overlap factor

$$O_{12} \equiv \int \chi_1^*(Q)\, \chi_2(Q)\, \mathrm{d}Q \tag{6.4.20}$$

which can be obtained either directly or from the obtained results, with the use of eq. (6.4.3) and the Franck–Condon principle. Only the square of the modulus will be needed, and this is given by, in the linear electron–lattice interaction,

$$|O_{12}(E)|^2 = \begin{cases} \dfrac{1}{4\pi}\left[\dfrac{\hbar\omega}{S(E_c-E)}\right]^{\frac{1}{2}} \exp\left[-2\Phi(E)\right] & (E < E_c) & (6.4.21) \\[2ex] \dfrac{1}{2\pi}\left[\dfrac{\hbar\omega}{S(E-E_c)}\right]^{\frac{1}{2}} & (E > E_c) & (6.4.22) \end{cases}$$

The averaged overlap factor at temperature T is then

$$\langle |O_{12}|^2\rangle = \frac{1}{(2\pi py)^{\frac{1}{2}}} \exp\left[p\left(\theta + \mu - x\cosh\theta - \log\frac{1+y}{x}\right) \right] \tag{6.4.23}$$

As discussed for the transition rate (6.4.18), this expression holds for all temperatures.

6.5 Transitions involving free carriers. The role of the Coulomb interaction

6.5.1 Multiphonon capture and emission of free charge carriers

The expressions for the multiphonon transition rate which were derived in sections 6.3 and 6.4 apply to transitions between discrete defect levels. We shall now turn to the multiphonon capture or emission of free electrons or holes whose energy falls within the continuum of band states.

Although the vibrational energy exchange which accompanies the transition can be handled along similar lines as for localized states, the CC diagram must take into account the multitude of free states. If the vibrational energy which corresponds to a free electron is written as $\frac{1}{2}\omega^2 Q$ (we set here $Q_2 = 0$), the potential curve $E_2(Q)$ which represents the total energy of the free electron is the sum of the vibrational energy and the electron energy $E(k)$ in state $\varphi_k(\mathbf{r})$ with wavevector $\mathbf{k}$:

$$E_2(Q, k) = E(k) + \tfrac{1}{2}\omega^2 Q \tag{6.5.1}$$

There is, therefore, a continuum of curves $E_2(Q, k)$ which corresponds to the continuum of wavevectors k. All these curves, however, have a vibrational equilibrium at the same point $Q = 0$.

The principally new element when dealing with the free states comes in the treatment of the electronic matrix elements, which connects the free and bound

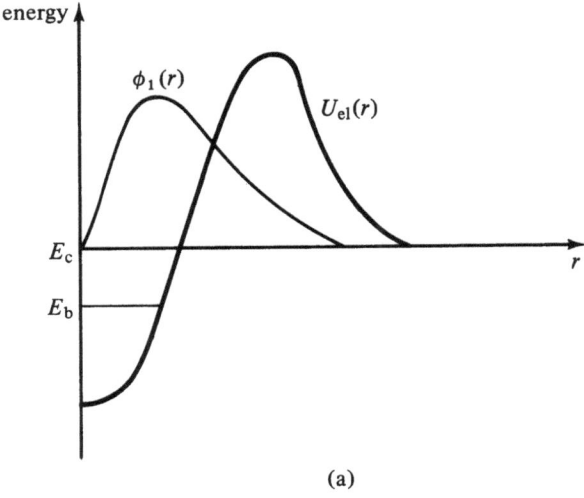

(a)

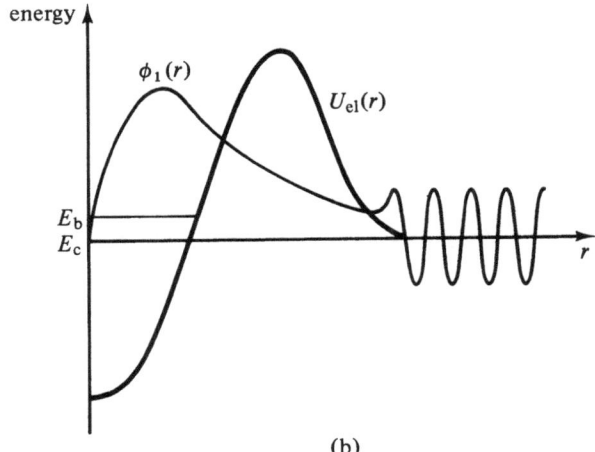

(b)

Fig. 6.5.1. The defect potential $U_{el}(\mathbf{r})$ which is used to illustrate the difference between a resonant and truly localized state at energy E_b. The wavefunctions $\varphi_1(r)$ shown in (a) and (b) correspond to a resonant state when $E_b > E_c$, and to $E_b < E_c$, where E_c denotes the energy of the conduction band edge.

state. It is known from the scattering theory [6.5.1]–[6.5.3] that a bound electron whose energy lies inside a band of allowed states in the crystal is not in a truly localized stationary state since a small perturbation will eject the electron into the band. Let us consider a defect potential $U_{el}(\mathbf{r})$ which corresponds to $U_{el}(\mathbf{r}, Q)$ of section 6.2 for a fixed Q and represents a potential well shown in Fig. 6.5.1(a) and (b). If the energy level of an electron in the well falls within the band gap, the electron cannot escape from the defect and the state will be strictly localized (Fig.

6.5.1a). A change of the defect potential, which can be brought about, for example, by the interaction with the lattice, can move the energy level into the allowed band (Fig. 6.5.1b). The bound electron can then escape from the well to infinity, with probability w per unit time, say. The wavefunction of this *resonant* state,

$$\varphi_n(\mathbf{r}, t) = \exp{(iEt/\hbar)}\,\varphi_n(r) \tag{6.5.2}$$

resembles an outgoing spherical wave outside the range of the defect potential. If $\varphi_n(\mathbf{r}, t)$ is normalized to give unit probability of finding the electron inside the well at $t = 0$, we have for the time independent part of the wavefunction far from the defect

$$\varphi_n(r) \simeq \frac{e^{i\alpha}}{r}\left[\frac{w}{4\pi v_e}\right]^{\frac{1}{2}} e^{ikr} \tag{6.5.3}$$

The wavefunction (6.5.3) corresponds to w particles escaping from a large sphere drawn around the defect per unit time. For simplicity, eq. (6.5.3) has been restricted to states with zero angular momentum, α is a phase factor which is independent of r, and v_e is the velocity of the emitted electron.

It can be shown [6.5.1], [6.5.2] that the wavefunction with the boundary condition (6.5.3) at infinity corresponds to the complex energy

$$E = E_b - i\Gamma/2 \tag{6.5.4}$$

where E_b is the energy of the bound state and $\Gamma = \hbar\omega$ is its width due to resonance. The probability of finding the particle inside the well is now

$$\int |\varphi_n(\mathbf{r}, t)|^2\,d\mathbf{r} = e^{-\Gamma t/\hbar} \tag{6.5.5}$$

where the integral extends over the interior of the well. Equation (6.5.5) shows that the resonant state has a lifetime

$$\tau = \hbar/\Gamma = 1/\omega \tag{6.5.6}$$

in accord with the definition of ω.

Assuming the applicability of perturbation theory, the emission rate w can be calculated using the Golden Rule (3.2.14):

$$w(k) = \frac{2\pi}{\hbar}|V_{nk}|^2\mathcal{N}_f \tag{6.5.7}$$

where the dependence of w on the wavevector k of the emitted electron has been shown explicitly, and the transition matrix element is given by

$$V_{nk} = \int \varphi_k^*(\mathbf{r})\,H_{int}(\mathbf{r}, Q)\,\varphi_n(\mathbf{r})\,d\mathbf{r} \tag{6.5.8}$$

The electronic wavefunctions in eq. (6.5.8) are taken within the spirit of the static coupling scheme as the eigenfunctions of the electron Hamiltonian $H_{el}(\mathbf{r}, Q)$ at the equilibrium point $Q = 0$ where the energy of the bound electron E_b is far removed from the energy of the free particle $E(k)$. The wavefunction φ_n is then strictly localized, and can be normalized to unity in the conventional manner. We assume, in addition, that the energy E_b lies far enough in the band gap for the wavefunction to be localized principally within the range of the defect potential, without a significant tail extending into the crystal. This somewhat idealized situation also ensures that the free-particle wavefunction constructed for the electron potential $H_{el}(\mathbf{r}, Q = 0)$ represents an electron elastically scattered by the defect without being affected significantly by the bound state (Fig. 6.5.2).

The free-particle wavefunction outside the range of the defect potential, normalized in a box of unit volume, has the form

$$\varphi_k(\mathbf{r}) = e^{i\mathbf{k}\cdot\mathbf{r}} + \frac{f(\theta)}{r} e^{ikr} \tag{6.5.9}$$

The first term in eq. (6.5.9) represents the incident wave, and the second term is an elastically scattered wave at an angle θ to the direction of incidence. The quantity $f(\theta)$ is the elastic scattering amplitude which is related to the differential elastic scattering cross section $d\sigma_{el}$ by

$$d\sigma_{el} = |f(\theta)|^2 \, d\Omega \tag{6.5.10}$$

where $d\Omega$ is an element of solid angle in the direction of θ.

The density of final states $\mathcal{N}_f$ is now set equal to $\frac{1}{2}\mathcal{N}(k)$, where $\mathcal{N}(k)$ is the number of free states in a unit energy range per unit volume, and the factor $\frac{1}{2}$ occurs because only states with the same direction of spin are accessible for emission:

$$\mathcal{N}_f = \frac{m^*k}{2\pi^2\hbar^2} \tag{6.5.11}$$

where the bottom of the band is assumed to lie at $\mathbf{k} = 0$, and m^* denotes the effective mass. Hence,

$$w(k) = \Gamma(k)/\hbar = \frac{m^*k}{\pi\hbar^2}|V_{nk}|^2 \tag{6.5.12}$$

Note that for small k, $w(k)$ and $\Gamma(k)$ are proportional to k on account of the density-of-states factor in eq. (6.5.7).

The transition rate of the multiphonon processes can be obtained analogously to eq. (6.5.7), but allowance must be made for the emission or absorption of vibrational energy, and for the vibrational overlap. If the vibrational energy after the emission lies in the range $E \to E + \delta E$, the emission rate (6.5.7) must be

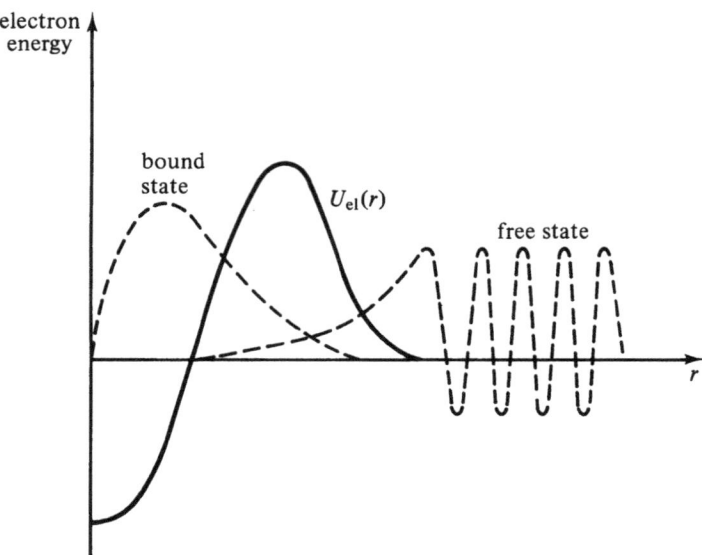

Fig. 6.5.2. The wavefunctions obtained by decomposing a resonant state into a free state φ_k and a bound state φ_n.

multiplied by the number of vibrational states in this energy range, equal to $dE/\hbar\omega$. Amending the matrix element (6.5.8) to include the total wavefunctions for the electronic as well as vibrational degrees of freedom by replacing $V_{nk} \to M_{nk}$, where

$$M_{nk} \equiv \int \chi_i(Q) \, V_{nk}(Q) \, \chi_f \, dQ = V(Q_{ck}) \, O_{if}(k, E) \qquad (6.5.13)$$

The second equality makes use of the Franck–Condon principle, as discussed in sections 6.2–6.4. In eq. (6.5.13), $O_{if}(k, E)$ is the vibrational overlap factor (6.4.20) which now depends on k through $E_2(Q, k)$, and $V(Q_{ck})$ is the electronic matrix element (6.5.8) evaluated at the crossing point Q_{ck} between the potential curve $E_2(Q, k)$ and the curve $E_1(Q)$ corresponding to the bound state. The emission rate then becomes

$$\delta W(k, E) = w(k) \, |O_{if}(k, E)|^2 \, \delta E/\hbar\omega \qquad (6.5.14)$$

where $w(k)$ denotes the emission rate (6.5.12) from the resonant state in the potential $H_{el}(\mathbf{r}, Q_{ck})$ if the lattice distortion by the bound electron is neglected.

The capture rate can be written down immediately by analogy with the transition rate between discrete states:

$$C(k, E) = \frac{2\pi}{\hbar} |M_{nk}|^2 \frac{1}{\hbar\omega} = \sigma(k, E) \, v_e \qquad (6.5.15)$$

where $\sigma(k, E)$ is the capture cross section,

$$\sigma(k, E) = \frac{2\pi^2}{k^2}|O_{\text{if}}(k, E)|^2 \frac{\Gamma(k)}{\hbar\omega} \qquad (6.5.16)$$

and $v_e = \hbar k/m^*$ is the incident velocity. Note that the capture cross section behaves as k^{-1} for small k but the capture rate tends to a k-independent constant as $k \to 0$.

Using eqs. (6.5.12) and (6.5.14)–(6.5.16), we obtain a relationship between the emission rate and capture cross section:

$$\delta W(k, E) = \tfrac{1}{2}C(k, E)\,\mathcal{N}(k)\,\delta E \qquad (6.5.17)$$

or

$$\sigma(k, E) = \frac{2\pi^2}{k^2}\hbar \frac{\delta W(k, E)}{\delta E} \qquad (6.5.18)$$

The thermally averaged total capture and emission rates per unit volume in a nondegenerate semiconductor which include the occupancy factors of the electron states are (see section 2.4.6)

$$\text{capture rate} = c_n\, n v_0 \qquad (6.5.19)$$

$$\text{emission rate} = e_n\, v_1$$

where v_0 and v_1 are the concentrations of empty and occupied defect levels, c_n and e_n are the capture and emission coefficients, and n is the electron concentration in the conduction band. Standard statistical mechanics now gives

$$c_n = \sum_{E'}\sum_k C(k, E')\,p(E')\,p_k \to \int \frac{dE'}{\hbar\omega} \int dE(k)\,\mathcal{N}(k)\,C(k, E')\,p(E')\,p_k \quad (6.5.20)$$

$$e_n = \sum_E \sigma W(k, E)\,p(E) \to \int \frac{dE}{\hbar\omega} \frac{\delta W(k, E)}{\delta E} p(E) \qquad (6.5.21)$$

where $p(E)$ is the occupation probability for a vibrational state at energy E which is given by eq. (6.3.24), and p_k is the occupation probability for an electron state with wavevector $\mathbf{k}$:

$$p_k = \exp\{(E_c - E(k))/kT)\}/N_c \qquad (6.5.22)$$

where N_c is the effective density of states per unit volume in the conduction band (1.6.6). Performing the integration, one can verify the detailed balance equation

$$c_n N_c = g\exp\{(E_b - E_c)/kT\}e_n \qquad (6.5.23)$$

where E_b refers to the defect level in vibrational equilibrium, and the degeneracy factor g for an occupied defect level is here equal to two on account of the two possible spin directions, and the degeneracy of an empty level is set equal to unity (see section 2.4).

The calculation of the emission and capture rate for a defect in vibrational equilibrium is hindered by the k-dependence of the vibrational overlap factor. How this can be tackled is indicated below, but it turns out that, to a reasonable approximation, this dependence can be neglected, at least for electrons at defects with weak coupling to the lattice. One then obtains for the capture coefficient

$$c_n = \sigma(T) v_t \tag{6.5.24}$$

$$\delta W = \frac{\Gamma(k_t)}{\hbar^2 \omega} \langle |O_{if}|^2 \rangle \, e^{-E(k)/kT} \, \delta E \tag{6.5.25}$$

$$e_n = \frac{\pi^2 N_c}{v_t} \frac{\Gamma(k_t)}{\hbar \omega} \langle |O_{if}|^2 \rangle \tag{6.5.26}$$

where $\langle |O_{if}|^2 \rangle$ is the thermally averaged overlap factor, given by eq. (6.4.23) for the case of linear electron–lattice interaction, v_t and k_t are the thermal velocity and wavevector,

$$v_t = (8kT/\pi m^*)^{\frac{1}{2}} \tag{6.5.27}$$

$$k_t = m^* v_t / \hbar \tag{6.5.28}$$

and the thermally averaged capture cross section is given by

$$\sigma(T) = \frac{2\pi^2}{k_t^2} \frac{\Gamma(k_t)}{\hbar \omega} \langle |O_{if}|^2 \rangle \tag{6.5.29}$$

The present method is based on the use of perturbation theory, but a nonperturbative method can also be used [6.5.4].

6.5.2 The role of the Coulomb interaction

Particularly important is the case of multiphonon capture or emission by a charged defect. We shall consider here the case when the electron is bound at the defect principally by a strong short-range defect potential $U_{el}(\mathbf{r})$ which is much stronger than the long-range Coulomb potential, as is frequently the case for deep level defects in covalent semiconductors. The Coulomb field can be allowed for by approximating the free-electron wavefunction $\varphi_k(\mathbf{r})$ as

$$\varphi_k(\mathbf{r}) = C_k(r) \, \varphi_{0k}(\mathbf{r}) \tag{6.5.30}$$

where $C_k(r)$ is the Coulomb wavefunction in the potential $1/\varepsilon_0 r$ for a particle with effective mass m^*, and $\varphi_{0k}(\mathbf{r})$ is the free-electron wavefunction for a neutral defect. If the defect potential acts over a distance much shorter than the Coulomb potential, we can approximate

$$V_{nk} \approx C_k(0) \, V_{nk}^0 \tag{6.5.31}$$

Here, V_{nk}^0 refers to the matrix element (6.5.8) for a neutral defect.

The transition rate is proportional to the square of the matrix element V_{nk}, and the effect of an attractive or repulsive Coulomb field is therefore expressed by the factor $|C_k(0)|^2$. This quantity which will be denoted by s, is usually called the Sommerfeld factor, and is well known from the quantum theory of a particle in the Coulomb field [6.5.1], [6.5.2], [6.5.5], [6.5.6]:

$$s \equiv |C_k(0)|^2 = \frac{K}{e^K - 1} \begin{cases} \simeq |K| \text{ for an attractive center, } K < 0 & (6.5.32) \\ \\ \simeq K e^{-K} \text{ for a repulsive center, } K > 0 & (6.5.33) \end{cases}$$

where

$$K = 2\pi Z/\tilde{r}_1 k \qquad (6.5.34)$$

Here, $\tilde{r}_1 = \hbar^2 \varepsilon_0 /(m^* q^2)$ is the 'Bohr radius' of a hydrogenic state of the charge carrier at an attractive Coulomb center, $Z = Q/q$, where Q and q are the charges of the defect and charge carrier, respectively, and the approximate equalities are valid for $k \to 0$. The Sommerfeld factor for the repulsive center (6.5.33) will be recognized as the Gamow probability for tunneling through the Coulomb barrier in the radioactive α-decay [6.5.7].

Other electric field distributions can be treated in a similar fashion. For example, Makram-Ebeid et al. [6.5.8] observed a field dependence of the emission rate in the electric field of the p–n junction which was explained theoretically in terms of electron tunneling through a potential barrier [6.5.9], [6.5.10].

If the k-dependence of the capture rate at a neutral center is neglected, the enhancement or reduction of the capture and emission rates at charged defects is given by the averaged Sommerfeld factors, and can be calculated without difficulty. For attractive centers we have, from eq. (6.5.32),

$$\langle s_a \rangle = \left[\frac{\hbar^2}{2\pi m^* k_B T} \right]^{\frac{3}{2}} \int_0^\infty \frac{2\pi |Z|}{a_B k} \exp\left[-\frac{\hbar^2 k^2}{2m^* k_B T} \right] 4\pi k^2 \, dk$$

$$= 4|Z| (\pi E_R/k_B T)^{\frac{1}{2}} \qquad (6.5.35)$$

where

$$E_R \equiv m^* q^4 /(2\hbar^2 \varepsilon^2) [= (m^*/\varepsilon^2 m) I_0] \qquad (6.5.36)$$

is the 'effective Rydberg', and the Boltzmann constant was furnished with a subscript B to avoid confusion with the wavevector k.

For repulsive Coulomb centers, using eq. (6.5.33),

$$\langle s_r \rangle = \left[\frac{\hbar^2}{2\pi m^* k_B T} \right]^{\frac{3}{2}} \int_0^\infty \frac{2\pi Z}{a_B k} \exp\left[-\frac{2\pi Z}{a_B k} - \frac{\hbar^2 k^2}{2m^* k_B T} \right] 4\pi k^2 \, dk \qquad (6.5.37)$$

The integral can easily be evaluated by the method of steepest descent with the result:

$$\langle s_\mathrm{r} \rangle = 8/\sqrt{3}(\pi^2 Z^2 E_\mathrm{R}/k_\mathrm{B} T)^{\frac{2}{3}} \exp\{-3(Z^2\pi^2 E_\mathrm{R}/k_\mathrm{B} T)^{\frac{1}{3}}\} \qquad (6.5.38)$$

The expressions (6.5.35) and (6.5.38) give the ratios between the transition rates at attractive and repulsive Coulomb centers, respectively, and the transition rate at a similar neutral center. Expression (6.5.38) was first obtained by Bonch-Bruevich and E.G. Landsberg [6.5.11].

Pässler [6.5.5], [6.5.12] discusses extensively these factors with allowance for the k-dependence of the multiphonon part of the capture cross section (6.5.16). This k-dependence can be taken into account by expanding the exponent of the overlap factor (which now depends on k through $E_2(Q, k)$, eq. (6.5.1)) in powers of k^2. For a defect with a weak coupling to the lattice, this results in the renormalization of temperature. For a defect strongly coupled to the lattice, however, the capture cross section increases with increasing k (this may be seen by an examination of the CC diagram, which shows that the nuclear barrier for the multiphonon process is reduced as k increases). This competition between thermal activation and electron tunneling can give rise to a preferred electron tunneling energy as a function of temperature – not unlike the nuclear tunneling at finite temperature which was discussed in section 6.4. A detailed discussion of this effect, however, will not be attempted.

6.6 Quantum mechanical treatment of the multiphonon transition rate

In sections 6.3–6.5, the multiphonon transition rate was calculated in the semiclassical approximation using the assumption of a single accepting mode. A full quantum mechanical treatment will now be given for a general spectrum of lattice vibrations.

Consider the function

$$F(E) = \sum_{n, m} |\langle f, n| H_\mathrm{int}|i, m\rangle|^2 p(E_{im}) \delta(E_{im} - E_{fn} + E) \qquad (6.6.1)$$

where f and i refer to the final and initial electronic states, n and m are the quantum numbers of vibrational states with energies E_{im} and E_{fn}, and $p(E_{im})$ is the Boltzmann factor

$$w_{im} = \mathrm{e}^{-\beta E_{im}}/Z \qquad (6.6.2)$$

where Z is a normalizing factor which, for a set of independent harmonic oscillators, is a product of factors (6.3.25) over all normal modes:

$$Z = \sum_m \mathrm{e}^{-\beta E_{im}} = \prod_k \{2\sinh(\beta\hbar\omega(k)/2)\}^{-1}; \quad \beta = 1/kT \qquad (6.6.3)$$

The interaction Hamiltonian H_int in eq. (6.6.1) is constructed using only the nondiagonal matrix elements (6.2.16), and the diagonal matrix elements of H_int (6.2.6) are included in the energies of the vibrational states. The analysis has been restricted to linear electron–phonon interaction by assuming that the frequencies $\omega(k)$ are independent of the electron state. A more general treatment can be found, for example, in the work of Kubo and Toyozawa [6.6.1].

To see the significance of the function $F(E)$ which is usually called the generating function we note that the multiphonon transition rate is obtained by setting $E = 0$:

$$W = \frac{2\pi}{\hbar} F(0) \tag{6.6.4}$$

Similarly, one can also obtain the radiative transition rate if one inserts for H_{int} the interaction matrix element with the radiative field $G^{(k)}$ in eq. (4.3.56). The energy E is then set equal to the photon energy $\hbar\Omega$.

The sum in eq. (6.6.1) can be simplified by defining the vibrational density matrix for each of the electron states i and f. In the basis formed by the vibrational eigenstates $|\alpha, 1\rangle$ ($\alpha = i, f$), this density matrix has the elements

$$\langle \alpha, 1 | \sigma(\tau) | \alpha', 1' \rangle = e^{-\tau E_{\alpha 1}} \delta_{11'} \delta_{\alpha\alpha'} \tag{6.6.5}$$

The Boltzmann factors (6.6.2) can therefore be written as

$$w_{im} = \langle i, m | \rho(\beta) | i, m \rangle / Z \tag{6.6.6}$$

It will be convenient to introduce the Fourier transform of $F(E)$,

$$f(t) = \int_{-\infty}^{\infty} F(E) e^{-iEt} \, dE \tag{6.6.7}$$

with the inverse

$$F(E) = \frac{1}{2\pi} \int_{-\infty}^{\infty} f(t) e^{iEt} \, dt \tag{6.6.8}$$

$$= \frac{1}{2\pi i} \int_{-i\infty}^{i\infty} f(\tau) e^{E\tau} \, d\tau$$

where $\tau = it$. The function $f(t)$ can now be written as

$$f(t) = \frac{1}{Z} \int_{-\infty}^{\infty} dE \, e^{-E\tau} \sum_{n,m} \langle i, m | H_{\text{int}} | f, n \rangle \langle f, n | H_{\text{int}} | i, m \rangle e^{-\beta E_{im}} \delta(E_{im} - E_{fn} + E)$$

$$= \frac{1}{Z} \sum_{n,m} \langle i, m | H_{\text{int}} | f, n \rangle e^{-\tau E_{fn}} \langle f, n | H_{\text{int}} | i, m \rangle e^{-(\beta-\tau)E_{im}}$$

$$= \frac{1}{Z} \text{Tr} \{ H_{\text{int}} \rho(\tau) H_{\text{int}} \rho(\beta - \tau) \} \tag{6.6.9}$$

where Tr denotes the trace of the matrix. Since the trace is independent of the choice of the basis, eq. (6.6.9) can be evaluated in any representation. We choose here the coordinate representation in terms of the real normal coordinates q_k and use the result for the matrix elements of the density matrix of a simple harmonic oscillator [6.6.1]:

$$\langle q | \rho(\lambda) | q' \rangle = \{ \omega/2\pi\hbar \sinh(\lambda\hbar\omega) \}^{\frac{1}{2}} \exp \{ -(\omega/4\hbar) \tanh(\lambda\hbar\omega/2)(q + q' - 2q_0)^2$$
$$- (\omega/4\hbar) \coth(\lambda\hbar\omega/2)(q - q')^2 \} \tag{6.6.10}$$

where q and q' are two values of the oscillator coordinate and q_0 is the equilibrium position. Hence, eq. (6.6.9) becomes

$$f(\tau) = \frac{1}{Z} \int_0^\infty dq_k \int_0^\infty dq_k' \prod_k \langle i,\{q_k\}| H_{int}|\{q_k\},f \rangle \langle f, q_k| \rho(\tau) |q_k',f \rangle$$

$$\langle f,\{q_k'\}| H_{int}|\{q_k'\}, i \rangle \langle i, q_k| \rho(\beta-\tau) |q_k',i \rangle \qquad (6.6.11)$$

$$= \frac{1}{Z_i} \int_{-\infty}^\infty dq_k \int_{-\infty}^\infty dq_k' \langle i,\{q_k\}| H_{int}|\{q_k\},f \rangle \langle f,\{q_k'\}| H_{int}|\{q_k'\}, i \rangle$$

$$Sh_k(\tau)\, Sh_k(\beta-\tau) \exp\{-\sum_k \{\tfrac{1}{2}T_k(\tau)(q_k+q_k'-2\delta q_k)^2 + \tfrac{1}{2}T_k(\beta-\tau)(q_k+q_k')^2$$

where

$$+\tfrac{1}{2}[C_k(\tau)+C_k(\beta-\tau)](q_k-q_k')^2\} + \tau\Delta E\} \qquad (6.6.12)$$

$$Sh_k(\lambda) \equiv \{\omega_k/2\pi\hbar \sinh(\lambda\hbar\omega_k)\}^{\frac{1}{2}}$$
$$T_k(\lambda) \equiv (\omega_k/2\hbar)\tanh(\lambda\hbar\omega_k/2) \qquad (6.6.13)$$
$$C_k(\lambda) \equiv (\omega_k/2\hbar)\coth(\lambda\hbar\omega_k/2)$$

In the second equality (6.6.12), the lattice coordinates were transformed as follows:

$$q_k \to q_k - q_{ki}$$
$$q_k' \to q_k' - q_{ki} \qquad (6.6.14)$$

and we denoted by δq_k the separation of the equilibrium displacements q_{ki} and q_{kf} (6.2.10) for mode k in the two electron states i and f:

$$\delta q_k = q_{kf} - q_{ki} \qquad (6.6.15)$$

The evaluation of the integral (6.6.12) over the coordinates q_k and q_k' represents a straightforward but tedious algebraic exercise which is carried out in Appendix G:

$$f(\tau) = g(x)\, G \exp[-ix\Delta E - \Phi(x)] \qquad (6.6.16)$$

where

$$x = \tau - i\beta/2$$

$$G = \exp\{-\sum_k S_k \coth(\beta\hbar\omega_k/2) - \beta\Delta E/2\} \qquad (6.6.17)$$

$$\Phi(x) = 2\sum_k S_k \cosh(x\hbar\omega_k)/\sinh(\beta\hbar\omega_k/2) \qquad (6.6.18)$$

$$g(x) = \tfrac{3}{4}\sum |B(k)|^2 \cosh(x\hbar\omega_k)/\sinh(\beta\hbar\omega_k/2)$$
$$+\tfrac{1}{4}|\sum B(k)\{1-\sinh(x\hbar\omega_k)/\sinh(\beta\hbar\omega_k/2)\}|^2 \qquad (6.6.19)$$

$B(k) \equiv B_{fi}(k)$ are given by eq. (6.2.16), and S_k are the Huang–Rhys factors (6.2.14) for each of the normal modes. The transition rate now becomes

$$W(E) = \frac{1}{i\hbar}\int_{-\infty i}^{i\infty} f(\tau)\,d\tau = \frac{1}{\hbar}\int_{-\infty-i\beta/2}^{+\infty-i\beta/2} f(x)\,dx$$

$$= G\int_{-\infty}^{+\infty} g(x)\exp[-ix\Delta E - \Phi(x)]\,dx \qquad (6.6.20)$$

where, in the last expression, the displacement of the integration contour is justified since no singularities are crossed while doing so. For a general dispersion law of ω_k, the integral (6.6.20) cannot be evaluated in a closed form without further approximations, and two solvable models will now be considered in more detail.

(i) The high-temperature limit

This can be obtained quite generally for any frequency spectrum, but the results are largely of academic interest since this limit is rarely attained in practice. We follow here the method of Holstein [6.6.2] who shows that, in this limit, the dominant contribution to the integral comes from the vicinity of the point $x = 0$. The function $\Phi(x)$ is then expanded in the powers of x up to the quadratic term, and the integral then becomes a Gaussian integral which can be evaluated in a straightforward manner. Keeping only the lowest order terms in β we obtain, after some algebra,

$$W(T) = |V(q_c)|^2 \left[\frac{\pi}{kTE_S}\right]^{\frac{1}{2}} \exp\left[-\frac{E_a}{kT}\right] \tag{6.6.21}$$

where

$$\begin{aligned} E_S &= \sum S_k \hbar\omega_k \\ E_a &= (\Delta E - E_S)^2/(4E_S) \end{aligned} \tag{6.6.22}$$

and the matrix element is evaluated in the configuration with

$$q_k = \delta q_k (E_a/E_S)^{\frac{1}{2}} \tag{6.6.23}$$

This is the configuration of the lowest total energy, equal to E_c, where the potential energy surfaces in the two electronic states intersect. Equations (6.6.21) and (6.6.22) generalize the result (6.3.27) to several normal modes.

(ii) Einstein model of the frequency spectrum

Here all frequencies ω_k are equal to a single frequency ω. The integration range in eq. (6.6.20) can be reduced to the interval $(-\pi, +\pi)$ by introducing new integration variables $y = x\hbar\omega - 2n\pi$, where $n = 0, \pm1, \ldots$, are integers. Noting that $g(y)$ and $\Phi(y)$ are periodic with period 2π, we obtain

$$W = (G/\hbar) \sum_{n=-\infty}^{\infty} \exp(2\pi i n\Delta E/\hbar\omega) \int_{-\pi}^{\pi} g(y) \exp[-iy\Delta E - \Phi(y)]\,dy \tag{6.6.24}$$

The sum in eq. (6.6.24) is equal to

$$\sum_{n=-\infty}^{\infty} \exp(2\pi i n\Delta E/\hbar\omega) = \sum_{n=-\infty}^{\infty} \delta(\Delta E/\hbar\omega - n) \tag{6.6.25}$$

which ensures that transitions occur only when ΔE is equal to an integral number of phonons. This condition, however, need not be taken too seriously as only a slight dispersion can ensure an exact match of the two energies. Henceforth, this factor will be omitted from eq. (6.6.24).

To calculate the integral the following properties of the modified Bessel functions will be needed (see, for example, [6.6.3])

$$I_n(z) = \frac{1}{2\pi} \int_{-\pi}^{\pi} \exp(z\cos y + iny)\,dy \tag{6.6.26}$$

$$\begin{aligned} I_{n-1}(z) - I_{n+1}(z) &= 2nI_n(z)/z \\ I_{n+1}(z) + I_{n-1}(z) &= 2I'_n(z) \end{aligned} \tag{6.6.27}$$

With the use of eqs. (6.6.26) and (6.6.27), we finally obtain

$$
\begin{aligned}
W = (1/\hbar)[|\sum B(k)\,\delta q_k|^2\{(p-S)^2/4S\}\,I_p(x) \\
+|\sum B(k)\,\delta q_k|^2 I_p'(x)/\{4S\sinh(\beta\hbar\omega/2)\} \\
+\{3\hbar/2\omega \sinh(\beta\hbar\omega/2)\}\sum |B(k)|^2 I_p'(x)] \\
\times \exp\{-S\coth(\beta\hbar\omega/2)-p(\beta\hbar\omega/2)\}
\end{aligned}
\tag{6.6.28}
$$

where

$$
S = \sum S_k
$$
$$
x = S\{p\sinh(\beta\hbar\omega/2)\}
$$

The first term in the pre-exponential factor in eq. (6.6.28) usually dominates, since, for large p, the second and third terms are smaller by a factor of the order p, as can be seen using an asymptotic expression for the Bessel functions [6.6.3]. It is interesting to note that the coefficient of I_p in the first term is equal to the matrix element $V(q)$ evaluated at the crossing point Q_c in the appropriate CC diagram. This again confirms the validity of the Franck–Condon principle. Using asymptotic expressions for the Bessel functions, expression (6.6.28) can be shown to reduce to eq. (6.4.18) in the limit $p \to \infty$ [6.6.4].

7

___

Recombination in low-dimensional
*semiconductor structures**

7.1 Introduction

In this chapter, an introduction to the theory of recombination in low-dimensional semiconductor structures is given. A low-dimensional semiconductor structure is one whose dimensions (i.e. layer thicknesses) are smaller than, or comparable to, the de Broglie wavelength of the carriers. An example of such a structure is a quantum well (QW), which consists of a small band-gap semiconductor, of width L, sandwiched between two larger band-gap semiconductors. If L is less than or equal to the de Broglie wavelength of the carriers, then the carriers are confined in the smaller band-gap material, although they are still free to move in the plane of the well. Thus, a two-dimensional electron (and hole) gas can be formed, whose density of states is significantly altered from the bulk (three-dimensional) density of states. The two-dimensional density of states has potential advantages for lasers utilizing such QW structures, as will be explained in the later sections of this chapter. The aim of this chapter is to give a simple description of the physics of QW structures, and to outline the reasons why devices employing such structures have possible advantages over conventional devices. The main emphasis will be on the physics of QWs for laser applications. The problems associated with long wavelength semiconductor lasers are outlined, and the use of QW lasers as a possible solution for the reduction of threshold currents and their temperature sensitivity in such systems is discussed.

Radiative and nonradiative recombination rates are then estimated for single, undoped QW structures, and a rough calculation is presented comparing the quantum efficiency of QW and double heterostructure (DH) lasers.

___

* This chapter has been written by R.I. Taylor, GEC-Marconi Materials Technology Ltd, Caswell.

Finally, the use of strained semiconductor QW lasers is considered, and it is shown that the changes that occur to the valence band structure of the active layer (due to the strain) can have advantageous effects in terms of lower, and less temperature dependent threshold currents for long wavelength lasers.

In order to obtain an idea of orders of magnitude, suppose that a typical inversion layer has a charge density of 10^{11} carriers per cm^2, i.e. 10^{-5} carriers per Å^2. Then a square box of side 10^3 Å contains only ten carriers, and this is reduced to a single carrier for a square box of side 316 Å. Quantum wells, quantum well wires, and quantum well boxes in particular thus lead one to the study of genuine few-electron systems, see for example [7.1.1] for a discussion of quantum well boxes.

These small numbers also lead to problems with the normal Fermi–Dirac statistics. We here deal with the average occupation numbers but do not discuss fluctuations.

For a one-particle system with two available states of energies E_1 and E_2 there can be as much as a 17% discrepancy between the canonical and grand canonical mean occupation probabilities for the lowest state. Let us put

$$b \equiv \exp\left(\frac{E_2 - E_1}{kT}\right) \tag{7.1.1}$$

Then the lowest state has a mean occupation probability [7.1.2]

$$n_1^g = \frac{1}{b^{-1/2} + 1} \quad \text{(grand canonical)} \tag{7.1.2}$$

$$n_1^c = \frac{1}{b^{-1} + 1} \quad \text{(canonical)} \tag{7.1.3}$$

with the maximum discrepancy, $(n_1^c - n_1^g)/n_1^c$, occurring for $E_2 - E_1 = 2.12\, kT$.

Equation (7.1.2) holds if, owing to occasional trapping of particles at the wall of the container, or due to some other process, there is in effect a particle reservoir which ensures that the mean total number of particles is unity. If the total number of particles is fixed at unity (and is not merely fixed on average), then eq. (7.1.3) holds. An experimental test of which (if either) of these alternatives hold would be of interest.

To prove eq. (7.1.2), the Fermi level equation

$$[x \exp(\eta_1) + 1]^{-1} + [x \exp(\eta_2) + 1]^{-1} = 1 \tag{7.1.4}$$

(with $x \equiv \exp(-\gamma)$, and $\gamma \equiv \mu/kT$), is solved for x to give

$$\gamma = \tfrac{1}{2}(\eta_1 + \eta_2) \tag{7.1.5}$$

where $\eta_i = E_i/kT$. Hence

$$n_1^g = \frac{1}{b^{-\frac{1}{2}} + 1} \tag{7.1.6}$$

To prove eq. (7.1.3), let $S(N; r, j)$ be the terms in the canonical partition function for N particles which assigns r particles to level j. Then

$$n_1^c = \frac{S(1; 1, 1)}{S(1; 0, 1) + S(1; 1, 1)}$$

$$= \frac{\exp(-\eta_1)}{\exp(-\eta_1) + \exp(-\eta_2)} = \frac{1}{b^{-1} + 1} \qquad (7.1.7)$$

as required.

7.2 Energy levels and density of states

In recent years, advances in epitaxial growth techniques, such as MBE (molecular beam epitaxy) and MOCVD (metal–organic chemical vapor deposition) have enabled the routine growth of semiconductor quantum well (QW) structures and superlattices. A QW consists of a semiconductor 'sandwich', in which a semiconductor of smaller band gap, and width L, lies between wider band gap cladding layers. If the QW width, L, is less than the de Broglie wavelength of the carriers, then the electrons and holes can be spatially confined, forming two-dimensional electron (and hole) gases.

If a series of QWs are stacked together, then the structure formed is termed either a multiple quantum well (MQW) or a superlattice. The distinction between the two depends upon whether or not carriers can easily tunnel from well to well (in which case the structure is termed a superlattice [7.2.1], [7.2.2]) or not (in which case the structure is an MQW).

Consider a MQW structure formed of alternating layers of semiconductors A and B (with different band gaps). Two possible structures can arise, depending on the relative line-up of the band edges. In a type I superlattice (or MQW), both types of carriers are spatially confined in the same layers (e.g. Fig. 7.2.1a). In a type II superlattice (or MQW), however, the electrons and holes are spatially separated (e.g. the electrons are in semiconductor A whereas the holes are in semiconductor B (see Fig. 7.2.1b). An example of a type I MQW is $GaAs/Ga_{1-x}Al_xAs$ (with $x < 0.4$) and a typical type II MQW is InAs–GaSb.

In the rest of this chapter, only type I MQW structures are considered, since the main emphasis is on lasers, and type II MQWs, with their spatially separated carriers, form much less efficient lasing structures than type I MQWs.

Thus, if a material with a small energy gap is sandwiched in between a material on either side of it whose energy gap is larger, a quantum well is created. This is shown in Fig. 7.2.2 for GaAs (typically of thickness 10–100 Å) and thicker layers of AlGaAs; these are frequently used materials for quantum well lasers. For a well with infinitely high potential wells, the wavefunctions have nodes at $z = 0$ and $z = L$. For the ground state, $j = 1$, there are no other nodes and the wavefunction is symmetrical with respect to the center of the well: $\psi_1(L/2 + z) = \psi_1(L/2 - z)$. For

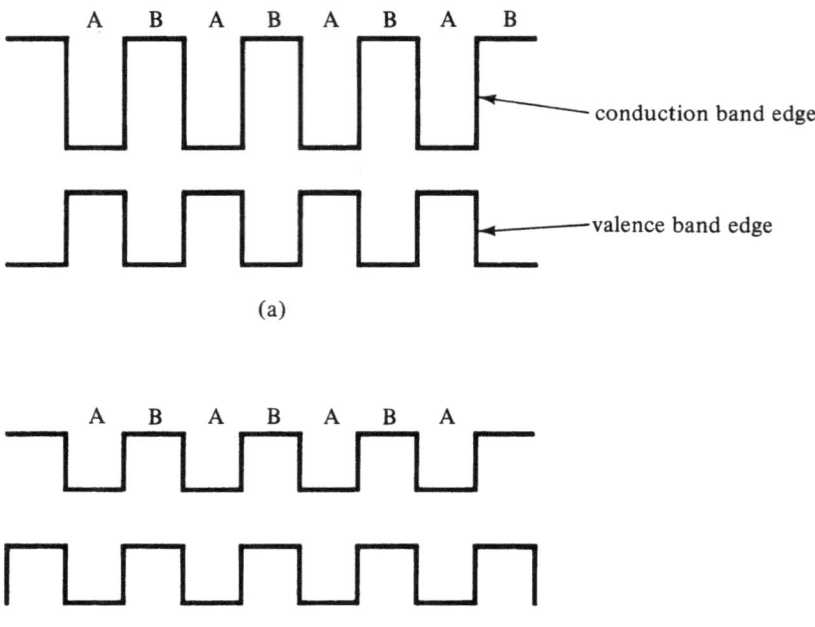

Fig. 7.2.1. (a) Type I superlattice. (b) Type II superlattice.

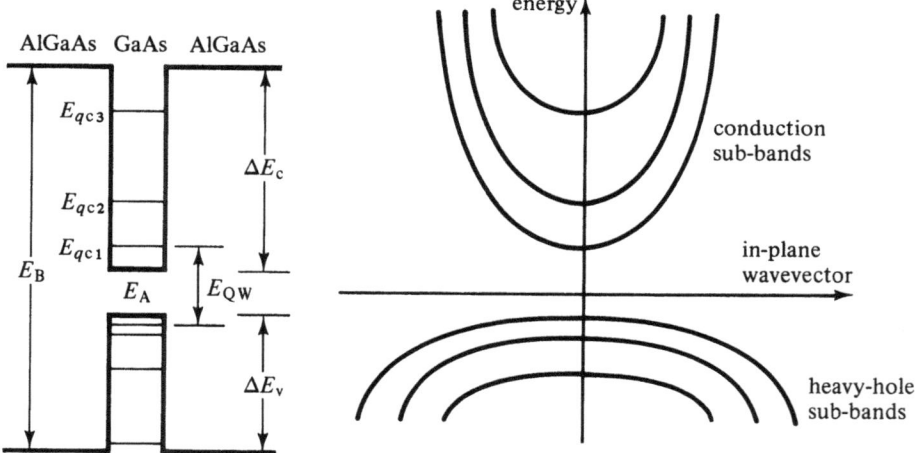

Fig. 7.2.2. The size of the confining barriers, ΔE_c and ΔE_v, determine the effectiveness of the wells in confining the carriers. The sum of ΔE_c and ΔE_v is equal to the difference in the band gaps of the materials A and B, but the ratio, $\Delta E_c : \Delta E_v$ has been found to be material dependent, and no *ab initio* model exists to predict this ratio. For $GaAs/Al_x Ga_{1-x} As$ ($x < 0.4$) one finds $\Delta E_c : \Delta E_v \simeq 0.6 : 0.4$.

the first excited state, $j = 2$, an additional node occurs at $z = L/2$, and the wavefunctions are antisymmetrical with respect to the center of the well, $\psi_2(L/2+z) = -\psi_1(L/2-z)$. For the second excited state, $j = 3$, there are four nodes (at $z = 0$, $L/3$, $2L/3$, and L), and $\psi_3(z)$ is again symmetrical with respect to the center of the well, and so on. One finds for the energy levels of a quantum well *box*, with infinite potential wells, and sides of length L_x, L_y, L_z,

$$E(n_x, n_y, n_z) = E_{r0} + \frac{h^2}{8\pi^2 m_r}(k_x^2 + k_y^2 + k_z^2) \tag{7.2.1}$$

$$(k_x, k_y, k_z) = \pi\left(\frac{n_x}{L_x}, \frac{n_y}{L_y}, \frac{n_z}{L_z}\right) \tag{7.2.2}$$

Here 'r' refers to the conduction or valence bands and m_r is the effective mass (which is negative for the valence bands), and E_{r0} is the energy of the extremum of the appropriate band in the limit of very large L_x, L_y, L_z. The wavefunction (or more accurately, the *envelope* function that modulates the underlying Bloch functions) is

$$\left(\frac{8}{L_x L_y L_z}\right)^{\frac{1}{2}} \sin\left(\frac{n_x \pi x}{L_x}\right) \sin\left(\frac{n_y \pi y}{L_y}\right) \sin\left(\frac{n_z \pi z}{L_z}\right) \tag{7.2.3}$$

If L_z is much smaller than L_x and L_y, then the energy spectrum can be written as

$$E_{ri}(n_{zri}, \mathbf{k}_\parallel) = E_{r0} + \frac{h^2}{8\pi^2 m_{ri}}\left[\mathbf{k}_{r\parallel}^2 + \left(\frac{\pi n_{zri}}{L_z}\right)^2\right] \tag{7.2.4}$$

$\mathbf{k}_{r\parallel}$ is a two-dimensional wavevector associated with free carrier motion in the (x, y)-plane. Also, r distinguishes the bands with edges at energies E_{r0}; $r = c, l, h, s$, for conduction band, light-hole, heavy-hole, and spin split-off bands, respectively. The subbands, or minibands, are labelled by $i = 1, 2, 3\ldots$. The subbands (7.2.4) are parabolae, with minima at

$$E_{qri} = E_{r0} + \frac{h^2 n_{zri}^2}{8m_{ri} L_z^2} \tag{7.2.5}$$

as illustrated in Fig. 7.2.2. The least energy difference between these levels, measured across the gap, for conduction and heavy-hole bands is the threshold energy for absorption and emission from this pair of bands. It is

$$E_{Gq} = E_G + \frac{h^2}{8m_{ch} L_z^2} \tag{7.2.6}$$

where $E_G = E_{c0} - E_{h0}$, and $1/m_{ch} = 1/|m_c| + 1/|m_h|$. One important advantage of the QWs can immediately be seen, namely that the fundamental frequency of the system can be 'tuned' by adjusting the value of L_z.

Other advantages of using QWs arise from the density of states, which is significantly modified from the bulk density of states since carriers are only free to move in two dimensions.

The density of states for the two-dimensional parabolae (7.2.4) arising from $\mathbf{k}_{r\parallel}$ is energy independent, in accordance with eq. (1.4.6). It has, therefore, the staircase appearance of Fig. 7.2.3. A step occurs each time n_z (the quantum number labeling the subband) changes by unity.

The concentration of electrons in the conduction band (r = c) is thus given by

$$n = \sum_i n_i = \sum_i \frac{4\pi m_{ci}}{h^2 L_z} \int_{E_{qci}}^{\infty} \frac{\mathrm{d}E}{1+\exp{(E-\mu_c/kT)}}$$

$$= \sum_i \frac{4\pi m_{ci} kT}{h^2 L_z} \ln\left(1+\exp\left(\frac{\mu_c - E_{qci}}{kT}\right)\right) \qquad (7.2.8)$$

where the factor in front of the integral gives the number of states, including spin, per unit volume per unit energy range, and is obtained from eq. (1.4.6) by putting $m_1 = m_2$, $g = 2$ and dividing by the full three-dimensional volume $V_2 L_z$. In the nondegenerate limit, eq. (7.2.8) yields

$$n = \sum_i \frac{4\pi m_{ci} kT}{h^2 L_z} \exp\left(\frac{\mu_c - E_{qci}}{kT}\right) \qquad (7.2.9)$$

The step-like density of states means that there are a large number of carriers at the QW band edges, in contrast to the bulk density of states, so that threshold current densities in QW lasers are expected to be lower. Also, since the temperature dependence of a typical concentration in r dimensions is, from eq. (1.4.4),

$$n_i = C \int_{E_{qci}}^{\infty} \frac{(E-E_{qci})^{r/2-1}\,\mathrm{d}E}{1+\exp{[(E-\mu_c)/kT]}}$$

$$= C(kT)^{r/2}\Gamma\left(\frac{r}{2}\right) F_{r/2-1}[(\mu_c - E_{qci})/kT] \qquad (7.2.10)$$

where C is independent of temperature and energy, it follows that threshold currents should be less temperature sensitive as r decreases.

In practice, the confining wells of a QW are not infinitely high, and carriers can 'leak' into the barrier layers. When this occurs, the results quoted above, for the energy levels of the QW, need to be corrected. The notion of envelope functions is still used, but in addition to an envelope function in the well, an envelope function is also used in the barrier. One has then to ensure that the envelope wavefunction inside the well matches the envelope function outside the well at the two interfaces, $z = 0$ and $z = L$. The boundary conditions on the derivative of the envelope functions are more controversial. However, the most widely used

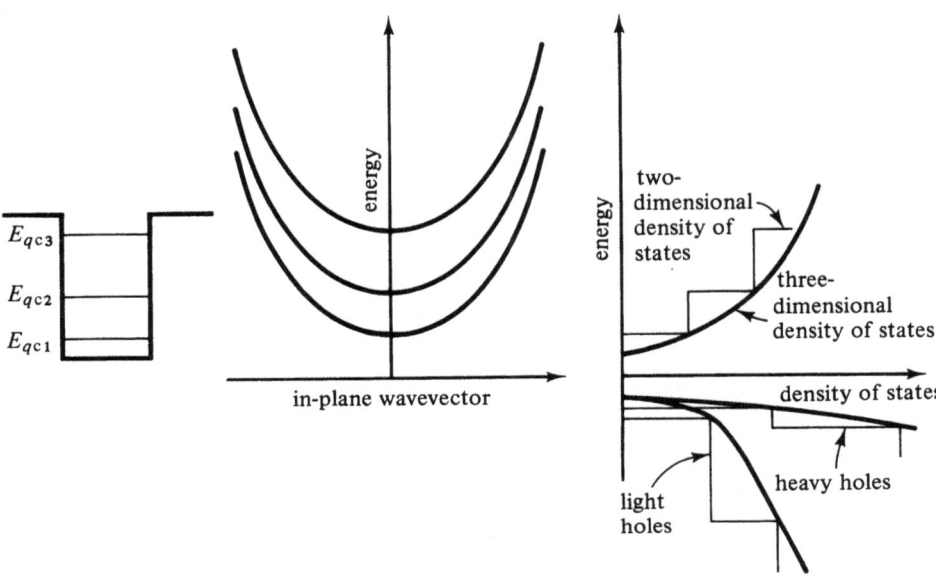

Fig. 7.2.3. The effects of combining quasi-continuous energies in two dimensions with quantization in the third direction.

boundary conditions on the derivative of the envelope functions F are that $(1/m^*)/(dF/dz)$ should be continuous across the QW interfaces [7.2.3], although this is only true if the underlying Bloch periodic functions are 'similar' in well and barrier materials (this is true for GaAs/AlGaAs). There are still a series of subbands associated with each confined QW level, but for a QW with finite barriers there are only a finite number of confined levels. In an infinite well, of course, there are an infinite number of energy levels. For further information on the theory of envelope functions, the reader is referred to eqs. (5.2.1) to (5.2.7) and references [7.2.4]–[7.2.6]. For wells with finite confining barriers, then the smaller m_{rt} and/or L_z, the further are the energy levels separated from their band at E_{r0} (in eq. (7.2.4)), and the greater the electron leakage into the barriers.

For the component of particle motion perpendicular to the (x, y)-planes (which constitute the potential wells) the energy levels, notably the lowest one, increase as L_z becomes smaller. This is the quantum size effect, also known as quantum confinement. Quantum confinement is responsible for raising the exciton binding energies in QWs, compared to their bulk values. If one treats the effective masses of electrons and holes in the z-direction as infinite, and assumes a QW with infinitely high barriers, one has, in fact, a two-dimensional H atom whose spectrum of bound states behaves as

$$\frac{-I_0}{(n-1/2)^2} \quad \text{where } n = 1, 2, 3, \ldots \tag{7.2.11}$$

For the normal three-dimensional case (see eq. (3.7.17)), the spectrum of bound states behaves as I_0/n^2, and so the confinement effect raises the binding energy of the ground state exciton from I_0 to $4I_0$ [7.2.7]. The same effect raises the binding energy of donors and acceptors, although these energies also depend on the precise position of the impurities in the well [7.2.8]. Of course, in practice, the wells are not infinitely deep, and this lowers the exciton binding energy relative to the extreme two-dimensional case [7.2.9]. The calculation of exciton binding energies for QWs of finite depth is carried out numerically, often using a variational approach [7.2.8]. Even for finite QWs though, the exciton binding energy is still enhanced over that of bulk excitons, and it is this that is responsible for the observation of free exciton recombination in single QWs *at room temperature* [7.2.10]. In addition, QW excitons still exist at room temperature when small electric fields are applied. The shift in absorption edge that occurs with an applied field is known as the quantum confined Start effect (QCSE) [7.2.11], and this effect can be used in the design of modulators. There is also the possibility of using QW excitonic effects for applications in optical bistability, four-wave mixing, and to obtain large electro-optic coefficients.

7.3 Radiative recombination in QWs

7.3.1 *Rough estimate of radiative recombination in a QW*

In this short section, a rough estimate of the temperature dependence of the radiative band–band recombination rate per unit volume, u_{cv}, in a QW is presented. As for bulk radiative band–band recombination, we expect

$$u_{cv} = B^s np \tag{7.3.1}$$

This expression arises from a relation of the following form (derived assuming **k**-conservation)

$$u_{cv} \sim \int_{E_{QW}}^{\infty} \mathcal{N}_{red} p_c (1 - p_v)\, d(E_c - E_v) \tag{7.3.2}$$

where $\mathcal{N}_{red}$ is the 'reduced' or 'joint' two-dimensional density of states, encountered in sections 1.5.2 and 5.4.3. Here $\mathcal{N}_{red}^{-1} = \frac{1}{2}(\mathcal{N}_e^{-1} + \mathcal{N}_h^{-1})$, where $\mathcal{N}_e$ and $\mathcal{N}_h$ are the electron and hole density of states. Since u_{cv} will thus, from expression (7.3.2), have the same temperature dependence as n and p, then, in r dimensions, we expect that

$$B_s = \frac{u_{cv}}{np} \sim \frac{T^{r/2}}{T^r} \sim T^{-r/2} \tag{7.3.3}$$

The exponential factor which occurs for nondegenerate materials will cancel. The total band–band radiative recombination rate per unit volume, u_{cv}, thus varies as $T^{r/2}$. Thus, in a bulk semiconductor, u_{cv} is expected to vary as $T^{3/2}$, whereas in a QW u_{cv} will vary as T. For a QW of width L, u_{cv} must approach the bulk value as L increases, and so it is expected that u_{cv} will vary as T^a, where a lies between 1 and $\frac{3}{2}$, and a should approach unity as L decreases, whereas a should approach $\frac{3}{2}$ as L increases.

If the band–band nonradiative time constant, τ_{nr}, is assumed to be large (a reasonable assumption for device quality GaAs/AlGaAs) then the luminescence decay time, τ, is given by

$$\frac{1}{\tau} = B^s n_0 + \frac{1}{\tau_{nr}} \sim B^s n_0 \qquad (7.3.4)$$

Thus, for a constant value of n_0 (the majority-carrier density), $1/\tau$ has the same temperature dependence as B^s. From the above discussion, it is expected that τ should vary as T^a. Leys et al. [7.3.1] have measured τ for GaAs/Al$_{0.33}$Ga$_{0.67}$As quantum wells, for various GaAs well thicknesses, L_z. Their results show that $\tau \sim T^a$, as expected, with the experimentally determined values of a presented in table 7.3.1. These experimental results provide confirmation of the simple qualitative prediction outlined above. As for absolute values, B^s has been estimated to be less than 7.0×10^{11} cm^3 s^{-1} for GaAs/AlGaAs QWs at room temperature [7.3.2].

In an ideal laser (i.e. one with no nonradiative recombination) the threshold current is proportional to u_{cv}, and so we expect the following temperature dependence:

$$i_{th} \sim T^a \qquad (7.3.5)$$

But, empirically, one finds that [7.3.3],

$$I_{th} \sim \exp\left(\frac{T}{T_0}\right) \qquad (7.3.6)$$

where T_0 is a constant (over a limited temperature range), and is a measure of the temperature sensitivity of the laser. If both the above laws hold, then

$$\frac{1}{I_{th}}\frac{dI_{th}}{dT} = \frac{a}{T} = \frac{1}{T_0} \qquad (7.3.7)$$

whence $T_0 = T/a$. Thus, for the GaAs/AlGaAs QW laser, at room temperature, the experimental results above [7.3.1] show that $T_0 \sim 300$ K, whereas for the conventional DH laser, $T_0 \sim 200$ K, indicating that, in the absence of nonradiative recombination, QW lasers are expected to be less temperature sensitive than their

Table 7.3.1

$L_z(\text{Å})$	65	130	300
a	1.07	1.22	1.41

bulk counterparts [7.3.4]. Arakawa and Sakaki [7.3.5] have also predicted that T_0 should increase as the dimensionality of the structure decreases, at least in the absence of nonradiative recombination, and they confirmed this by measuring T_0 for a conventional DH laser in zero magnetic field, and a strong magnetic field. The magnetic field has the effect of confining carriers in a plane. When this experiment was performed, T_0 was observed to increase from 144 K (at zero magnetic field) to 313 K (at a magnetic field of 24 T).

A further study of the temperature dependence of I_{th} should take into account lifetime broadening of the levels as well as broadening of the density of states by fluctuations [7.3.6].

7.3.2 Emission probabilities

In the preceding subsection, a simple approach to radiative band–band recombination suggested that QW lasers should be less temperature sensitive than bulk DH lasers, provided nonradiative recombination is negligible. In this subsection, calculations from first principles are carried out to verify this prediction.

The single-mode radiative transition probability between levels I and J per unit volume per unit time is, by eq. (4.4.3), for mode energy E_k

$$P_{IJ}(E_k) = \frac{2\pi e^2 h^2}{\mu^2 m^2 V^2 E_k} |M_{IJ}|^2_{av} p_I (1-p_J) \zeta \tag{7.3.8}$$

where

$$\zeta \equiv (N(E_k)+1)\,\delta(E_I - E_J - E_k) \quad \text{for emission}$$
$$\zeta \equiv N(E_k)\,\delta(E_I - E_J + E_k) \qquad \text{for absorption}$$

Multiplying by the number of modes $\mathcal{N}(E_k)$ in an energy range $(E_k, E_k + dE_k)$ one finds the multimode transition probability per unit volume per unit time per unit energy range

$$r_{IJ}^{sp}(E_k) = \frac{l}{2} \frac{16\pi^2 e^2 \mu E_k}{m^2 V c^3 h^2} |M_{IJ}|^2_{av} p_I (1-p_J)\,\delta(E_I - E_J - E_k) \tag{7.3.9}$$

where $l = 1$ for polarized light and $l = 2$ for unpolarized light. Also, $p_I \equiv p_I(\eta_I) = [1 + \exp(\eta_I - \gamma_i)]^{-1}$.

We have written the expression for spontaneous emission so as to include the occupation probability p_I of the uper state I, with quasi-Fermi level $\gamma_i kT$, and the probability $(1 - p_J)$ of a vacancy for the lower state J, with quasi-Fermi level $\gamma_j kT$. (For stimulated emission, the same expression applies, multiplied by the photon number $\mathcal{N}(E_k)$ in a mode of energy E_k.) The density of states in k-space for r dimensions, including spin, is

$$\frac{2V_r \, d^r k}{(2\pi)^r} \xrightarrow{r=2} \frac{2A \, d^2 k}{4\pi^2} \tag{7.3.10}$$

where A is the area of the material in the (x, y)-plane.

If state I is taken to be a state in the ith conduction miniband, and state J is taken to be a heavy-hole band state in its jth miniband, then an integration of eq. (7.3.9) with a fixed E_k (see Fig. 4.4.3) leads to

$$r^{sp}_{ci, hj}(E_k) = \frac{l}{2} \frac{16\pi^2 e^2 \mu E_k}{m^2 V c^3 h^2} 2\left(\frac{A}{(2\pi)^2}\right)^2$$

$$\times \int |M_{ci, hj}|^2_{av} p_c(E_{ci})[1 - p_h(E_{hj})] \, \delta(E_{ci} - E_{hj} - E_k) \, d^2 k_c \, d^2 k_h \tag{7.3.11}$$

The above expression is a general equation that may be applied to QWs with nonparabolic band structure, realistic matrix elements and Fermi–Dirac statistics. In such a calculation, however, only numerical results will be obtained. To gain an insight into spontaneous emission rates in QWs, it is useful to assume that transitions are vertical (i.e $\mathbf{k}$-conservation holds). With this assumption, eq. (7.3.11) yields algebraic results.

The condition of $\mathbf{k}$-conservation has been discussed at the end of section 4.4.2. In fact, if $|M|^2$ is the average matrix element for bulk Bloch states, then the average QW matrix element is [7.3.3]

$$|M_{ci, hj}|^2_{av} = |M|^2 \frac{4\pi^2}{A} \delta(\mathbf{k}_c - \mathbf{k}_h) \, \delta_{ij} \tag{7.3.12}$$

This expression can be derived by writing the total QW wavefunction as

$$\psi_{ci}(x, y, z) = \left(\frac{2V}{AL}\right)^{\frac{1}{2}} \sin\left(\frac{i\pi z}{L}\right) u_c(\mathbf{r}) \exp(i\mathbf{k}_\parallel \cdot \boldsymbol{\rho}) \tag{7.3.13}$$

$\boldsymbol{\rho} = (x, y)$, and $\mathbf{k}_\parallel$ is the two-dimensional wavevector. A similar expression is assumed for $\psi_{hj}(x, y, z)$, and then the matrix element for spontaneous emission can be calculated to give eq. (7.3.12). In deriving that equation, an additional assumption has been made, namely that L is much larger than the lattice constant

of the crystal, thus eq. (7.3.12) may not hold for extremely thin QWs, and then a
more detailed analysis is required. Equation (7.3.11) thus leads to

$$r^{sp}_{ci, hj}(E_k) = \frac{l}{2} \frac{16\pi^2 e^2 \mu E_k}{m^2 V c^3 h^2} 2 \left(\frac{A}{(2\pi)^2}\right)^2 \frac{4\pi^2}{A} |M|^2 \delta_{ij}$$

$$\times \int p_c(E_{ci})[1 - p_h(E_{hj})] \, \delta(\mathbf{k}_c - \mathbf{k}_h) \, \delta(E_{ci} - E_{hj} - E_k) \, d^2\mathbf{k}_c \, d^2\mathbf{k}_h \quad (7.3.14)$$

In the above equation, the integral contains delta-functions, and so algebraic
results may be obtained for the spontaneous emission rate. The result is

$$r^{sp}_{ci, hj}(E_k) = \frac{l}{2} \frac{64\pi^3 e^2 \mu |M|^2 m_{chi} E_k}{m^2 L_z c^3 h^4} p_c(1 - p_h) \quad (7.3.15)$$

where we have noted that $V = AL_z$ and used the 'reduced mass', m_{chi}, where
$1/m_{chi} = 1/m_{ci} + 1/|m_{hi}|$. Also,

$$p_c = p_c(E_{ci}) \quad \text{and} \quad 1 - p_h = 1 - p_h(E_{hi})$$

If we define the 'quantum well band gap' by

$$E_{QW} = E_G + \frac{h^2}{8L_z^2}\left(\frac{1}{m_{ci}} + \frac{1}{|m_{hi}|}\right) \quad (7.3.16)$$

then

$$E_{ci} = E_{QW} + \frac{m_{chi}}{m_{ci}}(E_{QW} - E_k) \quad \text{and} \quad E_{hi} = -\frac{m_{chi}}{|m_{hi}|}(E_{QW} - E_k)$$

To obtain the total spontaneous emission rate per unit volume, $R^{sp}_{ci, hi}$, eq. (7.3.15)
must be integrated over all values of E_k. Clearly, there can be no spontaneous
emission for energies less than E_{QW}, so that $R^{sp}_{ci, hi}$ is given by

$$R^{sp}_{ci, hi} = \frac{l}{2} \frac{64\pi^3 e^2 \mu |M|^2 m_{chi}}{m^2 L_z c^3 h^4} \int_{E_{QW}}^{\infty} E_k p_c(E_{ci})(1 - p_h(E_{hi})) \, dE_k \quad (7.3.17)$$

If Boltzmann statistics are used for the occupation probabilities, p_c and p_h, then the
above result is of the form $R^{sp}_{ci, hi} = B^s np$, where n and p are the electron and hole
carrier densities.

   To summarize the above results, the spontaneous emission rate per unit volume
is given by the above equation, provided the quantum numbers of the conduction
and heavy-hole subbands are equal. With a QW wavefunction of the form of eq.
(7.3.13) radiative band–band transitions between conduction and hole subbands
with different subband quantum numbers are forbidden. Thus, for an electron in
the ground state conduction subband, we only expect to see radiative band–band

transitions to light, heavy, and spin-split-off ground state hole subbands. Later, we shall discuss the reasons why 'forbidden' transitions (whereby electrons and holes in QW subbands of different quantum numbers recombine radiatively) have been seen experimentally [7.3.7].

A similar calculation may be carried out for the absorption coefficient, $\alpha(E_k)$. In order to calculate $\alpha(E_k)$, we start with the result for the absorption coefficient between two distinct states, I and J. After multiplying by the number of modes, $\mathcal{N}(E_k)$, in an energy range $(E_k, E_k + dE_k)$, the multi-mode absorption coefficient per unit volume per unit time per unit energy range may be derived. Then, after using the expression (7.3.10) for the two-dimensional density of electronic states, the absorption coefficient per unit volume at a particular energy E_k may be obtained. The result, if k-conservation and unpolarized incident light are assumed, is

$$\alpha(E_k) = \frac{16\pi^2 e^2 m_{chi} |M|^2}{m_0^2 c\mu h E_k} \frac{|M|^2}{L_z} (p_h(E_{hi}) - p_c(E_{ci})) \tag{7.3.18}$$

When this expression is integrated over all values of E_k, the total QW absorption coefficient is obtained. For high injected carrier densities, the absorption coefficient can become negative over an energy range which extends from E_{QW} to a higher energy E_{upper} (although E_{upper} is always less than $(\gamma_c - \gamma_h)kT$). In this energy range, gain occurs, and this forms the basis for laser operation. A gain coefficient $g(E_k)$ may be defined, which is simply equal to $-\alpha(E_k)$. It measures the extent to which stimulated emission exceeds absorption. One significant advantage of a QW laser is that the energy range over which $g(E_k)$ is positive is smaller than that of a bulk double heterostructure laser. This means that monomode laser action is more likely in QW lasers than in bulk lasers [7.3.8], since the possible laser modes are determined by the Fabry–Perot cavity, whereas the energy at which lasing occurs is determined by the energy region over which gain occurs. If the energy range over which gain occurs is smaller, then there is more likelihood of only one lasing mode occurring within that energy range. The physical reason for the above advantage that QWs have over bulk lasers arises because in an ideal bulk laser there are no carriers at the conduction and heavy-hole band edges, so the gain in a bulk laser must be zero at the band-gap energy and then rise to its maximum value before decreasing again. In contrast, for a QW, the steplike density of states means there are a large number of carriers at the QW band edges, and so the maximum gain occurs at E_{QW}. Burt [7.3.8] has shown that for a 100 Å QW at room temperature, with typical III–V semiconductor electron and hole effective masses, and an assumed threshold gain of 100 cm⁻¹, then gain only occurs for 2 meV above the QW band edge, whereas an equivalent bulk laser would have a gain spectrum that reaches its *maximum* value 4–5 meV above the bulk band edge. Thus, the QW gain spectrum is considerably narrower than that for a bulk laser.

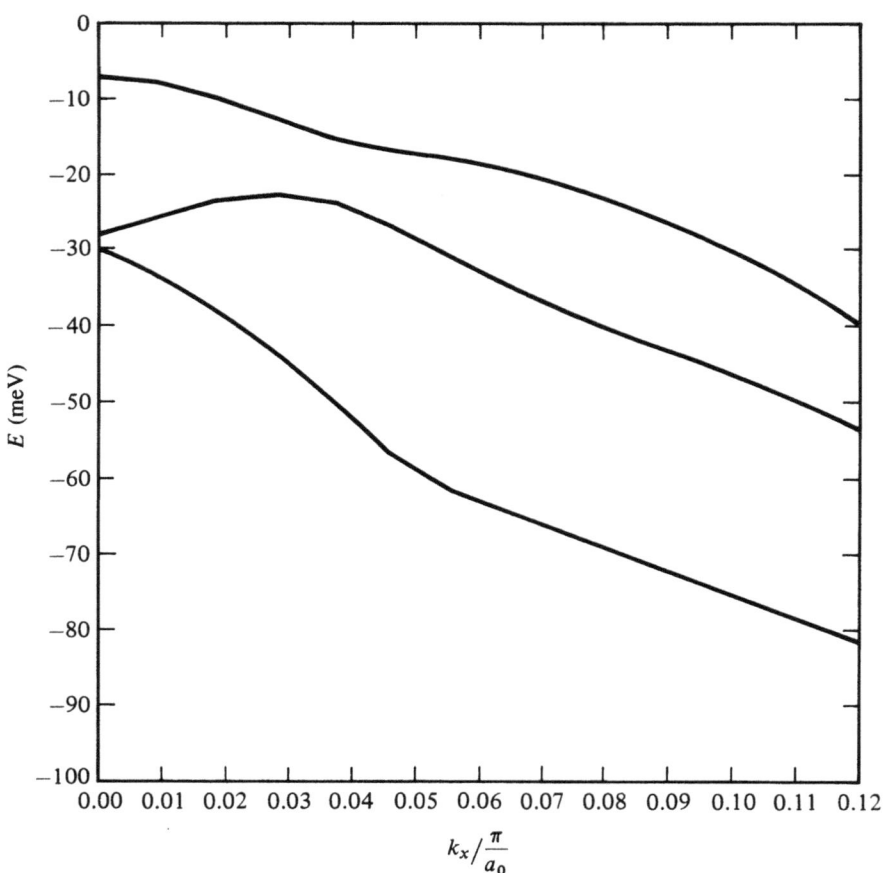

Fig. 7.3.1. The in-plane dispersion relation of a 100 Å GaAs/AlGaAs quantum well for the valence bands. (A.C.G. Wood, private communication, 1987.)

As discussed earlier, the QW wavefunction given by eq. (7.3.13) gave rise to a selection rule on radiative band–band transitions, namely that the quantum number of conduction and hole subbands should be identical, otherwise the transition is 'forbidden'. Experimentally, however, transitions which violate this selection rule have been observed [7.3.7]. The reason why experiment disagrees with the theory outlined above is that the QW wavefunction of eq. (7.3.13) is too simple. Only one bulk Bloch function has been associated with the confined state envelope function. In the valence bands of a QW, however, a considerable amount of 'band-mixing' occurs, as shown schematically in Fig. 7.3.1. Band-mixing is simply a phrase that indicates a number of bulk Bloch functions are required to describe the carrier wavefunction. The reason for the occurrence of band-mixing in the valence bands of a QW arises because holes that are heavy in the z-direction (i.e. with a small QW confinement energy) are relatively light in the (x, y)-plane.

Similarly, holes that are light in the z-direction are relatively heavy in the (x, y)-plane. This phenomenon can be understood on the basis of the Kane four-band $\mathbf{k} \cdot \mathbf{p}$ model, and has been succinctly explained by O'Reilly [7.3.9]. The ground state heavy-hole subband thus has a relatively light in-plane mass, whereas the ground state light-hole subband has a relatively heavy in-plane mass. Thus, on a parabolic subband picture, the ground state heavy- and light-hole subbands will intersect. In practice, however, an intersection does not occur, and an anticrossing is observed, with a subsequent change in the character of the subband. For example, the ground state heavy-hole subband has an in-plane effective mass that is relatively light close to the Γ point, and further away from Γ it becomes heavy. The important point to note is that, in general, the QW hole wavefunction comprises more than one bulk Bloch function. When eq. (7.3.13) is modified to take account of such band-mixing, the simplified selection rule (i.e. that only carriers in subbands of identical quantum number can recombine radiatively) is relaxed.

7.4 Nonradiative recombination in undoped QWs

7.4.1 Introduction

Intrinsic nonradiative recombination mechanisms provide an ultimate limit to the magnitude and temperature sensitivity of the threshold current of a laser, whether it is a bulk DH or a QW laser. Thus, an investigation of nonradiative recombination, whilst interesting in its own right, also gives an insight into how such mechanisms can be reduced in importance (e.g. by employing different materials systems or by changing the electronic band structure of the semiconductor by using strain effects). In section 7.3, band–band radiative recombination in single, undoped QWs was considered, and it was shown that an ideal QW should exhibit an improved, narrower gain spectrum [7.4.1], and lower, less temperature sensitive threshold currents [7.4.2]. This latter prediction, however, relied on the assumption that intrinsic nonradiative recombination was negligible. Whilst this is a reasonable approximation for short wavelength AlGaAs/GaAs lasers, it does not hold for the longer wavelength InGaAsP/InP lasers intended for use as sources in optical fiber communications systems. The radiative recombination rates for a QW were found to differ from bulk rates, and it is also expected that QW nonradiative rates will also differ from bulk values, and this is indeed found to be the case.

In lines 1, 2 and 4 of Table 3.5.1, the most important intrinsic Auger recombination rates are shown, and in Fig. 7.4.1, the most important intervalence band absorption processes are illustrated [7.4.3]. Auger recombination in bulk semiconductors has been studied thoroughly (see chapters 3 and 5). Auger recombination in a QW does, however, differ in detail from the bulk Auger

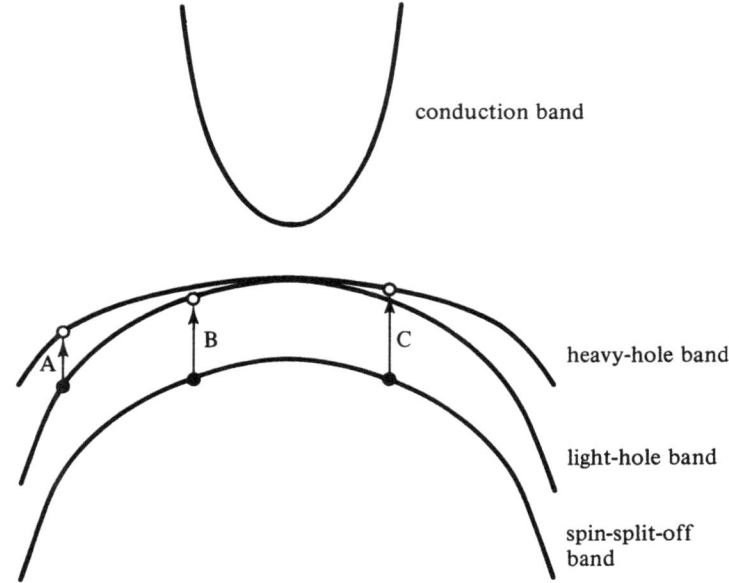

Fig. 7.4.1. The intervalence band absorption transition A between light- and heavy-hole bands, B between spin-split-off and light-hole bands, and C between spin-split-off and heavy-hole bands.

process. For example, carriers participating in a QW Auger process may occupy a number of possible subbands, and there is also the possibility that one of the carriers may be excited into an unbound state of the QW (i.e. a state which lies in the continuum of subbands above the confining barrier). The total QW Auger rate is thus the sum of all possible bound–bound Auger processes, together with the important bound–unbound Auger processes. The physical reason why these latter processes can be important is the reduction in the activation energy that can occur. The calculation of QW Auger rates is outlined in section 7.4.2, and intervalence band absorption is discussed in section 7.4.3. The results obtained are compared with the radiative recombination results obtained in section 7.3, and a rough estimate for the QW efficiency is given.

7.4.2 The theory of Auger recombination in a QW

Auger recombination in QWs includes

(a) bound–bound Auger processes; and
(b) bound–unbound Auger processes.

As in the case of bulk materials, bound–bound Auger processes are transitions in which all of the participating carriers start and finish in bound (confined) states of

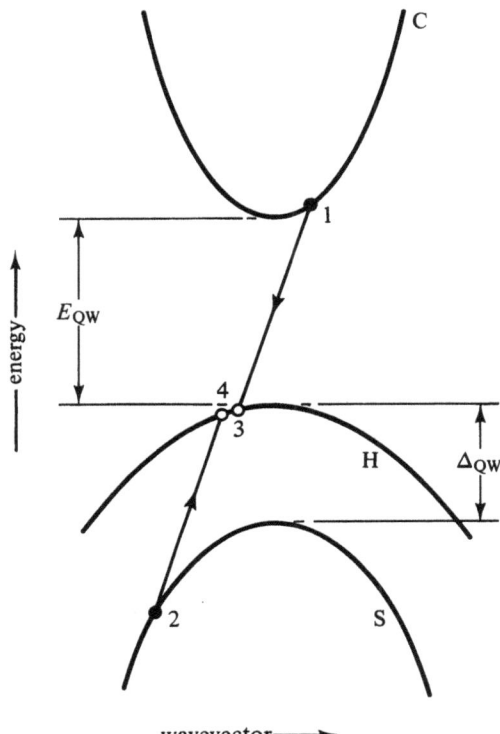

Fig. 7.4.2. A diagram illustrating the QW CHHS Auger process. Other subbands have been omitted for reasons of clarity.

the QW. The most important bound–unbound Auger process involves carriers which initially occupy bound states of the QW, but then one of the carriers is excited into an unbound state of the QW. The calculation of the QW bound–bound Auger recombination rate will be discussed first, since the bound–unbound rate may be obtained from this result. Also, the calculations will be presented for the so-called CHHS Auger rate (the notation used indicates the bands in which the carriers involved in the transition are to be found – see Fig. 7.4.2). The result for the CHHS Auger rate can be used to obtain all other important Auger rates.

The starting point for the calculation is Fermi's golden rule (3.2.13). The matrix element for the Auger transition can be calculated once the carrier wavefunctions and the perturbing potential are specified. The perturbing potential is taken to be the dynamically screened Coulomb potential [7.4.4], since, for typical transition energies of interest here (0.8–1.0 eV), carriers cannot respond quickly enough to effectively screen out the Coulomb potential. The wavefunctions used here are products of simple QW envelope wavefunctions and Bloch functions. The summation over all possible states which appears in Fermi's golden rule is converted to a phase space integral, and then the matrix element is removed from this integral and replaced by its most probable (or threshold) value, an approximation that is also often used in the case of bulk materials (see, for example, eq. (3.7.5)). The remaining phase space integral is readily evaluated if Boltzmann statistics and parabolic QW subbands are assumed. The effects of these latter two approximations on our results are discussed at the end of this section.

Thus, the QW CHHS bound–bound Auger recombination rate per unit volume, u, is given by:

$$u = \frac{1}{AL} \frac{2\pi}{\hbar} \sum_{\text{all states}} \Phi |M|^2 \delta(E) \tag{7.4.1}$$

where A is the area in the plane of the QW, L is the QW width, Φ is the statistical factor familiar from bulk Auger calculations (see chapter 3), and M is the matrix element for the Auger process. Also, $E = |E_1 + E_2 - E_3 - E_4|$, where E_i is the energy of the carrier in the ith state. Φ, the statistical factor, is given by

$$\Phi = p_{c,1}(\mathbf{k}_1) p_{v,3}(\mathbf{k}_3) p_{v,4}(\mathbf{k}_4)(1 - p_{v,2}(\mathbf{k}_2))$$
$$- (1 - p_{c,1}(\mathbf{k}_1))(1 - p_{v,3}(\mathbf{k}_3))(1 - p_{v,4}(\mathbf{k}_4)) p_{v,2}(\mathbf{k}_2) \tag{7.4.2}$$

where $p_{c,i}(\mathbf{k}_i)$ and $p_{v,i}(\mathbf{k}_i)$ are the electron and hole distribution functions for the ith state, respectively. If the assumption is made that Boltzmann statistics with quasi-Fermi levels are adequate to describe the distribution functions, then, as in section 3.8.1, the statistical factor simplifies to

$$\Phi = \exp(\eta_2 - F_h)(\exp(F_e - F_h) - 1) \tag{7.4.3}$$

Physically, the assumption of single quasi-Fermi levels for the conduction and valence subbands implies that carriers within the conduction subbands are in thermal equilibrium, as are carriers in the valence subbands, but the two sets of carriers are not in thermal equilibrium (i.e. intra-subband relaxation times are assumed to be much shorter than inter-subband relaxation times).

The matrix element for the Auger process can be written as

$$|M|^2 = \beta |M_D|^2 \tag{7.4.4}$$

where β lies between 2 and 4 [7.4.5], and where M_D denotes the direct matrix element, defined as in eq. (3.3.24) by

$$M_D = \int \int \varphi_3^*(\mathbf{r}_1) \varphi_4^*(\mathbf{r}_2) \frac{e^2}{\varepsilon |\mathbf{r}_1 - \mathbf{r}_2|} \varphi_1(\mathbf{r}_1) \varphi_2(\mathbf{r}_2) \, d\mathbf{r}_1 \, d\mathbf{r}_2 \tag{7.4.5}$$

$\varphi_i(\mathbf{r})$ denotes the wavefunction of the ith state, and ε is the relative permittivity. To obtain an explicit expression for the Auger recombination rate, the carrier wavefunctions need to be specified. If band-mixing effects are ignored, i.e. only one bulk Bloch periodic function is associated with the QW envelope function, then the wavefunction of a carrier in a bound state can be written as

$$\varphi_i(\mathbf{r}) = \left(\frac{2V}{AL}\right)^{\frac{1}{2}} u_i(\mathbf{r}) \sin\left(N_i \pi z / L\right) \exp\left(i\boldsymbol{\kappa}_i \cdot \boldsymbol{\rho}_i\right) \tag{7.4.6}$$

for $0 < z < L$, and, if wavefunction leakage out of the well is neglected, then $\varphi_i(\mathbf{r})$ is zero outside this range. The subscript i labels the band in which the carrier is to be found, and N_i labels the quantum number of the confined state. $u_i(\mathbf{r})$ is a bulk Bloch periodic function, and $\boldsymbol{\kappa}_i$, $\boldsymbol{\rho}_i$ are the two-dimensional wavevector and position vector, respectively. V is the total volume of the semiconductor. Equation (7.4.6) is an approximation since band-mixing has been ignored, but also the above expression assumes an infinite QW barrier height, since the evanescent parts of the wavefunction outside the well are ignored. These can, however, be included without difficulty, although the algebra is more complicated. Numerical results both with and without the evanescent parts of the QW wavefunction are presented later. By using the Fourier transform of $1/|\mathbf{r}|$, the expression for the direct matrix element can be rewritten as

$$M_{\mathrm{D}} = \frac{4\pi e^2}{\varepsilon} \frac{1}{(2\pi)^3} \iint \frac{I_{3,1}(\mathbf{q}) \, I_{4,2}(-\mathbf{q})}{|\mathbf{q}|^2} \, d\mathbf{q} \tag{7.4.7}$$

where

$$I_{m,n}(\mathbf{q}) = \int \varphi_m^*(\mathbf{r}) \, \varphi_n(\mathbf{r}) \exp\left(i\mathbf{q} \cdot \mathbf{r}\right) d\mathbf{q} \tag{7.4.8}$$

Using the wavefunctions above (eq. (7.4.6)), $I_{m,n}(\mathbf{q})$ may be calculated, although the expression is cumbersome [7.4.4]. The first important result arising from the analysis can be obtained by evaluating $I_{m,n}(\mathbf{q})$. Firstly, the assumption of a single bulk Bloch periodic function associated with the QW envelope function gives rise to a delta function, $\delta(\boldsymbol{\kappa}_1 + \boldsymbol{\kappa}_2 - \boldsymbol{\kappa}_3 - \boldsymbol{\kappa}_4)$. Thus, in addition to the energy conservation present in Fermi's golden rule, the wavefunctions used here give rise to momentum conservation in the plane of the well. The consequence of this is that carriers are, in general, situated away from the band edges, and so an activation energy is required. This will manifest itself in the final expression for the Auger rate having an activated form. Equation (7.4.8) can also be used to obtain a selection rule for the Auger process to occur. The selection rule is found to be $|N_1 + N_2 - N_3 - N_4| = 0$ or a positive or negative *even* integer. This result arises as a consequence of the confined state parity alone. When the matrix element is calculated from $I_{4,2}(\mathbf{q})$ and $I_{3,1}(\mathbf{q})$ the overlap integral of the bulk Bloch periodic functions in the conduction and heavy-hole band, and the heavy-hole and spin-split-off band, are required. This will be discussed in more detail later.

The expression for the Auger rate per unit volume is thus

$$u = C \int I^2(\kappa_0) \exp\left(\frac{E_2 - F_v}{kT}\right) \delta(E_1 + E_2 - E_3 - E_4)$$

$$\times \delta(\kappa_1 + \kappa_2 - \kappa_3 - \kappa_4) \, d\kappa_1 \, d\kappa_2 \, d\kappa_3 \, d\kappa_4 \qquad (7.4.9)$$

where C is a material dependent constant, $I(\kappa_0)$ represents the value of the matrix element evaluated at κ_0, where $\kappa_0 = |\kappa_1 - \kappa_3|$. If κ_0 is evaluated for the most probable, or threshold, process, then the factor $I^2(\kappa_0)$ may be removed from the above integral without appreciable loss of accuracy, since the statistical factor is highly peaked for wavevectors close to the threshold values, whereas the matrix element factor is slowly varying. Once $I^2(\kappa_0)$ has been removed from the phase space integral, the integral may be evaluated if the assumption of isotropic, parabolic subbands is used [7.4.5]. The final result for the QW CHHS Auger rate per unit volume for carriers whose confined state quantum numbers are N_1, N_2, N_3, N_4 is:

$$u(N_1, N_2, N_3, N_4) = \frac{1}{L}\left(\exp\left(\frac{F_c - F_v}{kT}\right) - 1\right) \exp\left(\frac{-\Delta_{QW} - F_v}{kT}\right) \frac{e^4 k^2 T^2}{\pi^2 \varepsilon^2 \hbar^7} I^2(\kappa_0)$$

$$\times \frac{m_c \, m_s \, m_h^2 (2m_h + m_c - m_s)}{(2m_h + m_c)^2}$$

$$\times \exp\left(-\frac{2m_h + m_c}{2m_h + m_c - m_s} \frac{E_{QW} - \Delta_{QW}}{kT}\right) \qquad (7.4.10)$$

E_{QW} is the difference between the conduction and heavy-hole subband edge energies, and Δ_{QW} is the difference between the heavy-hole and spin-split-off subband edge energies. Also,

$$I(\kappa_0) = M_{3,1} M_{4,2} \left(\frac{64L}{\pi^3}\right) N_1 N_2 N_3 N_4$$

$$\times \int_{-\infty}^{\infty} \frac{x^2 F(x)}{(x^2 + \kappa_0^2 L^2/\pi^2) \prod_{i=1,2} [x^2 - (N_{i+2} - N_i)^2][x^2 - (N_{i+2} + N_i)^2]} \, dx$$

$$(7.4.11)$$

where $F(x) = \sin^2(\pi x/2)$ if $(N_1 + N_3)$ is even, and $F(x) = \cos^2(\pi x/2)$ if $(N_1 + N_3)$ is odd.

In eq. (7.4.10), it has been assumed that all subbands associated with a particular band have the same effective masses (e.g. m_c for conduction subbands, m_h for heavy-hole subbands, and m_s for spin-split-off subbands). This is, however, only for algebraic convenience, and different effective masses for each of the four states can be included straightforwardly if required.

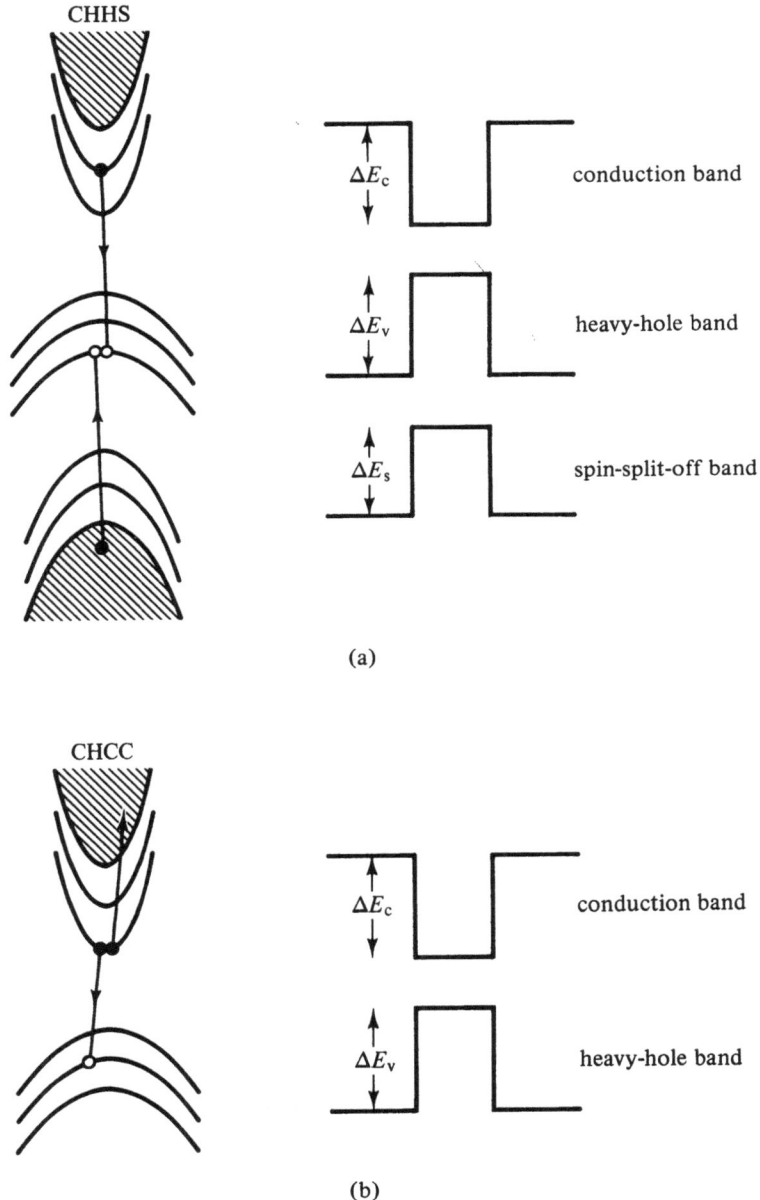

(a)

(b)

Fig. 7.4.3. (a) The CHHS bound–unbound Auger process in a quantum well.
(b) The CHCC bound–unbound Auger process in a quantum well.

The factors $M_{3,1}$ and $M_{4,2}$ are the overlap integrals mentioned earlier, and are defined by

$$M_{m,n} = \int_V u_m^*(\mathbf{r}) u_n(\mathbf{r}) \, d\mathbf{r} \qquad (7.4.12)$$

These overlap integrals have been discussed in section 3.6, where the more general notation $F_{nk}^{n'k'}$ was used.

As discussed earlier, the Auger rate is expected to be activated due to the combined constraints of energy and in-plane momentum conservation. Re-arranging eq. (7.4.10) we find

$$u(N_1, N_2, N_3, N_4) = C' I^2(\kappa_0) np^2 \exp\left(-\frac{m_\mathrm{s}}{2m_\mathrm{h}+m_\mathrm{c}-m_\mathrm{s}}\frac{\Delta E}{kT}\right) \qquad (7.4.13)$$

where C' is a material dependent constant and $\Delta E = E_\mathrm{QW} - \Delta_\mathrm{QW} = (E_\mathrm{s1} - E_\mathrm{s3}) - (E_\mathrm{s4} - E_\mathrm{s2})$, where the energies E_{si} refer to the subband edge energies for the ith state. Thus, the activation energy for the CHHS QW Auger process is $(m_\mathrm{s}/2m_\mathrm{h}+m_\mathrm{c}-m_\mathrm{s})\,\Delta E$, which is the same result as for the activation energy of the bulk Auger process [7.4.6], although the value of ΔE may differ since it refers to differences in QW band gaps as opposed to bulk band gaps. The above results hold for ΔE greater than zero, which is typically the case for III–V semiconductors. Haug [7.4.7] has given results for the case where $\Delta E < 0$.

Equation (7.4.13) illustrates the activated form of the Auger rate which is analogous to the bulk case (3.8.14), and also the strong dependence on carrier density. If Boltzmann statistics are used, then the Auger rate u is proportional to $p^2 n$, where n and p are the electron and hole densities, respectively. Such a dependence is physically reasonable since, for the CHHS Auger process to occur, two holes are required in the heavy-hole subband, and an electron is required in the conduction subband. For the CHCC Auger process, the carrier density dependence is $n^2 p$. Note, however, that such simple carrier density dependence is not obtained if Fermi–Dirac statistics are used.

Next, consider bound–unbound Auger processes (Fig. 7.4.3). Such an Auger process is formally the same as the bound–bound Auger transitions discussed above, except that the final state lies in a subband that forms part of the unbound state continuum above the top of the confining barrier. Thus, integration of an expression of the same form as the bound–bound QW Auger rate over the continuum of unbound states, weighted with the appropriate density of unbound states, will yield the total bound–unbound Auger rate.

For simplicity, the only bound–unbound Auger processes to be considered here are those in which the excited carrier is in the continuum of unbound states, and the other carriers are in their respective ground state subbands.

The physical reason why QW bound–unbound Auger processes are expected to be important lies in the fact that, for one of the continuum of unbound subbands, the activation energy of the Auger process can be zero. Auger transitions involving subbands close to this zero activation energy subband will thus have a considerably enhanced transition rate.

Consider the CHHS QW bound–unbound Auger process (Fig. 7.4.3a). If the bound–bound CHHS Auger rate per unit volume is written as $u(E_\perp)$, where $E_\perp$ is the bottom of the relevant spin-split-off continuum subband edge relative to the ground state subband edge, then the CHHS bound–unbound Auger rate can be written as:

$$u_{\text{unbound}} = \int_{\Delta E_s}^{\infty} u(E_\perp) \left(\frac{2m_s}{\hbar^2}\right)^{\frac{1}{2}} \frac{L_{\text{tot}}}{2\pi(E_\perp - \Delta E_s)^{\frac{1}{2}}} \, dE_\perp \qquad (7.4.14)$$

where $2L_{\text{tot}}$ is the total width of well and barrier regions (assumed to be $\gg L$), ΔE_s is the spin–orbit split-off barrier height, and m_s is the effective mass of the carrier in the spin-split-off band.

By using an appropriate choice for the unbound state wavefunction, the bound–bound Auger rate may be obtained by numerical integration of eq. (7.4.14) [7.4.4].

To illustrate the results of this section, numerical results are presented for both bound–bound and bound–unbound Auger transition rates in 1.3 μm InGaAsP/InP QWs (in a 1.3 μm InGaAsP/InP QW, the well material is $In_{1-x}Ga_xAs_yP_{1-y}$, lattice matched to the barrier material, InP, and the alloy composition is adjusted at each well width to ensure that the QW band gap corresponds to a wavelength of 1.3 μm). The parameters used in the calculations for InGaAsP (e.g. band gaps, effective masses etc.) are taken from [7.4.8]. 10^{18} carriers cm^{-3} are assumed to be injected into the QW at 300 K, and the quasi-Fermi levels and carrier density in any particular subband are calculated in the usual way [7.4.4]. Once the quasi-Fermi levels have been calculated, the expressions given above for the bound–bound and bound–unbound Auger transition rates can be used to obtain numerical estimates for the Auger transition rate. The overlap integrals discussed above have been taken from a calculation by Scharoch and Abram [7.4.9], which uses a modified four-band $\mathbf{k} \cdot \mathbf{p}$ approach, and gives results in excellent agreement with 15-band $\mathbf{k} \cdot \mathbf{p}$ and nonlocal pseudopotential calculations [7.4.10].

Using the above method, the QW CHCC bound–bound Auger rate has been calculated as a function of well width, and the results are shown in Fig. 7.4.4. A calculation which takes into account the evanescent parts of the QW wavefunction has also been performed, and the results are also shown in Fig. 7.4.4. The saw-tooth variation in the total bound–bound rate (obtained by summing over all possible inter-subband Auger processes) arises since new subbands become bound

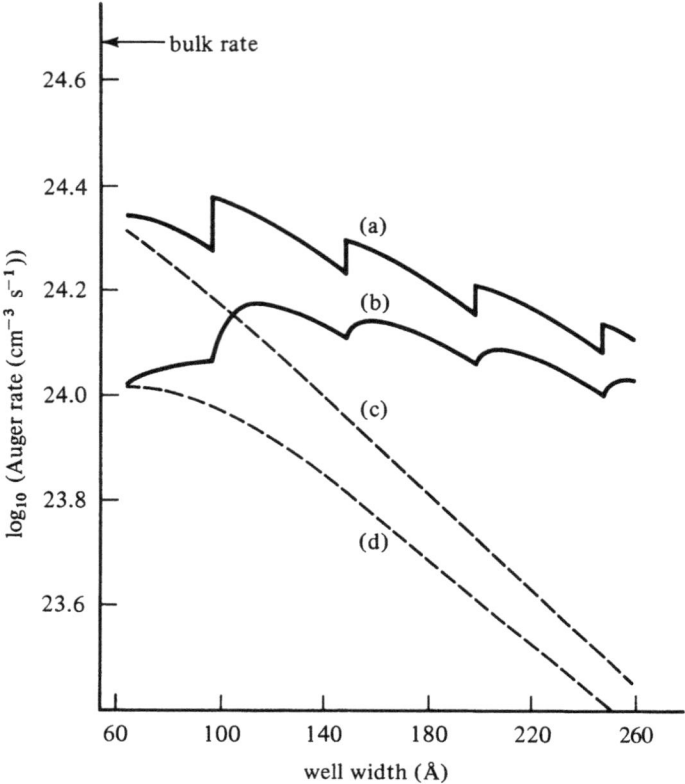

Fig. 7.4.4. Graph showing the variation of the QW CHCC bound–bound Auger rate as a function of quantum well width for a 1.3 μm InGaAsP/InP QW. The temperature is 300 K, and the injected carrier density is 10^{18} cm^{-3}. (a) and (c) assume no wavefunction leakage out of the wells, whereas this is taken into account in (b) and (d). The *total* bound–bound rates are shown in (a) and (b), whereas the ground state rate *only* is shown in (c) and (d). Overlap integrals are taken from [7.4.9].

by the well at certain well widths as the QW width is widened. Since the activation energy of a bound–bound Auger transition can be decreased significantly if the excited carrier occupies a state in a higher lying subband, these transitions can be favored, leading to discontinuities in the total bound–bound Auger rate. Also shown in Fig. 7.4.4 is the bound–bound Auger rate when all the participating carriers in the Auger process reside in their respective ground state subbands, and it can be seen that, for small well widths (i.e. < 100 Å), the total CHCC QW Auger rate is given, to a good approximation, by this 'ground state rate' alone. Also, at small well widths, it can be seen that neglect of wavefunction leakage out of the QW overestimates the Auger rate by a factor of two. At wider well widths, the two calculations are in better agreement, as expected. The ground state rate decreases

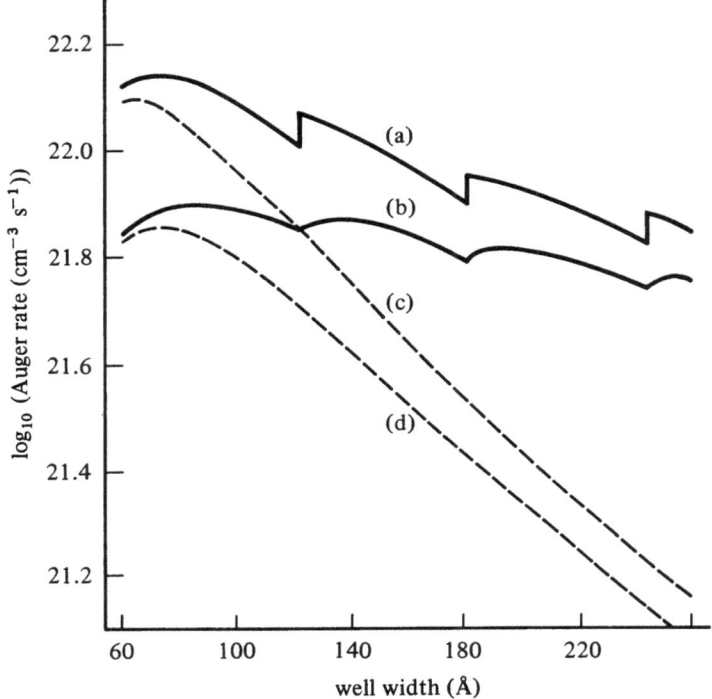

Fig. 7.4.5. Graph showing the variation of the QW CHHL bound–bound Auger rate as a function of quantum well width for a 1.3 μm InGaAsP/InP QW. (a)–(d) have the same meaning as in Fig. 7.4.4. The temperature is 300 K, and the injected carrier density is 10^{18} cm^{-3}. Overlap integrals are taken from [7.4.9].

with well width since the number of carriers occupying ground state subbands decreases with well width (since it is the number per unit *volume* which has been kept constant). The approximate Auger coefficient for the bound–bound CHCC QW transition is 0.2×10^{-29} cm^6 s^{-1}.

In Fig. 7.4.5, the same calculations have been performed for the QW CHHL bound–bound Auger transition rate, and the same general features are observed. The QW CHHL Auger coefficient is approximately two orders of magnitude less than the CHCC value though, because of the small value for the overlap integral between light- and heavy-hole bands. In fact, in the simplest four-band $\mathbf{k} \cdot \mathbf{p}$ model [7.4.11], this overlap integral vanishes for the case of collinear wavevectors which is appropriate here.

The CHHS bound–bound Auger rate cannot be calculated because, with the materials data used here [7.4.8], there is no spin-split-off confining potential.

Next, results for bound–unbound Auger rates are presented for the 1.3 μm InGaAsP/InP QW. In Fig. 7.4.6 the CHCC QW bound–unbound Auger rate is illustrated along with the ground state bound–bound QW CHCC Auger rate and, for comparison, the bulk rate. The oscillations in the bound–unbound Auger rate can be explained if it is assumed that the dominant contribution to the total bound–unbound rate arises from the zero activation energy process [7.4.4]. Fig. 7.4.6 also shows that the bound–unbound Auger rate will only be important for QWs of small width (< 100 Å).

Fig. 7.4.7(a) and (b) show the CHHL and CHHS bound–unbound QW Auger rates, respectively. The interpretation of these results is not as straightforward as for the bound–unbound CHCC QW Auger process since, for the zero activation energy transition, the Auger rate is zero because the overlap integral between the valence band states is zero. However, the absolute magnitude of the CHHL and CHHS bound–unbound Auger rates is approximately two orders of magnitude smaller than the corresponding CHCC bound–unbound transition rate.

Thus, the conclusion of the numerical work is that the CHCC Auger process provides the dominant contribution to the Auger coefficient in 1.3 μm InGaAsP/InP QWs. For well widths larger than about 100 Å, the total rate is given by the QW bound–bound CHCC Auger rate alone, and the Auger coefficient is approximately 0.2×10^{-29} cm^6 s^{-1}. This is approximately a factor of ten smaller than experimental measurements [7.4.12]–[7.4.14]. The results presented here are, however, expected to be underestimates since we have neglected band-mixing in the valence bands of the QW, which will tend to increase the value of the overlap integrals, and off-threshold contributions to the Auger rate have been neglected. If these effects are included, closer agreement between theory and experiment is expected.

The calculations above have assumed parabolic QW subbands and Boltzmann statistics with quasi-Fermi levels. These approximations seem at first sight to be fairly drastic, since the results are intended to apply to laser structures in which carrier densities are typically of the order of 10^{18} cm^{-3}, and the excited (Auger) carrier can be 1 eV, or more, from the band edge. Recent results [7.4.15]–[7.4.18] show that, in bulk semiconductors, the CHCC Auger rate is reduced by two orders of magnitude when accurate band structure is used. However, any reduction in the QW CHCC Auger rate is expected to be *less* because nonparabolicity is only important for carriers far from the subband edge, and, in the QW, large contributions to the CHCC Auger rate arise from inter-subband transitions involving higher lying subbands, because the activation energy is reduced, and thus the transition is more 'vertical', and so these carriers are all reasonably close to their respective subband edge. It has also been shown that the effects of using Boltzmann statistics are only likely to give errors in the Auger rate of about 20 %, [7.4.4], which is negligible in comparison with the uncertainties in overlap integrals, and neglect of off-threshold contributions to the Auger rate.

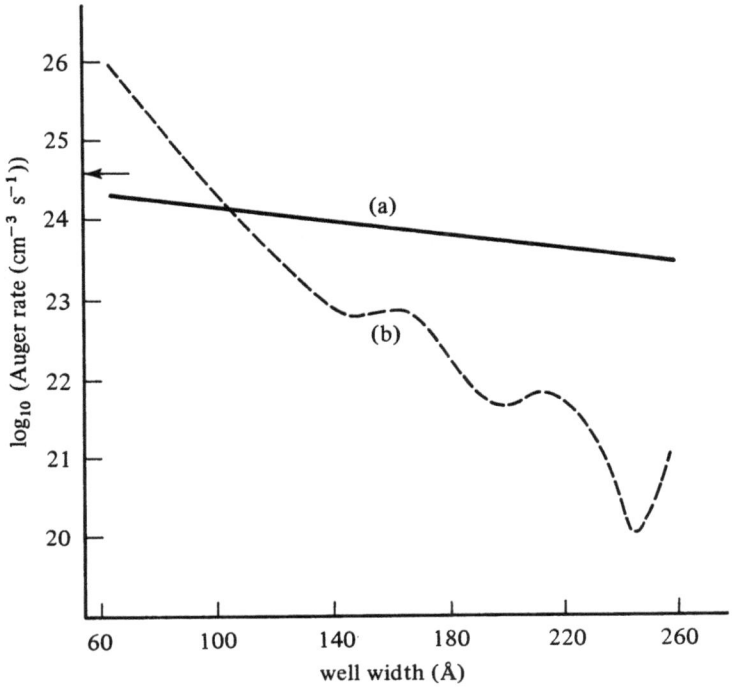

Fig. 7.4.6. Graph showing the variation with well width of the ground state bound–bound CHCC Auger rate (a) and the CHCC bound–unbound rate (b). The temperature and injected carrier densities are the same as for Figs 7.4.4 and 7.4.5. Overlap integrals are taken from [7.4.9].

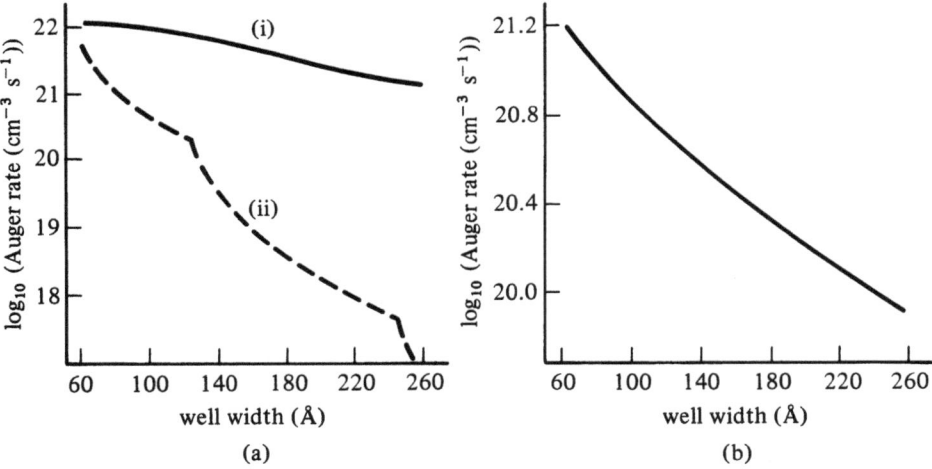

(a)   (b)

Fig. 7.4.7. (a) Graph showing the variation with well width of the ground state bound–bound CHHL Auger rate (i) and the CHHL bound–unbound rate (ii). All other parameters are as in Fig. 7.4.5. (b) Graph showing the variation with well width of the CHHS bound–unbound QW Auger rate. All other parameters are as in Fig. 7.4.5.

Thus, the results above are expected to provide a reasonable guide to Auger recombination in a QW. In order to compare Auger recombination in QW and bulk lasers, a 'rough and ready' calculation gives [7.4.19]

$$\frac{u_{\mathrm{Auger}}(\mathrm{bulk})}{u_{\mathrm{Auger}}(\mathrm{QW})} = \left(\frac{E_{\mathrm{a}}}{\pi k T}\right)^{\frac{1}{2}} \tag{7.4.15}$$

where it has been assumed that the same carrier densities are present in bulk and QW active regions. E_{a} is the activation energy for the dominant Auger process. The above result holds for wells larger than 50 Å, yet smaller than about 150 Å, so that the majority of carriers are in their ground state subbands. For the III–V materials of interest here, typical activation energies are of the order of 2–3 kT, and so Auger recombination rates are expected to be similar in QW and bulk semiconductors for equivalent carrier densities. The limited amount of experimental data [7.4.13], [7.4.14] is in agreement with this prediction.

The rough calculation above may be extended to show that the efficiency of a QW laser is expected to be the same as that of a bulk DH laser if the threshold carrier densities are equal [7.4.19], [7.4.20].

7.4.3 Intervalence band absorption

Intervalence band absorption (IVBA), shown schematically in Fig. 7.4.1, has been suggested as another possible cause of the high temperature sensitivity of long wavelength InGaAsP DH lasers [7.4.3].

Referring to Fig. 7.4.1, if only vertical transitions are considered, i.e. **k**-conservation is assumed, then the most important IVBA transition in InGaAsP is that between the spin-split-off and heavy-hole bands. The transition between light- and heavy-hole bands is unimportant at a wavelength of 1.3 μm since the energy difference between states in these two bands is much less than the InGaAsP band gap. Also, transitions between spin-split-off and light-hole bands are unimportant because of the small probability of hole occupancy in the light-hole band.

The expression for the IVBA coefficient for the spin-split-off to heavy hole transition can be obtained using Fermi's golden rule, to give, for bulk semiconductors [7.4.21]

$$\alpha(v) = \frac{e^2 h^2}{m^2 \mu \omega c \pi} \int |M|^2 (p_{\mathrm{H}} - p_{\mathrm{s}}) \, \delta(E_{\mathrm{HS}} - hv) \, \mathrm{d}\mathbf{k} \tag{7.4.16}$$

where p_{H} and p_{S} are the hole occupation factors for the heavy-hole and spin-split-off bands. $|M|^2$ is the squared matrix element, $\langle U_{\mathrm{s}} | \mathbf{e} \cdot \mathbf{p} | U_{\mathrm{h}} \rangle$, averaged over all polarization directions and spin states. m is the free electron mass, μ is the refractive index, c is the speed of light *in vacuo* and v is related to the energy of the transition, E, via $E = hv$. E_{HS} is the energy difference between the heavy-hole and spin-split-off bands at a particular value of wavevector.

If isotropic, parabolic bands are assumed, then

$$\alpha(v) = \frac{2e^2(2m^*)^{\frac{3}{2}}}{m^2 \mu h^2 cE}(hv - \Delta)^{\frac{1}{2}}|M|^2(p_H - p_S) \qquad (7.4.17)$$

where the transition with energy E ($= hv$) takes place at a wavevector

$$k = \left(\frac{2m^*(hv - \Delta)}{(h/2\pi)^2}\right)^{\frac{1}{2}} \qquad (7.4.18)$$

where $1/m^* = 1/m_s - 1/m_h$, and Δ is the spin–orbit splitting parameter. Note the negative sign that appears in the expression for m^*, which arises because IVBA is a radiative band–band transition involving energy bands with the same curvature.

Numerical calculations using accurate band structure have been performed [7.4.21], and typical IVBA coefficients for bulk InGaAsP, with injected carrier densities of the order of 10^{18} cm^{-3}, are found to be 10–20 cm^{-1}. In addition, the calculations show that the carrier dependence is linear, varying with p, the heavy-hole carrier density, as expected on physical grounds. No numerical calculations have been reported in the literature for QW IVBA coefficients. Note, though, that eq. (7.4.16) may be used to calculate the QW IVBA coefficient, and if isotropic, parabolic subbands are assumed, an expression similar to eq. (7.4.16) may be obtained for the QW IVBA coefficient, except that, for the QW, $\alpha(v)$ is proportional to $\theta(hv - \Delta)$, where $\theta(x)$ is the step function.

It is still unclear which of the two nonradiative recombination mechanisms discussed here, Auger recombination or IVBA, provides the dominant loss mechanism in long wavelength InGaAsP lasers. However, the different density dependence of the two mechanisms implies that the laser design plays a crucial role in determining which mechanism is the more important. In this chapter, only single QW structures have been considered, but in practical device structures MQWs are used to provide adequate carrier confinement, and wide cladding layers are used to provide confinement of electromagnetic radiation. The actual laser design thus determines the carrier density required for threshold, and so determines which of the two nonradiative recombination mechanisms is the most important. In other words, the question of which of the two, Auger recombination or IVBA, is the more important factor determining T_0 values, is somewhat meaningless unless the laser structure is also specified.

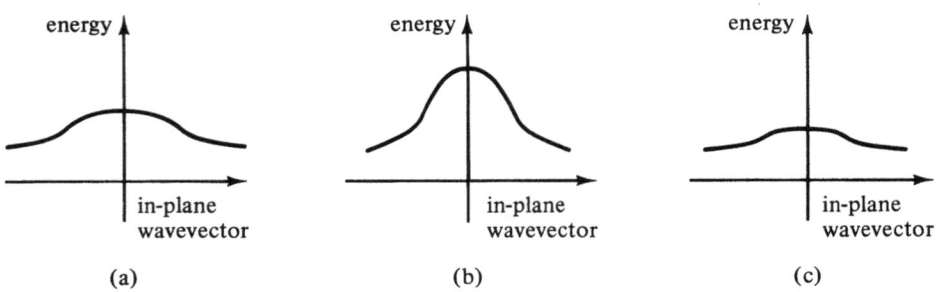

Fig. 7.5.1. Schematic diagram illustrating the effects of strain on the ground state hole band structure in a QW. (a) Unstrained QW (compare with Fig. 7.3.1). (b) The well material is in biaxial compression (e.g. InGaAs/GaAs QW). (c) The well material is in biaxial tension (e.g. GaAs/InP QW).

7.5 Ultra-low threshold current strained QW lasers

As we have seen in section 7.4, two intrinsic loss mechanisms, Auger recombination and intervalence band absorption, are responsible for the high temperature sensitivity of long wavelength InGaAsP bulk and QW lasers.

Recently, though, it has been suggested that strained QW laser structures may exhibit smaller and less temperature dependent threshold currents [7.5.1], since the nonradiative mechanisms discussed earlier will be of much less importance in such structures.

Adams proposed [7.5.1] the growth of an active layer on a substrate of smaller lattice constant (e.g. $In_x Ga_{1-x}As$, with $x > 0.53$, on InP). The elastic strain energy per unit area stored in the strained layer increases with layer thickness, and so, provided the layer is thinner than a certain critical thickness [7.5.2]–[7.5.4] it is energetically favorable for the layer to strain elastically so that its in-plane lattice constant matches that of the substrate, rather than strain plastically by the formation of misfit dislocations.

The resulting biaxial strain causes a tetragonal distortion of the active layer crystal structure, and this, in turn, affects the electronic band structure of the active layer. The valence band is particularly affected, and the uppermost valence band has a light in-plane effective mass, due to a light-hole 'cap' (see Fig. 7.5.1). Such a light-hole 'cap' ensures that IVBA is drastically reduced because there are no holes available in the heavy-hole band at the large wavevectors required for IVBA to occur. In addition, Auger recombination is expected to be reduced, since the activation energy is increased, because carriers taking part in Auger transitions are situated farther away from the valence band edge.

Thus, the use of strained QWs in laser structures offers the advantages of much reduced Auger recombination and IVBA, so possibly leading to lower and less temperature dependent threshold currents.

Appendix A

The delta function (section 3.2)

Consider the integral

$$I(\alpha, x - x') \equiv \frac{1}{2\pi} \int_{-\infty}^{\infty} e^{-\alpha|k| + ik(x - x')} \, dk \quad (\alpha > 0)$$

$$= \frac{1}{\pi} \int_{0}^{\infty} e^{-\alpha k} \cos k(x - x') \, dk \tag{A.1}$$

where we have put $e^{ik(x-x')} = \cos k(x - x') + i \sin k(x - x')$ and noted that the cosine is even and the sine is odd in its argument. The integral is the real part of

$$\int_{0}^{\infty} e^{-\beta k} \, dk = -\frac{1}{\beta} \equiv \frac{1}{\alpha - i(x - x')}$$

Hence

$$I(\alpha, x - x') = \frac{\alpha}{\pi[\alpha^2 + (x - x')^2]} \tag{A.2}$$

To study the properties of $I(\alpha)$, introduce a function $f(x)$ which is free of singularities.

The integral has the property that for $a < x < b$

$$\int_{a}^{b} f(x') I(\alpha, x - x') \, dx' = \frac{\alpha}{\pi} \left(\int_{a}^{x-\varepsilon} + \int_{x+\varepsilon}^{b} \right) \frac{f(x') \, dx'}{\alpha^2 + (x - x')^2}$$

$$+ \frac{1}{\pi} \left[f(x + \varepsilon) \tan^{-1} \frac{\varepsilon}{\alpha} - f(x - \varepsilon) \tan^{-1} \left(\frac{-\varepsilon}{\alpha} \right) \right]$$

$$+ \frac{1}{\pi} \left(\int_{-\varepsilon}^{0} + \int_{0}^{\varepsilon} \right) f'(z + x) \tan^{-1} \frac{z}{\alpha} \, dz \tag{A.3}$$

511

The last two terms arise from a partial integration of

$$\frac{\alpha}{\pi}\int_{-\varepsilon}^{\varepsilon}\frac{f(z+x)}{\alpha^2+z^2}\,dz\,(z\equiv x'-x;\ -\varepsilon\leqslant z\leqslant\varepsilon)$$

by regarding this integral as $\int uv'=uv|-\int u'v$ with

$$u=f(z+x),\quad v'=\alpha/\pi(\alpha^2+z^2),\quad v=(1/\pi)\tan^{-1}(z/\alpha)$$

One must let $\varepsilon\to 0$. The last two expressions then vanish and nothing interesting is found.

Let us now consider the case where the limit $\alpha\to 0$ is part of the definition of the integral (A.1). Thus consider

$$J(x-x')\equiv\lim_{\alpha\to 0}I(\alpha,x-x')=\frac{1}{2\pi}\lim_{\alpha\to 0}\left[\int_{-\infty}^{\infty}e^{-\alpha|k|+ik(x-x')}\,dk\right]\qquad (A.4)$$

and substitute $J(x-x')$ for $I(\alpha,x-x')$ in eq. (A.2). Then one must take the limit $\alpha\to 0$ prior to the limit $\varepsilon\to 0$. One then finds three interesting properties of eq. (A.4).

First property

From eq. (A.2)

$$J(x-x')=\begin{cases}0 & (x\neq x')\\ \infty & (x=x')\end{cases}\qquad (A.5)$$

Second property

Put $a=x-\varepsilon$, $b=x+\varepsilon$, when the first term in eq. (A.3) vanishes. If also $f(x)=1$, the third term in eq. (A.3) vanishes as well. The second term is

$$\frac{1}{\pi}\left[\frac{\pi}{2}-\left(-\frac{\pi}{2}\right)\right]=1$$

One finds, noting that the limit $\varepsilon\to 0$ is not needed,

$$\int_{x-\varepsilon}^{x+\varepsilon}J(x-x')\,dx'=\int_{x-\varepsilon}^{x+\varepsilon}\left[\lim_{\alpha\to 0}\frac{\alpha}{\pi}\frac{1}{\alpha^2+(x-x')^2}\right]dx'=1\qquad (A.6)$$

Third property

If the value x lies outside the range of integration, the second and third terms of eq. (A.3) do not arise and the first term yields zero since it contains α as a factor.

If the value x lies within the range of integration, the first term in eq. (A.3) still vanishes as $x' = x$ has been excluded and $\alpha \to 0$. The second term gives

$$\frac{1}{\pi}[f(x+\varepsilon)+f(x-\varepsilon)]\frac{\pi}{2} \xrightarrow{\varepsilon \to 0} f(x)$$

The third term gives

$$\frac{1}{\pi}\left[-\frac{\pi}{2}\int_{-\varepsilon}^{0} f'(z+x)\,dz + \frac{\pi}{2}\int_{0}^{\varepsilon} f'(z+x)\,dz\right] = \frac{1}{2}\left(\int_{0}^{\varepsilon} - \int_{-\varepsilon}^{0}\right) f'(z+x)\,dx$$

This vanishes in the limit $\varepsilon \to 0$. Hence

$$\int_{a}^{b} J(x-x')f(x')\,dx' = \begin{cases} 0 & \text{if } x \text{ is outside range } (a,b) \\ f(x) & \text{if } x \text{ is inside range } (a,b) \end{cases} \tag{A.7}$$

The properties (A.5)–(A.7) of $J(x-x')$ are those of the Dirac delta function. This is an 'improper' function which can be given a meaning only as a limit of an ordinary function. *One* such definition is

$$\delta(x-x') = \lim_{\alpha \to 0} \frac{1}{2\pi} \int_{-\infty}^{\infty} e^{-\alpha|k|+ik(x-x')}\,dk \quad (\alpha > 0) \tag{A.8}$$

A d-dimensional delta function is defined by

$$\delta(\mathbf{x} - \mathbf{x}') = \delta(x_1 - x_1')\,\delta(x_2 - x_2')\dots\delta(x_d - x_d')$$

where the vectors $\mathbf{x}$ are d-dimensional. It then follows that

$$\delta(\mathbf{x} - \mathbf{x}') = \lim_{\alpha \to 0} \frac{1}{(2\pi)^d} \int_{-\infty}^{\infty}\dots\int_{-\infty}^{\infty} e^{-\alpha(|k_1|+|k_2|+\dots+|k_d|)+i\mathbf{k}\cdot(\mathbf{x}-\mathbf{x}')}\,d\mathbf{k} \tag{A.9}$$

where $\mathbf{k} = (k_1, \dots, k_d)$ and $d\mathbf{k} = dk_1 \dots dk_d$.

It is usual to invert the order of integration and the taking of the limit, although this needs justification, so that we have finally

$$\delta(\mathbf{x} - \mathbf{x}') = \frac{1}{(2\pi)^d} \int_{-\infty}^{\infty}\dots\int e^{i\mathbf{k}\cdot(\mathbf{x}-\mathbf{x}')}\,d\mathbf{k} \tag{A.10}$$

where the integral extends over the whole of the space of $\mathbf{k}$. The dimension of the δ-function is that of $|k|^d$ or $|x|^{-d}$.

Note also that for any constant p

$$\int_{-\infty}^{\infty} d(px)\,d(px) = p\int_{-\infty}^{\infty} \delta(px)\,dx = 1 = \int_{-\infty}^{\infty} \delta(x)\,dx$$

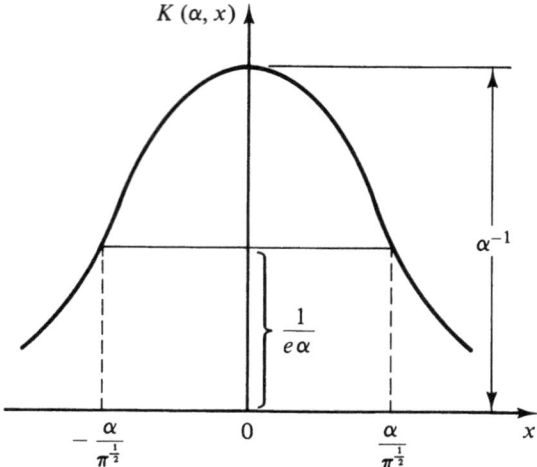

Fig. A.1. The function $K(\alpha, x)$.

so that

$$\delta(px) = \frac{1}{p}\delta(x) \tag{A.11}$$

A simple application yields the fundamental theorem of Fourier transforms. This states that

$$\text{if } f(\mathbf{x}) = \int_{-\infty}^{\infty} \dots \int g(\mathbf{k}') e^{i\mathbf{k}' \cdot \mathbf{x}} d\mathbf{k}',$$

then

$$g(\mathbf{k}) = \frac{1}{(2\pi)^d} \int_{-\infty}^{\infty} \dots \int f(\mathbf{x})^{-i\mathbf{k} \cdot \mathbf{x}} d\mathbf{x} \tag{A.12}$$

By way of proof compute the second integral of eqs. (A.12):

$$\frac{1}{(2\pi)^d} \int_{-\infty}^{\infty} \dots \int f(\mathbf{x}) e^{-i\mathbf{k} \cdot \mathbf{x}} d\mathbf{x} = \frac{1}{(2\pi)^d} \int_{-\infty}^{\infty} \dots \int g(\mathbf{k}') d\mathbf{k}' \int_{-\infty}^{\infty} \dots \int e^{i(\mathbf{k}'-\mathbf{k}) \cdot \mathbf{x}} d\mathbf{x}$$

$$= \int_{-\infty}^{\infty} \dots \int g(\mathbf{k}') \delta(\mathbf{k}'-\mathbf{k}) d\mathbf{k}' = g(\mathbf{k})$$

where we have used eqs. (A.10) and (A.7). One can also write the result (A.12) as

$$f(0) = \frac{1}{(2\pi)^d} \int_{-\infty}^{\infty} \dots \int d\mathbf{k} \int_{-\infty}^{\infty} \dots \int e^{i\mathbf{k} \cdot \mathbf{x}} f(\mathbf{x}) d\mathbf{x} \tag{A.13}$$

This is sometimes called the Fourier integral formula.

A slightly different representation of the δ-function is obtained from eq. (A.10) by the substitution $\mathbf{k} = 2\pi\mathbf{u}$ when

$$\delta(\mathbf{x} - \mathbf{x}') = \int_{-\infty}^{\infty} \cdots \int e^{2\pi i \mathbf{u} \cdot (\mathbf{x} - \mathbf{x}')} \, d\mathbf{u} \tag{A.14}$$

This gives the Fourier decomposition of the δ-function.

Another representation of the δ-function is (Fig. A.1)

$$\delta(x) = \lim_{\alpha \to 0} K(\alpha, x) \equiv \lim_{\alpha \to 0} [\alpha^{-1} e^{-\pi x^2/\alpha^2}] \tag{A.15}$$

To check this, note that the function on the right is even in x. Hence, putting

$$u \equiv \pi x^2/\alpha^2, \quad du = \frac{2\pi}{\alpha^2} \frac{\alpha u^{\frac{1}{2}}}{\pi^{\frac{1}{2}}} dx$$

$$\int_{-\infty}^{\infty} K(\alpha, x) \, dx = \frac{2}{\pi} \cdot \frac{\alpha}{2\pi^{\frac{1}{2}}} \int_{0}^{\infty} u^{-\frac{1}{2}} e^{-u} \, du = 1$$

which is the property (A.6) for all α. In addition, Fig. A.1 shows that $K(\alpha, x)$ becomes infinitely high and infinitely narrow as $\alpha \to 0$, which is the property (A.5). For an alternative definition of the δ-function see [A.1].

Appendix B

Useful identities arising from the periodic boundary condition (section 3.4)

In this appendix, periodic boundary conditions are assumed. They are applied to the volume $V = N\Omega$ of the crystal, called the fundamental domain. N is the number of unit cells, each of volume Ω, in this domain. We write for a vector joining lattice points in the direct lattice

$$\mathbf{R_n} = \sum_{l=1}^{3} n_l \, \mathbf{a}_l \tag{B.1}$$

where the n_l are integers, $\mathbf{n} = (n_1, n_2, n_3)$, and the $\mathbf{a}_l$ are the primitive translations which generate the direct lattice. In reciprocal space or $\mathbf{k}$-space we write

$$\mathbf{k}_n = \sum_{j=1}^{3} k_j \, \mathbf{b}_j \tag{B.2}$$

where the $\mathbf{b}_j$ are the primitive translations which generate the reciprocal lattice. The direct and the reciprocal lattice are connected by

$$\mathbf{b}_i \cdot \mathbf{a}_j = 2\pi\delta_{ij} \tag{B.3}$$

Crystallographers use a convention in which 2π is replaced by unity.

The usefulness of eq. (B.3) may be seen from the periodic boundary condition as applied to a plane wave. If the fundamental domain has N_1 cells along one crystallographic direction, N_2 along the second direction and N_3 along the third direction, then in $V = N\Omega$ one has

$$N = N_1 N_2 N_3 \tag{B.4}$$

For a plane wave, the periodic boundary condition gives

$$\exp[\mathbf{ik} \cdot (\mathbf{r} + N_i \mathbf{a}_i)] = \exp \mathbf{ik} \cdot \mathbf{r} \quad (i = 1, 2, 3)$$

516

It follows that:

$$\mathbf{k} \cdot \mathbf{a}_i = 2\pi t_i / N_i \tag{B.5}$$

where t_i is an integer. Thus $\mathbf{k}$ is quantized by the periodic boundary condition, as is expected from quantum mechanics. Equation (B.5) is satisfied if one uses eq. (B.3) and

$$k_j = t_j / N_j \tag{B.6}$$

The first Brillouin zone ('BZ') can be specified by any one of the following three ranges:

$$-\frac{N_i}{2} < t_i \leqslant \frac{N_i}{2}, \quad -\pi < \mathbf{k} \cdot \mathbf{a}_i \leqslant \pi, \quad -\frac{b_j}{2} < k_j \cdot \mathbf{b}_j \leqslant \frac{b_j}{2} \tag{B.7}$$

Note that the $\mathbf{a}_i$'s and the $\mathbf{b}_j$'s are not in general orthogonal or of unit length.

In order to manipulate matrix elements, one needs to know a little about Fourier expansions. The following are useful identities:

(i) $\displaystyle \sum_{\substack{\mathbf{k} \text{ in first} \\ \text{BZ}}} e^{i\mathbf{k} \cdot \mathbf{R}} = N\delta_{\mathbf{R}, 0}$

$$(\mathbf{R} \text{ is a lattice vector in fundamental domain}) \quad \text{(B.8)}$$

(ii) $\displaystyle \sum_{\substack{\mathbf{R} \\ (\text{in } V)}} e^{i\mathbf{k} \cdot \mathbf{R}} = N\delta_{\mathbf{k}, 0} \quad (\mathbf{k} \text{ in first BZ}) \tag{B.9}$

(iii) If $f(r)$ is periodic in the direct lattice,

$$\int_V e^{i\mathbf{k} \cdot \mathbf{r}} f(\mathbf{r}) \, d\mathbf{r} = N \sum_{\mathbf{m}} \left[\delta_{\mathbf{k}, \mathbf{K_m}} \int_\Omega e^{i\mathbf{K_m} \cdot \mathbf{r}} f(\mathbf{r}) \, d\mathbf{r} \right] \tag{B.10}$$

If $\mathbf{k}$ lies in the first BZ, then $\mathbf{K_m}$ can never equal $\mathbf{k}$ unless $\mathbf{K_m} = \mathbf{K_0} \equiv 0$. One then finds the special case

$$\int_V e^{i\mathbf{k} \cdot \mathbf{r}} f(\mathbf{r}) \, d\mathbf{r} = N \left(\int_\Omega f(\mathbf{r}) \, d\mathbf{r} \right) \delta_{\mathbf{k}, 0} \tag{B.10'}$$

(iv) $\displaystyle \int_V e^{i\mathbf{k} \cdot \mathbf{r}} \, d\mathbf{r} = V\delta_{\mathbf{k}, 0} \quad (\text{all } \mathbf{k}) \tag{B.11}$

$$\int_\Omega e^{i\mathbf{K} \cdot \mathbf{r}} = \Omega\delta_{\mathbf{K}, 0} \quad (\mathbf{K} \text{ is a lattice vector in } \mathbf{k}\text{-space}) \tag{B.12}$$

Proof of (i) *and* (ii)

$$\mathbf{k_t} \cdot \mathbf{R_n} = \sum_{j, l} \left[\frac{t_j}{N_j} \mathbf{b}_j \cdot (n_l \mathbf{a}_l) \right] = 2\pi \sum_{k=1}^{3} \frac{t_j n_j}{N_j}$$

$$e^{i\mathbf{k}_t \cdot \mathbf{R}_n} = \prod_{j=1}^{3} p_j^{n_j} = \sum_{j=1}^{3} q_j^{t_j}, \quad \left(p_j \equiv e^{2\pi i t_j/N_j}, \quad q_j \equiv e^{2\pi i n_j/N_j} \right)$$

Now

$$\sum_{n_j=0}^{(N_j-1)} p_j^{n_j} = \frac{1-p_j^{N_j}}{1-p_j}$$

$$\sum_{-\frac{1}{2}(N_j-1)}^{\frac{1}{2}(N_j-1)} q_j^{t_j} = q_j^{-\frac{1}{2}(N_j-1)}\left(1+q_j+q_j^2+\ldots+q_j^{N_j-1}\right) = q_j^{-\frac{1}{2}(N_j-1)}\frac{1-q_j^{N_j}}{1-q_j} = 0$$

provided $p_j \neq 1$, $q_j \neq 1$. The first sum extends over all N vectors $\mathbf{R}_n$ in V, the second over all *reduced* wave vectors (i.e. $\mathbf{k}$ in first BZ). If $p_j = 1$ or $q_j = 1$, the sums yield N. This happens if each t_j is a multiple of N_j, or if each n_j is a multiple of N_j. Thus

$$\sum_{t(BZ)} e^{i\mathbf{k}_t \cdot \mathbf{R}_n} = \begin{cases} N \text{ if each } n_j \text{ is an integral multiple of } N_j \\ 0 \text{ otherwise} \end{cases} \tag{B.13}$$

$$\sum_{n(V)} e^{i\mathbf{k}_t \cdot \mathbf{R}_n} = \begin{cases} N \text{ if } \mathbf{k}_t \text{ is a lattice vector in k-space} \\ 0 \text{ otherwise} \end{cases} \tag{B.14}$$

The range of the summation is indicated in brackets under the summation sign. The result N in expression (B.13) occurs if $\mathbf{R}_n = \mathbf{R}_0 \equiv 0$ and this is the case covered by eq. (B.8) which is a special case of (B.13). Similarly the special case when $\mathbf{k}_t$ is in the first BZ of expression (B.14) is covered in eq. (B.9).

Proof of (iii)

Let $\mathbf{R}$ be the position vector of the origin of a unit cell, these origins being chosen to lie in standard positions for all unit cells in V. Let $\boldsymbol{\rho}$ range over the unit cell $\mathbf{R}$. Then $\mathbf{r} = \mathbf{R} + \boldsymbol{\rho}$ and

$$\int_V e^{i\mathbf{k}\cdot\mathbf{r}} f(\mathbf{r})\, d\mathbf{r} = \sum_{\mathbf{R}(V)} \int_\Omega e^{i\mathbf{k}\cdot(\mathbf{R}+\boldsymbol{\rho})} f(\boldsymbol{\rho})\, d\boldsymbol{\rho}$$

$$= \left[\sum_{\mathbf{R}(V)} e^{i\mathbf{k}\cdot\mathbf{R}}\right]\left[\int_\Omega e^{i\mathbf{k}\cdot\boldsymbol{\rho}} f(\boldsymbol{\rho})\, d\boldsymbol{\rho}\right]$$

The sum vanishes by expression (B.14) unless $\mathbf{k}$ is of the form $\mathbf{K}_m$ (a lattice vector in k-space), when it is N. The sum is therefore $N\sum_m \delta_{\mathbf{k},\mathbf{K}_m}$, and eq. (B.10) follows.

Proof of (iv)

If $f(\mathbf{r}) \equiv 1$ in eq. (B.10), let $\mathbf{K_m} = \sum m_j \mathbf{b}_j$ and let $\mathbf{r} = \sum x_j \mathbf{a}_j$, then a typical integral on the right of eq. (B.10) is proportional to

$$\prod_j \left[\int_0^1 \exp\left(2\pi i m_j\, x_j\right) \mathrm{d}x_j \right]$$

and this vanishes if one of the m_j's is nonzero. The only nonzero case arises if $\mathbf{K_m} = \mathbf{0}$, and in that case eq. (B.10) yields Ω. Hence

$$\int_V e^{i\mathbf{k}\cdot\mathbf{r}}\,\mathrm{d}\mathbf{r} = N\Omega\delta_{\mathbf{k},0}$$

and this is eq. (B.11). Since

$$e^{i\mathbf{K}\cdot\mathbf{r}} \text{ is periodic in the direct lattice} \qquad\qquad (\text{B.15})$$

this yields

$$\int_\Omega e^{i\mathbf{K}\cdot\mathbf{r}}\,\mathrm{d}\mathbf{r} = \Omega\delta_{\mathbf{K},0}$$

as a special case. This is eq. (B.12).

Note that if $f(\mathbf{r})$ satisfies the periodic boundary condition, it can be Fourier-expanded

$$f(\mathbf{r}) = V^{-\frac{1}{2}} \sum_{\text{all } \mathbf{k}} f_{\mathbf{k}}\, e^{i\mathbf{k}\cdot\mathbf{r}}$$

and $f_{\mathbf{k}}$ identified using eq. (B.11), as discussed in Appendix C.

Notes

(i) We have applied the formula for a geometrical progression of complex numbers p_j, q_j. Is this valid? Yes, if one recalls the derivation

$$\left. \begin{aligned} s &\equiv 1 + a + \ldots + a^N \\ as &= a + a^2 + \ldots + a^{N+1} \end{aligned} \right\} \to (1-a)\,s = 1 - a^{N+1}$$

This holds if a is a complex number and even if a is a matrix.

(ii) We have considered in eqs. (B.8)–(B.14) only sums over $\mathbf{k}$ which are in the first BZ. Why not extend to sums over all $\mathbf{k}$-vectors? Such sums have an infinite number of terms, and t_j/N_j will be an integer an infinite number of times so as to contribute 1 to $\sum_t e^{i\mathbf{k}_t\cdot\mathbf{R_s}}$ an infinite number of terms. Such sums therefore diverge.

Appendix C

Fourier expansions (section 3.4)

Fourier expansions of lattice-periodic functions

In order to discuss Fourier expansions, the following results are useful:

(i) For orthonormal functions $\{\varphi_j(\mathbf{r})\}$ which form a complete set in some given space

$$\sum_n \varphi_n^*(\mathbf{r}') \varphi_n(\mathbf{r}) = \delta(\mathbf{r}-\mathbf{r}') \tag{C.1}$$

In particular, for plane waves

$$\sum_{\mathbf{k}} \exp(i\mathbf{k} \cdot \mathbf{r}) = V\delta(\mathbf{r}) \tag{C.2}$$

(ii) If $f(\mathbf{r})$ is periodic in the direct lattice,

$$f(\mathbf{r}) = \Omega^{-\frac{1}{2}} \sum_{\mathbf{K}} f_{\mathbf{K}} \exp(i\mathbf{K} \cdot \mathbf{r}), \quad f_{\mathbf{K}} = \Omega^{-\frac{1}{2}} \int_\Omega f(\mathbf{r}) \exp(-i\mathbf{K} \cdot \mathbf{r}) \, d\mathbf{r} \tag{C.3}$$

where $\mathbf{K}$ is a lattice vector in the reciprocal lattice and Ω is the volume of a unit cell in $\mathbf{k}$-space.

(iii) If $f(\mathbf{k})$ is periodic in the reciprocal lattice,

$$f(\mathbf{k}) = N^{-\frac{1}{2}} \sum_{\mathbf{R}(V)} f_{\mathbf{R}} \exp(i\mathbf{k} \cdot \mathbf{R}), \quad f_{\mathbf{R}} = N^{-\frac{1}{2}} \sum_{\substack{\mathbf{k} \\ (\mathrm{BZ})}} f(\mathbf{k}) \exp(-i\mathbf{k} \cdot \mathbf{R}) \tag{C.4}$$

The first sum goes over the lattice vectors in the fundamental domain V and the second sum goes over all the wavevectors in the first Brillouin zone. This is indicated in the sums. N is the number of unit cells in the fundamental domain.

(iv) In the passage to the limit $N \to \infty$, $V \to \infty$ for $\Omega = V/N$ fixed (corresponding to an infinite crystal), the following replacements are needed:

$$\frac{V}{8\pi^3}\delta_{\mathbf{kk'}} \to \delta(\mathbf{k}-\mathbf{k'})$$

$$V^{-1}\sum_{\mathbf{k}}g(\mathbf{k}) \to \frac{1}{8\pi^3}\int g(\mathbf{k})\,d\mathbf{k} \tag{C.5}$$

Proof of (i)

Any well-behaved function in the space considered may be expanded as

$$f(x) = \sum_n c_n\,\varphi_n(\mathbf{r}), \quad c_m \equiv \int f(\mathbf{r'})\,\varphi_m(\mathbf{r'})\,d\mathbf{r'} \tag{C.6}$$

The second result comes from

$$\int \varphi_m^*(\mathbf{r})\,\varphi_n(\mathbf{r})\,d\mathbf{r} = \delta_{mn} \tag{C.7}$$

From eqs. (C.6)

$$f(\mathbf{r}) = \sum_n \varphi_n(\mathbf{r})\cdot\int_V f(\mathbf{r'})\,\varphi_n^*(\mathbf{r'})\,d\mathbf{r'} = \int_V f(\mathbf{r'})\Big[\sum_n \varphi_n^*(\mathbf{r'})\,\varphi_n(\mathbf{r})\Big]\,d\mathbf{r}$$

The quantity in square brackets behaves therefore like a Dirac delta function, see eq. (A.7).

For plane waves,

$$\varphi(\mathbf{r}) = V^{-\frac{1}{2}}\exp(i\mathbf{k}\cdot\mathbf{r})$$

and eq. (C.7) takes the form

$$V^{-1}\sum_{\mathbf{k}}\exp[i\mathbf{k}\cdot(\mathbf{r}-\mathbf{r'})] = \delta(\mathbf{r}-\mathbf{r'})$$

which is eq. (C.2).

Proof of (ii)

Any well-behaved function $f(\mathbf{r})$ may be expanded as

$$f(\mathbf{r}) = A\sum_{\text{all }\mathbf{k}}f_{\mathbf{k}}\exp(i\mathbf{k}\cdot\mathbf{r}) \tag{C.8}$$

where A is a normalizing constant. If V is the volume of the fundamental domain,

$$\int_V f(\mathbf{r})\exp(-i\mathbf{k'}\cdot\mathbf{r})\,d\mathbf{r} = A\sum_{\text{all }\mathbf{k}}f_{\mathbf{k}}\int\exp[i(\mathbf{k}-\mathbf{k'})\cdot\mathbf{r}]\,d\mathbf{r} = AVf_{\mathbf{k'}} \tag{C.9}$$

by eq. (B.11). Suppose now that $f(\mathbf{r})$ is lattice periodic. Then if $\mathbf{R}$ is a vector joining lattice points (a 'lattice vector') in the direct lattice, then by eq. (C.8)

$$f(\mathbf{r}) = f(\mathbf{r}+\mathbf{R}) = A\sum_{\mathbf{k}}f_{\mathbf{k}}\exp[i\mathbf{k}\cdot(\mathbf{r}+\mathbf{R})]$$

It follows as in eq. (C.9) that

$$\int_V f(\mathbf{r}) \exp(-i\mathbf{k}' \cdot \mathbf{r}) d\mathbf{r} = AVf_{\mathbf{k}'} \exp(i\mathbf{k}' \cdot \mathbf{R}) \qquad (C.10)$$

Comparing eqs. (C.9) and (C.10),

$$f_{\mathbf{k}'} = f_{\mathbf{k}'} \exp(i\mathbf{k}' \cdot \mathbf{R}) \qquad (C.11)$$

Thus $f_{\mathbf{k}'} = 0$ or else $\mathbf{k}'$ is a lattice vector in the reciprocal lattice. We have arrived at

$$f(\mathbf{r}) = A \sum_{\mathbf{K}} f_{\mathbf{K}} e^{i\mathbf{K} \cdot \mathbf{r}}, \quad f_{\mathbf{K}} = \frac{1}{AV} \int_V f(\mathbf{r}) e^{-i\mathbf{K} \cdot \mathbf{r}} d\mathbf{r}$$

Since both $f(\mathbf{r})$ and $\exp(i\mathbf{K} \cdot \mathbf{r})$ are periodic in the direct lattice, the integral is N times an integral over a unit cell. Hence

$$f_{\mathbf{K}} = \frac{1}{A\Omega} \int_{\Omega} f(\mathbf{r}) e^{-i\mathbf{K} \cdot \mathbf{r}} d\mathbf{r}$$

This yields eq. (C.3) if the normalization $A = \Omega^{-\frac{1}{2}}$ is chosen. Sometimes one normalizes with $A = 1$.

Proof of (iii)

In analogy with eq. (C.3), put

$$f(\mathbf{k}) = B \sum_{\mathbf{R}'(V)} f_{\mathbf{R}'} e^{i\mathbf{k} \cdot \mathbf{R}'}$$

This is possible since $\exp(i\mathbf{k} \cdot \mathbf{R}')$ is periodic with the periodicity of the reciprocal lattice. If a term $\exp(i\mathbf{k} \cdot \mathbf{r})$ were present in the expansion for which $\mathbf{r}$ is not a lattice vector, the periodicity in reciprocal space would be violated. The Fourier coefficient is obtained by using eq. (B.8) in

$$\sum_{\mathbf{k}(BZ)} f(\mathbf{k}) e^{-i\mathbf{k} \cdot \mathbf{R}} = B \sum_{\mathbf{R}'(V)} f_{\mathbf{R}'} \sum_{\mathbf{k}(BZ)} e^{i\mathbf{k} \cdot (\mathbf{R}' - \mathbf{R})}$$

$$= BN \sum_{\mathbf{R}'} f_{\mathbf{R}'} \delta_{\mathbf{R}\mathbf{R}'} = BNf_{\mathbf{R}}$$

By choosing $B = N^{-\frac{1}{2}}$, the result (C.4) is found. Had one allowed all $\mathbf{r}$ in the expansion of $f(\mathbf{k})$, one would have found $f_{\mathbf{r}} = 0$ unless $\mathbf{r}$ is a lattice vector.

Proof of (iv)

Using eq. (B.11), $\int \exp(i\mathbf{k} \cdot \mathbf{r}) d\mathbf{r} = V\delta_{\mathbf{k}0}$, one has

$$\frac{V}{8\pi^3} \delta_{\mathbf{k}\mathbf{k}'} = \frac{1}{8\pi^3} \int_V \exp i(\mathbf{k} - \mathbf{k}') \cdot \mathbf{r} d\mathbf{r} = \delta(\mathbf{k} - \mathbf{k}')$$

where eq. (A.10) has been used. This requires that $V \to \infty$, but in order to maintain the unit cell, $N = V/\Omega$ has also to become infinitely large. This establishes expression (C.5).

Fourier expansions of some physically important functions

Consider now the Fourier expansions of functions in the direct lattice in a domain of volume V. The finiteness of the volume means that appendix B is relevant rather than appendix A. We have:

(i) *Dirac δ-functions*:

$$\delta(\mathbf{r}) = \sum_{\mathbf{k}} c_{\mathbf{k}} e^{i\mathbf{k}\cdot\mathbf{r}}, \quad c_{\mathbf{k}} = 1/V \tag{C.12}$$

(ii) *Bloch modulating functions*:

$$u_{\mathbf{k}}(\mathbf{r}) = \sum_{\mathbf{k}} v_{\mathbf{k}} e^{i\mathbf{k}\cdot\mathbf{r}}, \quad v_{\mathbf{k}} = 0 \text{ unless } \mathbf{k} = \mathbf{K}_{\mathbf{m}} \text{ for some integer triplet } \mathbf{m} \tag{C.13}$$

(iii) *Screened Coulomb potential*:

$$r^{-1}\exp(-\lambda r) = \sum_{\mathbf{k}} d_{\mathbf{k}} e^{i\mathbf{k}\cdot\mathbf{r}}, \quad d_{\mathbf{k}} = 4\pi/V(\lambda^2 + k^2) \tag{C.14}$$

Some explanatory remarks will be made about these results.

(i) To identify the Fourier coefficient $c_{\mathbf{k}}$ observe from the definition of $\delta(\mathbf{r})$ that

$$\int_V e^{-i\mathbf{k}'\cdot\mathbf{r}} \delta(\mathbf{r})\,d\mathbf{r} = 1$$

Using expressions (C.12) and (B.11), this is equivalent to

$$\sum_{\mathbf{k}} c_{\mathbf{k}} \int e^{i(\mathbf{k}-\mathbf{k}')\cdot\mathbf{r}}\,d\mathbf{r} = Vc_{\mathbf{k}'} = 1$$

Hence $c_{\mathbf{k}'} = 1/V$ as required.

(ii) The simple Bloch function is appropriate for a one-electron treatment of electrons in a perfect periodic structure. It has the form

$$\psi_{n\mathbf{k}}(\mathbf{r}) = u_n(\mathbf{k}, \mathbf{r}) e^{i\mathbf{k}\cdot\mathbf{r}}, \quad u_n(\mathbf{k}, \mathbf{r}) = u_n(\mathbf{k}, \mathbf{r}+\mathbf{R})$$

Its modulating part, being periodic in the direct lattice, is subject to eq. (C.3). Hence expression (C.13) results and has been used, for example in eq. (3.4.9).

(iii) If ε is an appropriate dielectric constant of the material and λ is a constant, we shall consider

$$U(r_1, r_2) \equiv \frac{e^2 e^{-\lambda|r_1-r_2|}}{\varepsilon|\mathbf{r}_1-\mathbf{r}_2|} \equiv \sum_{\mathbf{k}'} U_{\mathbf{k}'} e^{i\mathbf{k}'\cdot(r_1-r_2)}$$

Then

$$V^{-2} \int U(\mathbf{r}_1, \mathbf{r}_2) \, e^{i\mathbf{k} \cdot (\mathbf{r}_2 - \mathbf{r}_1)} \, d\mathbf{r}_1 \, d\mathbf{r}_2 = V^{-2} \sum_{\mathbf{k}'} U_{\mathbf{k}'} \int_V e^{i(\mathbf{k}' - \mathbf{k}) \cdot \mathbf{r}_1} \, d\mathbf{r}_1 \int_V e^{i(-\mathbf{k}' + \mathbf{k}) \cdot \mathbf{r}_2} \, d\mathbf{r}_2$$

The two integrals each have the value $V\delta_{\mathbf{k},\mathbf{k}'}$. It follows that:

$$U_{\mathbf{k}} = V^{-2} \int\int U(\mathbf{r}_1, \mathbf{r}_2) \, e^{-i\mathbf{k} \cdot (\mathbf{r}_1 - \mathbf{r}_2)} \, d\mathbf{r}_1 \, d\mathbf{r}_2$$

$$= V^{-1} \int U(\mathbf{r}) \, e^{-i\mathbf{k} \cdot \mathbf{r}} \, d\mathbf{r} = \frac{e^2}{\varepsilon V} \int r^{-1} e^{-(\lambda r + i\mathbf{k} \cdot \mathbf{r})} \, d\mathbf{r}$$

Now $d\mathbf{r} = r^2 \, dr \, dx \, d\varphi$, where x is the cosine of the angle between $\mathbf{k}$ and $\mathbf{r} = \mathbf{r}_1 - \mathbf{r}_2$, and φ is the azimuthal angle and can be integrated out at once. Hence

$$U_{\mathbf{k}} = \frac{2\pi e^2}{\varepsilon V} \int_0^\infty e^{-\lambda r} r \, dr \int_{-1}^1 e^{-ikrx} \, dx = \frac{2\pi e^2}{\varepsilon V k} i \int_0^\infty e^{-\lambda r} [e^{-ikr} - e^{ikr}] \, dr$$

$$= \frac{2\pi e^2}{\varepsilon V k} i \left[\frac{1}{\lambda + ik} - \frac{1}{\lambda - ik} \right] = \frac{4\pi e^2}{\varepsilon V (k^2 + \lambda^2)}$$

which is expression (C.14) and is used in eq. (3.4.19).

Appendix D

The effective mass sum rule and the dynamics of Bloch electrons (section 3.6)

The differentiation of an eigenvalue equation

Consider a Hermitian operator $H(\alpha)$ with eigenvalues and eigenfunctions specified by the following equation:

$$H(\alpha)\, u_{nl}(\alpha) = E_n(\alpha)\, u_{nl}(\alpha) \tag{D.1}$$

The suffix l numbers degenerate eigenfunctions due to spin or band overlap. The components $\alpha_1, \alpha_2, \ldots$ of α represent parameters entering into H. Rewrite this

$$A_n(\alpha)\, u_{nl}(\alpha) = 0, \quad A_n(\alpha) \equiv H(\alpha) - E_n(\alpha) \tag{D.2}$$

Differentiating with respect to α_i and taking matrix elements

$$\left[A_n(\alpha) \frac{\partial}{\partial \alpha_i} + \frac{\partial A_n}{\partial \alpha_i} \right] u_{nl}(\alpha) = 0 \tag{D.3}$$

so that with $|nl\rangle \equiv u_{nl}(r)$

$$\langle n'l' | \frac{\partial A_n}{\partial \alpha_i} | nl \rangle = - \langle n'l' | A_n(\alpha) \frac{\partial}{\partial \alpha_i} | nl \rangle$$

$$= [E_n(\alpha) - E_{n'}(\alpha)] \langle n'l' | \frac{\partial}{\partial \alpha_i} | nl \rangle \tag{D.4}$$

Differentiating eq. (D.3) again,

$$\left(\frac{\partial A_n}{\partial \alpha_j} \frac{\partial}{\partial \alpha_i} + A_n \frac{\partial^2}{\partial \alpha_i \partial \alpha_j} + \frac{\partial^2 A_n}{\partial \alpha_i \partial \alpha_j} + \frac{\partial A_n}{\partial \alpha_i} \frac{\partial}{\partial \alpha_j} \right) u_{nl} = 0 \tag{D.5}$$

Taking matrix elements,

$$-\langle nl'|\frac{\partial^2 A_n}{\partial\alpha_i\,\partial\alpha_j}|nl\rangle = \boxed{2}\sum_{\substack{n'',l''\\(n''\neq n)}} \langle nl'|\frac{\partial A_n}{\partial\alpha_i}|n''l''\rangle\langle n''l''|\frac{\partial}{\partial\alpha_j}|nl\rangle \qquad (D.6)$$

noting that the second term from eq. (D.5) does not contribute by eqs. (D.2), and that the diagonal term in eq. (D.6) can be omitted by eq. (D.4). The $\boxed{2}$ means that to each term which follows there should be added another term, not written down, obtained by interchanging i and j. Eliminating the $\partial/\partial\alpha_j$ term by eq. (D.4),

$$-\langle nl'|\frac{\partial^2 A_n}{\partial\alpha_i\,\partial\alpha_j}|nl\rangle =$$

$$\boxed{2}\sum_{\substack{n''(\neq n)\\l''}} \frac{\langle nl'|(\partial A_n/\partial\alpha_i)|n''l''\rangle\langle n''l''|(\partial A_n/\partial\alpha_j)|nl\rangle}{E_n(\alpha)-E_{n'}(\alpha)} \qquad (D.7)$$

The main results are eqs. (D.4) and (D.7), the latter being a sum rule.

To apply these results here, interpret α as the wavevector $\mathbf{k}$. Let H be the Hamiltonian of an electron in a periodic lattice, and $\psi_{nl} \equiv u_{nl}(\mathbf{k})\,e^{i\mathbf{k}\cdot\mathbf{r}}$ a Bloch function. The dependence of $u_{nl}(\mathbf{k})$ on $\mathbf{r}$ will not be indicated explicitly. The Schrödinger equation is

$$H\psi_{nl}(\mathbf{k}) = E_n(\mathbf{k})\,\psi_{nl}(\mathbf{k}) \rightarrow H_\mathbf{k}\,u_{nl}(\mathbf{k}) = E_n(\mathbf{k})\,u_{nl}(\mathbf{k}) \qquad (D.8)$$

where

$$H_\mathbf{k} \equiv e^{-i\mathbf{k}\cdot\mathbf{r}}\,H\,e^{i\mathbf{k}\cdot\mathbf{r}} \qquad (D.9)$$

Thus the theory based on eq. (D.1) applies to Bloch electrons. It can also be used to discuss virial theorems [D.1] and nondegenerate perturbation theory [D.2].

Application to Bloch electron

Now $E_n(\mathbf{k})$ is the nth eigenvalue (in ascending order of energies) at wavevector $\mathbf{k}$ and l allows for degeneracy of this eigenvalue at this point in k-space. Such degeneracy may be due to spin or it may be an r-fold 'essential degeneracy' due lattice symmetry in which each eigenvalue in a band is r-fold degenerate, or it can arise from bands which cross at $\mathbf{k}$ due to 'accidental' degeneracy. Thus

$$A_n(\alpha) \rightarrow A_n(\mathbf{k}) \equiv H_\mathbf{k} - E_n(\mathbf{k}) \qquad (D.10)$$

The transformation (D.9) has two useful properties which can be expressed in terms of a well-behaved function $g(\mathbf{k},\mathbf{r})$ and any operator $B(\mathbf{p},\mathbf{r})$ which is a power series in the electron momentum $\mathbf{p}$. The properties are, using eq. (D.9),

(i) $B_k(p, r) = B(p + \hbar k, r)$ (D.11)

(ii) $e^{-ik \cdot r} \nabla_p B(p, r) e^{ik \cdot r} g(k, r) = \hbar^{-1} [\nabla_k B_k(p, r)] g(k, r)$ (D.12)

where the square bracket means that ∇_k does not act on $g(k, r)$.

Proofs

(i) Observe that

$$p e^{ik \cdot r} = \frac{\hbar}{i} \nabla e^{ik \cdot r} = \frac{\hbar}{i} (ik \, e^{ik \cdot r} + e^{ik \cdot r} \nabla) = e^{ik \cdot r} (p + \hbar k)$$

Thus

$$(p^n)_k = (p + \hbar k)^n$$

holds for $n = 1$. Suppose it holds for some general integer n. Then it holds for $n + 1$, as can be seen from

$$(p^{n+1})_k = e^{-ik \cdot r} (p)^n e^{ik \cdot r} = (p + \hbar k)^n (p + \hbar k)$$

It follows that eq. (D.11) holds for any power series in p.

(ii) The operator ∇_p does not affect the transformation (D.9) since $e^{ik \cdot r}$ is independent of p. Hence the left-hand side of eq. (D.12) is, using eq. (D.11),

$$\nabla_p B_k(p, r) = \nabla_p B(p + \hbar k, r) = \hbar^{-1} \nabla_k B_k(p, r)$$

This is eq. (D.12).

We first apply the result (D.4) to Bloch electrons. If one takes $n = n'$, the right-hand side vanishes. It follows, using eq. (D.10) on the left, and $|nk\rangle$ as a notation for the modulating part $u_{nk}(r)$ of a Bloch function,

$$\langle nk | \frac{\partial H_k}{\partial k_i} | nk \rangle = \frac{\partial E_n(k)}{\partial k_i} \quad (i = 1, 2, 3)$$ (D.13)

This is the so-called Hellmann–Feynman theorem which has been used in several distinct areas of physics. For solid state theory eq. (D.7) is more important. One finds

$$\langle nkl | \frac{\partial^2 H_k}{\partial k_i \partial k_j} | nkl \rangle = \frac{\partial^2 E_n(k)}{\partial k_i \partial k_j}$$

$$+ \boxed{2} \sum_{\substack{n'(\neq n) \\ l'}} \frac{\langle nkl | \frac{\partial H_k}{\partial k_i} | n'kl' \rangle \langle n'kl' | \frac{\partial H_k}{\partial k_i} | nkl \rangle}{E_{n'}(k) - E_n(k)}$$ (D.14)

The terms $E_n(k)$ in $A_n(k)$ do not contribute to the sum, as the matrix elements which occur there are nondiagonal in n.

The velocity operator

We now turn to the velocity operator in order to interpret the terms in eq. (D.14). Quantum mechanics teaches that if H is the Hamiltonian and A is any operator which is not explicitly dependent on the time, then the operator representing the time-rate of change of A is

$$\dot{A} = \frac{i}{\hbar}[H, A], \quad \text{so that } v_j = \frac{i}{\hbar}[H(\mathbf{p}, \mathbf{r}), r_j] \quad (j = 1, 2, 3) \tag{D.15}$$

Now if $f(\mathbf{p})$ is a power series in $\mathbf{p}$, note that

$$\frac{i}{\hbar}[\mathbf{p}, x] = \frac{i}{\hbar}[p_1, x] = I$$

where p_1 is the x-component of the momentum. Also

$$\frac{i}{\hbar}[p_1^2, x] = \frac{i}{\hbar}\{p_1[p_1, x] - [x, p_1]p_1\} = 2p_1$$

and generally

$$\frac{i}{\hbar}[f(p_1), x] = df(p_1)/dp_1$$

Here the function is differentiated as if it were a classical function and the operator is substituted at the end. Lastly, if $\hat{\mathbf{X}}, \hat{\mathbf{Y}}, \hat{\mathbf{Z}}$ are unit vectors and $\mathbf{r} = (x, y, z)$,

$$\frac{i}{\hbar}[f(\mathbf{p}), \mathbf{r}] = \frac{i}{\hbar}\{[f(\mathbf{p}), x]\hat{\mathbf{X}} + [f(\mathbf{p}), y]\hat{\mathbf{Y}} + [f(\mathbf{p}), z]\hat{\mathbf{Z}}\} = \nabla_{\mathbf{p}}f(\mathbf{p})$$

We therefore see from expression (D.15) that

$$\mathbf{v} = \nabla_{\mathbf{p}}H(\mathbf{p}, \mathbf{r}) \tag{D.16}$$

Let us now make the transformation (D.9) using eq. (D.12):

$$\mathbf{v_k} = \hbar^{-1}\nabla_{\mathbf{k}}H_{\mathbf{k}}(\mathbf{p}, \mathbf{r}) \quad \text{or} \quad (\mathbf{v_k})_j = \hbar^{-1}\frac{\partial H_{\mathbf{k}}(\mathbf{p}, \mathbf{r})}{\partial k_j} \tag{D.17}$$

The expectation value of $(\mathbf{v_k})_j$ in a Bloch state $|n, \mathbf{k}\rangle$ (dropping the quantum number l for simplicity) is

$$(n\mathbf{k}|v_j|n\mathbf{k}) = \langle n\mathbf{k}|v_{\mathbf{k}j}|n\mathbf{k}\rangle = \hbar^{-1}\langle n\mathbf{k}|\frac{\partial H_{\mathbf{k}}}{\partial k_j}|n\mathbf{k}\rangle = \hbar^{-1}\frac{\partial E_n(\mathbf{k})}{\partial k_j} \tag{D.18}$$

We have here passed from a Bloch state representation $|\,)$ to a representation by the u's, $|n\mathbf{k}\rangle$, so that eq. (D.13) could be used. Equation (D.18) gives an identification of the matrix elements in the sum of eq. (D.14). Thus for Bloch state $|nk)$ one has rigorously

$$\mathbf{v}_{nk} \equiv (nk|\mathbf{v}|nk) = \hbar^{-1}\nabla_k E_n(\mathbf{k}) \tag{D.19}$$

The velocity (nk)-expectation value is the gradient in k-space of the energy surface n at wavevector $\mathbf{k}$ and can also be regarded as a group velocity. That an electron moves indefinitely without energy dissipation in a Bloch state (nk) in spite of its interaction with the ions (assumed here stationary) is a direct result of the periodic potential.

Forces and the rate of change of the wavevector

Suppose now a force $\mathbf{F}(\mathbf{r})$ acts on a Bloch electron in state (nk). The work done by it in time δt is, arguing classically,

$$\delta W = \mathbf{F}(\mathbf{r}) \cdot \mathbf{v}_{nk} \delta t \tag{D.20}$$

If as a result of this action the wavevector changes by δk, then the change in the energy of the electron is, using eq. (D.19),

$$\nabla_k E_n(\mathbf{k}) \cdot \delta k = \hbar \mathbf{v}_{nk} \cdot \delta k \tag{D.21}$$

Equating eqs. (D.20) and (D.21)

$$\mathbf{F}(\mathbf{r}) \cdot \mathbf{v}_{nk} = \hbar \mathbf{v}_{nk} \cdot \dot{\mathbf{k}}$$

This must hold for all $\mathbf{v}_{nk}$ so that

$$\hbar \dot{\mathbf{k}} = \mathbf{F}(\mathbf{r}) \tag{D.22}$$

Since the rate of change of $\hbar\mathbf{k}$ is given by the external forces only, the periodic field of the lattice being not involved, it is *not* the electron momentum in Bloch state (nk). It is therefore given a different name – it is called the *crystal momentum* of the electron.

One can also derive eq. (D.22) by a proper quantum mechanical argument. It utilizes an expansion over all bands for given $\mathbf{k}$ of the wavefunction (with superposition constants c_n):

$$|\mathbf{k}(t)\rangle = \sum_n c_n u_{nk}(\mathbf{r}) \exp[i\mathbf{k}(t) \cdot \mathbf{r}]$$

Using the translation operator $T(\mathbf{R})$ for any lattice vector $\mathbf{R}$,

$$T(\mathbf{R})|\mathbf{k}(t)\rangle \equiv \exp[i\mathbf{k}(t) \cdot \mathbf{R}]|\mathbf{k}(t)\rangle$$

since for any function of position $\mathbf{r}$

$$T(\mathbf{R})f(\mathbf{r}) = f(\mathbf{r}+\mathbf{R})$$

Hence

$$\langle \mathbf{k}(t)| \, T(\mathbf{R}) \, |\mathbf{k}(t)\rangle = \sum_{n,\,m} c_m^* \, c_n \, u_{m\mathbf{k}}^*(\mathbf{r}) \, u_{n\mathbf{k}}(\mathbf{r}) \exp\left[i\mathbf{k}(t) \cdot \mathbf{R}\right]$$

The time derivative of the operator is defined in the usual way by

$$b \equiv \langle \mathbf{k}(t)| \, \dot{T}(\mathbf{R}) \, |\mathbf{k}(t)\rangle = \frac{\mathrm{d}}{\mathrm{d}t} \langle \mathbf{k}(t)| \, T(\mathbf{R}) \, |\mathbf{k}(t)\rangle$$

$$= i\dot{\mathbf{k}}(t) \cdot \mathbf{R} \, \langle \mathbf{k}(t)| \, T(\mathbf{R}) \, |\mathbf{k}(t)\rangle$$

We next approach the matter via the Heisenberg equation of motion. The Hamiltonian consists of a lattice periodic part H_0 to which the effect of a constant force $\mathbf{F}$ has to be added

$$H = H_0 - \mathbf{F} \cdot \mathbf{r}, \quad [H_0, T(\mathbf{R})] = 0$$

so that

$$[H, T(\mathbf{R})] = -[\mathbf{F} \cdot \mathbf{r}, T(\mathbf{R})] = -\mathbf{F} \cdot \mathbf{r} T(\mathbf{R}) - \mathbf{F} \cdot (\mathbf{r} + \mathbf{R}) \, T(\mathbf{R}) = \mathbf{F} \cdot \mathbf{R} T(\mathbf{R})$$

Hence

$$\dot{T}(\mathbf{R}) = \frac{i}{\hbar}[H, T(\mathbf{R})] = \frac{i}{\hbar}\mathbf{F} \cdot \mathbf{R} T(\mathbf{R})$$

Taking the expectation value, one finds for the quantity

$$b = \frac{i}{\hbar}\mathbf{F} \cdot \mathbf{R} \, \langle \mathbf{k}(t)| \, T(\mathbf{R}) \, |\mathbf{k}(t)\rangle$$

Comparison of the two expressions for b now furnishes the required relation (D.22) with complete generality. This argument is developed from [D.3].

The effective mass tensor

Treating the acceleration $\mathbf{a}_{n\mathbf{k}}$ of an electron in Bloch state $(n\mathbf{k})$ not quantum mechanically by the procedure (D.15), but semiclassically, one has from eqs. (D.19) and (D.22),

$$(\mathbf{a}_{n\mathbf{k}})_i = \frac{\mathrm{d}}{\mathrm{d}t}(\mathbf{v}_{n\mathbf{k}})_i = \frac{\mathrm{d}}{\mathrm{d}t}\left[\hbar^{-1}\frac{\partial E_n(\mathbf{k})}{\partial k_i}\right] = \hbar^{-1}\sum_{j=1}^{3}\frac{\partial^2 E_n(\mathbf{k})}{\partial k_i \partial k_j}\frac{\mathrm{d}k_j}{\mathrm{d}t}$$

$$= \hbar^{-2}\sum_{j=1}^{3}\frac{\partial^2 E_n(\mathbf{k})}{\partial k_i \partial k_j}F_j \qquad\qquad\qquad\qquad (D.23)$$

Compare this with a free particle of mass m when

$$a_i = m^{-1}F_i \tag{D.24}$$

In a crystal the acceleration is clearly not always in the direction of the applied force so that m^{-1} is replaced by a 'reciprocal effective mass' *tensor*

$$\left(\frac{1}{m}\right)_{ij}^{nk} \equiv \hbar^{-2}\frac{\partial E_n(\mathbf{k})}{\partial k_i \partial k_j} \tag{D.25}$$

The equation of motion (D.24) is then

$$(a_{nk})_i = \sum_{j=1}^{3}\left(\frac{1}{m}\right)_{ij}^{nk} F_j \tag{D.26}$$

The first term on the right of eq. (D.14) receives a simple interpretation from eq. (D.25).

Turn now to an interpretation of the left-hand side of eq. (D.14). Suppose

$$H(\mathbf{p},\mathbf{r}) = \frac{p^2}{2m} + \sum_{j=1}^{3} a_j(\mathbf{r})p_j + b(\mathbf{r}) \tag{D.27}$$

where m is the free electron mass and a_j, b are functions of position. The a_j's, which allow for spin–orbit or other interactions, will not concern us here. Then by eq. (D.11)

$$H_{\mathbf{k}}(\mathbf{p},\mathbf{r}) = \frac{(\mathbf{p}+\hbar\mathbf{k})^2}{2m} + \sum_{j=1}^{3} a_j(\mathbf{r})(p_j + \hbar k_j) + b(\mathbf{r})$$

It follows that

$$\frac{\partial^2 H_{\mathbf{k}}}{\partial k_i \partial k_j} = \frac{\hbar^2}{m}\delta_{ij} \tag{D.28}$$

is an ordinary c-number.

Note that by eq. (D.25) for a diagonal reciprocal effective mass tensor

$$\frac{\partial^2 E_n(\mathbf{k})}{\partial k_i \partial k_j} = \hbar^2 \left(\frac{1}{m}\right)_{jj}^{mk} \delta_{ij} \tag{D.29}$$

If it is a constant, i.e. independent of k in addition, one can integrate eq. (D.29):

$$E_n(\mathbf{k}) - E_n(0) = \tfrac{1}{2}\sum_{j=1}^{3}\hbar^2 k_j^2 \left(\frac{1}{m}\right)_{jj}^2 \equiv \frac{1}{2}\sum_{j=1}^{3}\frac{\hbar^2 k_j^2}{2m_j} \tag{D.30}$$

where the m_j's may be called the principal effective masses.

The effective mass sum rule

Now multiply eq. (D.14) by $m/\hbar^2$ and substitute eqs. (D.28), (D.25) and (D.17) into eq. (D.14). One finds the effective mass sum rule

$$m\left(\frac{1}{m}\right)_{ij}^{n\mathbf{k}} + \sum_{n'(\neq n)} f_{\mathbf{k}ij}^{n'n} = \delta_{ij} \tag{D.31}$$

It is exact and not dependent on perturbation theory, as is sometimes suggested. A generalized oscillator strength

$$f_{\mathbf{k}ij}^{n'n} \equiv \boxed{2}\, m \frac{(n\mathbf{k}|v_i|n'\mathbf{k})(n'\mathbf{k}|v_j|n\mathbf{k})}{E_{n'}(\mathbf{k}) - E_n(\mathbf{k})} \tag{D.32}$$

has been introduced. It is a number, and in the simplest case, $i = j$ and $v_i = p_i/m$, one has

$$f_{\mathbf{k}ii}^{n'n} = \frac{2|(n\mathbf{k}|p_i|n'\mathbf{k})|^2}{m[E_{n'}(\mathbf{k}) - E_n(\mathbf{k})]} \tag{D.33}$$

The term *oscillator strength* is briefly discussed on p. 346.

For a localized electron the effective mass is infinite and one obtains from eq. (D.31) one of the original Thomas–Reiche–Kuhn sum rules $\sum_{n'(\neq n)} f_{ii}^{n'n} = 1$ [D.4]. Sum rules occur in different contexts and are often used to check approximately known functions.

For infinite band separations $f = 0$ and eq. (D.31) yields

$$\left(\frac{1}{m}\right)_{ij}^{n\mathbf{k}} = \frac{\delta_{ij}}{m}$$

The electron thus behaves like a free electron as one would expect. As the bands are brought together, one can imagine a curvature to be imposed on them by a mutual repulsion, which leads to negative effective masses for the valence band.

An alternative procedure is as follows. Consider only diagonal terms in an isotropic effective mass tensor

$$\left(\frac{1}{m}\right)_{ii}^{n\mathbf{k}} = \frac{1}{m_n}$$

The sum rule (D.31) is in this case with $|M_{nn'}^{(i)}| \equiv |(n\mathbf{k}|p_i|n'\mathbf{k})|$

$$\frac{1}{m}\sum_{\substack{n' \\ (\neq n)}} f_{\mathbf{k}ii}^{nn'} = \frac{2}{m^2}\sum_{n'(\neq n)}\frac{|M_{nn'}^{(i)}|^2}{E_n(\mathbf{k}) - E_{n'}(\mathbf{k})} = \frac{1}{m_n} - \frac{1}{m} \tag{D.34}$$

Summing over the Cartesian components $i = 1, 2, 3$

$$\frac{m}{6} \sum_{\substack{n' \\ (\neq n)}} \sum_{i=1}^{3} (f_{kii}^{nn'})_{av} = \sum_{n'(\neq n)} \frac{|M_{nn'}|_{av}^{2}}{E_{n}(\mathbf{k}) - E_{n'}(\mathbf{k})} = \frac{m^{2}}{2} \left(\frac{1}{m_{n}} - \frac{1}{m} \right) \tag{D.35}$$

where $|M_{nn'}|_{av}^{2} \equiv \frac{1}{3} \sum_{i} |M_{nn'}^{(i)}|^{2}$. For a typical III–V semiconductor one has to consider the conduction band C, the light-hole band L, the heavy-hole band H, and the split-off band S. Then in an obvious notation

$$\frac{|M_{CL}|_{av}^{2}}{E_{C} - E_{L}} + \frac{|M_{CH}|_{av}^{2}}{E_{C} - E_{H}} + \frac{|M_{CS}|_{av}^{2}}{E_{C} - E_{S}} = \frac{m}{2} \left(\frac{m}{m_{c}} - 1 \right) \tag{D.36}$$

Treating the three average matrix elements as approximately the same, with

$$E_{C} - E_{L} = E_{C} - E_{H} \equiv E_{G} \quad \text{and} \quad E_{C} - E_{S} \equiv E_{G} + \Delta$$

$$\frac{1}{2mE_{G}} |M|_{av}^{2} = \frac{E_{G} + \Delta}{12(E_{G} + 2\Delta/3)} \left(\frac{m}{m_{c}} - 1 \right) \tag{D.37}$$

Thus the energy derived from the momentum matrix element, expressed as a fraction of E_{G}, is then given very approximately by the above simple function of E_{G}, Δ and m_{c}. For GaAs one finds (with $E_{G} = 1.42$ eV, $\Delta = 0.33$ eV, $m_{c}/m = 0.067$) a value of 1.24 for the right-hand side of eq. (D.37). This means $|M|_{av}^{2}/2m_{c} \sim 26.3$ eV. One can infer roughly that the matrix element of the momentum operator, which determines the radiative transitions, increases with the energy gap.

There is now a considerable literature on various generalizations of these results [D.5]. We note here [D.6] that by summing eq. (D.31) over all n and l and denoting by d_{n} the degeneracy of band n,

$$\sum_{n} d_{n} \left[\left(\frac{1}{m^{*}} \right)_{ij}^{nk} - \frac{\delta_{ij}}{m} \right] = 0 \tag{D.38}$$

The sum must here extend over all bands likely to affect the given problem.

While sum rules are exact, the use of only selected terms is a procedure which must be used cautiously, as shown by the following example. For two bands, using the two substitutions $(n, n') = (c, v)$ or (v, c), eq. (D.35) takes the approximate form

$$(f_{kii}^{cv})_{av} = \frac{2|M_{cv}|_{av}^{2}}{mE_{G}} = \frac{m}{m_{c}} - 1 = 1 - \frac{m}{m_{v}} \tag{D.39}$$

from which one infers for $0 < m_{c} < m$ that $m_{v} < 0$ and

$$\frac{1}{2} \left[\frac{1}{m_{c}} - \frac{1}{|m_{v}|} \right] = \frac{1}{m} \tag{D.40}$$

It states that m is the harmonic mean of m_c and m_v. Only if this condition is broadly fulfilled can one use eq. (D.39). It has in fact been used for example to estimate previously unknown effective masses of lead polonide (PbPo) [D.7].

In recent years effective mass theory has been developed to cover the position dependence of effective masses [D.8].

Appendix E

Diagonalization and Jacobian for the threshold energy and impact ionization calculation (section 3.5.7)

In this appendix we shall carry out the change of variables from the parts $\mathbf{X}_1$ and $\mathbf{X}_2$ of the wavevectors $\mathbf{k}_1$ and $\mathbf{k}_2$ as given in eq. (3.5.22) to new variables $\mathbf{x}_1$ and $\mathbf{x}_2$. This change enables the energy–momentum conservation to be written as a diagonalized quadratic form, i.e. one which does not contain cross-terms of the variables $\mathbf{X}_1$ and $\mathbf{X}_2$.

The *first* step, using eq. (3.5.29), is to put

$$
E_i - E_t = \varepsilon + \frac{\hbar^2}{2m_{2'i}} X_{2'i}^2 + \frac{\hbar^2}{2} \sum_{i=1}^{3} \left[\frac{(X_{1i} + X_{2i} - X_{2'i} + Q_i)^2}{m_{1'i}} - \frac{X_{1i}^2}{m_{1i}} - \frac{X_{2i}^2}{m_{2i}} \right]
$$

$$
= \varepsilon + B(\mathbf{X}_{2'}) + \sum_{i=1}^{3} \sum_{j=1}^{2} a_{ji} X_{ji} + \sum_{i=1}^{3} \left[\frac{\hbar^2 X_{1i} X_{2i}}{m_{1'i}} + \frac{\hbar^2}{m_{1'i}} (X_{1i} + X_{2i})(Q_i - X_{2'i}) \right]
$$

$$\tag{E.1}$$

The underlinings will be explained later. They separate terms involving the variables $\mathbf{X}_1$, $\mathbf{X}_2$ from the constant terms, which include $\mathbf{X}_{2'}$.

Here we have put

$$
a_{ji} \equiv \frac{\hbar^2}{2} \left(\frac{1}{m_{1'i}} - \frac{1}{m_{ji}} \right) \quad (j = 1, 2; i = 1, 2, 3)
$$

$$\tag{E.2}$$

and

$$
B(\mathbf{X}_{2'}) \equiv \frac{\hbar^2}{2} \sum_{i=1}^{3} \left[\left(\frac{1}{m_{2'i}} + \frac{1}{m_{1'i}} \right) X_{2'i}^2 - \frac{1}{m_{1'i}} X_{2'i} Q_i + \frac{Q_i^2}{m_{1'i}} \right]
$$

$$
= \frac{\hbar^2}{2} \sum_{i=1}^{3} \left[\frac{X_{2'i}^2}{m_{2'i}} + \frac{(X_{2'i} - Q_i)^2}{m_{1'i}} \right] = \varphi_{2'}(\mathbf{X}_{2'}) + \sum_{i=1}^{3} \frac{\hbar^2 (Q_i - X_{2'i})^2}{2m_{1'i}}
$$

$$\tag{E.3}$$

535

As a *second* step we introduce the new variables formally with coefficients, u and v, to be determined shortly:

$$x_{1i} = u_{1i} X_{1i} + u_{2i} X_{2i} + u_{3i} \tag{E.4}$$
$$x_{2i} = v_{2i} X_{2i} + v_{3i}$$

Now, as a *third* step, we require the diagonalized form

$$E_i - E_f = A(\mathbf{X}_{2'}) - \sum_{i=1}^{3} (\sigma_{1i} x_{1i}^2 + \sigma_{2i} x_{2i}^2) \quad (\sigma_{1i}, \sigma_{2i} = \pm 1) \tag{E.5}$$

where $A(\mathbf{X}_{2'})$ is to be determined.

The rest is simply tedious manipulation. Introduce eq. (E.4) into eq. (E.5), multiply out, and compare the result with eq. (E.1). Equations (E.4) and (E.5) give

$$E_i - E_f = A(\mathbf{X}_{2'}) - \sum_{i=1}^{3} (\sigma_{1i} u_{3i}^2 + \sigma_{2i} v_{3i}^2)$$
$$\overline{-\sum_{i=1}^{3} [\sigma_{1i} u_{1i}^2 X_{1i}^2 + (\sigma_{1i} u_{2i}^2 + \sigma_{2i} v_{2i}^2) X_{2i}^2]}$$
$$\underline{-\sum_{i=1}^{3} [2\sigma_{1i} u_{1i} u_{2i} X_{1i} X_{2i} + 2\sigma_{1i} u_{1i} u_{3i} X_{1i} + 2(\sigma_{1i} u_{2i} u_{3i} + \sigma_{2i} v_{2i} v_{3i}) X_{2i}]} \tag{E.6}$$

Now we equate terms which have similar underlinings in eqs. (E.1) and (E.6). First we consider the solid underlinings and the coefficient of $X_{1i} X_{2i}$ to find

$$\frac{\hbar^2}{m_{1'i}} = -2\sigma_{1i} u_{1i} u_{2i}$$

Then we consider the coefficients of X_{1i} and X_{2i} to find

$$\frac{\hbar^2}{m_{1'i}} (Q_i - X_{2'i}) = -2\sigma_{1i} u_{1i} u_{3i} = -2(\sigma_{1i} u_{2i} u_{3i} + \sigma_{2i} v_{2i} v_{3i})$$

Lastly the coefficients of X_{1i}^2 and X_{2i}^2 are considered. This gives eventually

$$\left. \begin{array}{l} u_{1i} = (-\sigma_{1i} a_{1i})^{\frac{1}{2}}, \quad u_{2i} = -\dfrac{\hbar^2}{2m_{1'i}} \left(-\dfrac{\sigma_{1i}}{a_{1i}}\right)^{\frac{1}{2}} \\[4mm] u_{3i} = -\dfrac{\hbar^2}{2m_{1'i}} (Q_i - X_{2'i}) \left(-\dfrac{\sigma_{1i}}{a_{1i}}\right)^{\frac{1}{2}} \end{array} \right\} \tag{E.7}$$

$$v_{2i} = \left(\dfrac{\hbar^2}{2m_{2i}} \sigma_{2i} \dfrac{m_{1i} + m_{2i} - m_{1'i}}{m_{1i} - m_{1'i}}\right)^{\frac{1}{2}} \tag{E.8}$$

$$v_{3i} = (Q_i - X_{2'i}) \left[\sigma_{2i} \dfrac{\hbar^2 m_{2i}}{2(m_{1i} + m_{2i} - m_{1'i})(m_{1i} - m_{1'i})}\right]^{\frac{1}{2}} \tag{E.9}$$

Balancing the terms with broken underlining gives eventually

$$A(X_{2'}) = \varepsilon + \varphi_{2'}(X_{2'}) + \sum_{i=1}^{3} \frac{\hbar^2 (Q_i - X_{2'i})^2}{2(m_{1'i} - m_{1i} - m_{2i})} \tag{E.10}$$

To ensure the transformation (E.4) is real, eqs. (E.7) and (E.9) show that we need

$$\sigma_{1i} \text{ has the sign of } \frac{m_{1i} - m_{1'i}}{m_{1i} m_{1'i}} \tag{E.11}$$

$$\sigma_{2i} \text{ has the sign of } \frac{(m_{1i} + m_{2i} - m_{1'i})(m_{1i} - m_{1'i})}{m_{2i}} \tag{E.12}$$

This establishes eqs. (3.5.30)–(3.5.33).

The Jacobian of the transformation is

$$J = \prod_{i=1}^{3} J_i = \prod_i \begin{vmatrix} \left(\dfrac{\partial X_{1i}}{\partial x_{1i}}\right)_{x_{2i}} & \left(\dfrac{\partial X_{1i}}{\partial x_{2i}}\right)_{x_{1i}} \\ \left(\dfrac{\partial X_{2i}}{\partial x_{1i}}\right)_{x_{2i}} & \left(\dfrac{\partial X_{2i}}{\partial x_{2i}}\right)_{x_{1i}} \end{vmatrix} = \prod_i \begin{vmatrix} \dfrac{1}{u_{1i}} & -\dfrac{u_{2i}}{u_{1i} v_{2i}} \\ 0 & \dfrac{1}{v_{2i}} \end{vmatrix}$$

$$= \prod_{i=1}^{3} \left\{ \frac{2}{\hbar^2} \left[\frac{\sigma_{1i} \sigma_{2i}(m_{1'i} - m_{1i} - m_{2i})}{m_{1i} m_{2i} m_{1'i}} \right]^{-\frac{1}{2}} \right\} \tag{E.13}$$

Appendix F

The saddle-point method, and the method of stationary phase (section 6.3)

These methods are similar, and are discussed in detail by Jeffreys and Jeffreys [F.1]. The saddle-point method allows the evaluation of integrals of the form

$$I = \int_a^b f(x) \exp\left[\lambda g(x)\right] \tag{F.1}$$

in the limit $\lambda \to \infty$. In eq. (F.1), $f(x)$ and $g(x)$ are reasonably well-behaved functions of x, the derivative $dg(x)/dx$ has a zero at a point x_{sp} inside the interval (a, b), and the second derivative $g''(x_{sp})$ is negative. More general variants exist which deal with integrals in the complex plane, but the present formulation will suffice for the purposes of this book.

For large λ, the integral has a sharp maximum at the point x_{sp} (in a complex plane, this is a saddle point – hence the name of this method), as can be seen by expanding the function $g(x)$ in Taylor series about x_{sp}:

$$g(x) \simeq g(x_{sp}) - \tfrac{1}{2} g''(x_{sp})(x - x_{sp})^2 \tag{F.2}$$

since the first derivative vanishes at x_{sp}. For $\lambda \to \infty$, the contribution from the vicinity of x_{sp} will dominate the integral. The function $f(x)$ can then be evaluated at x_{sp} and taken outside the integral. The integral can be extended from $-\infty$ to ∞ without a significant error, giving

$$I = f(x_{sp}) \, e^{\lambda g(x_{sp})} \int_{-\infty}^{\infty} \exp\left[-\tfrac{1}{2} |g''(x_{sp})| (x - x_{sp})^2\right] dx$$

$$= \left[\frac{2\pi}{|g''(x_{sp})| \, \lambda}\right]^{\frac{1}{2}} f(x_{sp}) \, e^{\lambda g(x_{sp})} \tag{F.3}$$

where we used the result

$$\int_{-\infty}^{\infty} e^{-\sigma x^2} dx = (\pi/\alpha)^{\frac{1}{2}} \tag{F.4}$$

(see, for example, [F.2]).

The method of stationary phase concerns integrals (F.1) where the constant λ is a pure imaginary number:

$$\lambda = i\mu$$

where μ is real and positive, say. The point x_{sp} is then a point of stationary phase rather than a maximum: the integrand oscillates rapidly everywhere except in the vicinity of x_{sp} and this region therefore again gives the dominant contribution. Proceeding in a similar fashion as for the saddle-point method and using the result

$$\int_{-\infty}^{\infty} e^{\pm i\mu x^2} dx = (\pi/\mu)^{\frac{1}{2}} e^{\pm i\pi/4} \tag{F.5}$$

we obtain

$$I = \left[\frac{2\pi}{|g''(x_{sp})|\,\mu} \right]^{\frac{1}{2}} f(x_{sp}) \, e^{i[\mu g(x_{sp}) \pm \pi/4]} \tag{F.6}$$

where the signs $+$ and $-$ refer to the cases when the second derivative $g''(x_{sp})$ is positive and negative, respectively.

Appendix G

Evaluation of the integral (6.6.12)
(section 6.6)

It will be convenient to introduce the dimensionless variables x_k and y_k by

$$[\omega(k)/2\hbar]^{\frac{1}{2}} q_k = \tfrac{1}{2}(x_k + y_k) \tag{G.1}$$

$$[\omega(k)/2\hbar]^{\frac{1}{2}} q'_{k'} = \tfrac{1}{2}(x_k - y_k)$$

The product of the matrix elements of the Hamiltonian then becomes

$$\langle i, q | H_{\text{int}} | f, q \rangle \langle f, q' | H_{\text{int}} | i, q' \rangle$$

$$= \tfrac{1}{4} \sum_{k\ k'} S\{2\hbar/[\omega(k)\,\omega(k')]^{\frac{1}{2}}\} B(k)(x_k + y_k) B(k')(x_{k'} - y_{k'})$$

$$= \tfrac{1}{4} \sum_{k\ k'} S\{2\hbar/[\omega(k)\,\omega(k')]^{\frac{1}{2}}\} B(k) B(k')(x_k x_{k'} + x_{k'} y_k - x_k y_{k'} - y_k y_{k'}) \tag{G.2}$$

Not all the terms in the sum (G.2) contribute to the integral (6.6.12) where the exponent is a product of even functions of y_k. The second and third terms in the square brackets of eq. (G.2) give zero, and the last term contributes only for $k = k'$. Thus, inserting a factor $\tfrac{1}{2}$ for each pair of variables x_k, y_k which comes from the transformation (G.1), and noting the definition (6.6.13) of Sh_k,

$$f(\tau) = \tfrac{1}{4}(1/Z_i) \prod_k [Sh_k(\tau)\, Sh_k(\beta - \tau)/2]\, G$$

$$\times \int \cdots \int \{\sum_k |B(k)|^2 (x_k^2 - y_k^2) + \sum_{k \neq k'} B(k) B(k') x_k x_{k'}\}$$

$$\times \prod_{k'} \exp\{-\tfrac{1}{2}T_{k'}(x_{k'} - \zeta_{k'})^2\} \prod_{k'} \exp(-\tfrac{1}{2}C_{k'} y_{k'}^2)\, dx_1\, dy_1 \ldots \tag{G.3}$$

where

$$T_k = T_k(\tau) + T_k(\beta - \tau)$$
$$C_k = C_k(\tau) + C_k(\beta - \tau)$$
$$\zeta_k = 2\{T_k(\tau)/T_k\}\,\delta q_k\}$$
$$G = \exp[-\Sigma T_k(\beta - \tau)\,\zeta_k\,\delta q_k]$$

(G.4)

and $T_k(\tau)$, $C_k(\tau)$ are given by eq. (6.6.13). Equation (G.3) represents a product of Gaussian integrals which can be evaluated in a straightforward manner, giving

$$f(\tau) = \tfrac{1}{4}AG\{\sum |B(k)|^2\,(3/T_k - 3/C_k) + |\sum B(k)\,\zeta_k|^2\}$$

where

$$A = (1/Z_i)\prod\{\pi Sh_k(\tau)\,Sh_k(\beta - \tau)/(T_k\,C_k)^{\frac{1}{2}}\} = 1$$

Evaluating all the quantities with the use of eqs. (G.4) gives eq. (6.6.16).

References

Introduction

[1] H. Pauli. *Z. f. Phys.*, **31**, pp. 373 & 765 (1925).
R. Peierls. *Z. f. Phys.*, **53**, 255 (1929).
W. Heisenberg. *Ann. d. Phys.*, **10**, 888 (1931).

[2] A.H. Wilson. *Proc. Roy. Soc.*, **A133**, 458 (1931).

[3] N.F. Mott. *Proc. Roy. Soc.*, **A171**, 27 (1939).
W. Schottky. *Z. f. Phys.*, **118**, 539 (1942).

[4] N.F. Mott and R.W. Gurney. *Electronic Processes in Ionic Crystals* (Oxford University Press, 1940).

[5] H.K. Henisch. *Metal Rectifiers* (Oxford: Clarendon Press, 1949).
H.K. Henisch. *Semiconductor Contacts* (Oxford: Clarendon Press, 1984).

[6] W. Sweet. *Phys. Today*, May 1988, p. 87.

[7] R. Hilsch and R.W. Pohl. *Z. f. Phys.*, **111**, 399 (1938).

[8] W. Shockley. *Electrons and Holes in Semiconductors* (New York: van Nostrand, 1950).

[9] J. Bardeen and W.H. Brattain. *Phys. Rev.*, **74**, 230 (1948).

[10] W. Shockley, M. Sparks and G.K. Teal. *Phys. Rev.*, **83**, 151, footnote 9 (1951). See also [13].

[11] H.K. Henisch (ed.). *Semiconducting Materials* (London: Butterworths Scientific Publications Ltd, 1951).

[12] W. Shockley. *IEEE Trans. Electron Dev.*, **ED23**, 597 (1976).

[13] L. Esaki. *J. Res., Natl. Bur. Standards*, **86**, 565 (1981).

[14] H.J. Queisser. *The Conquest of the Microchip* (Harvard University Press, 1988).

Chapter 1

[1.1.1] J.R. Brews and C.J. Hwang. *J. Chem. Phys.*, **54**, 3263 (1971), discuss the limitation of eq. (1.1.5).

[1.1.2] P.T. Landsberg. *Thermodynamics and Statistical Mechanics* (Oxford University Press, 1978).

[1.2.1] T. Sakai. *Proc. Phys. Math. Soc., Japan*, **22**, 193 (1940).

[1.2.2] P.T. Landsberg. *Thermodynamics with Quantum Statistical Illustrations* (New York: Interscience, 1961).

[1.2.3] M.R.A. Shegelski. *Solid-State Commun.*, **58**, 351 (1986).
P.T. Landsberg and D.C. Browne. *Solid-State Commun.*, **62**, 207 (1987).

[1.2.4] E. Spenke. *Elektronische Halbleiter* (Berlin: Springer, 1955), translated as *Electronic Semiconductors* (New York: McGraw-Hill, 1958), pp. 293, 384.

[1.2.5] A. Pais, '*Subtle is the Lord...*'. *The Science and the Life of Albert Einstein* (Oxford: Clarendon Press, 1982).

[1.2.6] P.T. Landsberg. *Eur. J. Phys.*, **2**, 213 (1981).

[1.2.7] S.A. Hope, G. Féat and P.T. Landsberg. *J. Phys.*, **A14**, 2377 (1981).

[1.2.8] U. Weinert and E.A. Mason. *Phys. Rev.*, **A21**, 681 (1980).

[1.2.9] P.T. Landsberg and S.A. Hope. *Solid-State Electron.*, **20**, 421 (1977).
P.T. Landsberg. *J. App. Phys.* **56**, 1119 (1984).

[1.2.10] C. van den Broeck, F. Lostak and H.N.W. Lekkerkerker. *J. Chem. Phys.*, **74**, 2006 (1982).

[1.5.1] P.T. Landsberg and W.L. Wang. *J. App. Phys.*, **55**, 799 (1984).

[1.5.2] W. Jones and N.H. March. *Theoretical Solid State Physics*, vol. I (London: Wiley, 1973).

[1.5.3] F. Bassani and G. Pastori Parravicini. *Electronic States and Optical Transitions in Solids* (ed. R. Ballinger) (Oxford: Pergamon, 1975).

[1.5.4] P.T. Landsberg and D.J. Robbins. *J. Phys.*, **C10**, 2717 (1977).

[1.6.1] A.H. Wilson. *Proc. Roy. Soc.*, **A133**, 458 (1931).
A.H. Wilson. *Proc. Roy. Soc.*, **A134**, 277 (1931).

[1.6.2] N.F. Mott and R.W. Gurney. *Electronic Processes in Ionic Crystals* (Oxford University Press, 1940).

[1.6.3] W. Shockley. *Electrons and Holes in Semiconductors* (New York: Van Nostrand, 1950).

[1.6.4] E.O. Kane. *J. Phys. Chem. Sol.*, **1**, 249 (1957).

[1.6.5] P.I. Baranskii, V.V. Kolomoets and S.S. Korolyuk. *Phys. Stat. Sol.*, (b), **116**, 109 (1983).

[1.6.6] V.M. Bazovkin, G.L. Kurushev and V.G. Polovinkin. *Phys. Stat. Sol.*, (a), **74**, 297 (1982).

[1.6.7] Chhi-Chong Wu and Jensan Tsai. *J. Phys.*, **C15**, 4939 (1982).

[1.6.8] K.P. Ghatak, K.K. Gosh, H.M. Mukherjee and A.N. Chakravarti. *Phys. Stat. Sol.*, (b), **110**, 323 (1982).

[1.6.9] J.P. Leburton, J. Tang and K. Hess. *Solid-State Commun.*, **45**, 517 (1983).

[1.6.10] I.N. Dubrovskaya and Y.I. Ravich. *Sov. Phys. Solid State*, **8**, 1160 (1966).

[1.6.11] S. Chandrasekhar. *An Introduction to Stellar Structure* (University of Chicago Press, 1939), p. 385.

[1.6.12] P.T. Landsberg. *Phys. Rev.*, **B33**, 8321 (1986).

[1.6.13] W. Zawadzki, S. Klahn and U. Merkt. *Phys. Rev. Lett.*, **55**, 983 (1985).

[1.6.14] W. Zawadzki and J. Kołodziejczak. *Phys. Stat. Sol.*, **6**, 409 (1964).

[1.6.15] Y. Weissman. *J. Phys.*, **C9**, 2353 (1976).

[1.6.16] W. Ebeling. *Phys. Stat. Sol.*, (b), **46**, 243 (1971).

[1.6.17] G.D. Mahan. *Many-Particle Physics* (New York: Plenum, 1981).

[1.6.18] B. di Bartolo (ed.). *Collective Excitations in Solids* (New York: Plenum, 1981).

[1.6.19] P. Taylor. *A Quantum Approach to the Solid State* (Englewood Cliffs: Prentice-Hall, 1970).

[1.6.20] H. Smith. *Physica Scripta*, **28**, 287 (1983).

[1.6.21] M. de Llano. *Am. J. Phys.*, **51**, 247 (1983).

[1.6.22] J.S. Blakemore. *J. App. Phys.*, **53**, R123 (1982).

[1.6.23] H. Hazama, T. Sugimasa, T. Imachi and C. Hamaguchi. *J. Phys. Soc. Japan*, **55**, 1282 (1986).

[1.6.24] H.M. van Driel. *Appl. Phys. Lett.*, **44**, 617 (1984).

[1.6.25] A. Rogalski and K. Jóźwikowski. *Infrared Phys.*, **29**, 35 (1989).

[1.7.1] K.S. Shifrin. *J. Tech. Phys. USSR*, **14**, 43 (1944).

[1.7.2] P.T. Landsberg. *Proc. Phys. Soc.*, **B65**, 604 (1952).

[1.7.3] P.T. Landsberg. *Proc. Phys. Soc.*, **B69**, 1056 (1956).

[1.7.4] J.S. Blakemore. *Semiconductor Statistics* (Oxford: Pergamon, 1962).

[1.7.5] S.M. Ryvkin. *Photoelectric Effects in Semiconductors* (New York: Consultants Bureau, 1964).

[1.7.6] D.A. Evans. In *Solid State Theory: Methods and Applications* (ed. P.T. Landsberg) (London: Wiley, 1969).

[1.7.7] C.T. Sah and W. Shockley. *Phys. Rev.*, **109**, 1103 (1958).

[1.7.8] C.-H. Su and R.F. Bebrick. *J. Phys. Chem. Sol.*, **46**, 963 (1985).

[1.7.9] H. Brooks. In *Advances in Electronics and Electron Physics*, vol. 7 (1955), p. 117.

[1.7.10] E.W. Elcock and P.T. Landsberg. *Proc. Phys. Soc.*, **B70**, 161 (1957).
P.T. Landsberg. In *Solid State Physics in Electronics and Telecommunications*, Brussels, 1958, vol. 1 (London: Academic Press, 1960), p. 436.

[1.7.11] M.G. Clark. *J. Phys.*, **C13**, 2311 (1980).

[1.7.12] J.A. Burton. *Physica*, **20**, 845 (1954).

[1.7.13] C.O. Almbladh and G.J. Rees. *J. Phys.*, **C14**, 4576 (1981).

[1.7.14] C.D. Thurmond. *J. Electrochem. Soc.*, **122**, 1133 (1975).

[1.7.15] M.E. Foglio. *Phys. Stat. Sol.*, (*b*), **86**, 459 (1978).
M.E. Foglio. *Phys. Stat. Sol.*, (*b*), **87**, 87 (1978).

[1.7.16] O. von Roos. *J. App. Phys.*, **51**, 4523 (1980).

[1.7.17] M. Godlewski, D. Hommel, J.M. Langer and H. Przybylińska. *J. Luminescence*, **24/25**, 217 (1981).

[1.7.18] P.J. Dean. In *Handbook of Semiconductors* (ed. T.S. Moss), vol. 3 (ed. S.P. Keller) (Amsterdam: North-Holland, 1980), p. 113.

[1.7.19] B.H. Champness. *Proc. Phys, Soc.*, **B69**, 1335 (1956).

[1.7.20] P.T. Landsberg. In *Semiconductors and Phosphors* (Braunschweig: Vieweg, 1956), p. 45.

[1.7.21] W. Shockley and J.T. Last. *Phys. Rev.*, **107**, 392 (1957).

[1.7.22] S. Teitler and R.F. Wallis. *J. Phys. Chem. Sol.*, **16**, 71 (1960).

[1.7.23] M. Brown, C.L. Jones and A.F. Willoughby. *Solid-State Electron.*, **18**, 763 (1975).

[1.8.1] N.F. Mott. *Metal-Insulator Transitions* (London: Taylor and Francis, 1974).
P.A. Lee and T.V. Ramakrishnan. *Rev. Mod. Phys.*, **15**, 287 (1985).

[1.8.2] N.F. Mott. *Phys. Today*, November, 1978, p. 42.

[1.8.3] P.W. Anderson. *Phys. Rev.*, **109**, 1492 (1958).
P. Sheng (ed.) *Scattering and Localization of Classical Waves in Random Media* (Singapore: World Scientific, 1990).

[1.8.4] N.F. Mott and E.A. Davis. *Electron Processes in Non-Crystalline Materials* (Oxford: Clarendon Press, 1979).

[1.8.5]   E.O. Kane. *Phys. Rev.*, **131**, 79 (1963).

[1.8.6]   B.I. Halperin and M. Lax. *Phys. Rev.*, **148**, 722 (1966).

[1.8.7]   R.A. Abram, G.J. Rees and B.L.H. Wilson. *Adv. Phys.*, **27**, 799 (1978).

[1.8.8]   V. Sa-yakanit and H.R. Glyde. In *The Path Integral Method with Applications* (eds A. Ranfagni, V. Sa-yakanit and L.S. Shulman) (Singapore: World Scientific, 1988).

[1.8.9]   G.L. Pearson and J. Bardeen. *Phys. Rev.*, **75**, 865 (1949).

[1.8.10] P.P. Debye and E.M. Conwell. *Phys. Rev.*, **93**, 693 (1954).

[1.8.11] J. Leloup, H. Djerassi, J.H. Albany and J.B. Mullin. *J. App. Phys.*, **49**, 3359 (1978).

[1.8.12] N.V. Lien and B.I. Shklovskii. *Sov. Phys. Semicond.*, **13**, 1025 (1979).

[1.8.13] B.I. Shklovskii and A.L. Efros. *Electronic Properties of Doped Semiconductors* (Berlin: Springer, 1984). Translated by S. Luryi from a 1979 text in Russian.

[1.8.14] P.T. Landsberg (ed.). Proceedings of the International Conference 'Heavy Doping and the Metal–Insulator Transition in Semiconductors', *Solid-State Electron.*, **28**, no. 1/2 (1985).

[1.8.15] N.F. Mott. *Can. J. Phys.*, **34**, 1356 (1956).

[1.8.16] D.K. Roy and P.J. George. *Solid-State Electron.*, **18**, 757 (1975).

[1.8.17] F. Stern. *Phys. Rev.*, **148**, 186–94 (1966).

[1.8.18] D. Pines. *Elementary Excitations in Solids* (New York: Benjamin, 1963).

[1.8.19] D. Pines and P. Nozières. *The Theory of Quantum Liquids* (New York: Benjamin, 1966), p. 222.

[1.8.20] P.T. Landsberg. *Eur. J. Phys.*, **2**, 213 (1981).

[1.8.21] R.B. Dingle. *Phil. Mag.*, **46**, 831 (1955).

[1.8.22] H.C. Casey Jr. and F. Stern. *J. App. Phys.*, **47**, 631 (1976).

[1.8.23] W. Sritrakool, H.R. Glyde and V. Sa-yakanit. *Can. J. Phys.*, **60**, 373 (1982).

[1.8.24] N.H. March and M. Parinello. *Collective Effects in Solids and Liquids* (Bristol: Adam Hilger, 1982).

[1.8.25] J. Lindhard, Kgl. Danske Videnskab. *Selskab. Mat.-Fys. Medd.*, **28**, no. 1 (1954).

[1.8.26] P. Debye and E. Hückel. *Phys. Zeitschrift*, **24**, 185 (1923).

[1.8.27] N.F. Mott. *Proc. Camb. Phil. Soc.*, **32**, 281 (1936).

[1.8.28] N.F. Mott and H. Jones. *The Theory of the Properties of Metals and Alloys* (Oxford: Clarendon, 1936).

[1.8.29] P.T. Landsberg. *Proc. Phys. Soc.*, **A62**, 806 (1949).

[1.8.30] D. Pines and D. Bohm. *Phys. Rev.*, **85**, 338 (1952).

[1.8.31] G.D. Mahan. *Many-Particle Physics* (New York: Plenum, 1981).

[1.8.32] J. Hubbard. *Proc. Roy. Soc.*, **A276**, 238 (1963).

[1.8.33] P.W. Anderson. *Phys. Rev. Lett.*, **34**, 953 (1975).

[1.8.34] R. Car, P.J. Kelly, A. Oshiyama and S.T. Pantelides. *Phys. Rev. Lett.*, **52**, 1814 (1984).

[1.8.35] G.A. Baraff, E.O. Kane and M. Schlüter. *Phys. Rev.*, **B21**, 5662 (1980).

[1.8.36] J.R. Troxell and G.D. Watkins. *Phys. Rev.*, **B22**, 921 (1980).

[1.8.37] L.F. Makarenko, V.P. Markevich and L.I. Murin. *Sov. Phys. Semicond.*, **19**, 1192 (1985).

[1.8.38] H.J. Hoffmann. *Phys. Rev. Lett.*, **45**, 1733 (1980).

[1.8.39] H.J. Hoffmann. *Appl. Phys.*, **A27**, 39 (1982).

[1.9.1]   W. Shockley and J.L. Moll. *Phys. Rev.*, **119**, 1480 (1960).

[1.9.2]   M. Brown, C.L. Jones and A.F.W. Willoughby. *Solid-State Electron.*, **18**, 763 (1975).

[1.9.3]   T.A. O'Shaughnessy, H.D. Barber, D.A. Thompson and E.L. Heasell. *J. Electrochem. Soc.*, **121**, 1350 (1974).

[1.9.4]   R.K. Jain and R.J. van Overstraeten. *IEEE Trans. Electron Dev.*, **ED21**, 155 (1974).

[1.9.5]   P.T. Landsberg. *J. Phys.*, **D10**, 2467 (1978).

[1.9.6]   S. Teitler and R.F. Wallis. *J. Phys. Chem. Sol.*, **16**, 71 (1960).

[1.9.7]   K.L. Ashley, V. Jayakumar and R.T. Brown. *J. App. Phys.*, **42**, 1240 (1971).

[1.9.8]   E. Ohta and M. Sakata. *Solid-State Electron.*, **22**, 677a (1979).

[1.9.9]   T. Kunio, T. Nishino, E. Ohta and M. Sakata. *Solid-State Electron.*, **24**, 1087 (1981).

[1.9.10] P.T. Landsberg. In *Handbook of Semiconductors* (ed. T.S. Moss), vol. 1 (ed. W. Paul) (Amsterdam: North-Holland, 1982), p. 359.

[1.9.11] S.F. Cagnina. *J. Electrochem. Soc.*, **116**, 498 (1969).

[1.9.12] E.V.K. Rao, N. Duhamel and M. Gauneau. *J. App. Phys.*, **56**, 3413 (1984).

[1.9.13] D.G. Deppe, D.W. Nam, N. Holonyak, Jr., K.C. Hsieh, J.E. Baker, C.P. Kuo, R.M. Fletcher, T.D. Osentowski and M.G. Craford. *App. Phys. Lett.*, **52**, 1413 (1988).

[1.9.14] S. Froyen and A. Zunger. *Phys. Rev.*, **B34**, 7451 (1986).

[1.9.15] G.A. Baraff. In *Defects in Semiconductors*, Material Science Forum X (ed. H.J. von Bardeleben) (Aldermannsdorf, Switzerland: Trans. Tech. Publications, 1986), p. 377.

[1.10.1] E.A. Guggenheim. *Mixtures* (Oxford: Clarendon Press, 1952). Chapter 6 has the grand partition function for imperfections, which is rarely used in this context, in a somewhat different form.

[1.10.2] C. Kittel. *Am. J. Phys.*, **35**, 483 (1967).

[1.10.3] P.T. Landsberg. *Proc. Natl. Acad. Sci. Wash.*, **40**, 149 (1954).

[1.10.4] See, for example, N.N. Greenwood. *Ionic Crystals, Lattice Defects and Nonstoichiometry* (London: Butterworths, 1968).

[1.10.5] A. Chatterjea and J.R. Hauser. *Solid-State Electron.*, **31**, 1031 (1976).

[1.10.6] P.T. Landsberg and S. Canagaratna. *Phys. Stat. Sol.*, (*b*), **126**, 141 (1984).

[1.10.7] J.S. Anderson. *Proc. Roy. Soc.*, **A185**, 69 (1946).

[1.10.8] L.M. Atlas. In *The Chemistry of Extended Defects in Nonmetallic Crystals* (eds Le Roy Eyring and M. O'Keefe) (Amsterdam: North-Holland, 1970).

[1.10.9] G.W. Franti, D. Kahlmann-Wilsdorf, R.K. Mitchell, G.R. Proto and B.C. Wilson. *Crystal Lattice Defects*, **3**, 87 (1972).

[1.10.10] A.B. Lidiard. *Handbuch der Physik*, vol. 20 (Berlin: Springer, 1957), p. 258.

[1.10.11] R.W. Whitworth. *Adv. Physics*, **24**, 203 (1975).

[1.10.12] M. Yoshida. *Japan J. App. Phys.*, **15**, 2261 (1976).

[1.10.13] B. Henderson and A.E. Hughes (eds). *Defects and their Structure in Nonmetallic Solids* (New York: Plenum, 1976).

[1.10.14] J.A. van Vechten. In *Handbook of Semiconductors* (ed. T.S. Moss), vol. 3 (ed. S.P. Keller) (Amsterdam: North-Holland, 1980), p. 1.

[1.11.1] A.J. Rosenberg. *J. Chem. Phys.*, **33**, 665 (1960).

[1.11.2] C.J. Hwang and J.R. Brews. *J. Phys. Chem. Sol.*, **32**, 837 (1971).

[1.11.3] D.C. Herbert, D.T.J. Hurle and R.M. Logan. *J. Phys.*, **C18**, 3571 (1975).

[1.11.4] P.T. Landsberg. In *Semiconductors and Phosphors* (Braunschweig: Vieweg, 1958), p. 58.

[1.11.5] J.H. de Boer and W.C. van Geel. *Physica*, **2**, 186 (1935).

[1.11.6] J.S. Blakemore. *J. App. Phys.*, **51**, 1054 (1980).

[1.11.7] P.T. Landsberg and A.G. Guy. *Phys. Rev.*, **28**, 1187 (1983).

[1.11.8] K.M. van Vliet and A.H. Marshak. *Phys. Stat. Sol.*, (*b*), **101**, 525 (1980).

[1.11.9] P.T. Landsberg and H.C. Cheng. *Phys. Rev.*, **B32**, 8021 (1985).

[1.11.10] A.N. Chakravarti and B.R. Nag. *Int. J. Electron.*, **37**, 281 (1974).
A.N. Chakravarti, K.P. Ghatak, K.K. Gosh and G.B. Rao. *Phys. Stat. Sol.*, (*b*), **111**, K61 (1982).

[1.11.11] M.E. Orazem and J. Newman. *J. Electrochem. Soc.*, **131**, 2715 (1984).

[1.11.12] P.T. Landsberg. *Proc. Roy. Soc.*, **A213**, 226 (1952).

[1.11.13] R.K. Jain. *Phys. Stat. Sol.*, (*a*), **42**, 221 (1977).

[1.11.14] N.G. Nilsson. *Phys. Stat. Sol.*, (*a*), **19**, K75 (1973).

[1.11.15] A.N. Chakravarti and D.P. Parui. *Phys. Stat. Sol.*, (*a*), **14**, K55 (1972).

[1.11.16] K.P. Ghatak, A.K. Chowdhury, S. Gosh and A.N. Chakravarti. *Appl. Phys.*, **23**, 241 (1980).

[1.11.17] B.R. Nag and A.N. Chakravarti. *Phys. Stat. Sol.*, (*a*), **67**, K113 (1981).

[1.11.18] D. Tjapkin, V. Milanović and Z. Spasojević. *Phys. Stat. Sol.*, (*a*), **63**, 737 (1981).

[1.11.19] D. Tjapkin and V. Milanović. *Phys. Stat. Sol.*, (*b*), **116**, 653 (1983).

[1.11.20] B.R. Nag, A.N. Chakravarti and P.K. Basu. *Phys. Stat. Sol.*, (*a*), **68**, K75 (1981).

[1.11.21] K.M. van Vliet and A.H. Marshak. *Phys. Stat. Sol.*, (*b*), **78**, 501 (1976).

[1.11.22] P.T. Landsberg and S.A. Hope. *Solid-State Electron.*, **20**, 421 (1977).

[1.11.23] P.T. Landsberg. *J. App. Phys.*, **56**, 1119 (1984).

[1.11.24] K.M. van Vliet and A. van der Ziel. *Solid-State Electron.*, **20**, 931 (1977).

[1.11.25] H. Kroemer. *IEEE Trans. Electron Dev.*, **ED25**, 850 (1978).

[1.11.26] H. van Cong. *Phys. Stat. Sol.*, (*b*), **101**, K27 (1980).

[1.11.27] M. Mondal and K.P. Ghatak. *J. Phys.*, **C20**, 1671 (1987).

[1.12.1] See, for example, J.S. Blakemore. *Semiconductor Statistics* (Oxford: Pergamon Press, 1962).

[1.12.2] See, for example, A.G. Milnes. *Deep Impurities in Semiconductors* (New York: Wiley, 1973).

[1.12.3] P.T. Landsberg, R.W. Mackay and A.D. McRonald. *Proc. Phys. Soc.*, **A64**, 476 (1951).

[1.12.4] W. Ehrenberg. *Proc. Phys. Soc.*, **A63**, 75 (1950).

[1.12.5] J. McDougall and E.C. Stoner. *Phil. Trans. Roy. Soc.*, **A237**, 67 (1938).

[1.12.6] E. Mooser. *Z.A.M.P.*, **4**, 433 (1953).

[1.12.7] G. Busch. *J. Electron.*, **1**, 178 (1955).

[1.12.8] D. Bednarczyk and J. Bednarczyk. *Phys. Lett.*, **64A**, 409 (1978).

[1.12.9] W.B. Joyce and R.W. Dixon. *App. Phys. Lett.*, **31**, 354 (1977).

[1.12.10] W.B. Joyce. *App. Phys. Lett.*, **32**, 680 (1978).

[1.12.11] S.D. Jog. *Phys. Lett.*, **72A**, 303 (1979).

[1.12.12] M.A. Sobhan and S.N. Mohammad. *J. App. Phys.*, **58**, 2634 (1985).

[1.12.13] J.S. Blakemore. *Solid-State Electron.*, **25**, 1067 (1982).

[1.12.14] J.S. Blakemore. *Proc. Phys. Soc.*, **71**, 692 (1958).

[1.12.15] S.M. Sze. *Physics of Semiconductor Devices*, 2nd edn (New York: Wiley, 1981), p. 19.

[1.12.16] J.S. Blakemore. *J. Appl. Phys.*, **51**, 1054 (1980).

[1.12.17] D.S. Lee and J.G. Fossum. *IEEE Trans. Electron Dev.*, **ED30**, 626 (1983).

[1.12.18] P.T. Landsberg, A. Neugroschel, F.A. Lindholm and C.T. Sah. *Phys. Stat. Sol.*, (*b*), **130**, 255 (1985).

[1.12.19] R.A. Abram, G.J. Rees and B.L.H. Wilson. *Adv. Phys.*, **27**, 799 (1978).

[1.12.20] M.A. Shibib, F.A. Lindholm and F. Therez. *IEEE Trans. Electron Dev.*, **ED26**, 959 (1979).

[1.12.21] R.P. Mertens, R.J. van Overstraeten and H.J. de Man. *Adv. Electron. and Electron Phys.*, **55**, 77 (1981).

[1.12.22] A.H. Marshak and C.M. van Vliet. *Proc. IEEE*, **72**, 148 (1984).

[1.12.23] P.T. Landsberg (ed.). Proceedings of the International Conference 'Heavy doping and metal–insulator transition in semiconductors', *Solid-State Electron.*, **28**, no. 1/2 (1985).
Y. Pan and M. Kleefstra. *Semicond. Sci. Technol.*, **5**, 312 (1990).

[1.12.24] S. Selberherr. *Analysis and Simulation of Semiconductor Devices* (Wien: Springer, 1984), p. 39.

[1.12.25] R.P. Mertens, J.L. van Meerenbergen, J.F. Nijs and R.J. van Overstraeten. *IEEE Trans. Electron Dev.*, **ED27**, 949 (1980).

[1.12.26] J.W. Slotboom. *Solid-State Electron.*, **20**, 279 (1977).

[1.12.27] A.W. Wieder. *IEEE Trans. Electron. Dev.*, **ED27**, 1402 (1980).

[1.12.28] H.E.J. Wulms. *IEEE J. Solid-State Circuits*, **SC12**, 143 (1977).

[1.12.29] M.L. Mock. *Solid-State Electron.*, **16**, 1251 (1973).

[1.12.30] R.J. van Overstraeten, H.J. de Man and R.P. Mertens. *IEEE Trans. Electron. Dev.*, **ED20**, 290 (1973).

[1.12.31] V.L. Bonch-Bruevich. *The Electronic Theory of Heavily Doped Semiconductors* (Amsterdam: Elsevier Publishing Co., 1966).

[1.12.32] J.W. Slotboom and H.C. de Graaff. *Solid-State Electron.*, **19**, 857 (1976).

[1.12.33] D.D. Tang. *IEEE Trans. Electron. Dev.*, **ED27**, 563 (1980).

[1.12.34] G.E. Possin, M.S. Adler and B.J. Baliga. *IEEE Trans. Electron Dev.*, **ED31**, 3 (1984).

[1.12.35] A. Neugroschel, S.C. Pao and F.A. Lindholm. *IEEE Trans. Electron Dev.*, **ED29**, 894 (1982).

[1.12.36] J. del Alamo, S. Swirhun and R.M. Swanson. *Solid-State Electron.*, **28**, 47 (1985).

[1.12.37] K.-F. Berggren and B.E. Sernelius. *Phys. Rev.*, **B24**, 1971 (1981).
A. Selloni and S.T. Pantelides. *Phys. Rev. Lett.*, **49**, 586 (1982).

[1.12.38] G.D. Mahan. *J. App. Phys.*, **51**, 2634 (1980).

[1.12.39] P.A. Sterne and J.C. Inkson. *J. App. Phys.*, **52**, 6432 (1981).

[1.12.40] H.P.D. Lanyon and R.A. Tuft. *IEEE Trans. Electron Dev.*, **ED26**, 1014 (1979).

[1.12.41] P.T. Landsberg and G.S. Kousik. *J. App. Phys.*, **56**, 1696 (1984).

[1.12.42] J.G. Fossum and D.S. Lee. *Solid-State Electron.*, **25**, 741 (1982).

[1.12.43] B.I. Shklovskii and A.L. Efros. *Electronic Properties of Doped Semiconductors* (Berlin: Springer, 1984). Translated by S. Luryi from a 1979 text in Russian.

[1.12.44] M.G. Clark. *J. Phys.*, **C13**, 2311 (1980).

[1.12.45] G.G. Roberts, N. Apsley and R.W. Munn. *Phys. Reports*, **60**, 60 (1980).

[1.13.1] C.T. Sah, R.N. Noyce and W. Shockley. *Proc. Inst. Radio Eng.*, **45**, 9 (1957).

Chapter 2

[2.1.1] P.T. Landsberg. *Proc. IEE*, **B106**, 908 (1959).

[2.1.2] P.T. Landsberg. *Abh. Deutsch. Akad. Wiss., Berlin, Math-Phys. Kl.*, no. 7, p. 57 (1960).

[2.1.3] P.T. Landsberg. *Festkörperprobleme*, **6**, 174 (1967).

[2.1.4] P.J. Dean. *Prog. Solid-State Chem.*, **8**, 1 (1973).

[2.1.5] V.V. Vorob'ev, E.I. Tolpygo and M.K. Sheinkman. *Sov. Phys. Semicond.*, **19**, pp. 1313 & 1318 (1985).

[2.1.6] G. Bemski. *Proc. Inst. Radio Eng.*, **46**, 990 (1958).

[2.1.7] W. Schultz. *Festkörperprobleme*, **5**, 165 (1966).

[2.1.8] P.T. Landsberg. *Solid-State Electron.*, **10**, 513 (1967).

[2.1.9] R. Conradt. *Festkörperprobleme*, **12**, 449 (1972).

[2.1.10] P.T. Landsberg. *Phys. Stat. Sol.*, **41**, 457 (1970).

[2.1.11] B. Ross. In *Lifetime Factors in Silicon* (ed. R. Westbrook) (American Society for the Testing of Materials, Spec. Tech. Publ. 712, 1980), p. 14.

[2.1.12] G. Nimtz. *Phys. Reports*, **63**, 266 (1980).

[2.1.13] E. Yablonovitch and T.J. Gmitter. *Phys. Rev. Lett.*, **63**, 1950 (1989).
K.M. Ho, C.T. Chan and C.M. Soukoulis. *Phys. Rev. Lett.* **65**, 3152 (1990).
E.A. Hinds, *Adv. Atomic and Mol. Phys.* **28**, 237 (1990).

[2.2.1] A.G. Jordan, R.W. Lade and D.L. Scharfetter. *Am. J. Phys.*, **31**, 490 (1963).
P.T. Landsberg. *Festkörperprobleme*, **6**, 174 (1967).
H.K. Henisch. *Semiconductor Contacts* (Oxford University Press, 1984), p. 9.

[2.2.2] R.N. Hall. *Phys. Rev.*, **83**, 228 (1951).
R.N. Hall. *Phys. Rev.*, **87**, 387 (1952).
W. Shockley and W.T. Read. *Phys. Rev.*, **87**, 835 (1952).

[2.2.3] I.V. Karpova and S.G. Kalashnikov. *Proc. Intern. Conf. Physics of Semiconductors, Exeter* (London: Institute of Physics and Physical Society, 1962), p. 880.

[2.2.4] P.T. Landsberg, D.A. Evans and C. Rhys-Roberts. *Proc. Phys. Soc.*, **83**, 325 (1964).

[2.2.5] G.F. Neumark, D.J. DeBitetto, R.N. Bhargava and P.M. Harnack. *Phys. Rev.*, **15**, pp. B3147 & 3156 (1977).

[2.2.6] V.A. Zuev, V.G. Litovchenko and G.A. Sukach. *Sov. Phys. Semicond.*, **9**, 1083 (1976).

[2.2.7] J.M. Hermann III and C.T. Sah. *Solid-State Electron.*, **16**, 1133 (1973).

[2.2.8] G. Jones and A.R. Beattie. *Phys. Stat. Sol.*, (a), **4**, 193 (1971).
G. Jones and A.R. Beattie. *Phys. Stat. Sol.*, (a), **8**, 403 (1971).

[2.2.9] K.D. Glinchuk, A.V. Prokhorovich and V.T. Vovnenko. *Phys. Stat. Sol.*, (a), **54**, 121 (1979).

[2.2.10] A. Haug. *Phys. Stat. Sol.*, (b), **97**, 481 (1980).
M. Takeshima. *Phys. Rev.*, **B27**, 7524 (1983).

[2.2.11] D.J. Robbins and P.T. Landsberg. *J. Phys.*, **C13**, 2425 (1980).
V.B. Khalfin, M.V. Strikha and I.N. Yassievich. *Phys. Stat. Sol.*, (b), **131**, 203 (1985).

[2.2.12] R. Conradt. *Festkörperprobleme*, **12**, 449 (1972).

[2.2.13] M.H. Pilkuhn. *Proc. 13th International Conference on Physics of Semiconductors*, Rome, 1976, p. 61. No publisher given.

[2.2.14] P.T. Landsberg. *Phys. Stat. Sol.*, **41**, 457 (1970).

[2.2.15] D.J. Robbins. *Phys. Stat. Sol.*, **97**, pp. 9 & 387 (1980).
D.J. Robbins. *Phys. Stat. Sol.*, **98**, 11 (1980).
M.G. Burt, S. Brand, C. Smith and R.A. Abram. *J. Phys.*, **C17**, 6385 (1984).

[2.2.16] D. Hill and P.T. Landsberg. *Proc. Roy. Soc.*, **A347**, 547 (1976).

[2.2.17] D. Hill. *Proc. Roy. Soc.*, **A347**, 565 (1976).

[2.2.18] W. Lochmann. *Phys. Stat. Sol.*, (*a*), **45**, 423 (1978).

[2.2.19] F. Tröster. *Z. Naturf.*, **33a**, 1257 (1978).

[2.2.20] A. Haug. *Solid-State Commun.*, **28**, 291 (1978).

[2.2.21] A. Haug. *J. Luminescence*, **20**, 173 (1979).

[2.2.22] O. Ziep, M. Mocker, D. Genzow and K.H. Herrmann. *Phys. Stat. Sol.*, (*b*), **90**, 197 (1978).

[2.2.23] M. Takeshima. *J. App. Phys.*, **49**, 6118 (1978).
M. Takeshima. *Phys. Rev.*, **B28**, 2039 (1983).
M. Takeshima. *Phys. Rev.*, **B29**, 1993 (1984).

[2.2.24] A. Haug, D. Kerkhoff and W. Lochmann. *Phys. Stat. Sol.*, (*b*), **89**, 357 (1978).

[2.2.25] O. Ziep and M. Mocker. *Phys. Stat. Sol.*, (*b*), **98**, 133 (1980).

[2.2.26] T.N. Casselman and P.E. Peterson. *Solid-State Commun.*, **33**, 615 (1980).

[2.2.27] D.J. Robbins and A. Young. *Phys. Stat. Sol.*, (*b*), **102**, K143 (1980).

[2.2.28] M. Takeshima. *Phys. Rev.*, **B12**, 575 (1975).
D. Hill. *J. Phys.*, **C9**, 3527 (1976).

[2.2.29] M. Takeshima. *J. App. Phys.*, **44**, 4717 (1973).
R. Dornhaus, K.-H. Müller, G. Nimtz and M. Schifferdecker. *Phys. Rev. Lett.*. **37**, 710 (1976).

[2.2.30] A. Haug. *Solid-State Commun.*, **22**, 537 (1977).

[2.2.31] V. N. Abakumov and I.N. Yassievich. *Sov. Phys. Semicond.*, **11**, 766 (1977).

[2.2.32] A. Haug and W. Ekardt. *Solid-State Commun.*, **17**, 267 (1975).
M. Takeshima. *Phys. Rev.*, **B26**, 3192 (1982).
A. Das and R. Al-Jishi. *Phys. Lett.*, **A141**, 186 (1989).

[2.2.33] R.W. Martin and H.L. Störmer. *Solid-State Commun.*, **22**, 523 (1977).

[2.2.34] R.N. Silver and C.H. Aldrich. *Phys. Rev. Lett.*, **41**, 1249 (1978).

[2.2.35] L. Huldt, N.G. Nilsson and K.G. Svantesson. *App. Phys. Lett.*, **35**, 776 (1979).

[2.2.36] H. Bruhns and H. Kruse. *Phys. Stat. Sol.*, (*b*), **97**, 125 (1980).

[2.2.37] M.Y. Pines and O.M. Stafsudd. *Infrared Phys.*, **20**, 73 (1980).

[2.2.38] G. Nimtz. *Phys. Reports*, **63**, 265 (1980).

[2.2.39] S.M. Ryvkin. *Photoelectric Effects in Semiconductors* (New York: Consultants Bureau, 1964).

[2.2.40] V.L. Bonch-Bruevich and E.G. Landsberg. *Phys. Stat. Sol.*, **29**, 9 (1968).

[2.2.41] A.G. Milnes. *Deep Impurities in Semiconductors* (New York: Wiley, 1973).

[2.2.42] A.M. Stoneham. *Theory of Defects in Solids* (Oxford: Clarendon Press, 1975).

[2.2.43] P.T. Landsberg and A.F.W. Willoughby (eds). 'Recombination in semiconductors'. Special issue of *Solid-State Electron.*, **21**, 1275 (1978).

[2.2.44] G.K. Wertheim. *J. App. Phys.*, **30**, 1166 (1959).

[2.2.45] M.A. Habegger and H.Y. Fan. *Phys. Rev.*, **138**, 598 (1965).

[2.2.46] M. Lax. *Phys. Rev.*, **119**, 1502 (1960).

[2.2.47] E.F. Smith and P.T. Landsberg. *J. Phys. Chem. Sol.*, **27**, 1727 (1966).

[2.2.48] R.A. Brown and S. Rodriguez. *Phys. Rev.*, **153**, 890 (1967).

[2.2.49] R.M. Gibb, C.J. Rees, B.W. Thomas, B.L.H. Wilson, B. Hamilton, D.R. Wight and N.F. Mott. *Phil. Mag.*, **36**, 1021 (1977).

[2.2.50] C.H. Henry and D.V. Lang. *Phys. Rev.*, **15**, 989 (1977).

[2.2.51] J.-T. Rebsch. *Solid-State Commun.*, **31**, 377 (1979).

[2.2.52] D.J. Robbins. *J. Phys.*, **C13**, 1073 (1980).

[2.2.53] P.T. Landsberg and T.S. Moss. *Proc. Phys. Soc.*, **B69**, 661 (1956).
P.T. Landsberg. *Proc. Phys. Soc.*, **B70**, 282 (1957).

[2.2.54] O. von Roos. *Solid-State Electron.*, **21**, 633 (1978).

[2.2.55] F. Williams. *Phys. Stat. Sol.*, **25**, 493 (1968).

[2.2.56] V. Novotny. *Can. J. Phys.*, **47**, 1971 (1969).

[2.2.57] E.I. Tolpygo, K.B. Tolpygo and M.K. Sheinkman. *Sov. Phys. Semicond.*, **8**, 326 (1974).

[2.2.58] J.C. Tsang, P.J. Dean and P.T. Landsberg. *Phys. Rev.*, **173**, 814 (1968).

[2.2.59] K.P. Sinha and M. DiDomenico Jr. *Phys. Rev.*, **B1**, 2623 (1970).

[2.2.60] P.T. Landsberg and M.J. Adams. *Proc. Roy. Soc.*, **A334**, 523 (1973).

[2.2.61] A.T. Vink. *J. Luminescence*, **9**, 159 (1974).

[2.2.62] A.T. Vink, R.L.A. Van der Heijden and H.C. van Amstel. *J. Luminescence*, **9**, 180 (1974).

[2.2.63] F. Garcia-Moliner. *Catalysis Rev.*, **2**, 1 (1968).

[2.2.64] H.C. Casey Jr., B.I. Miller and E. Pinkas. *J. App. Phys.*, **44**, 1281 (1973).

[2.2.65] D.R. Wight. *J. Phys.*, **D10**, 431 (1977), Fig. 12.
M. Takeshima. *Phys. Rev.*, **B31**, 992 (1985).

[2.2.66] W. Schröter. *Phys. Stat. Sol.*, (*a*), **19**, 159 (1973).
C. Smith, R.A. Abram and M.G. Burt. *Superlattices and Microstructures*, **1**, 119 (1985).

[2.2.67] M.G. Burt and R.I. Taylor. *Electron. Lett.*, **21**, 733 (1985).

[2.2.68] P.T. Landsberg and M.J. Adams. *IEE Proc.*, **J 133**, 118 (1986).

[2.3.1] P.T. Landsberg and M.S. Abrahams. *J. App. Phys.*, **55**, 4284 (1984).

[2.3.2] W. Schmid and J. Reiner. *J. App. Phys.*, **53**, 6250 (1982).

[2.3.3] R.G. Pratt, J. Hewett, P. Capper, C.L. Jones and M.J. Quelch. *J. App. Phys.*, **54**, 5152 (1983).

[2.3.4] B. Ross. In *Lifetime Factors in Silicon* (ed. R. Westbrook) (Philadelphia: American Society for Testing and Materials, Special Technical Publication 712, 1980), p. 14.

[2.3.5] J.G. Fossum and D.S. Lee. *Solid-State Electron.*, **25**, 741 (1982).

[2.3.6] P.T. Landsberg and G.S. Kousik. *J. App. Phys.*, **56**, 1696 (1984).

[2.3.7] D.A. Evans and P.T. Landsberg. *Solid-State Electron.*, **6**, 169 (1963).

[2.3.8] G. Bemski. *Phys. Rev.*, **103**, 567 (1956).

[2.3.9] B. Ross and J.R. Madigan. *Phys. Rev.*, **108**, 1428 (1957).

[2.3.10] L. Elstner and W. Kamprath. *Phys. Stat. Sol.*, **22**, 541 (1967).

[2.3.11] G. Bemski and C.A. Dias. *J. App. Phys.*, **35**, 2983 (1964).

[2.3.12] M.L. Swanson. *Phys. Stat. Sol.*, **33**, 721 (1969).

[2.3.13] G. Swenson. *Phys. Stat. Sol.*, (*a*), **2**, 803 (1970).

[2.3.14] K.D. Glinchuk, N.M. Litovchenko, L.F. Linnik and R. Merker. *Phys. Stat. Sol.*, (*a*), **18**, 749 (1973).

[2.3.15] W. Leskochek, H. Feichtinger and G. Vidrich. *Phys. Stat. Sol.*, (*a*), **20**, 601 (1973).

[2.3.16] L.D. Yau and C.T. Sah. *Solid-State Electron.*, **17**, 193 (1974).

[2.3.17] K.D. Glinchuk, N.M. Litovchenko and R. Merker. *Phys. Stat. Sol.*, (*a*), **30**, K109 (1975).

[2.3.18] K.D. Glinchuk, N.M. Litovchenko and R. Merker. *Phys. Stat. Sol.*, (*a*), **33**, K87 (1976).

[2.3.19] K.D. Glinchuk, N.M. Litovchenko and R. Merker. *Phys. Stat. Sol.*, (*a*), **35**, K157 (1976).

[2.3.20] K.D. Glinchuk and N.M. Litovchenko. *Phys. Stat. Sol.*, (*a*), **58**, 549 (1980).

[2.3.21] G. Borchardt, E. Weber and N. Wiehl. *J. App. Phys.*, **52**, 1603 (1981).

[2.3.22] K. Wünstel and P. Wagner. *Solid-State Commun.*, **40**, 797 (1980).

[2.3.23] D.W. Ioannou. *Phys. Stat. Sol.*, (*a*), **72**, K33 (1982).

[2.3.24] A. Usami, Y. Fujii and K. Morioka. *J. Phys.*, **D10**, 899 (1977).

[2.3.25] L. Jastrezebski and P. Zanzucci. In *Semiconductor Silicon 1981* (ed. H.R. Huff) (Pennington, NJ: Electrochemical Society), p. 138.

[2.3.26] A.J.R. de Kock. In *Handbook of Semiconductors* (ed. T.S. Moss), vol. 3 (ed. S.P. Keller) (Amsterdam: North-Holland, 1980), p. 272.

[2.3.27] L. Passari and E. Susi. *J. App. Phys.*, **54**, 3935 (1983).

[2.3.28] M. Jaros. *Deep Levels in Semiconductors* (Bristol: Adam Hilger, 1982), p. 253.

[2.3.29] A. Rohatgi and P. Rai-Choudhury. In *Silicon Processing* (ed. D.C. Gupta) (American Society for Testing and Materials, Special Technical Publication 804, 1983), p. 383.

[2.3.30] P.T. Landsberg. In *Proceedings of the Flat Plate Solar Array Project Research Forum on High Efficiency Crystalline Silicon Solar Cells* (J.P.L. Publication 85-38, May 15, 1985), p. 13.

[2.3.31] P.T. Landsberg and D.C. Browne. *Semicond. Sci. Technol.*, **3**, 193 (1988). For discussion of additional features see W.D. Eades and R.M. Swanson. *J. App. Phys.*, **58**, 4267 (1985).

[2.3.32] E. Yablonovitch, D.L. Allara, C.C. Chang, T. Gmitter and T.B. Bright. *Phys. Rev. Lett.*, **57**, 249 (1986). E. Yablonovitch and T. Gmitter. *Appl. Phys. Lett.*, **49**, 587 (1986).

[2.3.33] P.T. Landsberg. *Appl. Phys. Lett.*, **50**, 745 (1987).

[2.3.34] P.T. Landsberg. 'Auger effects in Semiconductors' in *Proceedings of the International Workshop of Physics of Semiconductors* (eds S.C. Jain and S. Radhakrishna) (New Delhi: Wiley Eastern, 1982), p. 30.

[2.3.35] A. Haug. *Phys. Stat. Sol.*, (*b*), **108**, 443 (1981).

[2.3.36] C.H. Henry and D.V. Lang. *Phys. Rev.*, **B15**, 989 (1977).

[2.3.37] J. Dziewor and W. Schmid. *Appl. Phys. Lett.*, **31**, 346 (1977).

[2.3.38] O. von Roos and P.T. Landsberg. *J. App. Phys.*, **57**, 4746 (1985).

[2.3.39] P.T. Landsberg. In *Physical Limitations of Photovoltaic Solar Energy Conversion* (ed. A. Luque and F.L. Araújo) (Bristol: Adam Hilger, 1990), p. 134.

[2.3.40] R.A. Sinton and R.M. Swanson. *IEEE Trans. Electron Dev.*, **ED34**, 2116 (1987).

[2.4.1] M. Lax. *J. Phys. Chem. Sol.*, **8**, 66 (1959).

[2.4.2] M. Lax. *Phys. Rev.*, **119**, 1502 (1960).

[2.4.3] A.M. Stoneham. *Theory of Defects in Solids* (Oxford: Clarendon Press, 1975), p. 522.

[2.4.4] A.V. Rhzanov. *Sov. Phys. Solid State*, **3**, 2680 (1962).

[2.4.5] G.M. Guro and A.V. Rhzanov. *Sov. Phys. Solid State*, **4**, 2519 (1963).

[2.4.6] R.M. Gibb, G.J. Rees, B.W. Thomas, B.L.H. Wilson, B. Hamilton, D.R. Wight and N.F. Mott. *Phil. Mag.*, **36**, 1021 (1977).

[2.4.7] P. Lal and P.T. Landsberg. *Phys. Rev.*, **140**, A46 (1965).

[2.4.8] See, for example, D.E. Osterbrock. *Astrophysics of Gaseous Nebulae* (San Francisco: W.H. Freeman, 1974), section 4.2.

[2.4.9] S.R. Dhariwal, L.S. Kothari and S.C. Jain. *Solid-State Electron.*, **24**, 749 (1981).

[2.4.10] P.T. Landsberg and M.S. Abrahams. *Solid-State Electron.*, **26**, 841 (1983).

[2.4.11] P.T. Landsberg and S.R. Dhariwal. *Phys. Rev.*, **B39**, 91 (1989).

[2.4.12] S.R. Dhariwal and P.T. Landsberg. *J. Phys. Condensed Matter*, **1**, 569 (1989).

[2.4.13] P.T. Landsberg. *J. Luminescence*, **18–19**, 1 (1979).

[2.4.14] P.T. Landsberg. In *Handbook of Semiconductors* (ed. T.S. Moss), vol. 1 (ed. W. Paul) (Amsterdam: North-Holland, 1982), p. 359.

[2.4.15] G.J. Rees, H.G. Grimmeiss, E. Janzén and B. Skarstam. *J. Phys. C*, **13**, 6157 (1980).

[2.4.16] W. Pickin. *Solid-State Electron.*, **21**, 1299 (1978).

[2.4.17] V.L. Bonch-Bruevich and E.G. Landsberg. *Phys. Stat. Sol.*, **29**, 9 (1968).

[2.4.18] V.N. Abakumov, V.I. Perel' and I.N. Yassievich. *Sov. Phys. JETP*, **45**, 354 (1977).

[2.4.19] H.G. Grimmeiss, E. Janzén and B. Skarstam. *J. App. Phys.*, **51**, 4212 (1980).

[2.4.20] C.T. Sah, L. Forbes, L.L. Rosier and A.F. Tasch Jr. *Solid-State Electron.*, **13**, 759 (1970).

[2.4.21] E.F. Smith and P.T. Landsberg. *J. Phys. Chem. Sol.*, **27**, 1727 (1966).

[2.4.22] M.B. Chang, H. Tomokage, J.J. Shiau, R.H. Bube and J.C. Bravman. *J. App. Phys.*, **65**, 2734 (1989).

[2.4.23] V.N. Abakumov, L.N. Kreshchuk and I.N. Yassievich. *Sov. Phys. Semicond.*, **12**, 152 (1978).

[2.4.24] L.L. Rosier and C.T. Sah. *Solid-State Electron.*, **14**, 41 (1971).

[2.4.25] E. Fabre, M. Mautref and A. Mircea. *App. Phys. Lett.*, **27**, 239 (1975).

[2.4.26] A. Pogany. *Proceedings of 14th IEEE Photovoltaic Specialists Conference* (New York: IEEE, 1980), p. 410.

[2.4.27] V.G. Ivanov. *Sov. Phys. Solid State*, **8**, 1306 (1966).

[2.4.28] P.G. Wilson. *Solid-State Electron.*, **10**, 145 (1967).

[2.4.29] B.M. Ashkinadze, A.A. Patrin and I.D. Yaroshetskii. *Sov. Phys. Sem.*, **5**, 1471 (1972).

[2.4.30] K.D. Glinchuk, N.M. Litovchenko and L.F. Linnik. *Sov. Phys. Sem.*, **5**, 2088 (1972).

[2.4.31] W. Zimmerman. *Electron Lett.*, **9**, 378 (1973).

[2.4.32] V.L. Dalal and A.R. Moore. *J. App. Phys.*, **48**, 1244 (1977).

[2.4.33] C.T. Ho, R.O. Bell and F.V. Wald. *App. Phys. Lett.*, **31**, 463 (1977).

[2.4.34] C.T. Ho and J.D. Mathias. *J. App. Phys.*, **54**, 5993 (1983).

[2.4.35] I. Suemune, N. Uesugi, Y. Kan and M. Yamanishi. *Jap. J. App. Phys.*, **26**, L159 (1987).

[2.4.36] P. Spirito and G. Cocorullo. *IEEE Trans. Electron Dev.*, **ED34**, 2546 (1987).

[2.4.37] D.K. Schroder. *IEEE Trans. Electron Dev.*, **ED29**, 1336 (1982).

[2.4.38] K.J. Rawlings. *Nucl. Instrum. Meth. Phys. Res., Sec. A*, **260**, 201 (1987).

[2.4.39] W. Shockley and W.T. Read. *Phys. Rev.*, **87**, 835 (1952).

[2.4.40] C.T. Sah, R.N. Noyce and W. Shockley. *Proc. Inst. Radio Eng.*, **45**, 1228 (1958).

[2.4.41] C.T. Sah and W. Shockley. *Phys. Rev.*, **109**, 1103 (1958).

[2.4.42] W. Shockley. *Proc. Inst. Radio Eng.*, **46**, 973 (1958).

[2.4.43] P.T. Landsberg. In *Solid State Physics in Electronics and Telecommunications* (ed. M. Désirant), vol. I (London: Academic Press, 1958), p. 436.

[2.4.44] O. Engstrom and A. Alm. *Solid-State Electron.*, **21**, 1571 (1978).

[2.4.45] P.T. Landsberg. *J. App. Phys.*, **60**, 2189 (1986).

[2.4.46] S. Coffa, G. Calleri, L. Calcagno, S.U. Campisano and G. Ferla. *App. Phys. Lett.*, **52**, 558 (1988).

[2.4.47] S.D. Brotherton and J.E. Lowther. *Phys. Rev. Lett.*, **44**, 606 (1980).

[2.4.48] R. Kassing, L. Cohausz, P. van Staa, W. Mackert and H.J. Hoffman. *App. Phys. A*, **34**, 41 (1984).

[2.4.49] D.V. Lang, H.G. Grimmeiss, E. Meijer and M. Jaros. *Phys. Rev.*, **B22**, 3917 (1980).

[2.4.50] A.R. Peaker, U. Kaufmann, Z-G. Wang, R. Wörner, B. Hamilton and H.G. Grimmeiss. *J. Phys.*, **C17**, 6161 (1984).

[2.4.51] O. Engstrom and A. Alm. *J. App. Phys.*, **54**, 5240 (1983).

[2.4.52] O. Engstrom and J. Shivaraman. *App. Phys.*, **58**, 3929 (1985).

[2.4.53] M.J. Kirton and M.J. Uren. *App. Phys. Lett.*, **48**, 1270 (1986).

[2.4.54] Y.P. Varshni. *Physica*, **34**, 149 (1967).

[2.4.55] J.S. Blakemore. *J. App. Phys.*, **53**, 520 (1982).

[2.5.1] D.W. Greve, P.A. Potyraj and A.M. Guzman. *Solid-State Electron.*, **28**, 1255 (1985).

[2.5.2] S.J. Pearton, A.J. Tavendale and A.A. Williams. *Electron. Lett.*, **16**, 483 (1980).

[2.5.3] J. Frenkel. *Phys. Rev.*, **54**, 647 (1938).

[2.5.4] V. Dallacase and C. Paracchini. *IEEE Trans. Electrical Insulation*, **EI22**, 467 (1987).

[2.5.5] A.K. Jonscher. *Thin Solid Films*, **1**, 213 (1967).

[2.5.6] J.L. Hartke. *J. App. Phys.*, **39**, 4871 (1968).

[2.5.7] S.D. Brotherton and A. Gill. *App. Phys. Lett.*, **33**, 953 (1978).

[2.5.8] R.D. Harris, J.L. Newton and G.D. Watkins. *Phys. Rev. Lett.*, **48**, 1271 (1982).

[2.5.9] S.R. Dhariwal and P.T. Landsberg. *J. Phys. Chem. Sol.*, **50**, 363 (1989).

[2.5.10] E. Rosencher, V. Mosser and G. Vincent. *Phys. Rev.*, **B29**, 1135 (1984).

[2.5.11] T.H. Ning. *J. App. Phys.*, **47**, 3203 (1976).

[2.5.12] L.L. Rosier and C.T. Sah. *Solid-State Electron.*, **14**, 41 (1971).

[2.5.13] G.J. Rees, H.G. Grimmeiss, E. Janzén and B. Skarstam. *J. Phys.*, **C13**, 6157 (1980).

[2.5.14] A.F. Tasch and C.T. Sah. *Phys. Rev.*, **B1**, 800 (1970).

[2.5.15] G.A. Dussel and K.W. Böer. *Phys. Stat. Sol.*, **39**, 375 (1970).
         S.V. Bulyarskii, N.S. Grushko, A.A. Gutkin and D.N. Nasledov. *Sov. Phys. Semicond.*, **9**, 187 (1975).

[2.5.16] D.M. Pai. *J. App. Phys.*, **46**, 5122 (1975).

[2.5.17] G. Vincent, A. Chantre and D. Bois. *J. App. Phys.*, **50**, 5484 (1979).

[2.5.18] A.A. Grinberg. *Phys. Rev.*, **B33**, 7256 (1986).

[2.5.19] A. Sommerfeld and H.A. Bethe. In *Handbuch der Physik*, vol. 24/2, 2nd edn. (eds H. Geiger and K. Scheel) (Berlin: Springer, 1933), p. 440.
         M. Kleefstra and G.C. Herman. *J. App. Phys.*, **51**, 4923 (1980).
         A. Tugulea and D. Dascălu. *J. App. Phys.*, **56**, 2823 (1984).

[2.5.20] S.J. Fonash. *Solid-State Electron.*, **15**, 783 (1972).

[2.5.21] R. Stratton. *Phys. Rev.*, **125**, 67 (1962).
         R. Stratton. *Phys. Rev.*, **A135**, 794 (1964).

[2.5.22] R.V. Latham. *Vacuum*, **32**, 137 (1982).

[2.5.23] N.A. Cade, G.H. Cross, R.A. Lee, S. Bajic and R.V. Latham. *J. Phys.*, **D21**, 148 (1988).

[2.5.24] L. Onsager. *J. Chem. Phys.*, **2**, 599 (1934).

[2.5.25] L. Onsager. *Phys. Rev.*, **54**, 554 (1938).
         See also H. Sano and M. Tachiya. *J. Chem. Phys.*, **71**, 1276 (1979) for an alternative proof.

[2.5.26] C.L. Braun and R.R. Chance. In *Energy and Charge Transfer in Organic Semiconductors* (eds K. Masuda and M. Silver) (New York: Plenum Press, 1974), p. 17.

[2.5.27] R.H. Batt, C.L. Braun and J.F. Hornig. *App. Opt., Suppl.*, **3**, 20 (1969).

[2.5.28] R.H. Batt, C.L. Braun and J.F. Hornig. *J. Chem. Phys.*, **49**, 1967 (1968).

[2.5.29] M. Pope and C.E. Swenberg. *Electronic Processes in Organic Crystals* (Oxford University Press, 1982).

[2.5.30] R.R. Chance and C.L. Braun. *J. Chem. Phys.*, **59**, 2269 (1973).

[2.5.31] J. Ristein and G. Weiser. *Solid-State Commun.*, **66**, 361 (1988).

[2.5.32] G. Weiser and J. Ristein. *J. Non-Crystalline Sol.*, **97/98**, 1131 (1987).

[2.5.33] W. Fuhs and K. Jahn. In *Amorphous Silicon and Related Materials* (ed. H. Fritzsche) (Singapore: World Scientific, 1988), p. 767.

[2.5.34] D.F. Blossey. *Phys. Rev.*, **B9**, 5183 (1974).

[2.5.35] L.B. Loeb. *Basic Processes of Gaseous Electronics* (University of California Press, 1955), p. 525.

[2.6.1] J.J. Thomson. *Phil. Mag.*, **47**, 337 (1924).

[2.6.2] V.N. Abakumov and I.N. Yassievich. *Sov. Phys. JETP*, **44**, 345 (1976).

[2.6.3] V.N. Abakumov, V.I. Perel' and I.N. Yassievich. *Sov. Phys. Semicond.*, **12**, 1 (1978).

[2.6.4] S.A. Kaufman, K.M. Kulikov and N.P. Likhtman. *Sov. Phys. Semicond.*, **4**, 102 (1970).

[2.6.5] S.H. Koenig, R.D. Brown III and W. Schillinger. *Phys. Rev.*, **128**, 1668 (1962).

[2.6.6] K.D. Glinchuk, E.G. Miselynk and N.N. Fortunatova. *Ukr. Fis. Zh.*, **4**, 207 (1959).

[2.6.7] M. Lax. *Phys. Rev.*, **119**, 1502 (1960).

[2.6.8] V.N. Abakumov, L.N. Kreshchuk and I.N. Yassievich. *Sov. Phys. Semicond.*, **12**, 152 (1978).

[2.6.9] N. Sclar. *Prog. Quantum Electron.*, **9**, 222 (1984).

[2.6.10] E.F. Smith and P.T. Landsberg. *J. Phys. Chem. Sol.*, **27**, 1727 (1966).

[2.6.11] P. Lal and P.T. Landsberg. *Phys. Rev.*, **140**, A46 (1965).

[2.6.12] V.V. Antonov-Romanovskii. *Sov. Phys. Solid State*, **5**, 975 (1963).

[2.6.13] R. Stratton. *Progress in Dielectrics*, **3**, 235 (1961).

[2.6.14] D.R. Hamann and A.L. McWhorter. *Phys. Rev.*, **134**, A250 (1964).

[2.6.15] B.N. Brockhouse. *Phys. Rev. Lett.*, **2**, 256 (1959).

[2.6.16] A.M. Stoneham. *Theory of Defects in Solids* (Oxford: Clarendon Press, 1975), section 14.4.

[2.6.17] E.P. Pokatilov and M.M. Rusanov. *Sov. Phys. Semicond.*, **4**, 815 (1970).

[2.6.18] G. Güttler and H.J. Queisser. *Energy Conversion*, **10**, 51 (1970).

[2.6.19] G. Nimtz. *Phys. Reports*, **63**, 265 (1980).

[2.6.20] W. Pickin. *Phys. Stat. Sol.*, **96**, 617 (1979).

[2.6.21] W. Pickin. *Phys. Stat. Sol.*, **97**, 431 (1980).

[2.6.22] S.D. Baranovskii, V.G. Karpov and B.I. Shklovskii. *Sov. Phys. JETP*, **67**, 588 (1988).

[2.6.23] R.A. Street. *Adv. Phys.*, **30**, 593 (1981).

[2.7.1] D. Haneman. *Rep. Prog. Phys.*, **50**, 1045 (1987).

[2.7.2] S.P. Keller (ed.). *Handbook of Semiconductors*, vol. III (Amsterdam: North-Holland, 1980).

[2.7.3] P.T. Landsberg. *IEEE Trans. Electron Dev.*, **ED29**, 1284 (1982).

[2.7.4]   J.G. Fossum and F.A. Lindholm. *IEEE Trans. Electron Dev.*, **ED27**, 692 (1980).

[2.7.5]   J.G. Fossum and R. Sundaresen. *IEEE Trans. Electron Dev.*, **ED29**, 1185 (1982).
           P.T. Landsberg and M.S. Abrahams. *J. App. Phys.*, **55**, 4284 (1984).

[2.7.6]   P. Panayotatos and H.C. Card. *IEEE Electron Dev. Lett.*, **EDL1**, 263 (1980).

[2.7.7]   G. Baccarini, B. Ricco and G. Spadini. *J. App. Phys.*, **49**, 5565 (1978).

[2.7.8]   C.H. Seager. *J. App. Phys.*, **52**, 3960 (1981).

[2.7.9]   H.C. Card, J.G. Shaw, G.C. McGonigal, D.J. Thomson, A.W. de Groot and K.C. Cao. *Proc. 16th IEEE Photovoltaic Specialist Conference*, San Diego, California, 1982.

[2.7.10] M.S. Abrahams. Ph.D. Thesis, Faculty of Mathematical Studies, University of Southampton, 1984, p. 83.

[2.7.11] P.T. Landsberg and M.S. Abrahams. *Proc. 17th IEEE Photovoltaic Specialist Conference*, Orlando, Florida, 1984, p. 597.

[2.7.12] E.L. Heasell. *Semicond. Sci. Technol.*, **2**, 88 (1987).

[2.7.13] E. Schöll. *J. App. Phys.*, **60**, 1434 (1986).

[2.7.14] E. Schöll and P.T. Landsberg. *Z. Phys.*, **B72**, 515 (1988).

[2.7.15] A. Barhdadi, H. Amzil, J.C. Muller and P. Siffert. *App. Phys.* **A49**, 233 (1989).

[2.8.1]   W. Shockley. *Phys. Rev.*, **91**,. 228 (1953).

[2.8.2]   S.R. Morrison. *Phys. Rev.*, **104**, 619 (1956).

[2.8.3]   T. Figielski. *Solid-State Electron.* **21**, 1403 (1978).

[2.8.4]   Y.V. Gulyaev. *Sov. Phys. Solid State*, 3, pp. 279 & 796 (1961).

[2.8.5]   W.T. Read. *Phil. Mag.*, **46**, 111 (1955).

[2.8.6]   R.M. Broudy. *Adv. Phys.*, **12**, 135 (1963).

[2.8.7]   W. Schröter. *Phys. Stat. Sol.*, (a), **19**, 159 (1973).

[2.8.8]   R. Labusch. *J. de Phys. Colloque C6*, **40**, C6–81 (1979).

[2.8.9]   N.A. Drozdov, A.A. Patrin and V.D. Tkachev. *Sov. Phys. J. Expt. Theor. Phys. Lett.*, **23**, 597 (1976).

[2.8.10] L.A. Kazakevich, P.F. Lugakov and I.M. Filippov. *Phys. Stat. Sol.*, (a), **113**, 307 (1989).

[2.8.11] I.M. Filippov, L.A. Kazakevich, P.F. Lugakov and V.V. Shusha. *Phys. Stat. Sol.*, (a), **96**, 527 (1986).

[2.8.12] H.J. Leamy. *J. App. Phys.*, **53**, R51 (1982).

[2.8.13] I.E. Bondarenko, H. Blumtritt, J. Heydenreich, V.V. Kazmiruk and E.B. Yakimov. *Phys. Stat. Sol.*, (a), **95**, 173 (1988).

[2.8.14] H. Blumtritt, G.N. Panin, E.B. Yakimov and J. Heydenreich. *Phys. Stat. Sol.*, (a), **109**, K3 (1988).

[2.8.15] C. Donolato. *J. App. Phys.*, **54**, 1314 (1983).

[2.8.16] V.V. Aristov, I.E. Bondarenko, N.N. Drymova, V.V. Kazmiruk and E.B. Yakimov. *Phys. Stat. Sol.*, (a), **84**, K43 (1984).

[2.8.17] T.S. Fell and P.R. Wilshaw. International Conference on 'Structure and Properties of Dislocations in Semiconductors', *Inst. of Phys. Conf. Series no. 104* (Bristol: Adam Hilger, 1989), p. 227.

[2.8.18] B. Sieber. *Phil. Mag.*, **B55**, 585 (1987).

[2.8.19] C. Donolato. *Rev. de Phys. App.*, Colloque C6, p. 57 (1989).

Chapter 3

[3.1.1]   H.W.B. Skinner. *Phil. Trans. Roy. Soc.*, **A239**, 95 (1940).

[3.1.2] P.T. Landsberg. *Proc. Phys. Soc.*, **A62**, 806 (1949).
          P.T. Landsberg and D.J. Robbins. *Solid-State Electron.*, **28**, 137 (1985).
[3.1.3] See, for example, D. Pines. *Elementary Excitations in Solids* (New York: Benjamin, 1963).
[3.1.4] H. Fröhlich and J. O'Dwyer. *Proc. Phys. Soc.*, **A63**, 81 (1950).
          L. Pincherle. *Proc. Phys. Soc.*, **B68**, 319 (1955).
          M. Nagae. *Progr. Theor. Phys.*, **14**, 339 (1958).
          L. Bess. *Phys. Rev.*, **105**, 1469 (1959).
[3.1.5] A.R. Beattie and P.T. Landsberg. *Proc. Roy. Soc.*, **A249**, 16 (1958).
          P.T. Landsberg and A.R. Beattie. *J. Phys. Chem. Sol.*, **8**, 73 (1959).
[3.1.6] For example, M.G. Burt. *Electron. Lett.*, **18**, 806 (1982).
          A. Sugimura. *IEEE J. Quantum Electron.*, **QE19**, 932 (1983).
[3.1.7] A.A. Bergh and P.J. Dean. *Light-Emitting Diodes* (Oxford: Clarendon Press, 1976), p. 171.
[3.1.8] O. von Roos and P.T. Landsberg. *J. App. Phys.*, **57**, 4746 (1985).
[3.1.9] E. Schöll. *Non-equilibrium Phase Transitions in Semiconductors* (Berlin: Springer, 1987), p. 75.
[3.1.10] P.T. Landsberg. *Solid-State Electron.*, **30**, 1107 (1987).
[3.1.11] A.R. Beattie. *J. Phys. Chem. Sol.*, **49**, 589 (1988).
[3.1.12] P.T. Landsberg. *Proc. Flat Plate Solar Array Project Research Forum on High Efficiency Crystalline Silicon Solar Cells* (JPL Publication 85-38, 1985), pp. 13–35.
[3.1.13] P.T. Landsberg. *Phys. Stat. Sol.*, **41**, 457 (1970).
[3.1.14] M. Takeshima. *Phys. Rev.* **B28**, 2039 (1983).
[3.1.15] A. Das and R. Al-Jishi. *Phys. Lett.*, **A141**, 186 (1989).
          A. Das and R. Al-Jishi. *Phys. Rev.* **B41**, 3551 (1990).
[3.1.16] D.B. Laks, G.F. Neumark, A. Hangleiter and S.T. Pantelides. *Phys. Rev. Lett.*, **61**, 1229 (1988).

[3.4.1] S. Brand and R.A. Abram. *J. Phys.*, **C17**, L571 (1984).
[3.4.2] D.J. Robbins and P.T. Landsberg. *J. Phys.*, **C13**, 2425 (1980), equation (2.12).

[3.5.1] C.L. Anderson and C.R. Crowell. *Phys. Rev.*, **B5**, 2267 (1972).
[3.5.2] A.R. Beattie. *J. Phys. Chem. Sol.*, **24**, 1049 (1962).
[3.5.3] P.T. Landsberg. *Solid-State Commun.*, **10**, 479 (1972).
[3.5.4] O. Hildebrand, W. Kuebart, K.W. Benz and M.H. Pilkuhn. *IEEE J. Quantum Electron.*, **QE17**, 284 (1981).
[3.5.5] P.T. Landsberg and Y-J. Yu. *J. App. Phys.*, **63**, 1789 (1988).
          See also R.I. Taylor and R.A. Abram. *Semicond. Sci. Technol.*, **3**, 859 (1988).
[3.5.6] B.K. Ridley. *J. App. Phys.*, **48**, 754 (1977).
[3.5.7] F. Capasso. *Semiconductors and Semimetals*, **22D**, 1 (1985).
[3.5.8] P.T. Landsberg. In *Lectures in Theoretical Physics*, vol. 8A (Boulder: Colarado Press, 1966), p. 328.
[3.5.9] D.J. Robbins. *Phys. Stat. Sol.*, (b), **97**, 387 (1980).
[3.5.10] P.T. Landsberg and D.J. Robbins. *J. Phys.*, **C10**, 2717 (1977).
[3.5.11] W. Franz. In *Handbuch der Physik*, vol. 17 (Berlin: Springer, 1956), p. 190.
[3.5.12] E. Antončík. *Czech. J. Phys.*, **8**, 492 (1958).
          R.J. Hodgkinson. *Proc. Phys. Soc.*, **82**, 1010 (1963).
[3.5.13] S. Ahmad and W.S. Khokley. *J. Phys. Chem. Sol.*, **28**, 2499 (1967).
[3.5.14] R.A. Ballinger, K.G. Major and J.R. Mallinson. *J. Phys.*, **C6**, 2573 (1973).

[3.5.15] A.R. Beattie and P.T. Landsberg. *Proc. Roy. Soc.*, **A249**, 16 (1959).

[3.5.16] A.R. Beattie and G. Smith. *Phys. Stat. Sol.*, **19**, 577 (1967).

[3.5.17] P.T. Landsberg and M.J. Adams. *J. Luminescence*, **7**, 3 (1973).

[3.5.18] J.R. Hauser. *J. App. Phys.*, **37**, 507 (1966).

[3.5.19] D.L. Dexter. *Proc. Vth Intern. Conf. Semiconductors, Prague* (New York: Academic Press, 1960), p. 122.

[3.5.20] L. Huldt. *Phys. Stat. Sol.*, (a), **8**, 173 (1971).
L. Huldt. *Phys. Stat. Sol.*, (a), **24**, 221 (1974).

[3.5.21] D. Hill and P.T. Landsberg. *Proc. Roy. Soc.*, **A347**, 547 (1976).

[3.5.22] C. Shekhar and S.K. Sharma. *Phys. Lett.*, **A50**, 120 (1974).
C. Shekhar and S.K. Sharma. *Phys. Lett.*, **A51**, 339 (1975).

[3.5.23] T.P. Pearsall, R.E. Nahory and J.R. Chelikowsky. *Institute of Physics, Conf. Series No. 33b* (Bristol: Adam Hilger, 1977), p. 331.

[3.5.24] T.P. Pearsall, F. Capasso, R.E. Nahory, M.A. Pollack and J.R. Chelikowsky. *Solid-State Electron.*, **21**, 297 (1978).

[3.5.25] D.J. Robbins. *Phys. Stat. Sol.*, (b), **97**, pp. 9 & 387 (1980).
D.J. Robbins. *Phys. Stat. Sol.*, (b), **98**, 11 (1980).

[3.5.26] I.K. Czajkowski, J. Allam, M. Silver, A.R. Adams and M.A. Gell. *IEE Proc.*, **J137**, 79 (1990).

[3.6.1] E. Antončík and P.T. Landsberg. *Proc. Phys. Soc.*, **82**, 337 (1963).

[3.6.2] C. Kittel and A.H. Mitchell. *Phys. Rev.*, **96**, 1488 (1954).

[3.6.3] A.R. Beattie and G. Smith. *Phys. Stat. Sol.*, **19**, 577 (1967).

[3.6.4] P.T. Landsberg. In *Lectures in Theoretical Physics*, vol. 8A (Boulder: Colorado Press, 1966), p. 313.

[3.6.5] A.R. Beattie and P.T. Landsberg. *Proc. Roy. Soc.* **A258**, 486 (1960).

[3.6.6] G.A. Jones. *Bull. London Math. Soc.*, **16**, 241 (1984).

[3.6.7] G.A. Jones and P.T. Landsberg. *Phys. Stat. Sol.*, (b), **128**, 619 (1985).

[3.6.8] W. Brauer. *Phys. Stat. Sol.*, **5**, 139 (1964).

[3.6.9] V. Halpern. *J. Phys. Chem. Sol.*, **24**, 1495 (1963).

[3.6.10] A.R. Beattie and P.T. Landsberg. *Proc. Roy. Soc.*, **A249**, 16 (1958).

[3.6.11] P. Dzwig. *J. Phys.*, **C12**, 1809 (1979).

[3.6.12] W. Lochmann and A. Haug. *Solid-State Commun.*, **35**, 553 (1980).

[3.6.13] M.G. Burt, S. Brand, C. Smith and R.A. Abram. *J. Phys.*, C17, 6385 (1984).

[3.6.14] A.R. Beattie. *J. Phys.*, **C18**, 6501 (1983).

[3.6.15] P. Scharoch and R.A. Abram. *Semicond. Sci. Technol.*, **3**, 973 (1988).
A.R. Beattie. *Semicond. Sci. Technol.*, **3**, 48 (1988).
A.R. Beattie, P. Scharoch and R.A. Abram. *Semicond. Sci. Technol.*, **4**, 715 (1989).

[3.6.16] D.B. Laks, G.F. Neumark. A. Hangleiter and S.T. Pantelides. International Conference on Shallow Impurities in Semiconductors, Linköping, Sweden, 1988. *Int. Phys. Conf. Series 95* (ed. B. Monemar) (Bristol: Institute of Physics, 1989).

[3.6.17] D.M. Eagles. *Proc. Phys. Soc.*, **78**, 204 (1961).

[3.6.18] For example, A. Haug and W. Schmid. *Solid-State Electron.*, **25**, 665 (1982).

[3.7.1] P.T. Landsberg and D.J. Robbins. *Solid-State Electron.*, **21**, 1289 (1978).

[3.7.2] P.T. Landsberg. *Proc. Roy. Soc.*, **A331**, 103 (1972).

[3.7.3] P.T. Landsberg and W.L. Wang. *J. App. Phys.*, **55**, 799 (1984).

[3.7.4] B.K. Ridley. *Semiconductor Sci. Technol.*, **2**, 116 (1987).

[3.7.5] L.V. Keldysh. *Sov. Phys., JETP*, **37**, 509 (1960).

[3.7.6] A.R. Beattie. *Semicond. Sci. Technol.*, **3**, 48 (1988).

[3.7.7] M.G. Burt and S. McKenzie. *Physica*, **134B**, 247 (1985).
J.S. Marsland. *Solid-State Electron.*, **30**, 125 (1986).
B.K. Ridley. *Proc. Inst. Electron. Radio Eng.*, **57**, S75 (1987).

[3.7.8] R.C. Woods. *App. Phys. Lett.*, **52**, 65 (1988).

[3.7.9] G.F. Neumark, D.J. De Bitetto, R.N. Bhargava and P.M. Harnack. *Phys. Rev.*, **B15**, 3147 (1977).

[3.7.10] W. Schmid and P.J. Dean. *Phys. Stat. Sol.*, (*b*), **110**, 591 (1982).

[3.7.11] G. Jones and A.R. Beattie. *Phys. Stat. Sol.*, (*a*), **4**, 193 (1971).
G. Jones and A.R. Beattie. *Phys. Stat. Sol.*, (*a*), **8**, 403 (1971).

[3.7.12] Y.V. Vorob'ev, E.I. Tolpygo and M.K. Sheinkman. *Sov. Phys. Semicond.*, **19**, 1313 (1985).

[3.7.13] V.B. Khalfin, M.V. Strikha and I.N. Yassievich. *Phys. Stat. Sol.*, (*b*), **131**, 203 (1985).

[3.7.14] P.T. Landsberg and D.J. Robbins. *J. Phys.*, **C10**, 2717 (1977).

[3.8.1] A.R. Beattie and P.T. Landsberg. *Proc. Roy. Soc.*, **A249**, 16 (1958), equation (5.14).

[3.8.2] Information from A.R. Beattie, private communication.
See also A.R. Beattie and G. Smith. *Phys. Stat. Sol.*, **19**, 577 (1967).

[3.8.3] A.R. Beattie and P.T. Landsberg. *Proc. Roy. Soc.*, **A258**, 486 (1960), equation (4.1).

[3.8.4] J. Dziewor and W. Schmid. *App. Phys. Lett.*, **31**, 346 (1977).

[3.8.5] G. Krieger and R.M. Swanson. *J. App. Phys.*, **54**, 3456 (1983).

[3.8.6] D.B. Laks, G.F. Neumark, A. Hangleiter and S.T. Pantelides. *Phys. Rev. Lett.*, **61**, 1229 (1988).

[3.8.7] R. Conradt and A. Aengenheister. *Solid-State Commun.*, **10**, 321 (1972).

[3.8.8] N.G. Nilsson and K.G. Svantesson. *Proc. 11th Int. Conf. Phys. Semicond., Warsaw* 1972 (Warsaw: PWN Polish Scientific Publishers, 1972), p. 1105.

[3.8.9] L. Huldt. *Phys. Stat. Sol.*, (*a*), **24**, 221 (1974).

[3.8.10] A.R. Beattie. *J. Phys.*, **C18**, 6501 (1985).

[3.8.11] M. Young and D.R. Wight. *J. Phys.*, **D7**, 1824 (1974).

[3.8.12] J. Shah, E.F. Leheny, W.R. Harding and D.R. Wight. *Phys. Rev. Lett.*, **38**, 1164 (1977).

[3.8.13] P. Dzwig. *J. Phys.*, **C12**, 1809 (1979).

[3.8.14] D. Hill and P.T. Landsberg. *Proc. Roy. Soc.*, **A347**, 547 (1976).
D.J. Robbins and A. Young. *Phys. Stat. Sol.*, (*b*), **102**, K143 (1980).

[3.8.15] A. Haug. *J. Phys.*, **C16**, 4159 (1983).

[3.8.16] G. Benz and R. Conradt. *Phys. Rev.*, **B16**, 843 (1977).

[3.8.17] M. Takeshima. *Phys. Rev.*, **B29**, 1993 (1984).

[3.8.18] B. Sermage, H.J. Eichler, J.P. Heritage, R.J. Nelson and N.K. Dutta. *App. Phys. Lett.*, **42**, 259 (1983).

[3.8.19] A. Mozer, K.M. Romanek, W. Schmid, M.H. Pilkuhn and E. Schlosser. *App. Phys. Lett.*, **41**, 964 (1982).

[3.8.20] C.B. Su, J. Schlafer, J. Manning and R. Olshansky. *Electron. Lett.*, **18**, 595 (1982).

[3.8.21] A.R.E. Beattie. *Semicond. Sci. Technol.*, **2**, 281 (1987).

[3.8.22] D.L. Polla, S.P. Tobin, M.B. Reine and H.K. Sood. *J. App. Phys.*, **52**, 5182 (1981).

[3.8.23] B.L. Gelmont, Z.N. Sokolova and I.N. Yassievich. *Sov. Phys. Semicond.*, **16**, 382 (1982).

[3.8.24] K.H. Herrmann. *Solid-State Electron.*, **21**, 1487 (1978).

[3.8.25] P.R. Emtage. *J. App. Phys.*, **45**, 2563 (1976).

[3.8.26] M. Takeshima. *J. App. Phys.*, **58**, 3846 (1985).

[3.8.27] A. Haug. *Electron. Lett.*, **20**, 85 (1984).
W. Bardyszewski and D. Yevick. *J. App. Phys.*, **58**, 2713 (1985).
W. Bardyszewski and D. Yevick. *IEEE J. Quantum Electron.*, **QE21**, 1131 (1985).
D.Z. Garbuzov, V.V. Agaev, V.B. Khalfin and V.P. Chalyi. *Sov. Phys. Semicond.*, **17**, 992 (1983).
N.K. Dutta and R.J. Nelson. *J. App. Phys.*, **53**, 74 (1982).
B.L. Gelmont and Z.N. Sokolova. *Sov. Phys. Semicond.*, **16**, 1067 (1982).

[3.8.28] K. Lischka and W. Huber. *J. App. Phys.*, **48**, 2632 (1977).
K. Lischka, W. Huber and H. Heinrich. *Solid-State Commun.*, **20**, 929 (1976).
M. Mocker and M. Beiler. *Phys. Stat. Sol.*, (b), **116**, 205 (1983).
M. Mocker and O. Ziep. *Phys. Stat Sol.*, (b), **115**, 415 (1983).
B. Schlicht, R. Dornhaus, G. Nimtz, L.D. Haas and T. Jakobus. *Solid-State Electron.*, **21**, 1481 (1978).

[3.8.29] A.V. Voitsekhovskii and Y.V. Lilenko. *Phys. Stat. Sol.*, (a), **67**, 381 (1981).
R.R. Gerhardts, R. Dornhaus and G. Nimtz. *Solid-State Electron.*, **21**, 1467 (1978).
I.M. Baker, F.C. Capocci, D.E. Charlton and J.T.M. Wotherspoon. *Solid-State Electron.*, **21**, 1475 (1978).

[3.8.30] A.N. Titkov, G.V. Benemanskaya and G.N. Iluridze. *Sov. Phys. JETP Lett.*, **34**, 409 (1981).
A.N. Titkov, G.N. Iluridze, I.F. Mironov and V.A. Cheban. *Sov. Phys. Semicond.*, **20**, 14 (1986).
H. van Cong. *Solid-State Commun.*, **37**, 897 (1981).

[3.8.31] A. Rogalski and Z. Orman. *Infrared Phys.*, **25**, 551 (1985).

[3.8.32] A. Rogalski. *Infrared Phys.*, **27**, 353 (1987).

[3.8.33] M.I. D'yakonov and A.V. Khaetskii. *Sov. Phys. Semicond.*, **14**, 891 (1980).

[3.8.34] Y. Vaitkus and V. Grivitskas. *Sov. Phys. Semicond.*, **15**, 1102 (1981).
A. Haug and W. Schmid. *Solid-State Electron.*, **25**, 665 (1982).
C. Tanguy and M. Combescot. *Solid-State Commun.*, **57**, 539 (1986).

[3.8.35] T.S. Moss. *Proc. Phys. Soc.*, **66B**, 993 (1953).

[3.8.36] R. Conradt. *Festkörperprobleme*, **12**, 449 (1972).

[3.8.37] R.N. Zitter, A.J. Strauss and A.E. Attard. *Phys. Rev.*, **115**, 266 (1959).

[3.8.38] J.W. Ostrowski. *Acta Phys. Polonica*, **19**, 339 (1960).

[3.8.39] J.S. Blakemore. *Proc. Vth Intern. Conf. Physics of Semiconductors, Prague* 1960 (New York: Academic Press, 1960) p. 981.

[3.8.40] A. Haug. *J. Phys.*, **C21**, L287 (1988).

[3.8.41] L.M. Blinov, E.A. Bobrova, V.S. Vavilov and G.N. Galkin. *Sov. Phys. Solid State*, **9**, 2357 (1968).

[3.8.42] E. Wintner and E.P. Ippen, *App. Phys. Lett.*, **44**, 999 (1984).

[3.8.43] R. Olshansky, C.B. Su, J. Manning and W. Powazinik. *IEEE J. Quantum Electron.*, **QE20**, 838 (1984).

[3.8.44] K. Betzler, T. Weller and R. Conradt. *Phys. Rev.*, **B6**, 1394 (1972).

[3.8.45] A. Hangleiter. *Phys. Rev.*, **B35**, 9149 (1987).

[3.8.46] J.I. Pankove, L. Tomasetta and B.F. Williams. *Phys. Rev. Lett.*, **27**, 29 (1971).

[3.8.47] P.T. Landsberg. *J. Phys.*, **C9**, L111 (1976).

[3.8.48] O. Ziep. *Phys. Stat. Sol.*, (*b*), **115**, 161 (1983).
M. Beiler, M. Mocker and O. Ziep. *Phys. Stat. Sol.*, (*b*), **123**, 247 (1984).

[3.8.49] A. Pimpale. *J. Phys.*, **C11**, 1085 (1978).
A. Haug. *Solid-State Electron.*, **21**, 1281 (1978).

[3.8.50] I.S. Gradshteyn and I.M. Ryzik. *Table of Integrals Series and Products*, 4th edn (New York: Academic Press, 1988), p. 369.

Chapter 4

[4.1.1] J.R. Haynes and N.G. Nilsson. In *Radiative Recombination in Semiconductors*, Part of *Proc. VIIth Intern. Conf. Physics of Semiconductors, Paris* 1964 (Paris: Dunod, 1964), p. 21.

[4.1.2] C.J. Nuese, G.E. Stillman, M.D. Sirkis and N. Holonyak Jr. *Solid-State Electron.*, **9**, 735 (1966).

[4.1.3] P.T. Landsberg. *Solid-State Electron.*, **10**, 513 (1967).

[4.1.4] M.H. Pilkuhn. In *Handbook of Semiconductors* (ed. T.S. Moss), vol. 4 (ed. C. Hilsum) (Amsterdam: North Holland, 1981), p. 545.

[4.1.5] N. Holonyak Jr., C.J. Nuese, M.D. Sirkis and G.E. Stillman. *App. Phys. Lett.*, **8**, 83 (1966).

[4.1.6] E.J. Johnson. In *Semiconductors and Semimetals* (eds R.K. Willardson and A.C. Beer), vol. 3 (New York: Academic Press, 1967), p. 154.

[4.2.1] P.T. Landsberg. *J. Phys.*, **C14**, L1025 (1981)

[4.2.2] M.G.A. Bernard and B. Durrafourg. *Phys. Stat. Sol.*, **1**, 699 (1961).
P. Würfel. *J. Phys.*, **C15**, 3967 (1982).

[4.2.3] N.G. Basov, O.N. Krokhin and Y.M. Popov. *Sov. Phys. JETP*, **12**, 1033 (1961).
N.G. Basov, O.N. Krokhin and Y.M. Popov. *Sov. Phys. JETP*, **13**, 845 (1961).
C. Benoit à la Guillaume and C. Tric. *J. Phys. Rad., Paris*, **22**, 834 (1961).

[4.2.4] V.S. Mashkevich and V.L. Vinetskii. *Sov. Phys. Solid State*, **7**, 1605 (1966).

[4.2.5] M.J. Adams and P.T. Landsberg. *Proc. IXth Intern. Conf. Physics of Semiconductors, Moscow* 1988 (Leningrad: Akademiya Nauk, 1968), p. 619.

[4.2.6] A. Frova (ed.). *The Physics and Technology of Semiconductor Light Emitters and Detectors*, Pugnochiuso Symposium, 1972, (Amsterdam: North-Holland, 1973). See particularly the Panel Discussion on pp. 548–52. Reprinted in *J. Luminescence*, vol. **7** (1973).

[4.2.7] P.T. Landsberg. *Phys. Stat. Sol.*, **19**, 777 (1967).

[4.2.8] G.H.B. Thompson. *Physics of Semiconductor Laser Devices* (Chichester: Wiley, 1980).

[4.2.9] H. Kressel (ed.). *Semiconductor Devices for Optical Communication* (Berlin: Springer, 1982).

[4.2.10] J. Wilson and J.F.B. Hawkes. *Optoelectronics: An Introduction* (Englewood Cliffs, New Jersey: Prentice Hall, 1983).

[4.2.11] A. Pentzkofer. *Solid State Lasers. Progr. Quantum Electron.*, **12**, 294 (1988).

[4.2.12] E. Schöll and P.T. Landsberg. *J. Opt. Soc. Am.*, **73**, 1197 (1983).

[4.2.13] J. Parrott. *IEE Proc. J.*, **5**, 314 (1986).

[4.2.14] A. de Vos. *J. Phys.*, **D20**, 232 (1987).

[4.2.15] P.T. Landsberg and P. Baruch. *J. Phys.*, **A22**, 1911 (1989).

[4.3.1] A. Einstein. *Physik Zeitschrift*, **18**, 127 (1917).

[4.3.2] D. Bohm. *Quantum Theory* (New York: Prentice Hall, 1951).

[4.3.3] P.A.M. Dirac. *Proc. Roy. Soc.*, **A114**, 243 (1927).

[4.4.1] For example, L.I. Schiff. *Quantum Mechanics*, 3rd edn. (New York: McGraw-Hill, 1968), equation (45.22); 1st edn. (1949), equation (36.22).

[4.4.2] M.J. Adams and P.T. Landsberg. In *Gallium Arsenide Lasers* (ed. C.H. Gooch) (London: Wiley, 1969), p. 5.

[4.4.3] F. Bassani and G. Pastori Parravicini. *Electronic States and Optical Transitions in Solids* (Oxford: Pergamon Press, 1975).

[4.4.4] T.S. Moss (ed.). *Handbook of Semiconductors*, vol. 2 (ed. M. Balkanski) (Amsterdam: North Holland, 1981).

[4.4.5] P.T. Landsberg, M.S. Abrahams and M. Osiński. *IEEE J. Quantum Electron.*, **QE21**, 24 (1985).

[4.4.6] R. Nagarajan and T. Kamiya. *IEEE J. Quantum Electron.*, **QE25**, 1161 (1989).

[4.4.7] A. Mooradian and H.Y. Fan. *Phys. Rev.*, **148**, 873 (1966).

[4.4.8] J. Shah and R.C.C. Leite. *Phys. Rev. Lett.*, **22**, 1304 (1969).

[4.5.1] W. van Roosbroeck and W. Shockley. *Phys. Rev.*, **94**, 1558 (1954).

[4.5.2] Y.P. Varshni. *Phys. Stat. Sol.*, **19**, 459 (1967).
          Y.P. Varshni. *Phys. Stat. Sol.*, **20**, 9 (1967).

[4.5.3] J. Bardeen, F.J. Blatt and L.K. Hall. In *Atlantic City Photoconductivity Conference Report* (New York: Wiley, 1956), p. 146.

[4.5.4] G.J. Lasher and F. Stern. *Phys. Rev.*, **133**, A553 (1964).

[4.5.5] H.C. Casey Jr. and F. Stern. *J. App. Phys.*, **47**, 631 (1976).

[4.5.6] M.J. Adams and P.T. Landsberg. In *Gallium Arsenide Lasers* (ed. C.H. Gooch) (London: Wiley, 1969), p. 5.

[4.5.7] T.P. McLean. *Progress in Semiconductors*, **5**, 53 (1960).

[4.5.8] H.B. Bebb and E.W. Williams. In *Semiconductors and Semimetals* (eds R.K. Willardson and A.C. Beer), vol. 8 (New York: Academic Press, 1972), pp. 181 & 321.

[4.5.9] T.S. Moss, G.J. Burrell and B. Ellis, *Semiconductor Opto-Electronics* (London: Butterworths, 1973).

[4.5.10] R.A. Smith. *Semiconductors*, 2nd edn. (Cambridge University Press, 1978), chapter 10.

[4.5.11] F. Bassani and G. Pastori Parravicini. *Electronic States and Optical Transitions in Solids* (Oxford: Pergamon Press, 1975), chapter 5.

[4.5.12] T.P. McLean. In *Polarons and Excitons* (eds G.C. Kuper and G.D. Whitfield) (Edinburgh: Oliver and Boyd, 1963), p. 367.

[4.5.13] G.G. MacFarlane and V. Roberts. *Phys. Rev.*, **97**, 1714 (1955).

[4.5.14] G.G. MacFarlane and V. Roberts. *Phys. Rev.*, **98**, 1865 (1955).

[4.5.15] G.G. MacFarlane, T.P. McLean, J.E. Quarrington and V. Roberts. *Phys. Rev.*, **108**, 1377 (1957).
          G.G. MacFarlane, T.P. McLean, J.E. Quarrington and V. Roberts. *Phys. Rev.*, **111**, 1245 (1958).

[4.5.16] W.J. Choyke and L. Patrick. *Phys. Rev.*, **105**, 1721 (1957).

[4.5.17] G. Dresselhaus. *J. Phys. Chem. Sol.*, **1**, 14 (1956).

[4.5.18] N.O. Lipari. *Nuovo Cimento*, **23B**, 51 (1974).

[4.5.19] W. Kohn and J.M. Luttinger. *Phys. Rev.*, **98**, 915 (1955).

[4.5.20] W. Kohn and D. Schecter. *Phys. Rev.*, **99**, 1903 (1955).

[4.5.21] R.C. Casella. *J. App. Phys.*, **34**, 1703 (1963).

[4.5.22] S. Nikitine. In *Optical Properties of Solids* (eds S. Nudelman and S.S. Mitra) (New York: Plenum, 1969), p. 197.

[4.5.23] R.J. Elliott. *Phys. Rev.*, **108**, 1384 (1957).

[4.5.24] R.J. Elliott. In *Polarons and Excitons* (eds G.C. Kuper and G.D. Whitfield) (Edinburgh: Oliver and Boyd, 1963), p. 269.

[4.5.25] M.D. Sturge. *Phys. Rev.*, **127**, 768 (1962).

[4.5.26] F. Urbach. *Phys. Rev.*, **92**, 1324 (1953).
T.H. Keil. *Phys. Rev.*, **144**, 582 (1966).
T. Skettrup. *Phys. Rev.*, **B18**, 2622 (1978).
P. Bussemer. *Phys. Stat. Sol.*, (*b*), **94**, K77 (1979).
V. Sa-Yakanit and H.R. Glyde. *Comments Cond. Mat. Phys.*, **13**, 35 (1987).

[4.5.27] J.T. Pankove. *Phys. Rev.*, **140**, A2059 (1965).

[4.5.28] E. Burstein. *Phys. Rev.*, **93**, 632 (1954).

[4.5.29] T.S. Moss. *Proc. Phys. Soc.*, **B67**, 775 (1954).

[4.5.30] R.N. Hall. *Proc. IEE*, **B106**, Suppl. 17, p. 923 (1959).

[4.5.31] N.S. Baryshev. *Sov. Phys. Solid State*, **3**, 1037 (1961).

[4.6.1] R.E. Burgess. *Proc. Phys. Soc.*, **B69**, 1020 (1956).

[4.6.2] K.M. van Vliet and J.R. Fassett. In *Fluctuation Phenomena in Solids* (ed. R.E. Burgess) (New York: Academic Press, 1965).

[4.6.3] P.T. Landsberg. *Eur. J. Phys.*, **1**, 31 (1980).
E. Schöll. *Nonequilibrium Phase Transitions in Semiconductors* (Berlin: Springer, 1987).

[4.6.4] D.J. Robbins, P.T. Landsberg and E. Schöll. *Phys. Stat. Sol.*, (*a*), **65**, 353 (1981).

[4.6.5] D.A. Evans and P.T. Landsberg. *Proc. Roy. Soc.*, **A267**, 464 (1962).
P.T. Landsberg and E.A.B. Cole. *Proc. Phys. Soc.*, **87**, 229 (1966).

[4.6.6] A. van der Ziel. In *Semiconductors and Semimetals* (eds R.K. Willardson and A.C. Beer), vol. 14 (New York: Academic Press, 1979), p. 195.

[4.6.7] P.T. Landsberg, E. Schöll and P. Shukla. *Physica*, **D30**, 235 (1988).

[4.6.8] N.G. van Kampen. *Stochastic Processes in Physics and Chemistry* (Amsterdam: North-Holland, 1981), pp. 87 & 102.

[4.6.9] As [4.6.8], p. 140.

[4.6.10] E. Schöll. *App. Phys.*, **A48**, 95 (1989).
E. Schöll. *Physica Scripta*, **T29**, 152 (1989).

Chapter 5

[5.1.1] W. Shockley and J.T. Last. *Phys. Rev.*, **107**, 392 (1957).

[5.1.2] P.T. Landsberg. *Proc. Phys. Soc.*, **B69**, 1056 (1956).

[5.1.3] J.S. Blakemore and S. Rahimi. In *Semiconductors and Semimetals* (eds R.K. Willardson and A.C. Beer), vol. 20 (New York: Academic Press, 1984), p. 233.

[5.1.4] S.T. Pantelides. *Rev. Mod. Phys.*, **50**, 797 (1979).

[5.1.5] M. Jaros. *Deep Levels in Semiconductors* (Bristol: Adam Hilger, 1982).

[5.1.6] B.K. Ridley. *Quantum Processes in Semiconductors* (Oxford: Clarendon Press, 1982).

[5.1.7] G.F. Neumark and K. Kosai. In *Semiconductors and Semimetals* (eds R.K. Willardson and A.C. Beer), vol. 19 (New York: Academic Press, 1983), p. 1.

[5.1.8] S.T. Pantelides (ed.). *Deep Centers in Semiconductors* (New York: Gordon and Breach, 1986).

[5.1.9] A.G. Milnes. *Adv. Electron. Electron Phys.*, **61**, 63 (1983).

[5.1.10] D.V. Lang. *J. App. Phys.*, **45**, 3023 (1974).

[5.1.11] O. Madelung and M. Schulz (eds). *Impurities and Defects in Group IV Elements and III–V Compounds.* vol. 22b, Numerical Data and Functional Relationships in Science and Technology (new series) (Berlin: Springer, 1989).

[5.1.12] R. Englman and J. Jortner. *Mol. Phys.*, **18**, 145 (1970).

[5.1.13] L.A. Riseberg and H.W. Moos. *Phys. Rev.*, **174**, 429 (1968).

[5.1.14] D.L. Dexter, C.C. Klick and G.A. Russell. *Phys. Rev.*, **100**, 603 (1955).

[5.1.15] A.M. Stoneham and R.H. Bartram. *Solid-State Electron.*, **21**, 1325 (1978).

[5.2.1] F. Bassani and G. Pastori Parravicini. *Electron States and Optical Transitions in Solids* (Oxford: Pergamon Press, 1975), p. 237.

[5.2.2] P.T. Landsberg. In *Solid State Theory: Methods and Applications* (ed. P.T. Landsberg) (London: Wiley, 1969), p. 142.

[5.2.3] W. Kohn. *Solid-State Phys.*, **5**, 257 (1957).

[5.2.4] T.P. McLean. *Proc. Intern. School of Physics 'E. Fermi'* (New York: Academic Press, 1963), p. 479.

[5.2.5] W. Kohn. *Phys. Rev.*, **105**, 509 (1957).

[5.2.6] A. Miller. *J. Phys. Chem. Sol.*, **35**, 641 (1974).

[5.2.7] D.M. Eagles, *J. Phys. Chem. Sol.*, **16**, 76 (1960).

[5.2.8] W.P. Dumke. *Phys. Rev.*, **132**, 1948 (1963).

[5.2.9] P.T. Landsberg. *Thermodynamics with Quantum Statistical Illustrations* (New York: Interscience, 1961), p. 241.

[5.2.10] H.B. Bebb and E.W. Williams. In *Semiconductors and Semimetals* (eds R.K. Willardson and A.C. Beer), vol. 8 (New York: Academic Press, 1972), p. 181.

[5.2.11] T.S. Moss, G.J. Burrell and B. Ellis. *Semiconductors Opto-Electronics* (London: Butterworths, 1973), p. 208.

[5.2.12] H.C. Casey Jr. and F. Stern. *J. App. Phys.*, **47**, 631 (1976).

[5.2.13] A.P. Levanyuk and V.V. Osipov. *Sov. Phys. Usp.*, **24**, 187 (1981).

[5.2.14] B.I. Shklovskii and A.L. Efros. *Electronic Properties of Doped Semiconductors* (Berlin: Springer, 1984).

[5.2.15] D.Z. Garbuzov. In *Semiconductor Physics* (eds. V.M. Tuchkevich and V.Y. Frenkel) (New York: Consultants Bureau, 1986).

[5.2.16] D.Z. Garbuzov, V.B. Khalfin, M.K. Trukan, V.G. Agafonov and A. Abdullaev. *Sov. Phys. Semicond.*, **12**, 809 (1978).

[5.2.17] A. Erdélyi (ed.). *Bateman Manuscript Project*, vol. I. (New York: McGraw-Hill Book Co. Inc, 1954), p. 175.

[5.2.18] P.T. Landsberg, C. Rhys-Roberts and P. Lal. *Proc. Phys. Soc.*, **84**, 915 (1964).

[5.2.19] M.E. Cohen and P.T. Landsberg. *Phys. Rev.*, **154**, 683 (1967).

[5.2.20] M.E. Cohen and P. T. Landsberg. *Phys. Stat. Sol.*, (b), **64**, 39 (1974).

[5.2.21] M.J. Adams and P.T. Landsberg. Unpublished manuscript, 1969.

[5.2.22] N.F. Mott and E.A. Davis. *Electronic Processes in Non-Crystalline Materials* (Oxford: Clarendon Press, 1979).
M. Stutzmann. *Feskörperprobleme*, **28**, 1 (1988).

[5.2.23] M. Pollak and A.L. Efros (eds). *Electron–Electron Interactions in Disordered Systems* (Amsterdam: North-Holland, 1985).
B.L. Al'tshuler and P.A. Lee. *Physics Today*, December, 1988, p. 36.

[5.2.24] E.O. Kane. *Phys. Rev.*, **131**, 79 (1963).

[5.2.25] B.I. Halperin and M. Lax. *Phys. Rev.*, **148**, 722 (1966).
B.I. Halperin and M. Lax. *Phys. Rev.*, **153**, 802 (1967).

[5.2.26] V. Sa-yakanit and H.R. Glyde. In *The Path Integral Method with Applications*
(eds A. Ranfagni, V. Sa-yakanit and L.S. Schulman) (Singapore: World
Scientific, 1988).

[5.2.27] R.A. Abram, G.J. Rees and B.L.H. Wilson. *Adv. Phys.*, **27**, 799 (1978).
J.R. Lowney and J.C. Geist. *J. App. Phys.*, **55**, 3627 (1984).
H.S. Bennett. *Solid-State Electron.*, **28**, 193 (1985).

[5.2.28] E.O. Kane. *Solid-State Electron.*, **28**, 3 (1985).

[5.2.29] W. Sritrakool, V. Sa-yakanit and H.R. Glyde. *Phys. Rev. B*. **32**, 1090 (1985).

[5.2.30] J.C. Inkson. *J. Phys. C.*, **6**, 1350 (1973).
J.C. Inkson. *J. Phys. C.*, **9**, 1177 (1976).

[5.2.31] H.C. Casey Jr., D.D. Sell and K. W. Wecht. *J. App. Phys.*, **46**, 250 (1975).

[5.2.32] V. Sa-yakanit, W. Sritrakool and H.R. Glyde. *Phys. Rev. B*, **25**, 2776 (1982).

[5.2.33] G.D. Mahan and J.W. Conley. *App. Phys. Lett.*, **11**, 29 (1967).

[5.2.34] C. Benoit à la Guillaume and J. Cernogora. *Phys. Stat. Sol.*, **35**, 599 (1969).

[5.2.35] J.I. Pankove. *Proc. VIIIth Intern. Conf. on the Physics of Semiconductors
Kyoto*, 1966. *J. Phys. Soc. Jap.*, **21**, Supplement, p. 298, 1966.

[5.2.36] G.J. Burrell, T.S. Moss and A. Hetherington. *Solid-State Electron.*, **12**, 787
(1969).

[5.2.37] F. Stern. *Phys. Rev.*, **148**, 186 (1966).

[5.2.38] P.T. Landsberg. *Phys. Stat. Sol.*, **15**, 623 (1966).

[5.2.39] P.T. Landsberg. *Proc. Phys. Soc.*, **A62**, 806 (1949).

[5.2.40] P.T. Landsberg and D.J. Robbins. *Solid-State Electron.*, **28**, 137 (1985).

[5.2.41] G. Lucovski. *Solid-State Commun.*, **3**, 299 (1965).

[5.2.42] H.G. Grimmeiss and L.-Å. Ledebo. *J. Phys.*, **C8**, 2615 (1975).

[5.2.43] B.K. Ridley and M.A. Amato. *J. Phys.*, **C14**, 1255 (1981).

[5.2.44] M. Jaros. *Deep Levels in Semiconductors* (Bristol: Adam Hilger, 1982).

[5.2.45] B.K. Ridley. *Quantum Processes in Semiconductors* (Oxford: Clarendon Press,
1982).

[5.2.46] J.S. Blakemore and S. Rahimi. In *Semiconductors and Semimetals* (eds. R.K.
Willardson and A.C. Beer), vol. 40 (Orlando Academic Press, 1984), p. 233.
M. Godlewski. *Phys. Stat. Sol.*, (a), **90**, 11 (1985).

[5.2.47] A.H. Edwards and W.B. Fowler. *Phys. Rev.*, **B16**, 3613 (1977).

[5.2.48] W. van Roosbroeck and W. Shockley. *Phys. Rev.*, **94**, 1558 (1954).

[5.2.49] P.T. Landsberg. *Proc. Roy. Soc.*, **A331**, 103 (1972).

[5.2.50] E.A. Milne. *Phil. Mag.*, **47**, 209 (1924).

[5.2.51] D.H. Menzel and C.L. Perkins. *M.N. Roy. Ast. Soc.*, **96**, 77 (1935).

[5.2.52] P.T. Landsberg. In *Festkörperprobleme* (ed. O. Madelung), vol. 6,
(Braunschweig: Vieweg, 1967), p. 174.

[5.2.53] J.S. Blakemore. *Phys. Rev.*, **163**, 809 (1967).

[5.2.54] H.B. Bebb. *Phys. Rev.*, **B5**, 4201 (1972).

[5.2.55] H.J. Bowlden. *J. Phys. Chem. Sol.*, **3**, 115 (1957).

[5.2.56] J.M. Dishman. *Phys. Rev.*, **B3**, 2588 (1971).

[5.2.57] J.A.G. Slatter. *Phys. Stat. Sol.*, **40**, 31 (1970).

[5.2.58] F. Williams. *Phys. Stat. Sol.*, **25**, 493 (1968).

[5.2.59] R. Bowers and N.T. Melamed. *Phys. Rev.*, **99**, 1781 (1955).

[5.2.60] D.G. Thomas, J.J. Hopfield and K. Colbow. In *Radiative Recombination in
Semiconductors*. Part of the *Proc. VIIth Intern. Conf. Physics of
Semiconductors, Paris, 1964* (Paris: Dunod, 1965), p. 67.

[5.2.61] M. Gershenzon, R.A. Logan, D.F. Nelson and F.A. Trumbore. *Intern. Conf.
Luminescence, Budapest, 1968* (Budapest: Academia Kiado, 1968).

[5.2.62] A.T. Vink. *J. Luminescence*, **9**, 159 (1974).

[5.2.63] H.J. Zeiger. *J. App. Phys.*, **35**, 1657 (1964).

[5.2.64] P.T. Landsberg and M.J. Adams. *Proc. Roy. Soc.*, **A334**, 523 (1973).

[5.2.65] R. Dingle. *Phys. Rev.*, **184**, 788 (1969).

[5.2.66] K. Colbow. *Phys. Rev.*, **141**, 742 (1966).

[5.2.67] E. Zacks and A. Halperin. *Phys. Rev.*, **B6**, 3072 (1972).

[5.2.68] R. Binderman and K. Unger. *Phys. Stat. Sol.*, (*b*), **56**, 563 (1973).
         J.H. Jefferson, W.E. Hagston and H.H. Sutherland. *J. Phys.*, **C8**, 3457 (1975).

[5.2.69] G.F. Neumark. *Phys. Rev.*, **B29**, 1050 (1984).

[5.2.70] Y.E. Pokrovskii and K.I. Svistunova. *Sov. Phys. JETP Lett.*, **9**, 261 (1969).

[5.2.71] T.K. Lo. *Solid-State Commun.*, **15**, 1231 (1974).

[5.2.72] J.P. Wolfe, W.L. Hansen, E.E. Haller, R.S. Markiewicz, C. Kittel and C.D. Jeffries. *Phys. Rev. Lett.*, **34**, 1292 (1975).

[5.2.73] A. Forchel, B. Laurich, J. Wagner and W. Schmid. *Phys. Rev.*, **B25**, 2730 (1982).

[5.2.74] A.A. Rogachev. *Progr. Quantum Electron.*, **6**, 141 (1980).
         L.V. Keldysh. *Contemp. Phys.*, **27**, 395 (1986).
         S.G. Tikhodeev. *Sov. Phys. Usp.*, **28**, 1 (1985).
         J. Singh. *Solid-State Phys.*, **38**, 295 (1984).

[5.2.75] J.M. Blatt, K.W. Böer and W. Brandt. *Phys. Rev.*, **126**, 1691 (1962).

[5.2.76] J.M. Lévy-Leblond. *Phys. Rev.*, **178**, 1526 (1969).
         J.M. Lévy-Leblond. *Phys. Rev.*, **184**, 1006 (1969).

[5.2.77] J. Adamowski, S. Bednarek and M. Suffczynski. *Solid-State Commun.*, **9**, 2037 (1971).

[5.2.78] W. Gorzkowski and M. Suffczynski. *Phys. Lett.*, **29A**, 550 (1969).

[5.2.79] H. Nakayama, K. Ohnishi, H. Sawada, T. Nishino and Y. Hamakawa. *J. Phys. Soc. Jap.*, **46**, 553 (1979).
         K. Ohnishi, H. Nakayama, T. Nishino and Y. Hamakawa. *J. Phys. Soc. Jap.*, **49**, 1078 (1980).

[5.2.80] A.G. Steele, W.G. McMullen and M.L.W. Thewalt. *Phys. Rev. Lett.*, **59**, 2899 (1987).
         B. Monemar, U. Lindfelt and W.M. Chen. *Physica B*, **146**, 256 (1989).

[5.2.81] B.P. Zakharchenya. *Proc. 11th Int. Conf. Phys. Semiconductors Warsaw 1972* (Warszawa: PWN-Polish Scientific Publishers, 1972).

[5.2.82] C. Weisbuch and G. Lampel. *Solid-State Commun.*, **14**, 141 (1974).

[5.2.83] B.C. Cavenett. *Adv. Phys.*, **30**, 475 (1981).
         J.J. Davies. *Contemp. Phys.*, **17**, 275 (1976).

[5.2.84] F. Meier and B.P. Zakharchenya (eds). *Optical Orientation* (Amsterdam: North-Holland, 1984).

[5.2.85] M.C. Chen and D.V. Lang. *Phys. Rev. Lett.*, **51**, 427 (1983).

[5.2.86] Z. Kachwalla and D.J. Miller. *J. App. Phys.*, **62**, 2848 (1987).

[5.2.87] J. Tauc. In *Semiconductors and Semimetals* (eds R.K. Willardson and A.C. Beer), vol. 21B (New York: Academic Press, 1984), p. 299.

[5.2.88] J. Klafter, A. Blumen and G. Zumofen. *Phil. Mag. B Lett.*, **53**, L29 (1986).

[5.2.89] D.A. Evans and P.T. Landsberg. *J. Phys. Chem. Sol.*, **26**, 315 (1965).

[5.3.1] M.R.C. McDowell and J.R. Coleman. *Introduction to the Theory of Ion–Atom Collisions* (Amsterdam: North-Holland, 1970).

[5.3.2] D.J. Robbins and P.T. Landsberg. *J. Phys.*, **C13**, 2425 (1980).

[5.3.3] K. Omidvar. *Phys. Rev.*, **140**, pp. A26 & 38 (1965).

[5.3.4] W.L. Fite and R.T. Brackman. *Phys. Rev.*, **112**, 1141 (1958).

[5.3.5] E.W Rothe, L.L. Marino, R.H. Neynaber and S.M. Trujillo. *Phys. Rev.*, **125**, 582 (1962).

[5.3.6] G.D. Fletcher, M.J. Alguard, T.J. Gay, V.W. Hughes, P.F. Wainwright, M.S. Lubell and W. Raith. *Phys. Rev.*, **A31**, 2854 (1985).

[5.3.7] Y. Jiaqi and S. Yongrong. *J. Phys.*, **C21**, 3381 (1988).

[5.3.8] H.J. Queisser. *Solid-State Electron.*, **21**, 1495 (1978).

[5.3.9] G.F. Neumark and K. Kosai. In *Semiconductors and Semimetals* (eds. R.K. Willardson and A.C. Beer), vol. 19 (New York: Academic Press, 1983), p. 60.

[5.3.10] B.D. Belorusets and A.A. Grinberg. *Sov. Phys. Semicond.*, **12**, 345 (1978).

[5.3.11] P.T. Landsberg, C. Rhys-Roberts and P. Lal. *Proc. Phys. Soc.*, **84**, 915 (1964).

[5.3.12] M.E. Cohen and P.T. Landsberg. *Phys. Stat. Sol.*, (*b*), **64**, 39 (1974).
M.E. Cohen and P.T. Landsberg. *Phys. Stat. Sol.*, (*b*), **68**, 805 (1975).

[5.3.13] M.E. Cohen and P.T. Landsberg. *Phys. Rev.*, **154**, 683 (1967).

[5.3.14] M.S. Tyagi. *J. App. Phys.*, **54**, 2857 (1983).
M.S. Tyagi and R. van Overstraeten. *Solid-State Electron.*, **26**, 577 (1983).

[5.3.15] T.S. Moss. *Proc. Phys. Soc.*, **B66**, 993 (1953).

[5.3.16] J.R. Haynes and J.A. Hornbeck. *Phys. Rev.*, **150**, 606 (1955).

[5.3.17] L. Pincherle. *Proc. Phys. Soc.*, **B68**, 319 (1955).
L. Bess. *Phys. Rev.*, **105**, 1469 (1957).
M. Nagae. *Prog. Theor. Phys.*, **19**, 339 (1958).

[5.3.18] V.L. Bonch-Bruevich and Y.V. Gulyaev. *Sov. Phys. Solid State*, **2**, 431 (1960).

[5.3.19] S.H. Koenig, R.D. Brown III and W. Schillinger. *Phys. Rev.*, **128**, 1668 (1962).

[5.3.20] A. Haug. *Phys. Stat. Sol.*, (*b*), **97**, 481 (1980).

[5.3.21] A. Haug. *Phys. Stat. Sol.*, (*b*), **108**, 443 (1881).

[5.3.22] W. Schmid and J. Reiner. *J. App. Phys.*, **53**, 6250 (1982).

[5.3.23] J.M. Dishman. *Phys. Rev.*, **5**, 2258 (1972).

[5.3.24] I.V. Karpova and S.G. Kalashnikov. *Proc. VIth Int. Conf. on the Physics of Semiconductors, Exeter, 1962* (London: Institute of Physics and Physical Society, 1962) p. 880.

[5.3.25] P.T. Landsberg, D.A. Evans and C. Rhys-Roberts. *Proc. Phys. Soc.*, **83**, 325 (1964).

[5.3.26] J.M. Dishman, M. di Domenico and R. Caruso. *Phys. Rev.*, **82**, 1988 (1970).

[5.3.27] V.A. Zuev, V.G. Litovchenko and G.A. Sukach. *Sov. Phys. Semiconductors*, **9**, 1083 (1976).

[5.3.28] J.C. Tsang, P.J. Dean and P.T. Landsberg. *Phys. Rev.*, **173**, 814 (1968).

[5.3.29] J.W. Allen. *J. Phys. C*, **19**, 6287 (1986).
J.M. Langer. *J. Luminescence*, **40/41**, 589 (1988).

[5.3.30] J.M. Langer. In *Optoelectronic Materials and Devices* (ed. M.A. Herman, Warsaw: PWN, 1983), p. 303.
J.M. Langer. In *Electroluminescence* (eds S. Shionoya and H. Kobayashi) (Berlin: Springer, 1989), p. 16.
A. Suchocki and J.M. Langer. *Phys. Rev.*, **B39**, 7905 (1989).

[5.3.31] S.G. Ayling and J.W. Allen. *J. Phys. C*, **20**, 4251 (1987).

[5.3.32] J.M. Langer and L. van Hong. *J. Phys. C*, **17**, L923 (1984).

[5.3.33] G.F. Neumark, D.J. DeBitetto, R.N. Bhargava and P.M. Harnack. *Phys. Rev.*, **B15**, pp. 3147 & 3156 (1977).

[5.3.34] D.J. Robbins. *J. Phys. C*, **16**, 3825 (1983).

[5.3.35] M. Takeshima. *Phys. Rev. B23*, 771 (1981).

[5.3.36] C.H. Henry and D.V. Lang. *Phys. Rev.*, **B15**, 989 (1977).

[5.3.37] D.J. Robbins. *J. Phys. C*, **13**, L1073 (1980).
T. Markvart *J. Phys. C*, **14**, L 435 (1981).

[5.3.38] J.-T. Rebsch. *Solid-State Commun.*, **31**, 377 (1979).

[5.3.39] V.B. Khalfin, M.V. Strikha and I.N. Yassievich. *Phys. Stat. Sol.*, (*b*), **131**, 203 (1985).

[5.3.40] M.V. Strikha. *Sov. Phys. Semicond.*, **20**, 594 (1986).

[5.3.41] G. Jones and A.R. Beattie. *Phys. Stat. Sol.*, (*a*), **4**, 193 (1971).

[5.3.42] G.F. Neumark. *Phys. Rev.*, **B7**, 3802 (1973).

[5.3.43] F.A. Riddoch and M. Jaros. *J. Phys.*, **C13**, 6181 (1980).

[5.3.44] A. Haug. *J. Phys.*, **C21**, 6111 (1988).

[5.3.45] M. Kleverman, H.G. Grimmeiss, A. Litwin and E. Janzén. *Phys. Rev.*, **B31**, 3659 (1985).

[5.3.46] A.C. Wang, S.L. Lake and C.T. Sah. *Phys. Rev.*, **B30**, 5896 (1984).

[5.4.1] J. Singh. *Solid-state Phys.*, **38**, 295 (1984), Table IV.

[5.4.2] M. Trlifaj. *Czech. J. Phys.*, **B15**, 780 (1965).

[5.4.3] M. Trlifaj. *Czech. J. Phys.*, **B14**, 227 (1964).

[5.4.4] G.C. Osbourn and D.L. Smith. *Phys. Rev.*, **B16**, 5416 (1977).

[5.4.5] W. Schmid. *Phys. Stat. Sol.*, (*b*), **84**, 529 (1977).

[5.4.6] C. Richard and M. Dugue. *Phys. Stat. Sol.*, (*b*), **50**, 263 (1972).

[5.4.7] J. Singh and P.T. Landsberg. *J. Phys.*, **C9**, 3627 (1976).

[5.4.8] M. Trlifaj. *Czech. J. Phys.*, **9**, 446 (1959).

[5.4.9] Z. Khás. *Czech. J. Phys.*, **B15**, pp. 346 & 568 (1965).

[5.4.10] P.J. Dean, R.A. Faulkner, S. Kimura and M. Ilegems. *Phys. Rev.*, **B4**, 1926 (1971).

[5.4.11] D.F. Nelson, J.D. Cuthbert, P.J. Dean and D.G. Thomas. *Phys. Rev. Lett.*, **17**, 1262 (1966).

[5.4.12] M.J. Adams and P.T. Landsberg. *J. Luminescence*, **7**, 3 (1973).

[5.4.13] B.L. Gelmont, V.A. Kharchenko and I.N. Yassievich. *Sov. Phys. Solid State*, **29**, 1355 (1988).

[5.4.14] A. Hangleiter. *Phys. Rev.*, **B35**, 9149 (1987).
A. Hangleiter. *Phys. Rev.*, **B37**, 2594 (1988).

[5.4.15] M. Kleverman, J. Olajos and H.G. Grimmeiss. *Phys. Rev.*, **B37**, 2613 (1988).

[5.4.16] W. Czaja. *Festkörperprobleme*, **11**, 65 (1971).

[5.4.17] J.C. Tsang, P.J. Dean and P.T. Landsberg. *Phys. Rev.*, **173**, 814 (1968).

[5.4.18] P.D. Dapkus, W.H. Hackett, O.G. Lorimor and R.Z. Bachrach. *J. App. Phys.*, **45**, 4920 (1974).
B.L. Gelmont, N.N. Zinov'ev, D.I. Kovalev, V.A. Kharchenko, I.D. Yaroshetskii and I.N. Yassievich. *Sov. Phys. JETP*, **67**, 613 (1988).

[5.4.19] P.T. Landsberg and M.J. Adams. *Proc. Roy. Soc.*, **A334**, 523 (1973).

[5.4.20] P.J. Dean. *Proc. XIVth Intern. Conf. Physics of Semiconductors, Edinburgh*, 1978 (Bristol: Inst. of Physics 1979), p. 1259. Also catalogued as Inst. of Physics Conference Series No. 43.

[5.4.21] H. Przybylińska and M. Godlewski. *Phys. Rev.*, **B36**, 1677 (1987).

[5.4.22] J.M. Dishman. *Phys. Rev.*, **B3**, 2588 (1971).

[5.4.23] E.I. Tolpygo, K.B. Tolpygo and M.K. Sheinkman. *Sov. Phys. Semicond.*, **8**, 326 (1974).

[5.4.24] V.V. Diakin, E.A. Sal'kov, V.A. Khvostov and M.K. Sheinkman. *Sov. Phys. Semicond.*, **10**, 1357 (1976).

[5.4.25] N.E. Korsunskaya, I.B. Markevich, T.V. Torchinskaya and M.K. Sheinkman. *Sov. Phys. Semicond.*, **11**, 130 (1977).

[5.4.26] V.D. Bondar, O.B. Kushnir, A.B. Lyskovich, I.B. Markevich and M.K. Sheinkman. *Sov. Phys. Solid State*, **23**, 2024 (1981).

[5.4.27] E.I. Tolpygo, K.B. Tolpygo and M.K. Sheinkman. *Sov. Phys. Solid State*, **7**, 1442 (1965).

[5.4.28] M.K. Sheinkman. *Sov. Phys. Solid State*, **5**, 2035 (1964).
M.K. Sheinkman. *Sov. Phys. Solid State*, **7**, 18 (1965).

[5.4.29] Y.V. Vorob'ev, E.I. Tolpygo and M.K. Sheinkman. *Sov. Phys. Semicond.*, **19**, pp. 1313 & 1318 (1985).
Y.V. Vorob'ev, E.I. Tolpygo and M.K. Sheinkman. *Phys. Stat. Sol.*, (b), **123**, 295 (1984).

[5.4.30] V.A. Kovarskii, L.V. Chernysh and M.K. Sheinkman. *Phys. Stat. Sol.*, (b), **131**, 677 (1985).

[5.4.31] R. Sooryakumar, D.S. Chemla, A. Pinczuk, A.C. Grossard, W. Wiegmann and L.J. Sham. *Solid-State Commun.*, **54**, 859 (1985).

[5.4.32] L.J. Sham. *J. de Phys.*, **48**, Colloquium 5, p. 381 (1987).

[5.4.33] J.R. Troxell, A.P. Chatterjee, G.D. Watkins and L.C. Kimerling. *Phys. Rev.*, **B19**, 5336 (1979).

[5.4.34] L.C. Kimerling. *Solid-State Electron.*, **21**, 1391 (1978).

[5.4.35] A. Sibille. *J. Electron. Mat.*, **14a**, 1155 (1985).

[5.4.36] M. Levinson, C.D. Coombs and J.A. Kafalas. *Phys. Rev.*, **B34**, 4358 (1986).

[5.4.37] G.M. Martin. *App. Phys. Lett.*, **39**, 747 (1981).

[5.4.38] P.M. Petroff. *Semiconductors and Insulators* (New York: Gordon and Breach), vol. 5, p. 307 (1983).

[5.4.39] D.V. Lang. *Ann. Rev. Mater. Sci.*, **12**, 377 (1982).

[5.4.40] J.D. Weeks, J.C. Tully and L.C. Kimerling. *Phys. Rev.*, **B12**, 3286 (1975).

[5.4.41] F. Seitz and J.S. Koehler. *Solid-State Phys.*, **2**, 351 (1956).

[5.4.42] R.D. Gold and L.R. Weisberg. *Solid-State Electron.*, **7**, 811 (1964).

[5.4.43] T. Markvart and P.T. Landsberg. *Conference Radiation Physics of Semiconductors and Related Materials*, Tbilisi (eds G.P. Kerkelidze and V.I. Shakhovtsov) (Tbilisi State University Press, 1980), p. 510.

[5.4.44] J. Bourgoin and J.W. Corbett. *Phys. Lett.*, **38A**, 135 (1972).

[5.4.45] T. Markvart. *Mat. Sci. Forum*, **10/12**, 451 (1986).

[5.4.46] A.M. Stoneham. *Rep. Prog. Phys.*, **44**, 1251 (1981).

[5.4.47] D.V. Lang. *Ann. Rev. Mat. Sci.*, **12**, 377 (1982).

[5.4.48] G.F. Neumark and K. Kosai. In *Semiconductors and Semimetals* (eds R.K. Willardson and A.C. Beer), vol. 19 (New York: Acadenic Press, 1983), p. 1.

[5.4.49] J.A. van Vechten. *Phys. Rev.*, **B38**, 9913 (1988).

[5.4.50] H. Sumi. *J. Phys.*, **C17**, 6071 (1984).

Chapter 6

[6.1.1] J. Jortner. *Phil. Mag.*, **40**, 317 (1979).

[6.1.2] K. Huang and A. Rhys. *Proc. Roy. Soc.*, **A204**, 406 (1950).

[6.1.3] E. Sonder and W.A. Sibley. In *Point Defects in Solids* (eds J.H. Crawford and L.M. Slifkin), vol. 1, (New York: Plenum Press, 1975), p. 201.

[6.1.4] A.M. Stoneham. *Theory of Defects in Solids* (Oxford: Clarendon Press, 1975).

[6.1.5] M. Jaros. *Deep Levels in Semiconductors* (Bristol: Adam Hilger, 1982).

[6.1.6] C.H. Henry and D.V. Lang. *Phys. Rev.*, **B15**, 989 (1977).

[6.2.1] R. Englman. *Nonradiative Deacy of Ions and Molecules in Solids* (Amsterdam: North-Holland, 1979).
         A.M. Stoneham. *Rep. Prog. Phys.*, **44**, 1251 (1981).
[6.2.2] G. Vincent and D. Bois. *Solid-State Commun.*, **27**, 431 (1978).
         D. Bois and A. Chantre. *Rev. Phys. Appl.*, **15**, 631 (1980).
         H. Sumi. *J. Phys. Soc. Jap. Suppl. A*, **49**, 227 (1980).
         A. Chantre, G. Vincent and D. Bois. *Phys. Rev.*, **B23**, 5335 (1981).
[6.2.3] G.F. Neumark and K. Kosai. In *Semiconductors and Semimetals* (eds R.K. Williamson and A.C. Beer), vol. 19 (New York: Academic Press, 1983), p. 1.
[6.2.4] J. Franck. *Trans. Faraday Soc.*, **21**, 536 (1925).
[6.2.5] E.U. Condon. *Phys. Rev.*, **32**, 858 (1928).
[6.2.6] M. Lax. *J. Chem. Phys.*, **20**, 1752 (1952).
[6.2.7] J. Bardeen and W. Shockley. *Phys. Rev.*, **80**, 72 (1950).
[6.2.8] M. Jaros. *Deep Levels in Semiconductors* (Bristol: Adam Hilger, 1982).
[6.2.9] M. Born and K. Huang. *Dynamical Theory of Crystal Lattices* (Oxford: Clarendon Press, 1954).
[6.2.10] A.M. Stoneham. *Theory of Defects in Solids* (Oxford: Clarendon Press, 1982).

[6.3.1] C. Zener. *Proc. Roy. Soc.*, **A137**, 696 (1932).
[6.3.2] L.D. Landau. *Phys. Z. Sowjet.*, **1**, 88 (1932).
         L.D. Landau and E.M. Lifshits. *Quantum Mechanics* (Oxford: Pergamon Press, 1965).
[6.3.3] M. Born and J.R. Oppenheimer. *Ann. Phys. Leipzig*, **84**, 457 (1927).
[6.3.4] M. Born and K. Huang. *Dynamical Theory of Crystal Lattices* (Oxford: Clarendon Press, 1954).
[6.3.5] G. Helmis. *Ann. Phys. (Leipzig)*, **19**, 41 (1956).
[6.3.6] R. Pässler. *Czech. J. Phys.*, **B24**, 322 (1974).
         R. Pässler. *Czech. J. Phys.*, **B25**, 219 (1975).
[6.3.7] See, for example, M. Abramowitz and I.A. Stegun. *Handbook of Mathematical Functions* (New York: Dover, 1968), section 6.

[6.4.1] L.D. Landau and E.M. Lifshits. *Quantum Mechanics* (Oxford: Pergamon Press, 1965).
[6.4.2] T. Markvart. 'Multiphonon recombination in semiconductors' (unpublished).
[6.4.3] R. Englman and J. Jortner. *Mol. Phys.*, **18**, 145 (1970).

[6.5.1] L.I. Schiff. *Quantum Mechanics* (New York: Dover, 1968).
[6.5.2] L.D. Landau and E.M. Lifshits. *Quantum Mechanics* (Oxford: Pergamon Press, 1965).
[6.5.3] M. Jaros. *Deep Levels in Semiconductors* (Bristol: Adam Hilger, 1982).
[6.5.4] R.A. Abram. *J. Phys.*, **C13**, L753 (1980).
[6.5.5] R. Pässler. *Phys. Stat. Sol.*, (b), **78**, 625 (1976).
[6.5.6] B.K. Ridley. *Solid-State Electron.*, **21**, 1319 (1978).
[6.5.7] G. Gamow. *Z. Physik*, **51**, 204 (1928).
[6.5.8] S. Makram-Ebeid. *Appl. Phys. Lett.*, **37**, 464 (1980).
[6.5.9] D. Pons and S. Makram-Ebeid. *J. Phys. Appl.*, **40**, 1161 (1979).
[6.5.10] S. Makram-Ebeid and L. Lanoo. *Phys. Rev.*, **B25**, 6404 (1982).
[6.5.11] V.L. Bonch-Bruevich and E.G. Landsberg. *Phys. Stat. Sol.*, **29**, 9 (1968).
[6.5.12] R. Pässler. *Phys. Stat. Sol.*, (b), **85**, 203 (1978).

[6.6.1] R. Kubo and Y. Toyozawa. *Prog. Theor. Phys.*, **13**, 160 (1955).
[6.6.2] T. Holstein. *Ann. Phys.*, **8**, 343 (1959).
[6.6.3] M. Abramowitz and I.A. Stegun. *Handbook of Mathematical Functions* (New York: Dover, 1968).
[6.6.4] T. Markvart. *J. Phys. C*, **14**, L895 (1981).

Chapter 7

[7.1.1] G.W. Bryant. *Phys. Rev. Lett.*, **59**, 1140 (1987).
[7.1.2] P.T. Landsberg and P. Harshman. *J. Stat. Phys.*, **53**, 475 (1988).

[7.2.1] L. Esaki and R. Tsu. *IBM J. Res. Dev.*, **14**, 61 (1970).
[7.2.2] L. Esaki. *IEEE J. Quantum Electron.*, **QE22**, 1611 (1986).
[7.2.3] G. Bastard. *Phys. Rev. B*, **24**, 5693 (1981).
[7.2.4] R. Eppenga and M.F.H. Schuurmans. *Philips Tech. Rev.*, **44**, 137 (1988). G. Bastard, C. Delalande, Y. Guldner and P. Voisin. In *Advances in Electronics and Electron Physics*, vol. 72 (San Diego: Academic Press, 1988), p. 1.
[7.2.5] R. Dingle. *Festkörperprobleme*, **15**, 21 (1975).
[7.2.6] M. Altarelli. In *From Heterojunctions to Semiconductor Superlattices* (eds G. Allen and G. Bastard) (Berlin: Springer, 1986). A Les Houches Winter School.
[7.2.7] H.I. Ralph. *Solid-State Commun.*, **3**, 303 (1965).
[7.2.8] G. Bastard. *IEEE J. Quantum Electron.*, **QE22**, 1625 (1986).
[7.2.9] G.E.W Bauer and T. Ando. *Phys. Rev. B*, **3**, 6015 (1988).
[7.2.10] K. Fujiwara, N. Tsukada and T. Nakayama. *Appl. Phys. Lett.*, **53**, 675 (1988).
[7.2.11] A.J. Moseley, D.J. Robbins, A.C. Marshall, M.Q. Kearley and J.I. Davies. *Semicond. Sci. Technol.*, **4**, 184 (1989).

[7.3.1] M.R. Leys, M.P.A. Viegers and G.W. 't Hooft. *Philips Tech. Rev.*, **43**, 133 (1987).
[7.3.2] J.E. Fouquet and R.D. Burnham. *IEEE J. Quantum Electron.*, **QE22**, 1799 (1986).
[7.3.3] N. Holonyak, Jr., R.M. Kolbas, R.D. Dupuis and P.D. Dapkus. *IEEE J. Quantum Electron.*, **QE16**, 170 (1980).
[7.3.4] N.K. Dutta. *J. Appl. Phys.*, **53**, 7211 (1982).
[7.3.5] Y. Arakawa and H. Sakaki. *Appl. Phys. Lett.*, **40**, 939 (1982).
[7.3.6] P. Blood, S. Colak and A.I. Kucharska. *App. Phys. Lett.*, **52**, 599 (1988).
[7.3.7] R. Eppenga and M.F.H. Schuurmans. *Philips Tech. Rev.*, **44**, 137 (1988).
[7.3.8] M.G. Burt. *Electron. Lett.*, **19**, 182 (1983).
[7.3.9] E.P. O'Reilly. *Semicond. Sci. Technol.*, **1**, 128 (1986).

[7.4.1] M.G. Burt. *Electron. Lett.*, **19**, 210 (1983).
[7.4.2] Y. Arakawa and H. Sakaki. *Appl. Phys. Lett.*, **40**, 939 (1982).
[7.4.3] A.R. Adams, M. Asada, Y. Suematsu and S. Arai. *Japan J. Appl. Phys.*, **19**, L621 (1980).
[7.4.4] R.I. Taylor, R.A. Abram, M.G. Burt and C. Smith. *Semicond. Sci. Technol.*, **5**, 90 (1990).
[7.4.5] R.I. Taylor, R.A. Abram, M.G. Burt and C. Smith. *IEE Proc.*, **J132**, 364 (1985).
[7.4.6] P.T. Landsberg. *Solid-State Commun.*, **10**, 479 (1972).
[7.4.7] A. Haug. *J. Phys.*, **C20**, 1293 (1987).
[7.4.8] N.K. Dutta and R.J. Nelson. *J. App. Phys.*, **53**, 74 (1982).

[7.4.9] P. Scharoch and R.A. Abram. *Semicond. Sci. Technol.*, **3**, 973 (1988).

[7.4.10] M.G. Burt, S. Brand, C. Smith and R.A. Abram. *J. Phys.*, **C17**, 6385 (1984).

[7.4.11] E.O. Kane. *J. Phys. Chem. Sol.*, **1**, 249 (1957).

[7.4.12] C.B. Su, J. Schlafer, J. Manning and R. Olshansky. *Electron. Lett.*, **18**, 595 (1982).

[7.4.13] B. Sermage, D.S. Chemla, D. Sivco and A.Y. Cho. *Physica*, **134B**, 417 (1985).

[7.4.14] E. Zielinski, F. Keppler, S. Hausser, M.H. Pilkuhn, R. Sauer and W.T. Tsang. *IEEE J. Quantum Electron.*, **QE25**, 1407 (1989).

[7.4.15] T.P. Pearsall, R.E. Nahory and J.R. Chelikowsky. In *GaAs and Related Compounds, Institute of Physics Conf. Ser. No. 33b*, chapter 6, p. 331 (1977).

[7.4.16] A. Haug. *J. Phys. C*, **16**, 4159 (1983).

[7.4.17] P.T. Landsberg and Y-J. Yu. *J. Appl. Phys.*, **63**, 1789 (1988).

[7.4.18] R.I. Taylor and R.A. Abram. *Semicond. Sci. Technol.*, **3**, 859 (1988).

[7.4.19] M.G. Burt and R.I. Taylor. *Electron Lett.*, **21**, 733 (1985).

[7.4.20] P.T. Landsberg and M.J. Adams. *IEE Proc.*, **J 133**, 118 (1986).

[7.4.21] G.N. Childs, S. Brand, R.A. Abram. *Semicond. Sci. Technol.*, **1**, 116 (1986).

[7.5.1] A.R. Adams. *Electron. Lett.*, **22**, 249 (1986).

[7.5.2] F.C. Frank and J.H. van der Merwe. *Proc. Roy. Soc.*, **198**, 216 (1949).

[7.5.3] J.W. Matthews, A.E. Blakeslee and S. Mader. *Thin Solid Films*, **33**, 253 (1976).

[7.5.4] R. People and J.C. Bean. *Appl. Phys. Lett.*, **47**, 322 (1985).

Appendix A

[A.1] D. Zhang, Y. Ding and T. Ma. *Am. J. Phys.*, **57**, 281 (1989).

Appendix D

[D.1] D.J. Morgan and P.T. Landsberg. *Proc. Phys. Soc.*, **86**, 261 (1965).
       D.J. Morgan and J.A. Galloway. *Phys. Stat. Sol.*, **23**, 97 (1967).

[D.2] P.T. Landsberg and D.J. Morgan. *J. Math. Phys.*, **7**, 2271 (1966).

[D.3] H. Kroemer. *Am. J. Phys.*, **54**, 177 (1986).
       A. Manohar. *Phys. Rev.*, **B34**, 1287 (1986).

[D.4] W. Kuhn. *Z.f. Phys.*, **33**, 408 (1925).
       F. Reiche and W. Thomas. *Z.f. Phys.*, **34**, 510 (1925).

[D.5] See, for example, R.M. Wilcox. *J. Math. Phys.*, **8**, 962 (1967).

[D.6] A.R. Beattie and G. Smith. *Phys. Stat. Sol.*, **19**, 577 (1967).

[D.7] R. Dalven. *J. Phys.*, **C6**, 671 (1973).

[D.8] C.M. van Vliet and A.H. Marshak. *Phys. Rev.*, **B29**, 5960 (1984).
       P. Enders. *Phys. Stat. Sol.*, (*b*), **139**, K113 (1987).

Appendix F

[F.1] H. Jeffreys and B.S. Jeffreys. *Methods of Mathematical Physics*, 3rd edn. (Cambridge University Press, 1962).

[F.2] M. Abramowitz and I. Stegun. *Handbook of Mathematical Fucntions* (New York: Dover, 1968).

Index of names

* Gmitter changed initials from T. to T. J. in the late 1980s.

* van Vliet changed initials from C. M. to K. M. in the early 1980s.

Index of topics, concepts and materials